Scanning Electrochemical Microscopy

MONOGRAPHS IN ELECTROANALYTICAL CHEMISTRY AND ELECTROCHEMISTRY

consulting editor
Allen J. Bard
Department of Chemistry
University of Texas
Austin, Texas

Electrochemistry at Solid Electrodes, *Ralph N. Adams*

Electrochemical Reactions in Nonaqueous Systems, *Charles K. Mann and Karen Barnes*

Electrochemistry of Metals and Semiconductors: The Application of Solid State Science to Electrochemical Phenomena, *Ashok K. Vijh*

Modern Polarographic Methods in Analytical Chemistry, *A. M. Bond*

Laboratory Techniques in Electroanalytical Chemistry, *edited by Peter T. Kissinger and William R. Heineman*

Standard Potentials in Aqueous Solution, *edited by Allen J. Bard, Roger Parsons, and Joseph Jordan*

Physical Electrochemistry: Principles, Methods, and Applications, *edited by Israel Rubinstein*

Scanning Electrochemical Microscopy, *edited by Allen J. Bard and Michael V. Mirkin*

ADDITIONAL VOLUMES IN PREPARATION

Scanning Electrochemical Microscopy

edited by

ALLEN J. BARD
University of Texas at Austin
Austin, Texas

MICHAEL V. MIRKIN
Queens College–City University of New York
Flushing, New York

MARCEL DEKKER, INC. NEW YORK • BASEL

ISBN: 0-8247-0471-1

This book is printed on acid-free paper.

Headquarters
Marcel Dekker, Inc.
270 Madison Avenue, New York, NY 10016
tel: 212-696-9000; fax: 212-685-4540

Eastern Hemisphere Distribution
Marcel Dekker AG
Hutgasse 4, Postfach 812, CH-4001 Basel, Switzerland
tel: 41-61-261-8482; fax: 41-61-261-8896

World Wide Web
http://www.dekker.com

The publisher offers discounts on this book when ordered in bulk quantities. For more information, write to Special Sales/Professional Marketing at the headquarters address above.

Current printing (last digit):
10 9 8 7 6 5 4 3 2 1

PRINTED IN THE UNITED STATES OF AMERICA

PREFACE

A little more than ten years have elapsed since publication of the first papers describing the fundamentals of scanning electrochemical microscopy (SECM). During this decade, the field of SECM has evolved substantially. The technique has been used in a variety of ways, for example, as an electrochemical tool to study heterogeneous and homogeneous reactions, for high-resolution imaging of the chemical reactivity and topography of various interfaces, and for microfabrication. Quantitative theoretical models have been developed for different modes of the SECM operation. The first commercial SECM instrument was introduced in 1999. The SECM technique is now used by a number of research groups in many different countries. We think the time has come to publish the first monograph, providing comprehensive reviews of different aspects of SECM.

The first five chapters of this book contain experimental and theoretical background, which is essential for everyone working in this field: principles of SECM measurements (Chapter 1), instrumentation (Chapter 2), preparation of SECM ultramicroelectrodes (Chapter 3), imaging methodologies (Chapter 4), and theory (Chapter 5). Other chapters are dedicated to specific applications and are self-contained. Although some knowledge of electrochemistry and physical chemistry is assumed, the key ideas are discussed at the level suitable for beginning graduate students.

Through the addition of submicrometer-scale spatial resolution, SECM greatly increases the capacity of electrochemical techniques to characterize interfaces and measure local kinetics. In this way, it has proved useful for a broad range of interdisciplinary research. Various applications of SECM are discussed in this book, from studies of biological systems, to sensors, to probing reactions at the liquid/liquid interface. Although we did not intend to present even a brief survey of those diverse areas of research, each chapter provides sufficient details to allow a specialist to evaluate the applicability of the SECM methods for solving a specific problem. We hope it will be useful to all interested in learning about this technique and applying it.

We would like to thank our students, co-workers, and colleagues who have done so much to develop SECM. The future for this technique, which is unique among scanning probe methods in its quantitative rigor and its ability to study with ease samples in liquid environments, continues to be a bright one.

Allen J. Bard
Michael V. Mirkin

CONTENTS

CONTRIBUTORS

ALLEN J. BARD The University of Texas at Austin, Austin, Texas

BRADLEY D. BATH ALZA Corporation, Mountain View, California

KAI BORGWARTH Institute for Physical Chemistry, Albert Ludwig University of Freiburg, Freiburg, Germany

CHRISTOPHE DEMAILLE The University of Texas at Austin, Austin, Texas

GUY DENUAULT University of Southampton, Southampton, England

FU-REN F. FAN The University of Texas at Austin, Austin, Texas

JÜRGEN HEINZE Albert Ludwig University of Freiburg, Freiburg, Germany

BENJAMIN R. HORROCKS University of Newcastle upon Tyne, Newcastle upon Tyne, United Kingdom

JULIE V. MACPHERSON University of Warwick, Coventry, England

DANIEL MANDLER The Hebrew University of Jerusalem, Jerusalem, Israel

MICHAEL V. MIRKIN Queens College–City University of New York, Flushing, New York

GÉZA NAGY Janus Pannonius University, Pécs, Hungary

ERIK R. SCOTT Medtronic Corporation, Minneapolis, Minnesota

KLÁRA TÓTH Institute of General and Analytical Chemistry, Technical University of Budapest, Budapest, Hungary

MICHAEL TSIONSKY Gaithersburg, Maryland

PATRICK R. UNWIN University of Warwick, Coventry, England

HENRY S. WHITE University of Utah, Salt Lake City, Utah

DAVID O. WIPF Mississippi State University, Mississippi State, Mississippi

GUNTHER WITTSTOCK Wilhelm-Ostwald-Institute of Physical and Theoretical Chemistry, University of Leipzig, Leipzig, Germany

1

INTRODUCTION AND PRINCIPLES

Allen J. Bard

The University of Texas at Austin
Austin, Texas

I. BACKGROUND OF SCANNING ELECTROCHEMICAL MICROSCOPY

This volume is devoted to a complete and up-to-date treatment of scanning electrochemical microscopy (SECM). In this introductory chapter, we cover the historical background of the technique, the basic principles of SECM, and an overview of some of its applications (covered in more depth in later chapters). A number of reviews of this field have also been published (1–6).

SECM involves the measurement of the current through an ultramicroelectrode (UME) (an electrode with a radius, a, of the order of a few nm to 25 μm) when it is held or moved in a solution in the vicinity of a substrate. Substrates, which can be solid surfaces of different types (e.g., glass, metal, polymer, biological material) or liquids (e.g., mercury, immiscible oil), perturb the electrochemical response of the tip, and this perturbation provides information about the nature and properties of the substrate. The development of SECM depended on previous work on the use of ultramicroelectrodes in electrochemistry and the application of piezoelectric elements to position a tip, as in scanning tunneling microscopy (STM). Certain aspects of SECM behavior also have analogies in electrochemical thin-layer cells and arrays of interdigitated electrodes.

The movement of the tip is usually carried out by drivers based on piezoelectric elements, similar to those used in STM, as described in Chapter 2. Typically, inchworm drivers (Burleigh Instruments, Fishers, NY) are used, since they can move larger distances than simple piezoelectric tube scanners. However, where higher resolution is needed, piezoelectric pushers can be added, so that the inchworms provide coarse drives and the pushers nm-resolution drives. Generally the direction normal to the substrate is taken as the z direction, while x and y are those in the plane of the substrate.

There are several modes of operation of the SECM. In the tip generation–substrate collection (TG/SC) mode, the tip is used to generate a reactant that is detected at a substrate electrode. For example, the reaction O + ne → R occurs at the tip, and the reverse reaction occurs at the substrate. This mode of operation is similar to that at the rotating ring-disk electrode (7). Similar behavior is observed for a pair of side-by-side microband electrodes (8,9) and in thin-layer cells (10). In the SECM, TG/SC is usually used in studies of homogeneous chemical reactions, where the reaction of species R as it transits between tip and substrate causes a decrease in the substrate current (see Chapter 7). An alternative mode, where the substrate is the generator and tip the collector (SG/TC mode), can also be employed and is used in studies of reactions at a substrate surface (Chapters 6, 9, 11, and 12). The SG/TC mode was first used to study concentration profiles near an electrode surface without scanning and imaging (11–13).

The most frequent mode of operation of the SECM is the feedback mode, where only the tip current is monitored. As discussed in the next section, the tip current is perturbed by the presence of a substrate at close proximity by blockage of the diffusion of solution species to the tip (negative feedback) and by regeneration of O at the substrate (positive feedback). This effect allows investigation of both electrically insulating and conducting surfaces and makes possible imaging of surfaces and the reactions that occur there. This mode of operation with surface imaging was first described, along with the apparatus and theory, in a series of papers in 1989 (14–16).

II. PRINCIPLES OF SECM

A. Ultramicroelectrodes

An understanding of the operation of the SECM and an appreciation of the quantitative aspects of measurements with this instrument depends upon an understanding of electrochemistry at small electrodes. The behavior of ultramicroelectrodes in bulk solution (far from a substrate) has been the subject of a number of reviews (17–21). A simplified experimental setup for an electrochemical experiment is shown in Figure 1. The solution contains a species, O, at a concentration, c, and usually contains supporting electrolyte to decrease the solution resistance and insure that transport of O to the electrode occurs predominantly by diffusion. The electrochemical cell also contains an auxiliary electrode that completes the circuit via the power supply. As the power supply voltage is increased, a reduction reaction, O + ne → R, occurs at the tip, resulting in a current flow. An oxidation reaction will occur at the auxiliary electrode, but this reaction is usually not of interest in SECM, since this electrode is placed sufficiently far from the UME

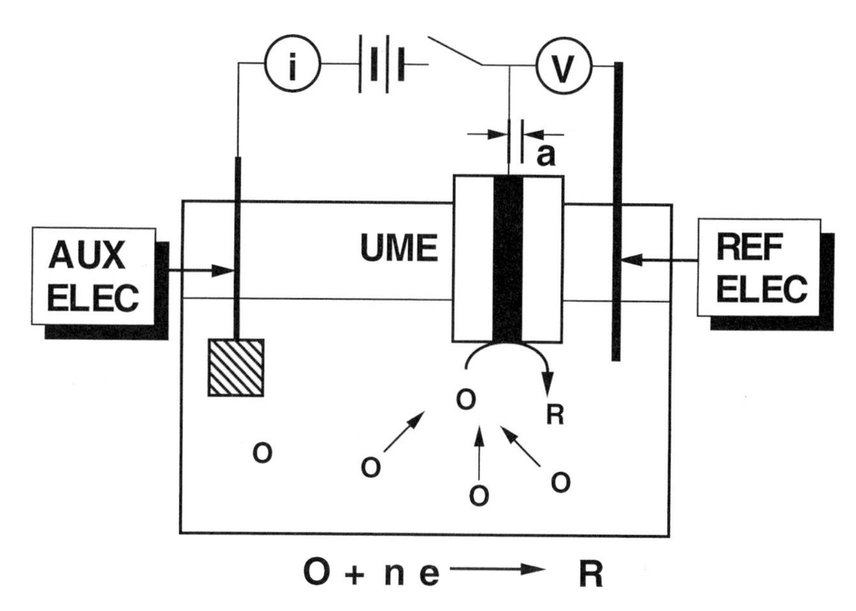

FIG. 1 Schematic diagram of a cell for ultramicroelectrode voltammetry.

that products formed at the auxiliary electrode do not reach the tip during the experiment. The potential of the tip electrode is monitored against a stable reference electrode, such as a silver/silver chloride electrode. A plot of the current flowing as a function of the potential of the UME is called a voltammogram; a typical one is shown in Figure 2. As shown, an S-shaped

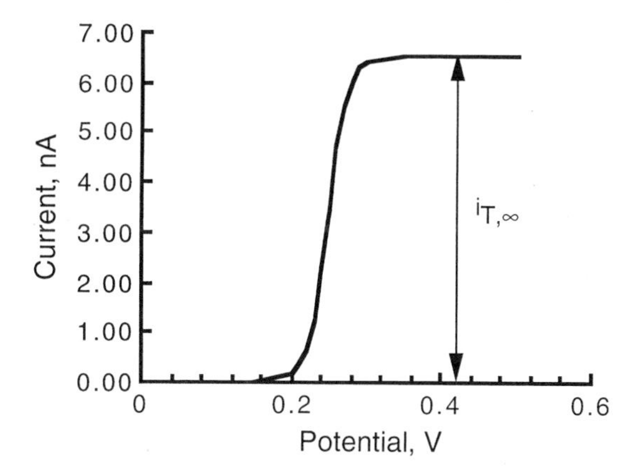

FIG. 2 Typical voltammogram for an ultramicroelectrde.

curve is produced. The current eventually limits to a value that is completely controlled by the rate of mass transfer by diffusion of O from the bulk solution to the electrode surface, where the electrochemical reaction has decreased its concentration to essentially zero. For a conductive disk of radius a in an insulating sheath, this steady-state diffusion-controlled current when the tip is far from a surface is given by:

$$i_{T,\infty} = 4nFDca \tag{1}$$

where D is the diffusion coefficient of species O, and F is the Faraday. The current at electrodes with other shapes, e.g., hemispheres or cones, can be expressed in a similar way, as discussed in Chapter 3, but almost all SECM experiments are carried out with disk-shaped electrodes, because they show the best sensitivity. The current is also relatively independent of the radius of the insulating sheath, r_g, often expressed in the SECM literature as $RG = r_g/a$. Moreover, because the flux of O to a small disk by diffusion ($\sim Dc/a$) is quite large, the current is relatively immune to convective effects like stirring in the solution. The current at a small disk also reaches steady state in a relatively short time ($\sim a^2/D$). For example, a 10 μm radius disk will attain steady state in a fraction of a second. These characteristics imply that an ultramicroelectrode used as a scanning tip and moved in a solution can be treated as a steady-state system. Finally, because of the small currents that characterize most experiments with ultramicroelectrode tips, generally pA to nA, resistive drops in the solution during passage of current are generally negligible.

B. Feedback Mode

The general principles of the feedback mode are shown in Figure 3. As shown in Eq. (1), the current, $i_{T,\infty}$, is measured at the ultramicroelectrode tip when it is far from any surface (A), the subscript, ∞, implying this long distance. In fact, as we shall see, this distance only has to be a few tip diameters. The current under these conditions is driven by the hemispherical flux of species O from the bulk solution to the tip (Fig. 3A). When the tip is brought near an electrically insulating substrate, like a piece of glass or plastic (Fig. 3C), the substrate blocks some of the diffusion of O to the tip and the current will decrease compared to $i_{T,\infty}$. The closer the tip gets to the substrate, the smaller i_T becomes. At the limit when the distance between tip and substrate, d, approaches zero, i_T also approaches zero. This decrease in current with distance is called *negative feedback*. When the tip is brought near an electrically conductive substrate, like a platinum electrode, while there is still blockage of diffusion of O to the tip by the substrate, there is also the oxidation of the product R back to O. This O generated at the

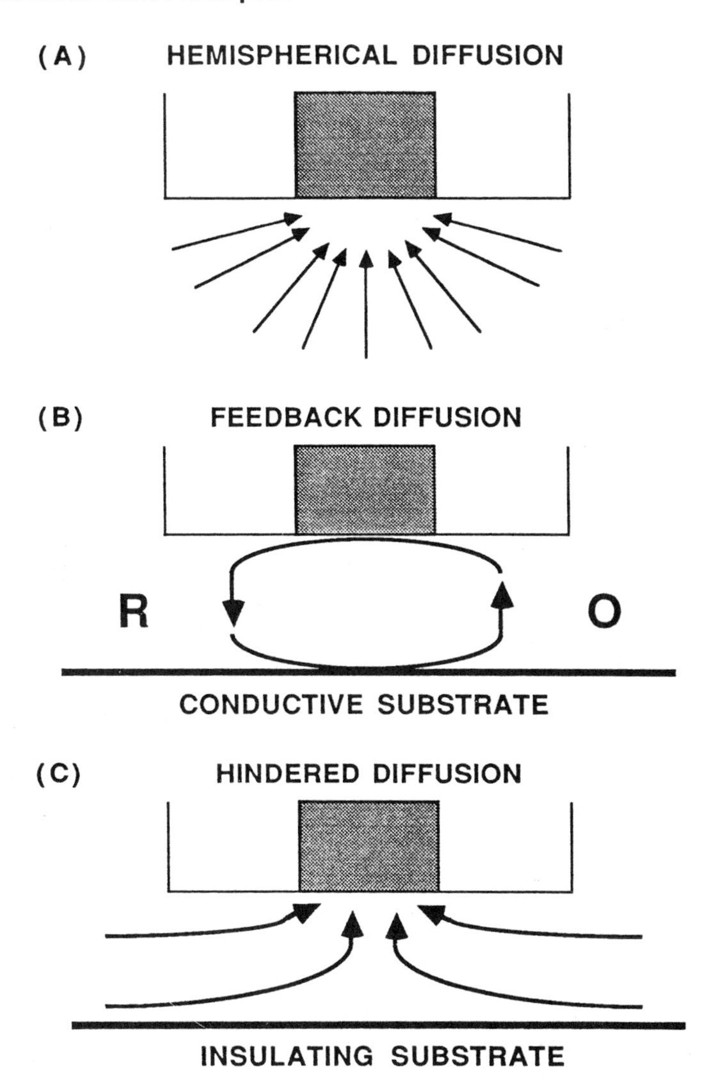

FIG. 3 Basic principles of scanning electrochemical microscopy (SECM): (A) far from the substrate, diffusion leads to a steady-state current, $i_{T,\infty}$; (B) near a conductive substrate, feedback diffusion leads to $i_T > i_{T,\infty}$; (C) near an insulating substrate, hindered diffusion leads to $i_T < i_{T,\infty}$. (Reprinted with permission from A. J. Bard, G. Denuault, C. Lee, D. Mandler, and D. O. Wipf, Acc. Chem. Res. *23*, 357 (1990). Copyright 1990 American Chemical Society.)

substrate diffuses to the tip and causes an increase in the flux of O compared with $i_{T,\infty}$. Thus with a conductive substrate $i_T > i_{T,\infty}$. In the limit as d approaches zero, the tip will move into a regime where electron tunneling can occur and the tip current will get very large. This increase of current with distance is called *positive feedback*. A plot of i_T versus d, as a tip is moved in the z direction, is called an *approach curve*.

A quantitative description of approach curves can be obtained by solving the diffusion equations for the situation of a disk electrode and a planar substrate (16), as discussed in Chapter 5. Typical approach curves for a conductive substrate (essentially infinite rate of regeneration of O from R) and an insulating substrate (zero rate of regeneration of O) are shown in Figure 4. These curves are given in dimensionless form by plotting $I_T = i_T/i_{T,\infty}$ (the tip current normalized by the current far from substrate) versus $L = d/a$ (the tip-substrate separation normalized by the tip radius). Since this plot involves only dimensionless variables, it does not depend upon the concentration or diffusion coefficient of O. From these curves one can readily find d from the measured I_T and a knowledge of a. The approach curves for an insulator actually depend upon r_g, since the sheath around the conducting portion of the electrode also blocks diffusion, but this effect is not usually important with most practical tips. If the rate constant for electron transfer at the substrate to species O is $k_{b,s}$, the limiting curves repesent $k_{b,s} \rightarrow 0$ (insulator) and $k_{b,s} \rightarrow \infty$ (conductor). The approach curves for intermediate values of $k_{b,s}$ can be found (Chapter 5) (Fig. 5). These are very useful in finding the rate of heterogeneous charge transfer at an interface (see Chapters 6 and 8).

C. Collection-Generation Modes

As discussed above, there are two modes of this type. In the TG/SC mode, the tip is held at a potential where an electrode reaction occurs and the substrate is held at a different potential where a product of the tip reaction will react and thus be collected. In most cases the substrate is considerably larger than the tip, so that the collection efficiency, given by i_S/i_T (where i_S is the substrate current), is essentially 1 (100%) for a stable tip-generated species, R. If R reacts on transit from tip to substrate, i_S/i_T becomes smaller, and its change with separation, d, allows determination of the rate constant of the homogeneous reaction (Chapter 7).

The alternative mode is the substrate generation–tip collection (SG/TC) mode. In this case the tip probes the reactions that are occurring on a substrate. For example, a scan in the z direction can produce the concentration profile, while a scan over the surface can identify hot spots, where reactions occur at a higher rate.

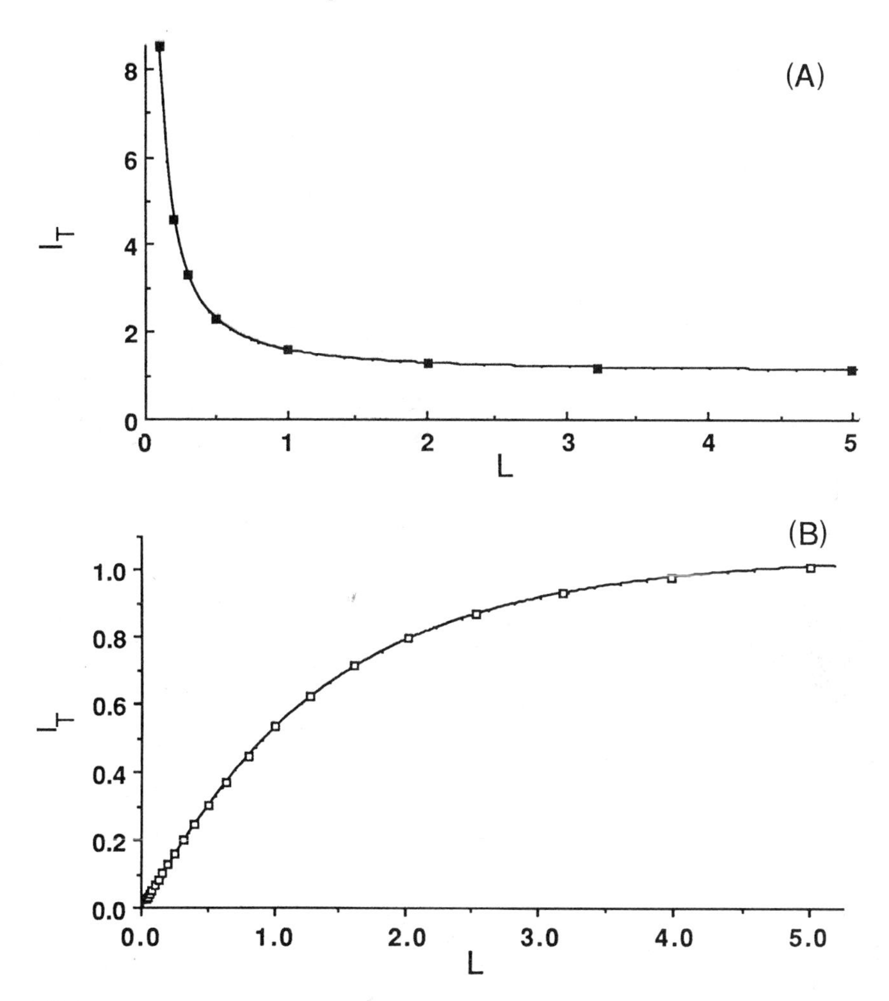

FIG. 4 Diffusion-controlled steady-state tip current as a function of tip-substrate separation. (A) Substrate is a conductor; (B) substrate is an insulator. (From Ref. 2.)

A related method involves the use of the tip reaction to perturb a reaction at a surface; an example of this approach is SECM-induced desorption (SECMID) (22). For example, the adsorption/desorption kinetics of protons on a hydrous metal oxide surface can be studied in an unbuffered solution by bringing the tip near the surface and reducing proton (to hydrogen) at the tip. This causes a local change in pH that results in proton desorption from the surface. The tip current can be used to study the kinetics of proton desorption and diffusion on the surface (Chapter 12).

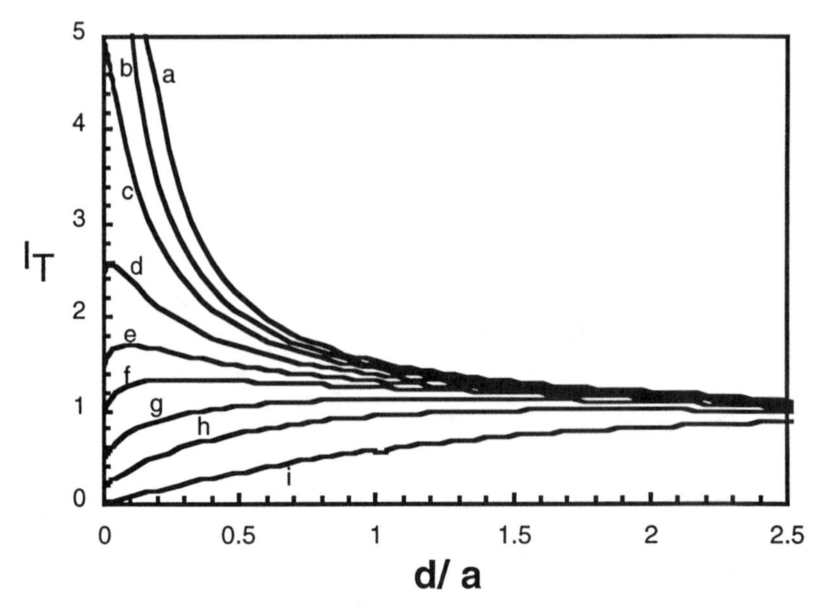

FIG. 5 Approach curves as a function of the heterogeneous reaction rate constant for electron transfer at the substrate, k, $I_T = i_T/i_{T,\infty}$. From top to bottom, k (cm/s) is (a) 1, (b) 0.5, (c) 0.1, (d) 0.025, (e) 0.015, (f) 0.01, (g) 0.005, (h) 0.002, (i) 0.0001. Curve (a) is identical to that for mass transfer control and curve (i) for an insulating substrate.

D. Transient Methods

Most SECM measurements involve steady-state current measurements. This can be a significant advantage in the measurement of kinetics, even for rapid processes, because factors like double-layer charging and adsorption do not contribute to the observed currents. However, one can also carry out transient measurements, recording i_T as a function of time. This can be of use in measurements of homogeneous kinetics (Chapter 7) and for systems that are changing with time. It can also be used to determine the diffusion coefficient, D, of a species without knowledge of the solution concentration or number of electrons transferred in the electrode reaction (23).

E. Fabrication

The SECM can also be used as a tool for modification of surfaces. For example, metals or semiconductors can be etched or metals deposited on a surface by passing the tip close to the surface and carrying out an appropriate electrochemical reaction. Two different modes are possible. In the *direct mode*, the tip acts as the counterelectrode and the desired electrochemical reaction occurs on the substrate. For example, Cu can be etched from a Cu substrate. Spatial resolution is determined by the current density distribution between tip and substrate. In the *feedback mode* a reactant is generated at the tip which promotes the reaction on the substrate. For example, Cu can be etched by bromine electrogenerated at the tip. In this case resolution is determined by the lateral (x-y) diffusion of reactant as it diffuses from tip to substrate. Details of fabrication using SECM are covered in Chapter 13.

III. APPLICATIONS OF SECM

The chapters that follow illustrate a wide range of applications of SECM that have appeared. Given below is an overview and some examples that might help put the technique in perspective before the detailed treatments.

A. Imaging

By scanning the tip in the x-y plane and measuring current changes (the *constant height mode*) (or, less frequently, by maintaining a constant current and measuring the changes in d in a *constant current mode*), one can obtain topographic images of conducting and insulating substrates (Chapter 4). The resolution of such images is governed by the tip radius, a, and d. However, by working in the thin film of water that condenses on a mica surface in humid air, it is possible to obtain higher resolution with a conical tip that is only slightly immersed in the water film. Of particular interest is the use of SECM to perform "chemical imaging," observing differences in reaction rates at different locations on the surface. This mode is useful in studying biological materials (e.g., enzyme sites) (Chapter 11) and surfaces that have active and passive sites.

B. Ultramicroelectrode Shape Characterization

It is frequently difficult to determine the actual shape of an ultramicroelectrode by examination using an optical or scanning electron microscope. For example, the conducting portion may be slightly recessed inside the glass mantle, or the shape may be that of a cone protruding from the insulator. Electrodes with radii of the order of 1 μm or less are particularly difficult to characterize. Simply determining a voltammogram with the tip in bulk solution is usually not useful in this regard, since almost all ultramicroelectrodes will produce a steady-state wave-shaped voltammogram characteristic of roughly hemispherical diffusion. However, by recording an approach curve, i_T versus d, one can frequently identify recessed tips (where i_T does not increase at small d when the insulator hits the substrate) or tips with shapes other than disks, which show different approach behavior (Chapter 5).

C. Heterogeneous Kinetics Measurements

As suggested above, by recording an approach curve or voltammogram with the tip close to a substrate, one can study the rates of electron transfer reactions at electrode surfaces (Chapter 6). Because mass transfer rates at the small tip electrodes are high, measurements of fast reactions without interference of mass transfer are possible. As a rule of thumb, one can measure $k°$ values (cm/s) that are of the order of D/d, where D is the diffusion coefficient (cm^2/s). For example, $k°$ for ferrocene oxidation at a Pt electrode in acetonitrile solution was measured at a 1 μm radius tip at a d of about 0.1 μm yielded a value of 3.7 cm/s (24). The use of small tips and small currents decreases any interference from uncompensated resistance effects.

D. Measurements of Homogeneous Kinetics

Rate constants for homogeneous reactions of tip-generated species as they transit between tip and conducting substrate can be determined from steady-state feedback current or TG/SC experiments or by transient measurements (Chapter 7). Generally rate constants can be measured if the lifetime of the species of interest is of the order of the diffusion time between tip and substrate, $d^2/2D$. Thus first-order reaction rate constants up to about 10^5 s^{-1} and second-order reaction rate constants up to about 10^8 M^{-1} s^{-1} are accessible.

E. Biological Systems

There have been a number of applications of SECM to biological systems (Chapter 11). These include imaging of cells, studies of enzymatic reactions, and oxygen evolution on leaf surfaces. SECM has also been applied in investigations of the transport of species through skin (Chapter 9). Because SECM is capable of monitoring a wide range of chemical species with good specificity and high spatial resolution, it should find wide application in studies of living organisms and isolated tissues and cells.

F. Liquid/Liquid Interfaces

There is considerable interest in ion and electron transfer processes at the interface between two immiscible electrolyte solutions (ITIES), e.g., water and 1,2-dichloroethane. SECM can be used to monitor such processes (Chapter 8). It allows one to separate ion transport from electron transfer and is relatively insensitive to the resistance effects often found with more conventional (four-electrode) electrochemical measurements.

G. Membranes and Thin Films

Different types of films on solid surfaces (e.g., polymers, AgBr) and membranes separating solutions have been examined by SECM (Chapters 6 and 9). SECM is a powerful technique for examining transport through membranes, with the ability to scan the surface to locate positions of different permeability. It has also been used with polymer films, e.g., polyelectrolytes or electronically conductive polymers, to probe the counterion (dopant) flux during redox processes. SECM can be particularly useful in probing film thickness as a film is grown on a surface (25). SECM is unique in its ability to probe inside some thin films and study species and electrochemical processes within the films (26,27). For example, the tip current versus z-displacement curve as a conical tip (30 nm radius, 30 nm height) was moved from a solution of 40 mM $NaClO_4$ into a nominally 2000 Å thick Nafion film containing $Os(bpy)_3^{2+}$ on a glass/ITO substrate (Fig. 6) (26). The tip was held at 0.80 V versus SCE, where $Os(bpy)_3^{2+}$ is oxidized to the 3+ form at a diffusion controlled rate. The different stages of penetration of the tip into the film, from initial contact to tunneling at the ITO can clearly be seen and the film thickness established. Moreover, with the tip at position c, a voltammogram can be recorded (Fig. 7). From such a voltammogram, one can determine the diffusion coefficient of $Os(bpy)_3^{2+}$ and information about the kinetics and thermodynamics of the reaction occurring in the film.

H. Surface Reactions

Measurements of the rates of surface reactions on insulator surfaces, such as dissolution, adsorption, and surface diffusion, are possible (Chapter 12). For example, proton adsorption on an oxide surface can be studied using the tip to reduce proton and induce a pH increase near the surface (22). Then, by following the tip current with time, information about proton desorption kinetics is obtained. Studies of corrosion reactions are also possible. Indeed, work has been reported where a tip-generated species has initiated localized corrosion and then SECM feedback imaging has been used to study it (28). In these types of studies, the tip is used both to perturb a surface and then to follow changes with time.

I. Semiconductor Surfaces

SECM has been used to probe heterogeneous electron transfer reaction kinetics on semiconductor electrodes, such as WSe_2 (29). In these studies, as

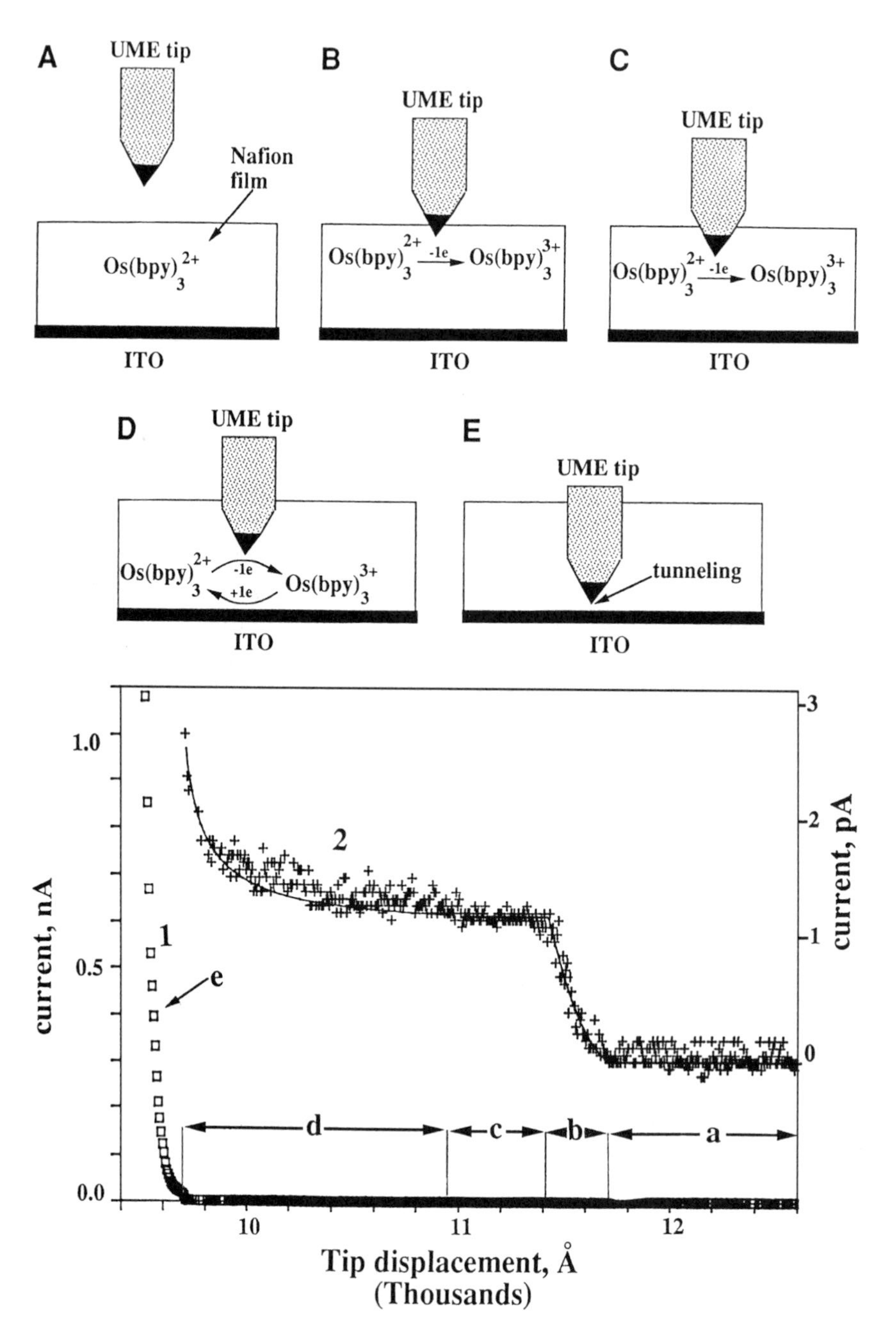
A
UME tip
Nafion film
Os(bpy)3 2+
ITO
B
UME tip
Os(bpy)3 2+ -1e Os(bpy)3 3+
ITO
C
UME tip
Os(bpy)3 2+ -1e Os(bpy)3 3+
ITO
D
UME tip
Os(bpy)3 2+ -1e +1e Os(bpy)3 3+
ITO
E
UME tip
tunneling
ITO
current, nA
current, pA
1
2
e
d
c
b
a
Tip displacement, Å
(Thousands)

in those at the liquid/liquid interface, the use of a separate metal probe electrode is useful in freeing the measured response from resistance effects. It also allows one to examine differences in behavior at different points on a surface. As discussed in Chapter 13 on applications to fabrication, SECM has also been used to etch semiconductor surfaces and study the nature of the etching reactions.

J. Electrochemistry in Small Volumes of Solution

Because of its ability to position an electrode tip with high spatial resolution in three dimensions, SECM can be used to probe electrochemistry in a small volume of liquid (e.g., on a conductive substrate that serves as a counter/reference electrode). For example, a solution volume of 3–20 μL was used to probe the adsorption isotherms on a mineral surface (30). Probing even smaller volumes, e.g., of liquids contained in pores, should be possible. Since electrochemical generation is an ideal method for producing small, controlled amounts of reactants, studies in which one wants to probe chemistry with very limited amounts of sample appear to be a good application. In such studies, means to maintain the sample volume and prevent evaporation, for example, by close control of the humidity or using an overlayer of an immiscible liquid, will be required.

K. Thin Liquid Layers

The SECM has been used to form thin liquid layers and probe electrochemical reactions in them. When the tip is pushed through the interface between

<

FIG. 6 (Top) A scheme representing five stages of the SECM current-distance experiment. (A) The tip is positioned in the solution close to the Nafion coating. (B) The tip has penetrated partially into Nafion and the oxidation of $OS(bpy)_3^{2+}$ occurs. The effective tip surface grows with penetration. (C) The entire tip electrode is in the film but is not close to the ITO substrate. (D) The tip is sufficiently close to the substrate to observe position SECM feedback. (E) The tunneling region. (Bottom) Dependence of the tip current versus distance. The letters a–e correspond to the five stages A–E described above. The displacement values are given with respect to an arbitrary zero point. The current observed during the stages a–d is much smaller than the tunneling current and therefore cannot be seen on the scale of curve 1 (the left-hand scale). Curve 2 is at higher current sensitivity to show the current-distant curve corresponding to stages a–d (the right-hand current scale). The solid line is computed for a conically shaped electrode with a height, h = 30 nm, and a radius, r_0 = 30 nm for zones a–c, and SECM theory for zone d. The tip was biased at 0.80 V vs. SCE, and the substrate at 0.20 V vs. SCE. The tip moved at a rate of 30 Å/s. (From Ref. 26.)

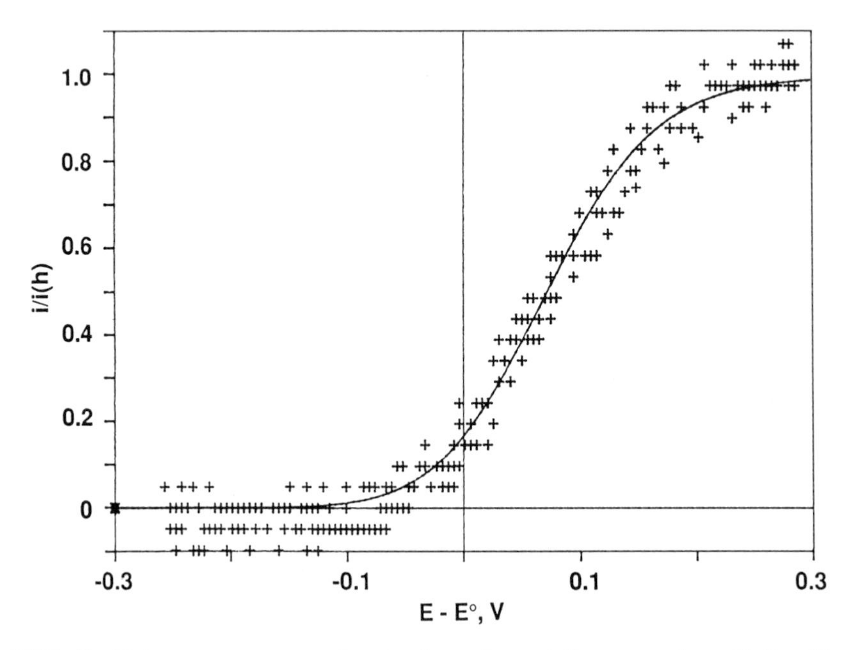

FIG. 7 Voltammogram at a microtip electrode partially penetrating a Nafion film containing 0.57 M $OS(bpy)_3^{2+}$. Scan rate, v = 5 mV/s. The substrate was biased at 0.2 V vs. SCE. The solid line is computed with a heterogeneous rate constant, $k° = 1.6 \times 10^{-4}$ cm/s and $D = 1.2 \times 10^{-9}$ cm^2/s. (From Ref. 26.)

two immiscible liquids, for example, through an aqueous layer above a layer of mercury or a layer of benzene above an aqueous layer, a thin film (several hundred nm to a few μm) of the top liquid layer is trapped on the surface of the tip (31–33). Electrochemical measurements can be used to probe reactions in this layer. Another type of thin layer that has been studied is the one that forms on a surface when exposed to humid air. In this case, a water layer that can be as thin as a few nanometers forms on a hydrophilic surface (e.g., mica). The SECM tip can probe into this layer, although studies have mainly been aimed so far at imaging rather than investigating the properties of the layer (Chapter 4).

L. Potentiometry

While most SECM studies are carried out with amperometric tips that drive faradaic (electron transfer) reactions, it is also possible to use potentiometric tips that produce a potential change in response to concentration changes of species. These are usually typical ion selective electrode tips, although other types, such as Sb tips for pH detection, have been described (34). Probes of this type and their applications are discussed in Chapter 10. They are particularly useful for studies of species that do not show electroactivity, like Na^+, K^+, and Ca^{2+}. Note, however, that ions of this type can be determined in an amperometric mode by the use of micropipet electrodes that respond to the transport of ions across an interface between two immiscible liquids (35,36).

M. Fabrication

A variety of studies have now been done that demonstrate that the SECM can carry out metal deposition, metal and semiconductor etching, polymer formation, and other surface modifications with high resolution. Such processes are discussed in Chapter 13. These SECM approaches have the advantage over analogous STM procedures in that the conditions of deposition or etching are usually known and well defined, based on electrochemical studies at larger electrodes.

REFERENCES

1. A. J. Bard, F.-R. F. Fan, D. T. Pierce, P. R. Unwin, D. O. Wipf, and F. Zhou, *Science 254*:68–74, 1991.
2. A. J. Bard, F.-R. F. Fan, and M. V. Mirkin, in *Electroanalytical Chemistry*, Vol. 18, A. J. Bard, ed., Marcel Dekker, New York, 1994, pp. 243–373.
3. M. Arca, A. J. Bard, B. R. Horrocks, T. C. Richards, and D. A. Treichel, *Analyst 119*:719–726, 1994.
4. M. V. Mirkin, *Mikrochim. Acta 130*:127–153, 1999.
5. A. J. Bard, F.-R. F. Fan, and M. V. Mirkin, in *The Handbook of Surface Imaging and Visualization*, A. T. Hubbard, ed., CRC, Boca Raton, Fl, 1995, pp. 667–679.
6. A. J. Bard, F.-R. F. Fan, and M. V. Mirkin, in *Physical Electrochemistry: Principles, Methods and Applications*, I. Rubinstein, ed., Marcel Dekker, New York, 1995, pp. 209–242.
7. A. J. Bard and L. R. Faulkner, *Electrochemical Methods*, Wiley, New York, 1980, p. 298.
8. C. Amatore, in *Physical Electrochemistry: Principles, Methods and Applications*, I. Rubinstein, ed., Marcel Dekker, New York, 1995, pp. 131–208.
9. A. J. Bard, J. A. Crayston, G. P. Kittlesen, T. V. Shea, and M. S. Wrighton, *Anal. Chem. 58*:2321, 1986.
10. A. T. Hubbard and F. C. Anson, in *Electroanalytical Chemistry*, Vol. 4, A. J. Bard, ed., Marcel Dekker, New York, 1970, pp. 129–214.
11. R. C. Engstrom, M. Weber, D. J. Wunder, R. Burgess, and S. Winquist, *Anal. Chem. 58*:844, 1986.
12. R. C. Engstrom, T. Meaney, R. Tople, and R. M. Wightman, *Anal. Chem. 59*: 2005, 1987.

13. R. C. Engstrom, R. M. Wightman, and E. W. Kristensen, *Anal. Chem. 60*:652, 1988.
14. A. J. Bard, F.-R. F. Fan, J. Kwak, and O. Lev, *Anal. Chem. 61*:132, 1989.
15. J. Kwak and A. J. Bard, *Anal. Chem. 61*:1221, 1989.
16. J. Kwak and A. J. Bard, *Anal. Chem. 61*:1794, 1989.
17. R. M. Wightman and D. O. Wipf, in *Electroanalytical Chemistry*, Vol. 15, A. J. Bard, ed., Marcel Dekker, New York, 1989, pp. 267–353.
18. M. I. Montenegro, M. A. Queirós, and J. L. Daschbach, eds., *Microelectrodes: Theory and Applications*, Kluwer Academic Publishers, Dordrecht, 1991.
19. J. Heinze, *Angew. Chem. Int. Ed. 32*:1268–1288, 1993.
20. R. J. Forster, *Chem. Soc. Rev.*, 289–297, 1994.
21. C. G. Zoski, in *Modern Techniques in Electroanalysis*, P. Vanysek, ed., Wiley-Interscience, New York, 1996, pp. 241–312.
22. P. R. Unwin and A. J. Bard, *J. Phys. Chem. 96*:5035, 1992.
23. A. J. Bard, G. Denuault, R. A. Friesner, B. C. Dornblaser, L. S. Tuckerman, *Anal. Chem. 63*:1282, 1991.
24. M. V. Mirkin, T. C. Richards, and A. J. Bard, *J. Phys. Chem. 97*:7672, 1993.
25. C. Wei and A. J. Bard, *J. Electrochem. Soc. 142*:2523, 1995.
26. M. V. Mirkin, F.-R. F. Fan, and A. J. Bard, *Science 257*:364, 1992.
27. M. Pyo and A. J. Bard, *Electrochim. Acta 42*:3077, 1997.
28. D. Wipf, *Colloids Surfaces A: Physicochem. Eng. Aspects 93*:251, 1994.
29. B. R. Horrocks, M. V. Mirkin, and A. J. Bard, *J. Phys. Chem. 98*:9106, 1994.
30. P. R. Unwin and A. J. Bard, *Anal. Chem. 64*:113, 1992.
31. M. V. Mirkin and A. J. Bard, *J. Electrochem. Soc. 139*:3535, 1992.
32. C. Wei, A. J. Bard, and M. V. Mirkin, *J. Phys. Chem. 99*:10633, 1995.
33. M. Tsionsky, A. J. Bard, and M. V. Mirkin, *J. Phys. Chem. 100*:17881, 1996.
34. B. R. Horrocks, M. V. Mirkin, D. T. Pierce, A. J. Bard, G. Nagy, and K. Toth, *Anal. Chem. 65*:1213, 1993.
35. T. Solomon and A. J. Bard, *Anal. Chem. 67*:2787, 1995.
36. Y. Shao and M. V. Mirkin, *Anal. Chem. 70*:3155, 1998.

2

INSTRUMENTATION

David O. Wipf

Mississippi State University
Mississippi State, Mississippi

I. INTRODUCTION

A scanning electrochemical microscope is a scanning probe microscope (SPM). The scanning electrochemical microscopy (SECM) instrument necessarily resembles other SPM instruments, but differences in the probe tip and signal lead to differences in design and capabilities. Differences also arise from the larger amount of research and engineering development in the commercially successful atomic force and scanning tunneling microscopes. The majority of the SECM instruments in use today are custom-built by the investigator. Although a commercial instrument dedicated to SECM has appeared on the market, both the commercial and "home-made" SECM instruments are less highly engineered than their other SPM cousins. Thus, there is still much opportunity for individual investigators to appreciate the design of SECM instruments and to make significant progress in SECM development.

This chapter discusses the components of the SECM instrument. Beginning with an overview of the major components, the discussion considers different choices in instrument construction and their effect on performance. The design of a commercial instrument is discussed in light of the range of choices presented in the overview. Several instrumental approaches to the important problem of "constant-current" imaging are discussed and evaluated. Further improvement in SECM instrumentation will likely involve use of ever-smaller tips in order to improve image resolution, and some practical problems related to the use of small imaging tips are discussed. Finally, some designs for construction of useful auxiliary equipment for SECM are presented.

II. OVERVIEW OF THE SECM APPARATUS

The illustration of an SECM instrument shown in Figure 1 outlines the discussion in this section. An important aspect of the SECM is the positioning system, which includes the positioning elements, translator stages, and motor controllers. Equally important is the data acquisition system, which begins with use of a potentiostat or electrometer to amplify the probe signal. After amplification, the signal is digitized with an analog-to-digital converter (ADC) and stored on a computer. Computer software is required to control the positioning and data acquisition system as well as to display and analyze the SECM data. Other important parts of an SECM are a probe mount system, video microscope, and vibration isolation.

A. Positioners and Translators

Accurate and reproducible positioning of the probe in three dimensions is an important design element in SECM. An SECM will typically allow movement in three orthogonal directions: x, y, and z. Ideally, the positioning elements for the SECM will allow a probe to move at desired scan rate (e.g., μm/s) over a given range. In addition, the positioner axes of motion are

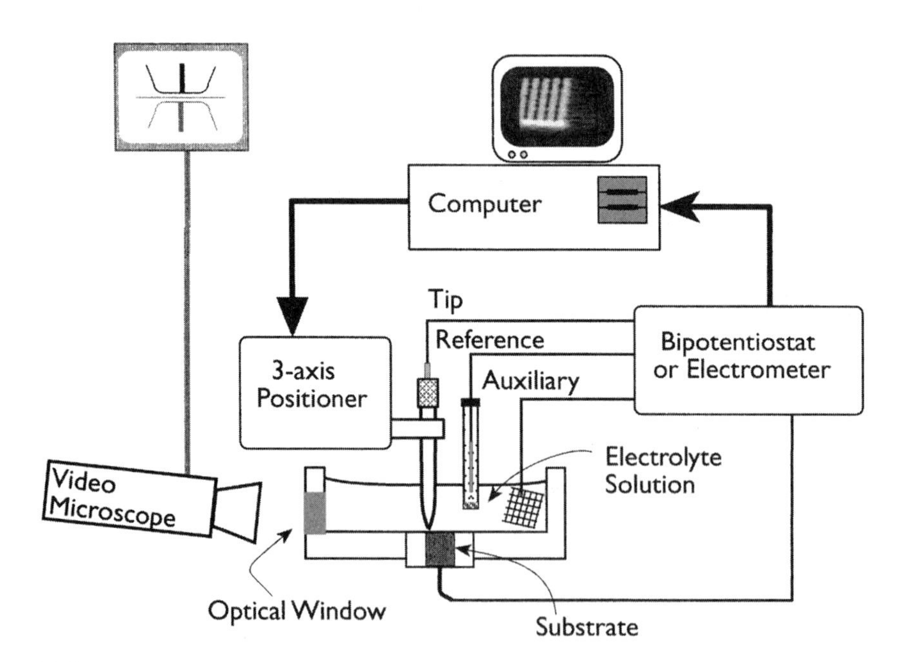

FIG. 1 An illustration of the SECM instrument.

ideally decoupled, and movement of one axis will not produce movement in the other axes. In practice, the positioning system used will only achieve these goals over a limited range of movement and scan speeds.

The smallest practical scan range is set by the SECM tip size. Since only image regions of tip size or larger will contain unique surface information, at least one image dimension should be significantly larger than the tip diameter. Thus, a minimum scan range is about 2 d (d = tip electrode diameter). A maximum scan range is set by physical limitations of the positioning device and, perhaps, by time or computer memory limitations. The maximum scan size is also limited by the maximum scan rate of the positioner, which again will depend on physical limitations of the positioner. The stability of the sample, tip, and solution as well as the patience of the operator will set the minimum scan rate. One hour is required to complete a 100 $d \times 100$ d image at a scan rate of 8.3 d/s, assuming that data points are collected at intervals of one-third d to avoid aliasing artifacts. Positioning accuracy and precision should also be considered. For most imaging experiments, a lateral position error of 0.1 d or less is sufficient. Vertical accuracy is more important than lateral accuracy, and errors of less than 0.01 d are desirable.

Most SECM experiments use tips with 1–25 μm diameters. Thus, a positioner for these probes should be able to scan regions of 100–1000 μm square at scan rates up to 50 μm/s. For these conditions, motorized positioners are suitable. With smaller probes, the greater accuracy of piezoelectric tube or tripod positioners of the type found in STM and AFM instruments is required.

1. Motorized Positioners

Many SECM designs employ Burleigh Instruments "Inchworm" motors to provide the large lateral scan range required for 1 μm and larger tips (1–3). The Inchworm positioner is a linear motor in which three piezoelectric elements act to move a central shaft (4). The two end elements are alternately clamped and disengaged from the shaft, while a center element expands and contracts. At the start of a movement cycle, the center element is fully contracted, the right element is clamped to the shaft, and the left element is unclamped. The center element is expanded by application of a staircase voltage ramp to propel the shaft to the right. When the center element is fully extended, the left element is clamped and the right element is unclamped. The center element is now contracted by a staircase ramp of opposite slope. At full contraction, the cycle repeats. Because of the staircase ramp, the Inchworm moves in discrete steps of about 4 nm. The center element can expand by 2 μm and so the clamping occurs at 2 μm intervals. The principal advantage of the Inchworm positioner is the lack of traditional magnetic motors, which eliminates rotating parts and gears. Other advantages include backlash elimination on reversal of motion and no power dissipation when the motor is stationary. Once stopped, the Inchworm shaft is firmly clamped so no creep or vibration is transmitted to the probe. The range of motion is 2.5 cm or more and is limited only by the shaft length, thus the Inchworm can act as both a "coarse" and "fine" positioning element. Disadvantages of the Inchworm are the higher cost due to a proprietary supplier and the specialized motor controllers. The Inchworm motors and controller also require some maintenance. The author's experience is that actively used motors will require factory service at 1- to 2-year intervals. A spare motor is recommended. Another quirk of the Inchworm motor is that clamping (called a click in some reports) produces two objectionable effects. The probe does not move smoothly at the clamp position, which produces an artifact in the probe signal (3). This is especially noticeable in current-distance curves. In addition, the end of the probe may move as much as ± 1 μm in a direction perpendicular to the Inchworm shaft at the clamp. Thus, directly mounting the probe to the Inchworm motor may cause problems and even tip crashes with the smaller probes. A factory modification is available to minimize this motion.

Stepper-motor positioners can be less expensive (but not always) than the Inchworm motors and are available in almost an infinite variety of choices. Nearly every optics supplier (such as Newport, Melles Griot, etc.) supplies positioning systems that are suitably precise for use with SECM. Several groups have used stepper motors successfully for SECM positioning (5–8). Stepper motor systems rely on gears to transfer rotary to linear motion. With any gear system, backlash limits the positioning accuracy, particularly for imaging applications, which requires many scan reversals. The motors consume power while stationary and a concern is the possibility of electrical interference at the tip.

2. *Continuous Positioners*

Continuous positioners are capable of high-resolution continuous motion. Most positioners of these types are based on lead zirconium titanate (PZT) piezoelectric ceramic (4). Two common configurations for piezoelectric SPM positioners are tripods or tube scanners. Tripod scanners are simply three orthogonal "sticks" of piezoelectric material. The probe is attached to the vertex of the tripod (9). Application of voltage to any one of the sticks moves the probe in that axis direction. It also moves the probe a small, but predictable, amount in the other two axes. Tube scanners use a piezoceramic tube. The tube has five electrodes, one coating the entire inner wall and four 90° segments on the outer wall (10). The probe is attached parallel to the tube axis. The tube bends by applying a voltage to opposing electrode segments: each of the opposing two electrodes are polarized either positively or negatively with respect to the inner electrode. As the tube bends, the probe tip is moved laterally. Vertical motion is produced by polarizing all outer segments simultaneously. Greater vertical motion is often achieved by using a composite scanner with separate piezoelectric elements for vertical and lateral motion (11,12). A disadvantage of the tube scanner is that lateral motion also generates some vertical movement. The vertical component of the motion is a complicated function of the lateral motion, but it is roughly arc shaped. The vertical motion can be minimized with careful design but is unavoidable. Because of their lower cost and greater vibration immunity, tube scanners are often chosen over tripod scanner for SPM positioners.

The distance resolution of both the tube and tripod scanners is limited by the precision of the driving voltage. A typical piezoelectric positioner may move 10 nm per applied volt; thus, subangstrom motions are feasible. In contrast, micrometer-scale movements will require hundreds of volts. This means that the practical maximum lateral movements are about 100 μm. Vertical movements of tube scanners are constrained to a few micrometers. The limited movement range is a handicap for using tips larger than a few micrometers in SECM because current-distance curves cannot be collected and the scan range is too small. Note that the scan range can be increased by use of levers to provide a mechanical advantage but resolution is proportionately degraded due to increased position noise.

Tube and tripod scanners have several nonideal properties. The behavior of piezoelectric scanners is well covered in the literature (4,13,14), and only a brief outline of these problems are given here. Hysteresis, linearity, and "creep" are the most critical. Hysteresis causes the piezo extension versus voltage curve to differ depending on whether the voltage is increased or decreased. In SPM applications, nonlinearity and hysteresis are compensated by calibration with a well-characterized sample (such as a diffraction grating) and by applying corrections during scanning or during data analysis (15,16). Because of hysteresis, data are only calibrated when acquired in the calibrated direction and scan rate. Because vertical movements cannot be confined to single directions and scan rates, they are uncalibrated and data interpretation is complicated. Creep is the slow relaxation to a steady-state extension following a rapid voltage change. Creep is apparent when a large change in driving voltage occurs. For example, when zooming from a large scan area to a small feature, creep causes a slow drift of the area scanned, and several minutes may be required to equilibrate the scanners.

Several SPM instrument manufacturers have corrected scanner nonideality by incorporating sensing devices in the piezoelectric elements. This modification allows real-time feedback control of the extension and eliminates many of the above problems. A number of different sensor types are available. Strain-gauge sensors indicate extensions by a resistance change, capacitance sensors detect the capacitance change between a fixed and moving element, and optical sensors use interferometry to detect a position change. The devices become desirable when large scan areas are used and when high linearity is required (as in metrology applications). Because the sensors ultimately limit the resolution of the piezoelectric positioner and can add noise, they are not often used in atomic resolution images in SPM. However, in SECM applications the larger distances usually used are not affected by the decreased resolution.

Micrometers are also useful positioning elements. Although manually operated micrometers are not suitable for most imaging applications, they provide an inexpensive way to acquire current-distance curves and perform substrate modifications (17,18). Submicron positioning is available with differential micrometers. Motorized micrometers are widely available, and micrometers with DC servo motor drives in closed-loop operation (see below) can give excellent results.

3. *Motor Controllers*

A motor controller contains the necessary electronics to drive a motor and provides a simplified electrical interface between the operator and the positioning elements. Piezoelectric and motorized positioners require large power levels and complicated drive signals that cannot be readily accommodated by the computer running the experiment. Motor controllers provide all the necessary electronics for driving the positioning elements, and the computer may only be required to supply signals to start and stop the positioning.

Controllers provide "closed-loop" or "open-loop" positioning. An open-loop controller does not verify that a movement has occurred at the speed or distance desired. Typically, a calibration scheme is used to determine the amount of movement produced by the positioner. As an example, four TTL signals are used to control each axis with the Burleigh Inchworm open-loop controller: axis select, forward/reverse, halt/run, and clock. To move a motor, the computer generates a set of TTL levels to select the axis and direction (i.e., z forward) and then provides a set of pulses to the clock input. The number and frequency of the pulses determines the amount of movement. Each axis must be individually calibrated to determine the average amount of movement per pulse as a function of both scan direction and speed. Calibration is accomplished either by imaging a well-characterized feature or by generating enough pulses to allow measurement of the movement with a caliper. Position is calculated with a computer program using the number of clock pulses and the calibration curve. More sophisticated open-loop controllers incorporate microprocessors to maintain position and calibration information. These controllers might accept higher level commands such as "move 100 micrometers" or "return to origin," but they still rely on a calibration procedure.

Closed-loop controllers use a feedback process that relies on information about the actual probe motion. As discussed above, some piezoelectric positioners use strain gauges for determining the actual extension of the piezoelectric element. Motorized positioners use other methods. A common approach is to incorporate a linear distance encoder in the translation stage. Usually the encoder consists of a pair of ruled patterns printed on a transparent strip. With the transparent strip attached to positioner, a pattern of light and dark is generated as the ruled pattern moves past a stationary photodetector. A microcomputer counts the number of transitions and generates an absolute movement. By arranging the patterns so that the marks of the paired patterns are 90° out of phase with each other (i.e., "in quadrature"), motion direction is also determined. The resolution of the encoder is determined by the precision in positioning the marks on the strip. Sophisticated closed-loop controllers used in micropositioning applications also use the phase differences of the photodetector pairs to increase the encoder resolution. Since the encoder "knows" the absolute position of the translation stage, any error between the programmed movement and the actual movement can be corrected by the controller.

Rotating motors, such as a stepper motor, use a similar encoding scheme but use rules marked on the rotating shaft or pairs of slotted disks (most computer mice use slotted disks). There are other methods of monitoring movement, such as optical interferometry, but all rely on instantly correcting the difference between the programmed and measured movements. The microprocessor in closed-loop controllers is able to understand high-level commands, thus, closed-loop controllers are usually much simpler to use for precision positioning. Some drawbacks to closed-loop controllers include expense, which may double or triple the cost of the open-loop system. In addition, the resolution of the controller is limited by the encoding method. For example, the Burleigh Inchworm system is available with resolutions from 0.02 to 0.1 μm, with cost greatly increased at higher resolution. Accuracy in closed-loop systems is related to encoder resolution with accuracy being at least 10–100 times worse, that is, an encoder with resolution of 0.05 μm may only have a guaranteed accuracy of ± 1 μm. Finally, the closed-loop system is a feedback system, and, like all feedback systems, it may become unstable and cause the positioner to oscillate or overshoot the desired position. While not particularly likely with higher quality controllers, imaging difficulty or tip crashes may result.

Controllers for piezoelectric tube or tripod scanners require high-stability, low-noise voltage amplifiers. Since piezoelectric materials have a large electrical capacitance, they also require amplifiers that produce the high power required for wide-bandwidth control of reactive loads. Closed-loop controllers mate an amplifier with a feedback loop and accept input from strain gauge or capacitive positioning sensors. For simple low-frequency open-loop positioning, commercial amplifiers may be more costly than necessary. An amplifier can be made using high-voltage operational amplifiers from companies such as Apex or Burr-Brown. Including the cost of a high-voltage power supply, a home-built amplifier can cost less than one tenth of a commercial unit. Apex, in particular, has helpful product literature on designing amplifiers for piezoelectric actuators (19).

B. Translation Stages

Most of the positioners mentioned above are coupled to a translation stage. Stages are primarily used to constrain motion from a positioner to a single axis and to provide a mount for the probe or sample. In considering a translation stage, the following sources of positioning errors must be considered. Runout error refers to the amount of "off-axis" linear motion; for example, an x-axis translator has out-of-plane motion (straightness) in the z-axis and in-plane motion (flatness) in the y-axis. Tilt and wobble are angular measures of the off-axis motion, and these are specified by the three orthogonal components: roll, pitch, and yaw. Abbe error is a consequence of tilt and wobble. It is produced by amplifying any translator angular error by the distance between the plane of travel and point of measurement. For example, the angular error in the stages is greatly amplified at a probe mounted several tens of centimeters above the x and y translation stages. Another significant error results from "play," which is a consequence of looseness of badly made or worn translator parts. Backlash is often the result of play in the translator. Friction and stiction (static friction) are also important since friction varies with motion velocity, leading to velocity-dependent errors. Stiction limits the translators ability to make small, incremental motions.

Several translation stage designs are available. Crossed-roller bearing stages provide high-quality, long-range positioning (at a high cost) and have low friction and stiction qualities. In comparison to the less expensive ball-bearing stages, crossed-roller stages have lower angular errors (100 μrad compared to 150 μrad) and are less susceptible to contamination. For smaller positioning ranges, flexure stages are ideal. The stage is attached to stiff flat springs, and motion is produced by elastic deformation of the spring. Since there is no friction or stiction, very small and rapid motions are possible. In addition, a mechanical advantage can be built into the stage, amplifying the positioner motion. Note that the two axes do not move independently in a flexure stage, and coupled motion between axes becomes objectionable at large excursions. For this reason, flexure stages are not suitable for movements much greater than 2000 μm.

C. Probe Mounting

Although many types of probes are used in SECM experiments, as a rule, vibration is minimized by making probes and probe mounts as small and rigid as possible. Careful design of the probe and mount is necessary for SPM experiments at nanometer resolution (14,20). Since most SECM experiments use probes with diameters of several micrometers, larger vibrations can be tolerated and probe mounting becomes less critical. The two mounts shown in Figure 2 are suitable for larger SECM tips. Figure 2A shows a probe holder composed of a section of aluminum angle bracket threaded to accept a Kel-F insert. Two attachment holes in the bracket permit the mount to be bolted to the vertical translation element. The Kel-F insert is a threaded section of rod that screws into the bracket. The insert contains a central hole that accepts the SECM probe and an intersecting horizontal hole tapped to accept a nylon screw, which is a clamp for the probe. Aside from its excellent chemical stability, the Kel-F insert is used to electrically isolate the probe and the translation stages. A thin electrolyte film on the probe body can easily wreak havoc if allowed to contact a nonisolated con-

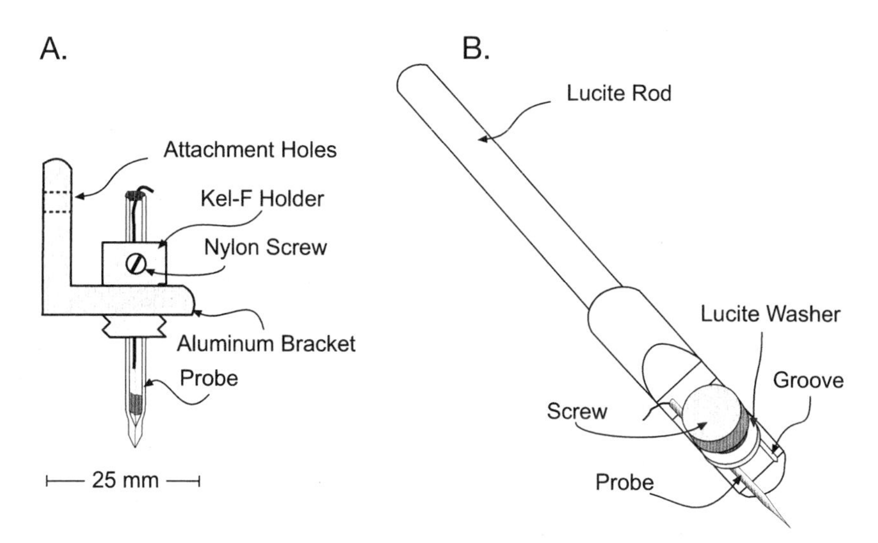

FIG. 2 Examples of two types of tip mounts: (A) home-built; (B) commercial micropipet holder.

ductive element. A nylon screw is preferred over a metal screw as the clamping element since it is more difficult to break an electrode if the clamp is overtightened. The clamp also allows the probe to slip up in case of a tip crash.

Commercially available probe holders are also useful. These mounts are sold as holders for micro-pipets and are available from electrophysiology suppliers. A pipet holder is illustrated in Figure 2B. The probe is laid in a groove and held by a screw and washer. This mount is not as rigid as the previous example, but it is convenient for use with fragile electrodes. For example, the holder can be used for probe preparation steps, such as polishing or etching, without removing the probe from the holder. This can minimize electrode breakage caused by manual handling.

The two mounts described above are too large for probes with nanometer-sized tips. In addition to vibration problems, the mounts would not fit onto piezoelectric tube or tripod scanners. Small SECM probes are very similar to electrochemical STM probes. These probes are often mounted by plugging the probe into a small electrical pin-socket glued to the tube scanner. The socket thus serves as a mechanical and electrical contact to the probe.

D. Vibration Isolation

Vibration increases noise and decreases lateral and vertical resolution. The most effective vibration isolation takes place in the instrument design. A goal is to keep vibrations resonances out of the band of frequencies used for data collection and feedback control, i.e., about 1–1000 Hz. Minimization of size and maximization of rigidity shifts vibration resonances to high frequencies (14,20). In addition, locating the SECM away from vibration sources such as air vents, pumps, or heavy machinery is advised. Note that bypassers are a source of vibrational and electrical interference and locations with foot traffic are inadvisable.

Additional vibration isolation may not be required for SECM. For SECM imaging with probes of 10 μm and larger, isolation is not critical. For smaller tips, some attempt should be made to determine if vibration isolation is needed. One simple check is to observe an increase in noise as the probe approaches the surface in a feedback experiment. Another check is to determine the frequency distribution of vibrations. Use a lock-in amplifier to examine frequencies from 10 mHz to 10 kHz with the probe positioned as close as possible to an electrode surface in a positive feedback configuration. Alternatively, perform a Fourier transform of the probe noise signal. Any peaks in the signal versus frequency plot are candidates for vibrational resonances. Discriminating vibrational and electrical noise is possible by moving the probe slightly away from the electrode surface. Vibrational peaks will decrease at larger probe-substrate separations.

Isolating the apparatus from the external environment is an effective means of reducing vibration, since energy from the environment is not available to excite mechanical resonances. For vibrations transmitted through air by acoustic or convective motion, the instrument is isolated by placing it in a closed box lined with acoustical foam or lead sheets. Energy from the building floor is transmitted to the instrument through the support. Transmission of frequencies larger than 10 Hz is effectively minimized by using pneumatic or spring supports. Elastic bands ("bungees") are often used as an inexpensive spring device. The instrument is placed on a metal plate supported by three or four bungee lines hanging from the ceiling or other support. About 10 kg of mass is added to the plate to extend the bungees to about one half of their maximum extension. Adjust the load and bungee length until the natural oscillation frequency of the suspended plate is about 1 Hz.

Air-tables are also effective isolation devices. These are available commercially and minimally consist of a rigid top supported by pneumatic cushions in the legs. The exceptional rigidity found in optical tables is unnecessary and less expensive workstation-type tables can be used. A simple homemade table often works as well as a commercial table. This consists of a motorcycle or bicycle inner tube supporting a 1 cm thick aluminum plate. The plate should be large enough to completely cover the inner tube with additional space on the edges for wooden support blocks, which support the plate during experiment set-up. Add mass for balance and adjust the tube pressure to give a 1–10 Hz resonant frequency.

High-frequency resonances can be excited by the movements of the translators. These can be damped by using rubber pads or foam under the instruments. Natural rubber or Sorbothane are often recommended for this purpose.

E. Signal Transduction and Amplification

The probe in SECM produces a signal that must be transduced and amplified prior to recording. At a voltammetric tip, electrolysis of either a mediator or a substrate-produced substance produces a faradaic current signal. At a potentiometric tip, the activity of a solution phase species generates a voltage signal.

The requirements for accurate recording of these two types of signals are quite different and lead to different equipment requirements. A sensitive potentiostat is required for a voltammetric probe and an electrometer is needed for a potentiometric probe.

1. Potentiostat and Current Transducer

A potentiostat is used for most voltammetric measurements. The potentiostat uses an electrical feedback loop to control the potential of the working electrode (the electrode at which the reaction of interest occurs) with respect to a reference electrode, even in the presence of ohmic drop. A third electrode, the auxiliary, is used to supply the current flowing at the working electrode. The use of a third electrode eliminates current flow through the reference electrode, permitting smaller reference electrodes and more accurate potential control. A simplified diagram of a potentiostatic circuit is given in Figure 3 (21). Most modern electrochemical equipment employ versatile high-input impedance differential amplifiers known as operational amplifiers (OAs) as circuit elements. In the potentiostat circuit, OA1 provides the auxiliary electrode with the required voltage and current to maintain the desired potential difference between the reference and working electrodes. OA2 is a buffer amplifier (voltage follower) that prevents significant current draw through the reference electrode and outputs a low-impedance measure of the reference electrode potential. In operation, the amplifier OA1 adjusts the current and voltage at the auxiliary electrode to whatever is necessary to minimize the voltage difference at the inputs. The negative feedback loop that exists between the output and input of OA1 includes OA2, the reference and auxiliary electrode, and the solution resistance between reference and auxiliary. Thus, the voltage at the auxiliary electrode will be the inverted sum of the reference electrode voltage (E_R), any externally applied voltage (E_W), and the ohmic voltage arising from current flow through the solution resistance. This means that, if the reference electrode is close to the working electrode, the effect of the ohmic drop is minimized and the potential difference between the working electrode and reference electrode is maintained or "clamped" close to E_W. Clamping the voltage at E_W is the basis for the

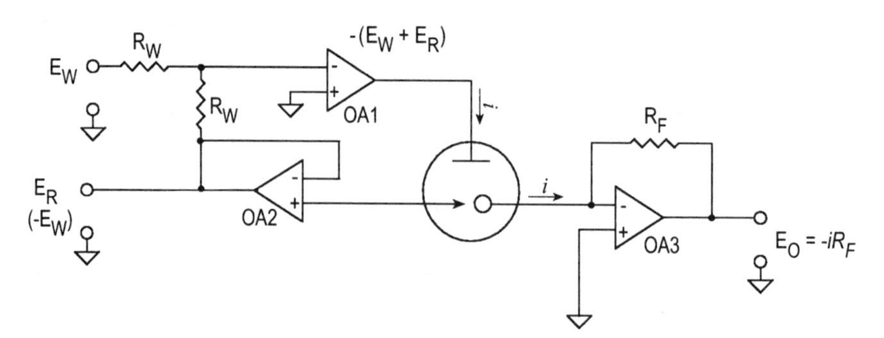

FIG. 3 Electrical schematic of a potentiostat circuit.

name "voltage clamp"; a term used outside the electrochemical literature as a more descriptive name for the potentiostat circuit.

The working electrode is maintained at the circuit common potential (which may not be earth ground) in the potentiostat circuit of Figure 3. The current-transducer amplifier, OA3, maintains the electrode potential at common potential and provides a voltage output proportional to the current input. As before, the amplifier uses negative feedback to minimize the potential difference between the inputs by adjusting the output. In the process, the inverting input (−) is set to be virtually the same as the potential of the noninverting input (+), and thus the inverting input is at "virtual ground." Note that the output voltage across the feedback resistor, R_F, produces a current to oppose the current flowing into the input and changing R_F changes the signal gain. An alternate method of maintaining the working electrode at the common potential is to connect a small-valued measuring resistor between the working electrode and common. By measuring the potential drop that develops across the resistor, the current flow can be calculated from Ohm's law.

Although the circuit in Figure 3 could be used in voltammetric experiments, most potentiostat circuits are considerably more sophisticated. Additional circuitry is added to provide damage protection from excessive signal inputs or outputs, additional current or voltage range (i.e., compliance) at the auxiliary electrode, and for variable gain and filtering of the input current. Some circuit designs maintain the reference electrode rather than the working electrode at the circuit common. A floating working electrode complicates the design of the working electrode amplifier slightly but allows a number of devices to share a reference electrode. For example, both a potentiometric and an amperometric working electrode can use the same reference electrode if the amperometric electrode is not set at ground potential. Some "potentiostats" eliminate the auxiliary electrode and apply E_W directly to the reference electrode. OA1 and OA2 are eliminated, but this requires that the total current flow through the cell is sufficiently small to avoid polarizing the reference electrode. Experiments with disk ultramicroelectrodes of 10 μm or smaller diameter can often use these "two-electrode" potentiostats.

2. *Bipotentiostat*

A bipotentiostat is simply a potentiostatic circuit designed to allow simultaneous potential control of two working electrodes in an electrochemical cell. A simplified schematic of a bipotentiostat circuit is illustrated in Figure 4. Note that connections in this schematic are always indicated by a dot at the intersection, and crossing lines without a dot are *not* connected. Part of this circuit is identical to the potentiostat circuit, with the addition of am-

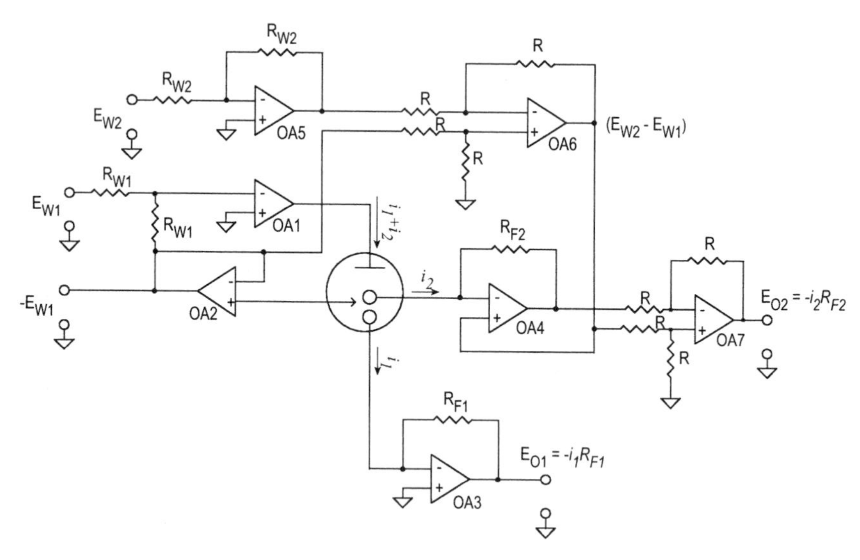

FIG. 4 Electrical schematic of a bi-potentiostat circuit.

plifiers OA4–OA7 to provide for the second working electrode. The bipotentiostat maintains working electrode 1 (WE1) at circuit common and while WE2 is lifted off circuit common by an amount equal to the potential difference between the two working electrodes (i.e., $E_{W2} - E_{W1}$). OA6 is the differential amplifier that generates this potential difference, which, when applied to the noninverting input of current-transducer amplifier OA4, shifts the potential of WE2 to a value equal to $E_{W2} - E_{W1}$. Current at WE2 is transduced by amplifier OA4. Since the voltage output of OA4 is not referenced to circuit common, differential amplifier OA7 is used to restore the output to a circuit common reference. In the bipotentiostat, the auxiliary electrode supplies the total current required at WE1 and WE2; however, the potential difference between WE1 and the reference electrode is always maintained at E_{W1} by OA2 and OA1. If desired, additional working electrodes can be added to this circuit by duplicating the circuits of OA4–OA7 as many times as necessary.

As an alternative to a bipotentiostat, two separate potentiostats can be used to control two working electrodes. For conventional potentiostats, both working electrodes would be held at circuit common, which requires that one of the potentiostats be operated in a floating mode. With potentiostats that allow a floating working electrode, the circuit common can be shared at the reference electrode and no isolation is required. Whether to use a single bipotentiostat or dual potentiostats in a SECM experiment is up to the investigator to decide. The bipotentiostat often allows a simpler experimental setup; however, there are few commercial bipotentiostats, which can limit an experiment if a bipotentiostat does not supply a desired feature.

Although bipotentiostats are recommended for SECM, dual potentiostats may be preferred in some situations. An example arises when using SECM for corrosion studies. In this case, a large corrosion sample requires a stable potentiostat with high power output to passivate the sample or to run potentiodynamic curves. In this case, the experimenter may feel justified in using the same potentiostat for SECM studies as is used in other corrosion experiments. However, the probe potentiostat may need to measure currents in the pA range. Bipotentiostats are often designed with the assumption of similar signal levels at both working electrodes. Thus, it may be difficult to either build or buy a bipotentiostat with the required power-handling ability for the corrosion sample and the high sensitivity needed at the probe. In this case, using two potentiostats is preferable.

A bipotentiostat must be stable to accurately control both working electrodes. Although the feedback loop (see OA1 and OA2 in Fig. 4) maintains the potential of the working electrodes, experimental artifacts can be introduced. These arise from the fact that the loop does not instantaneously respond to changes in current at the working electrode. In particular, potential steps at either working electrode will introduce spurious current transients at the other electrode due to the finite response time of the feedback loop. The transients not only obscure the current signal but also indicate a brief loss of potential control. For example, these spikes can limit the ability of the SECM to perform diffusional transit time experiments at very short time scales. Using dual potentiostats is not necessarily a solution to this problem, since large currents flowing from either of the two auxiliary electrodes to closely spaced working electrodes will cause a similar loss of potential control. A second consideration in using a bipotentiostat is that the potential at WE2 is less stable than the potential at WE1 since the potential of WE2 is always referenced to the potential of WE1 rather than to circuit common. The measured potential of WE1 cannot be more stable than circuit common, thus potential control of WE2 will be compromised, leading to increased current noise at WE2.

3. S/N and Dynamic Range Requirements

The ability of the current transducer to amplify low-current signals must be considered when choosing a potentiostat for SECM experiments. As a tip is miniaturized to nanometer dimensions, the accurate and stable measurement of SECM probe current becomes challenging. A 10 μm diameter disk electrode produces about 2 nA of steady-state limiting current in a 1 mM solution of mediator, but a 10 nm disk electrode would produce only 2 pA. In the feedback mode, low tip currents can be partially compensated by using higher mediator concentrations (although high mediator concentrations may cause other problems). In addition, tip current will not generally change by more than a factor of 10 from the baseline current (i.e., $i_{T,\infty}$). However, imaging in generation-collection mode does not allow a direct choice of the mediator concentration, and the mediator concentration may vary 100-fold during an image acquisition. Thus, using submicrometer tip electrode requires accurately measuring currents in the femtoampere range.

Many modern commercial potentiostats have built-in or optional ability to measure signals in the pA range. When choosing or designing a potentiostat for use with micrometer or smaller electrodes, ensure that the current transducer has a minimum of 10 pA/V gain ratio and that the resolution is at least 10 fA. Older potentiostats, designed for larger electrodes, are often too noisy to support measurements of nA or smaller currents. In such cases, a current preamplifier can be added that will allow low current measurements. A design for a simple preamplifier is in the literature (22). In addition to conventional electrochemical suppliers, one should consider suppliers of electrophysiology equipment. The amplifiers used for micropipet techniques can make excellent low-noise, high-bandwidth potentiostats for currents in the fA range (23).

4. *Electrometer*

Electrically, potentiometric electrodes are voltage sources with a high source resistance, R_s. The voltage produced is typically a function of the logarithm of the activity of the desired ion. A pH electrode, for example, may produce a voltage change of 59 mV per decade change in the activity of hydrogen ion. For 1% accuracy in activity measurement, a measurement accuracy of 0.6 mV is required. A difficulty occurs because micrometer-diameter potentiometric electrodes can have source resistances (e.g., a membrane resistance), R_s, in the range of 10^9–10^{11} Ω (24,25). To avoid measurement errors, the input resistance, R_{in}, must be much larger than R_s. In addition, bias or offset currents from the measuring device must not depolarize the electrode. Commercial electrometers are specifically designed to accurately measure high-resistance voltage sources. A modern high-quality electrometer might have R_{in} of 10^{14} Ω and input offset current of 50 fA. Such an instrument would produce less than 0.5 mV of polarization in a probe with an $R_s < 10^{10}$ Ω (i.e., 10^{10} $\Omega/5.0 \times 10^{-14}$ A). In contrast, a typical digital multimeter has a R_{in} of 10^7 ohm.

The high resistance of the potentiometric probe is an effective noise generator. The Johnson noise produced by a resistor at room temperature is 1.3×10^{-10} $\Omega^{-1/2}$V/Hz$^{1/2}$ (26,27). The 10^{10} Ω R_{in} of a micropotentiometric probe will produce 41 μV rms of noise in a 10 Hz measurement bandwidth, reducing accuracy to 0.01% in our theoretical micro-pH sensor. In addition, the high impedance of the potentiometric micro-probe makes it susceptible to external noise sources, and electrical shielding is required to reduce noise to manageable levels (28). Of concern for SECM measurements are voltages induced at the potentiometric tip due to current flow through the solution, especially when the tip is near a substrate surface.

Although somewhat dependent on the method of electrode construction, the R_s of a potentiometric probe depends on the active tip area (24,25). One limiting factor for the ultimate size of the probe used in potentiometric measurements is the electrical characteristics. As R_s approaches 10^{13} Ω, the noise level of the measurement will not allow greater than 1% accuracy, assuming very naively that the only noise source is the resistor Johnson noise. The noise level of small tips suggests that potentiometric tips of 0.05 μm diameter are close to the smallest useful size.

Excellent electrometers are available from commercial sources (e.g., Keithley), and the experimenter is unlikely to improve substantially on these devices in terms of stability and versatility. However, an experimenter can build a suitable device that costs substantially less than a commercial unit. For example, the Analog Devices OA AD515KH has an input impedance greater than 10^{13} Ω and an input offset current of less than 150 fA at a price of less than $15 (29). In a voltage follower configuration using a battery power supply, this OA makes a serviceable low-noise electrometer. However, laying out a circuit to reduce leakage currents to fA levels is a nontrivial problem (30,31).

5. *Floating Measurements*

Making G/C SECM measurements using a potentiometric probe and a potentiostatic substrate electrode is often desirable. In this case, isolation of the electrometer and potentiostat is required to avoid polarizing the potentiometric probe. Home-built and commercial electrometers use isolated power supplies and instrumentation amplifiers to isolate both inputs from the output common line. Unfortunately, the isolation is often defeated when the electrometer and potentiostat outputs are recorded on a computer. One might think that using differential inputs on a PC-based analog-to-digital card (ADC) would be sufficient. However, the differential inputs of the ADC are not truly isolated. The ADC "low" input is often connected to the computer ground through a drain resistor of several kΩs. Thus, the electrometer and potentiostat common are in electrical contact and μA-level currents can flow into the potentiometric electrode. Removing the drain resistor on the ADC board input is usually not recommended by the manufacturer since it leads to poor stability and increased noise and poses a slight safety risk. A simple solution is the incorporation of true galvanic isolation at the electrometer output. The required isolation is available with a simple and inexpensive "one-chip" device described later in this chapter.

F. Software

A SECM instrument requires a significant amount of software to control probe movement and to collect, display, and interpret data. Additional software is required to operate a distance control circuit in a constant current mode or to provide transformation of the data. In some cases, prewritten software is available either as gifts from fellow SPM users, freeware, or from commercial sources. In many cases, especially for control of positioners or data acquisition, software must be developed. In the sections below, general guidelines for software development in SECM are given. Because of the wide range of hardware and software available for SECM instrument development, it is not feasible to illustrate any specific software code. However, the following suggestions should be considered if a SECM software development is undertaken. All probe movements should be monitored on a "real-time" basis to avoid damage to probes or samples if a crash occurs. Data should be displayed graphically and numerically as soon as they are collected. Data should be protected at all times to permit recovery in case of a computer crash or operator error. For example, individual scan lines of a image should be saved to disk as they are collected rather than waiting for the entire image to be collected. Data files should be self-documenting. All experimental information, such as the date, hardware gain, scan speed, etc., should be saved with the data file. If necessary, conversion programs can be used to produce data files for other applications, but the primary data file should remain inviolate.

Software development will likely occur on Windows or Macintosh platforms. It is becoming difficult to find MSDOS-mode support for devices and other operating systems are often not well supported by vendors. Windows and Macintosh are GUI (graphical user interfaces) operating systems, and they provide a familiar interface to users of SECM software, but they have some significant drawbacks. One is "latency"—the inability of the operating system (OS) to guarantee a response to a user or hardware request within a specified interval. Unpredictable delays occur between receipt of a hardware signal and a software-generated response or vice versa. All OSs have latency. Some OSs, known as real-time operating systems (RTOSs), guarantee that the latency will not exceed a certain, small time interval. DOS, Windows, and Macintosh are not RTOSs, and software must account for possible delays. Restrictions in directly accessing the hardware in a modern operating system require that drivers for hardware devices be written to rigid specifications. Thus, because of the difficulty in dealing with latency issues and driver development, most SECM developers will rely on hardware vendors to supply prewritten drivers for hardware control. Drivers for hardware devices can be accessed through traditional programming languages. C, C++, and Basic programming languages are popular in the Windows OS, and all vendors will supply drivers and interface functions for these languages. One should always inquire about latency issues for any time critical hardware devices.

Writing software requires some thought as to the user interface. A clumsy interface can make experiments tedious and may promote operator errors, such as tip crashes. Fortunately, most modern languages have prewritten templates for menus, user input, etc., to make the programmer's job easier. Despite this, most programming in GUI interfaces is made more difficult by the complexity of the operating system and the fact that GUI OSs are "event driven." In writing event-driven programs, the programmer must be prepared for random user input, i.e., mouse clicks, keystrokes, diskette ejections, etc. Of course, the programmer can ignore many events, but at the cost of diminishing the utility of the GUI interface.

Some of the programming difficulty in a GUI OS is relieved by the use of function libraries. These libraries contain routines to implement prebuilt functions, for example, emulation of a strip-chart recorder. Libraries can also include sophisticated function blocks to perform Fourier transforms or digital filtering on data sets. A well-known example is LabView (National Instruments). LabView is actually a sophisticated object-oriented visual programming language with a rich array of functions for data collection and transformation. HP VEE (Hewlett Packard) is a similar product. Programs are constructed in LabView by linking together function blocks to produce the desired effect. Since each function block can react to GUI events, the programmer's job is simplified. A further advantage is that these function libraries are modular and cross-platform. Thus, a data acquisition routine written in LabView on the Macintosh can be ported to Windows simply by changing the driver for the hardware device. (At least in principle. Speed-sensitive routines may need to be "tweaked" and allowances made for different hardware specifications.) More significantly, cross-platform compatibility allows code to run on newer hardware and operating systems, unlike, for example, DOS programs that must be completely rewritten to work on modern GUI systems. A drawback is that an additional software driver must be written specifically for the function library. Thus, only the National Instrument data acquisition and I/O boards are natively supported under LabView. Some hardware vendors supply a "wrapper" library, which provides an interface between LabView and their own hardware drivers.

1. Probe Movement

Probe movements in three dimensions (axes) are required for SECM. Fortunately, all probe movements required in SECM can be made by moving a single axis at a time. Typical movements are a single line scan, a raster, a spiral, or an arbitrary set of movements. All these cases can be handled by preparing a movement list. Each list item contains the axis, direction, speed, and whether data acquisition will occur. Pauses can be handled by insertion of a dummy axis. Processing the list sequentially will produce the desired motion.

Probe movement is closely tied to the data acquisition process. Knowledge of the probe position at the time of data acquisition is vital. The start of data collection and probe movement should be synchronized with a hardware trigger. The alternative, initiating probe movement and data collection with a software command would entail unacceptable latency in the data collection.

Programming probe movements is dependent on whether an open-loop or closed-loop controller is used. In open-loop control, a calibration factor is used to calculate a movement. Once the movement begins, there is no feedback to indicate the actual movement or velocity. Thus, a position is calculated based on the calibration curve. A computer program might send an open-loop positioner a specific number of digital clock pulses at a specific frequency to generate a movement. The Inchworm controller, for example, moves the Inchworm motor about 10 nm for every clock pulse received. However, the actual amount of movement depends on the individual motor, direction of movement, velocity, and load. For example, a movement of 10 μm in the forward direction at 10 μm/s with motor x requires 1010 pulses (10.1 nm/pulse) at a frequency of 1010 Hz, while a movement of 10 μm in the reverse direction at 10 μm/s with motor y requires 980 pulses (9.80 nm/pulse) at a frequency of 980 Hz. Once the calibration factor is known, the required number of pulses and pulse rate can be calculated. For a piezoelectric scanner, the calibration would be used to program the output of a DAC for a given voltage range and voltage step. A good overview of open-loop software control for SECM is given by Wittstock and coworkers (3). A pulse sequence can be generated with a counter/timer board or by using the clock on the ADC/DAC board. A counter is programmed to produce the required number of counts at the desired frequency. Provision should be made for the finite resolution of the clock when calculating the probe movement. Although a 10 μm/s movement is desired, the resolution of the clock may dictate a 9.98 μm/s movement. Also, user intervention must be taken into account. If the user stops a scan in the middle of an acquisition, the counter can be queried to determine the actual number of pulses made, permitting the actual probe position to be calculated. DAC-based movements are similar. Again, the finite resolution of the DAC and its clock should be considered in calculating probe positions and speed.

Closed-loop control is often considerably more simple than open-loop control. The programmer usually must send a command, perhaps over a serial port, telling the motor controller what movement to make. The probe motion and data acquisition must be synchronized. The programmer is advised to not rely on simply computing the probe position based on the requested motion, but instead to acquire the position information directly from the controller. One must beware of latencies between the reported position and the actual position during scanning.

2. *Tip/Substrate Signal Collection*

Data acquisition during SECM will generally involve measuring a voltage in the range of ±10 to 0.001 V, i.e., a dynamic range of 10,000 or 80 dB. Voltage signals arise from the output of a potentiostat current transducer or an electrometer output. In these cases, the analog signal must be digitized. In some cases, the signal might already be digitized, such as when a signal is sent from an instrument over a serial or IEEE-488 (GPIB) bus digital bus. However, most difficulties in data acquisition arise during the digitization step, and this section will focus on digitization of analog voltages.

SECM data is usually acquired through an ADC board. At minimum, the board should have at least four input channels, 12-bit resolution, and 100 kHz aggregate sampling rate. Four channels are preferred in order to be able to simultaneously acquire tip and substrate signals as well as two auxiliary channels. For example, signals for tip and substrate current, vertical position, and error or phase signals are collected in constant-current mode operation. A board with 12-bit resolution has insufficient dynamic range to capture the expected dynamic range of the input data (2^{12} levels or 72 dB). This is partially offset by using on-board amplification to allow mV level inputs. 16-bit boards are recommended for SECM since they easily cover the entire input dynamic range. A high dynamic range also allows a "zoom" into the data to extract small signals from a large baseline. A sampling rate of over 100 kHz is recommended for SECM, which allows good performance during voltammetric experiments and improves SECM data by allowing averaging.

In programming a data acquisition routine, one should attempt to maximize the signal-to-noise ratio (SNR) and to minimize distortion due to clock skew and aliasing. Maximizing SNR requires that the input range of the ADC be matched to the input range. For example, an input signal of between ±1 V presented to a 12-bit ADC with a ±10 V input range will be digitized with only 9 bits of resolution, limiting the ultimate precision to 0.2% of the full scale. Amplifying the input signal 10-fold makes use of all 12 bits of input range, increasing the ultimate precision to 0.02%. Since it is often not possible to predetermine the input signal levels, it is better to use a 16-bit (or better) ADC that can retain precision of 0.01% over a ± 10 V range.

Another strategy to increase SNR and digital resolution is to employ sample averaging. Rather than acquire a single sample at sample rate f, n samples are acquired at sample rate $n \times f$. The samples are then averaged to retain the effective sampling rate f. Averaging has a number of advantages. If a signal contains only white noise, SNR is improved by a factor of $n^{1/2}$. In addition, the effective resolution in the digitized sample is increased by a factor of $0.5 \times \log_2(n)$. Averaging 16 12-bit samples produces 1 sample with an effective 14-bit resolution. A point of diminishing returns occurs with this method since all ADC cards show a loss in their effective number of bits (ENOB) as the sampling rate increases. High sampling frequencies do not allow input circuitry to completely settle and resolution is lost. However, sample rates of greater than 10 kHz are required before the ENOB drops significantly.

Aliasing arises by sampling too infrequently. The Nyquist criterion states that an input signal can be recovered if it is sampled at at least twice the highest frequency of the signal. Sampling slower than the Nyquist rate leads to aliasing, where the undersampled signal appears as an artifact at lower frequencies. Aliasing can be avoided by using a low-pass filter on the input to the ADC and sampling at twice the filter cut-off frequency or higher. Note that the filter will attenuate the signal power at frequencies higher than the filter cutoff, but strong noise sources above the cutoff frequency may produce aliased signals. Failure to prevent aliasing will place the noise present at twice the Nyquist rate and higher into the digitized data samples.

Channels are acquired sequentially in most ADC boards. If four channels are being acquired at 10 Hz each, the difference in acquisition time for each sample is 100 ms because of clock skew. For a 10 μm/s scan in SECM, the clock skew corresponds to a 3 μm skew in the data of the first and last channels. Clock skew can be minimized with multiple acquisition and averaging. In the above example, 8 samples of each of the 4 channels can be acquired at 80 Hz and then averaged 8-to-1. Since the four channels are acquired sequentially, averaging interleaves the acquisition times of the four samples. Clock skew is minimized in proportion to the number of averages and skew between the first and last samples is reduced to 0.375 μm. ADC boards that use sample-and-hold (S/H) circuitry or sigma-delta converters allow true simultaneous acquisition, avoiding clock skew, but are significantly more expensive than sequential acquisition boards.

Given the above information, a reasonable data acquisition rate can be calculated. As an example, a 1 μm diameter electrode is scanned at 10 μm/s.

Data should be sampled with a period of two times the electrode diameter to avoid aliasing. Using a sample rate of 50 Hz will ensure a faithful data period, i.e., 0.2 μm/pt. Recording four channels at this rate produces a shift of 0.6 μm between the first and fourth channel because of clock skew. A four-channel average will reduce the shift to 0.15 μm. In this example, image data is low-pass filtered at frequencies less than 20 Hz to satisfy the Nyquist rate. Although four sample averages are sufficient to avoid clock skew, increasing the number of averages will improve SNR and resolution. The aggregate sample rate should be less than 10 kHz to maintain a high ENOB. Averaging 40 samples per channel requires a 2 kHz per channel and 8 kHz aggregate sample rate, reduces the clock skew to 0.1 of the sample interval, increases SNR by 6.3 times, and increases the effective resolution of the data by 2.7 bits.

3. *Auxiliary Routines*

The main SECM program is involved in acquiring data as a function of position. However, the SECM program is enhanced by the ability to perform voltammetry (i vs. E) or other time-based routines at a stationary electrode. A DAC on the ADC board can generate a voltage sweep or step. Alternately, a trigger pulse starts an external generator.

A tip-approach function is useful. A simple approach is to move the probe relatively slowly towards the surface while monitoring the tip current. A 10% change from $i_{T,\infty}$ indicates the electrode is at the surface. This simple method is slow and susceptible to noise. The Model CH900 SECM uses a PID (proportional, integral, derivative) function to allow a rapid approach while minimizing noise.

4. *Display and Analysis*

Data may be displayed during acquisition in a number of ways. Common methods are using the computer display to emulate a strip chart recorder or digital multimeter. Simple two-dimensional plots are easily generated by encoding the magnitude of the data point as a color or gray level. By mapping the tip position to an *x*, *y* pixel screen coordinate, a two-dimensional color display can be generated while imaging. However, the computer display has a very small dynamic range, and some care is needed to produce usable images during acquisition. A strip-chart trace on a computer screen is limited to several hundred pixels or about 55 dB of dynamic range. Color maps on a monitor have wide dynamic ranges due to a wide array of colors, but a quantitative relationship between color and magnitude is complicated by a nonlinear human visual system (HVS) response and the presence of color deficiencies in many people. The HVS can only distinguish about 10^3 intensity levels simultaneously (a range of 10^{10} is available by allowing the

HVS to adapt) (32). Simultaneous use of numeric displays, color-scale plots, and strip-chart displays will provide sufficient information to the experimenter.

Once data is acquired, more sophisticated data analysis and displays can be used. With a range of commercial and free graphical software packages available to analyze images, it is difficult to recommend that programmers develop their own routines. However, routines such as tilt correction, baseline correction, and interpolation make initial data interpretation easier and are readily programmed. Tilt correction is used to minimize the effect of sample tilt on an image. The simplest method calculates a mean intensity value and a least-squares plane from the image. The least-squares plane is subtracted from the image and the mean value is added. Baseline-correction removes scan-line to scan-line variances in an image. A mean value for the entire image is calculated as well as the least-squares line for each scan line. The least-squares line is subtracted from each scan-line and the mean value is added back. Interpolation routines are usually necessary when displaying SECM images. SECM images are often small (less than 200×200 pixels) when compared to images on a monitor or printer, and the data pixels are usually not "square." A SECM pixel might represent a 0.1×3 μm region, for example. To display these data properly on a high-resolution output device, an interpolation program is required to convert the data pixels to square image pixels and to increase the number of pixels. A bilinear interpolation function is straightforward to implement, but an interpolation using polynomial or spline functions can produce better (smoother) results. An algorithm for these functions is available from Press et al. (33). Note that all image manipulation functions can add artifacts and mislead the viewer as to the quality of the original data. Carefully consider the benefits of using these or other filter functions.

Some commercial software packages, such as Sigmaplot (SPSS), TecPlot (Amtech Engineering), and IGOR Pro (WaveMetrics), are relatively inexpensive to academic users and provide excellent graphical displays of two- and three-dimensional image data. Built-in macro languages simplify data import and graphing operations. Some software packages are designed specifically for scanning probe microscopy. SPIP (Image Metrology) and Image SXF (Steve Barrett) are available free. The free version of SPIP is a demo that limits the image size. Both programs include many image manipulation routines including Fourier filtering and analysis, image measurements, linearity and baseline correction, and flexible data import. Other "freeware" two-dimensional image manipulation and display routines are also available. A well-known program is NIH Image, which runs on Macintosh computers, but a Windows port is available. Another very good program is Image Tool, available from the University of Texas Health Science

Center, San Antonio. SECM data files can also be converted into files that can be read by other commercial SPM instruments. The author wrote a program (available upon request) to convert SECM data files to Nanoscope (Digital Instruments) format, making all the data analysis and display routines of the commercial software available.

G. Video Microscope

A video microscope is not required for an SECM instrument, but it is very useful and is a highly recommended addition. A video microscope is preferred over a normal optical microscope because it allows the probe to be continuously observed while operating the instrument. A video record of the experiment is also available. The video microscope aids in positioning the probe in generation/collection experiments, where the lack of a feedback response makes accurate distance control difficult. In addition, video microscopy helps in positioning the probe near features of interest on the substrate.

Microscopes can be arranged to look down or at a grazing angle at the substrate. The down orientation is only suitable for low-magnification use since the region under the probe is obscured by the probe itself and the image is distorted by the meniscus formed by the probe entering the solution. Glare is also a problem, but it can be minimized with use of polarizing filter on the microscope. High-magnification microscopes should use a grazing orientation. This allows a good view of the probe-substrate separation and, by adjusting the grazing angle, a view of the substrate surface. A microscope in this orientation can easily be used to position a probe to within several micrometers of the substrate surface.

Consider the following factors when choosing a video microscope. The working distance should not be so close that it interferes with the experiment or is subject to damage by electrolyte. Alternately, realize that some video microscopes are essentially telescopes and require inconveniently long working distances. The depth of field should be large enough to include the probe-scanning region (i.e., up to several hundred micrometers). Do not be seduced by high magnifications; lower magnifications give a more complete view of the substrate and a larger depth of field. A resolution of 2–10 μm per video display line is enough. Color cameras can distinguish colored by-products of surface modification reactions, but black-and-white cameras are usually more sensitive. In either case, a high-intensity illumination is required. A fiber-optic source minimizes heating and allows flexible light-source arrangements. In the United States, Edmund Scientific sells complete video microscope systems and their online catalog is a useful resource for choosing an appropriate system (34).

Video records of experiments can be made by video recorders or video digitizers. Recent consumer demand has made computer-based video capture equipment and high-resolution (S-VHS) recorders widely available.

III. COMMERCIAL SECM INSTRUMENT IMPLEMENTATION

The model 900 SECM (CH Instruments) is the first commercial instrument specifically designed for SECM. Although STM instruments equipped with electrochemical accessories can be adapted for some SECM experiments, they cannot replace a general purpose SECM instrument. This section will describe the features of the model 900. The model 900 includes the cell and probe positioner illustrated in Figure 5 as well as a computer, motor controller, and bipotentiostat (35).

The probe positioning system is influenced by designs originating in the Bard laboratory (1–3). Lateral probe scans as large as 2.5 cm are accomplished by two Inchworm motors driving crossed-roller–bearing translator stages. Vertical motion is provided by mounting the probe directly to shaft of the third motor. The electrochemical cell is stationary and mounted on an adjustable tilting platform to compensate for any substrate tilt. Open-loop motor control is used, and software calibration corrects scan-rate, axis, and direction dependent effects on the probe motion. The CH Instruments motor controller uses motor step sizes of less than 1 nm during scanning, which should minimize the effect of the clamping step on the shaft movement and provide smoother movements than the standard Burleigh motor controller does.

The data acquisition system and bipotentiostat of the CH SECM are microprocessor based and communicate with the personal computer controller via a serial data link. The bipotentiostat is suitable for analytical voltammetry and is able to amplify signals in the pA range. Adjustable second-order Bessel filters provide noise reduction and prevent aliasing by the ADC. The compliance voltage is ± 12 V and the maximum current is ± 10 mA. An advantage is an internal ADC with at least 20 bits of resolution, which implies a very large 120 dB dynamic range. However, the dynamic range is ultimately limited by the analog circuitry to less than 100 dB. The frequency response of the potentiostat extends to 100 kHz. However, the ADC limits the maximum data acquisition rate to about 1 kHz.

The software for the model 900 SECM runs on the Windows operating system with a Pentium class machine. The software provides all the controls necessary for positioning the probe in three dimensions and setting parameters for electrochemical experiments. The software also supports a wide range of electrochemical experiments, graphical displays, data processing,

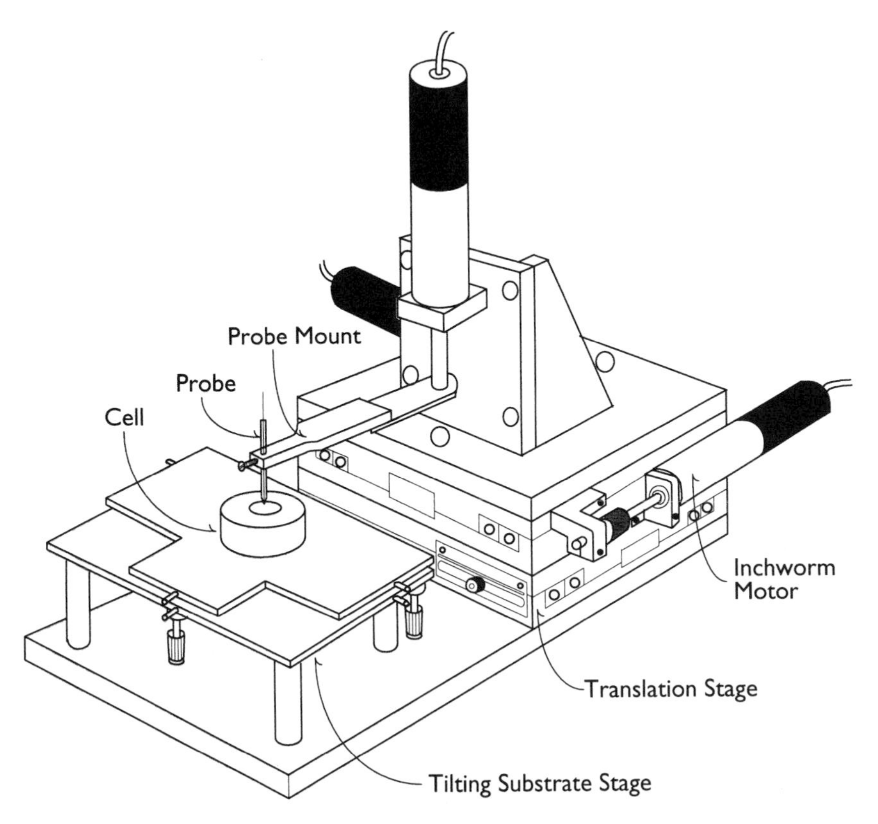

FIG. 5 An illustration showing translators and tip and cell mounts for the commercially available CH model 900 SECM.

and even digital simulation of electrochemical events, which adds value to the device for use simply as a versatile electrochemical workstation. During probe scanning, only amperometric or potentiometric signals can be recorded. Other software features include a probe approach curve routine that automatically stops the probe when the current reaches a set level and that automatically slows the speed of approach as the probe approaches the substrate surface. Graphics output during scanning and post-scan is by grayscale or color-coded plots. A three-dimensional "mesh" plot is also available, but other data manipulation must be done outside the main program.

The CH Instrument appears to be a quite useful implementation of an SECM. The design lends it to use in feedback imaging and surface modification experiments. However, some modifications by the purchaser or the manufacturer can be considered. A desirable modification would be a "constant-current" imaging mode. There appears to be no barrier to the experimentalist adding this capability. Both the probe and cell mounting could be modified to add additional vibration isolation and the use of a video microscope to assist in probe positioning. Particularly lacking are sophisticated graphical output and data manipulation routines for the SECM image. However, this is not a serious problem since data analysis and display software is available (see above).

IV. TIP POSITION MODULATION INSTRUMENTATION

Some aspects of feedback mode imaging can be improved by employing the tip position modulation (TPM) mode of the SECM (36). Images made with TPM SECM show an apparent increase in lateral resolution and improved vertical sensitivity for images of insulating materials. In TPM mode, the tip position is modulated with a small-amplitude motion perpendicular to the substrate surface. In a typical experiment, the modulation is sinusoidal with a frequency of 100 Hz or greater and has amplitude, δ, equal to 1–10% of the tip electrode radius, a. The modulation is supplied by a piezoelectric stacked-disk element. A schematic diagram of the instrumental setup is shown in Figure 6. After the tip current signal is transduced to voltage and amplified by the potentiostat, the tip signal is isolated by a low-pass filter into a dc and a ac component. The V_{DC} signal is effectively identical to the signal observed in normal SECM feedback imaging as long as the modulation amplitude is less than 10% of the disk radius (36,37). The modulated component, V_{AC}, is passed to a lock-in amplifier that measures the rms (root-mean-square) amplitude and phase of V_{AC} relative to the probe oscillation. Because the lock-in amplifier effectively narrows the bandwidth of the measurement, the signal-to-noise ratio of V_{AC} is greatly increased (38).

The advantages of TPM compared to standard feedback SECM are (1) TPM SECM is more sensitive to tip-substrate separation, (2) the signal baseline is zero, and (3) the signals for conducting and insulating samples have opposite arithmetic signs (i.e., the phase difference between the two signals is 180°). The phase difference is important, because it is the basis for a method of constant-current imaging with the SECM (see below). Although the TPM experiment is slightly more complicated than the standard SECM experiment, no other disadvantages exist. The V_{DC} signal is effectively the same as the standard SECM feedback signal, and the TPM signal is useful for feedback imaging with very small tips (see below).

V. CONSTANT-CURRENT MODE INSTRUMENTATION

Nearly all of the published SECM images are made in the "constant-height" mode, where the probe is scanned at a constant reference plane above the

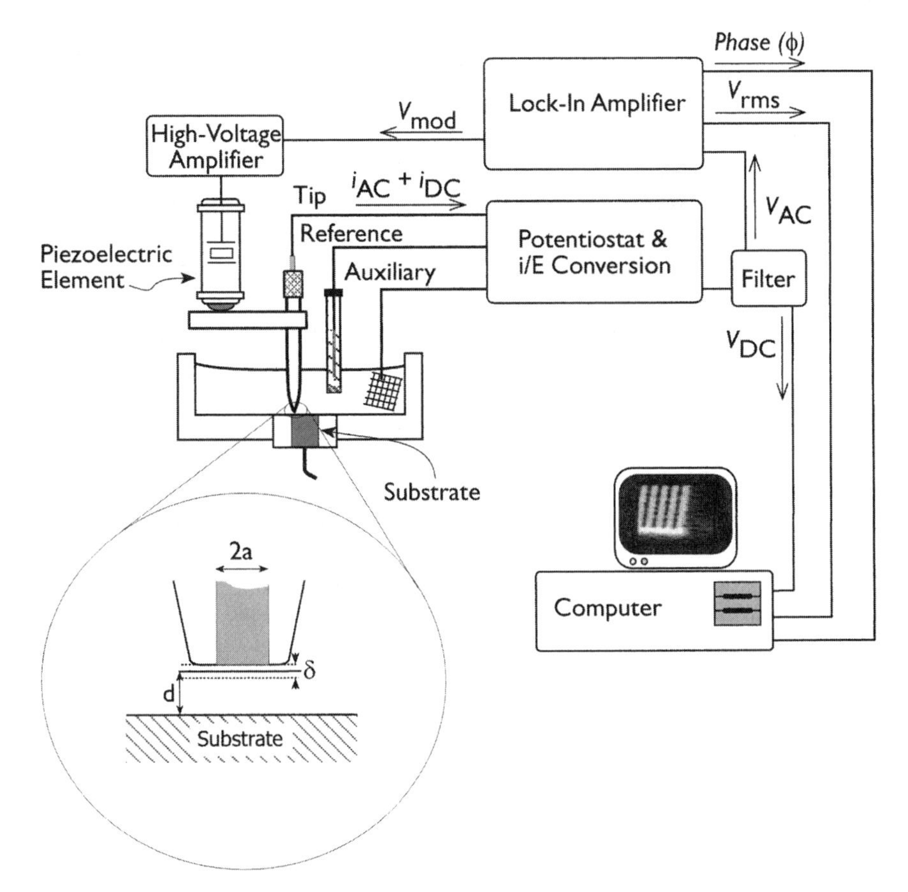

FIG. 6 An illustration of a SECM configured in the tip-position modulation (TPM) mode. The magnified portion of the image illustrates the tip, with radius *a*, positioned at an average distance *d* above the substrate and undergoing a 0.2*a*(p-p) modulation.

substrate surface. The simplicity of the constant-height mode and the ability to mathematically convert constant-height feedback images to true topographic images is an advantage (39,40). In contrast, "constant-current" scans maintain a constant separation between the tip and substrate. The constant-current mode makes collisions between the tip and substrate (tip crashes) less likely and, in particular, is important for imaging with small tips. Small tips are difficult to use when scanning in the constant height mode because the required close tip-substrate separation (i.e., about one tip radius) is likely to promote tip crashes. Despite these benefits, a completely satisfactory method for constant current imaging has yet to appear.

A. Control Loops

Constant-current imaging in SPM uses a servo feedback loop to maintain a desired tip-substrate separation. As the tip moves away from the desired position, an error signal is generated. The conditioned and amplified error signal drives a positioner to move the tip back to the desired position. As the tip approaches the desired position, the error signal diminishes. This type of feedback loop is in the same class of control loops used to control temperature, motor speeds, chemical process variables, etc., and is similar to the negative feedback loops used to stabilize electronic circuitry (41). As such, a wide literature is available for consultation to produce an optimized method of using the error signal to return the tip to the desired position (42). The real difficulty in SECM is producing the error signal—knowing when the tip is not at the desired position.

To illustrate a control loop, consider a system in which a SECM probe is controlled at a set position over a conducting sample (Fig. 7). To maintain the tip position, the probe is mounted on a piezoelectric positioner, which produces precise vertical movement by applying a control voltage to the piezoelectric element. An error signal is generated by electronically amplifying the difference in the tip signal from a reference level. The difference is sent to a high-voltage piezoelectric amplifier. Since the piezoelectric el-

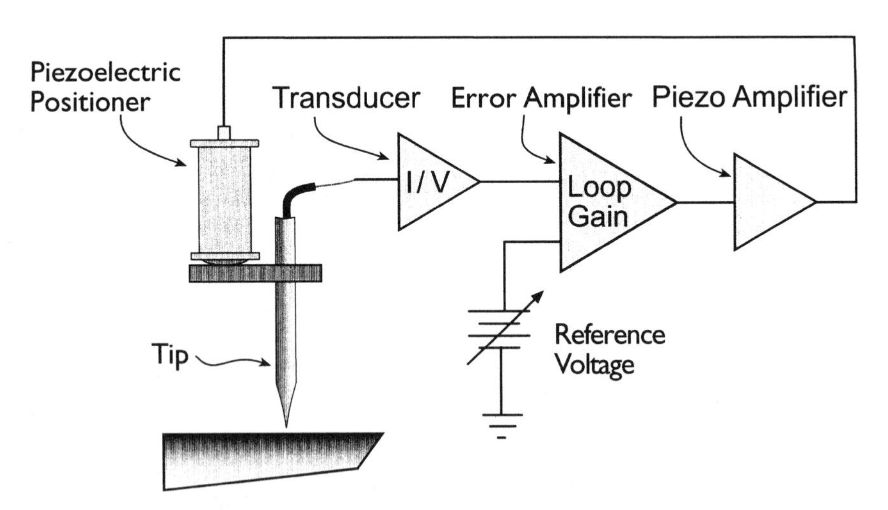

FIG. 7 An illustration of a simple control-loop used for tip-position control.

ement produces a roughly linear movement with applied voltage, such as 10 nm/V, the control voltage can be used directly as an indication of the tip position. The control loop operates in a negative feedback mode; the output (the piezoelectric voltage) is used to minimize the input (the error signal). The degree of amplification of the error signal, the "loop-gain," is an important parameter. An infinite loop-gain is required to ensure that the tip current and reference current become equal (since otherwise an infinitesimal error signal would not be amplified sufficiently to cause a voltage sufficient to move the piezoelectric element). However, large loop gains are not especially desirable, since the loop becomes very sensitive to noise in the signal and reference. More importantly, the probe movement system is subject to mechanical restraints, which limit the ability to accelerate and decelerate the tip. A delay is always present between the application of the tip control voltage and the probe movement. In other words, a frequency-dependent phase shift develops in the control loop. Once the phase shift becomes large (>180°), the loop will have positive feedback at higher frequencies, leading to oscillation of the piezoelectric device. The loop gain must be low enough to prevent oscillation but still sufficiently large to rapidly move the tip to the desired position.

The simple control loop described above is a proportional control because of the linear relationship between the error signal and control voltage. The low loop gain required for stability in proportional control leads to practical problems. Low gain makes the loop respond slowly, slowing scan speeds. Since the error signal is never truly minimized, the tip position also tends to "hunt" for the desired position, wandering above and below the desired minimum in the error signal. A considerable performance improvement is possible by adding integral and differential control elements to the proportional control loop. Applying a correction proportional to the integral the residual error allows a higher proportional loop gain and reduces "hunting." The differential element applies a correction proportional to the time derivative of the error signal. Thus, large changes in the error signal, produced, for example, from scanning over a sharp edge, can be rapidly corrected. The proportional, integral, and differential elements (i.e., PID) can be made with analog elements, such as operational amplifier circuitry, or by digitizing the signal and using a digital signal processor (DSP) to implement digital equivalents of the PID control (20,43).

The PID controller loop must be "tuned" to give optimum performance. The gain of each of the PID elements and the integration time must be adjusted to give good transient response while maintaining stability. In addition, changing the tip diameter or substrate type (conductive or insulating) requires adjusting the tuning parameters. Finding the optimum PID parameters is not trivial and is the subject of much research in control loop optimization. Commercial STM and AFM systems use analog or digital control loops that can be optimized through software control (43). SECM instrumentation is not yet at that level, and constant-current operation is not offered with the existing commercial SECM instrument.

B. Constant-Current Mode Designs

The tip current can be used to generate an error signal when operating in either the negative or positive feedback mode. Because the tip current has a definite functional dependence on distance, it can easily be incorporated in the feedback loop. Note that the feedback loop must be modified when scanning in the negative or positive feedback mode. (An unfortunate aspect of SECM terminology is the use of positive and negative feedback as descriptions of scanning in the feedback mode over a conducting or insulating surface. Feedback refers, in this context, to the shuttling of the solution mediator between the substrate and tip. One should not assume the use of the word feedback in this context implies use of a constant-current mode—quite the contrary, feedback is usually used to indicate a constant height mode. In addition, negative feedback is used here to indicate a condition of a stable feedback loop. The meaning of "feedback" should be clear from the context.) In negative feedback operation, the tip current must be compared to a reference signal below $i_{T,\infty}$ and the control loop must move the tip towards the surface when the tip current is larger than the reference current. In positive feedback operation, the tip current is be compared to a reference signal above $i_{T,\infty}$ and the control loop must move the tip away from the surface when the tip current is larger than the reference current. The similarity of STM and positive feedback SECM imaging allows published designs for STM control loops to be adapted for use in SECM (11,20,44).

Using the feedback current in the control loop can lead to tip crashes. Consider the situation where the tip is operating in constant-current positive-feedback mode. If the sample surface has an inactive region in the scan area, a tip crash can occur as the feedback loop interprets the decrease in tip current over the inactive region as an increase in the tip-substrate separation. A crash also occurs when operating in constant-current negative feedback mode if a conducting region is encountered. The addition of some intelligence to the feedback loop will guard against some of these problems. For example, an "emergency" circuit could detect a tip signal above or below $i_{T,\infty}$ in negative or positive feedback mode, respectively, and shut off the feedback loop during these "impossible" signal levels.

Several SECM researchers have suggested alternate methods in providing constant current imaging. They hope to improve the simple approach outlined above by providing an additional source of information about the tip-substrate distance. The additional information used in all the proven designs is gathered by a tip oscillation. Interestingly, the oscillation range in amplitude from hundreds of micrometers to a few nanometers, indicating the designs are still evolving towards an optimal solution. Three varieties of constant-current mode devices are discussed below.

1. Tip Position Modulation

The tip position modulation (TPM) mode of SECM suggests a method for constant-current imaging where the surface type (insulating or conducting) is automatically detected. As described above, TPM uses a small-amplitude vertical modulation of the tip position. A lock-in amplifier detects the modulated tip current, i_{AC}, and generates an output proportional to the rms magnitude and phase of i_{AC}. A low-pass filter removes the ac component of the tip current from the feedback-mode signal, i_{DC}. Comparing i_{DC} to a reference level produces an error signal. The phase of i_{AC} determines the type of surface because of the 180° phase shift of i_{AC} between insulating and conducting surface. For example, a phase range of 30–150° might be allowed for positive feedback signals while a phase range of 210–330° might indicate a negative feedback situation. A dual-window comparator detects the phase signal and sets the loop gain and reference current level as necessary for the type of surface. An important part of the circuit is detection of out-of-range or emergency conditions. If the phase signal is not in the proper range or if the tip signal is above or below $i_{T,\infty}$ in negative or positive feedback mode, respectively, the feedback circuit stops the piezo motion until the phase signal $i_{T,\infty}$ are both consistent. Using a simple proportional control loop, this circuit was able to image a surface containing a mixture of insulating and conducting phases with a 2 μm diameter tip while automatically maintaining the desired tip position without operator input (45).

2. Picking

In the "picking" mode, a surface-induced convective current is caused when the tip rapidly approaches from a large tip-substrate separation (8). Borgwarth and coworkers showed that when a tip approaches a surface at a speed of 50 μm/s from an initial height of 200 μm above that surface, a significant tip current increase occurs as the tip-substrate separation drops below 1 tip radius. Importantly, the increase occurs at both conducting and insulating substrates. In operation, the picking mode instrument repeatedly approaches the surface at high speed until the tip current exceeds a set value. After halting the tip and allowing the tip current to achieve steady state, tip current is measured. The tip is then withdrawn 200 μm, and the tip position is incremented in a lateral direction to accomplish a scan. This mode allows better than 2 μm vertical accuracy in positioning with 5–25 μm diameter tips over surfaces of unknown type. Better accuracy is found when the surface is known to be insulating or conducting. The approach from large separations minimizes the chance of tip-substrate crashes during lateral motion, and the picking mode has an advantage in imaging high-relief (rough) samples. The picking mode does not require a conventional feedback loop as described above and can be readily implemented using equipment normally present in a standard SECM setup. A potential disadvantage to this mode is that sufficient vertical accuracy might be unavailable when using micrometer or smaller tips. In addition, the method can only be used with the SECM feedback mode since addition of a redox mediator is required to provide the tip-positioning signal.

3. Shear Mode

The shear mode was developed for use with near-field scanning optical microscopy (NSOM) (47–49). In NSOM imaging the probe is scanned at a distance of about 25 nm from a surface. A number of designs have been implemented to provide the desired subnanometer positioning accuracy in NSOM, but the shear force method has emerged as the most popular. In the shear force method, a "dither" piezoelectric element is used to vibrate the NSOM probe tip into a mechanical resonance at a frequency from 10 to 50 kHz. The undamped probe oscillates with a 10–50 nm amplitude parallel to the surface. When the probe tip is less than 25 nm from the surface, an interaction force, commonly called the shear force, damps the oscillation amplitude and changes the resonance frequency of the probe. The basis for a control loop is the change in tip-vibration amplitude. The tip-vibration amplitude is monitored at the undamped probe resonance frequency with a lock-in amplifier. The control-loop error signal is the deviation of the tip-vibration amplitude from a reference vibration amplitude.

The success of the shear mode in NSOM has led to several instruments incorporating variants of the shear mode for SECM. The differences in NSOM and SECM designs primarily arise from the method of sensing the vibration amplitude of the probe and the difference between NSOM and SECM probes. The active participation of the probe in shear mode requires new SECM probe designs. It is likely that only lightweight and fragile electrodes with small tip diameters will be suitable for use with shear mode, i.e., "disposable" probes. These probes can be made using metal-filled pulled-glass capillaries and wax- or polymer-coated etched metal (or carbon) fibers. Metal sputtered optical fiber probes can be used in a hybrid electrochemical and optical scanning instrument (50).

An instrument employing an optical detection of the probe vibration amplitude was reported by Schuhmann and coworkers (37). A piezoelectric element is used to induce a probe vibration while a focused laser illuminates the probe electrode near its end. By detecting the rate at which a diffraction line oscillates between the two segments of a split photodiode, the probe vibration frequency is determined. The probe remains stationary to maintain the laser and photodiode alignment, while the substrate is scanned to make images. The probe for this initial experiment was a 1.5 cm long section of a pulled glass capillary containing a 25 μm diameter Pt disk electrode. The long pulled section is necessary to produce the flexibility required for oscillation. Several resonance peaks of the electrode were noted, and the lowest frequency (about 2 kHz) was used in imaging. Modulation amplitudes in this experiment are about 1–2 μm, which is significantly larger than commonly employed in NSOM. The large amplitudes produce a proportionately larger shear mode distance so that the sample probe separation was in the micrometer range when shear mode damping was observed. The authors also note that the large amplitude tip oscillations increased the steady-state tip current.

Use of optical methods in NSOM is often objectionable because of the interference of stray light on the image signal (a moot point in SECM), difficulty with liquid operation, and difficulty with miniaturization. Karraï and Grober developed a nonoptical method, which employs a miniature tuning fork as the vibration amplitude detector (51,52). Figure 8A is an illus-

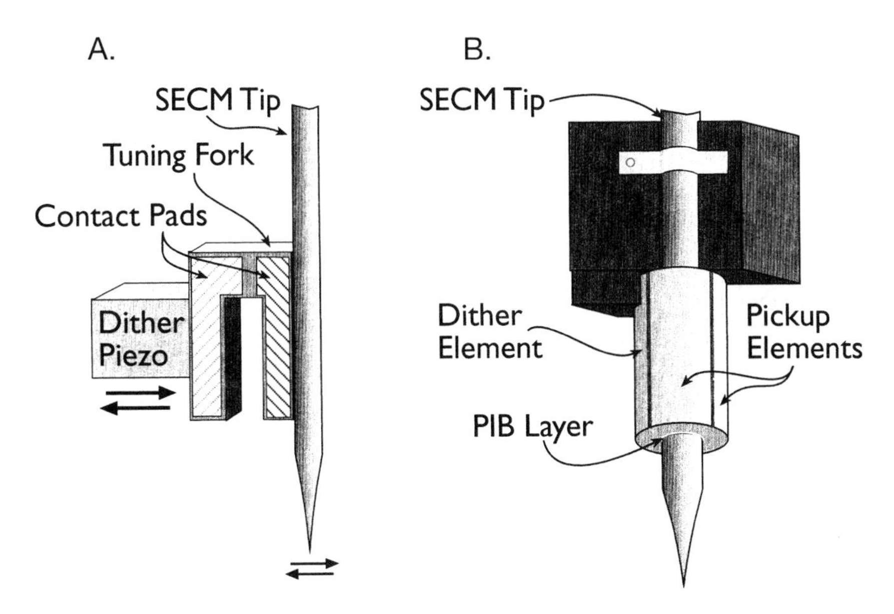

FIG. 8 Tip mounts used for implementing shear-force feedback in SECM: (A) the tuning fork mount; (B) proposed piezo tube mount.

tration of the essential components. The tuning fork is an inexpensive piezoelectric device used in quartz clocks. The dither piezoelectric element is rigidly mounted to one of the tuning fork prongs. The probe is glued to the other prong. Excitation of the fork with the dither piezo produces a vibrating motion of the probe monitored by detecting the induced piezoelectric potential produced between the contact pads of the tuning fork. As in other designs, the resonance frequency changes as the tip approaches to within the shear force boundary. An important advantage of the tuning fork is that the probe need not be flexible since the tuning fork provides the motion to drive the probe oscillation.

Smyrl and coworkers adapted the tuning-fork based feedback loop and x, y, z positioning found in a commercial NSOM instrument (the Topometrix Aurora) to produce SECM images (53). In their initial report, an etched tungsten probe insulated with polymer was used as the SECM probe. SECM and AFM imaging gave similar topographic profiles, (the AFM exhibiting higher lateral resolution), suggesting that the tuning fork method worked properly. The exact tip surface distance was not determined but was estimated to be less than 50 nm during SECM imaging with distance feedback.

An alternate method for piezoelectric detection is illustrated in Figure 8B. Barenz and coworkers use a four-segment piezoelectric tube to provide the dither and detection elements for NSOM (54). The dither element is one segment firmly attached to the mount. Electrical excitation of this segment produces bending motion in the piezoelectric tube and, hence, in the probe. The other three segments are connected in parallel and used to detect the vibration amplitude by monitoring the induced piezoelectricity. Two advantages recommend this design for possible use with SECM probes. The first is the relatively low Q factor of the device. The Q (quality) factor is the sharpness of the resonance peak and is calculated by dividing the peak frequency by the peak width at −3 dB. High Q factors lead to less stable and slower scanning because sharp resonance peaks make it more difficult to stay at the desired frequency. Dropping completely off the resonance peak leads to loss of feedback control and a crash. In addition, high Q resonances tend to ring, which requires longer settling times. A second advantage is an innovative mounting method. A small spring holds the probe in place, but the probe is rigidly coupled to the piezoelectric element by a polyisobutylene layer. This viscous fluid has a high Young's modulus at frequencies above 10 kHz, which produces the required rigid coupling at operating frequencies but permits rapid replacement of the probe. The probe length protruding from the piezoelectric element can also be adjusted easily, maintaining a nearly constant resonance frequency from probe to probe.

VI. EXPERIMENTAL DIFFICULTIES IN DATA ACQUISITION

Acquiring SECM data requires optimization of three simultaneous factors: signal-to-noise ratio (SNR), experiment duration, and image resolution. Improved SNR is accomplished by isolating external noise sources, designing proper grounds, and employing electrical shielding with Faraday cages (28,55,56). Ultimately, though, increasing SNR requires decreasing the measurement bandwidth, *f*, since most noise sources are proportional to $f^{1/2}$. Measurement bandwidth is decreased by filtering, ensemble (signal) averaging, lock-in amplification, or box-car averaging (38,57). For most SECM experiments, the signal is at sufficiently low frequency to be considered a dc level signal and low-pass filtering will be the primary means of reducing the signal bandwidth.

Reducing the experiment duration is desirable. For example, a $100 \times 100\ \mu m^2$ image requires 16 minutes to acquire with 1 μm resolution at a scan rate of 10 μm/s. That may be too long to resolve chemical or electrochemical changes on a substrate surface. Increasing the probe scan rate will decrease the acquisition time, but this may not be feasible.

Decreasing the tip size can increase image resolution. However, difficulties arise in simultaneously improving resolution, experiment duration, and SNR in the SECM experiment. At least one goal must be compromised to improve the others, and improving image resolution will be most challenging.

A. Low-Pass Filtering of *I-L* Curves

As tip electrodes become smaller, the tip current will decrease, resulting in decreased SNR. Thus, the temptation exists to "improve" the data by filtering. Filtering can increase SNR, and most potentiostats do include a low-pass filter to help reduce noise on the working electrode signal. Some instruments offer only one or two fixed settings, while others offer a variable frequency filter. However, filtering causes distortion of the data, which leads to problems in data interpretation. The effect of low-pass filtering a SECM signal is illustrated for two important SECM experiments shown in Figures 9 and 10. Figure 9 shows theoretical positive feedback I_T versus L plots for a 1 and 5 μm radius disk-shaped tip over a conducting substrate. Here, I_T is $i_T/i_{T,\infty}$, the normalized tip current, and L is d/a, where d is the tip-substrate separation and a is the electrode radius. The tip is initially located at a separation of 0.1 L from the substrate surface in the calculated curves. A filtered version of the theoretical curves is also plotted on the same graph. The filtered response is generated by convolution of the theoretical SECM

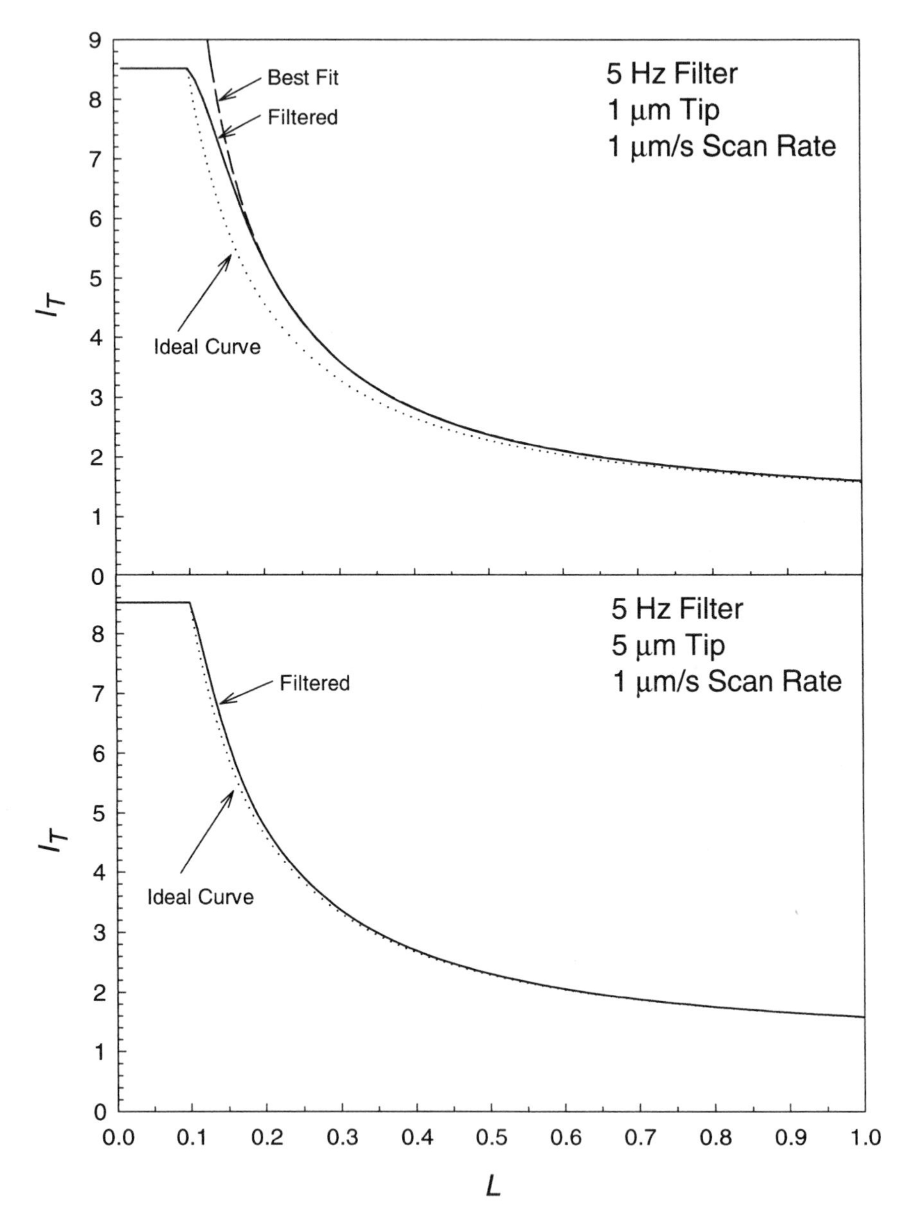

FIG. 9 Theoretical $I_T - L$ plots for 1 μm (top) and 5 μm (bottom) tips over a conducting substrate. The ideal curves are illustrated with calculated curves showing the effect of a 5 Hz low-pass filter on the curve shape as the tip is withdrawn from the surface at a rate of 1 μm/s.

signal with the time response of a single-pole low-pass filter with a 5 Hz cutoff frequency (58). The time scale for the curves is fixed by the scan rate. Note that the simulated time scale starts sufficiently prior to probe motion to avoid a discontinuity at $t = 0$. The model assumes that the tip is initially located near surface and moves continuously away at $t > 0$. As the lower curve in Figure 9 illustrates, the effect of filtering is minimal at the 5 μm radius tip. For the identical filter and probe scan rate, filtering the 1 μm radius tip signal causes significant distortion. The extent of the distortion is clearer when a second undistorted *I-L* curve with an offset of 0.034 *L* is overlaid on the filtered response. A good fit to the distorted results is observed for *I* values up to about 5.4, but the remaining data do not fit the theoretical curve.

Low-pass filtering introduces significant distortion *I-L* curves generated by recording the tip current as the probe is scanned to or from the substrate. A particular danger is noted when *I-L* curves are used to analyze heterogeneous kinetic parameters. The shape of the filtered curves in Figure 9 could be confused with the effect of heterogeneous kinetics, especially as the curve approaches higher I_T values. In addition, the distorted curves can lead to a misunderstanding of the actual separation between tip and substrate. Filtering distortion is distinguished from kinetic perturbation by its dependence on the direction of tip motion. The curve is shifted towards larger or smaller *L* as the tip moves to smaller or larger *L* values, respectively, when filtering occurs. Filtering distortion also affects negative feedback curves but with a smaller shift due to the less steep approach curve. The actual distortion observed on the curves will be dependent on the tip radius, the direction and speed of approach, the conductivity of the surface, and the point at which the scan starts. By applying the following rule of thumb, a shift in *L* of 0.01 or less will occur in *I-L* curves.

$$af/v > 13 \tag{1}$$

where a is the electrode radius in μm, f is the frequency in s^{-1}, and v is the scan rate in μm/s.

B. Low-Pass Filtering While Imaging

Use of low-pass filtering while acquiring an image also produces distortion in the observed data. Figure 10 illustrates the effect of a single-pole low-pass filter with a 5 or 15 Hz cutoff frequency on a single line scan across an artificial substrate. The artificial data set consists of three conductive regions 16, 1, and 4 μm wide surrounded by insulating substrate (see Fig. 10A). The calculation assumes the tip is scanned at 20 μm/s at a tip-substrate distance of 1.0 *L*. In order to more closely approximate a real image; the

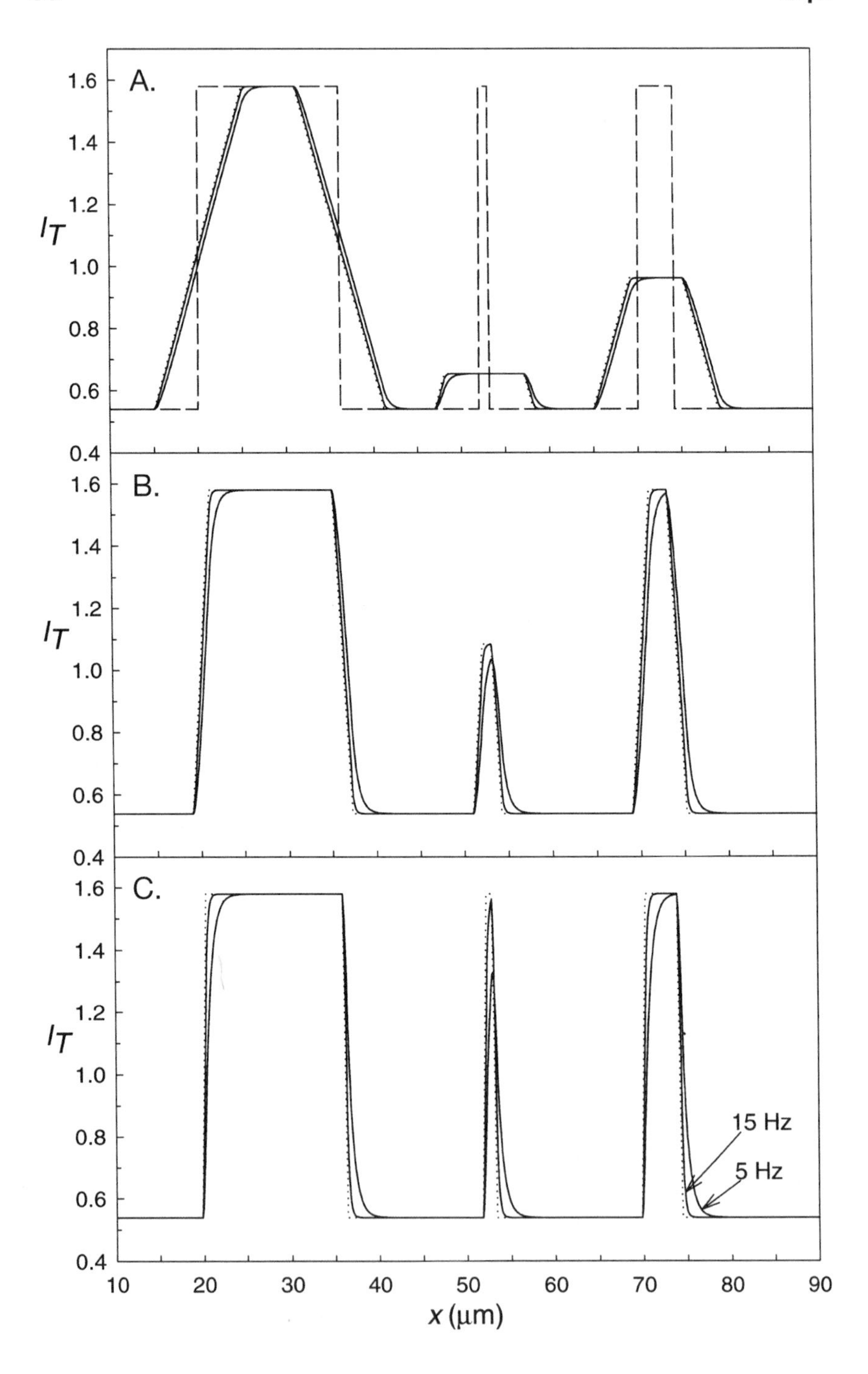
A.
B.
C.
1.6
1.4
1.2
1.0
0.8
0.6
0.4
I_T
10
20
30
40
50
60
70
80
90
x (μm)
15 Hz
5 Hz

data are first convoluted with a tip function to simulate the effect of the tip size on image resolution. The tip function is a unit area pulse with a width equal to the tip diameter. The tip resolution effect is plotted as the dotted line in the figure and is most dramatically illustrated in Figure 10A, which models a 5 μm radius tip. The additional distortion due to diffusion is not modeled in the figures (58). As predicted by the results of the previous section, the distortion due to low-pass filtering increases as the tip radius decreases. In Figure 10A, the 5 μm radius tip signal is not seriously degraded at either the 5 or the 15 Hz filter cutoff. However, the slower filter slightly distorts the data at a 1 μm radius tip and seriously distorts the 0.2 μm radius tip data, particularly for the 1 μm feature (Fig. 10B, C). A 1 μm feature should be resolvable with a 0.2 μm radius tip, but the filtering process significantly reduces the apparent magnitude of the current. A rule of thumb for filter cutoff frequencies during image acquisition is

$$af/v > 1 \tag{2}$$

This is a relaxed requirement compared to *I-L* data acquisition since the distortion due to tip-size function diminishes the effect of low-pass filtering.

C. Improving Image Resolution with Small Diameter Tips

A number of advantages occur when the tip electrode size is reduced. Lateral resolution during imaging is improved and diffusional transit times for the mediator between the tip and substrate are very short at close tip-substrate separations, which allows rapid tip scan rates and produces larger mass-transfer coefficients in heterogeneous and homogeneous rate measurements. However, the faradaic tip signal is proportional to tip size. A 1 nA signal at a 5 μm radius tip is reduced to a 10 pA signal at a 50 nm tip. The unfortunate conclusion of both Eqs. (1) and (2) is that decreasing the electrode radius requires a proportionate increase in filter frequency or a decrease in probe scan rate—exactly the opposite of what may be desired. Simply stated, use of a smaller tip effectively shifts the *I-L* curve or image features to higher frequencies. Since imaging occurs at higher scan rates than collection of

FIG. 10 Simulated SECM line-scan data over a substrate containing conductive regions of 16, 1, and 4 μm wide embedded in an insulating matrix. The ideal response (neglecting diffusional distortion) is shown by the dashed line (---) in A. (···) shows the ideal response when including the effect of tip-size on the data. (——) are curves calculated to include the effect of low-pass filters of 5 and 15 Hz when the tip is scanned across the sample at 20 μm/s. (A) 5 μm radius tip; (B) 1.0 μm radius tip; (C) 0.2 μm radius tip.

I-L curves, providing adequate filtering will be difficult when imaging with nanometer-sized tips. For example, scanning a 0.1 μm radius tip at 10 μm/s will require a filter frequency of about 100 Hz, permitting power-line noise to interfere. Using a filter type other than a single-pole low-pass filter will somewhat reduce the distortion, but many multipole filter types introduce nonlinear phase distortion in the passed signal (59), and this may prove to be even more objectionable. Similarly, postacquisition digital filters may be applied to the data to provide better S/N ratios. Clearly, greatly increased attention to electrical shielding will be required with small tips.

For critical results, the author suggests that continuous scans be avoided in collecting *I-L* curves. Instead, use a stepped method to move the tip to the desired *L* value, and, after a delay of 5 RC time constants to allow the filter to settle, record the *I* value. For imaging with tips smaller than 1 μm, the author suggests use of ac-modulated signals to improve the probe scanning speed while maintaining SNR. For example, tip-position modulation SECM provides a narrow bandwidth signal that is virtually unaffected by the rate of probe scanning (36).

D. Effect of Tip Vibration

In addition to electrical noise sources, the tip signal experiences noise due to vibration of the tip near the substrate surface. As the tip vibrates normal to the surface in the feedback mode, positive and negative deviations from the mean signal will be generated. Vibrations with frequencies in the measurement bandwidth will appear as noise in the data. Since the current-distance relationship is nonlinear in the feedback mode, vibration-induced signals at frequencies above the measurement bandwidth will become rectified (36). The rectification effect increases as the tip-substrate separation decreases. When the current noise due to vibration is integrated (by the low-pass filter in the measuring circuit), an additional DC component due to the rectified component is added to the tip signal. The effect of an oscillating tip in the positive feedback mode has been examined and a vertical oscillation of 0.4 *d/a* peak-to-peak amplitude will increase the DC signal by 10%. Nanometer-sized tips are vulnerable to tip vibration since even a 1 nm p-p vibration will begin to be troublesome at a 25 nm radius tip. In this case, attention must be paid to the tip-substrate mount to provide a small and stiff structure with high resonant frequencies, similar to those used for AFM and STM experiments.

In some experiments, such as TPM SECM or shear-force mode, the probe is intentionally vibrated. The TPM mode is a vertical oscillation that can be treated as outlined above. Experimenters should be wary of intentional vibration applied to the probe to avoid excitation of additional mechanical vibrations in the probe mount. Nonfeedback SECM experiments will be less effected by noise since the signal is only weakly related to the tip-substrate distance.

VII. ACCESSORY EQUIPMENT FOR SECM

A. SECM Cells

SECM experiments use a variety of samples and a universal sample cell is not feasible. However, the three cell designs sketched in Figure 11A–C are able to accommodate a variety of samples. Figure 11A is a versatile cell easily dissembled for cleaning and substrate polishing or replacement. The cell is circular in cross section and machined out of a Teflon rod. The bottom section seats into a $\frac{1}{4}$-inch-thick aluminum plate, acting as a cell mount. Cylindrical substrates are introduced into the cell via a circular hole in the cell bottom. The lubricity and cold-flow properties of the Teflon are often sufficient to provide a leak-free seal simply by press-fitting the sample into the cell. If required, a better seal is achieved by wrapping the substrate with Teflon tape before insertion. Small-diameter Teflon or Tygon tubing can be inserted into a hole drilled through a plexi-glass cover plate. Other holes are drilled for reference, auxiliary, and tip electrodes. Using the cell cover while sparging with inert gas will reduce but not completely exclude oxygen from the solution. The traditional practice of maintaining a flow of inert gas over the solution will generally produce an objectionable amount of noise during imaging scans. For experiments sensitive to oxygen or moisture, the entire cell and probe positioner must be enclosed in a controlled atmosphere chamber.

Substrates for use in this cell can be machined to fit into the bottom opening. Alternately, the sample can be potted in epoxy. A rubber septum makes a simple and reusable mold for the epoxy. Epon 828 resin (Shell Chemicals) with triethylene tetraamine (TETA) curing agent makes a highly chemically resistant and hard potting agent (60). Other epoxies should be checked for chemical attack in nonaqueous or extreme pH solutions. Commercially available electrodes are also accommodated in this cell by changing the bottom hole size to accept the electrode. Alternately, several different diameters of substrate can be accommodated by preparing Teflon bushings to fit the cell and substrate (see Fig. 11B).

Figure 11B is a modification incorporating an optical window. The window allows monitoring of tip-substrate separation with a video microscope. The window is simply a section of a microscope slide sealed against the cell using a clamp and O-ring. The cell has a square external shape to accommodate thin brass bar clamps held in place by four screws tapped into

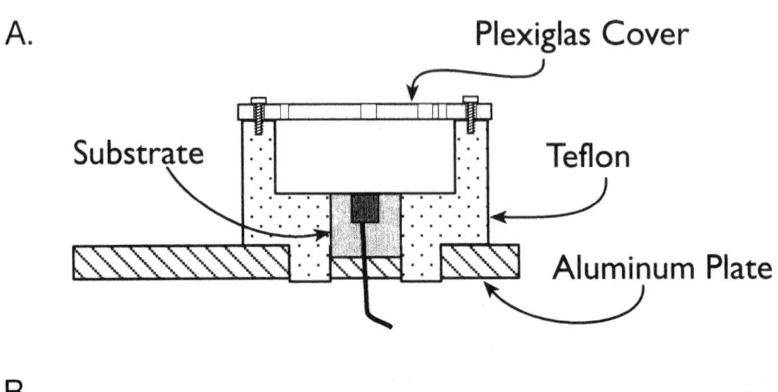

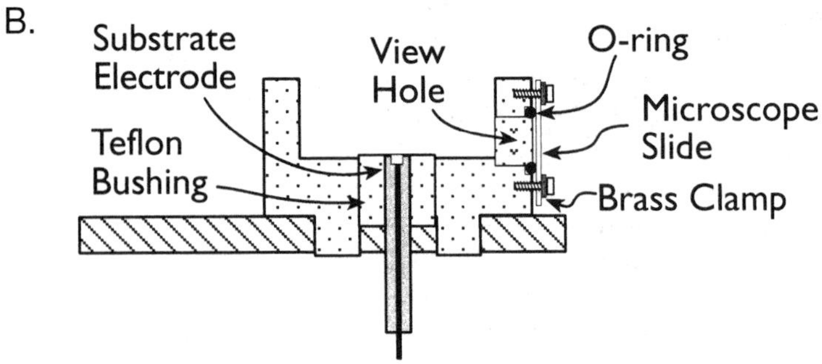

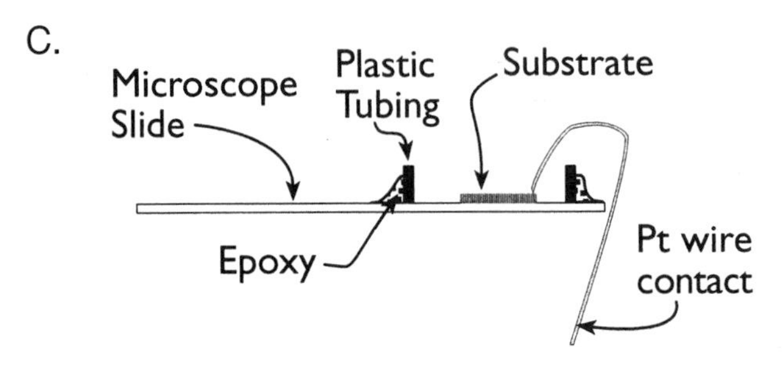

FIG. 11 Sample cells for SECM.

the face of the cell. Threaded brass or steel inserts provide a secure attachment for the clamping screws and prevent loosening due to Teflon cold flow. The O-ring and microscope slide assembly is inexpensive, easy to disassemble for cleaning, and stable in most solvents. Teflon-based O-rings can be used for aggressive organic solvents.

Figure 11C is a very inexpensive, disposable cell design, suitable for flat specimens. A short (5–10 mm) section of plastic or glass tubing is cut and mounted to a glass microscope slide with epoxy cement or hot-melt glue to make a shallow cell. Mounting can be as simple as a dab of vacuum grease applied to the backside of the sample. If required, silicon adhesive, epoxy, or hot-melt glue can also be used. Note that vapors from curing adhesives can significantly contaminate a sample surface. In some cases, the glass or plastic tube is eliminated by using just epoxy or holt-melt glue to form the cell on the microscope slide. Even simpler, a ring of vacuum grease can be used to confine a bubble of electrolyte solution While hot-melt glue works well as an adhesive in aqueous solution, it usually leaks in non-aqueous solutions. Figure 11C illustrates a method for making electrical contact to a substrate. A "fish-hook" of 0.5 mm diameter Pt wire is prepared. The slight downward pressure exerted by the working electrode lead is sufficient to hold the wire in place pressed against the surface of the sample. Since the area of the Pt lead exposed to solution is small compared to the substrate electrode, the effect on the electrochemistry can often be ignored. Note that external means are necessary to mount reference and auxiliary electrodes for the cell in Figure 11C.

B. Nano-Current Source

A nano-current source, easily constructed by the experimenter, allows using one of the working electrodes of the SECM bipotentiostat in a galvanostatic (constant current) mode while maintaining potential control over the second electrode. The electrode can be switched between galvanostatic and potentiostatic mode without changing the electrode connections. This is useful in some types of micro-fabrication modes and in iontophoretic experiments. A block diagram of the device is shown in Figure 12. The two essential parts of the device are the current source and a set of relays to switch modes. The device allows using the same auxiliary electrode as the potentiostatic electrode while in galvanostatic mode. Alternately, two auxiliary electrodes can be used if interference is noted during operation of the constant-current source. Remotely enabling the current source is accomplished by grounding the input or applying a low TTL logic signal. The device can accurately supply currents over a 7-decade range: 1 mA to less than 100 pA.

A detailed schematic of the circuit is shown in Figure 13. The upper part of the circuit is the current source. Voltage reference LT1004 and a potentiometer provide a voltage to program the current through the load (i.e., the electrochemical cell). The operational amplifier U1a attempts to produce a current equal to Vin/R_R, where R_R (i.e., R1–R7) is the current scaling resistor. A set of relays (K2-K4, Hamlin HE721C0510) is used to switch the current through the load. As drawn, the circuit shows the position of the switches when the current source is not supplying current to the external load. In this case, the bipotentiostat working electrode lead is con-

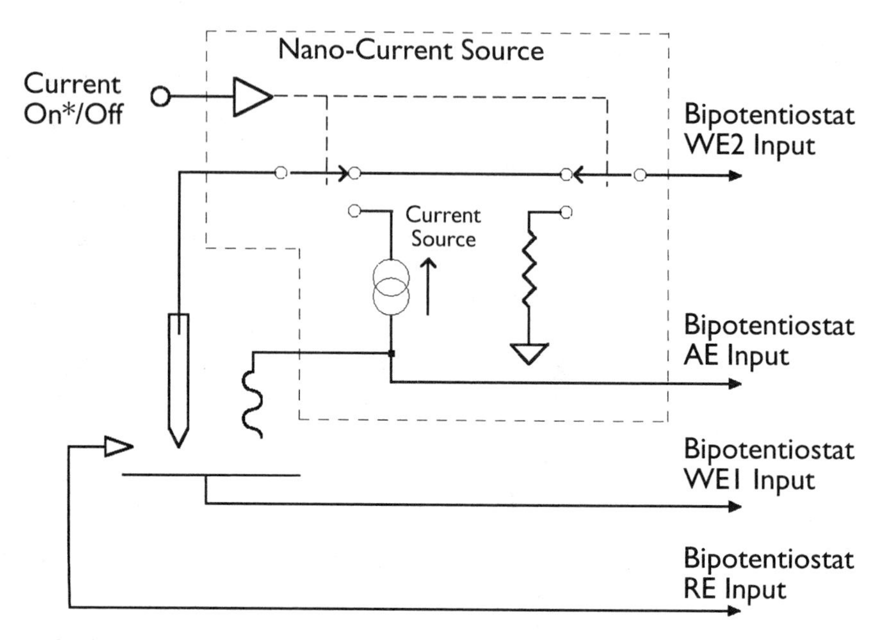

FIG. 12 Block schematic of a nanocurrent source for use in SECM. WE1, WE2, AE, and RE refer to working, auxiliary, and reference electrodes.

nected directly to the tip electrode. When the current source is switched on, the bipotentiostat working electrode lead is switched via relay K4 to a dummy cell (R13), and the current source is routed through the external load. The polarity of the current is switched by S2. Depending on the load, the OA may not have sufficient compliance voltage to produce the required current, and an overload lamp, D1, will light to indicate that condition.

The current source uses a 9 V alkaline battery to permit "floating," low-noise operation. The battery lifetime is in excess of 1 month with the use of micropower devices U1a and LT1004-1.2. Switching the device to 1 nA range when not in use will preserve the battery life. Before use, the battery can be checked by using a voltmeter to measure the voltage produced by applying 1 mA of current to a 1 to 5 kΩ load resistance. The voltage should be that predicted by Ohm's law, $V = iR$. Using a larger load resistance permits calibration of the output at lower currents. At the low current scales, a high-input impedance electrometer is required to accurately measure the voltage developed across the load resistor.

The lower part of the diagram is the switching circuit. Applying a ground potential to the input causes relay K1 (Hamlin HE721A0510) to become energized, lighting LED D2. K1 acts as a power switch for relays

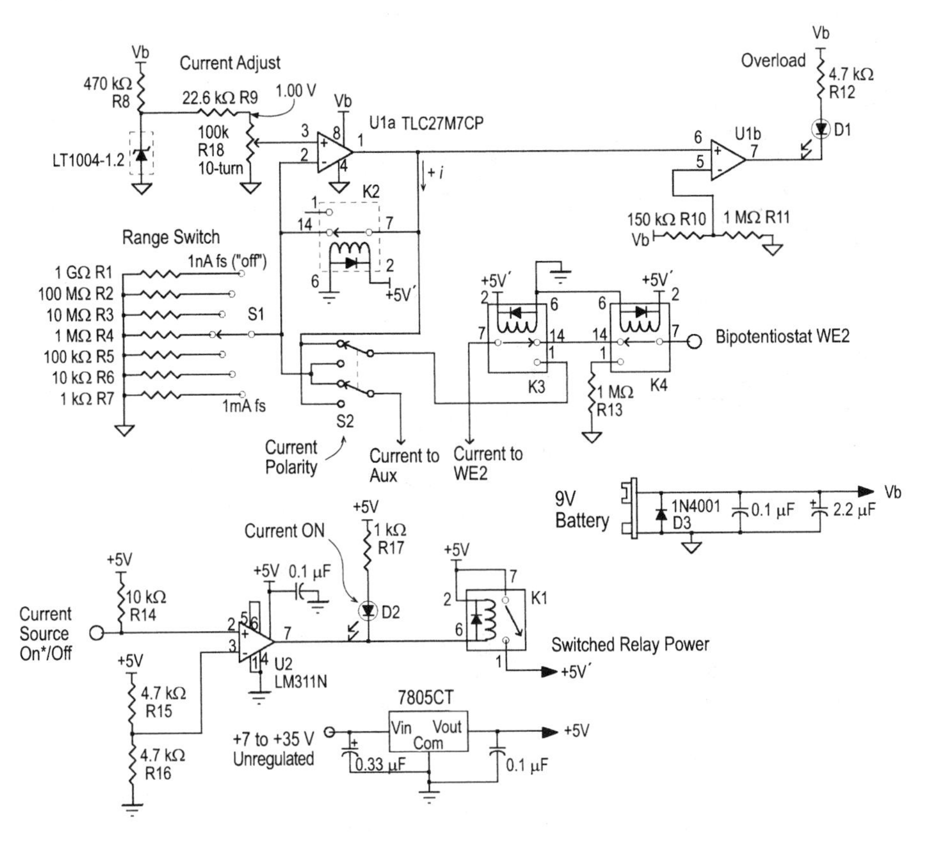

FIG. 13 Detailed electrical schematic of the nanocurrent source.

K2-4. A separate relay power supply is needed to supply the large currents required by the relays. The relay power supply can be an inexpensive "wall-wart." Note that the relay power supply and ground lines are completely isolated from the current supply section of the circuit.

In constructing the current source, resistors R1–R7 and R9–R11 should be at least 1% tolerance, and a counting type dial should be used on potentiometer R18. Also, ensure that the common of the current source is not connected at any point to the earth ground of the switching logic and relays. It may be difficult to purchase small quantities of high-value resistors; most vendors require a purchase of 50 or more units. Small quantities of precision gigaohm range resistors can be purchased from Axon, a neurophysiology instrumentation company. The basic design of the current source is easily modified to provide additional functions. For example, the current supply may be switched to supply constant current between two working electrodes. With a display to show the voltage between the WE and REF electrodes during constant current operation, the device operates as a nano-galvanostat. If provision for an external current-programming voltage is made, current sweeps or other functions can be produced. In the above cases, the programming voltage or output voltage must be galvanically isolated from earth ground.

C. Squawk Box

Another useful device is a "squawk" box. Connected to the WE output of a potentiostat or current transducer, the device provides an audible tone with a frequency that is proportional to the absolute value of an input voltage. A detailed schematic is shown in Figure 14. A tone output is helpful in a number of experimental situations. For example, during initial setup or approach of the tip to the substrate, the experimenter not only hears the approach but can also see the tip approach with an optical magnifier or video. In addition, a small change in signal on the computer screen can be difficult to observe of the limited dynamic range of the display. However, most people can detect very small tone shifts, and listening for a tone shift can aid in finding a target location on a surface. Finally, the tone allows an experimenter to remotely monitor a scan while engaged in other activities such as analyzing a previous scan. The experimenter can listen for the end of a scan or a tip crash.

The squawk box accepts voltage inputs of ± 10 V and emits tones up to 10 kHz. The dynamic range of the tone generator is about 10,000. Thus, a 1 mV signal change generates a 1 Hz output change. In addition, the device has pitch, volume, and on/off controls to better accommodate varying signal levels. With output frequencies of less than 80 Hz, a series of clicks rather than a continuous tone is heard. At the input, the voltage signal is rectified by a precision absolute value circuit. The rectified signal is applied to an AD537 voltage-to-frequency converter, which generates the tone. The AD537 chip is chosen for its excellent linearity and 80 dB dynamic range. The square wave output is amplified by the LM386 and sent to a $2\frac{1}{2}$ inch diameter speaker. The completed device is mounted in a small plastic box. A set of small (1/8″) holes drilled in the case serves as a sound port by simply gluing the speaker to the inside of the box.

D. Isolation Amplifier

Some SECM experiments require an isolation amplifier to avoid electrical interference. One example occurs with a potentiometric tip and a potentiostatic electrode, such as a pH-sensitive tip at a polarized corrosion test sam-

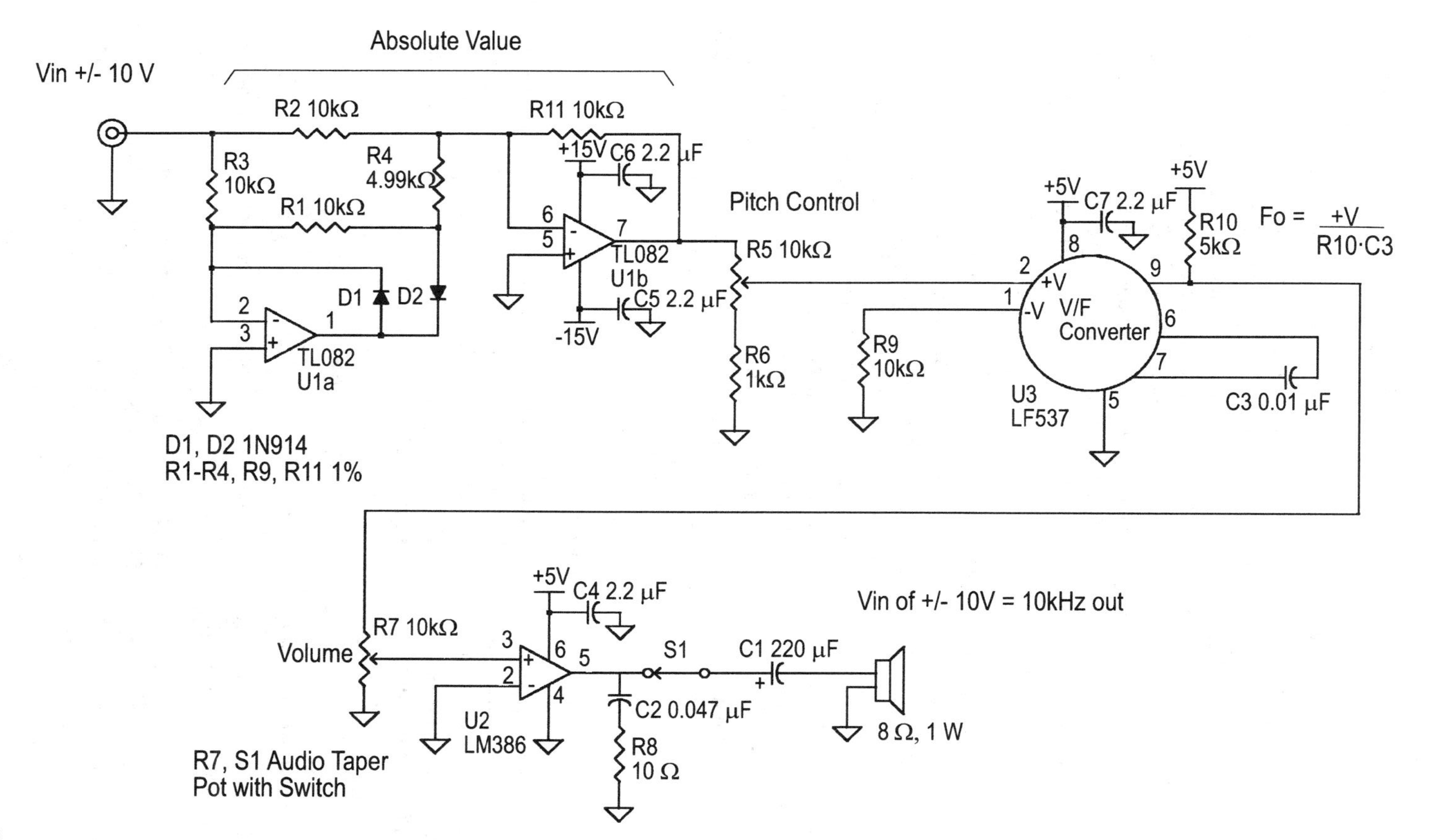

FIG. 14 Detailed electrical schematic of the squawk box.

ple. When the potentiometric and the potentiostatic electrode use the same reference electrode, polarization of the reference or potentiometric electrode is likely even if the electrometer is floating. The problem often is due to the recording process. If the ADC (analog-to-digital converter) in the computer has "single-ended" inputs, the return lines (i.e., common) of the electrometer and potentiostat output are set to ground potential, grounding the reference or potentiometric electrode. If the A/D converter has "differential" inputs, the return lines are often only isolated by 10–100 kΩ from the earth ground to reduce noise and provide a safety feature. Thus, differential inputs are effectively no better than single-ended inputs in isolating the reference electrode from ground.

Three solutions are suggested. The first is the use of a separate reference electrode for the potentiometric electrode, which may be unfeasible given the cell geometry. The second is to employ a potentiostat with a grounded reference electrode. A third option is the use of an isolation amplifier between the output of the electrometer and the A/D input on the computer. A low-cost "single-chip" solution is available with a Burr-Brown ISO113 isolation amplifier module (61). Capacitive coupling between the inputs and outputs provides true galvanic isolation of over 10^{14} Ω. Using the module involves connecting a few passive components and a power supply, as illustrated on the data sheet. Note that use of the isolation amplifier still requires that the electrometer be operated in a floating or isolated mode.

APPENDIX: SUPPLIERS

The following is a listing of addresses and web pages for companies mentioned by name in the text. Also included are suppliers of material or equipment useful in constructing or maintaining a SECM. It is not an exhaustive list, and emphasis is on North American suppliers. Addresses are for the United States unless otherwise noted.

A. Image Processing

1. General Scientific (Freeware)

The following are free and can be used for two-dimensional image analysis and publication quality display.

The Department of Dental Diagnostic Science provides Image Tool at the University of Texas Health Science Center, San Antonio, TX (http://www.uthscsa.edu/dig/itdesc.html). A full-featured image display and analysis program designed to run on the Windows operating system.

NIH Image is a powerful image manipulation and display program that runs on Macintosh Computers. Available from the NIH (http://rsb.info.nih.gov/nih-image). A Windows clone is available from the Scion Corporation (Frederick, MD, http://www.scioncorp.com).

Osiris is available from the Digital Imaging Unit at the University Hospital of Geneva and is designed for display of medical images. (http://www.expasy.ch/www/UIN/html1/projects/osiris/osiris.html).

2. Graphics Libraries (Freeware)

The following are freely available sets of functions that one can use in a script or a C or C++ program to do image manipulation and file I/O (i.e., gif or PostScript images).

ImageMagic is for the X Windows shell, but ports and shells are available for Windows. (http://www.wizards.dupont.com/cristy/ImageMagick.html).

Magick++ is a C++ library available at Magick++ (http://www.simplesystems.org/Magick++/).

3. SPM Analysis and Display

These programs are designed specifically to analyze and display SPM image data.

Image SXM runs as a script on NIH Image. Available at no cost from Steve Barrett, S.D.Barrett@liv.ac.uk (http://reality.caltech.edu/imagesxm.htm).

SPIP is available in a free version for images of 64 × 64 pixels. The commercial version is identical but allows larger images. Available from Image Metrology ApS, Anker Engelundsvej 1, Bldg. 307, DK-2800 Lyngby, DENMARK, +45 4525 5885 (http://www.imagemet.com).

4. Commercial Graphing Programs

These programs can be used to produce two- and three-dimensional (surface, grid, or contour) plots.

Igor Pro, WaveMetrics, Inc., P.O. Box 2088, Lake Oswego, OR 97035, 503-620-3001 (http://www.wavemetrics.com).

SigmaPlot, SPSS Inc., 233 S. Wacker Drive, 11th floor, Chicago, IL 60606-6307, 800-543-2185 (http://www.spss.com).

TecPlot, 13920 SE Eastgate Way Suite 220, Bellevue, WA 98005, 425-653-1200 (http://www.amtec.com). High-quality three-dimensional shaded surface-plots. Steep discount for academic users.

B. Positioning Equipment

These companies provide complete positioning solutions, motors, translation stages, and vibration isolation equipment.

Burleigh Instruments, Inc., Burleigh Park, P.O. Box E, Fishers, NY 14453, 716-924-9355 (http://www.burleigh.com). Supplies Inchworm positioners and inexpensive SPMs.

Edmund Scientific, 101 East Gloucester Pike, Barrington, NJ 08007, 609-573-6250 (http://www.edmundscientific.com). Also supplies video microscopy equipment.

Mad City Labs, Inc., 6666 Odana Rd. PMB 198, Madison, WI 53719, 608-298-0855 (http://madcitylabs.com/).

Melles Griot, 1770 Kettering Street, Irvine, CA 92614, 714-261-5600 (http://www.mellesgriot.com).

New Focus, Inc., 2630 Walsh Ave., Santa Clara, CA 95051, 408-980-8088 (http://www.newfocus.com).

Newport Corporation, 1791 Deere Ave., Irvine, CA 92606, 714-863-3144 (http://www.newport.com).

Piezosystem Jena Präzisionsjustierelemente GmbH, Wildenbruchstr. 15, D-07745 Jena/Thür., GERMANY, 49-3641-675390 (http://www.piezojena.com/index.html).

Polytec PI, Inc., Suite 212, 23 Midstate Drive, Auburn, MA 01501, 508-832-3456 (http://www.polytecpi.com).

C. Micromanipulation Equipment and Supplies

These companies supply equipment for microscopy and electrophysiology including amplifiers, micromanipulators, electrode holders, reference electrodes, pipet pullers, and microelectrodes.

FHC, Inc., 9 Main Street, Bowdoinham, ME 04008, 207-666-8190 (http://www.fh-co.com).

Ted Pella, Inc., P.O. Box 492477, Mountain Lakes Blvd., Redding, CA 96003, 530-243-2200 (http://www.tedpella.com). A leading supplier of microscopy supplies.

World Precision Instruments, International Trade Center, 175 Sarasota Center Boulevard, Sarasota, FL 34240, 941-371-1003 (http://www.wpiinc.com).

Sutter Instrument Co., 51 Digital Drive, Novato, CA 94949, 415-883-0128 (http://www.sutter.com). Makes a wide range of precision pipet pullers including the laser-based model P-2000.

D. Piezoelectric Materials and Actuators

These companies can supply piezoelectric material and devices such as tubes, stacked assemblies, unimorphs, and bimorphs.

DynaOptic Motion, 23561 Ridge Route, Suite U, Laguna Hills, CA 92653, 949-770-9911 (http://www.dynaoptics.com).

Piezo Kinetics, Inc., PO Box 756, Mill Road & Pine Street, Bellefonte, PA 16823, 814-355-1593 (http://www.piezo-kinetics.com).

Piezomechanik GmbH, Berg am Laim Str. 64, D-81673 München/Munich, GERMANY, (0)89-431-5583 (http://www.piezomechanik.com).

Stavely Sensors, Inc., 91 Prestige Park Circle, East Hartford, CT 06108, 860-289-3189.

Sensor Technology, P.O. Box 97, 20 Stewart Rd., Collingwood, Ont., CANADA, L9Y 3Z4, 705-444-1440 (http://www.sensortech.ca).

E. SPM Instrumentation

CH Instruments, Inc., 3700 Tennison Hill Drive, Austin, TX 78733, 512-402-0176 (http://chinstruments.com). Sole commercial supplier of a production SECM.

Digital Instruments, 112 Robin Hill Road, Santa Barbara, CA 93117, 805-967-1400 (http://www.di.com).

Molecular Imaging Co., 9830 S. 51st Street, Suite A124 Phoenix, AZ 85044, 480-753-4311 (http://www.molec.com).

Pacific Scanning Co., 2038 Foothill Blvd., Pasadena, CA 91107, 626-796-2626 (http://www.pacificscanning.com). Supplies SPM subsystems.

ThermoMicroscopes, 1171 Borregas Avenue, Sunnyvale, CA 94089, 408-747-1600 (http://www.thermomicro.com). Also supplies the Aurora NSOM with tuning-fork shear-force distance control.

Quesant Instrument Co., 29397 Agoura Road, Suite 104, Agoura Hills, CA 91301, 818-597-0311 (http://www.quesant.com). Inexpensive SPM systems.

F. Electrochemical Equipment

Makers of potentiostats, bipotentiostats, electrodes, and electrochemical cells are included.

Axon Instruments, Inc., 1101 Chess Drive, Foster City, CA 94401, 415-571-9500 (http://www.axon.com). A neurophysiology company with very high quality low-noise amplifiers and "voltage clamps" (i.e., potentiostats).

Bioanalytical Systems, Inc., 701 Kent Avenue, West Lafayette, IN 47906, 765-463-4527 (http://www.bioanalytical.com).

Cypress Systems, Inc., 2449 Iowa St # KL, Lawrence, KS 66046, 785-842-2565 (http://www.cypresshome.com). Supplies the EI-400 bi-potentiostat.

Perkin Elmer Instruments, Inc., Attn: Princeton Applied Research, 801 S. Illinois Ave., Oak Ridge, TN 37831, 865-481-2442 (http://www.par-online.com).

Gamry Instruments, Inc., 734 Louis Drive, Warminster, PA 18974, 215-682-9330 (http://www.gamry.com).

Pine Instrument Company, 101 Industrial Drive, Grove City, PA 16127-1091, 724-458-6391 (http://www.pineinst.com).

Solartron, Victoria Road, Farnborough, Hampshire, UK, GU14 7PW, +44 (0) 1252 376666 (http://www.solartron.com).

G. Electronics Suppliers or Manufacturers

Analog Devices, One Technology Way, P.O. Box 9106, Norwood, MA 02062, 781-329-4700 (http://www.analog.com). IC supplier.

Burr-Brown Co., PO Box 11400, Tucson, AZ 85734, 520-746-1111 (http://www.burr-brown.com). IC supplier.

Apex Microtechnology Co., 5980 N. Shannon Road, Tucson, AZ 85741-5230, 520-690-8600 (http://www.apexmicrotech.com). Specializes in high-power operational amplifiers.

Data Translation, Inc., 100 Locke Drive, Marlboro, MA 01752, 508-481-3700 (http://www.datx.com). Data capture and I/O boards.

Keithley Instruments, Inc., 28775 Aurora Road, Cleveland, OH 44139, 440-248-0400 (http://www.keithley.com). Data capture boards, I/O, picoammeters, and electrometers.

National Instruments Co., 11500 N Mopac Expwy, Austin, TX 78759, 512-794-0100 (http://www.natinst.com). Data capture boards, I/O, and LabView software.

H. Useful Web Pages

The Homebrew STM Page (July 28, 1999) (http://www.metanet.org/mnt/lib/homebrew_stm.html). A useful discussion of how to build your own STM.

SPM Links (July 28, 1999) (http://www.ifm.liu.se/Applphys/spm/links.html). Thorough collection of links of interest to the SPM investigator.

NCSU SPM Links (July 28, 1999) (http://spm.aif.ncsu.edu/spmlinks.htm).

I. General Supplies and Materials

McMaster-Carr, P.O. Box 4355, Chicago, IL 60680, 630-833-0300 (http://www.mcmaster.com/). The ultimate hardware store.

REFERENCES

1. J Kwak, AJ Bard. Scanning electrochemical microscopy—apparatus and 2-dimensional scans of conductive and insulating substrates. Anal Chem 61:1794–1799, 1989.
2. DO Wipf, AJ Bard. Scanning electrochemical microscopy 7. Effect of heterogeneous electron-transfer rate at the substrate on the tip feedback current. J Electrochem Soc 138:469–474, 1991.
3. G Wittstock, H Emons, TH Ridgway, EA Blubaugh, WR Heineman. Development and experimental evaluation of a simple system for scanning electrochemical microscopy. Anal Chim Acta 298:285–302, 1994.
4. The Piezo Book. Fishers, NY: Burleigh Instruments.
5. Y Selzer, I Turyan, D Mandler. Studying heterogeneous catalysis by the scanning electrochemical microscope (SECM): the reduction of protons by methyl viologen catalyzed by a platinum surface. J Phys Chem B 103:1509–1517, 1999.
6. RC Engstrom, T Meany, R Tople, RM Wightman. Spatiotemporal description of the diffusion layer with a microelectrode probe. Anal Chem 59:2005–2010, 1987.
7. I Kapui, RE Gyurcsanyi, G Nagy, K Toth, M Arca, E Arca. Investigation of styrene-methacrylic acid block copolymer micelle doped polypyrrole films by scanning electrochemical microscopy. J Phys Chem B 102:9934–9939, 1998.
8. K Borgwarth, DG Ebling, J Heinze. Scanning electrochemical microscopy—a new scanning-mode based on convective effects. Ber Bunsen Ges Phys Chem 98:1317–1321, 1994.
9. C Gerber, R Binnig, H Fuchs, O Marti, H Rohrer. Scanning tunneling microscope combined with a scanning electron microscope. Rev Sci Instrum 57:221–224, 1986.
10. G Binnig, H Rohrer. Scanning tunneling microscopy. IBM J Res Develop 30:355–369, 1986.
11. S Kliendiek, KH Hermann. A minaturized scanning tunneling microscope with large operation range. Rev Sci Instrum 64:692–693, 1992.
12. F Besenbacher, K Laegsgaard, K Mortensen, U Nielsen, I Stensgaard. Compact, high-stability, "thimble-size" scanning tunneling microscope. Rev Sci Instrum 59:1035–1038, 1988.
13. R Howland, L Benetar. A Practical Guide to Scanning Probe Microscopy. Park Scientific Instruments, 1996.
14. DW Pohl. Some design criteria in scanning tunneling microscopy. IBM J Res Develop 30:417–427, 1986.
15. GE Poirier, JM White. Diffraction grating calibration of scanning tunneling microscope piezoscanners. Rev Sci Instrum 61:3917–3918, 1990.

16. J Akila, SS Wadhwa. Correction for nonlinear behavior of piezoelectric tube scanners used in scanning tunneling and atomic force microscopy. Rev Sci Instrum 66:2517–2519, 1995.
17. RC Engstrom, M Weber, DJ Wunder, R Burgess, S Winguist. Measurements within the diffusion layer using a microelectrode probe. Anal Chem 58:844–848, 1986.
18. RD Martin, PR Unwin. Scanning electrochemical microscopy. Kinetics of chemical reactions following electron-transfer measured with the substrate-generation-tip-collection mode. J Chem Soc, Faraday Trans 94:753–759, 1998.
19. Apex Microtechnology. http://www.apexmicrotech.com/.
20. S-I Park, CF Quate. Theories of the feedback and vibration isolation systems for the scanning tunneling microscope. Rev Sci Instrum 58:2004–2009, 1987.
21. AJ Bard, LR Faulkner. Electrochemical Methods: Fundamentals and Applications. New York: John Wiley & Sons, 2000, Chapter 14.
22. HJ Huang, P He, LR Faulkner. Current multiplier for use with ultramicroelectrodes. Anal Chem 58:2889–2891, 1986.
23. Cellular Neuroscience Product Line. Axon Instruments, Inc. http://www.axon.com/CN_Neuroscience.html.
24. UE Spichiger, A Fakler. Potentiometric microelectrodes as sensors and detectors. Magnesium-selective electrodes as sensors, and Hofmeister electrodes as detectors for histamine in capillary electrophoresis. Electrochim Acta 42:3137–3145, 1997.
25. WE Morf, NF Derooij. Micro-adaptation of chemical sensor materials. Sensor Actuator A Phys 51:89–95, 1995.
26. P Horowitz, W Hill. The Art of Electronics. Cambridge: Cambridge University Press, 1989, p. 431.
27. Low Level Measurements. Cleveland, OH: Keithley Instruments, 1998, Chapter 3.
28. Low Level Measurements. Cleveland, OH: Keithley Instruments, 1998, Chapter 4.
29. AD515A. Analog Devices. http://www.analog.com/pdf/1138_a.pdf.
30. P Horowitz, W Hill. The Art of Electronics. Cambridge: Cambridge University Press, 1989, pp. 391–466.
31. RA Pease. Troubleshooting Analog Circuits. Boston: Newnes, 1993, pp. 50–64.
32. RC Dorf. Electrical Engineering Handbook. Boca Raton, FL: CRC Press, 1997, p. 416.
33. WH Press, BP Flannery, SA Teukolsky, WT Vetterling. Numerical Recipes in C. Cambridge: Cambridge University Press, 1988.
34. Welcome to Edmund Scientific. http://www.edmundscientific.com/.
35. CHI 900 SECM. CH Instruments. http://chinstruments.com/secm.htm.
36. DO Wipf, AJ Bard. Scanning electrochemical microscopy 15. Improvements in imaging via tip-position modulation and lock-in detection. Anal Chem 64: 1362–1367, 1992.

37. M Ludwig, C Kranz, W Schuhmann, HE Gaub. Topography feedback mechanism for the scanning electrochemical microscope based on hydrodynamic-forces between tip and sample. Rev Sci Instrum 66:2857–2860, 1995.
38. GM Hieftje. Signal-to-noise enhancement through instrumental techniques. Anal Chem 44:81A–88A, 1972.
39. AJ Bard, F-RF Fan, DT Pierce, PR Unwin, DO Wipf, FM Zhou. Chemical imaging of surfaces with the scanning electrochemical microscope. Science 254:68–74, 1991.
40. JV Macpherson, PR Unwin. Scanning electrochemical microscopy as a probe of silver-chloride dissolution kinetics in aqueous solutions. J Phys Chem 99: 14824–14831, 1995.
41. HV Malmstadt, GG Enke, SR Crouch. Electronics and Instrumentation for Scientists. Reading, MA: Benjamin/Cummings Publishing Co, 1981, pp. 409–455.
42. RC Dorf. Electrical Engineering Handbook. Boca Raton, FL: CRC Press, 1997, pp. 2258–2331.
43. NanoScope® IIIa Technical Specifications. http://www.di.com/Products/NS3/a/techspecs.html.
44. S-I Park, CF Quate. Scanning tunneling microscope. Rev Sci Instrum 58:2010–2017, 1987.
45. DO Wipf, AJ Bard, DE Tallman. Scanning electrochemical microscopy 21. Constant-current imaging with an autoswitching controller. Anal Chem 65: 1373–1377, 1993.
46. G Binnig, DPE Smith. Single-tube three-dimensional scanner for scanning tunneling microscopy. Rev Sci Instrum 57:1688–1689, 1986.
47. H Shiku, RC Dunn. Near-field scanning optical microscopy. Anal Chem 71: 23A–29A, 1999.
48. DW Pohl, W Denk, M Lanz. Optical stethoscopy: Image recording with resolution $\lambda/2$. Appl Phys Lett 44:651–653, 1984.
49. H Heinzelmann, DW Pohl. Scanning near-field optical microscopy. Appl Phys A 59:89–101, 1994.
50. G Shi, LF Garfias-Mesias, WH Smyrl. Preparation of a gold-sputtered optical fiber as a microelectrode for electrochemical microscopy. J Electrochem Soc 145:2011–2016, 1998.
51. K Karraï, RD Grober. Piezo-electric tuning fork tip-sample distance control for near field optical microscopes. Ultramicroscopy 61:197–205, 1995.
52. K Karraï, RD Grober. Piezoelectric tip-sample distance control for near field optical microscopes. Appl Phys Lett 66:1842–1844, 1995.
53. PI James, LF Garfias-Mesias, PJ Moyer, WH Smyrl. Scanning electrochemical microscopy with simultaneous independent topography. J Electrochem Soc 145:L64–L66, 1998.
54. J Barenz, O Hollricher, O Marti. An easy-to-use non-optical shear-force distance control for near-field optical microscopes. Rev Sci Instrum 67:1912–1916, 1996.
55. RC Dorf. Electrical Engineering Handbook. Boca Raton, FL: CRC Press, 1997, pp. 1003–1014.

56. P Horowitz, W Hill. The Art of Electronics. Cambridge: Cambridge University Press, 1989, pp. 454–466.
57. P Horowitz, W Hill. The Art of Electronics. Cambridge: Cambridge University Press, 1989, pp. 1026–1034.
58. DO Wipf, EW Kristensen, MR Deakin, RM Wightman. Fast-scan cyclic voltammetry as a method to measure rapid heterogeneous electron-transfer kinetics. Anal Chem 60:306–310, 1988.
59. P Horowitz, W Hill. The Art of Electronics. Cambridge: Cambridge University Press, 1989, pp. 263–284.
60. Shell Chemical Products—EPON (TM) Resins. Shell Chemicals Ltd. http://www2.shellchemical.com/CMM/WEB/GLOBCHEM.NSF/Products/EPONS.
61. ISO113.PDF. Burr-Brown Corp. http://www.burr-brown.com/download/DataSheets/ISO113.pdf.

3

THE PREPARATION OF TIPS FOR SCANNING ELECTROCHEMICAL MICROSCOPY

Fu-Ren F. Fan and Christophe Demaille

The University of Texas at Austin
Austin, Texas

I. INTRODUCTION

In this chapter, several methods for the fabrication of different types of amperometric tips suitable for SECM are described. We have also suggested some methods for microelectrode fabrication, which have not yet been tested for SECM but may provide alternative ways for its tip preparation. Section II.A describes the techniques for the preparation of various metallic microelectrodes, including Pt, Ir-Pt, Au, Hg, and W. The manufacture of carbon microelectrodes is presented in Section II.B. Most of these tips are encapsulated in or supported with glass capillaries. Other coating materials and techniques are treated in Section II.C.

While most of the SECM work has been carried out with amperometric tips for measuring feedback current or for use in the generation/collection mode, other types of tips such as potentiometric and enzymatic tips are also possible but are only briefly described in Section II.D. The readers who are interested in these tips are encouraged to refer to the appropriate chapters in this monograph. Finally, Section III deals with an approach to determining the shape of an ultramicroelectrode from its SECM response.

II. PREPARATION TECHNIQUES

A. Metal Microelectrodes

1. *Disk-in-Glass Microelectrodes*

The conventional technique frequently used for the preparation of SECM tips is largely based on the fabrication technique for disk-shaped microelectrodes, which have been described in the literature (1–3). The procedures for the manufacture of this type of SECM tips have been described in detail previously (4) and are only presented briefly here.

A fine wire of Pt, Au or a carbon fiber of the desired radius ($\geq$5 μm, commercially available) is placed in a 10-cm-long, 1-mm i.d. Pyrex (or soft glass for Au) tube sealed at one end. The open end of the tube is connected to a vacuum line and heated with a nichrome wire helix for approximately half an hour to desorb any impurities or moisture on the wire and glass tube. One end of the wire is then sealed in the glass at the closed end of the tube by increasing the heating coil temperature. The glass should melt around the wire for at least 1–2 mm at the tip of assembly. The whole glass tube, including the part that seals the wire, should be straight. After the glass has cooled, the electrode is inspected under a microscope to see whether the wire is completely sealed at the tip and to make sure that there are no trapped air bubbles. The sealed end is polished with coarse sandpaper until the cross section of the wire is exposed, and then successively with 15, 6, 3, 1, and 0.25 μm diamond paste (Buehler, Lake Bluff, IL). Electrical connection to the unsealed end of the wire is made with silver paint (Ekote, No. 3030, Acme Chemicals and Insulation Co., New Haven, CT) or colloidal silver (Cat. # 16031, Ted Pella, Inc., P.O. Box 510, Tustin, CA) to a Cu wire. After this, a small amount of Torr Seal Epoxy (Varian Associates, Palo Alto, CA) is packed into the open end of the glass capillary. This seals the barrel and provides strain relief for the contact wire. The glass wall surrounding the conducting disk is conically sharpened with emery paper (grit 600) and 6 μm diamond paste, with frequent checking with an optical microscope until the diameter of the flat glass section is less than 10 times that of the conducting disk (i.e., RG $<$ 10). A small RG decreases the possibility of contact between glass and substrate because of any slight deviation in the axial alignment of the tip as the tip approaches the substrate. So far, this technique for the sharpening of the glass sheath remains to be improved. This technique has been successfully used to prepare tips of diameter larger than 1 μm. The manufacture of smaller tips from Wollaston wire (Goodfellow Metals, Cambridge, UK) based on this technique is also possible, but it takes a lot of practice and patience (4). Very sturdy, polishable, and long lifetime tips can be prepared by this technique, although the procedures are tedious and time-consuming. Preparation of submicron SECM tips by this technique is very challenging.

Reduction of the size of SECM probes is important, since the resolution of SECM depends strongly on the size of tip used. In order to achieve this goal, one has to thin down the diameter of commercially available metal wires by electrochemical etching or by pulling in glass capillary as described below. Further coating and exposing steps are then necessary, as in the case of unetched wires, to expose only the electrode surface. One method for the preparation of submicron Pt disks in glass was developed by Lee et al. (5). A 1 to 2 cm long Pt wire (25 μm diameter) is sharpened by electrochemical etching in a solution containing saturated $CaCl_2$ (60% by volume), H_2O (36%), and concentrated HCl (4%) at 2 V rms ac applied with a Variac transformer (6,7), as described below. A Pyrex capillary with a tip diameter greater than 100 μm is drawn from a piece of Pyrex tubing (2 mm o.d., and 1 mm i.d.) by using a microelectrode puller (Cat. No. 51217, Stoelting Co., Chicago, IL) with heat setting at 65 and pull setting at 100. The tip of the capillary is sealed by heating in a gas flame for 1 s. The sharpened wire is transferred inside the glass capillary, and the open end of the tube is connected to a vacuum line. The capillary tip is then heated with four loops (1/4 inch i.d.) of nichrome wire (1.3 mm thick) at an ac voltage of 3.8 V (temperature ~ 800°C). Excess glass near the tip of the Pt wire is removed by heating the electrode in the resistive heating coil, which pulls the glass toward the Pt wire. Heating is halted just before the Pt wire is exposed. The diameter of the exposed Pt disk can be controlled from 0.2 to 25 μm by controlling the extent of the final polishing. Further heating for 0.5–1 minutes causes the glass to pull away from the Pt wire so that the small exposed Pt disk protrudes from the sealed glass. Electrical connection to the unsealed end of the wire and the packing of Torr Seal Epoxy into the open end of the glass capillary are made as described above.

2. Submicrometer Glass-Encapsulated Microelectrodes

An alternative preparation procedure for submicron glass-encapsulated SECM tips involved pulling metal wires with the glass capillary, with a pipet puller (8,9). This technique is quite attractive and is under development by Fish et al. (10). Platinum or silver micro-wire (50 μm diameter) and borosilicate tubes (1.2 mm o.d.) were used as starting materials. Several different parameters of the pulling program can be varied in order to control the shape and size of the micropipet. A more detailed description of the pulling procedure can be found elsewhere (10). Immediately after pulling, the metal core is entirely cover with glass, and the next step is to expose the tip of the wire. This was done by micro-polishing with a micro-pipet beveler (Model BV-10, Sutter Instrument Company, 40 Leveroni Court, Novato, CA) equipped with a micro-manipulator and a working distance stereomicroscope. The micromanipulator was used to move the pipet toward the slowly rotating abrasive disk covered with 0.05 μm alumina (Sutter). Initially, the axis of the pipet is perpendicular to the plane of the disk. When the pipet touches the disk, the small change in the angle can be detected microscopically. The size of the prepared tip increases with the length of polishing. It has been reported that only a fraction of the complete turn of the polishing wheel is required to produce a nanometer-sized polished electrode. The electrode resistance between the tip and an InGa alloy liquid was used to estimate the surface area of the electrode during preparation. When the tip is completely covered with glass, the resistance is extremely high, and it decreases sharply when the tip becomes exposed to solution by polishing. By this technique, electrodes with an effective radius as small as 2 nm and as large as a few micrometers can be manufactured.

Figure 1A shows a steady-state voltammogram of 10 mM $Ru(NH_3)_6^{3+}$ at a polished Pt electrode, whose effective radius is found to be 84 nm. At a scan rate of 20 mV/s, the voltammogram is well shaped with a flat plateau. At a higher scan rate, e.g., 2 V/s, no peaks characteristic for non–steady-state diffusion could be detected, and the limiting current plateau remained flat. This indicates that the size of the electrode is significantly smaller than 1 μm. A modest contribution of the charging current to the total current was consistent with the apparent radius of about 100 nm. The response of a smaller tip (radius ~ 22 nm, thick line in Fig. 1B) is in good agreement with the theory (thin line). The current at an even smaller (radius ~ 2 nm) electrode (Fig. 1C) is slightly noisy, but its voltammogram remains sigmoidal and retraceable at scan rates as high as several volts per second. This indicates that the true area of the electrode is very small. The low background currents observed at these polished electrodes indicate that there is no apparent solution leakage through the cracks in the glass insulator.

An SECM approach curve under favorable condition for a polished Pt tip to a Pt substrate is shown in Figure 2. Good agreement between the experimental data and the theoretical curve for a microdisk electrode and a relatively large maximum value of the normalized tip current suggest that the tip is a planar electrode with an effective radius of ~380 nm. It is quite unlikely that such an approach curve can be obtained with an electrode having the conducting disk deeply recessed behind the insulating glass sheath.

3. *Electrochemical Etching of Metal Wires*

The electrochemical etching procedures usually involve the anodic dissolution of the metals. There are two ways in which this can be done: an alternating current (ac) etch or direct current (dc) etch according to the applied potential. Each procedure gives a different tip shape; the ac-etched tips have a conical shape and much larger cone angles than the dc-etched tips. The dc-etched tips, on the other hand, have the shape of a hyperboloid and are much sharper than ac-etched tips. The etching procedures used for several metal wires are described in the following sections.

Pt, Ir, or Pt-Ir Alloy. A 1 to 2 cm long Pt or Pt-Ir (80-20%) wire (125 or 250 μm diameter) is sharpened by electrochemical etching in a solution containing saturated $CaCl_2$ (60% by volume), H_2O (36%), and concentrated

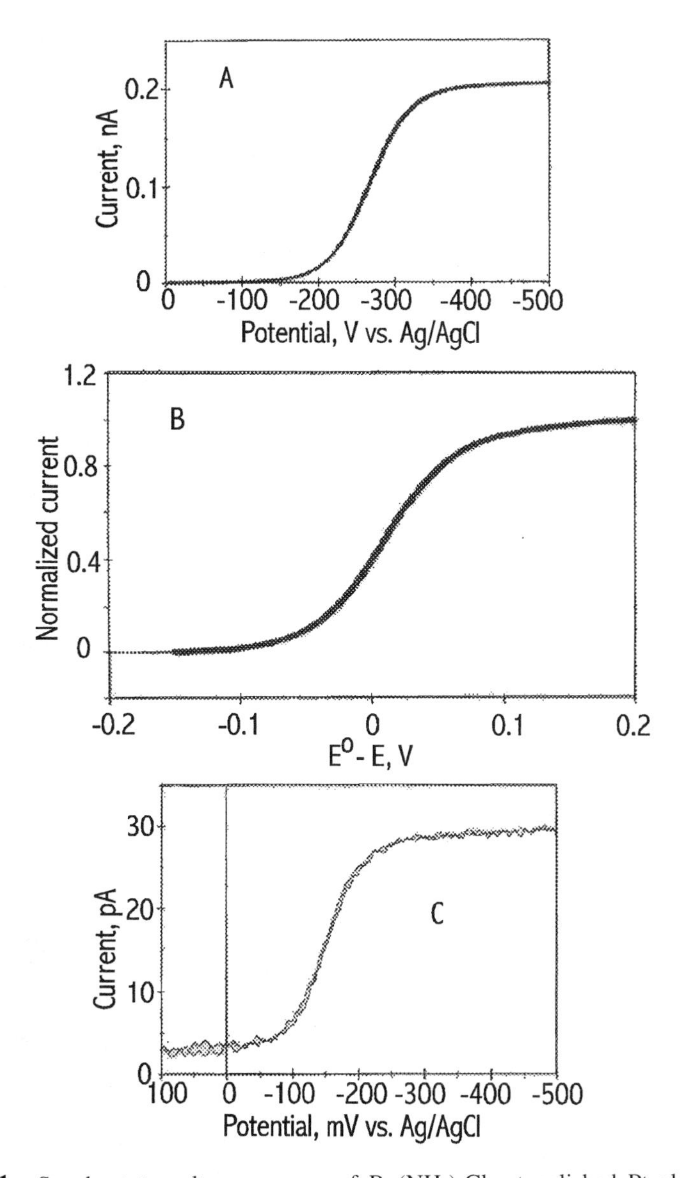

FIG. 1 Steady-state voltammograms of $Ru(NH_3)_6Cl_3$ at polished Pt electrodes. Solution contained (A) 10 mM $Ru(NH_3)_6Cl_3$ and 1 M KCl; (B) 10 mM $Ru(NH_3)_6Cl_3$ and 0.5 M KNO_3; and (C) 50 mM $Ru(NH_3)_6Cl_3$ and 0.5 M KCl. The effective electrode radius (assuming disk geometry) and the potential sweep rate were (A) a = 84 nm, v = 20 mV/s; (B) a = 22 nm, v = 5 mV/s; and (C) a = 2 nm, v = 20 mV/s. The thin line in (B) is the theory for a 22 nm microdisk. (Reprinted with permission from Ref. 9. Copyright © 1997 American Chemical Society.)

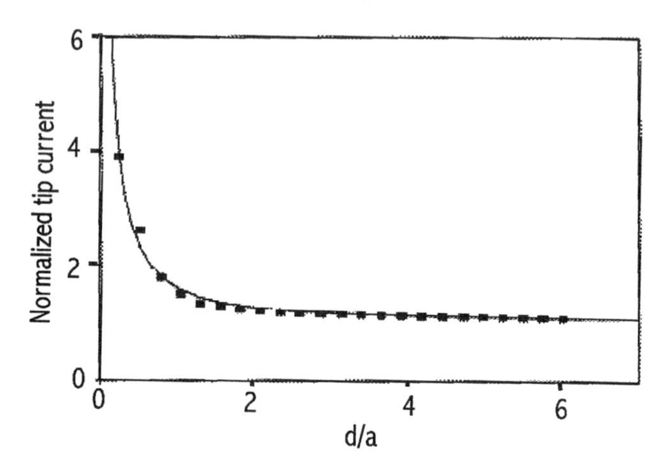

FIG. 2 The SECM approach curve obtained with a polished submicrometer-sized tip over a polished Pt substrate. The thin line shows the theoretical curve computed for disk-shaped 380 nm radius tip. (Reprinted with permission from Ref. 9. Copyright © 1997 American Chemical Society.)

HCl (4%) at ~20 V rms ac applied with a Variac transformer (6,7). Lower voltage (e.g., 2 V rms ac) is used for thinner wires (e.g., 25 μm diameter). A carbon rod or plate serves as the counterelectrode in a two-electrode etching cell. The etching current, which depends on the area of immersed wire and the applied voltage, was adjusted to an initial value of ~0.5 A (for 250 μm diameter Pt-Ir rod) by varying the immersed depth of the metal wire. The etching proceeds uniformly over the immersed wire except at the extreme lower end and the air/solution interface where the etching rate is faster. This process produces a neck shape near the interface. As the etching proceeds, the neck-like region becomes thinner and thinner, and eventually the lower portion falls off. The current shows a uniform decrease during etching. However, it decreases abruptly as the lower portion falls off. To get a very sharp tip with a small protrusion at the end, one needs to switch off the circuit as soon as the current abruptly drops. Failing to switch off the circuit as soon as the lower portion falls off, the upper part of the wire will continue to etch as long as it remains in the electrolyte under an applied potential and will end up with a rounded tip end. The switching off can be best achieved by means of an electronic circuit. Several circuits have been designed to monitor these electrochemical etching or polishing processes (11–13). It has been found that the cutoff time of the etch circuit has a significant effect on the radius of curvature and cone angle of the etched tip; i.e., the faster the cutoff time, the sharper the tip (13).

Au, Ag, or W. These metals are etched by a similar technique but with different etching solutions. The commonly used etching solution for Au and Ag is 3 M NaCN in 1 M NaOH, while 3 M NaOH solution is used for W wire etching. The cyanide facilitated the complexation and dissolution of metal from the immersed wire, and the NaOH inhibited the formation of the toxic HCN. All electrochemical-etching processes were performed in a fume hood. It should be mentioned that placing the tungsten tip in an aqueous solution or exposing it to air results in the formation of a surface oxide. There are several cleaning procedures for the removal of the surface oxide (14–19). The ease of oxide formation prevents W from being a useful electrode material for SECM amperometric tips in aqueous solutions. However, it could be beneficial as a tip used for imaging in humid air as described in the chapter on SECM imaging in this monograph.

4. Self-Assembled Spherical Gold Microelectrodes

Nano-Gold-Particle Self-Assembly Process. This technique, developed by Bard and coworkers (20), relies on the finding of Schiffrin and coworkers, who reported that nanometer sized gold nano-particles could be self-assembled, using thiols cross-linkers, to form conductive bulk materials (21) and multilayer thin films (22). The fabrication technique itself originated from the idea of confining the dithiol linker (1,9-nonane dithiol) inside the micrometer-sized tip of a pulled capillary that was then immersed into a solution of octyl ammonium protected 8-nm gold nanoparticles (20), forcing the encounter of the gold nanoparticles and of the dithiol linker to occur only at the very end of the tip. Smooth spherical shiny structures were observed to form at the end of the tip (Fig. 3). The growth of such spherical structures raised numerous questions about their growth mechanism. The growth process was, however, qualitatively represented, as shown in Figure 4. Nanoparticles diffused toward the tip, where they were cross-linked by dithiol molecules. A diffusion-limited growth would, however, have resulted in a dendritic structure. The fact that smooth spherical structures were obtained was indicative of some mobility of the individual particles within the assembly. The particles can accommodate themselves into a sphere, which is the most favorable geometry to minimize the surface energy. That the nanoparticles maintained their individualities inside the material was demonstrated by growing a similar material out of porous glass and studying it by x-ray diffraction (23). The width of the diffraction peak showed an average domain size corresponding to the nanoparticle size. Moreover, the high degree of self-organization of the particles within the self-assembled structure was suggested by the presence of a peak at low diffraction angle corresponding to a lattice parameter of the order of the nanoparticle. This is consistent with previous TEM observations that solutions of similar nano-

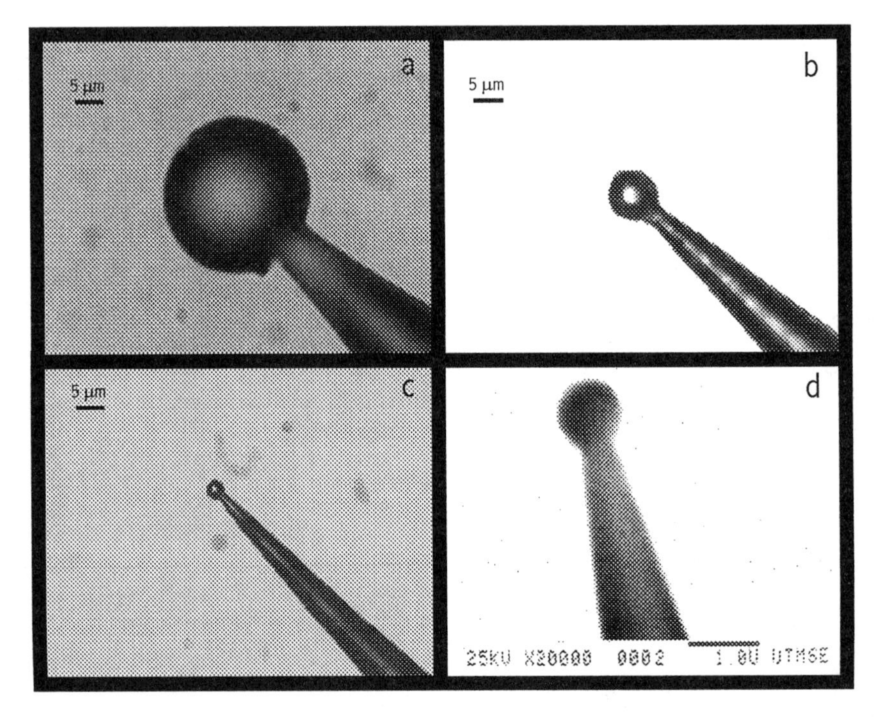

FIG. 3 Optical (a–c) and SEM (d) pictures of the self-assembled spherical gold microelectrodes. The horizontal bar at the lower corner of (d) represents 1 μm. Electrode diameter: (a) 26, (b) 8, (c) 3, and (d) 0.9 μm. (Reprinted with permission from Ref. 20. Copyright © 1997 American Chemical Society.)

particles, when reacted with dithiols, form ordered globular aggregates on the 100 nm scale rather than random dendritic structures (21). Similar preparations were also observed to lead to the formation of two-dimensional superlattices of gold nanoparticles (24). The fact that the spheres have diameters greater than the capillary tip size indicates that the sphere growth continues even though the tip opening is already closed by numerous layers of cross-linked nanoparticles. This indicates that the dithiol molecules are able to move within the growing spheres possibly, as shown in Figure 4, by replacement reactions. The sphere then grows until all the dithiol is consumed. By varying the capillary size, spherical electrodes having diameters from 1 to 30 μm could be easily obtained. The sphere diameter ranged from 2 to 20 times the capillary tip diameter, depending on the amount of dithiol inside the tip.

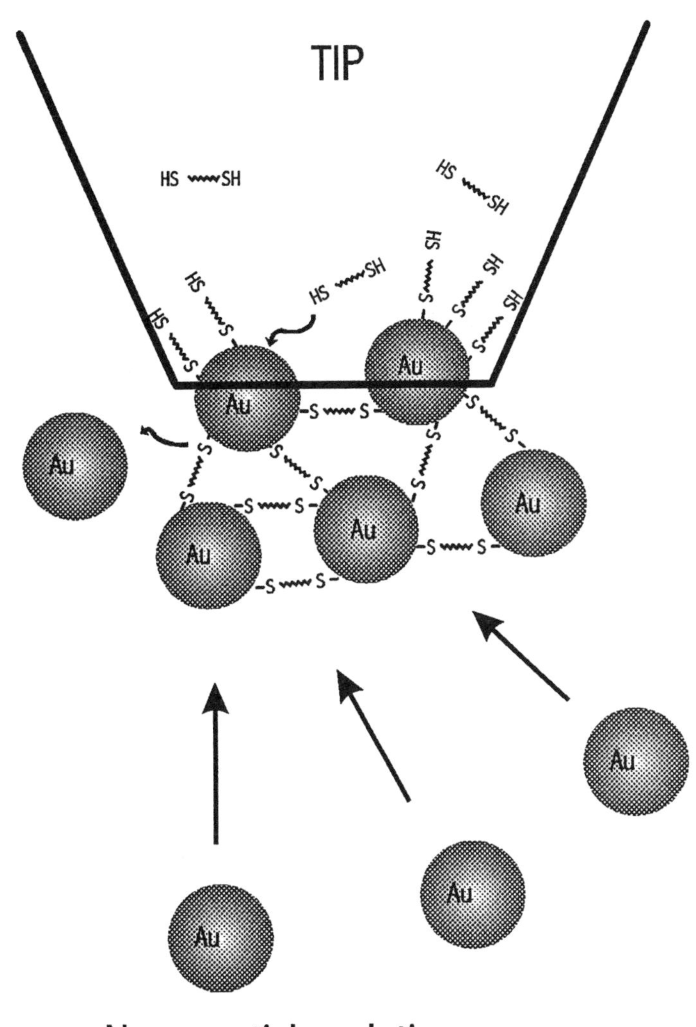

FIG. 4 Schematic picture of the self-assembly process leading to the formation of gold microspheres at the end of the micropipet tip. (Reprinted with permission from Ref. 20. Copyright © 1997 American Chemical Society.)

The well-defined spherical shape of these structures and their smoothness made them very attractive to be used as microelectrodes and as SECM tips. In order to establish an electrical contact with these spherical assemblies, the quartz capillary carbonization technique (described later) was used. This time it was desirable not to close the capillary but to still obtain conductive carbon coverage on the inner walls of the capillary. This could be achieved using the described carbonizing set-up and introducing the capillary into the heated zone only once. This carbon-coated capillary was then used for the preparation of spherical gold microelectrode based on the procedures as summarized in Figure 5 and resulted in self-assembled spherical gold microelectrodes such as those presented in Figure 3.

Electrochemical Behavior of Self-Assembled Spherical Gold Microelectrodes. The electrochemical behavior of these spherical microelectrodes

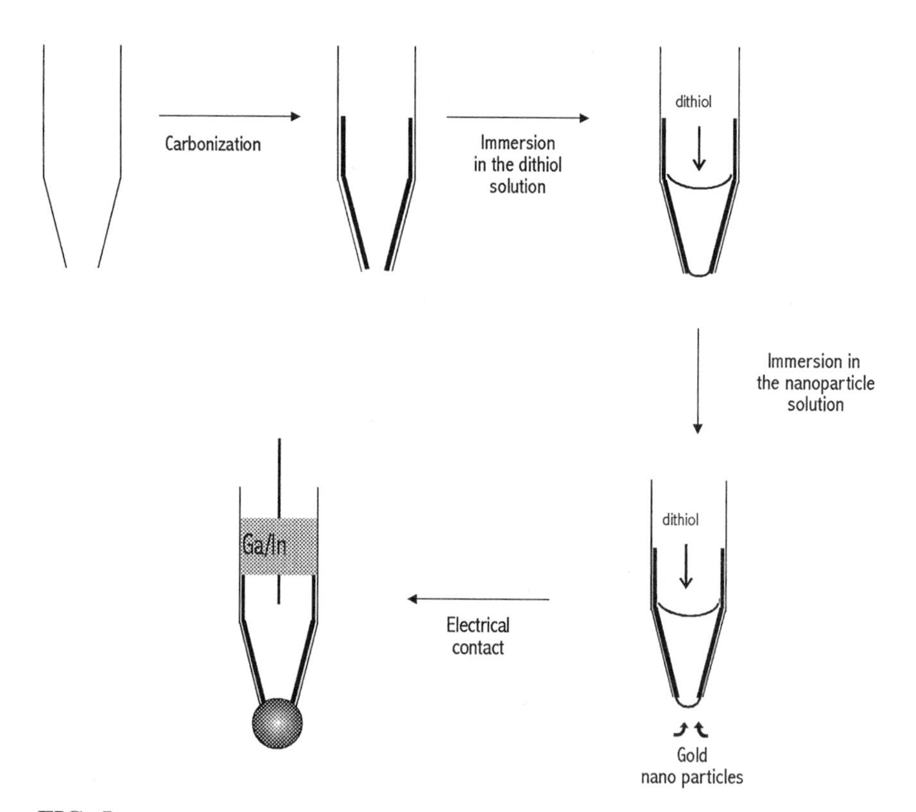

FIG. 5 Schematic diagram for the preparation of the self-assembled spherical gold microelectrodes. (Reprinted with permission from Ref. 20. Copyright © 1997 American Chemical Society.)

was first characterized by recording the gold oxide film formation/reduction in acidic conditions. The voltammogram obtained, presented in Figure 6, is very similar to the one expected for a bulk gold surface. Moreover, the electrode surface as deduced from the area under the oxide peak (25) was found to correspond to the geometrical surface as calculated from that measured optically. This was indicative of the high smoothness and the absence of porosity in the self-assembled structures. The use of the self-assembled spherical structures as microelectrodes was further tested by cyclic voltammetry both in aqueous (Fig. 7a) and in nonaqueous media (Fig. 7b). In every case, nernstian behavior was observed. The steady-state current (26) corresponded well with the one expected from the size of the electrode measured optically. This ideal behavior was maintained even at high scan rates (up to 10 V/s), especially in nonaqueous solvents. As the scan rate was raised, any porosity or a bad seal between the glass and the sphere would cause thin-layer cell behavior in voltammogram (as observed in the case of pyrolytic carbon microelectrodes, described below). It was therefore concluded that the self-assembled spherical gold microelectrodes, although consisting of individual gold nanoparticles, have an electrochemical behavior very similar to that of bulk gold microelectrodes. Although the resistivity of the material is expected to be higher than that of bulk gold, it is still too small to have

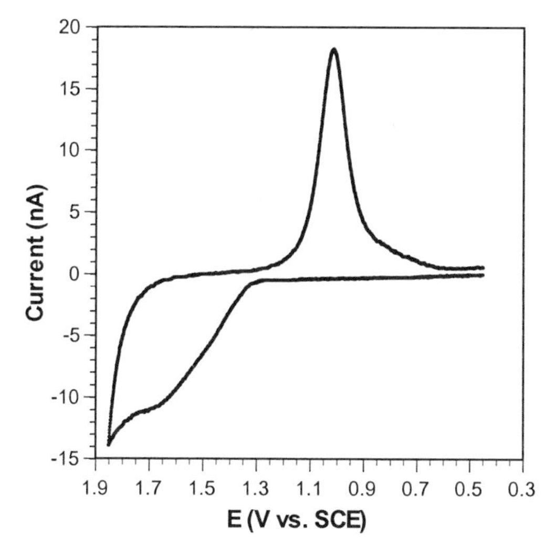

FIG. 6 Cyclic voltammogram of a self-assembled spherical gold microelectrode in 0.5 M H_2SO_4. Scan rate 1 V/s. (Reprinted with permission from Ref. 20. Copyright © 1997 American Chemical Society.)

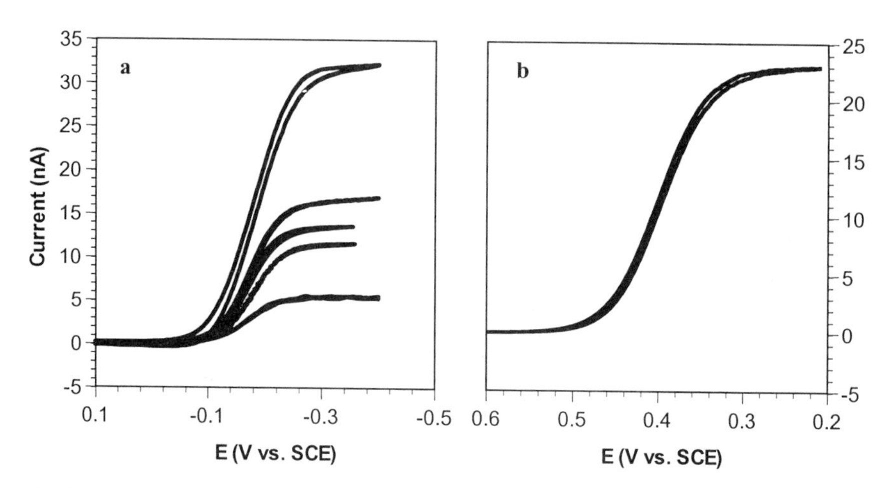

FIG. 7 Cyclic voltammogram of a self-assembled spherical gold microelectrode. (a) In a 7.5 mM aqueous solution of hexaamineruthenium(III) chloride and 1 M KCl electrolyte. Electrode diameters, from top to bottom, 10, 5, 4, 3, 1.7 μm. Scan rate 0.1 V/s. (b) In a 6.5 mM solution of tetracyanoquinodimethane in acetonitrile. Supporting electrolyte: tetrabutylammonium tetrafluoroborate (0.1 M). Electrode diameter: 6 μm. Scan rate 0.1 V/s. (Reprinted with permission from Ref. 20. Copyright © 1997 American Chemical Society.)

any consequences on the voltammetric behavior, since the currents flowing through microelectrodes are usually very small.

Size Limitations. The size of the spherical electrodes prepared by this technique could be controlled by the dithiol concentration or by the size of the pulled capillary. These two parameters determined the amount of dithiol present in the pipette tip. In order to obtain fully developed spheres, the self-assembly process was left to proceed until completion. It was observed that spheres having a diameter greater than 15–20 μm were very fragile and had a tendency to split open. Although it was possible to grow spherical structures a few hundreds of nanometers in diameter from capillary tips measuring a few tens of nanometers, the growth of such submicrometer structures was more difficult than micrometer-size tips. The fabrication of submicrometer self-assembled spherical electrodes was difficult to reproduce, although a few were obtained and characterized by cyclic voltammetry, such as the one presented in Figure 8.

Use of Self-Assembled Spherical Gold Microelectrodes as SECM Tips. The great smoothness, well-defined spherical shape, and good voltammetric behavior of the self-assembled electrodes made them very attractive to be used as SECM tips. One of the major advantages of these elec-

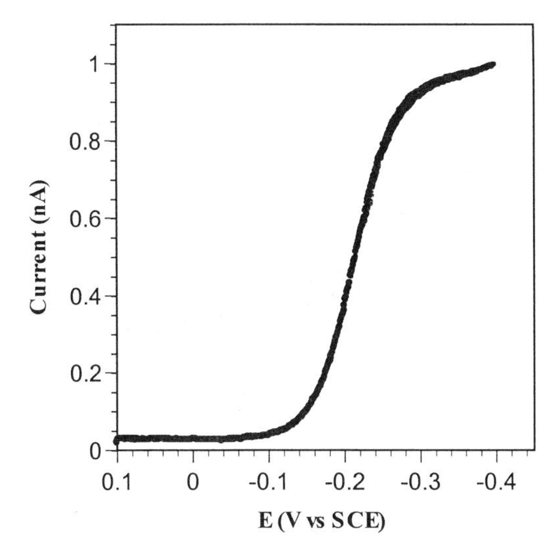

FIG. 8 Cyclic voltammogram of a 0.35 μm diameter self-assembled spherical gold microelectrode in a 6.5 mM aqueous solution of hexaamineruthenium(III) chloride and 1 M KCl electrolyte. Scan rate is 0.1 V/s.

trodes was their accessibility to the substrate surface. The approach of disk-shaped SECM tips to a planar substrate is mainly limited by the tip misalignment that results in an early contact between the insulating sheath of the probe and the substrate. The spherical shape and the absence of an insulating sheath render self-assembled microelectrodes much less sensitive to such alignment errors. At much shorter distances the tip roughness might become the limiting factor for the approach. The smoothness of the surface of the self-assembled microelectrodes may therefore allow very small tip-substrate separation to be obtained. Not only must the shape of the very end of a SECM tip be compatible for a close approach to the substrate, but its geometry has to be well defined in order to extract information from the approach curves. The theoretical approach curve corresponding to the tip geometry must also be known. The theoretical approach curve for a spherical electrode was derived from the curve calculated for a hemispherical electrode (27) by assuming that the upper part of the spherical electrode does not participate in the feedback process (see also Sec. IV for the SECM approach curves for nondisk tips). Thus, it only contributes to the tip current by a constant current corresponding to the unperturbed diffusion of the mediator to the upper half of the spherical electrode (see Fig. 9). This constant current therefore equals half the current measured at infinite tip-substrate separation. This assumption results in a simple expression that relates the

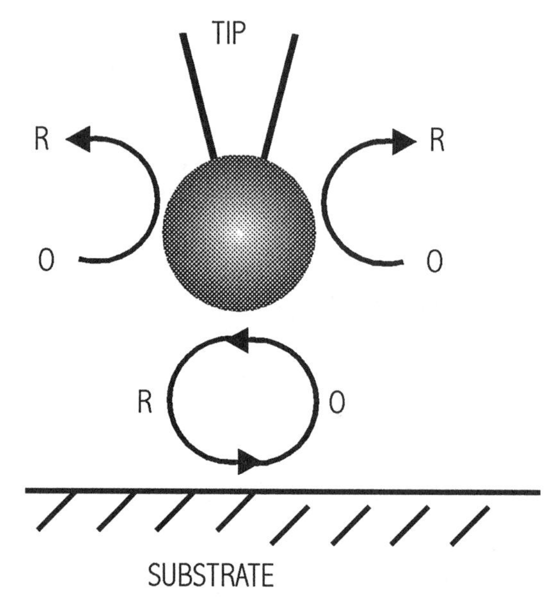

FIG. 9 Schematic picture of the diffusional processes taking place at a spherical electrode approaching a conductive substrate showing how positive feedback occurs at the lower hemisphere while the contribution of the upper hemisphere to the tip current remains constant. (Reprinted with permission from Ref. 20. Copyright © 1997 American Chemical Society.)

feedback current for a spherical probe (I_s) to the one for a hemispherical probe (I_h): $I_s = (I_h + 1)/2$. Experimental approach curve for a conducting or an insulating substrate fits this expression reasonably well (see Fig. 10).

When a self-assembled spherical electrode approaches a conducting substrate (Fig. 10, curve a), the current increase due to the positive feedback process is expectedly not as sharp as the one for a disk-shaped electrode but is sufficiently high to sense the surface. An interesting feature of the experimental approach curve obtained in this case was that a very small normalized distance ($L \sim 0.05$) could easily be achieved when a disk-shaped SECM tip only allows for $L \sim 4$ times greater. When a self-assembled spherical electrode approaches an insulating substrate (Fig. 10, curve b), only a small current decrease is observed, as expected for spherical electrodes. Thus, spherical electrodes have a reduced ability to sense an insulating substrate. The agreement between the theoretical and experimental results in this case is not as good as in the case of conducting substrate. This can be explained by the fact that the original model for the hemispherical electrode

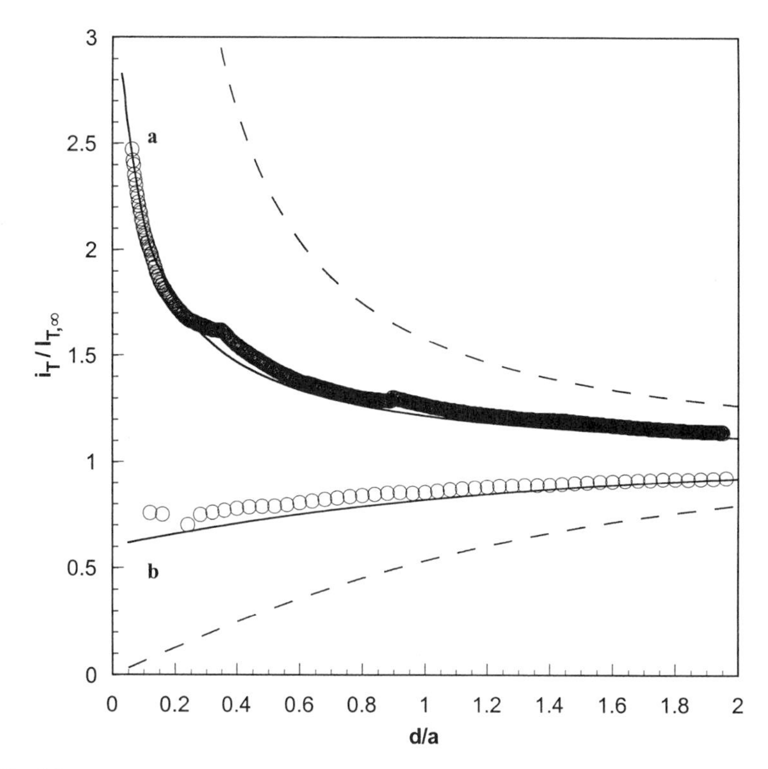

FIG. 10 Experimental approach curve obtained with a 4 μm diameter spherical self-assembled ultra-microelectrode: variation of the normalized current with the dimensionless tip-substrate separation d/a (d is the tip-substrate separation, a is the electrode effective radius). (———) Theoretical approach curve for a spherical electrode calculated as described in the text. (– – – –) Theoretical approach curve for a disk electrode. (a) Conductive platinum substrate. (b) Insulating glass substrate. Aqueous solution of hexaamineruthenium(III) chloride and 1 M KCl electrolyte. (Reprinted with permission from Ref. 20. Copyright © 1997 American Chemical Society.)

was calculated from data corresponding to planar electrodes having an insulating sheath, whose influence is negligibly small for positive feedback but significant for negative feedback. The closest distance attained in the negative feedback mode seems greater than that obtained in the positive feedback mode, which might also just be limited by an early contact due to the roughness of the glass slide used as the insulating substrate.

It can be concluded from the above discussion that the improvements gained by utilizing self-assembled spherical microelectrodes as SECM probes strongly depends on the operational mode of the SECM used. For surface imaging, the use of a small electrode has the obvious advantage of increasing the lateral resolution. In this mode, the improvement brought by the self-assembly technique is simply the ease of making relatively small probes as compared to conventional techniques. The main drawback of self-assembled electrodes for surface imaging derives from their reduced sensitivity to insulating substrates. When SECM is used as a tool for the kinetic studies of fast homogeneous reactions following electron transfer (28), the capability to bring a self-assembled spherical electrode very close to the substrate surface can be a major improvement. In this kinetic mode, short-lived intermediates generated at the tip can be collected and detected at the substrate if the tip-to-substrate distance can be made small enough that the intermediates can reach the substrate before being consumed by the fast chemical reactions. In other words, the diffusional tip-to-substrate transit time, $\tau \sim d^2/D$ (d is the tip-substrate separation and D the diffusion coefficient of the species), must be of the order of the lifetime of the intermediate. This would mean that for an 1 μm diameter spherical electrode that could be approached 25 nm away from the substrate, intermediates having lifetimes of $<\mu$s can be detected and characterized.

As explained above, the upper part of a spherical tip always gives rise to a permanent and distance-independent component of the tip current. In the feedback mode, this feature can limit the sensitivity of the feedback detection of small amounts of intermediate species. However, in the tip-generation/substrate-collection (TG/SC) mode, no current would theoretically flow through the substrate until it ultimately detects the intermediate species. This characteristic allows for the measurement of a very small substrate current due to the redox reaction of intermediate species. However, this would require minimizing the background current flowing through the substrate by limiting its size to an upper limit of an order of magnitude greater than the probe size. As mentioned above, for the probe to be able to closely approach the substrate, a smooth substrate must be used. Good candidates for the required small and smooth substrates would be the biggest self-assembled spherical electrodes obtainable if one can solve the problem associated with the alignment of two micrometer-sized spheres.

5. *Mercury Microelectrodes*

Together with carbon, mercury is one of the most attractive electrode materials. It has high proton reduction overpotential and a very well-defined and smooth surface. For these reasons it has been widely used as electrode material in organic electrochemistry. The preparation of mercury microelectrodes has thus been an interesting and challenging problem that has been addressed by several groups (29–31). The technique of making mercury microelectrodes consists in the reduction of mercury ions at the surface of a disk-shaped metallic or carbon microelectrode. The choice of substrate material for mercury deposition is critical. Mercury should properly wet the substrate surface but not form an amalgam with it. It was found that iridium is the most suitable material, while platinum and carbon are relatively acceptable but worse substrates for mercury deposition. The growth of the mercury electrode can be monitored by recording the time variation in the reduction current of the mercury ion. From the amount of mercury deposited, assuming a hemispherical growth, the radius of the microelectrode can be calculated. It was found that the radius corresponded quite well with that deduced from the diffusion-limited current expected for the reduction of an electroactive species at the same microelectrode, assuming a hemispherical shape. From these observations, it was suggested that mercury microelectrodes are shaped like hemispheres a few μm in diameter (29). In one case, the growth of a mercury microelectrode of a few tens of μm in size was followed by optical microscopy. It was observed to grow to an almost hemispherical microelectrode (30). In all cases, the radii of the mercury microelectrodes seem to be equal to or larger than that of the supporting disk-shaped electrode.

Even though it was stated that mercury microelectrodes having radii much larger than the supporting disk-shaped surface would in fact more resemble to shielded spheres than to actual hemispheres (31), all these microelectrodes share an interesting feature: they behave as very smooth, hemispherical microelectrodes. As mentioned previously and discussed in Section IV, although the approach curves of spherical-segment-shaped microelectrodes are not as steep as a microdisk, they are still quite suitable for SECM applications. Another merit of such mercury probes is their inherent smoothness. Finally, one could also consider that a controlled deposition of mercury could be used to grow spherical-shaped mercury tips on recessed tips such as the small but often recessed micropipet-based metallic microelectrodes, without dramatically increasing their diameter.

B. Carbon Microelectrodes

In the last 10 years, a lot of work has been done on the design of carbon microelectrodes, and several reviews have been devoted to this large amount of work (for reviews see Refs. 32 and 33). The purpose of this section is therefore only to review the techniques useful for the preparation of SECM tips based on the criteria presented earlier defining a good SECM tip. Some of our own work about the use of pyrolytic carbon electrodes as SECM probes will also be discussed.

1. Carbon Fiber Microelectrodes

Glass Encapsulation of Carbon Fiber Microelectrodes. The electrochemical behavior of carbon electrodes makes carbon a very interesting substitute for metals as microelectrode material. Unlike metallic electrodes, carbon electrodes do not form oxide films. They also have a much higher proton reduction overpotential than most metallic electrodes. As a result, they allow the exploration of a larger potential window. These characteristics have made carbon a versatile electrode material widely used in organic electrochemistry. Neurochemists also take advantage of this versatility to use carbon as the electrode material for the detection of neurotransmitters such as dopamine. Their requirement of ultimately working in vivo, i.e., in a confined environment, led to the initial development of amperometric micron-sized carbon fiber microelectrodes (34). These early carbon fiber microelectrodes consisted of short bare fibers held by an insulating support. This cylindrical geometry was not ideal for purely electrochemical investigations and is even less suitable for SECM experiments. More interesting for this purpose is the development of disk-shaped carbon microelectrodes that followed the initial papers by Wightman et al. (35,36). In these early reports and many of the following ones, the fabrication of disk-shaped carbon microelectrodes consisted in sealing the carbon fiber into a glass tube either by pulling and melting the glass around the fiber or by filling the space between the fiber and the glass taper with very fluid epoxy resins that filled the electrode end by capillary action. Mercury or conducting epoxy was used as the electrical contact to the fiber. The surface of the disk was then exposed by cutting the end of the electrode with a scalpel. Depending on the initial capillary tip size, this resulted in a disk-shaped electrode having a diameter of a few μm located somewhere in the plane of a larger (or much larger) insulating disk. Starting from small glass capillaries, microelectrodes having a RG (the ratio of the total microelectrode tip diameter over the diameter of the active carbon electrode) as small as 10 could be obtained. Microelectrodes to be useful as SECM probes should not have an RG exceeding 10 in order to be able to approach closely to the surface of the substrate. Exposing the surface of a disk shaped electrode by cutting the end of the capillary was, however, shown to result in fragile electrodes having quite rough, slightly protruding electrode surfaces. In spite of this drawback, this fabrication technique could still be useful for making SECM tips because it is fast and can directly produce electrodes with a small RG, although one needs to consider how to cut the electrodes perpendicular to their axes.

Tips of smaller RGs were later produced in order to be beveled (37). The beveling technique consisted in using a microelectrode beveler, i.e., a polishing setup consisting of a slowly rotating abrasive wheel to which a microelectrode mounted on a micro-manipulator is approached at a certain angle (e.g., 45°). This resulted in carbon fiber microelectrodes having small RGs and much smoother electrode surfaces. The beveled tips were well suited for their intended use, i.e., for easy penetration of the biological tissues. However, since the surface of a beveled electrode in most cases is not a disk but an ellipse and because the electroactive plane is not perpendicular to the tip axis, these tips are unsuitable for SECM. However, by placing these tips perpendicular to the polishing wheel instead of using a 45° angle one could make disk-shaped smooth carbon microelectrodes that would be usable as SECM probes. A few attempts of using a beveling apparatus in such a way to polish the very end of glass capillary-based SECM probes have been recently reported (38) with some success.

When the diameter of the capillary used to seal the carbon fiber is of the order of a few millimeters, much more robust disk-shaped electrodes were produced (3). The active electrode surface can then be easily exposed and smoothened by a regular polishing procedure. This kind of microelectrode can then be changed into a SECM tip by following the glass-sheath grinding procedure described above. Most of the carbon fiber SECM probes reported in the literature were made by this technique (39). As a result, these carbon tips present the same limitations encountered in the metallic ones: a tedious and time-consuming preparation procedure and a minimum size of a few μm, corresponding to the smallest carbon fiber tips available. One could think of addressing these two drawbacks, respectively, by using other sealing techniques and by thinning down those existing carbon fibers with different etching techniques. Several techniques allowing such improvements are reported in the literature and are examined below.

Carbon Fiber Etching. It is worth examining various different techniques that were developed to make carbon microelectrodes smaller than the lower limit imposed by the diameter of the commercially available carbon fibers (a few μm). In order to achieve this goal, the diameter of commercial carbon fibers was decreased by different etching procedures. These techniques generally result in the etched part of the fiber having the shape of a truncated cone. The tip diameter and its length depend on the etching conditions. Further coating and exposing steps are then necessary, as in the case of unetched fibers, in order to expose only the electrode surface.

Taking advantage of the fact that carbon fibers can be etched down to a size of approximately 100 nm by passing the fiber through a flame, Strein and Ewing (40) constructed polymer coated submicrometer disk electrodes from flame-etched carbon fibers. Electrodes with an apparent diameter as small as 124 nm were obtained. The flame-etched fibers had a very rough aspect (4).

Another technique based on electrochemical etching a carbon fiber in a half-saturated solution of sodium nitrite with an alternating current (50 Hz 2–6 V) has been proposed (41). Even though the final conically etched carbon fiber diameter was ~0.4–1.5 μm, it had the advantage of being embedded in glass. The embedding technique consisted in introducing the etched tip in a capillary that was later pulled and melted around the etched conical carbon fiber end. This resulted in a microelectrode having an overall conical shape. Although it was not clear how well the very tip of the electrode was characterized, further cutting or micro-polishing of the produced electrode could produce an interesting SECM probe.

Using a different electrochemical etching technique, a 480-nm diameter carbon microelectrode was fabricated (42). A carbon fiber was first etched by applying to it a 2 V amplitude and 1–50 Hz square wave in a 3 M KOH solution. The etched fiber was initially insulated with a very thin layer of sputtered SiO_2 and then embedded in epoxy glue. More recently a technique of carbon fiber etching by an argon ion beam thinner was proposed (43). The unetched part of the fiber was then embedded in glass, resulting in a very long conical microelectrode. This geometry is unsuitable for SECM applications. However, this technique could provide etched carbon fibers having a diameter as small as 50 nm, which are produced in much milder conditions than the flame-etching technique. The surfaces of these fibers were also much smoother than flame-etched ones.

All etched fibers have submicrometer diameters, making them attractive to use as SECM probes. However, they share the same constraints as the unetched fibers in terms of designing a coating technique that would make them suitable for SECM probes. In the case of etched tips, an extra problem arises. If disk-shaped electrodes are desired, the etched part has to be long enough so that the technique used to expose the carbon disk surface (e.g., micro-polishing or cutting) exposes only the very end of the fiber. Alternatively, one could think of using these conically shaped etched carbon fibers to make conical SECM probes. This would require coating the conical fiber down to its very end, leaving only a very short cone exposed at the end as described in other sections. However, one should keep in mind the limitations in terms of resolution and sensitivity inherent to the use of such conical geometry for SECM probes (27).

2. *Pyrolytic Carbon Microelectrodes*

Survey of Previously Reported Carbonization Techniques. The use of pyrolytic carbon deposited as a thin film resulting from the pyrolysis of methane as an electrode material has been reported by Blaedel and Mabbot (44). The pyrolysis of methane on a heated quartz cylinder under an inert atmosphere resulted in the deposition of a carbon film onto the quartz cylinder. The reasonably good electrochemical behavior of this pyrolytic carbon film was demonstrated. This technique was latter applied for the deposition of carbon inside heated quartz capillaries. At first the deposition of a thin carbon film on the inside walls of micrometer-sized capillaries was reported (45). In order to be usable the capillary had then to be back-filled with epoxy glue and cut with a razor blade. This procedure resulted in ring-shaped carbon microelectrodes a few μm in size. More recently it was reported that this technique could also result in disk-shaped carbon microelectrodes if the carbonization time was increased; the end of the capillary was then supposedly ultimately closed by a carbon cap (46). The resulting electrodes displayed, at least at slow scan rates reported, remarkably good electrochemical behavior, both in water and in organic solutions. Carbon microelectrodes apparently a few tenths of μm could be obtained. This microelectrode fabrication technique consisted of introducing an open quartz capillary filled by methane under high pressure (about 130 PSI) into a larger heated quartz tube in which an inert gas flows. A high methane pressure is required in order to ensure the presence of methane at the very end of the capillary. The inert gas flow avoids the combustion of the carbon film deposited. The temperatures required for the deposition being around 1000°C, the use of quartz capillaries is required. Since commercially available laser-based pullers allow for quartz capillaries having a tip diameter of a few tens of nanometers to be pulled, this carbonization technique seems particularly attractive for the preparation of submicrometer-sized carbon microelectrodes.

Modification of Carbonization Techniques. The feasibility of using pyrolytic carbon-filled capillary microelectrodes as SECM tips was explored by Bard and coworkers (47). Quartz capillaries of 0.5 mm i.d. and 1 mm o.d. were pulled using a laser puller in order to obtain capillaries of a tip size ranging from a few μm down to a few tens of nanometers. The size of samples of capillaries was estimated by scanning electron microscopy or by conductivity measurements (48). In order to acquire a better control on the carbonization conditions, the set-up presented in Figure 11 was built. This set-up was modified from the ones previously described in the literature (44–46). To have better control of the carbonizing temperature, the heat source consists of a nichrome wire coil instead of an unstable Bunsen burner flame. The introduction of the capillary into the heated quartz tube is controlled via the use of a mechanical Z-positioner. By mounting the pulled capillary to the positioner's arm, this set-up allows for a perfect control of the introduction and the withdrawing rate of the capillary from the heated zone. The carbonization time of the tip inside the hot zone can also be well controlled. The carbonizing temperature used was typically about 1000°C. The methane pressure ranged from 240 PSI for the largest capillaries to 500 PSI for the smallest ones. The use of a gas chromatography line and a ferrule-based

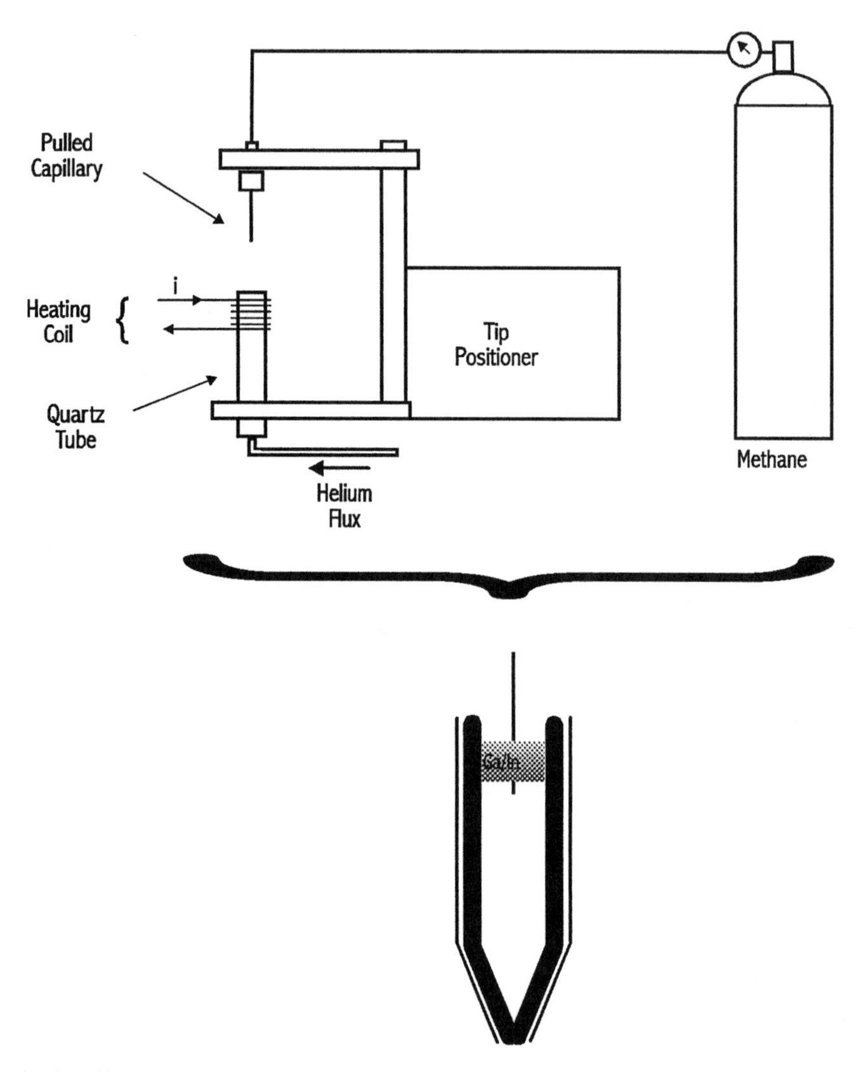

FIG. 11 Top: Schematic diagram of the capillary carbonization setup. Bottom: Schematic picture of the pyrolytic carbon microelectrode made from pulled quartz capillaries using this setup.

capillary holder allowed for the flow of high-pressure methane. For a given size of pulled capillary, the minimum pressure required was determined as the one at which bubbles started to come out from the capillary through a water bubbler. However, the methane flux had to be kept below a certain level, because an excessive flux always resulted in the absence of carbonization. It was observed that removal of the traces of oxygen remaining inside the system greatly improved the apparent quality of the carbon film. This was achieved by connecting the methane line to vacuum before introducing the methane. Once the interior of the tip was carbonized, electrical contact to the carbon film was established by introducing liquid gallium indium eutectic from the back of the capillary. A schematic picture of the tips obtained with this set-up is presented in Figure 11.

The voltammetric characteristics of the fabricated pyrolytic carbon micro-electrodes in water can be very good at slow scan rates (Fig. 12a). However, at high scan rates and/or in organic solvents, they show thin-layer-cell behavior (Fig. 12b), indicating that the tips are open. A longer carbonization time does not help to close the opening. The only way to close the end was by slowly introducing the capillary back and forth into the heated zone several times. By this technique, pyrolytic carbon microelectrodes of a size ranging from a few tens to a few hundreds of nanometer, showing a nice voltammetric behavior in both water and organic solvent at various scan rates, could be obtained.

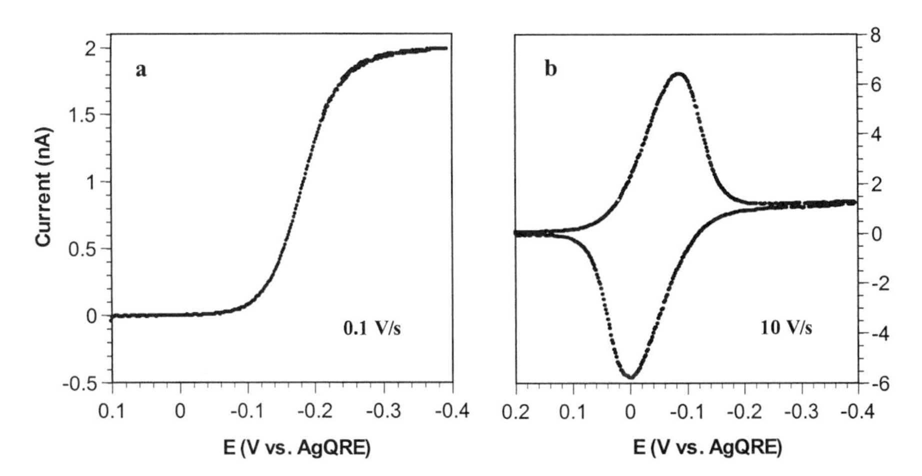

FIG. 12 Cyclic voltammogram of a pyrolytic carbon microelectrode in a 5 mM aqueous solution of hexaamineruthenium(III) chloride and 1 M KCl electrolyte. (a) Scan rate 0.1 V/s. (b) Scan rate 10 V/s.

SECM Behavior of Pyrolytic Carbon Microelectrodes. A characteristic approach curve for a 0.4 μm pyrolytic carbon micro-electrode to a platinum substrate, is presented in Fig. 13. As can be seen, instead of an expected monotonic increase in the current for the approach to a conducting surface (positive feedback, upper dashed curve), the current initially decreases (following roughly the negative feedback behavior, lower dashed curve) before rising again at close tip substrate separation. This behavior is characteristic of a recessed electrode, i.e., an electrode with the conductive surface recessed from the glass rim of the insulating sheath [49].

The biggest pyrolytic carbon micro-electrodes obtained from this technique have a diameter of a few microns. Their exact size could be measured using an optical microscope and compared with the apparent size derived from the steady state current obtained by cyclic voltammetry. This apparent

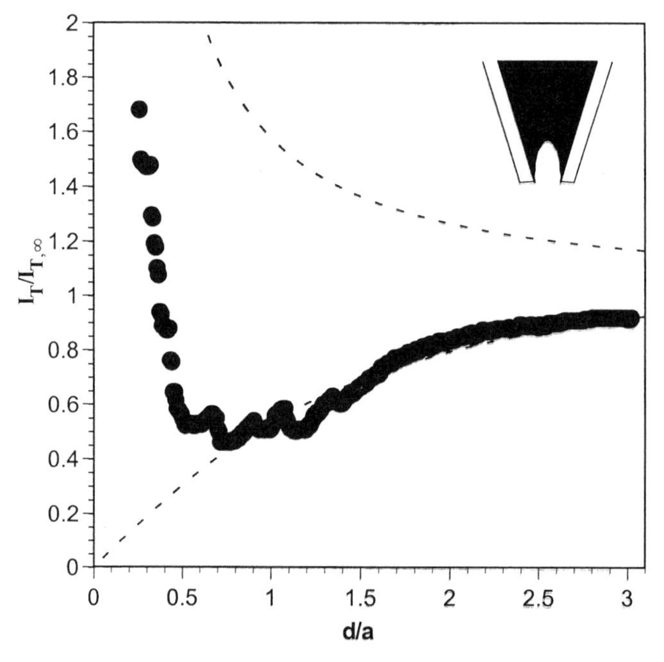

FIG. 13 Experimental approach curve obtained with a 0.4 μm apparent diameter pyrolytic carbon microelectrode in a 5 mM aqueous solution of hexaamineruthenium(III) chloride and 1 M KCl electrolyte. The substrate is a 100 μm platinum disk. (Upper dashed curve) Theoretical positive feedback approach curve. (Lower dashed curve) Theoretical negative feedback approach curve. (Inset) Schematic picture of the very end of the carbonized capillary compatible with the experimental approach curve.

size is expected to decrease with the recessed depth of the electrode [49]. In the case of pyrolytic carbon microelectrodes, the sizes of electrodes measured optically and by cyclic voltammetry were very close, indicating that the recessed depth was less than 1 or 2 electrode diameters. The structure of the carbon micro-electrodes obtained by the carbonization technique can therefore be schematically presented as shown in the inset of Fig. 13. A very thin conducting carbon film is deposited on the inside of the capillary walls down to the very end of the tip and the capillary is eventually closed by carbon a few electrode diameters away from the capillary tip. We therefore conclude that the carbonization technique can easily produce slightly recessed sub-micron carbon tips, however, their recessed nature renders them not useful as SECM tips. Attempts to polish the very end of the micron sized tip using a pipette beveler were unsuccessful due to the very small thickness of the carbon film closing the tip. However, the carbonization technique allows one to coat the inside walls of very small capillaries with a stable conducting carbon film, which can be used as an electrical contact to the macroscopic end of the electrode as described previously.

C. Development of Other Sealing Techniques

As mentioned above, the radius of the insulating sheath surrounding the electrode surface is a parameter needed to be considered if a microelectrode is used as an SECM probe. Microelectrodes with large RGs are very robust but can only be useful as SECM probes after a tedious and delicate grinding procedure. On the other hand, the use of tips with too small RGs may result in deviation from the theoretical behavior, especially when an insulating substrate is used. A large insulating sheath, resulting in a large overall diameter of the microelectrode tip, is also detrimental when the microelectrode is used in a confined biological environment. Thus, much effort has been devoted to develop alternative coating techniques that would produce microelectrodes having smaller overall diameters than the ones obtained from the conventional glass-encapsulating technique. Several alternative methods are described as following.

1. Deposition of an Organic Insulating Polymer

Electropolymerization. One coating method consists in the oxidative electropolymerization onto the surface of the electrode. Films of poly(oxyphenylene) were initially deposited from a 1 : 1 (by volume) water : methanol solution containing 0.25 M 2-allylphenol, 0.2 M 2-butoxyethanol, and 0.4 M ammonium hydroxide or allylamine by biasing the wires at a constant voltage of 4 volts for half an hour (50). Under these conditions the phenolic groups of the monomers are oxidized, generating a free radical, which initiates the growth of polymer. Later a phenol-allyphenol copolymer was preferred to the simple allylphenol polymer (40). In all cases, the deposited thin polymer film was cured at 150°C for 1/2–1 hour. This technique resulted in a fiber coated by a few μm of insulating polymer film. The disk-shaped microelectrode was then exposed either by cutting the insulated wire with a sharp scalpel or razor blade or by inducing an arc of current at the very end of the electrode, which locally desorbed the polymer. The thickness of the coating could be varied from 0.1 to a few μm by changing the deposition conditions such as the electrodeposition time and the concentration of the monomer solution.

Electrophoretic Deposition. Recently, an alternative approach consisting in the anodic electrophoretic deposition of paint was proposed for insulating carbon fibers with organic polymers (51). In this technique, negatively charged micelles of poly(acrylic-carboxylic acid) are electrophoretically attracted toward the anodically biased fiber. Electrolysis at the fiber surface decreases the pH locally, resulting in the precipitation of the micelles as a thin and uniform film on the fiber surface. Heat curing is then necessary to obtain carbon fibers covered with a few μm of thick insulating films.

An interesting aspect of these two techniques is that they result in disk-shaped electrode surfaces located exactly at the center of the insulating polymer plane. Moreover, the small RG of these electrodes is useful for SECM applications. However, the electroactive surface exposed by the cutting procedure can be quite rough. Polishing of the electrode surface for submicrometer wires, if possible, may be difficult. Finally, an unsupported electrode remains so slender and flexible that it may be difficult to align its surface with the substrate. However, the high aspect ratio of these polymer-coated electrodes could be beneficial as an SECM probe for imaging rough surfaces.

Coating with Molten Polymers or Apiezon Wax. The use of poly(α-methylstyrene) as an insulating material has previously been reported (52). Other promising insulating materials are Apiezon wax and low-melting-point polyethylene (53,54), although they are unstable and tend to dissolve or swell in most organic solvents. For insulation based on the procedures reported in Ref. 54, the sharp-etched fibers are mounted on a manipulator. A copper plate (1.5 mm thick), held as shown in Figure 14, is heated using a soldering iron element and is used to melt the polymers or Apiezon wax. A 1 mm wide rectangular slit extends from one side to the center of the copper plate, providing a temperature gradient (colder at the open end and hotter at the closed end) for the melted polymer or wax. The tip is pushed through the molten wax or polymer and allowed to break the top surface of the melt. If the tip breaks the surface at a too hot region, it is mostly bare. If the tip breaks the surface at the colder region of the melt, it raises a bulb of wax

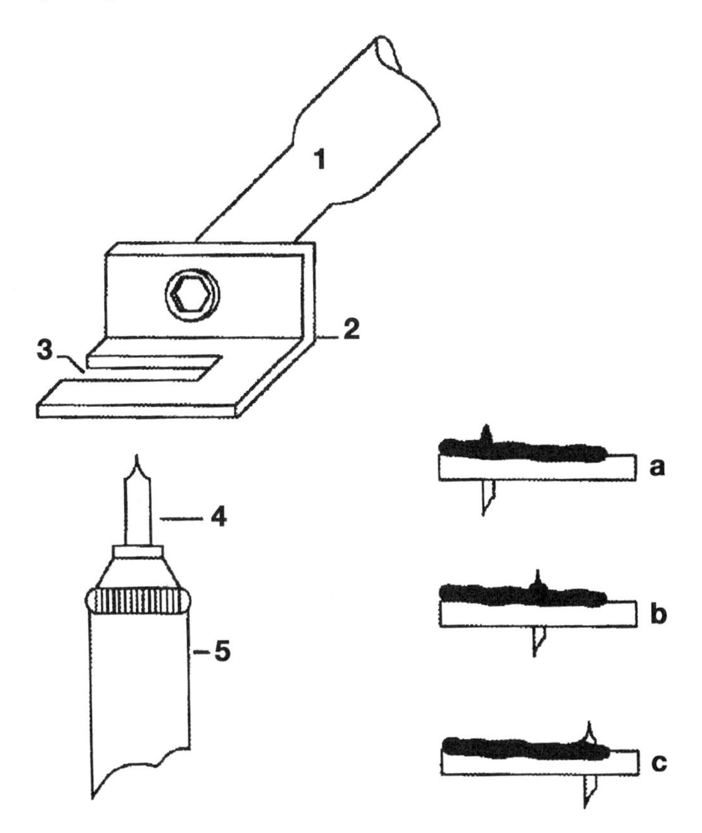

FIG. 14 Illustration of the apparatus/technique used for tip insulation. (1) Soldering iron, (2) copper plate, (3) 1 mm wide slit, (4) STM tip, and (5) pin socket for the tip. (a) At too cold a region, the tip is completely covered with polymer or Apiezon wax. (b) Optimum point allows only the extreme end of the tip to be exposed. (c) At too hot a region, the tip receives little insulation and is thus mostly bare. (Reprinted with permission from Ref. 54. Copyright © 1989 American Institute of Physics.)

or polymer above it, and the tip must be reentered into a hot region to remove the excess wax or polymer. Between these two regions exists an optimum zone where the wax or polymer coats the tip well. The insulated tip is moved sideways out of the groove so as to leave the very end of the tip unperturbed. Several coatings are usually required to insulate the tip completely or nearly completely. The degree of insulation of a tip can be checked by carrying out cyclic voltammetry in a solution containing a redox species. A completely insulated tip will not show detectable voltammetric waves. For a completely insulated tip, the very end of the tip is exposed by placing it in a scanning tunneling microscope (STM) with bias potentials on the tip and a conductive substrate [e.g., a Pt disk or an indium tin oxide (ITO) plate] and the STM set in the constant current mode. The onset of current flow (e.g., 0.1 nA) produces a hole in the tip insulation at the point of closest approach of tip to substrate, leaving most of the tip still insulated. The amount of exposed area of the tip can be controlled by the bias voltage and the onset current flow. A further treatment of "micro-polishing" these tips, by continuously scanning them over the substrate surface, can enlarge the exposed area relative to that of tips not receiving this treatment. These electrodes are very fragile and much more difficult, if possible at all, to repolish as compared to the glass-encapsulated ones and are discarded after 1 or 2 days of experiments.

Penner et al. used a similar coating technique for polystyrene and its derivatives (52). A procedure utilizing a soldering iron with an "omega"-shaped wire (diameter = 4 mm) was employed. The polymer bead was supported by this wire and was heated by the soldering iron. It was found that the rate of tip translation during the polymer coating step did not appear to be an important variable in determining the final coating morphology. The quantity of exposed metal at the coated tip varied reproducibly with the polymer melt temperature. During coating, the polymer melt was adjusted to a temperature of 190–205°C. Within this temperature range, the cone/hemisphere radius decreased from 10 to 0.5 μm as the temperature was increased.

2. *Deposition of an Inorganic Insulating Film*

Glass-coated tips were early developed by Heben et al. (55) and are also commercially available (56). However, glass cracks easily, resulting in an increased faradaic current. Interesting techniques for insulating tips by deposition of SiO_2 films have also been reported. The original technique involved covering the electrode surface with SiO_2 thin films by RF sputtering (42). In later works (57), the SiO_2 layer was deposited onto ohmically heated electrodes by a chemical vapor deposition (CVD) technique starting from a gaseous mixture of SiH_2Cl_2, $SiOEt_4$, H_2, and O_2. This CVD technique has the advantage of producing a very smooth SiO_2 coating with a thickness gradually decreasing along the wire from a few μm down to zero (Fig. 15). This allows one to make electrodes having a desired overall o.d. and thus a desired RG simply by cutting the coated wire at the corresponding position. The coated wire is cut with a diamond optical fiber cleaver. The electrochemical characteristics of the produced microelectrodes are very good and are indicative of a very good seal between the carbon and the silica coating. Moreover, as in the case of the electropolymerized coating, the disk surface

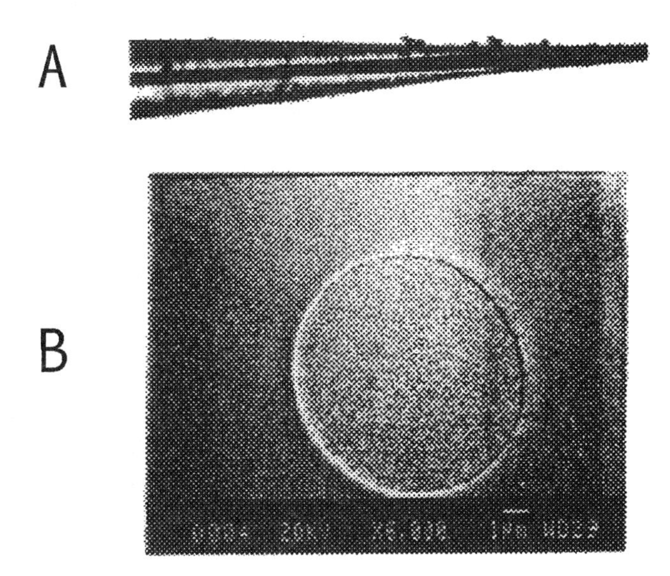

FIG. 15 (A) Optical micrograph of a silica-coated 10 μm carbon fiber at a magnification of 210. (B) Scanning electron micrograph of a polished silica-coated carbon fiber. Film thickness, 100 μm; magnification, 6000×. (Reprinted with permission from Ref. 57a. Copyright © 1994 Elsevier.)

is positioned exactly at the center of the microelectrode. Finally, the conical SiO_2 coating provides rigidity and durability to the otherwise fragile and flexible wire. All these characteristics make these electrodes very attractive for SECM application.

3. *Other Sealing Materials*

Other sealing materials recommended previously include Torr-seal (Varian Associates, 611 Hansen Way, Palo Alto, CA) (58), EPON 828 (Miller-Stephenson Chemical Co., Washington Highway, Danbury, CT) (59), and Epoxylite 6001 varnish (Epoxylite Corp., Irvine, CA) (7). Various sealing materials reported early for electrochemical STM (60) may be also interesting for SECM. The proper choice of the sealing materials depends on the intended use of the microelectrodes. For example, one should pay attention to the solubility, chemical stability, and swelling effect of a sealant in organic solvents if a long-term exposure is required.

D. Micropipet and Other Tips

Pipet-based electrodes have been used for a wide range of applications, e.g., micropipet ion-selective electrodes (ISEs) for potentiometric determination of intracellular and extracellular activities of different ions in living cells (61). Several scanning probe techniques based on micropipet electrodes have been developed to obtain topographic and chemical information about a sample substrate. For example, scanning ion conductance microscopy (62), which uses an electrolyte-filled Ag/AgCl micropipet electrode as the probe, was used to image the topography of surfaces, and scanning ion-selective potentiometric microscopy (48,63) was developed to map local concentration profiles of various ions. A micropipet tip has also been utilized to form a micro-ITIES to study ion-transfer and electron-transfer processes and also for SECM imaging (64). Miniaturization of the electrode body has made possible the fabrication of pipet-based ISEs with tip diameters of micrometer and submicrometer dimensions. Nanometer-sized pipets can be fabricated by the use of microprocessor-controlled pipet pullers (Laser puller Model P-2000, Sutter Instruments, Novato, CA) with laser heating of quartz tubes to produce orifice diameters of a few tens of nanometer, and nanopipet electrodes can be constructed using these and an electrolyte filling (63).

III. NONDISK TIPS AND TIP SHAPE CHARACTERIZATION

Two aspects of the use of nondisk micro-tips in SECM experiments are the evaluation of the shape of microelectrodes and the utilization of advantages associated with a specific tip geometry. For very small electrodes (nm-sized electrodes), one cannot obtain good images by electron microscopic techniques, and SECM approach curves rather than the voltammetric measurements are used to evaluate ultra-microelectrode (UME) geometry. In an early report for the steady-state current in a thin layer cell formed by two electrodes, it was shown that the normalized steady-state diffusion-limited current is fairly sensitive to the geometry of the electrode (65). However, the thin-layer theory does not describe accurately the steady-state current between a small tip and a larger planar substrate because the tip steady-state current, $i_{T,\infty}$, was not included in that approximate model. A more realistic approximate theory for SECM with a tip shaped as a cone or spherical segment versus a planar substrate was later presented (27). The surface of the nonplanar electrode was considered to be a series of thin circular strips, each of which is parallel to the planar electrode (Fig. 16). The diffusional flux to each strip was calculated from Eq. (1) for a conductive substrate or Eq. (2) for an insulating substrate.

$$I_{disk}(\Im) = 0.68 + 0.78377/\Im + 0.3315 \exp(-1.0672/\Im) \quad (1)$$

$$I_{disk}(\Im) = 1/[0.15 + 1.5385/\Im + 0.58 \exp(-1.14/\Im) + 0.0908 \exp\{(\Im - 6.3)/(1.017\Im)\}] \quad (2)$$

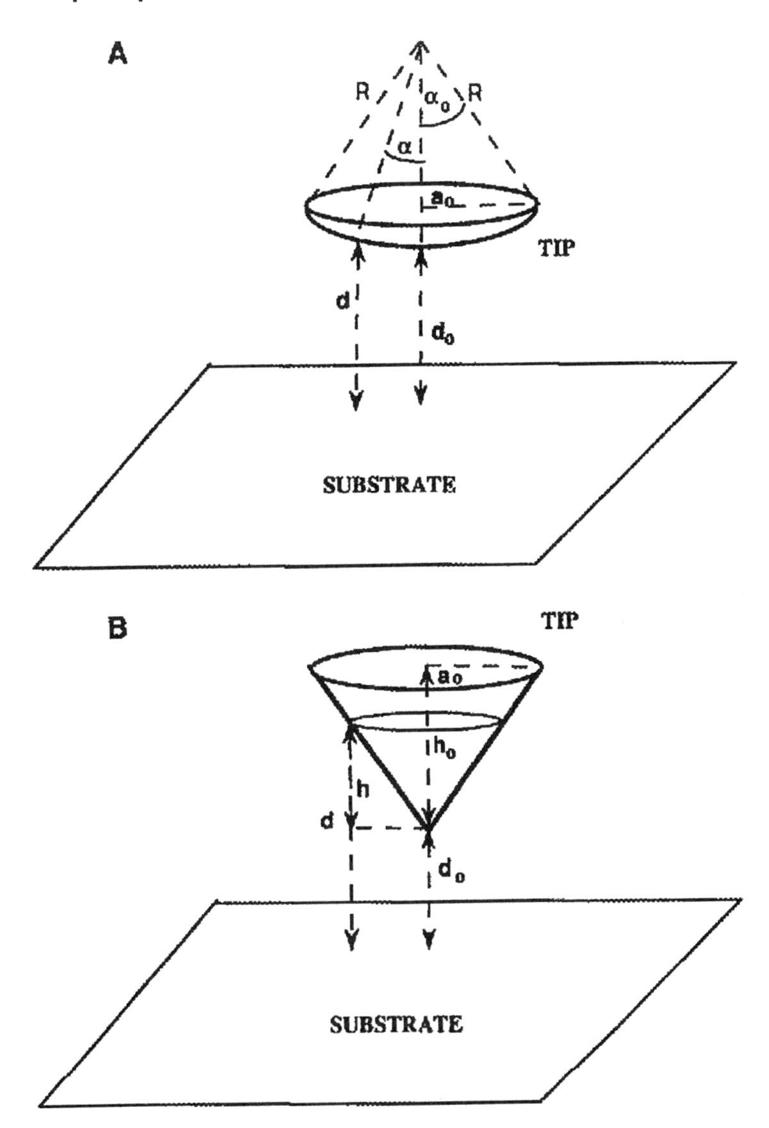

FIG. 16 Schematic diagrams of thin layer cells formed by a spherical segment tip and a planar substrate (A) and by a conical tip and a planar substrate (B). (Reprinted with permission from Ref. 27. Copyright © 1992 Elsevier.)

in which $\Im$ is z for a spherical tip or x for a conical tip. The normalized current to the whole surface of the segment can be expressed as

$$I_{sp}(L) = [2/\sin^2\alpha_0] \int_{\cos\alpha_0}^{1} I_{disk}(z)y\,dy \qquad (3)$$

for a spherical tip and

$$I_{cone}(L) = [2/k^2] \int_{0}^{k} I_{disk}(x)y\,dy \qquad (4)$$

for a conical one, where L is the distance between the substrate and the point of tip closest to it normalized by the tip radius (i.e., $L = d_0/a_0$), $z = L + (1 - y)/\sin\alpha_0$, $x = L + y$, $I_{disk}(z)$ is the SECM current function given by Eq. (1) or Eq. (2), and k is the ratio of the cone height to its base radius; other parameters are explained in Figure 16. The families of working curves, as shown in Figure 17, for conductive and insulating substrates were computed from Eqs. (1)–(4) for spherical and conical tip shapes, respectively.

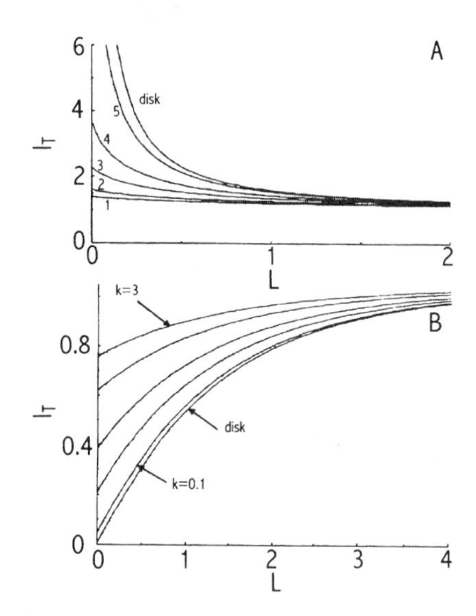

FIG. 17 Steady-state current-distance curve for a conical tip over conductive (A) and insulating (B) substrates corresponding to different values of the parameter $k = h_0/a_0$. (A) k = 3 (curve 1), 2 (curve 2), 1 (curve 3), 0.5 (curve 4), and 0.1 (curve 5). The upper curve was computed for a disk-shaped tip from Eq. (1). (B) From top to bottom, k = 3, 2, 0.5, and 0.1. The lower curve was computed for a disk-shaped tip from Eq. (2). (Reprinted with permission from Ref. 27. Copyright © 1992 Elsevier.)

One can see that the different working curves have substantially different curvatures, thus a unique curve can be found to obtain the best fit with the experimental data. One can also see that the current for a conical tip tends to reach some limiting values as $L \to 0$ and as $k \to 0$, the working curves for a conical tip approach that computed for a disk-shaped tip.

IV. CONCLUSIONS

Since the resolution of SECM depends strongly on the size of tip, future work on tip preparation will probably be directed toward the fabrication of sturdy nm-size tips. It would also be useful to devise a broader range of tip materials, including synthetic tips such as chemically modified ones, and to develop multiple tips for both sensing and detection, e.g., SECM-ECL (electrogenerated chemiluminescence), SECM-AFM (atomic force microscope), and SECM-NSOM (near-field scanning optical microscope) tips. The fabrication of potentiometric and enzymatic tips with ultra-high-resolution and fast response would also be very valuable.

ACKNOWLEDGMENTS

F.-R. F. Fan would like to thank Prof. Allen J. Bard for his constant support and discussion. He also wishes to thank his colleagues for their input and discussion during the development of SECM. The support of the research in SECM by the Robert A. Welch Foundation, the National Science Foundation, and the Texas Advanced Research Program is also gratefully acknowledged.

REFERENCES

1. M. Fleischmann, S. Pons, D. R. Rolison, and P. P. Schmidt, *Ultramicroelectrodes*, Datatech Systems, Morgantown, NC, 1987.
2. J. Kwak, Ph.D. dissertation, The University of Texas at Austin, 1989.
3. R. M. Wightman and D. O. Wipf, in *Electroanalytical Chemistry*, Vol. 15 (A. J. Bard, ed.), Marcel Dekker, New York, 1988, p. 267.
4. A. J. Bard, F.-R. F. Fan, and M. V. Mirkin, in *Electroanalytical Chemistry*, Vol. 18 (A. J. Bard, ed.), Marcel Dekker, New York, 1994, p. 243.
5. C. Lee, C. J. Miller, and A. J. Bard, *Anal. Chem. 63*:78 (1991).
6. H.-Y. Liu, F.-R. F. Fan, C. W. Lin, and A. J. Bard, *J. Am. Chem. Soc. 108*: 3838 (1986).
7. A. A. Gewirth, D. H. Craston, and A. J. Bard, *J. Electroanal. Chem. 261*:477 (1989).
8. B. D. Pendley and H. D. Abruna, *Anal. Chem. 62*:782 (1990).

9. Y. Shao, M. V. Mirkin, G. Fish, S. Kokotov, D. Palanker, and A. Lewis, *Anal. Chem. 69*:1627 (1997).
10. (a) G. Fish, O. Bouevitch, S. Kokotov, K. Lieberman, D. Palanker, I. Turovets, and A. Lewis, *Rev. Sci. Instrum. 66*:3300 (1995).
11. R. Morgan, *J. Sci. Instrum. 4*:808 (1967).
12. Y. Chen, W. Xu, and J. Huang, *J. Phys. E22*:455 (1989).
13. J. P. Ibe, P. P. Bey, Jr., S. L. Brandow, R. A. Brizzolara, N. A. Burnham, D. P. DiLella, K. P. Lee, C. R. K. Marrian, and R. J. Colton, *J. Vac. Sci. Technol. A8*:3570 (1990).
14. P. K. Hansma and J. Tersoff, *J. Appl. Phys. 61*:R1 (1987).
15. J. A. Stroscio, R. M. Feenstra, and A. P. Fein, *Phys. Rev. Lett. 58:*1668 (1987).
16. H. Neddermeyer and M. Drechsler, *J. Microsc. 152*:459 (1988).
17. V. T. Binh, *J. Microsc. 152*:355 (1988).
18. D. K. Biegelsen, F. A. Ponce, J. C. Tramontana, and S. M. Koch, *Appl. Phys. Lett. 50*:696 (1987).
19. D. K. Biegelsen, F. A. Ponce and J. C. Tramontana, *Appl. Phys. Lett. 54*:1223 (1989).
20. C. Demaille, M. Brust, M. Tsionsky, and A. J. Bard, *Anal. Chem. 69*:2323 (1997).
21. M. Brust, D. Bethell, D. J. Schiffrin, and C. J. Kiely, *Adv. Mater. 7*:795 (1995).
22. D. Bethell, M. Brust, D. J. Schiffrin, and C. J. Kiely, *J. Electroanal. Chem. 409*:137 (1996).
23. C. Demaille, M. Brust, M. Tsionsky, and A. J. Bard, unpublished results.
24. (a) R. L. Whetten, J. T. Khoury, M. M. Alvarez, S. Murthy, I. Vezmar, Z. L. Wang, P. W. Stephens, C. L. Cleveland, W. D. Luedtke, and U. Landman, *Adv. Mater. 8*:428 (1996). (b) R. P. Andres, J. D. Bielefeld, J. I. Henderson, D. B. Janes, V. R. Kolagunta, C. P. Kubiak, W. J. Mahoney, and R. G. Osifchin, *Science 273*:1690 (1996).
25. D. A. J. Rand and R. Woods, *J. Electroanal. Chem. 31*:29 (1971).
26. M. I. Montenegro, M. A. Queiros, and J. L. Daschbach (eds.), *Microelectrodes: Theory and Applications*, NATO ASI Ser. Appl. Sci. Vol. 197, Kluwer Acad. Publ., Dordrecht, 1991.
27. M. V. Mirkin, F.-R. F. Fan, and A. J. Bard, *J. Electroanal. Chem. 328*:47 (1992).
28. (a) P. R. Unwin and A. J. Bard, *J. Phys. Chem. 95*:7814 (1991). (b) F. Zhou, P. R. Unwin, and A. J. Bard, *J. Phys. Chem. 96*:4317 (1992). (c) P. R. Unwin and A. J. Bard, *J. Phys. Chem. 96*:5035 (1992). (d) D. A. Treichel, M. V. Mirkin, and A. J. Bard, *J. Phys. Chem. 98*:5751 (1994). (e) F. Zhou and A. J. Bard, *J. Am. Chem. Soc. 116*:393 (1994). (f) C. Demaille, P. R. Unwin, and A. J. Bard, *J. Phys. Chem. 100*:14137 (1996).
29. K. R. Wehmeyer and R. M. Wightman, *Anal. Chem. 57*:1989 (1985).
30. J. Golas, Z. Galus, and J. Osteryoung, *Anal. Chem. 59*:389 (1987).
31. Z. Stojek and J. Osteryoung, *Anal. Chem. 61*:1305 (1989).
32. T. E. Edmonds, *Anal. Chim. Acta 175*:1 (1985).
33. J. O. Besenhard, A. Schulte, K. Schur, P. P. Jannakoudakis, in *Microelectrodes: Theory and Applications*, M. I. Montenegro, M. A. Queiros, and J. L. Dasch-

bach (eds.), NATO ASI Ser. Aool. Sci. Vol. 197, Kluwer Acad. Publ., Dordrecht, 1991.
34. J. L. Ponchon, R. Cespuglio, F. Gonon, M. Jouvet, and J.-F. Pujol, *Anal. Chem. 51*:1487 (1979).
35. M. A. Dayton, J. C. Brown, K. J. Stutts, and R. M. Wightman, *Anal. Chem. 52*:846 (1980).
36. R. M. Wightman, *Anal. Chem. 53*:1125A (1981).
37. R. S. Kelly and R. M. Wightman, *Anal. Chim. Acta 187*:79 (1986).
38. (a) D. T. Miles, A. Knedlik, and D. O. Wipf, *Anal. Chem. 69*:1240 (1997). (b) M. Tsionsky, C. Demaille, and A. J. Bard, unpublished results.
39. A. J. Bard, F.-R. F. Fan, J. Kwak, and O. Lev, *Anal. Chem. 61*:132 (1989). (b) D. T. Pierce, P. R. Unwin, and A. J. Bard, *Anal. Chem. 64*:1795 (1992).
40. T. G. Strein and A. G. Ewing, *Anal. Chem. 64*:1368 (1992).
41. A. Meulemans, B. Poulain, G. Baux, L. Tau, and D. Henzel, *Anal. Chem. 58*: 2091 (1986).
42. T. Abe, K. Itaya, and I. Uchida, *Chem. Lett.* 399 (1988).
43. X. Zhang, W. Zhang, X. Zhou, and B. Ogorevc, *Anal. Chem. 68*:3339 (1996).
44. W. J. Blaedel and G. A. Mabbot, *Anal. Chem. 50*:933 (1978).
45. Y.-T. Kim, D. M. Scanulis, and A. G. Ewing, *Anal. Chem. 58*:1782 (1986).
46. D. K. Y. Wong and L. Y. F. Xu, *Anal. Chem. 67*:4086 (1995).
47. A. J. Bard, C. Demaille, and M. Tsionsky, unpublished results.
48. C. Wei, A. J. Bard, G. Nagy, and K. Toth, *Anal. Chem. 67*:1346 (1995).
49. F.-R. F. Fan, J. Kwak, and A. J. Bard, *J. Am. Chem. Soc. 118*:9669 (1996).
50. K. Pojte-Kamloth, J. Janata, and M. Josowicz, *Ber. Bunsen-ges. Phys. Chem. 93*:1480 (1990).
51. (a) A. Schulte and R. H. Chow, *Anal. Chem. 68*:3054 (1996). (b) A. Schulte and R. H. Chow, *Anal. Chem. 70*:985 (1998). (c) C. E. Bach, R. J. Nichols, H. Meyer, and J. O. Besenhard, *Surf. Coat. 67*:139 (1994). (d) C. E. Bach, R. J. Nichols, W. Beckman, H. Meyer, A. Schulte, and J. O. Besenhard, *J. Electrochem. Soc. 140*:1281 (1993). (e) C. J. Slevin, N. J. Gray, J. V. Macpherson, M. A. Webb, and P. R. Unwin, *Electrochem. Commun. 1*:282 (1999).
52. R. M. Penner, M. J. Heben, and N. S. Lewis, *Anal. Chem. 61*:1630 (1989).
53. J. Wiechers, T. Toomey, D. M. Kolb, and R. J. Behm, *J. Electroanal. Chem. 248*:451 (1988).
54. L. A. Nagahara, T. Thundat, and S. M. Lindsay, *Rev. Sci. Instrum. 60*:3128 (1989).
55. M. J. Heben, M. M. Dovek, N. S. Lewis, R. M. Penner, and C. F. Quate, *J. Microsc. 152*:651 (1988).
56. Longreach Scientific Resources, Orr's Island, ME.
57. (a) G. Zhao, D. M. Giolando, and J. R. Kirchhoff, *J. Electroanal. Chem. 379*: 505 (1994). (b) G. Zhao, D. M. Giolando, and J. R. Kirchhoff, *Anal. Chem. 67*:2592 (1995).
58. W. Thormann and A. M. Bond, *J. Electroanal. Chem. 218*:187 (1987).
59. P. M. Kovach, M. R. Deakin, and R. M. Wightman, *J. Phys. Chem. 90*:4612 (1986).

60. A. J. Bard and F.-R. F. Fan, in *Scanning Tunneling Microscopy and Spectroscopy: Theory, Techniques, and Applications*, D. A. Bonnell (ed.), VCH Publishers, Inc., New York, 1993.
61. D. Ammann, *Ion Selective Microelectrodes: Principles, Design and Application*, Springer-Verlag, New York, 1986.
62. (a) P. K. Hansma, O. Marti, S. A. C. Gould, and C. B. Prater, *Science 243*:641 (1989). (b) C. B. Prater, P. K. Hansma, M. Tortonese, and C. F. Quate, *Rev. Sci. Instrum. 62*:2634 (1991).
63. (a) K. Toth, N. Nagy, C. Wei, and A. J. Bard, *Electroanalysis 7*:801 (1995). (b) B. R. Horrocks, M. V. Mirkin, D. T. Pierce, A. J. Bard, N. Nagy, and K. Toth, *Anal. Chem. 65*:1213 (1993). (c) C. Wei, A. J. Bard, and S. W. Feldberg, *Anal. Chem. 69*:4627 (1997).
64. (a) A. A. Stewart, G. Taylor, and H. H. J. Girault, *J. Electroanal. Chem. 296*: 491 (1990). (b) T. Solomon and A. J. Bard, *Anal. Chem. 67*:2787 (1995). (c) Y. Shao and M. V. Mirkin, *J. Am. Chem. Soc. 119*:8103 (1997).
65. J. M. Davis, F.-R. F. Fan, and A. J. Bard, *J. Electroanal. Chem. 238*:9 (1987).

4

SECM IMAGING

Fu-Ren F. Fan

The University of Texas at Austin
Austin, Texas

I. INTRODUCTION

This chapter is intended to provide the reader with an up-to-date account of scanning electrochemical microscopy (SECM) imaging in chemical and biological applications. The operational principle and methodology of SECM imaging is described in Sec. II. Some examples of SECM imaging in bulk liquid environments are demonstrated in Sec. III, including membranes, leaves, artificially patterned biological systems, and corrosion systems. In Sec. IV a developing imaging technique involving an ultra-thin liquid layer on a mica substrate is described. The images of an ionic conductive polymer, e.g., Nafion, and several macrobiological molecules are shown to demonstrate the applicability of this technique for imaging. Since the main focus of this chapter is on the developments related to imaging, other applications of SECM and a detailed account of the instrumentation are presented in other chapters in this monograph.

II. PRINCIPLE AND METHODOLOGY OF SECM IMAGING

A three-dimensional SECM image is obtained by scanning the tip in the x-y plane and monitoring the tip current, i_T, as a function of tip location, the so-called constant height mode. The image can be converted to a plot of z-height (distance between tip and substrate), d, versus x-y position via an i_T versus d calibration plot, or it can be plotted as a gray-scale image, where high values of i_T are shown in light colors and small values as dark colors. Figure 1A shows a constant height SECM image of a 80-μm × 80-μm portion of an interdigitated electrode array (IDA), which consists of 3-μm-wide bands of Pt separated by 5 μm of SiO_2. For this image, a 1-μm-radius Pt disk served as the tip, and this tip and the IDA were immersed in

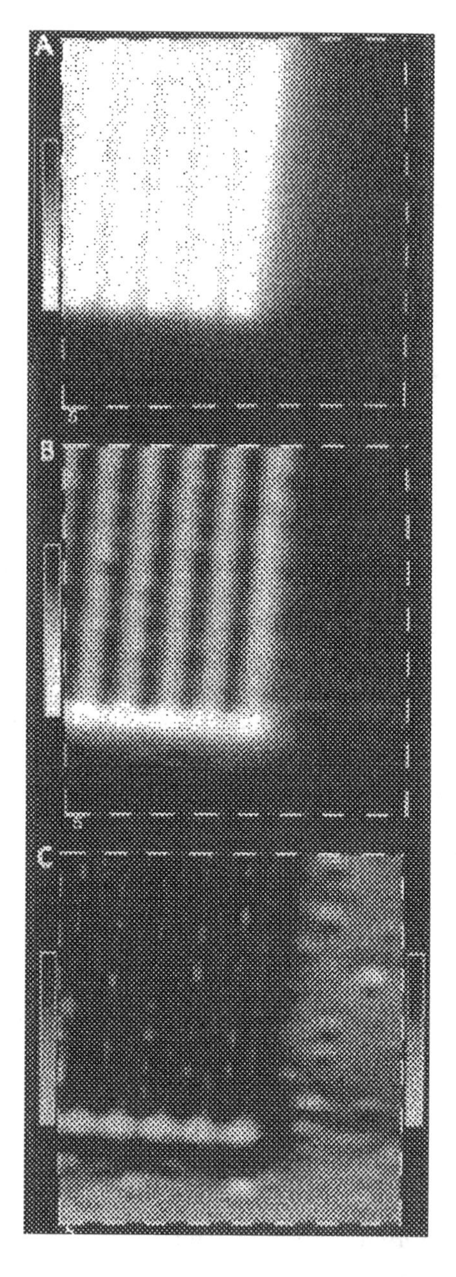
A
B
C

a solution containing 1.5 mM $Ru(NH_3)_6^{3+}$ in pH 4.0 buffer. The imaging tip current was 200–850 pA and $i_{T,\infty}$ = 470 pA.

The resolution attainable with SECM is largely governed by the tip size and the distance between tip and sample. With a very small diameter tip (e.g., diameter < 100 nm), scanning the tip in close proximity to the substrate surface (e.g., 100 nm above the surface) and measuring the tip current become very difficult because stray vibrations, irregularities in the sample surface, or tilt of the sample can cause a "tip crash." Thus, for high resolution, SECM must be carried out in the constant current mode, as is often used with the scanning tunneling microscope (STM) (1), where the distance is adjusted by a feedback loop to the z-piezo to maintain i_T constant. This is straightforward when the sample is either all conductive or all insulating, since the piezo feedback can be set to counter a decrease in tip current by either moving the tip closer (conductor) or farther away (insulator).

For samples that contain both types of regions, a method of recognizing the nature of the substrate must be available for designing the feedback loop. One approach is to modulate the motion of the tip normal to the sample surface and record di_T/dz, the so-called tip position modulation (TPM) technique (2). The use of TPM can substantially improve the sensitivity and resolution of the SECM image and provide a method of distinguishing between conductive and insulating areas on the substrate surface being examined. Figure 1B and C show the SECM images of the same IDA structure used in Fig. 1A acquired by TPM technique. As shown, the images acquired with the TPM signal are significantly more detailed than the dc image (Fig. 1A). In particular, note the improved detail in the insulating region of the image due to the better sensitivity of the TPM signal at insulators. Another interesting topography feedback mechanism for SECM imaging is the one based on shear force modulation reported by Ludwig et al. (3) and by James et al. (4), in which a microelectrode is vibrated laterally and the damping

←

FIG. 1 (A) A constant height SECM image of a 80-μm × 80-μm portion of an interdigitated electrode array (IDA) in a solution containing 1.5 mM $Ru(NH_3)_6^{3+}$ in pH 4.0 buffer. The IDA structure consists of 3-μm-wide bands of Pt separated by 5 μm of SiO_2. For this image, a 1-μm-radius Pt disk served as the tip. The imaging tip current was 200–850 pA and $i_{T,\infty}$ = 470 pA. d ~ 0.9 μm. (B) and (C) The SECM images of the same IDA structure as used in A acquired by TPM technique. Modulation frequency = 160 Hz. (B) TPM SECM image, scale −7 to 15 pA_{rms}. (C) Absolute value TPM SECM image, left scale = 0–15; right scale = 0–7 pA_{rms}. (Reprinted with permission from Ref. 2, Copyright © 1992 American Chemical Society.)

of the tip vibration amplitude upon its approach to the sample surface is measured. In the former a laser/split-photodiode assembly was used as the vibration amplitude detector, whereas in the latter a standard tuning fork resonator was employed. Such an approach allows for controlling the vertical position of the electrode independent of electrochemical parameters. Figure 2 shows a series of scanning probe images (35 μm × 35 μm) obtained for a Pt ultramicroelectrode (UME) imbedded in glass (4). Figure 2A and 2B show the topography in air of the sample obtained by atomic force microscope (AFM) and by shear-force technique, respectively. The shear-force

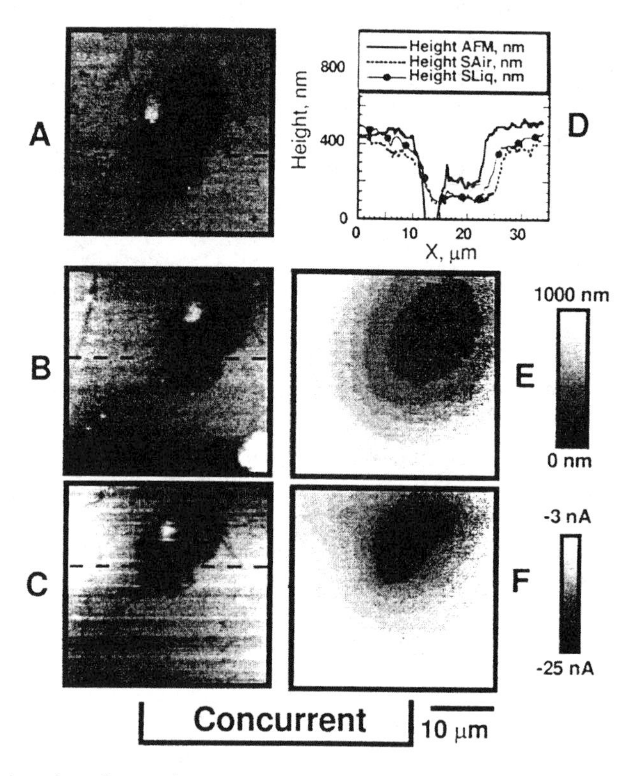

FIG. 2 A series of scanning probe images (35 μm × 35 μm) with corresponding surface profiles obtained for the sample (Pt UME imbedded in glass). The SECM images were obtained in 0.1 M $K_4Fe(CN)_6$ and images C and F were obtained concurrently. (A) AFM image in air; (B) shear-force topography in air; (C) shear-force topography in liquid; (D) overlay of surface profiles from topography images A, B, and C; (E) SECM image without feedback at an approximate distance between the sample and the probe ca. 4 μm; (F) SECM image with feedback. (Reprinted with permission from Ref. 4, Copyright © 1998 The Electrochemical Society, Inc.)

topography obtained by tuning-fork method in liquid is shown in Fig. 2C. Figure 2D shows one line scan in the same area of each of the topography images A, B, and C. Figure 2E is a constant height SECM image obtained in 0.1 M $K_4Fe(CN)_6$ with a tungsten tip of radius ~1 μm, located at an approximated distance of 4 μm from the sample surface. Figure 2F is the SECM image obtained with shear-force feedback and is obtained concurrently with Fig. 2C. As demonstrated, SECM and scanning force imaging, based on tuning fork feedback mechanism, gave similar topographic profiles. Readers interested in a more detailed description of these and related techniques should consult Chapter 2 in this monograph.

III. IMAGES IN SOLUTIONS

Since the SECM response is a function of the rate of the heterogeneous reaction at the substrate, it can be used to image the local chemical and electrochemical reactivity of surface features. A technique called reaction-rate imaging, which is unique to SECM, is particularly useful in imaging the areas on a surface where reactions occur. Membranes (5–7), leaves (8–10), polymers (11,12), surface films (13–17), and artificially patterned biological systems (7,18–20) have been imaged with SECM.

A. Membranes

Images of the molecular flux across porous membranes, e.g., track-etched mica membranes (5) and excised mouse skins (6), have been obtained by monitoring the iontophoretic current through the pores. The ability to detect ~1-μm-radius pores separated by 50–100 μm has been demonstrated. The contribution of the electrophoretic flux of individual types of ions to the total conductivity through individual pores was also studied. As stated above, the resolution attainable with the SECM is largely governed by the tip size and the distance between tip and sample. Figure 3 shows a series of constant height SECM images of polycarbonate filtration membranes having a nominal average pore size of ~14 μm. These images were taken with a commercially available SECM (21) at Pt disk tips immersed in a $KCl/K_4Fe(CN)_6$ solution. As shown in Fig. 3A–C for a 10-μm-diameter Pt disk, the resolution of SECM images can be improved by decreasing the height at which the tip is rastered above the surface. The image taken at a normalized distance of 2.4 (Fig. 3A) shows significant overlap between the diffusion fields of the electroactive mediator around individual pores. Decreasing the tip/substrate spacing to a normalized distance of 0.26 (Fig. 3C) resolves individual pores almost completely, although the apparent pore size is substantially larger than the nominal value. Figure 3D shows one SECM image of

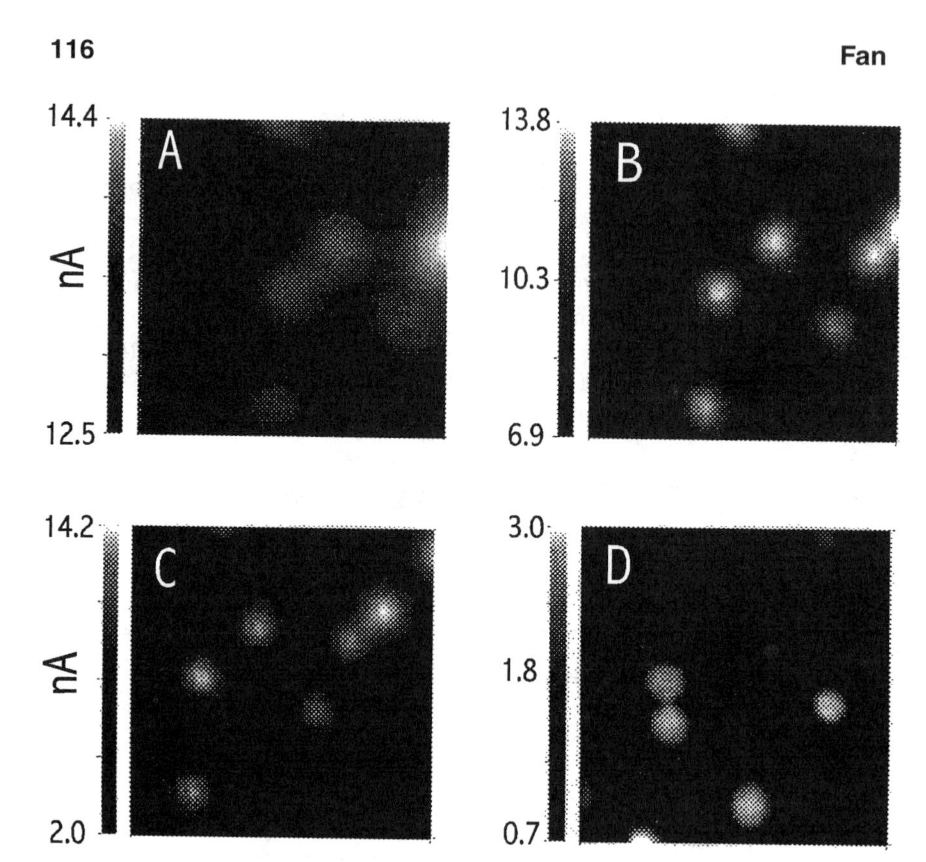

FIG. 3 A series of constant-height SECM images of polycarbonate filtration membranes having a nominal average pore size of ca. 14 μm. The scan size of each image is 100 μm $\times$ 100 μm. These images were taken with a commercially available SECM unit at Pt disk tips immersed in a solution containing $K_4Fe(CN)_6$ as the redox mediator and KCl as the supporting electrolyte. Images A–C for a 10-μm diameter Pt disk were taken at different normalized distances, L. L = 2.4 (A); 1.0 (B); 0.25 (C). Image D shows one SECM image of the same membrane taken with a smaller tip (2 μm diameter) at a normalized distance of 1.

the same membrane with a smaller tip (2 μm diameter) at a normalized distance of 1. As can be seen, not only are individual pores well resolved, but the apparent pore size is fairly close to the nominal value.

B. Leaves

In a previous work, SECM was used to obtain topographic information about biological samples immersed in an electrolyte solution either by using the feedback mode or by detecting a substrate-generated electroactive species (e.g., oxygen in the photosynthesis process) at the tip (8). For example, the upper surface of a *Ligustrum sinensis* leaf that was immersed in an aqueous solution of $Fe(CN)_6^{4-}$ could be imaged by SECM (Fig. 4A). The image in gray scale shows the parallel venation pattern characteristic of monocot leaves. An SECM image over the bottom surface of the leaf is shown in Fig. 4B. Several open stoma structures can be seen, showing that the guard cells (pairs of specialized epidermal cells surrounding each stoma) are swollen and protrude above the surrounding epidermal cells. The image of Fig. 4B after image processing by the use of the Laplacian of Gaussian (LOG) filter (22) is shown in Fig. 4C. A large, crater-like feature at middle left shows finer structure, and some stomata at the upper right and middle right edges show their structures more clearly. The tip can also be used to detect electroactive products generated at a biological substrate. For example, the detection of oxygen during illumination of an Elodea leaf has been carried out in a 10 mM NaCl solution saturated with CO_2 (8).

SECM has also been used to examine in vivo topography and photosynthetic electron transport of individual guard cells in *Tradescantia fluminensis* (10). A topographic image by SECM of a 200 μm $\times$ 200 μm leaf area shows three stoma complexes (Fig. 5). Prior to taking the image, the stomata were opened by applying 0.1 M KCl, and SECM clearly detects the open stoma pores. Figure 5A shows the current image; the dark areas represent bulges on the leaf surface and correspond to low tip current. Figure 5B shows the proper topographic image by converting the tip current to distance. As shown, a typical *T. fluminensis* stoma complex protrudes approximately 4 μm from the surrounding epidermal region of the leaf. In addition to topographic images, SECM has also been used to monitor oxygen evolution from an illuminated leaf for an individual stoma region. Figure 6A shows three peaks in the tip current corresponding to oxygen evolution from three stomata. i_T values for these peaks are at least twice background i_T for the leaf surface as a whole. If the tip is held in position above an individual stoma, the response of the tip current to the change in light intensity can be examined. As shown in Fig. 6B, when the light source is eliminated, i_T decreases as oxygen evolution from the stoma slows. When the light source is restored, i_T increases as photosynthesis and O_2 evolution recommence.

C. Artificially Patterned Biological Systems

As stated above, SECM can be used to image the reactivity of surface features. A feedback detection scheme was used to observe the localized reaction of glucose oxidase and mitochondria-bound NADH–cytochrome c reductase (7). The spatial resolution of the imaging is high for enzymatic

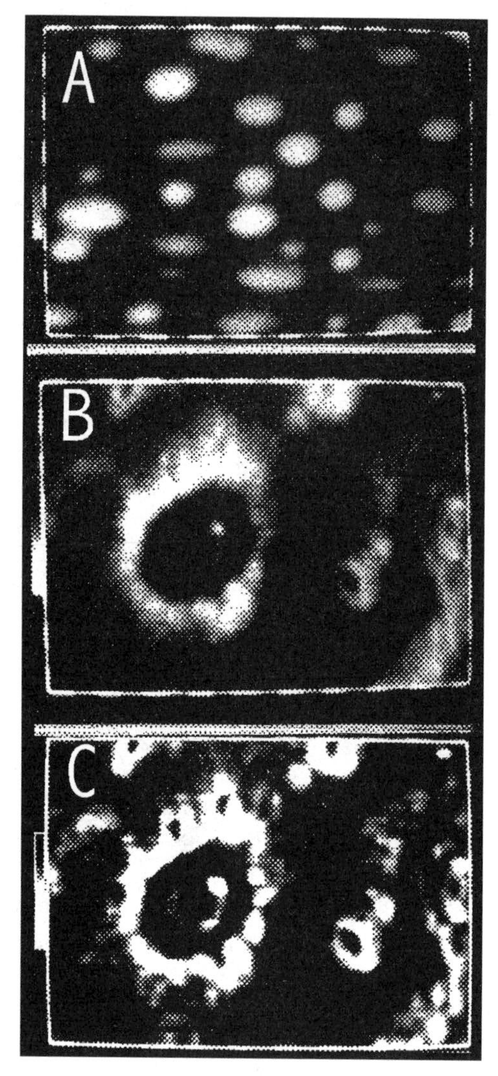

FIG. 4 Gray-scaled images of leaves in a 20 mM $K_4Fe(CN)_6$ and 0.1 M KCl solution scanned with a 1-μm-radius Pt tip: scan area 188 μm × 142 μm (white markers around edges = 10 μm). (A). Top surface of a grass leaf. (B). Bottom surface of a *L. sinensis* leaf. (C). The image in (B) after image processing by the use of the Laplacian of Gaussian (LOG) filter (22). (Reprinted with permission from (A, B) Ref. 8, Copyright © 1990 the National Academy of Science, and (C) Ref. 22, Copyright © 1991 American Chemical Society.)

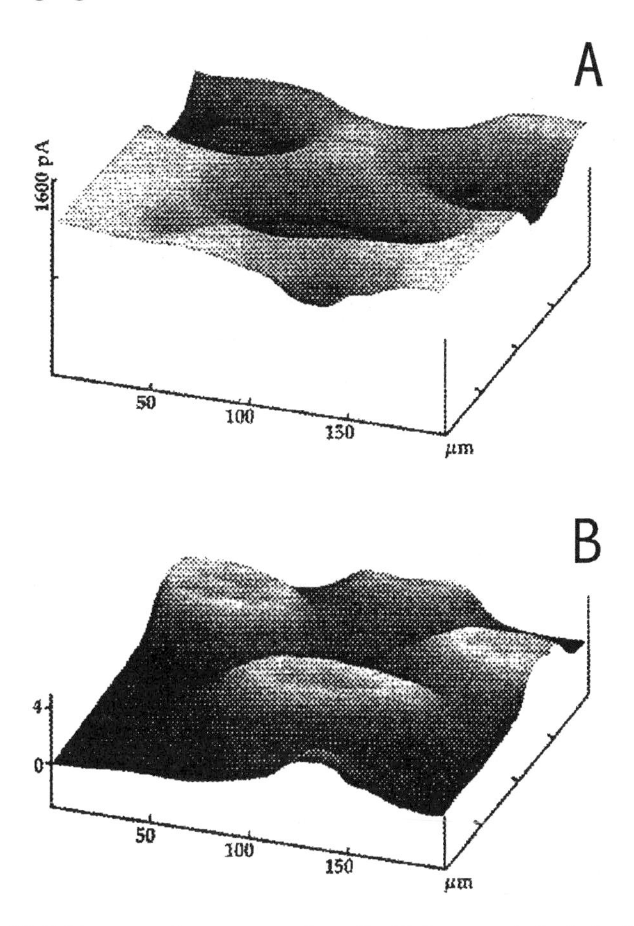

FIG. 5 SECM images of a green portion of a *T. fluminensis* leaf obtained with a 7-μm-diameter carbon tip in an air-saturated solution in the dark. (A) i_T across the leaf surface. The image is inverted, i.e., the dark areas represent bulges on the surface and correspond to low i_T. (B) The image in A returned to the proper physical orientation (y axis in μm). The images were obtained 30 minutes after addition of 0.1 M KCl to the buffer solution to open stomata. (Courtesy from Ref. 10, Copyright © 1997 American Society of Plant Physiologists.)

systems with a fast catalytic turnover. For slower reactions, however, sensitivity constraints need to be considered and image resolution must be sacrificed in favor of adequate detection level.

The enzyme-linked immunosorbent assay (ELISA) method (23) has recently been used for the preparation of ultrasmall biological structures for

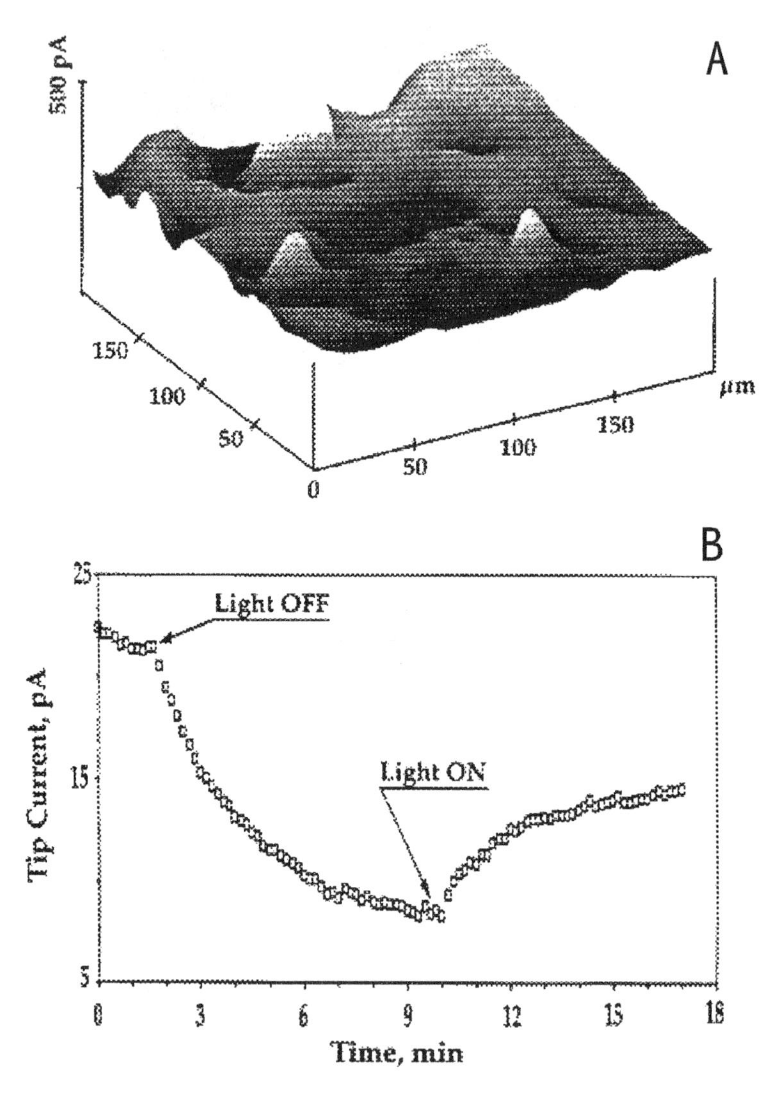

FIG. 6 (A) O_2 evolution from three stomata of an illuminated green leaf region (200 μm × 200 μm). The x and y axes represent position on the leaf; the z axis is i_T measured in feedback mode. (B) The response of i_T to step changes in light availability when held over a stoma in a green leaf region. Measured current tracks O_2 diffusion and decreases after the leaf is placed in the dark and increases after reillumination (800 μE m^{-2} s^{-1}). Measurements for A and B were made with a 7-μm-diameter carbon microelectrode. (Courtesy from Ref. 10, Copyright © 1997 American Society of Plant Physiologists.)

SECM measurements. For example, microspots of carbinoembryonic antigen (CEA) on glass substrates were imaged by SECM (20). CEA was immobilized via a sandwich method using horseradish peroxidase (HRP)–labeled anti-CEA. The reduction current of the oxidized form of ferrocenylmethanol generated by the HRP reaction was monitored to view SECM images. Figure 7 shows an SECM image of the substrate with CEA/HRP-labeled anti-CEA spots. The appearance of circular spots with large reduction currents is attributed to the HRP-catalyzed reactions at the spots, which indicates the localized presence of CEA. The reduction currents depend on the CEA concentration of the solution for making a CEA spot. In Figure 7, the five spots on the left side were made using a 2.0 μg/mL CEA solution, while the five

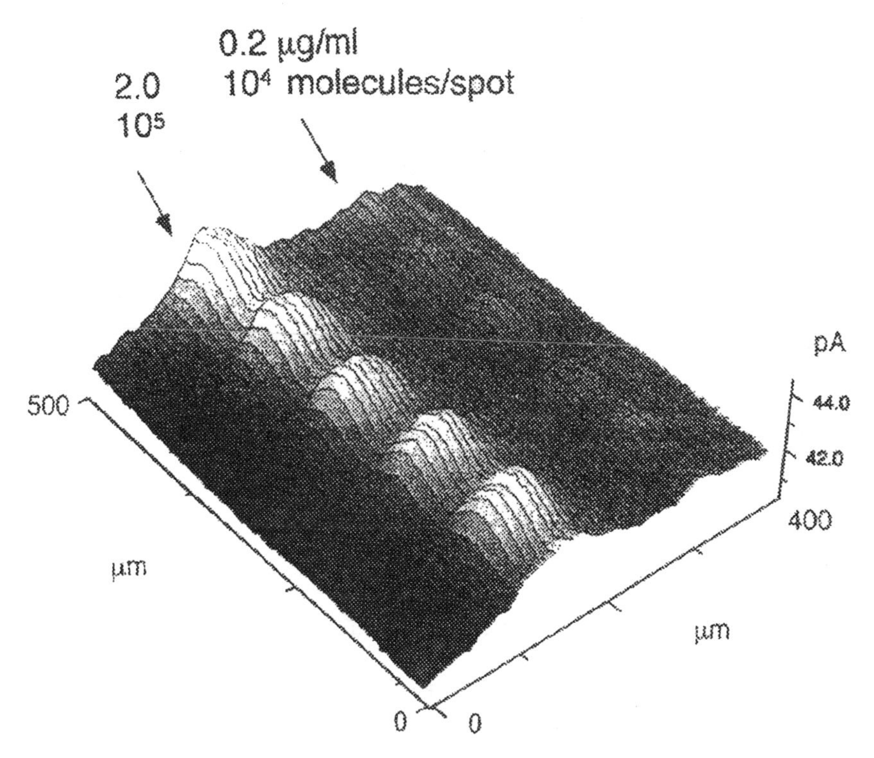

FIG. 7 SECM image of a series of antigen-antibody immobilized spots in a 1.0 mM ferrocenylmethanol/0.5 mM H_2O_2/0.1 M phosphate buffer solution (pH 7.0). Carcinoembryonic antigen concentration of the solution for making spots, 2.0 (left) and 0.2 μg/mL (right); distance, d = 10 μm. (Reprinted with permission from Ref. 20, Copyright © 1996 American Chemical Society.)

spots on the right were for a 0.2 μg/mL CEA solution. The number of CEA molecules in a single spot ($\sim$20-μm radius) on the right was $\sim 10^4$. Recently, Wittstock et al. (19) also demonstrated the imaging of immobilized antibody layers. Detailed discussions on the bioanalytical applications of SECM are given in another chapter in this monograph.

D. Corrosion Systems

Using SECM, Casillas et al. (15) demonstrated the first direct experimental observation of precursor sites for pitting on the surface of a TiO_2-converted Ti substrate. Figure 8 shows the voltammetric response of a Ti/TiO_2 (5 nm thick oxide) electrode immersed in a 1 M KBr/0.05 M H_2SO_4 solution open to the ambient atmosphere. The broad wave between 1.1 and 2.0 V corresponds to Br^- oxidation. The anodic wave decays to a background current

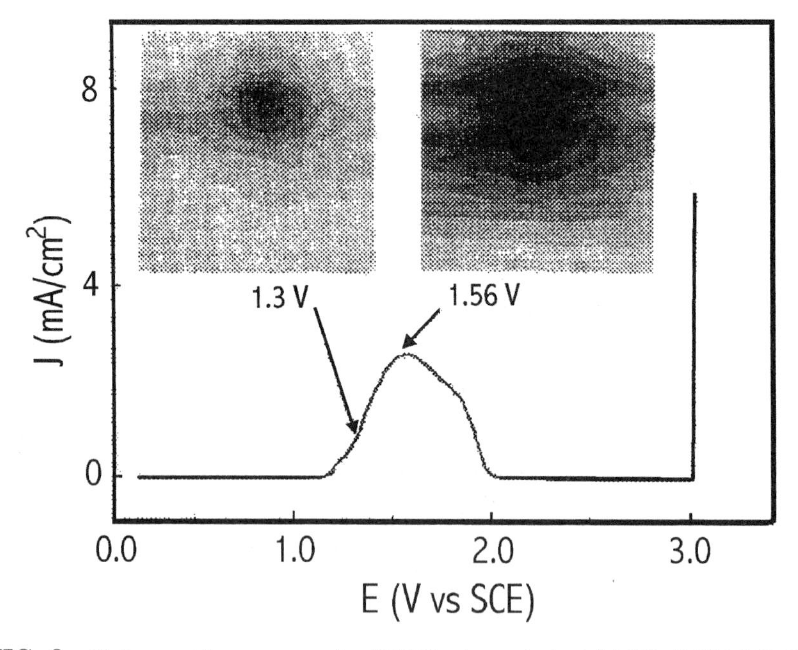

FIG. 8 Voltammetric response of a Ti/TiO_2 electrode in 1 M KBr/0.05 M H_2SO_4 at 1 mV/s. Insets: SECM images of a 400 μm $\times$ 400 μm region of the Ti/TiO_2 surface at 1.3 and 1.56 V vs. SCE. SECM tip potential: 0.6 V vs. SCE. (Reprinted with permission from Ref. 15, Copyright © 1993 The Electrochemical Society, Inc.)

level when the potential is more positive than ~1.6 V versus SCE due to the passivation of the surface sites. The sudden and large increase in current at 3.0 V is due to the onset of pitting corrosion, as readily verified by visual inspection of the electrode surface. The images shown in the insets of Fig. 8 are constant-height (~20 μm) SECM images of a 400 μm × 400 μm region of the Ti/TiO_2 surface recorded at the tip potential of 0.6 V versus SCE, where Br_2 generated at the Ti/TiO_2 surface is reduced to Br^-.

When the Ti/TiO_2 surface was poised at potentials below 1.0 V, SECM images of the surface were essentially featureless and the current was small (~0.1 nA). At potentials between 1.0 and 2.0 V, SECM images revealed a few microscopic sites, such as the one shown in Fig. 8, where intense Br_2 generation occurred. The maximum current measured at the tip above an electroactive site was typically between 0.2 and 2 nA. SECM images were stable and reproducible at Ti/TiO_2 potentials between 1.0 and 1.5 V. At slightly more positive potentials, the current at the tip decayed to background levels over a period of a few minutes as a result of the passivation of the active sites. The number density of active sites for Br_2 electrogeneration on the Ti/TiO_2 surface is estimated to be on the order of 30 sites/cm^2. The dimensions of the surface-active sites varied considerably from site to site but appear to be on the order of 10–100 μm. As depicted in the SECM images of Fig. 8, the apparent electrochemical activity of the surface sites at any particular potential correlates very well with the magnitude of the faradaic current observed in the voltammogram. The intensity of Br_2 generation in the image taken at 1.56 V is clearly much greater than that at 1.3 V. Other examples of electrochemically active sites are imaged on tantalum/tantalum oxide electrodes (24) and inclusions on aluminum alloy surfaces (25) and stainless steel (26). (For a detailed discussion on imaging molecular transport across membranes, readers should consult individual chapters, e.g., Chapter 9, in this monograph.)

Related to the corrosion problems was a recent SECM study, which demonstrated the possibility of eliminating typical experimental problems encountered in the measurements of heterogeneous electron transfer at semiconductor electrodes (27). In this experiment, the redox reaction of interest (e.g., reduction of $Ru(NH_3)_6^{3+}$) is driven at a diffusion-controlled rate at the tip. The rate of reaction at the semiconductor substrate is probed by measuring the feedback current as a function of substrate potential. By holding the substrate at a potential where no other species than the tip-generated one would react at the substrate, most irreversible parasitic processes, such as corrosion, did not contribute to the tip current. Thus, separation of the redox reaction of interest from parallel processes at the semiconductor electrode was achieved.

E. Other Systems and Potentiometric Imaging

SECM images of other systems, including mixed metal-insulator surfaces (28) and substrates composed of two conductors with different reactivity [e.g., gold and glassy carbon (29) or Pt and carbon (30)] have been reviewed recently (31). Imaging of Prussian blue films in the constant current mode has also been described (32). Here the SECM feedback occurred inside the solid phase. The resulting image depended strongly on the substrate potential, which in turn governed the film reactivity. SECM imaging can also be carried out with potentiometric probes instead of amperometric tips (31). For example, Horrocks et al. (33) used antimony tips as pH sensors to image local pH changes around a disk of immobilized urease-hydrolyzing urea and a disk of immobilized yeast cells in glucose solution. Other types of potentiometric imaging are also possible. Readers who are interested in detailed information on this imaging mode should refer to the proper chapter in this monograph.

IV. IMAGES IN HUMID AIR

Most SECM measurements are carried out with the sample under a thick liquid layer, and thus the tip must be sheathed in an insulator to achieve high resolution. SECM measurements can also be carried out within a thin layer of water that forms on the surface of a sample in air. In this case very high resolution can be attained using tips without insulation (e.g., the usual W or Pt-Ir STM-type tips) because the tip area is defined by the small part of the tip that touches the liquid layer (34,35). Studies of mica surfaces, polymer films, and some biological samples as described below are possible by this technique. With this mode it is also possible to fabricate small metal structures in polymer films as demonstrated previously (36). High-resolution electrochemical deposition of silver nanostructures on mica surfaces in humid air was also achieved (35). For detailed discussion on SECM applications for fabrication, see Chapter 13.

A. Instrumentation and Sample Preparation

The SECM/STM instrument employed in the following studies has a current sensitivity of <0.1 pA for imaging (37). Electrochemically etched tips were prepared from either 250-μm-diameter Pt-Ir (80%–20%) wires [etching solution: saturated $CaCl_2$: concentrated HCl : water (60% : 4% : 36% by volume)] or 500-μm W rods [etching solution: 3 M NaOH]. The scanning head was contained in a Faraday cage and a Plexiglas box in which the relative humidity (RH) of the atmosphere was controlled by various humidity-limiting solutions (38,39). A nearly constant humidity inside the box was usually attained in about an hour as determined by monitoring the surface conductance of the mica sample or by measuring the RH with a humidity probe (Model 6517-RH, Keithley Instruments, Inc., Cleveland, OH). The tip-sample arrangement is shown in Fig. 9.

Freshly cleaved mica substrates (Spruce Pine Mica Co., Spruce, NC) were vacuum-coated with an approximately 500 Å thick layer of gold at one end; this served as the reference/counter electrode contact. In all experiments the tip was laterally positioned approximately 1 mm from the Au contact and the imaging process was operated in the constant current mode. In ac-admittance measurements, a small ac voltage (10 mV peak-to-peak amplitude, 10 kHz) was superimposed on the dc bias and the admittance was measured with a phase-sensitive detector. If not otherwise mentioned, the mica substrate was treated with one small drop ($\sim$10 μL) of a phosphate

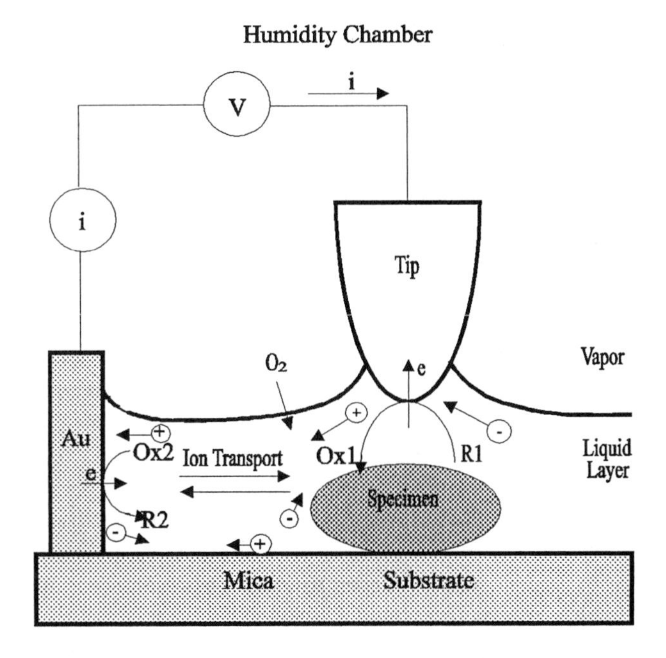

FIG. 9 Schematic diagram for the SECM chamber with controlled humidity and the electrochemical processes that control the current. The tip was located $\sim$1–2 mm from the Au contact. V = voltage bias between the tip and Au contact. i = current flow through the tip. R and Ox represent the reduced and oxidized forms of an electroactive species. ⊕ and ⊖ represent cations and anions in the water film. (Reprinted with permission from Ref. 34, Copyright © 1995 American Association for the Advancement of Science.)

buffer solution (0.12 mM NaH_2PO_3, 0.84 mM NaCl, and 0.046 mM NaN_3, pH 7.4) containing the specimen for approximately 10 min. The sample was then mounted on the SECM/STM and allowed to equilibrate with the atmosphere within the humidity chamber for approximately 1 hour before the experiments were carried out. In some experiments, the mica sample was treated with one small drop ($\sim$10 μL) of a solution containing the specimen in Tris-EDTA buffer solution [10 mM Tris buffer and 1 mM ethylenediaminetetraacetic acid (EDTA) (disodium salt), pH 7.6] for 5 minutes. The sample was then dipped for $\sim$1 second in water, dried for 5 minutes in air, and dipped again for $\sim$1 second in water. Excess water on the surface was carefully removed with filter paper, and the sample was mounted on the SECM/STM and allowed to equilibrate with the atmosphere within the box before the experiments were carried out.

B. Electrochemical and SECM Behavior

1. *Electrochemical Measurements*

Information about the mechanism of current flow was obtained from current-voltage (i-V) curves under different conditions. In these experiments, a rather blunt tip (radius of curvature $\sim$ 10 μm) was brought into contact with the sample and the piezo-feedback was switched off to avoid changes in the tip-substrate gap during the potential scan. Typical i-V curves are shown in Fig. 10. The curve for a Nafion film with a Pt-Ir tip (Fig. 10A) shows some hysteresis in the range of +1 and −1 V where pA currents flow. Outside this range the curves become linear and are largely independent of scan direction. The shape of the curves is consistent with electrochemical reactions, presumably mainly water electrolysis, at the tip and Au contact. Above this region the current is limited by the resistance of the thin film ($\sim 7.5 \times 10^{11}$ ohms). The i-V curves with a blunt Pt-Ir tip on a mica substrate treated with Tris-EDTA buffer solution are more complex and depend upon scan history (Fig. 10B). Thus, for scans from 0 V to a negative tip potential (C1 to C3), the i-V relation is fairly linear with the current increasing slightly on each scan. An initial scan of the tip towards positive potentials following these scans is also almost linear (A1). However, on successive scans (A2 to A4) a definite peak appears; the current decreases and the peak shifts to less positive potentials on each scan. The voltammetric scan is restored if the tip is biased to negative potentials for about 2 minutes, suggesting that a species, probably H_2, builds up around the tip at negative bias that is depleted during successive anodic tip scans. After a series of scans of the tip to positive potentials, an initial cathodic tip scan shows decreased current (C4) but gradually attains the linear behavior shown in successive cathodic scans. For

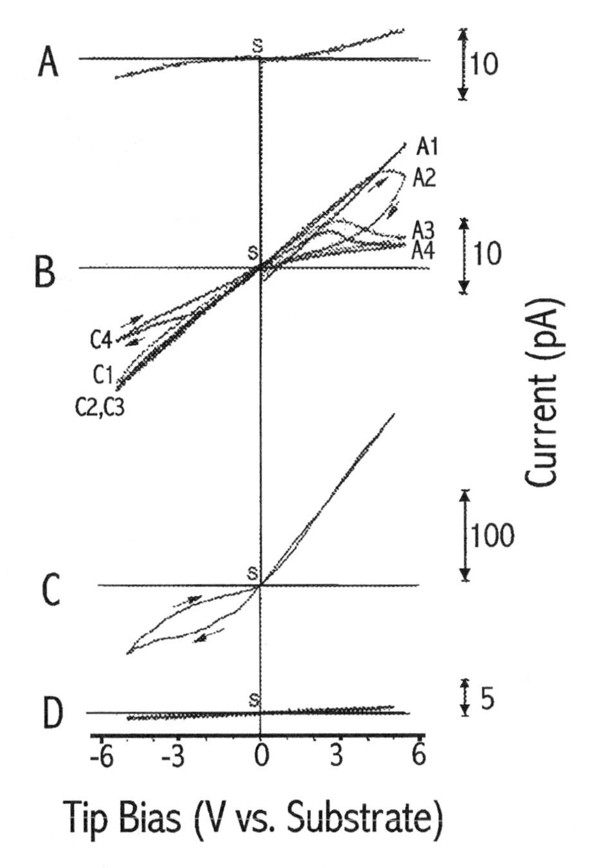

FIG. 10 Typical voltammetric curves in the humidity chamber at 100% RH, 25°C. Radius of tip curvature = 10 μm. In all cases, the voltage was scanned from 0 V (point S) in either direction and then returned to 0 V; the scan rate was 0.2 V/s. (A) Pt-Ir tip, Nafion film (~200 nm thick) on mica. (B) Pt-Ir tip, mica substrate treated with Tris-EDTA buffer solution. Curves C1–C3: first, second, and third negative scans. Curves A1–A4: first, third, fourth, and fifth positive scans after C3. Curve C4: first negative scan after A4. (C) W tip, mica substrate treated with Tris-EDTA buffer solution. (D) W tip, mica substrate treated only with water. (Reprinted with permission from Ref. 34a, Copyright © 1995 American Association for the Advancement of Science.)

purposes of imaging with a Pt-Ir tip, the current is fairly stable for negative tip bias values; with positive tip bias, oscillations frequently occurred.

As expected for an electrochemical process, the i-V curves also depend upon the tip material. For a blunt W tip on mica treated with Tris-EDTA buffer solution (Fig. 10C), the current is stable and increases almost linearly for positive tip bias but tends to saturate and shows considerable hysteresis on scan reversal in the negative tip bias region. Thus, with a W tip, imaging at positive bias is appropriate. The observed electrochemical behavior also depends on the pretreatment of the insulating substrate like mica. For example, the voltammetric behavior is very different when the mica is simply immersed in water for a few hours rather than treated with Tris-EDTA buffer solution. Here an i-V curve at 100% RH at a W tip (Fig. 10D) shows only very small currents across a region of ± 5 V. The current at a Pt-Ir tip at 100% RH is at least one order of magnitude smaller for a water-treated sample compared to one treated with Tris-EDTA buffer solution. The current under these conditions, however, was steady and showed little hysteresis. At low RH (e.g., 33%), no appreciable current (<0.1 pA) was observed in the region of ± 5 V. Thus, pretreatment of the insulating substrate to provide ions in the water layer and thus increase its conductivity appears to be useful. Presumably following treatment with Tris-EDTA buffer solution, a sufficient number of ions remain adsorbed on the mica surface, even after washing, to yield some conductance in the water film that forms on exposure to humid air.

The electrochemical signal observed here could arise from both capacitive charging and faradaic processes. However, charging processes are transient and cannot sustain a true steady direct current. Because the observed i-V curves are perturbed by resistive drops in the solution, one cannot identify with certainty the nature of the faradaic processes at the tip and Au contact. At Pt-Ir, candidate reactions are oxidation of water to O_2, reduction of water or protons to H_2, reduction of dissolved O_2, and processes involving absorbed species, e.g., oxidation of EDTA or adventitious impurities. At W, in addition to these processes, oxidation of the W to form the oxide and transiently, reduction of native or electrochemically generated oxide, are possible. The large and stable currents with a W tip at positive tip bias compared to Pt-Ir (see also the following discussion) suggest that the W oxidation reaction occurs at the tip. Note that the faradaic processes also generate ions that can contribute to the solution conductivity. Ions in the water layer also play an important role in establishing the double layers at both electrodes and providing charge compensation for electrogenerated species. They may also affect the hydrophilicity of the substrate surface and the thickness and structure of the water layer that forms on it.

It is interesting to estimate the effective tip radius immersed in the water layer, which is responsible for a tip current of ~1 pA at 1.5 V bias. As shown in Fig. 11, a polyurethane-coated W tip behaves as a microelectrode. A sigmoidal diffusion-limited current superimposed on the linear background current was obtained for the reduction of 1 mM $Ru(NH_3)_6^{3+}$ in 10 mM $NaClO_4$ solution. An effective radius estimated from the nearly steady-state current is ~3 μm. Also shown in Fig. 11 is the anodic background current due to the oxidation of W at potentials positive of ~0.4 V versus SCE (curve b). From the data shown in curve c of Fig. 10 and curve b of Fig. 11, if one assumes that similar effective tip radius is responsible for both anodic and cathodic redox processes, an estimated effective contact radius of ~3 nm can be obtained for a background current flow of 1 pA at a bias voltage of 1.5 V.

2. Conductance Measurement

In this experiment, two 50-nm-thick layers of Au separated by ~1 mm were deposited on 1 cm × 0.8 cm pieces of mica. One small drop (~10 μL) of

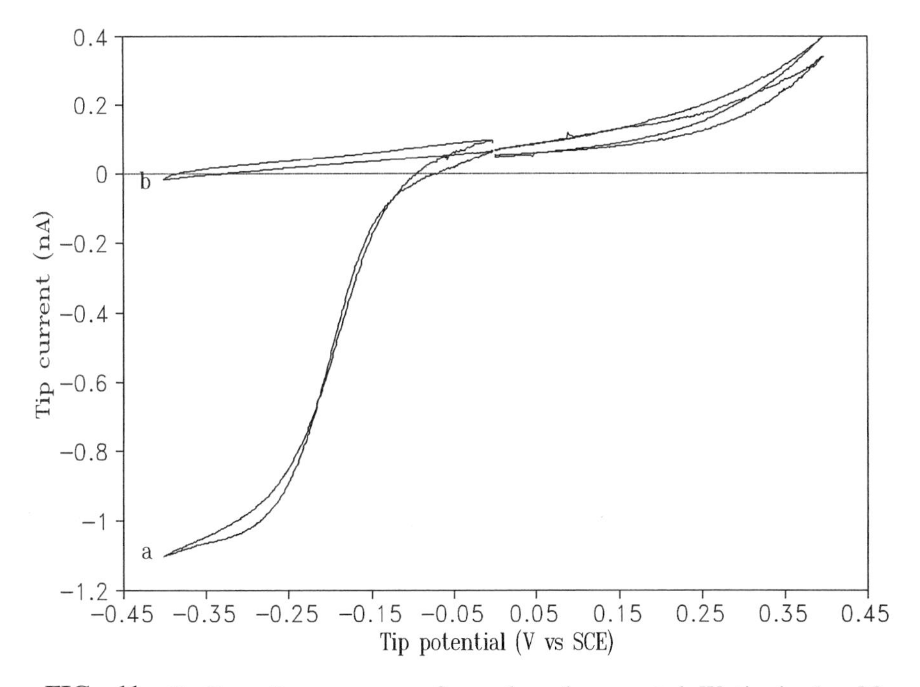

FIG. 11 Cyclic voltammograms of a polyurethane-coated W tip in 1 mM $Ru(NH_3)_6^{3+}$/10 mM $NaClO_4$ solution (curve a) and 10 mM $NaClO_4$ only (curve b). Potential scan rate is 5 mV/s.

a buffer solution was spread over the gap to cover an area of approximately 0.3 cm × 0.8 cm. After being partially dried in the ambient for ~10 minutes, the treated (or untreated) mica substrate was then allowed to equilibrate with the atmosphere within the humidity chamber for about one hour before the voltammetric experiments were carried out. Figure 12 shows a series of semi-log plots of current-voltage curves for a bare mica substrate at different RH. The measured current was very small (~20 fA at 2 V bias) in a desiccated (anhydrous $MgClO_4$ as the desiccant, RH < 10%) chamber, while it increased at least five orders of magnitude to ~6.9 nA at 93% RH. Figure 13 summarizes the conductance values (measured as the slope of the current-voltage curve in the bias range of 1.5–2.0 V) at various RH for the same mica sheet before and after the treatment with phosphate or Tris-EDTA buffer solution. As shown, the absolute conductance values of the bare mica surface increased slowly at low RH and more rapidly at high RH. Similar behavior was reported previously (40) for the adsorption isotherm measured by ellipsometry, suggesting that the observed conductance is closely related to the amount of water adsorbed on the mica surface. The adsorption iso-

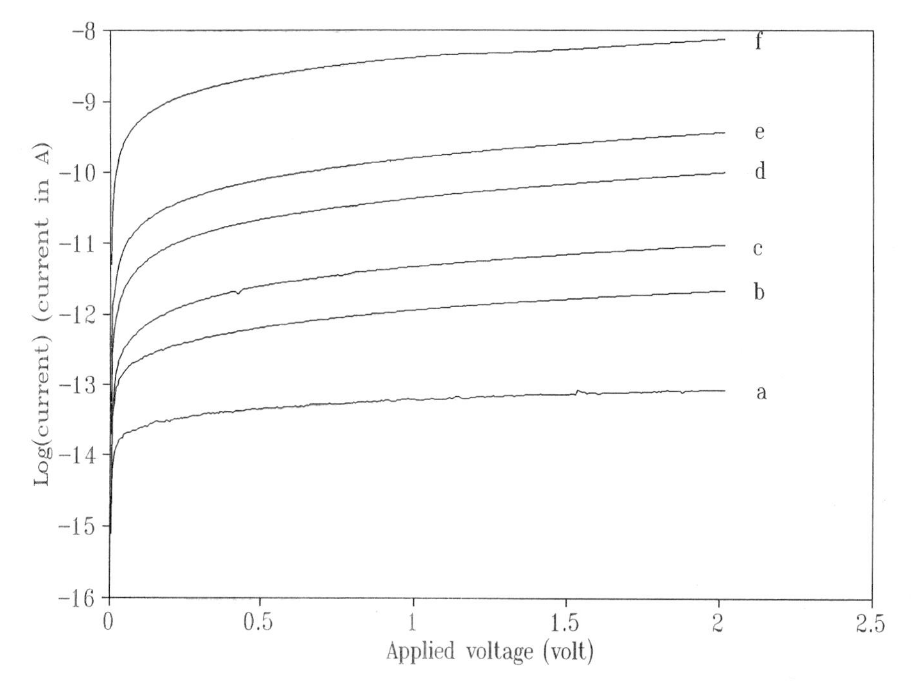

FIG. 12 A series of semilog plots of current-voltage curves for a bare mica substrate at different RH: 33% (a); 58% (b); 65% (c); 74% (d); 81% (e); 93% (f).

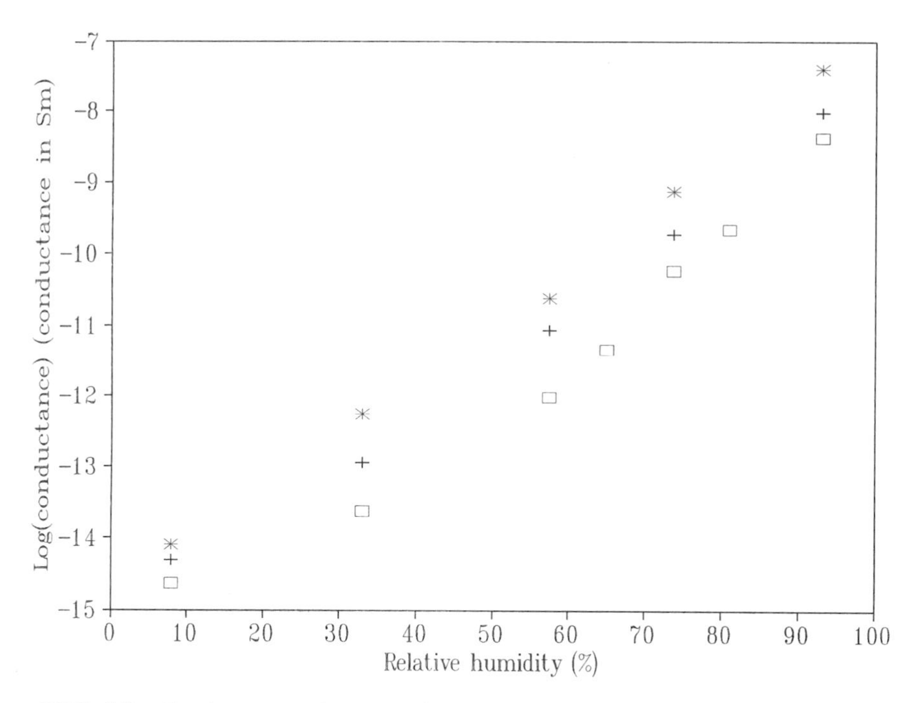

FIG. 13 Conductance values at various RH for the same mica sheet before (rectangles) and after treatment of Tris-EDTA buffer solution (asterisks) or phosphate (pluses) buffer. The conductance was measured as the slope of the current-voltage curve in the bias range of 1.5–2.0 V.

therm indicates initial weak interaction between water molecules and the mica surface followed by growth of layers of water. The average thickness of the water layer on a bare mica surface is ~0.4 nm at 70% RH. It is interesting to notice that, although the surface conductance of the mica substrate was substantially enhanced (one to two orders of magnitude) by the treatment with phosphate or Tris-EDTA buffer solution, similar behavior was observed for the RH dependence of the conductance. This enhanced conductance could be attributed to the increased ion concentration in the thin electrolyte film on the mica substrate, which in turn can affect the thickness and structure of the water layer that forms on it.

3. *Approach Curves*

When a W or Pt-Ir tip (biased at +3 V) approached the sample at 3 nm/s, the current remained at zero until it contacted the liquid layer, where it increased sharply, showing several orders of magnitude increase over a distance of a few nm. Accompanying this sharp increase in the tip current when it contacted the liquid layer, both parallel tip conductance and parallel tip capacitance also increased very rapidly (see Fig. 14). This is characteristic of an electrochemical process, and it provides the necessary feedback mechanism for controlling the tip position in high-resolution imaging. As reported previously, this arrangement can be used for imaging (34) and fabrication (36,41).

C. Imaging

As stated above, on the specimen and the mica substrate, there is normally a thin film of water (a few nm or less) at moderate RH. As the tip, biased at a certain voltage, is brought into contact with the film containing a sufficient concentration of ions, a measurable current occurs. When the constant-current mode is used, the feedback mechanism will keep the tip moving

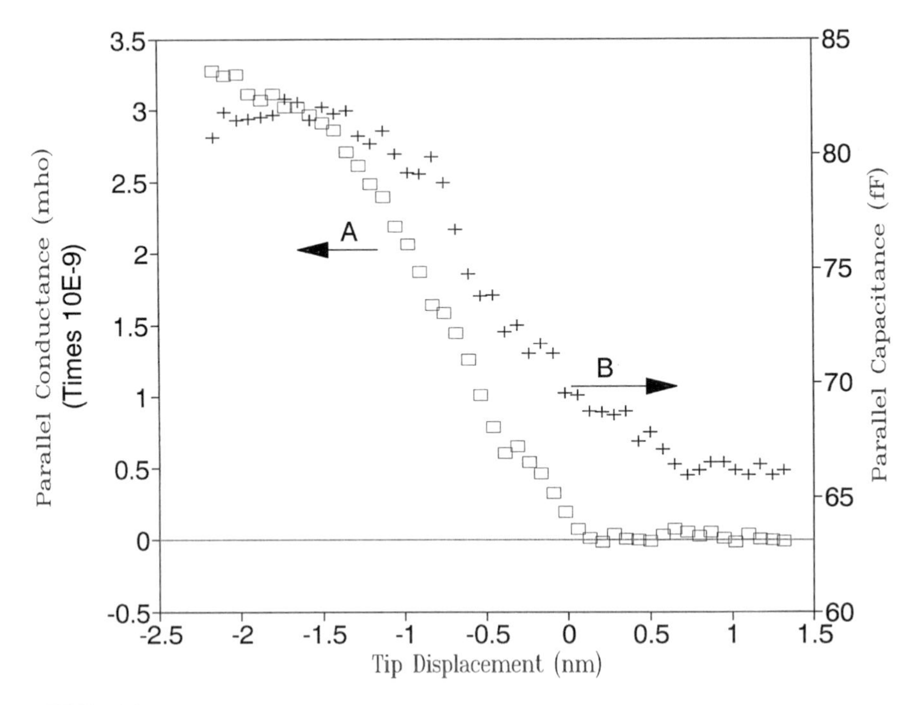

FIG. 14 Typical parallel conductance (A) and parallel capacitance (B) vs. tip displacement curves for a blunt W tip at 100% RH and 25°C for a mica substrate treated with Tris-EDTA buffer solution. Tip bias is 3 V with respect to the Au counterelectrode. The tip approaches the substrate surface at a rate of 3 nm/s.

up and down to maintain contact in the thin electrolyte layer. Therefore, the tip position provides some information about the surface topology of the specimen.

1. Mica Surface and Nafion Thin Film

The image of the mica surface obtained in the constant current mode at a reference tip current of 0.3 pA is shown in Fig. 15A. The step running diagonally from the lower left corner represents a cleavage plane for a single mica layer, ~1 nm high, and the small pieces seen on the lower plane probably represent salt residue or debris left after cleavage. A thin film (~200 nm) of Nafion could also be imaged by this technique. The film was prepared on the mica substrate by spin-coating at 3000 rpm with a photoresist spinner (Headway Research, Garland, TX) from an isopropanol-ethanol (4:1 by volume) solution containing 5% Nafion. The image of this film (Fig. 15B) shows smaller circular domain structures, 1–2 nm in diameter, consisting of a more conductive central zone surrounded by a less conductive region. These structures may correspond to those proposed in Gierke's cluster model (Fig. 15C) for a Nafion membrane (42) in which a central hydrophilic domain consisting of backbone ions and water is surrounded by a Teflon-like hydrophobic zone. This image demonstrates that SECM can distinguish between zones of different ionic conductivity in a sample. It also suggests that such an imaging mode does not pose serious limitation on the specimen thickness as long as the surface water film provides a closed circuit for ion conduction.

2. DNA

As first reported by Guckenberger et al. (43), images of DNA on mica could be obtained by this technique. DNA on mica was prepared as follows: A small drop (10 μL) of the DNA specimen (2.96 kbp, 21.75 μg/mL) in Tris-EDTA buffer solution was placed on mica to cover ~1 cm^2 and left to adsorb for ~5 minutes. The sample was then dipped for ~1 second in water, dried for 5 minutes in air, and dipped again for ~1 second in water. Excess water on the surface was carefully removed with filter paper, and the sample was then mounted on the SECM/STM and allowed to equilibrate with the atmosphere within the box before the imaging was carried out. Figure 16 shows corresponding images of several supercoiled DNA molecules on mica. In general, imaging is more stable at positive tip bias if a W tip is used. There is a tendency to lose the image after repeated scans at the same area, perhaps because of the depletion of electroactive species near the tip. Note that the magnitude of the signal can be partially recovered by waiting for a short period of time between different frames of imaging. The signal can also be recovered by imaging at different areas. As shown in Fig. 16, the

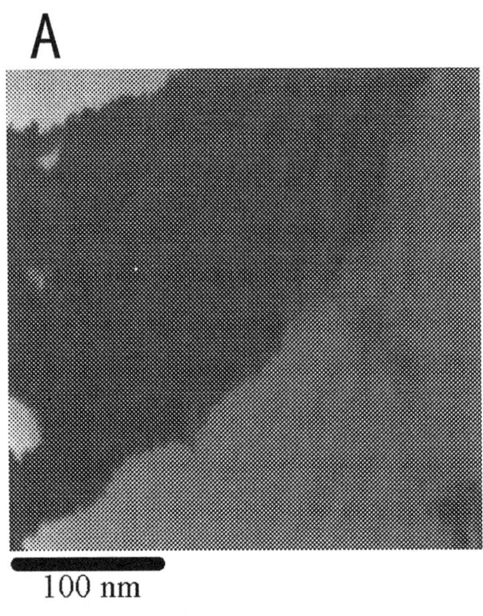

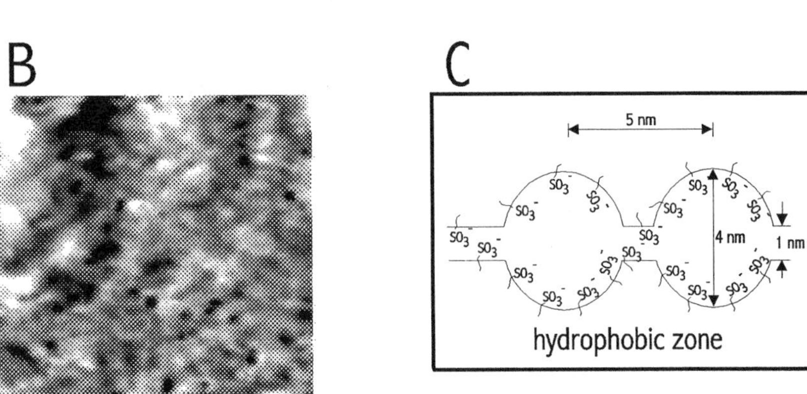

FIG. 15 (A) Image of a mica surface treated with Tris-EDTA buffer solution taken in humid air (80% RH at 25°C) with a sharp W tip at a reference current of 0.3 pA and a tip bias of 3 V. The tip rastering rate was 0.25 Hz. The total z-range is 3 nm. (B) Image of a Nafion film on mica taken in humid air (100% RH at 25°C) with a sharp Pt-Ir tip at a reference current of 3 pA and a tip bias of 3 V with respect to the Au counterelectrode. The tip rastering rate was 0.25 Hz. The image is inverted to enhance the visual effect of the domain structure; the dark region has a higher transient current than the lighter region and is corresponding to the hydrophilic domain. (C) Gierke's cluster model for Nafion membrane (42). (Reprinted with permission from (A, B) Ref. 34, Copyright © 1995 American Association for the Advancement of Science.)

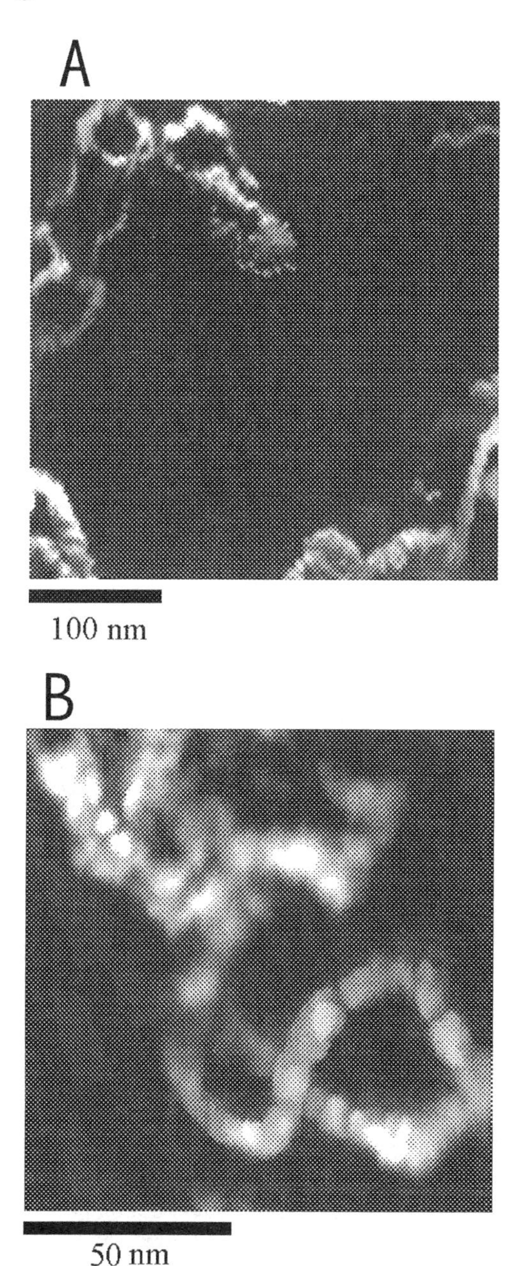

FIG. 16 Constant current images of fragments of DNA specimen (A: supercoiled; B: "Bluescript," 2.96 kbp) on mica substrates in humid air (80% RH at 25°C). Images were taken with a sharp W tip at a reference current of 0.3 pA and a tip bias of 3 V. The tip rastering rate was 0.25 Hz.

lateral dimension obtained (4–8 nm) is considerably larger than the expected diameter of DNA (~2.5 nm). Information in the z-axis is limited, due to the lack of a quantitative relation between the charge transfer rate and the tip-sample distance for this system.

3. Proteins

Besides DNA, antibody (e.g., a mouse monoclonal IgG), enzyme (e.g., glucose oxidase, GOD), and hemocyanin (e.g., keyhole limpet hemocyanin, KLH) molecules could also be imaged by the same technique. All of these protein molecules have characteristic three-dimensional structures and the majority of their surfaces is hydrophilic and so allow one to test the applicability of this technique for imaging. A small drop of 10-μL protein solution to cover ~1 cm^2 of the mica surface was used for the experiment. The stock solutions of IgG (15 μg/mL) and GOD (22.5 μg/mL) were prepared in a phosphate buffer solution. It is important to use low concentrations of protein to minimize aggregate formation. However, for imaging purposes the protein concentration should be high enough to have a reasonable coverage of protein molecules on the mica surface. The images were recorded by using a W tip. The reference current was 1 pA at a tip bias of 1.5 V (vs. the Au contact). Figure 17 shows a region of the mica surface that contains several mouse IgG (MW = 150,000 daltons) molecules. Some of them are of similar size and have a very characteristic quaternary structure (two arms and one body perhaps corresponding to the two Fab and one Fc fragments of the IgG molecule). The image of native IgG obtained by this technique is similar to those obtained from x-ray crystallography (44,45) and TEM (46). However, the apparent dimensions (~12 nm × 9 nm) are significantly larger than those (8.5 nm × 6 nm) determined from x-ray crystallographic studies. The estimated height of the molecule is ~5 nm, which is somewhat higher than the TEM and x-ray data (~4 nm).

Figure 18 shows the image of a group of GOD (MW = 160,000 daltons, from *Aspergillus niger*) molecules on a mica surface at 81% RH. Most of the molecules show a dimeric structure perhaps corresponding to the folded form of the two identical polypeptide chains of the GOD molecules. The monomeric unit is apparently a compact spheroid with approximate dimensions of 8 nm × 4 nm. The top-view images of some of the molecules are observed as nearly as circles with diameters of ~7 nm, which might represent the third dimension of the spheroidal monomeric unit. Contacts between two monomeric units forming the dimer are confined to a long, narrow stretch. The overall dimensions of the GOD molecules are thus approximately 8 nm × 7 nm × 8 nm, significantly larger than those (6.0 nm × 5.2 nm × 7.7 nm) determined from x-ray crystallographic data on the partially deglycosylated enzyme (47).

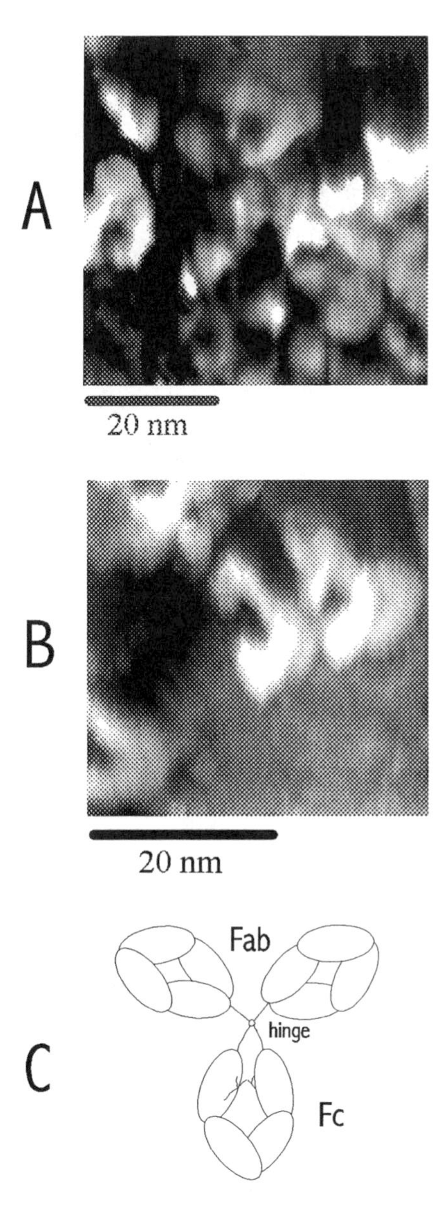

FIG. 17 (A) and (B). Constant current images showing several mouse IgG (MW = 150,000 daltons) molecules on mica substrates. The images were recorded using a W tip in humid air (80% RH at 25°C). The reference current was 1 pA at a tip bias of 1.5 V (vs. the Au contact). (C) Schematic drawing of an IgG molecule showing two Fab arms and the Fc portion (57).

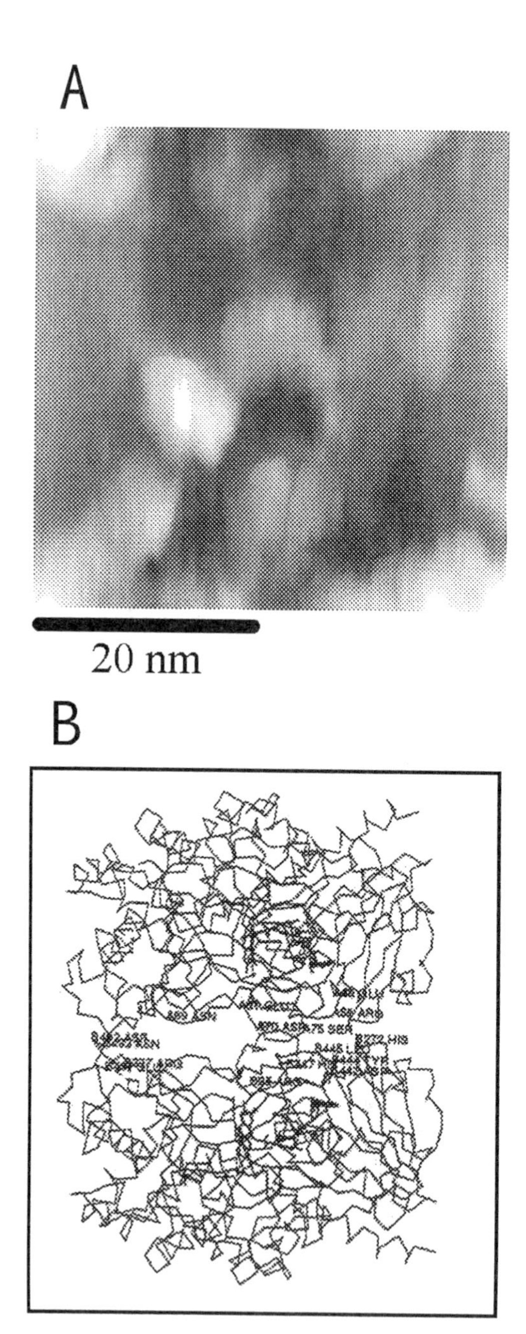

FIG. 18 (A) Constant current image showing several glucose oxidase (MW = 160,000 daltons, from *A. niger*) molecules on mica substrates. The images were recorded using a W tip in humid air (80% RH at 25°C). The reference current was 1 pA at a tip bias of 1.5 V vs. the Au counterelectrode. (B) C^{α} tracing of the dimer structure of a glucose oxidase molecule (47). Contacts between molecules forming the dimer are confined to a long, narrow stretch.

Finally, we chose hemocyanin molecules from Gastropods (e.g., KLH) as a testing specimen for this imaging technique, since it has a distinctive cylindrical profile having 6, 9, 12, etc. parallel rows, depending on its size (48). Images for KLH molecules (MW $\sim$ 7–8 $\times$ 10^6) were taken at 65% RH with a W tip biased at 1.7 V with a reference current of 1 pA. Figure 19 suggests a cylindrically shaped molecule having nine parallel sections. It has a rectangular profile with one dimension of 30–35 nm and the other dimension of 60–65 nm, which are quite close to those observed by TEM. The shape of the molecule is similar to the model for the structure of a six parallel-rows Gastropod hemocyanin molecule proposed by Mellema and Klug (49) based on a computer-processed three-dimensional reconstruction of the TEM data. The reconstruction showed a hollow cylindrical molecule partly closed at both ends by a collar and possibly a central cap. The walls are made from 60 morphological units of similar size, shape, and orientation, which were correlated with 120 oxygen-binding sites. This result suggests that such an imaging mode may provide some information about the surface topology of the specimen as long as the surface water film provides a closed circuit for ion conduction.

The significant increases in size of the DNA and small protein molecules determined by this technique as compared with other techniques are most likely due to the presence of water and salt around the molecules, and the convolution of the nonnegligible tip size. They could also be caused by the deformation of the molecules by the tip during imaging, as frequently encountered in AFM or STM images for biological samples (50–55). It is conceivable that when the dimensions of the specimen are large as compared with the tip curvature, the relative deviation due to tip geometry is small. It is also possible to deconvolute the images if one knows the exact tip dimensions and the detailed mechanisms of imaging. One can also minimize the size deviation by optimizing the RH and the salt concentration. In spite of this discrepancy, the overall shapes of the DNA and protein molecules are well reproduced in this experiment and are similar to those determined from other more established techniques. These results give us confidence that this technique accurately reflects the native conformation of the DNA and protein molecules. They further indicate that the use of a mica surface, which is atomically flat and can easily be chemically modified, is a suitable substrate for studying protein samples as suggested previously.

V. CONCLUSIONS AND FUTURE PROJECTIONS

Complementary with other imaging techniques, such as electron microscopy and other scanning probe microscopies (SPMs), SECM routinely offers μm resolution of the surface structures in bulk electrolyte solutions. SECM also

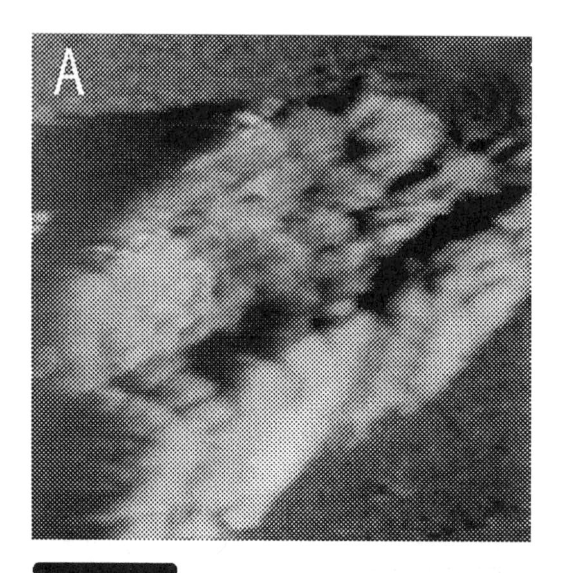

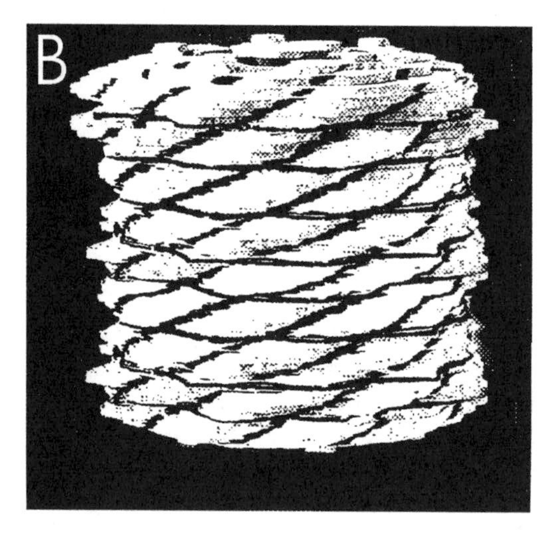

FIG. 19 (A) Constant current image (A) of a keyhole limpet hemocyanin (MW $\sim$ 7–8 $\times$ 10^6 daltons) molecule on a mica surface. The image was taken at 65% RH with a W tip biased at 1.7 V with a reference current of 1 pA. (B) An analogous Mellema and Klug's model (49) for a nine-parallel-row Gastropod hemocyanin.

provides the local chemical and electrochemical reactivity of surface features. The latter application is unique to SECM and is particularly useful in identifying and understanding the functional aspects of these surface structures. Those images of biomacromolecules taken in humid air further demonstrate that the resolution can reach a few nm, although it is still not as high as that seen in STM of ordered conductive surfaces. Improved resolution should result from lower RH conditions to decrease the thickness of the water film but still keep the ionic conductivity of the mica substrate and the individual molecules high enough for detection. As stated above, the resolution is also a function of the exposed tip area and perhaps could be improved by treatment of the tip except at the very end with a hydrophobic agent to decrease the wetted area above the contact point. The potential of this technique for ultra-high-resolution chemical imaging as successfully applied in bulk liquid phase studies is very attractive. Yet the problems of getting better control of the electrochemical reactions within the film have not been resolved. Since the resolution of SECM images depends strongly on the tip size and the distance between tip and substrate, it is also interesting to continue developing some universal technique for the feedback mechanism for near-to-surface imaging. Results reported previously (3,4,56) as well as those obtained most recently based on force-feedback technique seem promising. Continuous effort in the development and improvement of various feedback-control and tip-preparation techniques may allow SECM to image surface structures routinely with sub-μm resolution in bulk electrolyte solutions. Although the potentiometric imaging technique is not discussed here, it should be very valuable to devise small ion-selective and enzyme electrode tips with fast response for high-resolution chemical imaging.

ACKNOWLEDGMENTS

We would like to thank Professor B. L. Iverson and Dr. G. Chen for their generous donation of DNA and hemocyanin samples. We would also like to thank Drs. X.-H. Xu and C. Demaille for their helpful technical assistance, suggestions, and discussions. Particularly, we would like to thank Professor Allen J. Bard for his constant support and discussion. The support of the research in SECM by the Robert A. Welch Foundation, the National Science Foundation, and the Texas Advanced Research Program is also gratefully acknowledged.

REFERENCES

1. G Binnig, H Rohrer. Helv Phys Acta 55:726, 1982.
2. DO Wipf, AJ Bard. Anal Chem 64:1362, 1992.

3. M Ludwig, C Kranz, W Schuhmann, HE Gaub. Rev Sci Instrum 66:2857, 1995.
4. PJ James, LF Garfias-Mesias, PJ Moyer, WH Smyrl. J Electrochem Soc 145: L64, 1998.
5. ER Scott, HS White, JB Phipps. J Membr Sci 58:71, 1991.
6. ER Scott, HS White, JB Phipps. Solid State Ionics 53:176, 1992.
7. DT Pierce, AJ Bard. Anal Chem 65:3598, 1993.
8. CM Lee, JY Kwak, AJ Bard. Proc Natl Acad Sci USA 87:1740, 1990.
9. RB Jackson, M Tsionsky, ZG Cardon, AJ Bard. Plant Physiol 111:354, 1996.
10. M Tsionsky, ZG Cardon, AJ Bard, RB Jackson. Plant Physiol 113:895, 1997.
11. CM Lee, FC Anson. Anal Chem 64:528, 1992.
12. IC Jeon, FC Anson. Anal Chem 64:2021, 1992.
13. JY Kwak, CM Lee, AJ Bard. J Electrochem Soc 137:1481, 1990.
14. CM Lee, AJ Bard. Anal Chem 62:1906, 1990.
15. N Casillas, S Charlebois, W Smyrl, HS White. J Electrochem Soc 140:L142, 1993.
16. N Casillas, S Charlebois, W Smyrl, HS White. J Electrochem Soc 141:636, 1994.
17. P James, N Casillas, W Smyrl. J Electrochem Soc 143:3853, 1996.
18. H Shiku, T Takeda, H Yamoda, T Matsue, I Uchida. Anal Chem 67:312, 1995.
19. G Wittstock, K-J Yu, HB Halsall, TH Ridgway, WE Heineman. Anal Chem 67:3578, 1995.
20. H Shiku, T Matsue, I Uchida. Anal Chem 68:1276, 1996.
21. Model CHI900 SECM, CH Instruments, Inc., Austin, TX. JF Zhou, F-RF Fan, AJ Bard, unpublished results.
22. CM Lee, DO Wipf, AJ Bard, K Bartels, AC Bovik. Anal Chem 63:2442, 1991.
23. DS Hage. Anal Chem 67:455R, 1995.
24. SB Basame, HS White. Anal Chem 71:3166, 1999.
25. JPH Sukamto, WH Smyrl, N Casillas, M Al-Odan, P James, W Jin, L Douglas. Mat Sci Eng A 198:177, 1995.
26. YY Zhu, DE Williams. J Electrochem Soc 144:L43, 1997.
27. BR Horrocks, MV Mirkin, AJ Bard. J Phys Chem 98:9106, 1994.
28. RC Engstrom, M Weber, DJ Wunder, R Burgess, S Winquist. Anal Chem 58: 844, 1986.
29. DO Wipf, AJ Bard. J Electrochem Soc 138:L4, 1991.
30. RC Engstrom, B Small, L Kattan. Anal Chem 64:241, 1992.
31. AJ Bard, F-RF Fan, MV Mirkin. In: AJ Bard, ed. Electroanalytical Chemistry. Vol. 18. New York: Marcel Dekker, 1994, p 243.
32. H Sugimura, N Shimo, N Kitamura, H Masuhara. J Electroanal Chem 346: 146, 1993.
33. BR Horrocks, MV Mirkin, DT Pierce, AJ Bard, G Nagy, K Toth. Anal Chem 65:1213, 1993.
34. (a) F-RF Fan, AJ Bard. Science 270:1849, 1995; (b) Proc Natl Acad Sci USA 96:14222, 1999.
35. F Forouzan, AJ Bard. J Phys Chem B 101:10876, 1997.
36. DH Craston, CW Lin, AJ Bard. J Electrochem Soc 135:785, 1988.

37. F-RF Fan, AJ Bard. J Electrochem Soc 136:3216, 1989.
38. JA Dean, ed. Lange's Handbook of Chemistry. 13th ed. New York: McGraw-Hill, 1985.
39. AN Kirgintsev, AV Luk'yanov. Zh Neor Khim 12:2032, 1967.
40. D Beaglehole, EZ Radlinska, BW Ninham, HK Christenson. Phys Rev Lett 66:2084, 1991.
41. RL McCarley, SA Hendricks, AJ Bard. J Phys Chem 96:10089, 1992.
42. WY Hsu, TD Gierke. J Membr Sci 13:307, 1983.
43. R Guckenberger, M Heim, G Ceve, H Knapp, W Wiegräbe, A Hillebrand. Science 266:1538, 1994.
44. TNC Wells, M Stedman, RJ Leatherbarrow. Ultramicroscopy 42:44, 1992.
45. M Marquart, J Deisenhofer, R Huber. J Mol Biol 141:369, 1980.
46. RC Valentine, NM Green. J Mol Biol 27:615, 1967.
47. HJ Hecht, HM Kalisz, J Hendle, RD Schmid, D Schomburg. J Mol Biol 229: 153, 1993.
48. EFJ Van Bruggen. In: EJ Wood, ed. Structure and Function of Invertebrate Respiratory Proteins. London: Harwood Academic Publishers, 1982, p. 1.
49. JE Mellema, A Klug. Nature 239:146, 1972.
50. RJ Leatherbarrow, M Stedmann, TNC Wells. J Mol Biol 221:361, 1991.
51. GJ Leggett, MC Davies, DE Jackson, CJ Roberts, SJB Tendler, PM Williams. Phys Chem 97:8852, 1993.
52. JN Lin, B Drake, AS Lea, PK Hansma, JD Andrade. Langmuir 6:509, 1990.
53. Q Chi, J Zhang, S Dong, E Wang. J Chem Soc Faraday Trans 90:2057, 1994.
54. SL Tang, AJ McGhie. Langmuir 12:1088, 1996.
55. H Hansma, J Hoh. Annu Rev Biophys Biochem Struct 23:115, 1994.
56. JV Macpherson, PR Unwin, AC Hillier, AJ Bard. J Am Chem Soc 118:6445, 1996.
57. DR Burton. In: F Calabi, MS Neuberger, eds. Molecular Genetics of Immunoglobulin. Amsterdam: Elsevier, 1987, p. 1.

5

THEORY

Michael V. Mirkin

Queens College–City University of New York
Flushing, New York

I. INTRODUCTION

The quantitative SECM theory was developed during the last decade for various regimes of measurements and electrochemical mechanisms. Different operating modes of the SECM, e.g., feedback and generation/collection modes, steady-state and transient measurements, require significantly different theoretical descriptions. Quantitative treatments are available for both diffusion-controlled processes and finite kinetics as well as for more complicated mechanisms involving adsorption and homogeneous reactions in the gap. The focus of this chapter is on fundamentals of the SECM theory, i.e., steady-state and transient measurements, feedback mode, and generation/collection experiments. While the theory for a one-step heterogeneous reaction will be covered in detail, only a brief discussion of other processes (e.g., adsorption, charge transfers at the liquid/liquid interface) will be included to avoid overlap with other chapters.

II. FEEDBACK MODE OF SECM OPERATION

Feedback theory has been the basis for most quantitative SECM applications reported to date. Historically, the first theoretical treatment of the feedback response was the finite-element simulation of a diffusion-controlled process by Kwak and Bard (1), but we will start from a more general formulation for a quasi-reversible process under non-steady-state conditions and then consider some important special cases.

A. General Theory of SECM Feedback for a First-Order Heterogeneous Reaction at the Tip/Substrate

In the feedback mode experiment, both the tip and the substrate are immersed in a solution containing an electrolyte and a redox species (e.g., a reducible species, O). This species is reduced at the microdisk tip electrode, and the product, R, can be reoxidized at the substrate:

$$O + ne^- \rightarrow R \qquad \text{(tip electrode)} \tag{1}$$

$$R - ne^- \rightarrow O \qquad \text{(substrate)} \tag{2}$$

The considerable complexity of SECM theory is due to the combination of a cylindrical diffusion to the ultramicroelectrode (UME) tip and a thin-layer–type diffusion space. The time-dependent diffusion problem for a simple quasireversible reaction in cylindrical coordinates is as follows (2,3):

$$\frac{\partial C_i}{\partial T} = \frac{\partial^2 C_i}{\partial z^2} + \frac{\partial^2 C_i}{\partial R^2} + \frac{1}{R}\frac{\partial C_i}{\partial R} \qquad 0 < T,\ 0 \leq R,\ 0 < Z < L \tag{3}$$

where L is the normalized tip-substrate distance (d/a) and other dimensionless variables are

$$R = r/a \tag{4}$$

$$Z = z/a \tag{5}$$

$$C_i = c_i/c_O^\circ \tag{6}$$

$$T = tD_O/a^2 \tag{7}$$

where r and z are, respectively, the coordinates in the directions radial and normal to the electrode surface; D_i and c_i are the diffusion coefficient and concentration of the species i (i = O, R), c_O° is the bulk concentration of O, a is the tip electrode radius, and t is time.

The assumption of equal diffusion coefficients ($D_O = D_R = D$) allows the problem to be described in terms of a single species (O). The boundary conditions are of the form:

$0 < T$, $0 \leq R < 1$, $Z = 0$ (tip electrode surface):

$$\frac{\partial C_O}{\partial Z} = K_{f,T}C_O - K_{b,T}(1 - C_O) = J_T \tag{8}$$

$0 < T$, $1 \leq R \leq RG$, $Z = 0$ (glass insulating sheath):

$$\frac{\partial C_O}{\partial Z} = 0 \tag{9}$$

$0 < T$, $0 \leq R \leq h$, $Z = L$ (substrate surface):

$$\frac{\partial C_O}{\partial Z} = K_{b,s}C_O - K_{f,S}(1 - C_O) = J_S \tag{10}$$

$R > RG$, $0 \leq Z \leq L$:

$$C_O = 1 \tag{11}$$

where J_T and J_S are the normalized fluxes of species at the tip and the substrate,

$$K_{f/b,S} = k_{f/b,S}\, a/D \tag{12}$$

$$K_{f/b,T} = k_{f/b,T}\, a/D \tag{13}$$

$$RG = rg/a \tag{14}$$

$$h = a_S/a \tag{15}$$

where rg is the radius of the tip-insulating material and a_S is the substrate radius. The initial condition, completing the definition of the problem, is

$$T = 0,\ 0 \leq R,\ 0 \leq Z \leq L; \qquad C_O = 1 \tag{16}$$

If the substrate is an insulator, $J_S \equiv 0$. If both tip and substrate reactions are electrochemical processes, the rate constants for reduction (k_f) and oxidation (k_b) are given by the Butler-Volmer relations:

$$k_f = k^\circ \exp[-\alpha f(E - E^{\circ\prime})] \tag{17}$$

$$k_b = k^\circ \exp[(1 - \alpha) f(E - E^{\circ\prime})] \tag{18}$$

where k° is the standard rate constant, E is the electrode potential, $E^{\circ\prime}$ is the formal potential, α is the transfer coefficient, the number of electrons transferred per redox event is taken as 1, and $f = F/RT$ (here F is the Faraday, R is the gas constant, and T is the temperature). The solution of the problem can be obtained in terms of the dimensionless currents $I_T(T)$ and $I_S(T)$:

$$I_T(T) = -\frac{\pi}{2}\int_0^1 J_T(T, R) R\, dR$$

$$I_S(T) = \frac{\pi}{2}\int_0^h J_S(T, R) R\, dR \tag{19}$$

where I_T and I_S are equal to the physical currents normalized by the limiting diffusion tip current, $I = i/i_{T,\infty}$, where

$$i_{T,\infty} = 4nFDc^\circ a \tag{20}$$

The above formulation is somewhat overly general, since it includes the possibility of mixed diffusion/kinetic control of both the tip and substrate processes. In practice, at least one of those electrodes is held under diffusion control, so either

$$0 < T,\ 0 \leq R < 1,\ Z = 0; \qquad C_O = 0 \tag{21a}$$

can be used instead of Eq. (8), or Eq. (10) can be replaced with a much simpler boundary condition:

$$0 < T, 0 \leq R \leq h, Z = L; \qquad C_O = 1 \tag{21b}$$

The time-dependent SECM problem was solved semi-analytically in terms of two-dimensional integral equations (2,3) and numerically by using Krylov integrator (4) and the alternating-direction implicit finite difference (ADI) method (3,5). Potentiostatic transients were computed for two limiting cases: a diffusion-controlled process and totally irreversible kinetics (3–5). The analysis of the simulation results (4) revealed several time regions typical of SECM transients (Fig. 1). In the short-time region ($T \leq 0.001$) an SECM tip current follows closely the microdisk transient, after which it starts to deviate and finally levels out at a constant value of i_T. For a conductive substrate, $i_T \geq i_{T,\infty}$, and it is always larger than the value calculated for the same time from thin-layer cell (TLC) theory. For an insulating substrate, $i_T < i_{T,\infty}$. The smaller the tip/substrate separation is, the earlier the deviations between the SECM and microdisk transients occur. The time when the SECM undergoes a transition from the microdisk regime to the TLC regime is related to the time needed by the species to diffuse across the gap between the tip and the substrate. This time can be determined experimentally and can be used to evaluate the diffusion coefficient (4). The transients computed for different rates of an irreversible heterogeneous reaction (5) also showed a microdisk-type behavior at short times. The current magnitude at longer times and its eventual steady-state value are determined by the value of the dimensionless heterogeneous rate constant, $K_{b,S}$ (Fig. 2). The substrate behaves as a conductor as $K_{b,S} \rightarrow \infty$ and as an insulator as $K_{b,S} \rightarrow 0$. A few transients and non–steady-state CVs for a quasireversible process were computed for both insulating and conducting substrates (2).

More recently, Martin and Unwin simulated chronoamperometric feedback allowing for unequal diffusion coefficients of the oxidized and reduced forms of the redox mediator (6). Unlike steady-state SECM response, the shape of the tip current transients is sensitive to the ratio of the diffusion coefficients, $\gamma = D_O/D_R$ (Fig. 3). When $D_O = D_R$, the tip current attains a steady-state value much faster than for any $\gamma \neq 1$. At $\gamma < 1$, a characteristic minimum appears in the short-time region, which is quite unusual for potentiostatic transients.

FIG. 1 Comparison of simulated SECM transients with transients corresponding to different electrode geometries (all processes are diffusion-controlled). (A) SECM transient for a conductive substrate; (B) two-electrode thin-layer cell; (C) microdisk; (D) planar electrode; (E) SECM with an insulating substrate; (F) one-electrode thin-layer cell. Curves A, B, E, and F were computed with $L = d/a = 0.1$. (Adapted with permission from Ref. 4. Copyright 1991 American Chemical Society.)

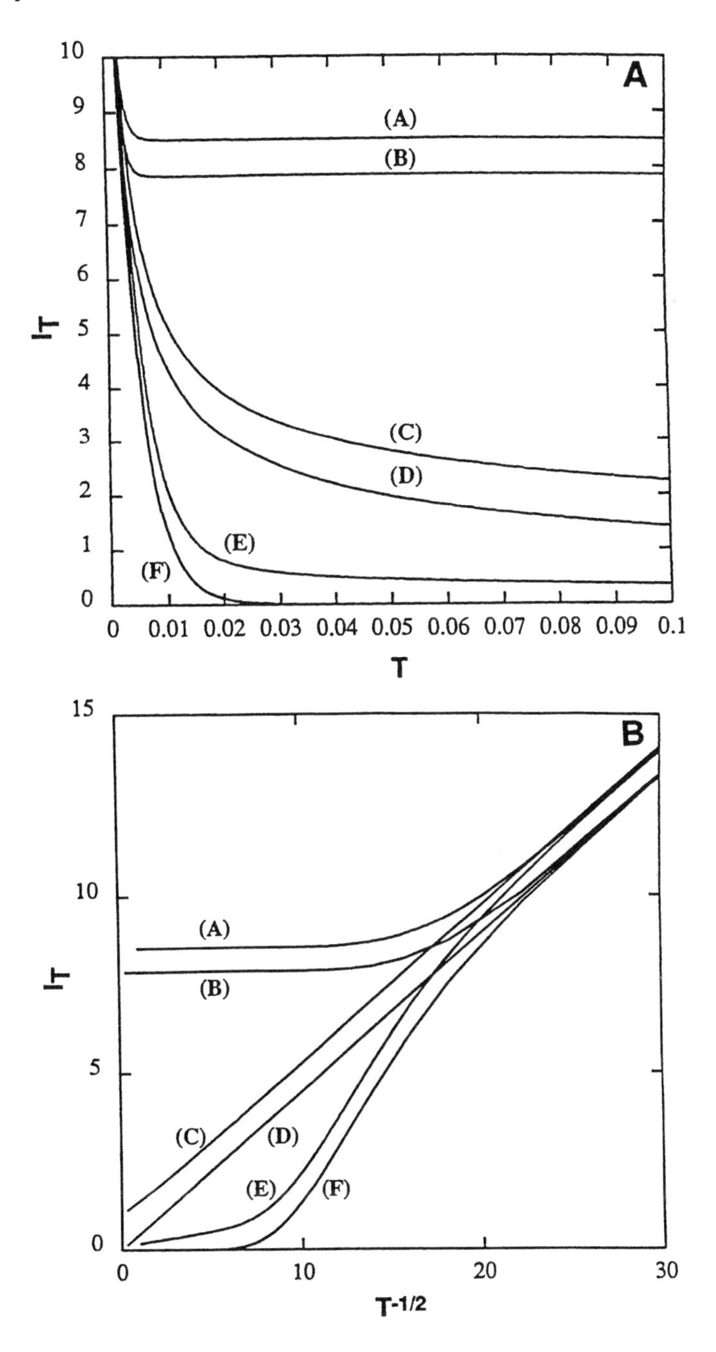
A
(A)
(B)
(C)
(D)
(E)
(F)
I_T
T
0 0.01 0.02 0.03 0.04 0.05 0.06 0.07 0.08 0.09 0.1
0 1 2 3 4 5 6 7 8 9 10
B
(A)
(B)
(C)
(D)
(E)
(F)
I_T
$T^{-1/2}$
0 10 20 30
0 5 10 15

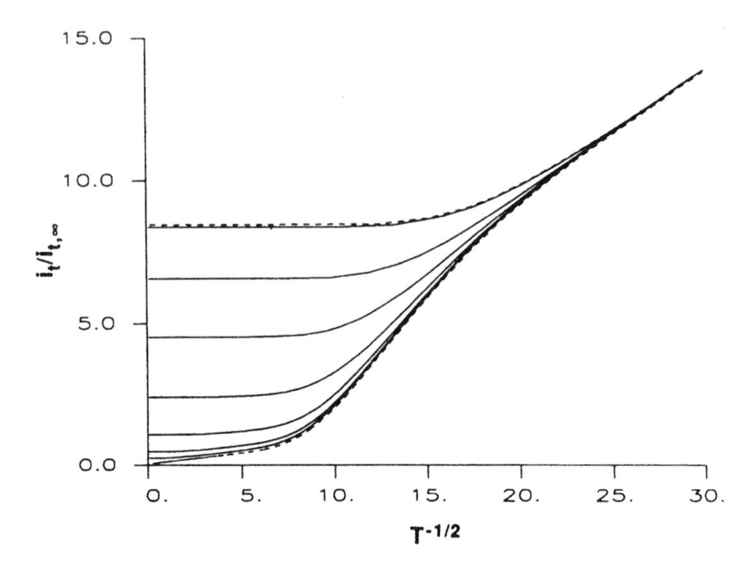

FIG. 2 Simulated feedback SECM transients with various rate constants for an irreversible heterogeneous process at the substrate. The upper and lower dashed curves correspond respectively to the limits $K_{b,S} \rightarrow \infty$ and $K_{b,S} = 0$. The solid curves (from top to bottom) represent log $K_{b,S}$ = 3.0, 1.5, 1.0, 0.5, 0, −0.5, and −1.0. RG = 10. L = d/a = 0.1. (Reprinted with permission from Ref. 3. Copyright 1992 American Chemical Society.)

One should notice that Eq. (9) implies that concentration of the mediator is equal to its bulk value everywhere beyond the limits of the tip-insulating sheath ($R \geq RG$). This approximation is applicable only if $RG \gg 1$, and even for RG = 10 it leads to about 2% error, which is present in almost all published SECM simulations.

B. Steady-State Conditions

The formulation of the steady-state SECM problem is significantly simpler (8). It includes a single Laplace equation

$$\frac{\partial^2 C}{\partial Z^2} + \frac{\partial^2 C}{\partial R^2} + \frac{1}{R}\frac{\partial C}{\partial R} = 0 \tag{22}$$

for oxidized (or reduced) form of the mediator with boundary conditions (9) and (11) and one of two pairs of boundary conditions for the tip and substrate surfaces, i.e., either Eqs. (10) and (21a) (diffusion-controlled tip reaction and quasireversible substrate kinetics) or Eqs. (8) and (21b) (diffusion-controlled substrate reaction and finite kinetics tip at the tip). All variables in these

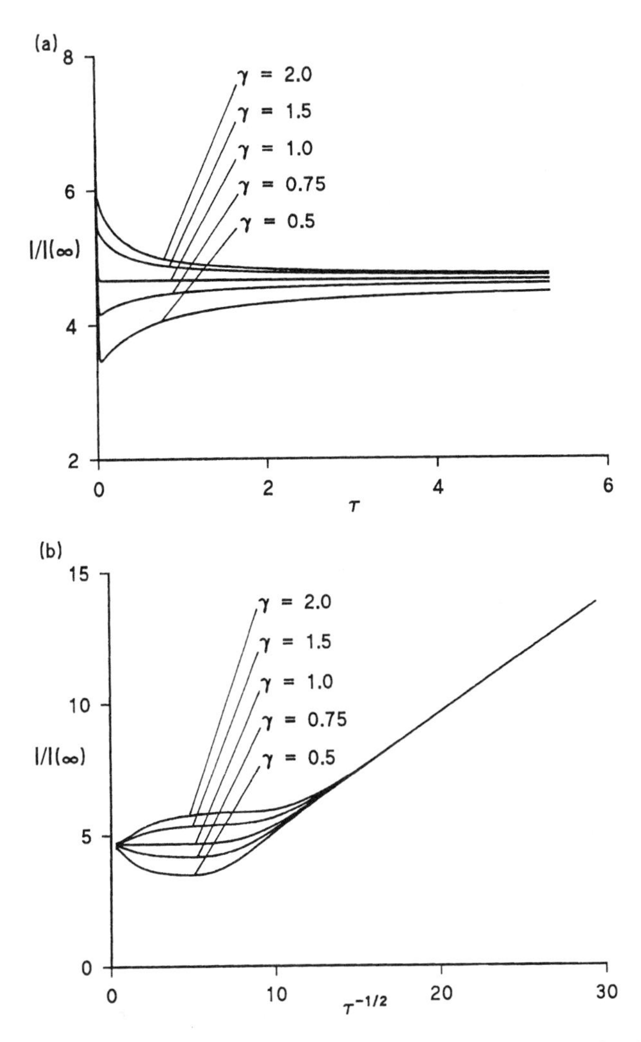

FIG. 3 The effect of γ on the chronoamperometric characteristics for the positive feedback process at a tip/substrate separation, L = 0.2. The reduced form initially present in solution is oxidized at the tip and regenerated at the substrate. Normalized tip current is plotted as a function of dimensionless time, $\tau = tD_R/a^2$. (Reprinted with permission from Ref. 6. Copyright 1997 Elsevier Science S.A.)

equations are now time-independent. The assumption of equal diffusion coefficients is unnecessary under steady state, because the concentrations of O and R at steady state are interrelated as follows

$$c_O + c_R D_R/D_O = c_O^\circ + c_R^\circ D_R/D_O \tag{23}$$

The solution for the finite kinetics at the tip and diffusion-limited substrate reaction is (8)

$$\frac{1 - \pi J_T(R)/4\kappa'}{\theta} = \int_0^1 u J_T(u) du \int_0^\infty J_0(pR) J_0(pu) \tanh(pL) dp \tag{24}$$

where J_0 is the Bessel function of the first kind of order zero,

$$\kappa' = \pi a k^\circ \exp[-\alpha f(E - E^{\circ\prime})]/(4D_O) \tag{25a}$$

and

$$\theta = 1 + \exp[nf(E - E^{\circ\prime})] D_O/D_R \tag{25b}$$

However, Eq. (24) has not been utilized in any calculations. Most published computational results represent the long-time limit of the data obtained by solving time-dependent equations. By fitting these results, several analytical approximations have been obtained for different situations including diffusion-controlled, irreversible, and quasireversible reactions at both tip and substrate surfaces.

1. *Diffusion-Controlled and Nernstian Steady-State Processes*

The knowledge of the shape of the I_T-L curve for a diffusion-controlled process is critical for both imaging and quantitative kinetic measurements because it allows one to establish the distance scale. The equations describing a diffusion-controlled SECM process represent two limiting cases of the general steady-state problem discussed above, i.e., (1) "pure positive feedback" produced by rapid regeneration of the reactant species at the substrate and (2) "pure negative feedback" observed when no mediator regeneration occurs at the substrate. Both problems were treated numerically by Kwak and Bard (1). The dimensionless current-distance curves were tabulated for both insulating and conductive substrates, assuming the equality of the diffusion coefficients and an infinitely large substrate. Later (9), analytical approximations were obtained for both working curves. Both numerical results in Ref. 1 and analytical expressions in Ref. 9 are suitable only for large values of RG (i.e., RG $\geq$ 10). However, tips with much thinner insulating sheathes have often been employed because they allow one to obtain closer tip/substrate separation. A typical RG for a pipet-based tip is ~1.1 (7).

A more recent treatment of this problem (7) is suitable for 1.1 $\leq$ RG $\leq$ 10. The diffusion problem in dimensionless form is

$$\frac{\partial^2 C}{\partial Z^2} + \frac{\partial^2 C}{\partial R^2} + \frac{1}{R}\frac{\partial C}{\partial R} = 0 \qquad 0 \le R,\ L1 < Z < L \tag{26}$$

where $L1 = \ell_1/a$ shows how far behind the tip surface ($Z = 0$) the simulation goes (Fig. 4); other dimensionless variables are defined above. The boundary conditions are:

$0 \le R < 1$, $Z = 0$ (tip surface):

$$C = 0 \tag{27a}$$

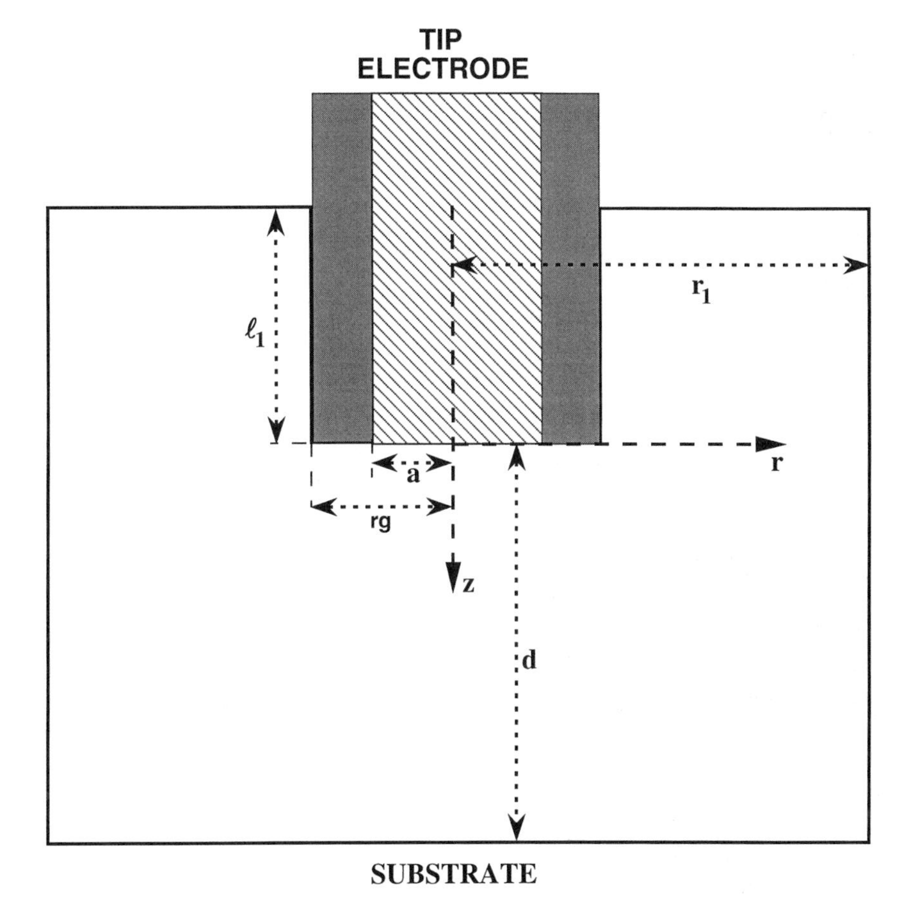

FIG. 4 Geometry of the simulation domain and the parameters defining the diffusion problem for SECM. (Adapted with permission from Ref. 7. Copyright 1998 American Chemical Society.)

$1 \leq R \leq RG$, $Z = 0$ and $R = RG$, $L1 < Z < 0$ (insulating glass):

$$\frac{\partial C}{\partial n} = 0 \tag{27b}$$

where $\partial C/\partial n$ is the normal derivative on the surface of insulating glass surrounding the conductive core of the tip.

$0 \leq R \leq RMAX$, $Z = L$ (substrate);

$$C = 1 \text{ (positive feedback)} \tag{28a}$$

or

$$\frac{\partial C}{\partial Z} = 0 \text{ (negative feedback)} \tag{28b}$$

$RG \leq R < RMAX$, $Z = L1$ or $R = RMAX$, $L1 < Z < L$ (simulation space limits):

$$C = 1 \tag{29}$$

The main difference between Eqs. (26)–(29) and the previous simulation (1) is in problem geometry. The values of $L1 = 0$ and $RMAX = RG$ were used in Ref. 1, i.e., the concentration of the mediator was assumed to equal its bulk value everywhere beyond the limits of the tip-insulating sheath ($R \geq RG$). However, for $RG \ll 10$ satisfactory simulation accuracy can only be achieved by taking into account mediator diffusion from behind the shield and extending the simulation domain much further than $R = RG$. The values of $L1 \leq -20$ and $R1 = 100$ were sufficient to ensure the relative error of less than 1% for all L and RG values (7).

Several normalized current vs. distance curves for pure positive (A) and negative (B) feedback and different RG values are presented in Fig. 5. To fit an experimental approach curve to the theory, one usually normalizes the measured tip current by the experimental $i_{T,\infty}$ value. For $RG \geq 10$ the experimental $i_{T,\infty}$ is close to the 4nFaDc value given by Eq. (20). In earlier SECM studies, the tip current was normalized by this quantity. But at $RG \ll 10$ $i_{T,\infty}$ becomes larger than 4nFaDc due to the mediator diffusion from the back of the tip (10). The correction for the finite insulator thickness can be introduced by replacing the factor 4 in Eq. (20) by one suitable for a given RG, e.g., 4.06 ($RG = 10$), 4.43 ($RG = 2$), 4.64 ($RG = 1.5$), and 5.14 ($RG = 1.1$). All tip current values in Fig. 5 are normalized by the $i_{T,\infty}$ corresponding to a specified RG.

For positive SECM feedback the effect of the RG on the shape of the current-distance curve is relatively small. But when the substrate is an insulator, a smaller RG results in a significantly higher current at any tip/

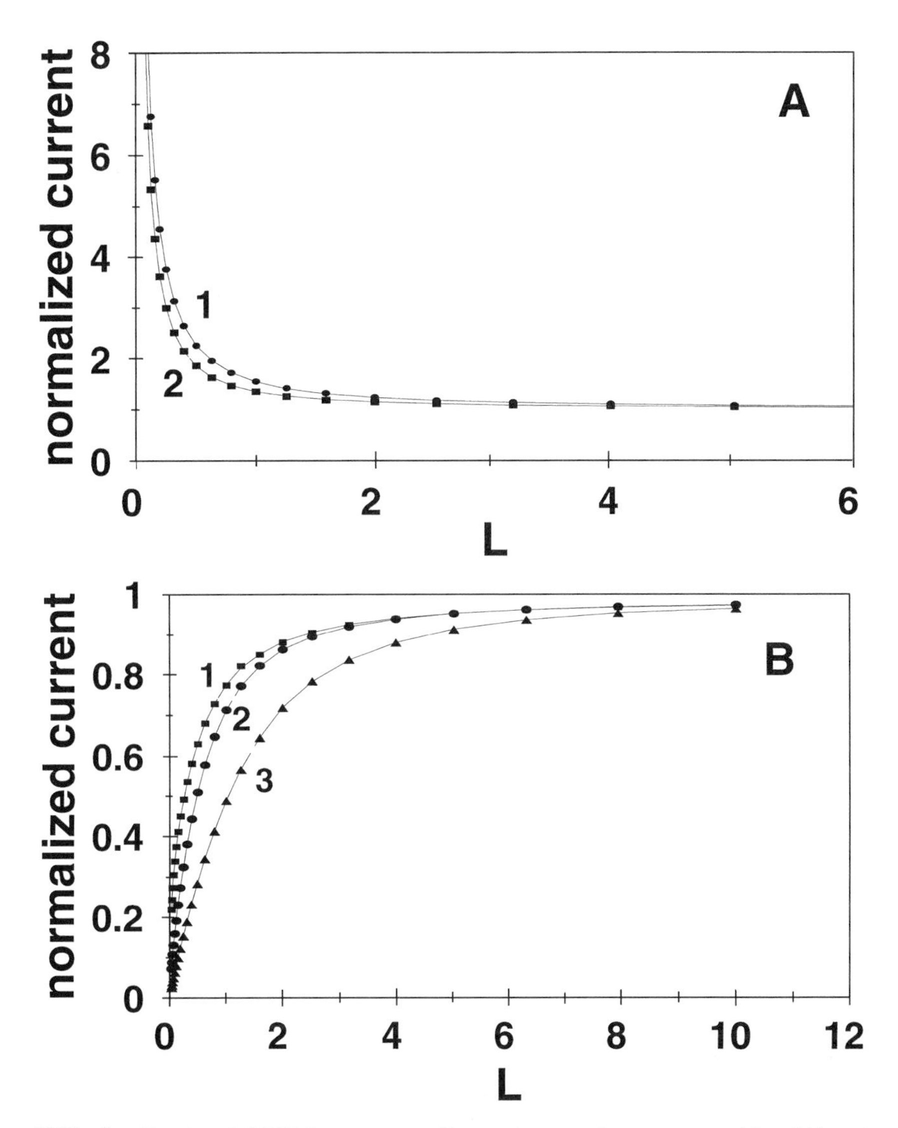

FIG. 5 Simulated SECM current vs. distance curves for a pure positive (A) and negative (B) feedback and different RG values. Symbols, simulation; solid line in (A), Eq. (30); solid line in (B), Eq. (31). (A) RG = 10 (1), 1.1 (2). (B) RG = 1.1 (1), 2 (2), and 10 (3). (Reprinted with permission from Ref. 7. Copyright 1998 American Chemical Society.)

TABLE 1 Parameter Values for Eq. (30)

RG	A	B	C	D
1.1	0.5882629	0.6007009	0.3872741	−0.869822
1.5	0.6368360	0.6677381	0.3581836	−1.496865
2.0	0.6686604	0.6973984	0.3218171	−1.744691
10	0.7449932	0.7582943	0.2353042	−1.683087
5.1[a]	0.72035	0.75128	0.26651	−1.62091

[a]From Ref. 11.

substrate separation. Analytical approximations can be obtained for both positive and negative feedback working curves (solid lines in Fig. 5). Equation (30) with the parameter values listed in Table 1 fits within 1% the simulated I_T-L data for a conductive substrate over the L interval from 0.04 to 10 and for RG values between 1.1 and 10:

$$I_T(L) = A + B/L + C \exp(D/L) \tag{30}$$

For an insulating substrate, Eq. (31) with the parameter values listed in Table 2 is similarly accurate over the same ranges of L and RG values:

$$I_T(L) = 1/[A + B/L + C \exp(D/L)] + E \times L/(F + L) \tag{31}$$

Amphlett and Denuault (11) have formulated a time-dependent SECM problem based on the same ideas (i.e., the simulation space is expanded beyond the edge of the insulating sheath and diffusion from behind the shield is taken into account). The steady-state responses were calculated as a long-time limit of the tip transient currents. These authors also obtained two equations describing SECM approach curves for a pure positive and negative feedback. The equation for a diffusion-controlled positive feedback is identical to Eq. (30) (this is not surprising because both equations are based on the same approximate expression from Ref. 9). The parameters reported in Ref. 11 for RG = 10.2 and 1.51 are quite close to those obtained in Ref. 7

TABLE 2 Parameter Values for Eq. (31)

RG	A	B	C	D	E	F
1.1	1.1675164	1.0309985	0.3800855	−1.701797	0.3463761	0.0367416
1.5	1.0035959	0.9294275	0.4022603	−1.788572	0.2832628	0.1401598
2.0	0.7838573	0.877792	0.424816	−1.743799	0.1638432	0.1993907
10	0.4571825	1.4604238	0.4312735	−2.350667	−0.145437	5.5768952

TABLE 3 Parameter Values for Eq. (32)

RG	k_1	k_2	k_3	k_4	% error	Validity range
1002	0.13219	3.37167	0.8218	−2.34719	<1%	0.3–20
100	0.27997	3.05419	0.68612	−2.7596	<1%	0.4–20
50.9	0.30512	2.6208	0.66724	−2.6698	<1%	0.4–20
20.1	0.35541	2.0259	0.62832	−2.55622	<1%	0.4–20
15.2	0.37377	1.85113	0.61385	−2.49554	<1%	0.4–20
10.2	0.40472	1.60185	0.58819	−2.37294	<1%	0.4–20
8.13	0.42676	1.46081	0.56874	−2.28548	<1%	0.4–20
5.09	0.48678	1.17706	0.51241	−2.07873	<1%	0.2–20
3.04	0.60478	0.86083	0.39569	−1.89455	<0.2%	0.2–20
2.03	0.76179	0.60983	0.23866	−2.03267	<0.15%	0.2–20
1.51	0.90404	0.42761	0.09743	−3.23064	<0.7%	0.2–20
1.11	−1.46539	0.27293	2.45648	8.995E-7	<1%	2–20

for RG = 10 and 1.5, respectively. The last row in Table 1 gives the parameters for RG = 5 (11). The approximate expression for the negative feedback in Ref. 11 is of the same form as an approximation for RG = 10 in Ref. 9:

$$I_T(L) = 1/[k_1 + k_2/L + k_3 \exp(k_4/L)] \qquad (32)$$

with the parameter values listed in Table 3. Although Eq. (32) is not valid for small tip/substrate distances ($L \geq 0.2$ for all RG values, and $L \geq 2$ for RG = 1.11), it can be useful for 2 < RG < 10 and for larger RG values unavailable in Table 2.

The reversible (nernstian) steady-state voltammogram for any electrode geometry can be calculated from the following equation:

$$i(E) = i_{dif}/\theta \qquad (33a)$$

where i_{dif} is the diffusion limiting current and θ was defined in Eq. (25b). Thus, for the SECM

$$i_T(E, L) = I_T(L)/\theta \qquad (33b)$$

where $I_T(L)$ is given by Eqs. (30) and (31) for a conductive and an insulating substrate, respectively.

2. *Irreversible Substrate Kinetics*

A special case of totally irreversible substrate kinetics ($K_{f,S} = 0$) was treated numerically (3). Later, the entire family of theoretical working curves (I_T vs. $K_{b,S}$) calculated for various L was fit to Eq. (34) (12):

$$I_T(L) = I_S(1 - I_T^{ins}/I_T^c) + I_T^{ins} \tag{34a}$$

$$I_S = 0.78377/L(1 + 1/\Lambda) + [0.68 + 0.3315 \exp(-1.0672/L)]/[1 + F(L, \Lambda)] \tag{34b}$$

where I_T^c and I_T^{ins} are given by Eqs. (30) and (31) with RG = 10, and I_S is the kinetically controlled substrate current; $F(L, \Lambda) = (11/\Lambda + 7.3)/(110 - 40L)$, and $\Lambda = K_{b,S}d/D$. Figure 6 shows a family of working curves for different values of L along with the simulated data from Ref. 3. The numerical results (triangles) fit Eq. 34 (squares) over an L interval from 0.1 to 1.5 and $-2 \leq \log\Lambda \leq 3$ within ~1–2%. These working curves were calculated for RG = 10 and should not be used for the analysis of data obtained with very sharp tips (e.g., RG < 2).

The radius of the portion of the substrate surface participating in the SECM feedback loop can be evaluated as $h \cong 1 + 1.5$ L (3). Thus at small tip/substrate separations (e.g., $L \leq 2$), a large substrate behaves as a virtual

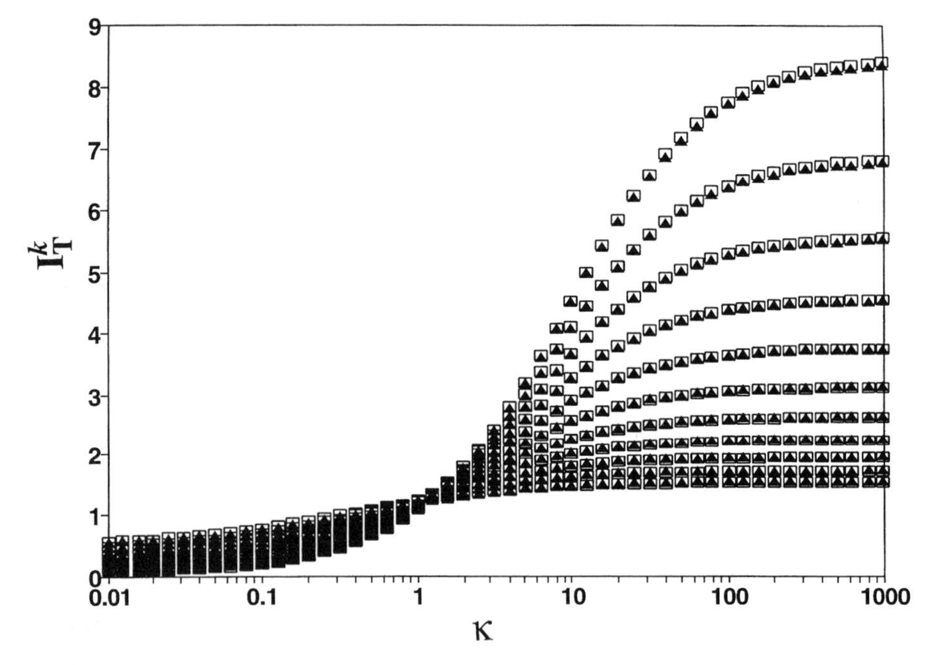

FIG. 6 Working curves of I_T vs. $K_{b,S}$ for different values of L. Open squares, Eq. (34); filled triangles, simulation in Ref. 3. From top to bottom, log (d/a) = −1.0, −0.9, −0.8, −0.7, −0.6, −0.5, −0.4, −0.3, −0.2, −0.1, and 0.0. (Reprinted with permission from Ref. 12. Copyright 1995 American Chemical Society.

UME of a size comparable with that of the tip electrode. The SECM allows probing local kinetics at a small portion of the macroscopic substrate with all of the advantages of microelectrode measurements.

Equation (34) is equally applicable to finite electrochemical and chemical kinetics at the substrate. The only difference is that an electrochemical rate constant is a function of electrode potential, while for a one-step chemical process ($R \xrightarrow{k_b} O$) k_b is potential-independent. For more complicated surface reactions, the theory has to be modified.

3. Quasireversible Processes

For quasireversible (Butler-Volmer) kinetics at the tip and a diffusion-controlled mediator regeneration at the substrate, one can obtain an approximate equation for the tip current at any potential assuming uniform accessibility of the tip surface (i.e., the current density is uniform over the tip surface) (13):

$$I_T(E, L) = [0.68 + 0.78377/L + 0.3315 \exp(-1.0672/L)]/(\theta + 1/\kappa) \tag{35}$$

where $\kappa = k^\circ \exp[-\alpha f(E - E^{\circ\prime})]/m_O$, θ is defined in Eq. (25b), and the effective mass-transfer coefficient for SECM is

$$\begin{aligned} m_O &= 4D_O[0.68 + 0.78377/L + 0.3315 \exp(-1.0672/L)]/(\pi a) \\ &= i_T(L)/(\pi a^2 nFc) \end{aligned} \tag{36}$$

One can see from Eq. (36) that at $L \gg 1$, $m_O \sim D/a$ (as for a microdisk electrode alone), but at $L \ll 1$, $m_O \sim D/d$, i.e., a thin-layer cell (TLC)–type behavior. This suggests that the SECM should be useful for studying rapid heterogeneous electron transfer kinetics. By decreasing the tip/substrate distance, the mass-transport rate can be increased sufficiently for quantitative characterization of the electron-transfer kinetics, preserving the advantages of steady-state methods, i.e., the absence of problems associated with ohmic drop, adsorption, and charging current.

At constant L, Eq. (35) describes a quasireversible steady-state tip voltammogram. Such a curve can be obtained by scanning the potential of the tip while the substrate potential is held constant. Unlike Eq. (35), Eq. (37) does not assume uniform accessibility of the tip and may be somewhat more accurate [14]:

$$I_T(E, L) = \frac{0.78377}{L(\theta + 1/\kappa_{TLC})} + \frac{0.68 + 0.3315 \exp(-1.0672/L)}{\theta\left[1 + \dfrac{\pi}{\kappa'\theta}\dfrac{2\kappa'\theta + 3\pi}{4\kappa'\theta + 3\pi^2}\right]} \tag{37}$$

where $\kappa_{TLC} = 2\kappa' d(D_O + D_R)/(\pi a D_R)$ and κ' is given by Eq. (25a). Both equations (35) and (37) yielded essentially the same values of α and $E^{\circ\prime}$ for rapid oxidation of ferrocene at a Pt tip in acetonitrile (14). The differences between extracted k° values were about 20%.

The computational results in Ref. 2 (cyclic voltammograms) and Ref. 3 (steady-state current-distance curves) illustrate the essential features of the SECM response when substrate reaction is quasireversible. In this case the assumption of uniformly accessible substrate surface is inaccurate, and no simple approximation similar to Eq. (35) is available. The non–steady-state response depends on too many parameters to allow presentation of a complete set of working curves. Although an extensive table in Ref. 3 contains values of the normalized steady-state current for various values of the dimensionless rate constant and L, it may be hard to use for analysis of experimental results.

C. Microscopic or Partially Blocked Substrates

The discussion in the previous sections was based on assumption of a macroscopic (i.e., infinitely large compared to the tip) uniformly reactive substrate. This assumption works well for most kinetic SECM experiments, but it is not appropriate when the substrate is small [e.g., another microdisk electrode (3)], contains small three-dimensional features, or consists of small patches of different reactivities. It was shown in Ref. 3 that the magnitude of either positive or negative feedback produced by a microscopic substrate is lower than the effect that would be produced under the same conditions by the large substrate. The current-distance curves in Fig. 7 computed for various values of h using both numerical solution of integral equations and the ADI simulations (3). Apparently, the size of the substrate is most important when $h \leq 1$.

At $h \geq h^{\infty} = 1 + 1.5L$, the substrate behaves essentially as an infinite one. The value of h^{∞} defines the surface area that is actually seen in an SECM feedback experiment and thus has important implications in terms of the resolution of SECM images (see Ref. 15 for a detailed discussion). The difference between the steady-state feedback current for a finite conductive substrate and that for an insulating one becomes small at $h < 0.05$–0.1. Thus it should be possible to identify particles (or other objects) 10–20 times smaller than the tip by SECM (if these particles are sufficiently well separated). With a 1-μm-diameter tip, this corresponds to particles 50–100 nm in size.

Particular caution is necessary for interpretation of images containing small conductive spots ($0 < h < h^{\infty}$); from Fig. 7, one can see that such features may appear as conductors at smaller L and as insulators at larger L. These are typical for surfaces modified with self-assembled molecular monolayers (SAMs). The theory was recently developed for a substrate cov-

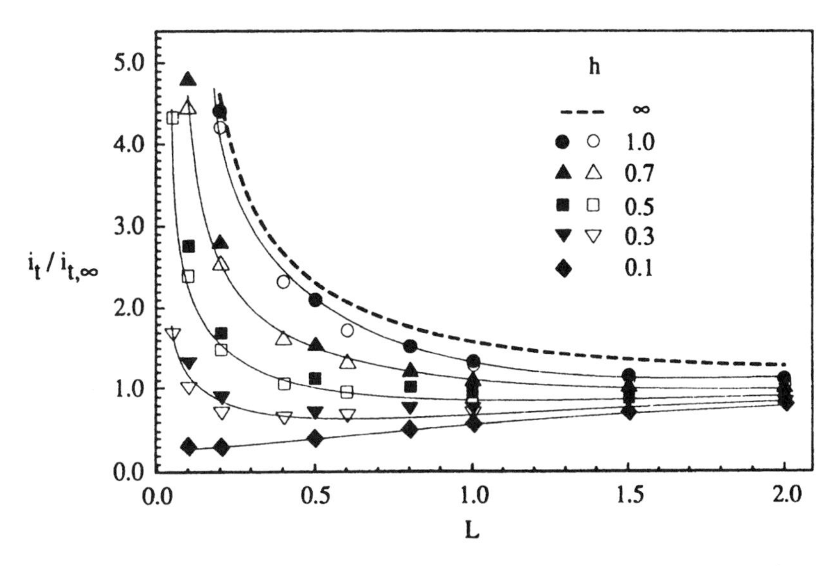

FIG. 7 I_T vs. L curves for finite disk-shaped substrates. Both tip and substrate reactions are diffusion-controlled. Filled symbols calculated using multidimensional integral equations; open symbols are from ADI simulation. The h (a_s/a) values are as indicated. Dashed line is simulation for H = ∞ from Ref. 1. The lines through the symbols are drawn as a guide. (Reprinted with permission from Ref. 3. Copyright 1992 American Chemical Society.)

ered with a blocking film, which contains microscopic conductive disk-shaped defects (16). Use was made of the effective medium approach developed by Zwanzig and Szabo (17). It was shown that the steady-state diffusion-limiting current to a uniformly accessible (e.g., spherical) inert surface partially covered by randomly distributed active disks is equivalent to kinetically controlled current at the same surface with an effective rate constant

$$k_{eff} = 4(1 - \theta)D/\pi R_d \tag{38}$$

where θ is the fraction of surface that is insulating, i.e., free from active disk-shaped defects, and R_d is the defect radius. Thus the SECM current-distance curve at a partially covered substrate should have the same shape as the kinetically controlled $i_T - d$ curve at an uncovered substrate described by Eq. (34) with the rate constant given by Eq. (38).

In a compact SAM, where both the defect size and density are small, the local current density at defect sites is very high, and the overall redox process may be kinetically controlled. When the process at the microdisks is governed by finite irreversible heterogeneous kinetics and $\theta \to 1$, the

SECM approach curve at a partially covered substrate should be equivalent to that at the uniform substrate with

$$k_{eff} = \frac{4(1 - \theta)D}{\pi R_d \left[1 + \frac{\pi}{\kappa}\frac{2\kappa + 3\pi}{4\kappa + 2\pi^2}\right]} \tag{39}$$

where $\kappa = \pi k_d R_d/4D$ and k_d is the heterogeneous constant for all reactive disks. By fitting experimental current-distance curves to Eq. (34), one can obtain the k_{eff} value and evaluate θ and R_d using Eq. (38) or Eq. (39). It is not possible to distinguish between the diffusion-controlled process at the disk-shaped defects [Eq. (38)] and finite heterogeneous kinetics [Eq. (39)] from a single approach curve. However, by shifting the substrate potential to more extreme values, the diffusion limit where k_{eff} becomes potential-independent can be reached.

D. Nondisk Tips and Tip Shape Characterization

Several advantages of the inlaid disk-shaped tips (e.g., well-defined thin-layer geometry and high feedback at short tip/substrate distances) make them most useful for SECM measurements. However, the preparation of submicrometer-sized disk-shaped tips is difficult, and some applications may require nondisk microprobes [e.g., conical tips are useful for penetrating thin polymer films (18)]. Two aspects of the related theory are the calculation of the current-distance curves for a specific tip geometry and the evaluation of the UME shape. Approximate expressions were obtained for the steady-state current in a thin-layer cell formed by two electrodes, for example, one a plane and the second a cone or hemisphere (19). It was shown that the normalized steady-state, diffusion-limited current, as a function of the normalized separation for thin-layer electrochemical cells, is fairly sensitive to the geometry of the electrodes. However, the thin-layer theory does not describe accurately the steady-state current between a small disk tip and a planar substrate because the tip steady-state current $i_{T,\infty}$ was not included in the approximate model (19).

A more realistic approximate theory for the SECM with a tip shaped as a cone or spherical segment was presented in Ref. 9. The surface of the nonplanar tip electrode was considered to be a series of thin circular strips, each of which is parallel to the planar substrate. The diffusional flux to each strip was calculated using approximate equations for a disk-shaped tip over a conductive or an insulating substrate. The normalized current to the nonplanar tip was obtained by integrating the current over the entire tip surface. Two families of working curves for conical tips over conductive (Fig. 8A) and insulating substrates (Fig. 8B) illustrate the effect of the tip geometry.

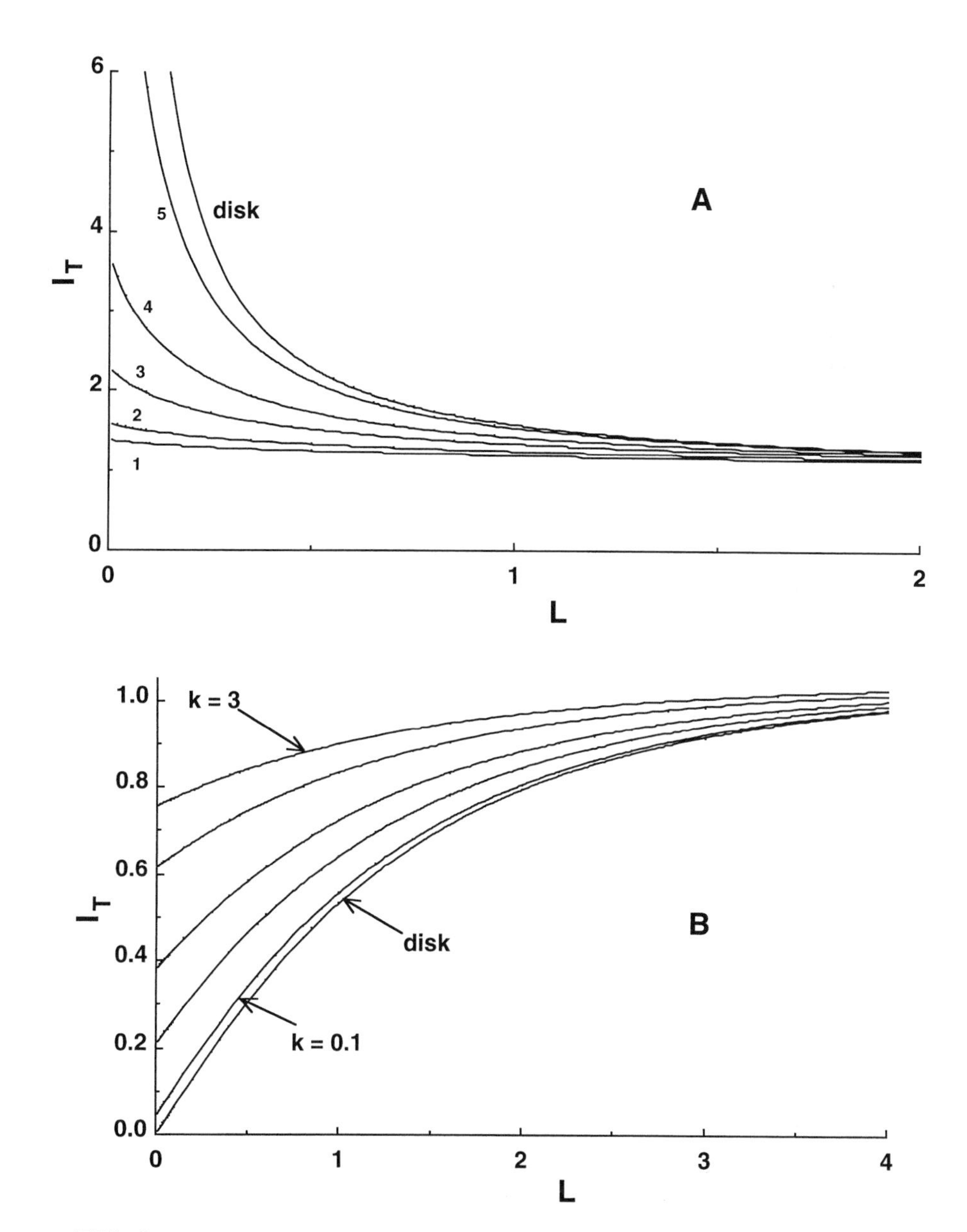

FIG. 8 Steady-state current-distance curves for a conical tip over conductive (A) and insulating (B) substrates corresponding to different values of the parameter k = height/base radius. L is the distance between the substrate and the point of tip closest to it normalized by the base radius. (A) k = 3 (curve 1), 2 (curve 2), 1 (curve 3), 0.5 (curve 4), and 0.1 (curve 5). The upper curve was computed for a disk-shaped tip. (B) From top to bottom, k = 3, 2, 0.5, and 0.1 The lower curve was computed for a disk-shaped tip. (Reprinted with permission from Ref. 9. Copyright 1992 Elsevier.)

One can see that the different working curves possess substantially different curvatures, thus a unique curve can be found to obtain the best fit with the experimental data. The upper curve in Fig. 8A and the lower one in Fig. 8B represent the theory for a microdisk tip (1). The current in a cone/plane cell (unlike that in a disk/plane cell) tends to some limiting value as $L \rightarrow 0$. The same is true for spherical microprobes (20), which also exhibit significantly lower feedback than disk-shaped tips.*

When a submicrometer-sized UME is used as an SECM tip, it is essential to distinguish between planar and recessed geometries. In the latter case, the metal core of an UME is recessed into the solution-filled microcavity inside the glass insulator. Such electrodes are usually unsuitable for kinetic measurements (21). On the other hand, a nm-thick solution layer trapped between such a nm-sized recessed tip and the substrate can be used for single-molecule detection (22). Fan et al. (22) simulated a special case of the recessed electrode with the radius of the circular hole in the insulator (r_a) equal to the disk radius. They also derived an approximate expression for the $i_{T,\infty}$ at a recessed tip with $r_a = a$:

$$i_{T,\infty} = \frac{4nFDc^{\circ}a}{1 + (4/\pi)\ell/a} \tag{40}$$

where ℓ is the depth of the metal recession. The response of a recessed tip with orifice larger than the conductive disk radius is controlled by the size of the conductive core rather than the orifice. This type of geometry can be diagnosed from the lack of a significant positive SECM feedback (large ℓ/a) or the absence of tunneling (smaller ℓ/a) until the breakage of glass insulator.

An expression for a dimensionless current-distance curve for a recessed microdisk tip with the orifice radius, r_a, significantly smaller than the disk radius, a, over a conductive substrate was derived in Ref. 23:

$$I = \frac{I_T[1 + (1.2/\pi)\ell/a]}{1 + (1.2/\pi)I_T\ell/a} \tag{41}$$

where I_T is the normalized current for an r_a-radius inlaid disk tip given by Eq. (30). The maximum normalized tip current observed when the substrate contacts the insulating sheath and plugs the orifice is

$$I_{max} = \frac{1 + (1.2/\pi)\ell/a}{(1.2/\pi)\ell/a} = 1 + \frac{\pi a}{1.2\ell} \tag{42}$$

Thus the deeply recessed tips (i.e., $\ell \gg a$) do not show any significant positive feedback and can be easily distinguished from nonrecessed elec-

*More exact numerical theory for SECM with a hemispherical tip was published recently by Selzer and Mandler (Anal Chem 2000; 72:2383).

trodes. For smaller ℓ, the approach curve analysis is problematic because it is not easy to obtain normalized feedback current higher than 3 or 4 using a small tip of any shape.

Since the shape of a nonplanar tip is typically imperfect, one is not motivated to carry out extensive simulations for these complicated systems. All equations describing the approach curves under diffusion control are highly approximate, and no theory has been developed for finite kinetics at either tip or substrate electrode.

III. GENERATION/COLLECTION (G/C) MODE OF SECM OPERATION

In different publications the term "generation/collection mode" has been applied to several substantially different SECM experiments. The two main types are tip generation/substrate collection (TG/SC) and substrate generation/tip collection (SG/TC) modes. The modifications of the latter type of G/C experiments include potentiometric G/C mode, where the tip is a potentiometric sensor, and two amperometric modes. In the first the tip/substrate separation is so large that no feedback effect is observed, and the amperometric tip only detects the species produced at the substrate. At much shorter distances the species reoxidized (or rereduced) at the tip can return to the substrate producing a feedback effect.

A. Potentiometric G/C Mode

Unlike the feedback mode of the SECM operation where the overall redox process is essentially confined to the thin layer between the tip and the substrate, in SG/TC experiments the tip travels within a thick diffusion layer produced by the large substrate. The theoretical treatment is easier when the tip is a potentiometric sensor. Such a passive sensor does not change the concentration profile of electroactive species generated (or consumed) chemically or electrochemically at the substrate. Still, a consistent theoretical treatment was proposed only for a steady-state situation when a small substrate (a microdisk or a spherical cap) generates stable species. The concentration of such species can be measured by an ion-selective microtip as a function of the tip position. The concentration at any point can be related to that at the source surface. For a microdisk substrate the dimensionless expression is (24,25)

$$c(R, Z)/c(0, L) = \frac{2}{\pi} \tan^{-1} \cdot \left[\frac{2}{R^2 + Z^2 - 1 + \sqrt{(R^2 + Z^2 - 1)^2 + 4Z^2}} \right]^{1/2} \tag{43}$$

[A somewhat similar expression was derived for a disk-shaped sink (26).]

Sometimes it is more useful to relate the concentration distribution in the diffusion layer to the flux at the source surface. This relation was obtained in Ref. 27:

$$C(R, Z) = \frac{2}{\pi} \int_0^1 \sqrt{u/R} J_S(u) p K(p)\, du \tag{44}$$

for a disk-shaped substrate, where K(p) is the complete elliptic integral of the first kind (28) and $p = 2\sqrt{Ru}/\sqrt{Z^2 + (R + u)^2}$. Note that R and Z in Eqs. (43) and (44) are normalized by the substrate radius a_S, rather than the tip radius, a. The flux at the substrate, $J_S(R)$, can be defined using either a constant current or a constant concentration approximation (29). Both assumptions lead to fairly complicated integral expressions. The dimensionless concentration profiles were obtained numerically from those expressions and tabulated (27).

The potentiometric SECM experiment yields the potential of the tip electrode, E, as a function of the tip position. To establish the correspondence between these data and the above theory, one needs to use a calibration curve, i.e, a Nernstian E vs. c plot. Using such a calibration, one can transform the experimental results to the c vs. (z, r) dependence and fit them to the theory in order to find J_S and establish the distance scale.

The above theory implies that the products generated on the substrate do not participate in any chemical reaction in solution. Otherwise [e.g., when the tip is a pH sensor used to monitor proton concentration in a buffered aqueous solution (27)], a more complicated treatment may be necessary.

One can also evaluate the relative change in rate of heterogeneous reaction at the substrate by measuring concentration of the reaction product at the tip. In this setup, the tip is positioned at fixed distance from the substrate, and the time dependence of concentration is measured. This simpler approach is based on the proportionality between the heterogeneous reaction rate and the product concentration. It is most useful when the substrate flux cannot be measured directly (e.g., the substrate reaction is not an electrochemical process) [30,31].

B. Amperometric G/C Modes

1. *Substrate Generation/Tip Collection Mode*

In the feedback mode of the SECM operation, the overall redox process is essentially confined to the thin layer between the tip and the substrate. In a SG/TC experiment there are two different distance scales. One scale is determined by the tip size, the other by the substrate size. If $a_S/a \gg 1$, the diffusion layer generated by the substrate is much thicker than that at the tip electrode. The theory assuming no feedback (i.e., the substrate current unaffected by the tip process) is applicable either at $d/a \gg 1$ or when the product of the tip process does not react at the substrate. Rigorous theoretical description is difficult because (1) the moving tip stirs the substrate diffusion layer—disturbances are especially significant when the tip is an amperometric sensor and has its own diffusion layer: (2) when the substrate is large, no true steady state can be achieved; and (3) the tip blocks the diffusion to the substrate surface, and this screening effect is hard to take into account because of the imperfect geometry of the tip-insulating sheath.

The substrate generation/tip collection (SG/TC) mode with an amperometric tip was historically the first SECM-type measurement performed (32). The aim of such experiments was to probe the diffusion layer generated by the large substrate electrode with a much smaller amperometric sensor. A simple approximate theory (32a,b) using the well-known $c(z, t)$ function for a potentiostatic transient at a planar electrode (33) was developed to predict the evolution of the concentration profile following the substrate potential perturbation. A more complicated theory was based on the concept of the "impulse response function" (32c). While these theories have been successful in calculating concentration profiles, the prediction of the time-dependent tip current response is not straightforward because it is a complex function of the concentration distribution. Moreover, these theories do not account for distortions caused by interference of the tip and substrate diffusion layers and feedback effects.

When the separation distance is sufficiently large to eliminate feedback to the substrate, the diffusion of species between two electrodes can be probed by transient SG/TC measurements. The tip current transient following the application of a short potential pulse to the substrate electrode is peak-shaped. The time corresponding to the maximum tip current, t_{max}, is independent of the tip size and can be expressed as (34)

$$t_{max} = 0.11d^2/D \tag{45}$$

If the tip/substrate distance, d, is known, one can use Eq. (45) to evaluate the diffusion coefficient. Alternatively, using a mediator with an established D value, one can evaluate d [finding d is not straightforward when the tip is very small (23)].

Few amperometric SG/TC experiments with the separation distance of the same order of magnitude as the tip radius have been reported. The applications of this mode have been limited by the shortcomings listed in the beginning of this section and by low collection efficiency (i_T/i_S), significant IR-drop caused by a large substrate current, and the lack of adequate theoretical treatments. Recently, Martin and Unwin developed the first quantitative treatment for amperometric SG/TC experiments (35). Although a chronoamperometric current at a macroscopic substrate does not exhibit any steady state, the collection current at the tip electrode at long times reaches a stable, time-independent value. The time-dependent theory for a diffusion-controlled process was illustrated by a number of three-dimensional concentration profiles and working curves covering a range of experimental conditions.

The steady-state SG/TC measurements were shown to be complimentary to the feedback mode experiment in the same solution (35). In a feedback experiment (assuming that only oxidized form, O, is initially present in solution), the tip and the substrate reactions are given by Eqs. (1) and (2), respectively. In a SG/TC experiment the tip and the substrate reactions are transposed:

$$O + ne^- \rightarrow R \qquad \text{(substrate)} \tag{46}$$

$$R - ne^- \rightarrow O \qquad \text{(tip electrode)} \tag{47}$$

At long times the O form becomes depleted within the thick diffusion layer of the substrate electrode. The tip probes only a micrometer-thick layer adjacent to the substrate surface, in which $c_O \cong 0$ and $c_R = \gamma c_O^\circ$ according to Eq. (23). When the tip is a few radii away from the substrate, the feedback effects are negligible and the tip current is

$$i_{collection,\infty} = 4nFDc_O^\circ a\gamma = \gamma i_{T,\infty} \tag{48}$$

The shape of the SG/TC current-distance curves is very similar to that of the feedback mode $i - d$ curves, and the ratio of the collection current to the corresponding feedback current is the same at any separation distance (Fig. 9):

$$i_{collection}/i_{feedback} = \gamma \tag{49}$$

Thus the diffusion coefficient ratio of the oxidized and reduced forms (γ) can be found by measuring the ratio of the feedback and collection currents at the same d (35). No prior knowledge of a or d is required for this determination. One should also notice that the transient current flowing at the macroscopic substrate electrode vanishes at long times. The steady-state current is confined to the microscopic area of the substrate surface facing the tip. This largely eliminates the IR-drop problem.

2. *Tip Generation/Substrate Collection Mode*

The tip generation/substrate collection (TG/SC) mode is somewhat similar to the feedback mode of the SECM. In both cases an electroactive species is generated at the tip and collected at the substrate. The TG/SC experiment

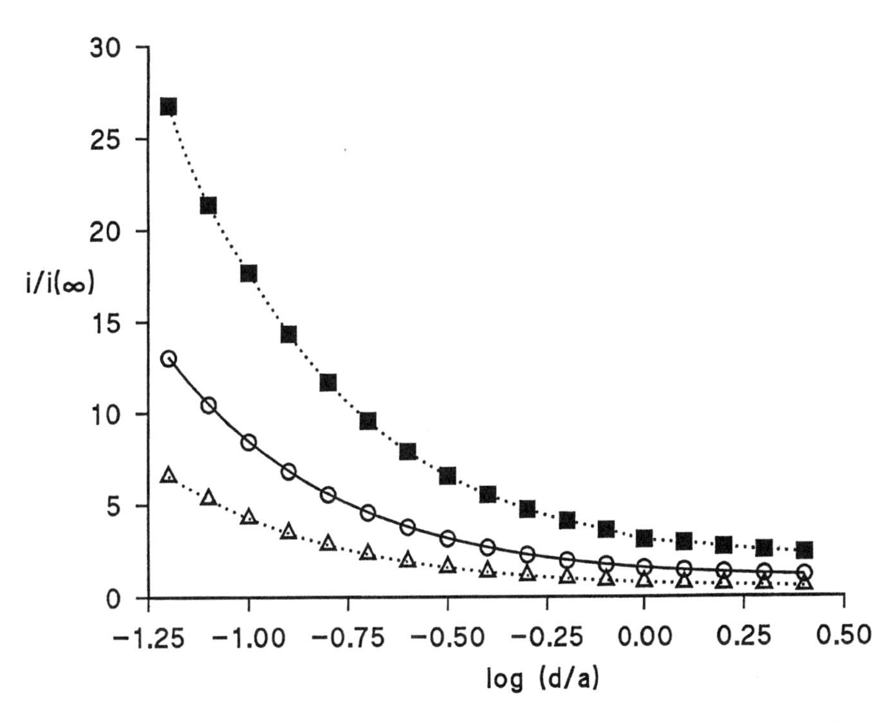

FIG. 9 Simulated steady-state tip current vs. distance curves showing the effect of γ on SG/TC responses. $\gamma = 1$ (○), 0.5 (△), and 2 (■). The solid and dotted lines are calculated from Eq. (49). For $\gamma = 1$ the SG/TS approach curve is indistinguishable from the corresponding feedback mode approach curve. (Reprinted with permission from Ref. 35. Copyright 1998 American Chemical Society.)

includes simultaneous measurements of both tip and substrate currents. Immediately after the application of a potential step to the tip electrode, the substrate current is close to zero. The i_S grows with time while the tip current decreases. For an uncomplicated process, the difference between i_T and i_S vanishes gradually. At steady state, these quantities are almost identical, if L is not very large (36,37) [the collection efficiency, i_S/i_T, is more than 0.99 at $L \leq 2$, and $i_S/i_T \cong 0.8$ at $L = 5$ (36)]. Under these conditions, the tip-generated species predominantly diffuses to the large substrate rather than escaping from the tip/substrate gap. For a process with a coupled chemical reaction, the collection efficiency may be much less than one, and important kinetic information can be extracted from the i_S/i_T vs. L dependencies (see Sec. IV.A).

IV. SECM OF MORE COMPLICATED CHEMICAL SYSTEMS

A. Processes Involving Homogeneous Reactions

A homogeneous chemical reaction occurring in the gap between the tip and substrate electrodes causes a change in i_T, therefore its rate can be determined from SECM measurements. If both heterogeneous processes at the tip and substrate electrodes are rapid (at extreme potentials of both working electrodes) and the chemical reaction (rate constant, k_c) is irreversible, the SECM response is a function of a single kinetic parameter $K = \text{const} \times k_c/D$, and its value can be extracted from I_T vs. L dependencies.

SECM theory has been developed for four mechanisms with homogeneous chemical reactions coupled with electron transfer, i.e., a first-order irreversible reaction (E_rC_i mechanism) (5), a second-order irreversible dimerization (E_rC_{2i} mechanism) (36), ECE and DISP1 reactions (38). [The solution obtained for a E_qC_r mechanism in terms of multidimensional integral equations (2) has not been utilized in any calculations.] While for E_rC_i and E_rC_{2i} mechanisms analytical approximations are available (39), only numerical solutions have been reported for more complicated ECE and DISP1 reactions (38).

Three approaches to kinetic analysis were proposed: (1) steady-state measurements in a feedback mode, (2) generation/collection experiments, and (3) analysis of the chronoamperometric SECM response. Unlike the feedback mode, the generation/collection measurements included simultaneous analysis of both I_T-L and I_S-L curves or the use of the collection efficiency parameter (I_S/I_T when the tip is a generator and the substrate is a collector). The chronoamperometric measurements were found to be less reliable (5), so only steady-state theory will be discussed here.

1. *First-Order Following Reaction*

For the E_rC_i mechanism

$$O + e^- \rightarrow R \qquad \text{(tip)} \tag{50a}$$

$$R \xrightarrow{k_c} \text{products} \qquad \text{(gap)} \tag{50b}$$

$$R - e^- \rightarrow O \qquad \text{(substrate)} \tag{50c}$$

the SECM diffusion problem is represented by the system of differential equations

$$\frac{\partial c_O}{\partial t} = D_O \left[\frac{\partial^2 c_O}{\partial z^2} + \frac{\partial^2 c_O}{\partial r^2} + \frac{1}{r} \frac{\partial c_O}{\partial r} \right] \tag{51a}$$

$$\frac{\partial c_R}{\partial t} = D_R \left[\frac{\partial^2 c_R}{\partial z^2} + \frac{\partial^2 c_R}{\partial r^2} + \frac{1}{r}\frac{\partial c_R}{\partial r} \right] - k_c c_R \tag{51b}$$

with initial and boundary conditions

$$t = 0,\ 0 \leq r,\ 0 < z < d; \qquad c_O = c_O^\circ; \qquad c_R = 0 \tag{52}$$

$$0 \leq r \leq a,\ z = 0; \qquad c_O = 0; \qquad D_O \left[\frac{\partial c_O}{\partial z} \right] = -D_R \left[\frac{\partial c_R}{\partial z} \right] \tag{53}$$

$$a \leq r \leq rg,\ z = 0; \qquad D_O \left[\frac{\partial c_O}{\partial z} \right] = D_R \left[\frac{\partial c_R}{\partial z} \right] = 0 \tag{54}$$

$$0 \leq r \leq rg,\ z = d; \qquad c_R = 0; \qquad D_O \left[\frac{\partial c_O}{\partial z} \right] = -D_R \left[\frac{\partial c_R}{\partial z} \right] \tag{55}$$

$$rg < r,\ 0 < z < d; \qquad c_O = c_O^\circ; \qquad c_R = 0 \tag{56}$$

$$r = 0,\ 0 < z < d; \qquad D_O \left[\frac{\partial c_O}{\partial z} \right] = D_R \left[\frac{\partial c_R}{\partial z} \right] = 0 \tag{57}$$

It was solved numerically using the alternating-direction implicit (ADI) finite difference method (5). The steady-state results were obtained as a long time limit and presented in the form of two-parameter families of working curves (5). These represent steady-state tip current or collection efficiency as functions of $K = ak_c/D$ and L.

It was shown later (39) that the theory for E_rC_i process under SECM conditions can be reduced to a single working curve. To understand this approach, it is useful to first consider a positive feedback situation with a simple redox mediator (i.e., without homogeneous chemistry involved) and with both tip and substrate processes under diffusion control. The normalized steady-state tip current can be presented as the sum of two terms:

$$I_T = I_f + I_T^{ins} \tag{58}$$

where I_f is the feedback current coming from the substrate and I_T^{ins} is the current due to the hindered diffusion of the electroactive species to the tip from the bulk of solution given by Eq. (31); all variables are normalized by $i_{T,\infty}$. The substrate current is

$$I_S = I_f + I_d \tag{59}$$

where I_f is the same quantity as in Eq. (58), representing the oxidized species, which eventually arrive at the tip as a feedback current, and I_d is the dissipation current, i.e., the flux of species not reaching the tip. It was shown in Ref. 36 that I_S/I_T is more than 0.99 at $0 < L \leq 2$, i.e., for any L within this interval the tip and the substrate currents are essentially equal to each other. Thus, from Eqs. (58) and (59)

$$I_d = I_T^{ins} \quad (60)$$

or

$$I_d/I_S = I_T^{ins}/I_T = f(L) \quad (61)$$

where f(L) can be computed for any L as the ratio of the right-hand side of Eq. (31) to that of Eq. (30) assuming RG = 10.

Analogously, for an electrochemical process followed by an irreversible homogeneous reaction of any order, one can write

$$I'_T = I'_f + I_T^{ins} \quad (62)$$

$$I'_S = I'_f + I'_d \quad (63)$$

where the variables labeled with the prime are analogous to unlabeled variables in Eqs. (58) and (59), and I_T^{ins} is unaffected by the occurrence of the homogeneous reaction in Eq. (50b). Since the species O are stable, the fraction of these species arriving at the tip from the substrate should also be unaffected by the reaction in Eq. (50b), i.e., the relation $I'_d/I'_S = f(L)$ holds true. Consequently,

$$I'_T = I_T^{ins} + I'_S - I'_d$$
$$= I_T^{ins} + I'_S(1 - f(L)) \quad (64)$$

i.e., for an SECM process with a following homogeneous chemical reaction of any order, the dependence I'_T vs. I'_S at any given L should be linear with a slope equal to 1 − f(L) and an intercept equal to $I_T^{ins}(L)$. Thus, the generation/collection mode of the SECM (with the tip electrode serving as a generator) for these mechanisms is completely equivalent to the feedback mode, and any quantity, I'_T, I'_S, or I'_S/I'_T, can be calculated from Eq. (64) for a given L if any other of these quantities is known.

For mechanisms with irreversible reactions, one can expect the collection efficiency, I'_S/I'_T, to be a function a single kinetic parameter κ. If this parameter is known, the SECM theory for this mechanism can be reduced to a single working curve. After the function $\kappa = F(I'_S/I'_T)$ is specified, one can immediately evaluate the rate constant from I'_S/I'_T vs. L or I'_T vs. L experimental curves. If only tip current has been measured, the collection efficiency can be calculated as

$$I'_S/I'_T = \frac{1 - I_T^{ins}/I'_T}{1 - f(L)} \quad (65)$$

For the E_rC_i mechanism $\kappa = k_c d^2/D$. Figure 10A represents the working curve, κ vs. I'_S/I'_T, along with the simulated data from Ref. 5. The numerical results fit the analytical approximation

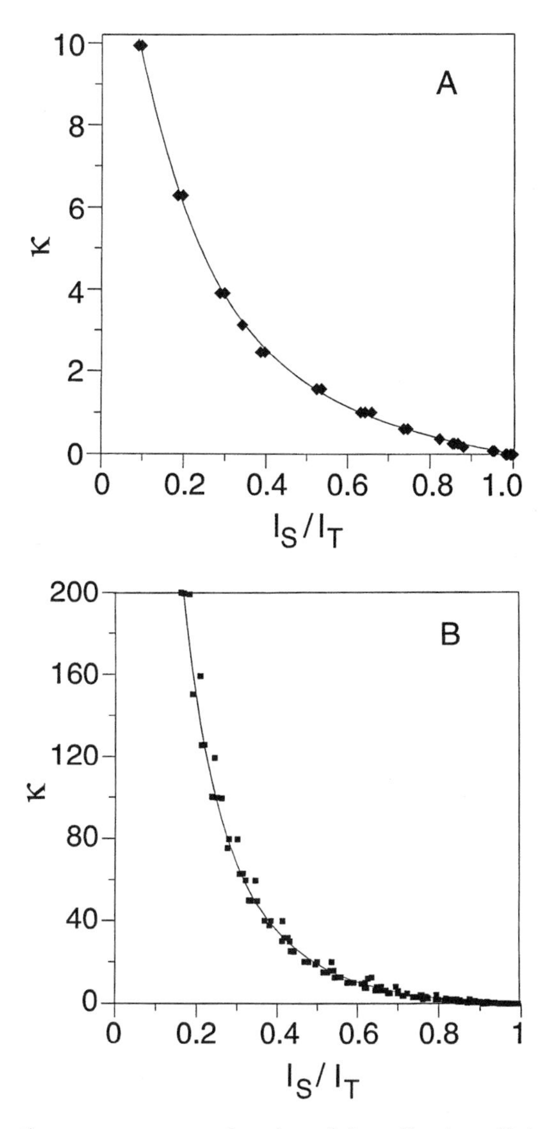

FIG. 10 Kinetic parameter κ as a function of the collection efficiency (I'_S/I'_T). (A) E_rC_i mechanism; $\kappa = k_c d^2/D$, solid line was computed from Eq. (66), triangles are simulated data taken from Ref. 5. (B) E_rC_{2i} mechanism; $\kappa' = c^\circ k'_c d^3/aD$, solid line was computed from Eq. (69), squares are simulated data taken from Ref. 36. (Reprinted with permission from Ref. 39. Copyright 1994 American Chemical Society.)

$$\kappa = F(x) = 5.608 + 9.347 \exp(-7.527x) - 7.616 \exp(-0.307/x) \tag{66}$$

(solid curve in Figure 10A), where $x = I'_S/I'_T$, within about 1%. First-order rate constants in excess of $2 \times 10^4\ s^{-1}$ should be accessible to SECM measurements under steady-state conditions.

The latest contribution to the theory of the EC processes in SECM was the modeling of the substrate generation/tip collection (SG/TC) situation by Martin and Unwin (40). Both the tip and substrate chronoamperometric responses to the potential step applied to the substrate were calculated. From the tip current transient one can extract the value of the first-order homogeneous rate constant and (if necessary) determine the tip/substrate distance. However, according to the authors, this technique is unlikely to match the TG/SC mode with its high collection efficiency under steady-state conditions.

2. *Second-Order Following Reaction*

The detailed TG/SC theory was developed for an electrode process with a following dimerization reaction (E_rC_{2i} mechanism) (36):

$$\text{tip:} \qquad O + ne^- \rightarrow R \tag{67a}$$

$$\text{gap:} \qquad 2R \xrightarrow{k'_c} \text{products} \tag{67b}$$

$$\text{substrate:} \qquad R - ne^- \rightarrow O \tag{67c}$$

This diffusion problem is similar to one for E_rC_i mechanism [Eqs. (50)] except for the different Fick's equation for c_R:

$$\frac{\partial c_R}{\partial t} = D_R \left[\frac{\partial^2 c_R}{\partial z^2} + \frac{\partial^2 c_R}{\partial r^2} + \frac{1}{r}\frac{\partial c_R}{\partial r} \right] - k'_c c_R^2 \tag{68}$$

Both chronoamperometric and steady-state responses were calculated by solving the related equations numerically. An analytical approach discussed in the previous section is equally applicable to E_rC_{2i} mechanism under steady-state conditions. Equation (64) was verified using the data simulated from Ref. 36, and I'_T vs. I'_S dependencies were plotted for different values of L (39). Although for the E_rC_{2i} mechanism the choice of κ is less straightforward than for E_rC_i, an acceptable fit for all the data points computed in Ref. 36 was obtained using $\kappa' = c°k'_c d^3/aD$ (Fig. 10B). These data were fit to Eq. (69):

$$\kappa' = 104.87 - 9.948x - 185.89/\sqrt{x} + 90.199/x + 0.389/x^2 \tag{69}$$

Although this approximation is less accurate than Eq. (66), its use would not lead to an error of more than about 5–10%, which is within the usual range of experimental error. The invariability of k'_c computed from different experimental points would assure the validity of the results. The upper limit for the rate constant accessible to steady-state TG/SC SECM measurements is about $5 \times 10^8\ M^{-1}$.

3. ECE and DISP1 Mechanisms

These reactions involving two electron-transfer steps are shown schematically in Fig. 11 (38). The product of homogeneous chemical reaction (C) often is easier to reduce than A. This second reduction can either occur at the tip electrode (ECE mechanism in Fig. 11a) or via the disproportionation (DISP1 mechanism in Fig. 11B):

$$C + B \xrightarrow{k} A + D \tag{70}$$

With the substrate biased at a potential slightly more positive than E° of A/B couple, B is oxidized to form A for both DISP1 and ECE mechanisms. However, in the latter case the reduction of C also occurs at the substrate. The numerical solution of corresponding diffusion problems (see Ref. 38 for problem formulations) yielded several families of working curves shown in Fig. 12 (DISP1 pathway) and Fig. 13 (ECE pathway). In both cases the tip and the substrate currents are functions of the dimensionless kinetic parameter, $K = ka^2/D$. The normalization of the i_T and i_S for two-electron processes is somewhat problematic. In Ref. 38 both quantities are normalized with respect to the one-electrode steady-state current, which flows at infinite tip/substrate separation ($i_{T,1e,\infty} = 4FDac_a^\circ$). However, this value is not equal to experimentally measured tip current at $d \to \infty$, which

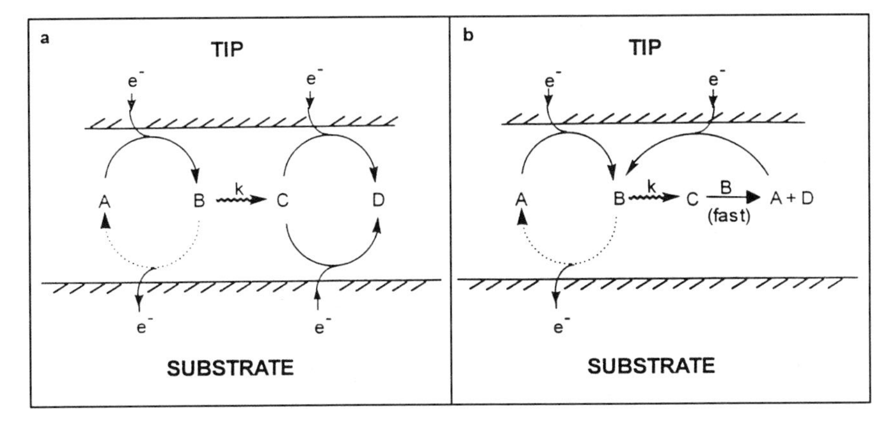

FIG. 11 Diffusional and chemical processes occurring within the tip/substrate domain in the case of (a) an ECE pathway and (b) a DISP1 pathway. (Reprinted with permission from Ref. 38. Copyright 1996 American Chemical Society.)

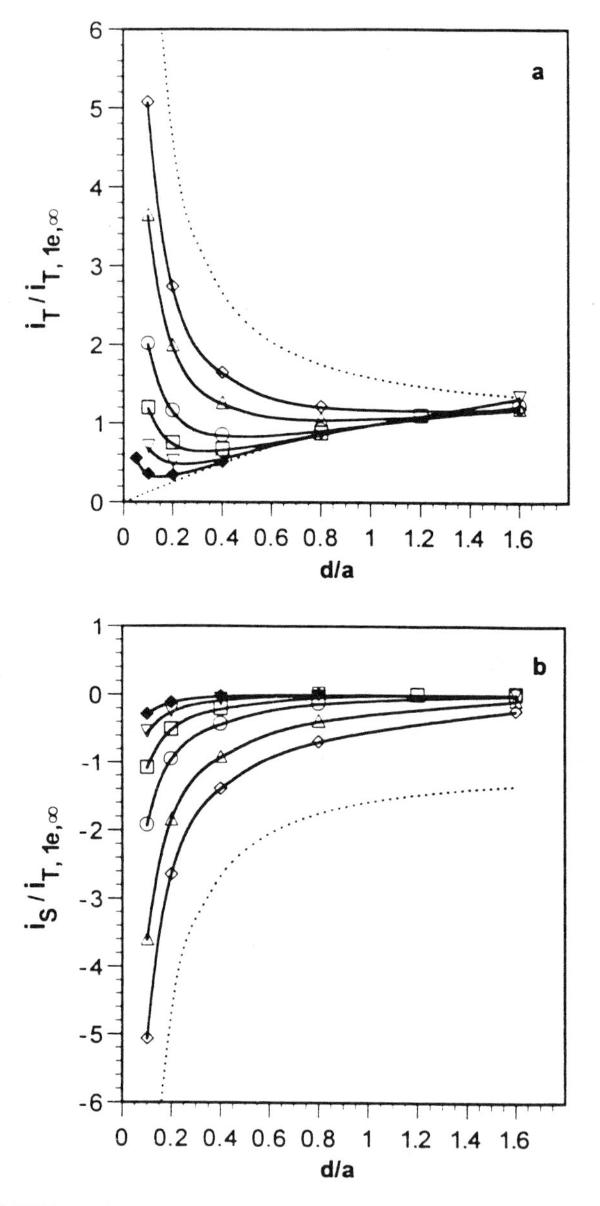

FIG. 12 DISP1 pathway. Theoretical current-distance curves for several values of $K = ka^2/D$: 1 (◇), 2 (△), 5 (○), 10 (□), 20 (▽), and 50 (♦). The upper dashed line in a represents the one-electron pure positive feedback ($K = 0$). The lower dashed line is the two-electron pure negative feedback ($K = \infty$). In part b the dashed line represents the one-electron pure positive feedback. (Reprinted with permission from Ref. 38. Copyright 1996 American Chemical Society.)

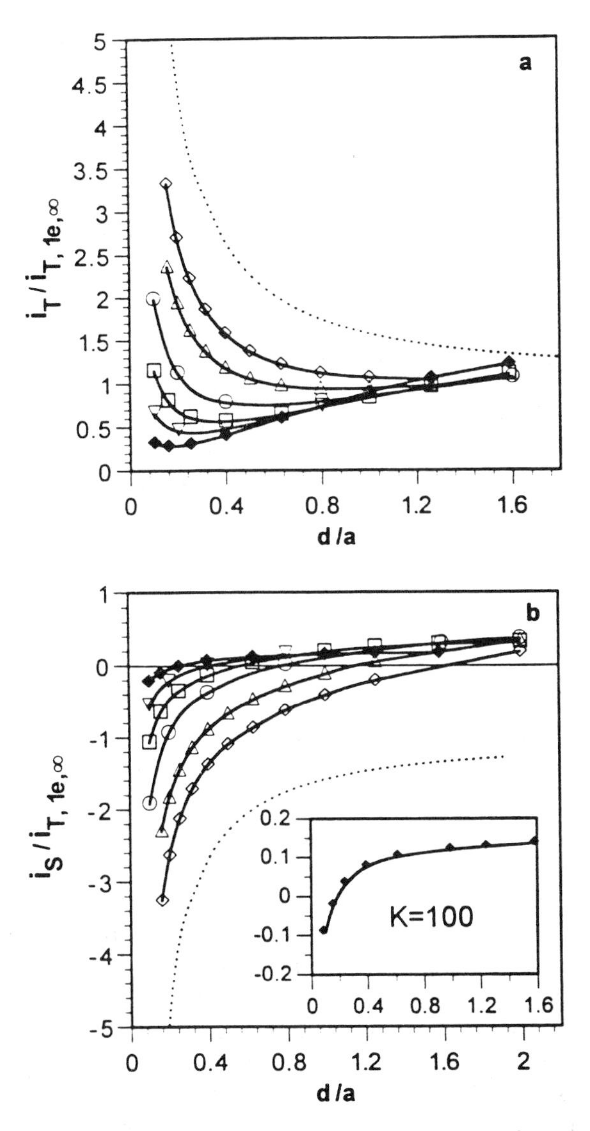

FIG. 13 ECE pathway. Theoretical current-distance curves for several values of $K = ka^2/D$: 1 (◇), 2 (△), 5 (○), 10 (□), 20 (▽), and 50 (◆). The upper dashed line in a represents the one-electron pure positive feedback (K = 0). In part b the dashed line represents the one-electron pure positive feedback. The inset shows the substrate current for K = 100. (Reprinted with permission from Ref. 38. Copyright 1996 American Chemical Society.)

also includes the contribution from the second electron transfer. Nevertheless, by comparing experimental current-distance curves to the theory one can distinguish between DISP1 and ECE pathways and evaluate the k value (38).

B. Surface Reactions and Interfacial Transfer Processes

Many different types of interfacial boundaries can be probed by SECM. The use of the SECM for studies of surface reactions and phase transfer processes is based on its abilities to perturb the local equilibrium and measure the resulting flux of species across the phase boundary. This may be a flux of electrons or ions across the liquid/liquid interface, a flux of species desorbing from the substrate surface, etc. Furthermore, as long as the mediator is regenerated by a first-order irreversible heterogeneous reaction at the substrate, the current-distance curves are described by the same Eqs. (34) regardless of the nature of the interfacial process. When the regeneration kinetics are more complicated, the theory has to be modified. A rather complete discussion of the theory of adsorption/desorption reactions, crystal dissolution by SECM, and a description of the liquid/liquid interface under SECM conditions can be found in other chapters of this book. In this section we consider only some basic ideas and list the key references.

1. *Adsorption/Desorption Processes*

Two opposite reactions, adsorption on and desorption from the substrate surface, can be probed by the SECM microtip (41,42). In the former case, the SECM functioned only as a micromanipulator aimed to bring the microtip electrode inside the small drop of liquid (3.5–20 μL volume). The adsorption was determined through the decrease in the tip current caused by decreasing concentration of adsorbing species in a liquid phase. To study the desorption process the UME tip is positioned close to the surface of the substrate covered with an adsorbate (42). Following the application of the potential step to the tip electrode, the reduction (or oxidation) begins leading to the depletion of adsorbing species in the tip-substrate gap. The depletion in turn results in two competing processes: the desorption from the substrate and surface diffusion driven by the surface concentration gradient. The desorbed species can diffuse to the tip and react at its surface, thus contributing to the measured tip current. The rates of both desorption and surface diffusion processes can be deduced from the chronoamperometric SECM response (42).

One should notice that only small amount of species is adsorbed on a microscopic portion of the substrate facing the tip. Thus the desorption kinetics can be obtained from the short-time transient behavior when the effect of surface diffusion is negligible. At times sufficiently long for a true steady state to prevail, I_T always attains the value for an inert insulating substrate, and the adsorption/desorption kinetics are not accessible.

Surface diffusion provides an additional path for the transport of the adsorbed mediator into the tip/substrate domain. Its main effect is to increase the magnitude of the current flowing in the longer time region of the transient and, in particular, to enhance the final steady-state current as compared to that at an inert substrate. The adsorption/desorption process is at equilibrium, and thus the tip current depends on the solution and surface diffusion rates. The larger the surface diffusion coefficient, as compared to that in solution, the larger the steady-state current at the tip UME. The steady-state current becomes increasingly sensitive to the surface diffusion process as L is minimized. The related theory is complicated and involves a large numbers of empirical parameters. Several families of chronoamperometric transients and steady-state approach curves can be found in Ref. 42 (see also Chapter 12 in this book).

2. *Dissolution of Ionic Crystals*

Macpherson and Unwin (43) developed the theory for dissolution processes at the substrate induced by depleting of electroactive species at the SECM tip. The UME tip can oxidize or reduce the species of interest in solution at the crystal surface. If this species is one of the crystal components, the depletion of its concentration in the solution gap between the tip and substrate induces crystal dissolution. This process produces additional flux of electroactive species to the tip similarly to positive feedback situation discussed in previous sections. Unlike the desorption reaction, where only a small amount of adsorbed species can contribute to the tip current, the dissolution of a macroscopic crystal is not limited by surface diffusion. Accordingly, the developed theory is somewhat similar to that for finite heterogeneous kinetics at the substrate. Several models developed in Ref. 43a–d use different forms of the dissolution rate law applicable to different experimental systems. In general, the rate of the substrate process is (43a):

$$j = k_n \sigma^n \qquad (71)$$

where j is the flux of dissolving species, k_n is the rate constant, n is the reaction order, and σ is the undersaturation produced by the tip reaction. The SECM theory for the first-order process was developed (43a) and applied to dissolution of the (100) face of copper sulfate single crystal, while the second-order kinetic model was shown to describe well the dissolution of potassium ferrocyanide trihydrate (43c). By considering a dislocation-free crystal surface on which dissolution sites are only nucleated above a certain critical value of the undersaturation one can also model an oscillatory dissolution process (43b). In all cases the change in geometry caused by dissolution of the substrate and electrodeposition at the tip were neglected.

The developed theory was extended to the case of salt dissolution in solution containing no supporting electrolyte (43e). Although the theory describing mass transport via diffusion/migration in such systems is more complicated, it was possible to fit the experimental current-distance curves for AgCl dissolution and to demonstrate the applicability of the second-order rate law to this process. A more complete discussion of the theory of the SECM induced dissolution can be found in Chapter 12.

3. *Charge-Transfer Processes at the Liquid/Liquid Interface*

With regard to SECM theory, the main difference between charge transfers at the liquid/liquid boundary and heterogeneous electron-transfer (ET) reactions considered in the above sections is in the presence of the second liquid phase. The charge-transfer reactions occurring at the interfacial boundary include ET between redox species confined to different liquid phases and ion transfer (IT) reactions (see Chapter 8 for more detailed discussion). All these processes can be studied by the SECM (7,12,44,45). ET can be probed using a feedback mode of the SECM. A tip UME is placed in the upper liquid phase (e.g., organic solvent) containing one form of the redox species (e.g., the reduced form, R). When the tip is held at a positive potential, R reacts at the tip surface to produce the oxidized form of the species, O. When the tip approaches the ITIES, the mediator can be regenerated at the interface via the biomolecular redox reaction between O in the organic phase and a reduced form of aqueous redox species (Fig. 14A). In addition to the ET step, the overall interfacial process includes the transfer of a common ion between two liquid phases and the mass transfer in the bottom phase. If these steps are rapid, the current-distance curves are described by Eqs. (34). Otherwise a more complicated theory may be required (12,45).

IT at the liquid/liquid interface can be probed using a micropipet tip (7,46). Two types of IT processes, i.e., facilitated and simple IT, require different treatments. In the first case, an aqueous solution in the pipet contains some ion, e.g., a metal cation (M^+). The micropipet is immersed in organic solvent containing dissolved ionophore, R, capable of complexing the ion of interest. The facilitated IT reaction at the micropipet tip is

$$M^+(w) + R(o) = MR^+(o) \qquad \text{(at the pipet tip)} \tag{72a}$$

When the tip approaches the bottom (aqueous) layer, M^+ is released from the complex and transferred to the aqueous solution, and R, which serves as a mediator, is regenerated (Fig. 14B):

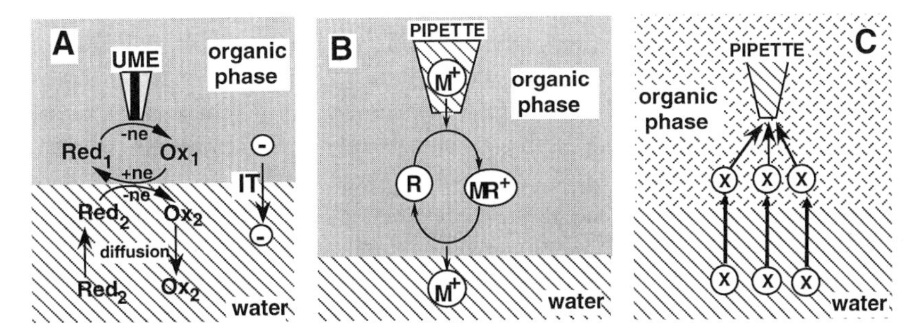

FIG. 14 Studying charge transfer processes at the liquid/liquid interface by SECM. (A) Probing interfacial electron transfer. The SECM operates in the feedback mode with a reversible redox mediator. (B) Probing facilitated IT with the SECM operating in the feedback mode. A cation is transferred from the aqueous solution inside the pipet into the organic phase by interfacial complexation and from organic phase into the bottom aqueous layer by interfacial dissociation mechanism. (C) Probing simple IT with the SECM. The tip depletes the concentration of species X at the phase boundary inducing the transfer of X across the ITIES or liquid/membrane interface.

$$MR^+(o) = M^+(w) + R(o) \qquad \text{(at the liquid/liquid interface)} \qquad (72b)$$

It was shown in Ref. 46 that the overall process in Eqs. (72a) and (72b) is mathematically equivalent to conventional feedback situation. If regeneration of the mediator at the interface [reaction (72b)] is rapid, the SECM theory for a conductive substrate is applicable; with no regeneration, the insulating behavior is observed.

Unlike conventional feedback mode and facilitated IT experiments (in which R acts as a mediator), no mediator species is involved in a simple IT process (Fig. 14C):

$$M^+(o) = M^+(w) \qquad \text{(at the pipet tip)} \qquad (73a)$$

$$M^+(w) = M^+(o) \qquad \text{(at the liquid/liquid interface)} \qquad (73b)$$

In this case, the top and the bottom liquid phases contain the same ion M^+ at equilibrium. A micropipet tip is used to deplete concentration of this ion in the top solvent near the phase boundary. This depletion results in ion transfer across the liquid/liquid interface. Reaction (73b) can produce positive feedback if concentration of M^+ in the bottom phase is sufficiently high. Any solid surface (or a liquid phase containing no specific ion) acts as an insulator in this experiment. The interface between the top and the bottom layers is nonpolarizable, and the potential drop is determined by the ratio of concentrations of the common ion (i.e., M^+) in two phases.

Unlike interfacial ET and facilitated IT, the simple IT reaction (3) is initially at equilibrium and has to be treated as a quasi-reversible process (7). The theory for interfacial transfer processes in two-phase systems developed in Ref. 45a,b is applicable to both steady-state conditions and transient experiments. The model accounts for reversibility of the transfer reaction and allows for diffusion limitations in both liquid phases. The possibility of different diffusion coefficients in two phases is also included. The steady-state situation is defined by three dimensionless parameters, i.e., $K_e = c_o/c_w$ (the ratio of bulk concentrations in organic and aqueous phases), $\gamma = D_O/D_w$ (the ratio of diffusion coefficients), and $K = k°a/D_O$ (normalized rate constant for the transfer from organic to water). The tip current (at a given distance) increases strongly with both K and γK_e values but is largely independent of individual values of K_e and γ as long as their product is constant. The diffusion limitations in the second phase (i.e., the liquid phase containing no electrodes) are noticeable only at $\gamma K_e \lesssim 10$. Separating the effects of all three parameters on the shape of current-distance curves may be difficult.

V. NUMERICAL SOLUTION OF SECM DIFFUSION PROBLEMS USING PDEase2 PROGRAM PACKAGE

The SECM theory for most systems reviewed in previous sections is rather complicated. Analytical approximations are available only for a limiting number of relatively simple processes. The generation of theoretical data requires numerical solution of partial differential equations, which makes it harder for an experimentalist to carry out quantitative SECM studies. A similar situation existed in macroelectrode electrochemistry before the development of DigiSim program (47), which allows one to treat various mechanisms (described by one-dimensional diffusion equations) without doing any calculus or computer programming. A commercially available and relatively inexpensive PDEase2 software package (48) can be equally useful for two-dimensional microelectrode simulations. This program employs the Galerkin finite element method and creates a triangular adaptive grid to solve problems involving singularities with high accuracy and computational efficiency. To our knowledge, the application of this software to solving SECM-related problems was reported only once (7), and no discussion of programming and computations was presented.

Unlike the DigiSim package, PDEase2 is a general solver of partial differential equations rather than electrochemical software. Thus it is suitable for modeling a broad range of physicochemical problems involving mass transfer, fluid dynamics, electrostatics, etc. However, unlike DigiSim, the user has to write pertinent differential equations and specify the geometry of his problem. The examples of the input files given in this section illustrate the solution of three progressively more complicated steady-state problems, i.e., diffusion to an embedded microdisk, pure positive (or negative) feedback at the SECM tip, and the E_rC_i mechanism. Solving time-dependent SECM problems with PDEase2 will be discussed elsewhere.

A. Diffusion to an Embedded Microdisk

The input files for PDEase2 are regular text files, which can be created using any word processor. There is no case sensitivity, and spaces are not important. An input file consists of a number of sections (e.g., "Title," "Select," "Equations," and so on). Each section contains one or more statements. The following descriptor file, **disk.pde**, contains the formulation required for PDEase2 to compute steady-state diffusion flux to an embedded microdisk and generate the graphical and numerical output. The statements in this file are either self-explanatory or clarified by short comments written in curly brackets.

The geometry of the problem and boundary conditions are specified in the section "Boundaries." This section uses a hierarchial structure consisting of one or more subsections called "regions." In each region, the statement "start" defines the starting point from which the boundary of that region is drawn. Each boundary condition precedes a geometrical statement (e.g., "line") defining the segment (or segments) to which this condition applies. The first coordinate in each bracket is "R," and second is "Z." Region 1 is an all-inclusive region, which describes the total extent of the problem. Thus the simulation domain (region 1) in the **disk.pde** program is a rectangle whose vertices have the coordinates (0,0), (100,0), (100,100), and (0,100). Regions 2 to 4 in **disk.pde** are pseudoregions. They represent the areas where the concentration and/or flux of redox species change sharply. By creating a higher density of nodes in these areas, one can decrease the number of regrids. These regions are optional and are used only to reduce computation time. In more complicated (e.g., multiphase) simulations the problem geometry can be defined in more than one region.

The diffusion equation in **disk.pde** looks somewhat different from the corresponding Eq. (26). Nevertheless, these equations are completely equivalent, and eliminating the 1/R singularity makes the program work better. Symbol "dR" means the first derivative with respect to R, and "dZZ" represents the second derivative with respect to Z. Similarly, the boundary condition written in the form "natural(C) = −M*C*R" (where natural = inward diffusion flux × R and M is a sufficiently large number) is equivalent to Eq. (27a) but results in a significantly better computational efficiency. All variables in **disk.pde** are normalized according to Eqs. (4) to (6) and (14). The computed diffusion limited current is normalized by $i_{T,\infty}$ [Eq. (20)].

Figure 15 contains some graphical output. The triangular grid generated by the program (Fig. 15A) is dense near the electrode surface (especially at the edge of the disk where the current density is very large) and sparse far from it. An equal contour map (Fig. 15B) shows how the concentration of redox species changes from zero at the disk to one in the bulk solution. The flux distribution over the disk surface is shown in Fig. 15C. The normalized current computed by integration of the flux ($i/4nFac^{\circ}D = 0.995$) is within 0.5% from the exact value (1.000). If required, the accuracy can be improved by decreasing the "errlim" number at the expense of longer computation time.

```
{disk.pde}

Title "Diffusion to an embedded disk"

Select
  errlim = 5E-7           {acceptable error}
  elevationgrid = 1000 {the number of integration intervals
                          on the disk surface}
  gridlimit = 20          {maximum number of regrids}

Coordinates
  ycylinder(R,Z)    {cylindrical coordinates}

Variables
  C              {define C as the system variable}

Definitions      {assign values to constant parameters}
  RG = 100        {RG = maximum value of R; thick insulating
                   sheath}
  ZMAX = 100     {the extent of the system in Z-direction}
  M = 10000       {how strongly to drive boundary
                   concentrations to zero}
  flux = dZ(C)   {normalized flux, dC/dZ}
  curr = flux*PI*2*R  {normalized current density to be
                        integrated over the disk surface}

Initial values
  C = 1          {tell PDEase2 the rough scale of the answer}

Equations        {define the diffusion equation}
    dR(R*dR(C)) + R*dZZ(C) = 0

Boundaries       {define the problem domain}

region 1
    start(0,0)
```

```
  natural(C) = -M*C*R  {at the disk, drive C strongly to
                         zero}
    line to (1,0)          {0<R<1, Z=0 correspond to the disk
                            surface}
  natural(C) = 0
    line to (RG,0)       {1<R<RG, Z=0 insulator surface
                          surrounding the disk}
  natural(C) = M*(1-C)*R    {at the outer boundary of the
                              problem domain, drive C strongly
                              to 1}
    line to (RG,ZMAX) to (0,ZMAX)   {far from the disk}
  natural(C) = 0        {axis of symmetry}
    line to finish        {back to (0,0)}
region 2
  start(0,0)
  line to (2,0) to (2,1) to (0,1) to finish
region 3
  start (0,0)
  line to (1.2,0) to (1.2,0.2) to (0,0.2) to finish
region 4
  start (0,0)
  line to (1.05,0) to (1.05,0.02) to (0,0.02) to finish

Plots
  grid(R,Z)  pause       {show final grid}
  contour(C) zoom(0,0,12,12)    pause
  elevation(C) from (0,0) to (10,0)   {show C on R axis} pause
  elevation(C) from (0,0) to (0,ZMAX)   {show C on Z axis}
                                          pause
  elevation(flux) from (0.0) to (1.5,0)   {distribution of
   flux over the disk surface} pause
  elevation(curr) from (0,0) to (1,0)   {the "area" value =
   total normalized current to the disk is printed}

End                     {end of descriptor file}
```

B. Diffusion-Controlled SECM Feedback

One would only need to change a few lines in the **disk.pde** descriptor file to switch from the microdisk to the SECM problem. Somewhat more extensive changes are required to simulate the SECM feedback situation taking into account mediator diffusion from behind the shield. The following input file (**secm.pde**) was used in Ref. 7 to simulate negative feedback current (insulating substrate) by solving diffusion problem from Sec. II.B [Eq. (26) with boundary conditions given by Eqs. (27), (28b) and (29)]. The presented simulation uses the RG = 1.1 and L = 1, but they can be replaced by any

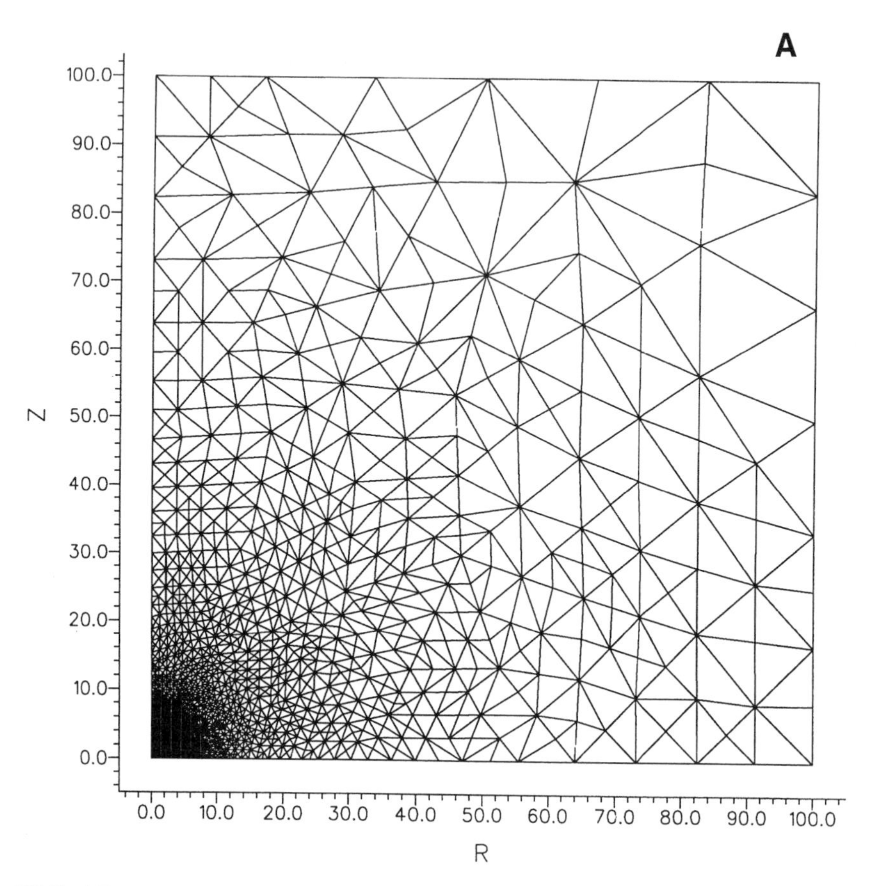

FIG. 15 Simulation of steady-state diffusion-limited current at an embedded microdisk electrode by PDEase2. (A) The final grid generated by the program after 11 regrids. (B) An equal contour map of concentration of redox species. The concentration value for each contour line is given in the column in the right side of the figure and designated by the corresponding letter. (C) The flux distribution over the disk surface. For parameters see **disk.pde** file.

other values. Figure 16 shows the final grid for the SECM problem (A) and the distributions of concentration (B) and of the flux (C) over the R-axis. One should notice the "negative" flux from the back of the tip in Fig. 16C. The computed dimensionless tip current ($i/4nFac^{\circ}D$) was 1.017.

To simulate positive feedback situation one has to replace the boundary condition "natural(C)=0" for the last segment of region 1 representing the substrate surface with "natural(C)=M*(1-C)*R." Simulating positive feedback takes significantly less time, and larger "errlim" values can be used

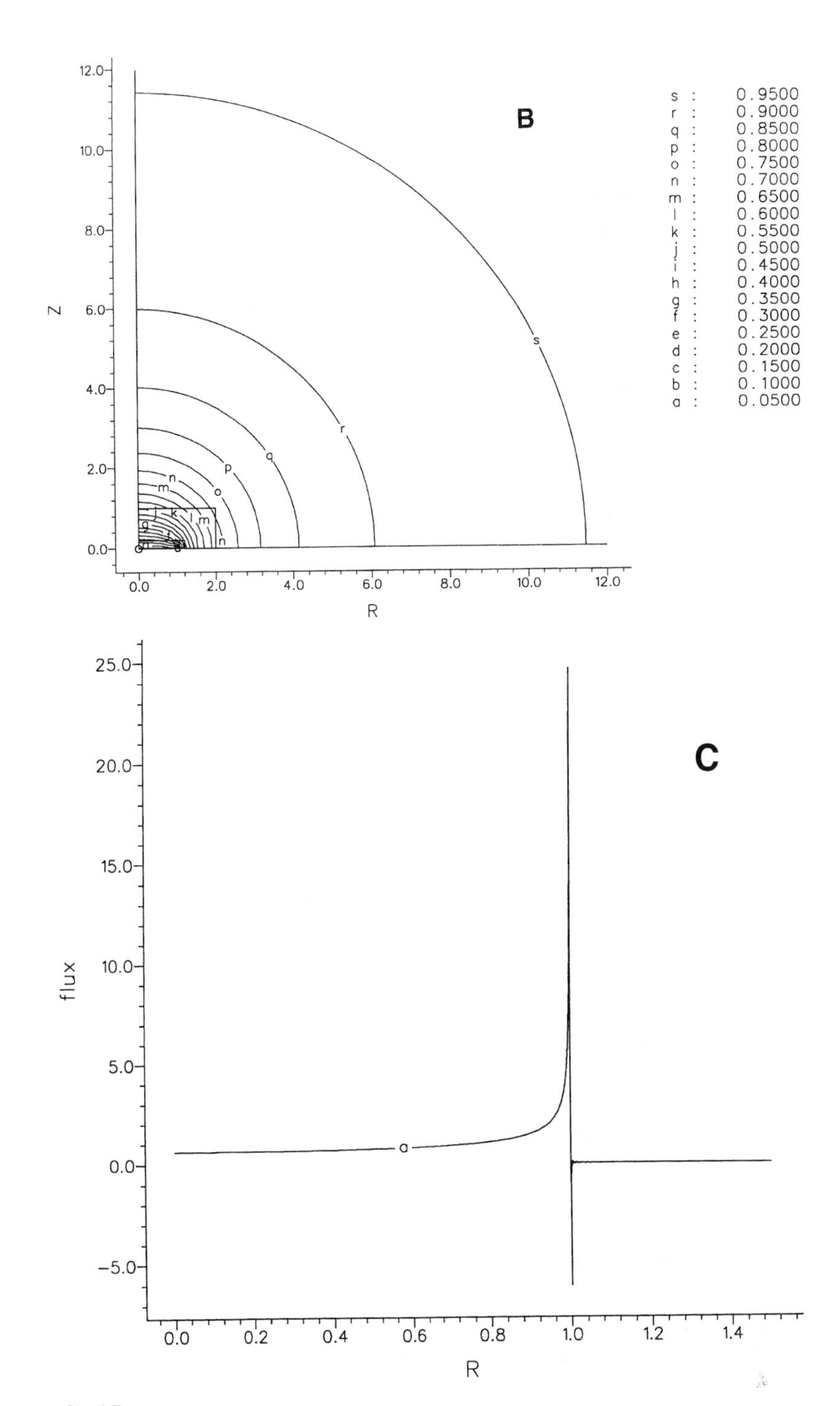

FIG. 15 Continued

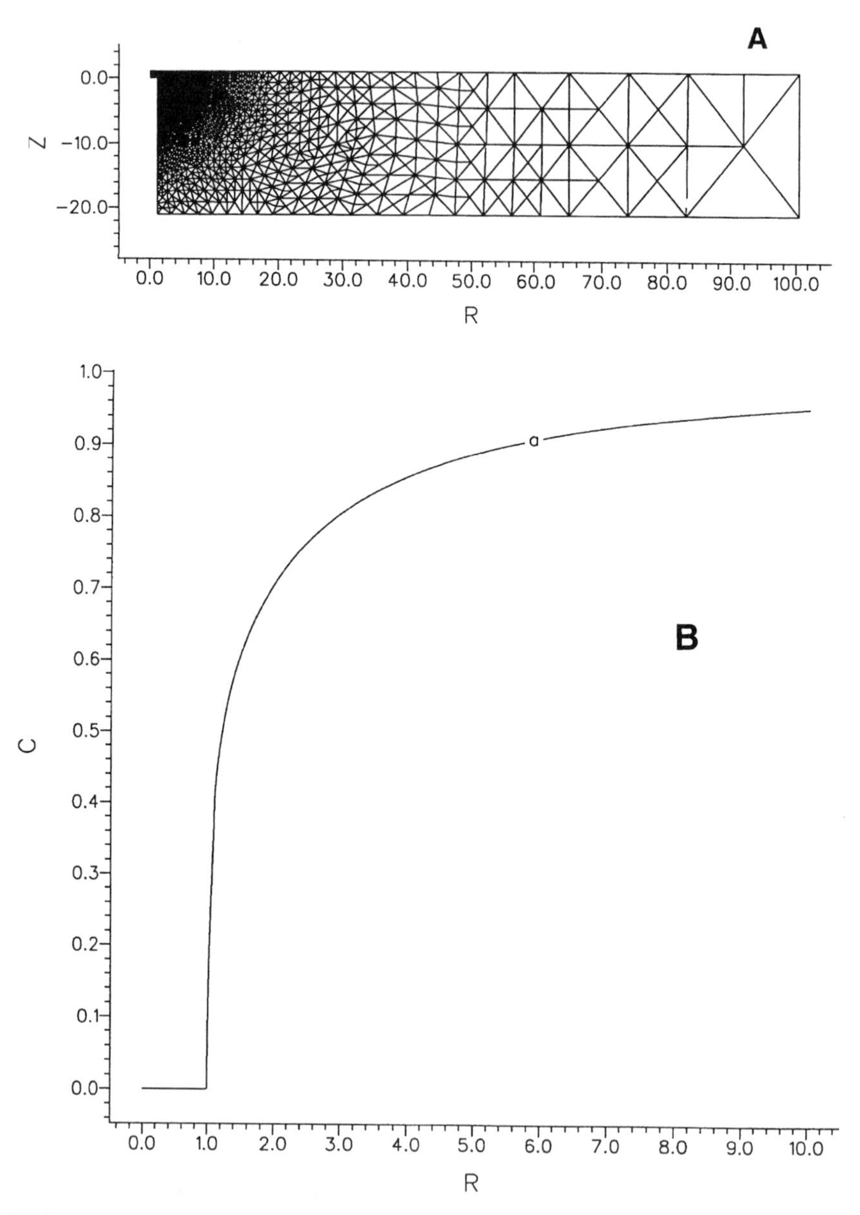

FIG. 16 Simulation of the diffusion-controlled SECM process at an insulating substrate by PDEase2. (A) The final grid generated by the program consists of 8659 cells and 17588 nodes. (B) Distribution of concentration of redox species over R-axis. (C) Distribution of the flux over R-axis. For parameters see **secm.pde** file.

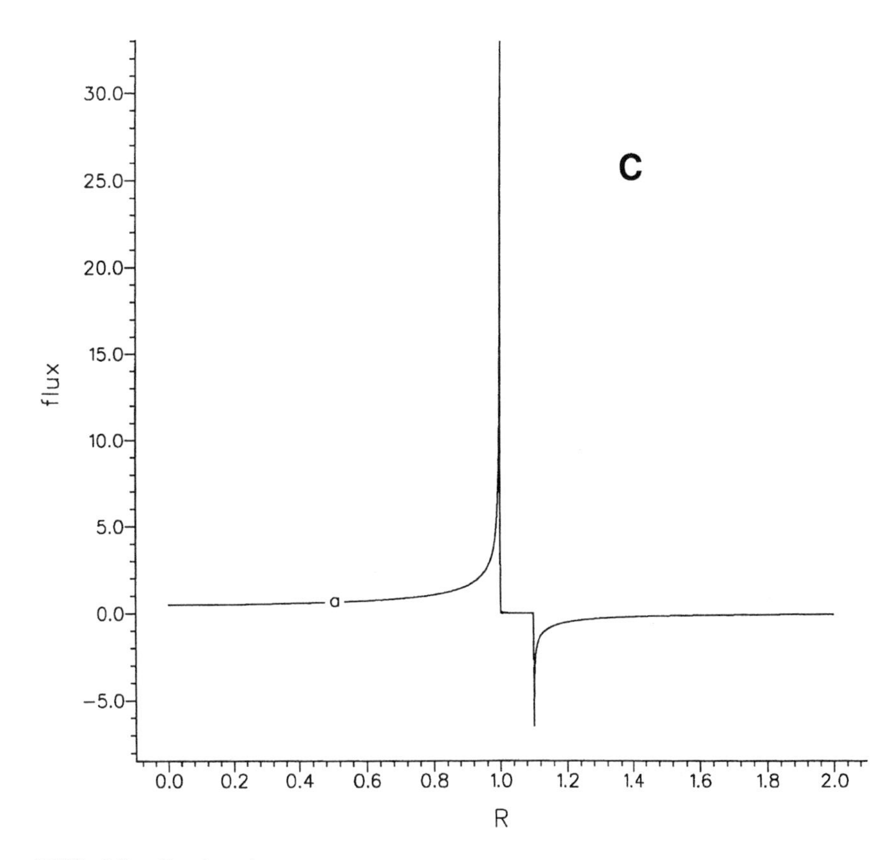

FIG. 16 Continued

without sacrificing accuracy. Minor modifications of the input file allow the simulation of the SECM responses for recessed or protruding tips, finite heterogeneous kinetics at the tip and/or substrate, etc.

```
{secm.pde}

Title "SECM with an insulating substrate"

Select
  errlim = 5E-7
  elevationgrid = 1000
  gridlimit = 25

Coordinates
  ycylinder(R,Z)
```

```
Variables
  C

Definitions
  RG = 1.1
  L = 1
  L1 = 20   {show how far behind the tip surface (Z = 0) the
             simulation goes}
  RMAX = 100   {the extent of the system in R-direction}
  flux = dZ(C)
  curr = flux*PI*0.5*R
  M = 10000

Initial values
  C = 1

Equations
  dR(R*dR(C)) + R*dZZ(C) = 0

Boundaries
region 1
    start(0,0)
  natural(C) = -M*C*R
     line to (1,0)    {the disk surface}
  natural(C) = 0
     line to (RG,0) to (RG,-L1-L)   {insulating plane and
        cylindrical inert surface surrounding the tip}
  natural(C) = M*(1-C)*R    {C tends to 1 far from the disk
                              surface}
     line to (RMAX,-L1-L) to (RMAX,L)
  natural(C) = 0
     line to (0,L) to finish     {axis of symmetry}
region 2
  start (0,0)
  line to (RG,0) to (RG,-5) to (RG+5,-5) to (RG+5,L) to (0,L)
   to finish
region 3
  start (0,0)
  line to (RG,0) to (RG,-1) to (RG+1,-1) to (RG+1,L) to (0,L)
   to finish
region 4
  start (0,0)
  line to (RG,0) to (RG,-0.1) to (RG+0.1,-0.1) to (RG+0.1,L)
   to (0,L) to finish
region 5
  start(0.98,0)
  line to (1.02,0) to (1.02,0.02) to (0.98,0.02) to finish
```

```
Plots
  grid(R,Z) pause
  elevation(C) from (0,0) to (10,0) pause
  elevation(C) from (0,0) to (0,L) pause
  elevation(flux) from (0,0) to (1.5,0) pause
  elevation(curr) from (0,0) to (1,0)

End
```

C. E_rC_i Mechanism

The diffusion problem solved in this section is a steady-state version of Eqs. (51) to (57), in which $\partial c_i/\partial t = 0$, all variables are time-independent, and the initial condition [Eq. (52)] is omitted. The normalized variables employed are the same as above besides the dimensionless homogeneous rate constant, $K = k_c a^2/D_R$. The main difference between the input file **erci.pde** and the ones previously considered is that concentrations of both forms of redox species (COX and CR) have to be computed by solving two partial differential equations. To save space the diffusion from behind the shield has been neglected. It can be taken into account by introducing additional boundary segments as it was done in **secm.pde**. An implicit assumption of equal diffusion coefficients ($D_{Ox} = D_R$) is not essential and can easily be eliminated. Finding a proper value of "errlim" (i.e., such that making it 2–5 times smaller does not change the result within the required precision level) requires some experimentation. Clearly this number should be as high as feasible. Optimizing pseudoregions may also save computer resources.

```
{erci.pde}

Title "ErCi mechanism under SECM conditions"

Select
  errlim = 1E-5
  elevationgrid = 1000
  gridlimit = 20

Coordinates
  ycylinder(R,Z)

Variables   {COX and CR are concentrations of the oxidized
  COX         and reduced forms}
  CR

Definitions
  RG = 10
  L = 0.25
  fluxo = dZ(COX)     {flux of the oxidized form}
```

```
  fluxr = -dZ(CR)      {flux of the reduced form}
  K = 1    {dimensionless homogeneous rate constant}
  curr = fluxo*PI*0.5*R
  M = 10000

Initial values
  COX = 1
  CR = 0

Equations
  dR(R*dR(COX)) + R*dZZ(COX) = 0
  dR(R*dR(CR)) + R*dZZ(CR) - K*CR*R = 0

Boundaries

region 1
    start(0,0)
  natural(COX) = -M*COX*R   {at the disk, drive COX strongly
                              to zero}
  natural(CR) = M*COX*R     {flux at R at the tip = -flux of Ox}
    line to (1,0)
  natural(COX) = 0
  natural(CR) = 0
    line to (RG,0)   {1<R<RG, Z=0 inert surface surrounding
                       the disk}
  value(COX) = 1
  value(CR) = 0
    line to (RG,L)            {the outer boundary of the
                                problem domain}
  natural(CR) =-M*CR*R      {at the substrate, drive CR
                              strongly to zero}
  natural(COX) = M*CR*R   {flux of Ox at the substrate = -flux
                            of R}
    line to (0,L)
  natural(COX) = 0
  natural(CR) = 0
    line to finish

region 2
    start(0,0)
    line to (2,0) to (2,L/5) to (0,L/5) to finish

region 3
    start(0,0)
    line to (1.1,0) to (1.1,0.1*L) to (0,0.1*L) to finish

region 4
    start(0,0)
    line to (1,0) to (1,L/100) to (0,L/100) to finish
```

```
region 5
    start(0.98,0)
    line to (1.02,0) to (1.02,0.02) to (0.98,0.02) to finish

region 6
    start(0,0.9*L)
    line to (1.+1.5*L,0.9*L) to (1.+1.5*L,L) to (0,L) to
     finish

Plots
  grid(R,Z)
  contour(COX) zoom(0,0,1+2*L,L) pause
  contour(CR) zoom(0,0,1+2*L,L) pause
  elevation(COX) from (0,0) to (10,0) pause
  elevation(COX) from (0,0) to (0,L) pause
  elevation(fluxo) from (0,0) to (2,0) pause
  elevation(fluxr) from (0,L) to (2,L) pause
  elevation(curr) from (0,0) to (1,0)

End
```

Computational results in Fig. 17 include a concentration profile of Ox (A) and distributions of fluxes of Ox across the tip surface (B) and R over the substrate surface (C) obtained with L = 0.25, K = 1, and RG = 10.

ACKNOWLEDGMENTS

The support of our research in SECM by grants from the Petroleum Research Fund administrated by the American Chemical Society and PSC-CUNY Research Award Program is gratefully acknowledged.

LIST OF SYMBOLS

a	radius of electrode
a_S	radius of substrate
c	concentration of redox species
c°	bulk concentration of electroactive species
C	dimensionless variable equal to c/c_O°
d	distance between tip and substrate
D	diffusion coefficient of redox species

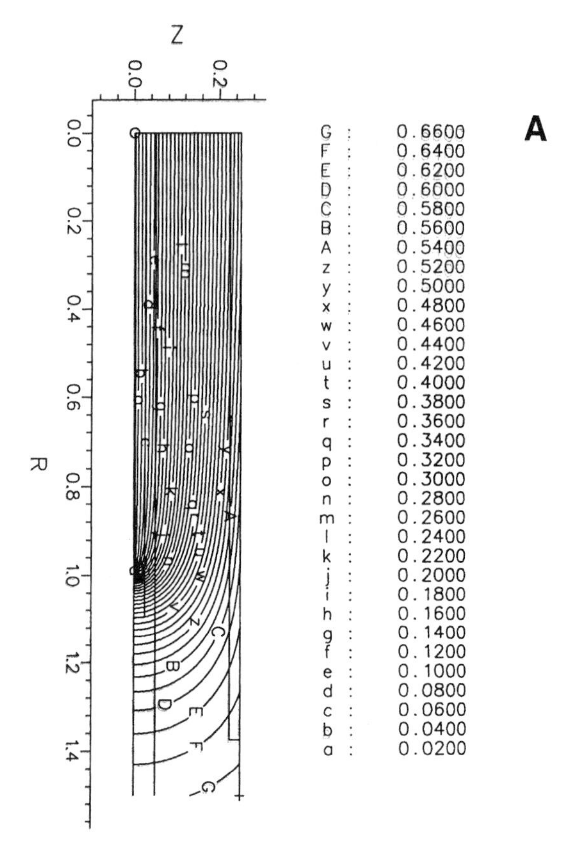

FIG. 17 Simulation of the E_rC_i mechanism by PDEase2. (A) An equal contour map of concentration of oxidized species. (B) Distribution of the flux of O across the tip surface. (C) Distribution of the flux of R over the substrate surface. For parameters see **erci.pde** file.

E	electrode potential
$E^{\circ\prime}$	formal potential
f	parameter equal to F/RT (F is the Faraday, R is the gas constant, and T is temperature)
F	the Faraday
h	ratio of substrate radius to tip radius (a_S/a)
h	the value of h (equal to 1 + 1.5L) at which the substrate behaves as an infinite one

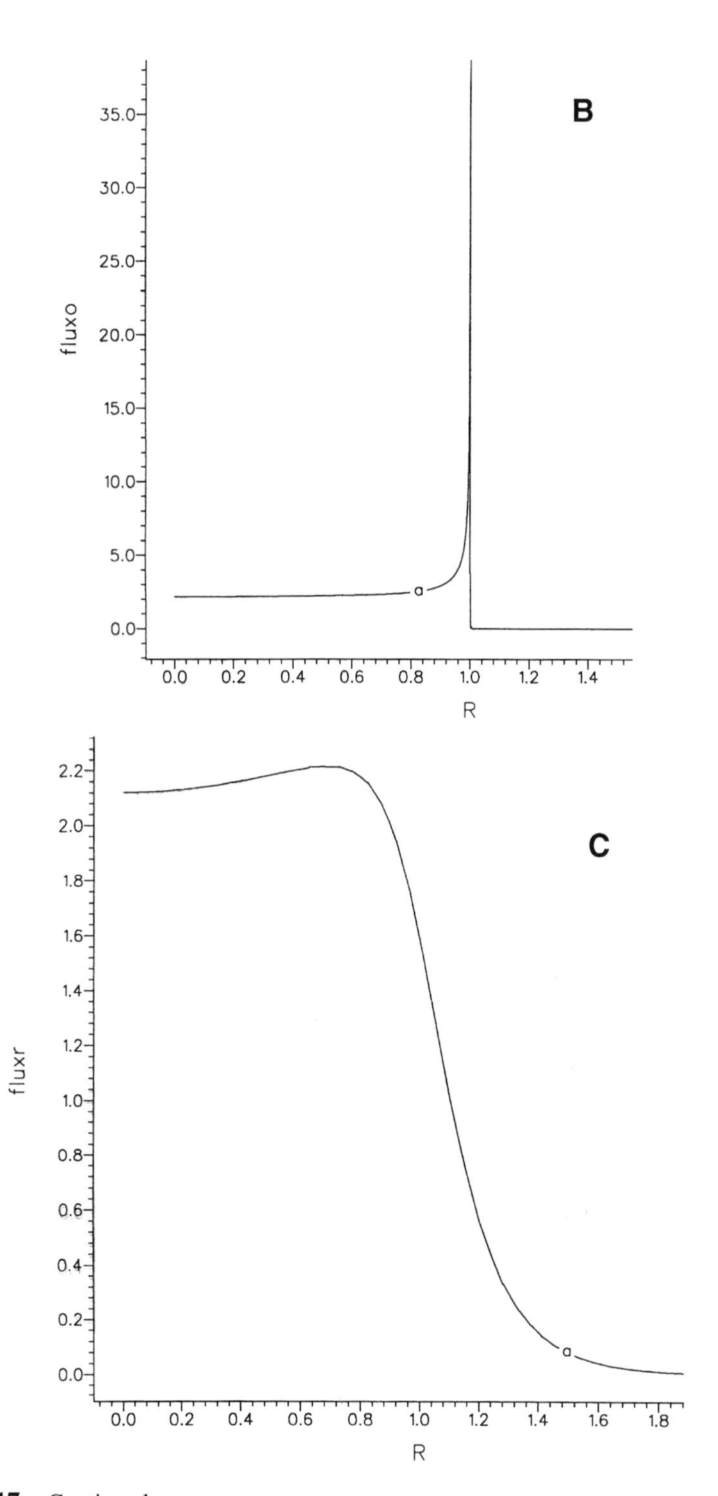

FIG. 17 Continued

i_S	current flowing through substrate
i_T	current flowing through tip
i_T^c, i_T^{ins}	tip current at conductor and insulator
$i_{T,\infty}$	steady-state current at a tip electrode when the tip is far from the substrate
i_S/i_T	collection efficiency of substrate
i_T/i_S	collection efficency of tip
i_{dif}	diffusion-limited current for any electrode geometry
I_d	dissipation current in Eq. (59)
I_f	feedback current coming from the substrate in Eq. (59)
I_T	normalized tip current equal to $i_T/i_{T,\infty}$
I_0	modified Bessel function of the first kind of order zero
$I_T(L)$	diffusion-limiting tip current at a normalized tip/substrate separation equal to L
$I_S(T)$	dimensionless substrate current given by Eq. (19)
j	diffusion flux density
J_T, J_S	dimensionless variable equal to $ja/(Dc^\circ)$
J_0	the Bessel function of the first kind of order zero
k°	standard rate constant
$k_{f/b}$	heterogeneous rate constants for oxidation and reduction
$K_{f/b,S}$	dimensionless rate constant equal to $k_{f/b,S}a/D$
k	homogeneous rate constant in an ECE reaction
k_c	homogeneous rate constant in a first-order chemical reaction
k_c'	homogeneous rate constant in a dimerization reaction
k_n	dissolution rate constant
k_{eff}	kinetic parameter given by Eq. (38)
K(p)	complete elliptical integral of the first kind
K_e	ratio of bulk concentrations in organic and aqueous phases
ℓ	the depth of the metal recession

L	normalized tip/substrate distance equal to d/a
L1, RMAX	simulation space limits
m_O	effective mass transfer coefficient
n	number of electrons transferred per redox event
n	reaction order in Eq. (71)
O, R, S, T	when used as subscripts, refer to oxidized form, reduced form, substrate, and tip, respectively
r, z	spatial variables
R	dimensionless variable equal to r/a
R_d	defect radius in the surface film
rg	radius of the insulating ring around a microtip
RG	dimensionless sheath radius equal to rg/a
t	time
t_{max}	time corresponding to the maximum tip current
T	dimensionless variable equal to tD/a^2
v	scan rate
Z	dimensionless variable equal to z/a
α	transfer coefficient
γ	equal to D_O/D_R
γ	ratio of diffusion coefficients in organic and aqueous phases
κ	parameter equal to $k^\circ \exp[-\alpha n f(E - E^{\circ\prime})]/m_O$
κ	kinetic parameter in Eq. (66) equal to $k_c d^2/D$
κ'	parameter equal to $ak^\circ \exp[-\alpha n f(E - E^{\circ\prime})]/(4D_O)$
κ'	dimensionless parameter equal to $k_c' a^2 c_O^\circ/D$
κ_d	kinetic parameter equal to $\pi k_d R_d/4D$
κ_{TLC}	parameter equal to $2\kappa' d(D_O + D_R)/(\pi a D_R)$
Λ	variable equal to dk°/D
σ	undersaturation produced by the tip reaction

θ parameter equal to $1 + \exp[nf(E - E^{\circ\prime})]D_O/D_R$

θ fractional surface coverage

REFERENCES

1. Kwak J, Bard AJ. Anal Chem 1989; 61:1221.
2. Mirkin MV, Bard AJ. J Electroanal Chem 1992; 323:29.
3. Bard AJ, Mirkin MV, Unwin PR, Wipf DO. J Phys Chem 1992; 96:1861.
4. Bard AJ, Denuault G, Friesner RA, Dornblaser BC, Tuckerman LS. Anal Chem 1991; 63:1282.
5. Unwin PR, Bard AJ. J Phys Chem 1991; 95:7814.
6. Martin RD, Unwin PR. J Electroanal Chem 1997; 439:123.
7. Shao Y, Mirkin MV. J Phys Chem B 1998; 102:9915.
8. Bard AJ, Fan, F-RF, Mirkin MV. In: AJ Bard, ed. Electroanalytical Chemistry. Vol. 18, New York: Marcel Dekker, 1993, p 243.
9. Mirkin MV, Fan F-RF, Bard AJ. J Electroanal Chem 1992; 328:47.
10. (a) Shoup D, Szabo A. J Electroanal Chem 1984; 160:27; (b) Fang Y, Leddy. J Anal Chem 1995; 67:1259.
11. Amphlett JL, Denuault G. J Phys Chem B 1998; 102:9946.
12. Wei C, Bard AJ, Mirkin MV. J Phys Chem 1995; 99:16033.
13. Mirkin MV, Bard AJ. Anal Chem 1992; 64:2293.
14. Mirkin MV, Richards TC, Bard AJ. J Phys Chem 1993; 97:7672.
15. Borgwarth K, Ricken C, Ebling DG, Heinze J. Fresenius J Anal Chem 1996; 356:288.
16. Forouzan F, Bard AJ, Mirkin MV. Isr J Chem 1997; 37:155.
17. (a) Zwanzig R. Proc Natl Acad Sci USA 1990; 87:5856; (b) Zwanzig R, Szabo A. Biophys J 1991; 60:671; (c) Szabo A, Zwanzig R. J Electroanal Chem 1991; 314:307.
18. (a) Mirkin MV, Fan F-RF, Bard AJ. Science 1992; 257:364; (b) Fan F-RF, Mirkin MV, Bard AJ. J Phys Chem 1994; 98:1475.
19. Davis JM, Fan F-RF, Bard AJ. J Electroanal Chem 1987; 238:9.
20. Demaille C, Brust M, Tsionsky M, Bard AJ. Anal Chem 1997; 69:2323.
21. Baranski AS. J Electroanal Chem 1991; 307:287; (b) Oldham KB. Anal Chem 1992; 64:646.
22. Fan F-RF, Bard AJ. Science 1995; 267:871; (b) Fan F-RF, Kwak J, Bard AJ. J Am Chem Soc 1996; 118:9669.
23. Shao Y, Mirkin MV, Fish G, Kokotov S, Palanker D, Lewis A. Anal Chem 1997; 69:1627.
24. Saito Y. Rev Polarogr 1968; 15:177.
25. (a) Scott ER, White HS. Anal Chem 1993; 65:1537; (b) Bath BD, Lee RD, White HS, Scott ER. Anal Chem 1998; 70:1047.
26. Bond AM, Oldham KB, Zoski CG. J Electroanal Chem 1988; 245:71.
27. Horrocks BR, Mirkin MV, Pierce DT, Bard AJ, Nagy G, Toth K. Anal Chem 1993; 65:1213.

28. Abramowitz M, Stegun I, eds. Handbook of Mathematical Functions. New York: Dover, 1965.
29. Fleischmann M, Pons S. J Electroanal Chem 1988; 250:257.
30. (a) Troise Frank MH, Denuault G. J Electroanal Chem 1993; 354:331; (b) Troise Frank MH, Denuault G. J Electroanal Chem 1994; 379:405.
31. Horrocks BR, Mirkin MV. J Chem Soc Faraday Trans 1998; 94:1115.
32. (a) Engstrom RC, Weber M, Wunder DJ, Burgess R, Winquist S. Anal Chem 1986; 58:844; (b) Engstrom RC, Meaney T, Tople R, Wightman RM. Anal Chem 1987; 59:2005; (c) Engstrom RC, Wightman RM, Kristensen EW. Anal Chem 1988; 60:652.
33. Bard AJ, Faulkner LR. Electrochemical Methods, Fundamentals and Applications. New York: Wiley, 1980.
34. Mirkin MV, Arca M, Bard AJ. J Phys Chem 1993; 97:10790.
35. Martin RD, Unwin PR. Anal Chem 1998; 70:276.
36. Zhou F, Unwin PR, Bard AJ. J Phys Chem 1992; 96:4917.
37. Lee C, Kwak J, Anson FC. Anal Chem 1991; 63:1501.
38. Demaille C, Unwin PR, Bard AJ, J Phys Chem 1996; 100:14137.
39. Treichel DA, Mirkin MV, Bard AJ. J Phys Chem 1994; 98:5751.
40. Martin RD, Unwin PR. J Chem Soc Faraday trans 1998; 94:753.
41. Unwin PR, Bard AJ. Anal Chem 1992; 64:113.
42. Unwin PR, Bard AJ. J Phys Chem 1992; 96:5035.
43. (a) Macpherson JV, Unwin PR. J Phys Chem 1994; 98:1704; (b) Macpherson JV, Unwin PR. J Phys Chem 1994; 98:11764; (c) Macpherson JV, Unwin PR. J Phys Chem 1995; 99:3338; (d) Macpherson JV, Unwin PR. J Phys Chem 1995; 99:14824; (e) Macpherson JV, Unwin PR. J Phys Chem 1996; 100: 19475.
44. (a) Tsionsky M, Bard AJ, Mirkin MV. J Phys Chem 1996; 100:17881; (b) Tsionsky M, Bard AJ, Mirkin MV. J Am Chem Soc 1997; 199:10785.
45. (a) Slevin CJ, Macpherson JV, Unwin PR. J Phys Chem 1997; 101:10851; (b) Barker AL, Macpherson JV, Slevin CJ, Unwin PR. J Phys Chem 1998; 102: 1586.
46. Shao Y, Mirkin MV. J Electroanal Chem 1997; 439:137.
47. Rudolph M, Reddy DP, Feldberg SW. Anal Chem 1994; 66:589A.
48. PDEase2. Version 2.5.2. Reference Manual and Application Notes. Bass Lake: SPDE, Inc., 1995.

6

HETEROGENEOUS ELECTRON TRANSFER REACTIONS

Kai Borgwarth and Jürgen Heinze

Albert Ludwig University of Freiburg, Freiburg, Germany

I. INTRODUCTION

The scanning electrochemical microscopy (SECM) concept of bringing a probe close to a phase boundary makes it possible to measure rates of transfer processes across interfaces in a wide range of applications (1). For scientists, ultramicroelectrodes are the ideal probe, since they can be used to simultaneously initiate transfer reactions of electrons and chemical species across phase boundaries by driving chemical reactions and to monitor the response. Such SECM experiments elegantly combine the advantages of ultramicroelectrodes (UME) with those of thin-layer cells (TLC) (2). Thus, the analysis of heterogeneous transfer processes, primarily electron transfer, represents a popular application of the SECM, and further applications are being developed.

On the one hand, lateral differences in heterogeneous rate constants can be exploited to obtain images of reactivity by moving the tip parallel to a sample's surface at a constant distance. A number of studies reported lateral differences in electrical conductivity, such as the interdigitated array consisting of platinum and glass shown in Figure 1 (3). Moreover, lateral variations in chemical reactivity, even those involving complex kinetics of heterogeneous catalysis and enzymes, become visible using the high chemical specificity of the method. This makes it possible to compare directly properties of different parts of a surface and to identify "hot spots," the locations of reactions as shown in the example of a commercial microelectrode array for amperometric determination of heavy metals in blood in Figure 2. On the other hand, quantitative data can be obtained from steady-state voltammetry and from current-distance curves at fixed locations, normally using the feedback mode. This opens up considerable possibilities, since the SECM method is capable of measuring very fast heterogeneous rate constants of up to about 100 cm/s. In multiphase systems the mechanism of complex reactions can be studied and the role of phases and interfaces as

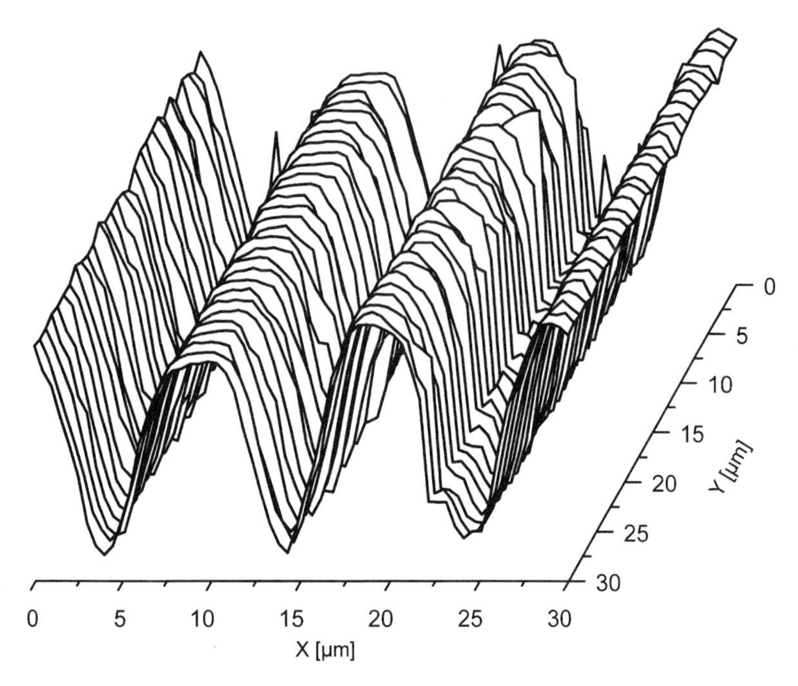

FIG. 1 SECM image of an interdigitated array (IDA) with a period of 3 μm Pt and 7 μm glass stripes, using a 1 μm platinum tip, $RG = 30$, $d = 300$ nm at a scan speed $v = 50$ μm/s. Conditions: mediator 10 mM $K_3[Fe(CN)_6]$ in aq. 100 μM KSCN, 500 mM KCl, current scale (-5 to -8) nA. (From Ref. 3.)

conductors or locations of transfer reaction of electrons or ions can be analyzed. To extract electrochemical parameters quantitatively, the experimental data have to be fitted to numerical simulations taking into account the mass transfer and the heterogeneous and, if applicable, homogeneous reactions.

This chapter reviews in detail the principles and applications of heterogeneous electron transfer reaction analysis at tip and sample electrodes. The first section summarizes the basic principles and concepts. It is followed by sections dedicated to one class of sample material: glassy carbon, metals and semiconductors, thin layers, ion-conducting polymers, and electrically conducting polymers. A separate section is devoted to practical applications, in essence the study of heterogeneous catalysis and in situ characterization of sensors. The final section deals with the experiments defining the state of the art in this field and the outlook for some future activities. Aspects of heterogeneous electron transfer reactions in more complex systems, such as

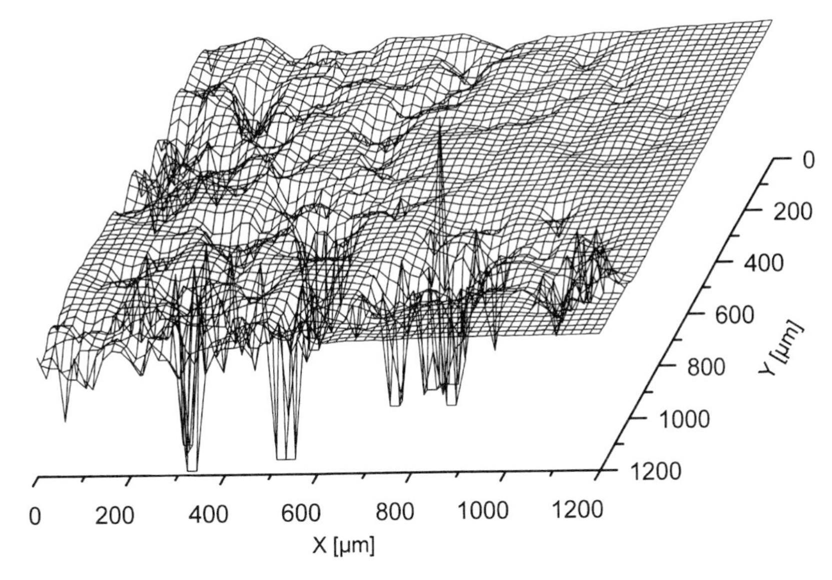

FIG. 2 Hot spots on a commercial graphite electrode to determine heavy metals in blood. Imaged by a 10 μm Pt tip in aq. 20 mM $K_3[Fe(CN)_6]$, 500 mM Na_2SO_4 at v = 10 μm/s. The sample is courtesy of Metrohm AG, Switzerland.

biological systems, and liquid/liquid and liquid/solid interfaces involving chemical reactions, such as corrosion and dissolution of salts, can be found in Chapters 8, 11, and 12.

II. PRINCIPLES

A. Basic Concepts

In general, the kinetics of heterogeneous electron transfer can be determined with high lateral resolution while scanning a tip parallel to the surface. Distance-dependent measurements provide quantitative information on sample properties. They are characterized by more than one parameter and improve the quality of data. This aspect gains significant importance with increasing sample complexity, e.g., in studies of multiphase systems with phase boundaries parallel to the surface, the location and the rate of electron transfer can be determined. For example, the location of charge transfer of dissolved molecules with redox active films or conducting polymers on metal electrodes has been studied. This powerful technique has proven to be suitable for the determination of reaction mechanisms and even of the positions of boundaries of stacked phases. Further important objectives of studies are the film thickness and conductivity.

Principally, all voltammetric techniques can be carried out in SECM equipment, additionally benefiting from the high lateral resolution. Because transient techniques provide quite complex data, steady-state methods are preferable. All experiments to obtain quantitative data for heterogeneous electron transfer reactions were performed amperometrically in the feedback mode (4). Since this method does not require electrical contact with the sample, there are no restrictions on the chemical nature of the sample. Therefore, chemical surface reactions can be included, which considerably extends the range of applications to ion conductors and nonconducting samples. At large distances the current can be reliably attributed to faradaic processes. However, at small distances in the low nanometer range, electron tunneling also makes a contribution with conductive substrates. The SECM response with insulating substrates with thin water films has been the subject of several papers (5).

Recently, as an alternative to metal-based ultramicroelectrodes, micropipettes in the ion-transfer mode have been used (6). They broaden the SECM application to the monitoring of redox inactive charged species and to the study of transfer processes of these species across phase boundaries. In cases of samples that may undergo surface-modifying processes under the influence of tip-generated species, any contributions to the detected signal have to be separated from the pure electrochemical feedback. This can be accomplished by varying the time scale of the experiment or by employing further in situ techniques such as quartz crystal microbalance (QMB) or atomic force microscopy (AFM).

The theory of SECM and computation of the tip and substrate current are discussed in detail in Chapter 5. Here, we briefly summarize the main correlations and focus on the concept of the experiments.

The use of ultramicroelectrodes in electrochemical microscopy has been the subject of several books and reviews (7,8). These electrodes provide very high rates of mass transport, since the effective mass transfer coefficient m (9) is inversely proportional to the active radius a (10). For a microdisk electrode m can be approximated as

$$m = \frac{\pi}{2}\frac{D}{a} \tag{1}$$

where D represents the diffusion coefficient of the species that is converted at the electrode. As a consequence of the fast mass transfer rate at very small electrodes, the heterogeneous charge transfer may become the rate-determining factor and can be determined quantitatively from the shape and shift of voltammetric curves (7,11). These applications have the advantage of small solution volumes, negligible capacitive charging currents, and small iR drops. The latter enables the determination of rate constants in highly resistive solvents such as liquid ammonia and benzene.

When the UME is part of an SECM apparatus and is close to a substrate, the rapid regeneration (positive feedback) increases the mass transfer coefficient even more and the mass transfer coefficient becomes inversely proportional to the distance d from the sample (12,13):

$$m = \frac{D}{d} \quad \text{for} \quad d \ll a \tag{2}$$

In practice, it is much easier to bring a flat-ended tip within very small distances, as shown in Figure 3, than to make electrodes with a well-defined geometry and a similar value as tip radius. Thus, the SECM strongly facilitates the determination of fast rate constants. Bringing the tip down to the present technical limit of about 10 nm distance and assuming a typical diffusion coefficient $D = 10^{-5}$ cm^2/s, mass transfer coefficients of 10 cm/s can be achieved, which provides an upper limit of determinable rate constants of about 100 cm/s.

In general, the electrode reaction rate is governed by the rates of steps such as mass transfer, electron transfer, and, occasionally, chemical reactions of the species involved (9), as shown in Figure 4 for a species Ox being

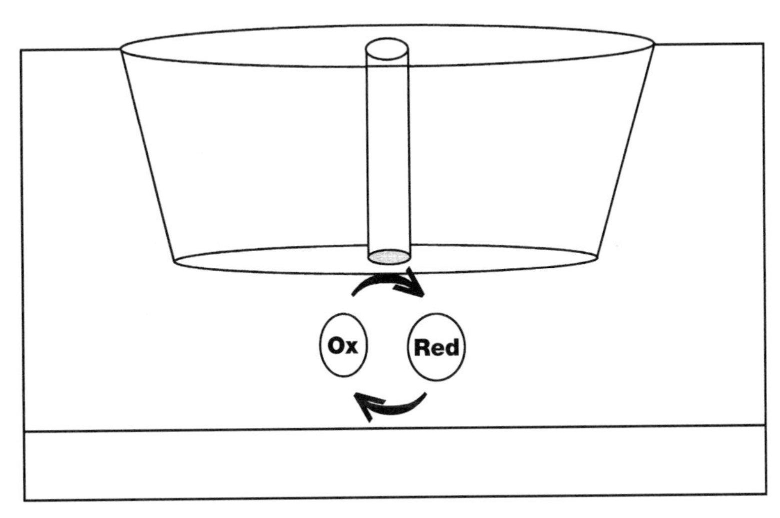

FIG. 3 Scheme of redox recycling in the feedback mode enabling very high rates of mass transport at small distances.

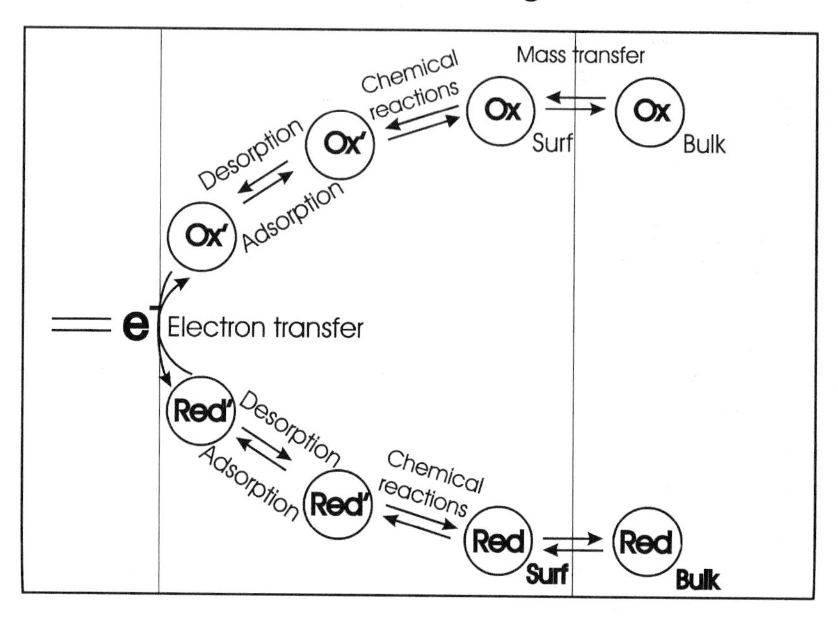

FIG. 4 Pathway of general electrode reaction. (Adapted from Ref. 9.)

reduced to Red by a transfer of n electrons. On the electrode area ($r \leq a$) of a geometry shown in Figure 3, the faradaic current density j is

$$j(t, r) = nF[k_f c_O(t, r) - k_b c_R(t, r)] \tag{3}$$

where F is the Faraday constant and C_O and C_R represent the concentration of the redoxactive species. The rate constants k_f and k_b are given by the well-known Butler-Volmer relations:

$$k_f = k^0 \exp[-\alpha nf(E - E^{0'})] \tag{4a}$$

$$k_b = k^0 \exp[(1 - \alpha)f(E - E^{0'})] \tag{4b}$$

where k^0 is the standard rate constant, α is the transfer coefficient, E is the electrode potential, $E^{0'}$ is the formal standard potential, $f = F/RT$, R is the gas constant, and T is the temperature.

Far above any sample the current of a disk-shaped tip electrode in laboratory units results as

$$i_{T,\infty} = 4nFDca \tag{5}$$

The value of $i_{T,\infty}$ gives the normalizing factor for currents: $I = i/i_{T,\infty}$. The general solution of equations involving heterogeneous kinetics with respect to a tip above a conducting substrate can be obtained under reasonable boundary conditions in the form of two-dimensional integral equations (see Chapter 5).

1. Transients

The special case of totally irreversible substrate kinetics ($k_{f,S} = 0$) has been simulated by Bard et al. (14). When a potential pulse is applied with diffusion-controlled turnover to a tip electrode positioned above a sample surface, the tip current decreases from initially very high values according to the Cottrell equation. After a distance-dependent time, the turnover remains at a constant steady-state level if the sample can regenerate the consumed species at a high rate of electron transfer. If, however, the sample kinetics are slow, the signal continuously decreases to small values. Such simulated transients for a disk-shaped electrode of a shielding ratio $RG = 10$ and a distance $L = 0.1$ are shown in Figure 2 in Chapter 5 with a series of dimensionless heterogeneous rate constants $K_{b,S}$:

$$K_{b,S} = k_{b,S}a/D \tag{6}$$

The upper and lower dashed curves correspond respectively to the limits of a conductor ($K_{b,S} \rightarrow \infty$) and of an insulator ($K_{b,S} = 0$) Equation (6) gives a quantitative estimate of the effect of tip size and rate constant on the transient. The faster the electron transfer, the smaller the electrode has to be to perform a transient measurement of the rate constant.

2. Steady-State Conditions

It is a characteristic of ultramicroelectrodes that the tip current generally reaches stationary values within very short times after changing any parameter of the experiment. The use of these steady-state values strongly benefits from the applicability of working curves. Figure 5 shows simulated curves of the normalized tip current I_T—still in the case of totally irreversible substrate kinetics—as a function of the sample heterogeneous rate $K_{b,S}$ at various distances L, which were obtained by extrapolating the transients shown above to large normalized times ($T \rightarrow \infty$). These theoretical working curves provide a direct means of extracting the heterogeneous rate constant of sample reactions from the tip current and the tip-sample distance. Especially when using small electrodes, the exact correspondence of the real tip geometry and the one applied in the model is of crucial importance. The range of measurable rate constants is a function of the dimensionless tip-substrate

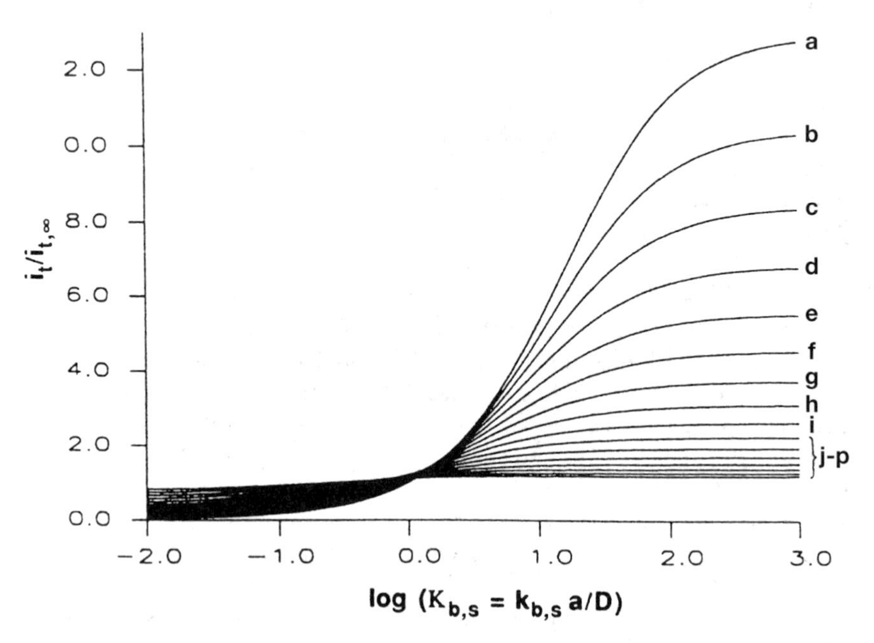

FIG. 5 Working curves of IT vs. log $K_{b,S}$ log $L = -1.2, -1.1, -1.0, \ldots, 0.2$ (a-p). (From Ref. 14. Copyright 1992 American Chemical Society.)

separation and can be determined from the kinetic zone diagram, as shown in Figure 6.

Table 1 contains the theoretical results for SECM with a quasi-reversible electron transfer reaction at the substrate (sample) under steady-state conditions. It contains values of the dimensionless steady-state current as a function of the dimensionless rate constant $\Lambda_S = k_{b,S}\, a/D$ and the dimensionless tip-sample distance $L = d/a$. Each row in Table 1 corresponds to a given value of the normalized tip-sample distance. Seven groups of columns (three columns in each group) correspond to various values of Λ_S spanning the range from virtually reversible ($\Lambda_S = 25$) to irreversible kinetics ($\Lambda_S = 0.001$). In the calculations α was assumed to be 0.5. Three contrasting values of the dimensionless substrate potential $E_1 = (E_S - E^{0'})nf$ were considered for each L. These correspond to the formal potential, shown in Figure 7, to a potential where the substrate feedback is almost diffusion controlled, and to a potential between the two. Table 1 indicates that an increase in overpotential (while keeping all other parameters constant) leads to a higher feedback current. At sufficiently high overpotentials the substrate behaves as a conductor with diffusion-controlled feedback, i.e., with L, I_T decreases monotonically to unity. At $E_1 = 0$ and small Λ_S, the rate of the feedback

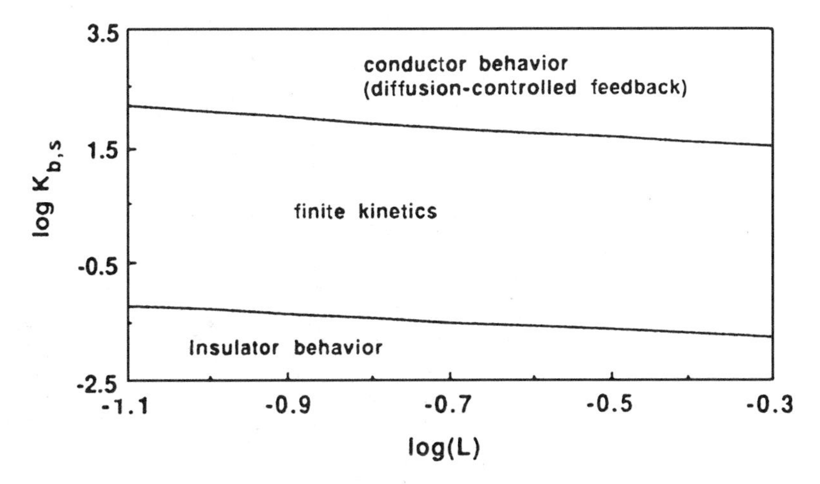

FIG. 6 Kinetic zone diagram illustrating the regions of finite irreversible kinetics, diffusion-controlled feedback, and insulating substrate behavior. (From Ref. 14. Copyright 1992 American Chemical Society.)

process is negligible. However, at $E_1 = 0$ and large Λ_S, a significant positive feedback is observed with a small L. The working curves, I_T vs. Λ_S were calculated for a few values of L and E_S and are shown in Figure 8. They reveal that a significant difference in tip current of about 10%, corresponding to $\Lambda_S = 25$ and $\Lambda_S = 50$, can be detected at close tip-substrate separations ($L = 0.1$).

In cases of complex sample behavior it is advisable to vary the sample potential at slow rates and to monitor the tip current. This method was named T/S (tip/substrate) cyclic voltammetry.

The dependence of near-steady-state voltammograms with finite tip kinetics is similar to that shown for a microdisk electrode (7,8). However, because of a higher mass-transfer coefficient in the case of a positive feedback, the half-wave potential $E_{1/2}$ becomes more sensitive to the charge-transfer rate. When the tip is at small distances, the mass transfer becomes accelerated due to the feedback effect evoked by the substrate electrode. Thus, restrictions in current are increasingly governed by tip kinetics, and the dimensionless parameter Λ_T characterizing the heterogeneous transfer kinetics changes from ak^0/D to dk^0/D. Studies of rapid electron transfer kinetics benefit by the high mass transfer rates at small tip-sample separation. Manufacturing microelectrodes of the same size represents a significantly higher obstacle.

The feedback response discussed in the preceding sections assumes a substrate that is much larger than the tip. Bard et al. described the effects

TABLE 1 SECM of Normalized Steady-State Current Computed for Various Values of the Kinetic Parameter, Λ, Normalized Tip-Substrate Distance, L, and Dimensionless Substrate Potential, E_1[a]

	$\Lambda = 25$			$\Lambda = 5$			$\Lambda = 1$			$\Lambda = 0.5$			$\Lambda = 0.1$			$\Lambda = 0.05$			$\Lambda = 0.001$		
	E_1			E_1			E_1			E_1			E_1			E_1			E_1		
L	0.0	0.585	2.926	0.0	1.171	4.682	0.0	2.341	7.803	0.0	3.902	9.754	0.0	7.803	11.71	0.0	7.803	13.66	0.0	15.61	21.4
0.1	3.71	4.79	7.66	2.40	3.84	7.38	1.00	2.42	7.39	0.68	2.62	7.67	0.21	3.25	6.95	0.17	2.07	7.31	0.11	2.04	7.30
0.2	2.12	2.74	4.20	1.66	2.62	4.22	0.93	2.05	4.24	0.72	2.22	4032	0.24	2.59	4.10	0.19	1.89	4.22	0.14	1.86	4.21
0.5	1.16	1.48	2.20	1.07	1.62	2.26	0.84	1.61	2.28	0.69	1.72	2.29	0.49	1.87	2.25	0.43	1.60	2.27	0.36	1.60	2.27
0.8	0.89	1.14	1.68	0.85	1.30	1.74	0.78	1.39	1.76	0.68	1.49	1.76	0.60	1.58	1.75	0.57	1.44	1.76	0.50	1.44	1.76
1.0	0.81	1.03	1.51	0.79	1.18	1.57	0.72	1.31	1.58	0.67	1.40	1.58	0.61	1.47	1.57	0.60	1.37	1.58	0.55	1.37	1.58
1.5	0.70	0.89	1.32	0.69	1.03	1.35	0.68	1.18	1.36	0.67	1.27	1.36	0.66	1.31	1.35	0.66	1.26	1.36	0.67	1.26	1.36
2.0	0.66	0.82	1.21	0.65	0.96	1.24	0.66	1.12	1.25	0.66	1.19	1.25	0.70	1.22	1.25	0.73	1.20	1.285	0.78	1.20	1.25
5.0	0.63	0.78	1.05	0.64	0.90	1.08	0.65	1.01	1.08	0.66	1.06	1.08	0.74	1.07	1.08	0.79	1.07	1.08	0.92	1.07	1.08

[a] $L = d/a$, $E_1 = (E_S - E^{0'})nF/RT$, $\Lambda = ak_S^0/D$, and $\alpha = 0.5$ for all cases.

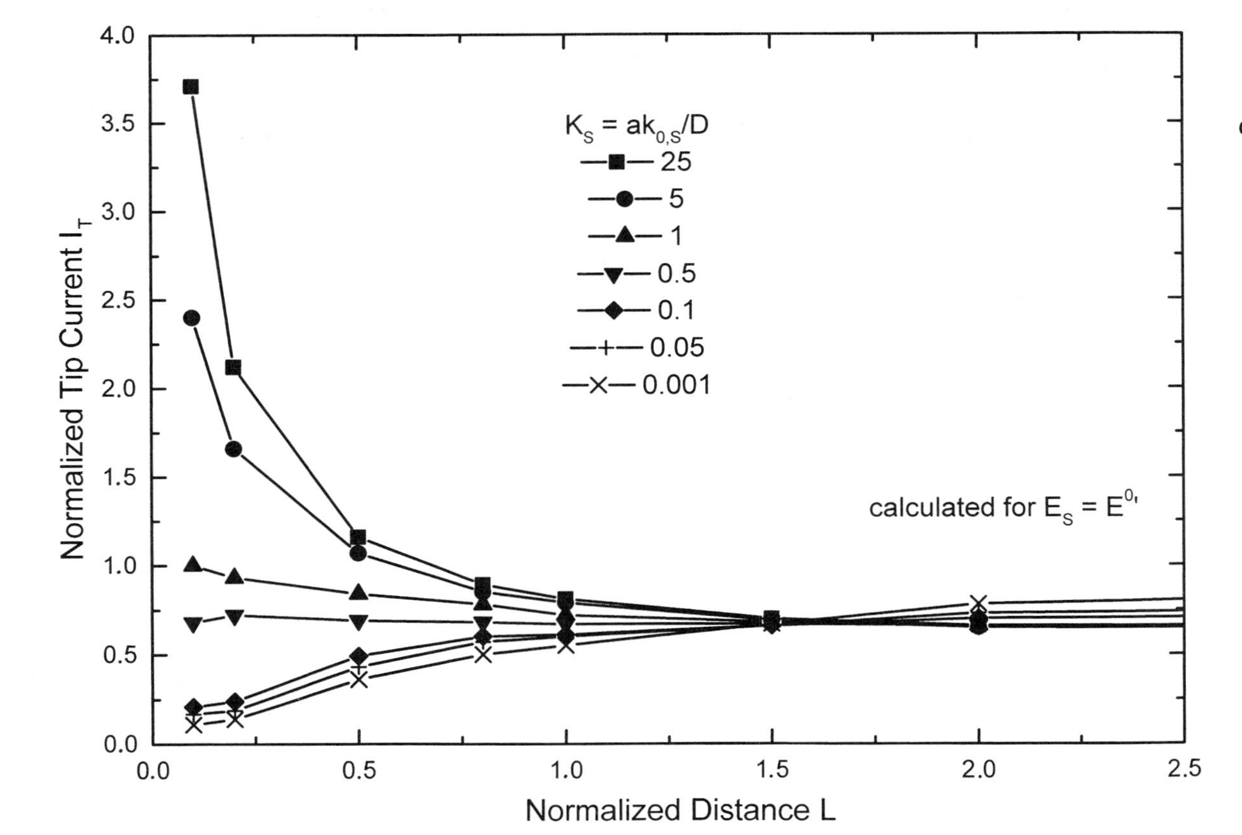

FIG. 7 Current-distance dependence simulated at different normalized rate constant k_S. The sample electrode potential E_S was held constant at the formal potential $E^{0'}$. (Data from Ref. 14.)

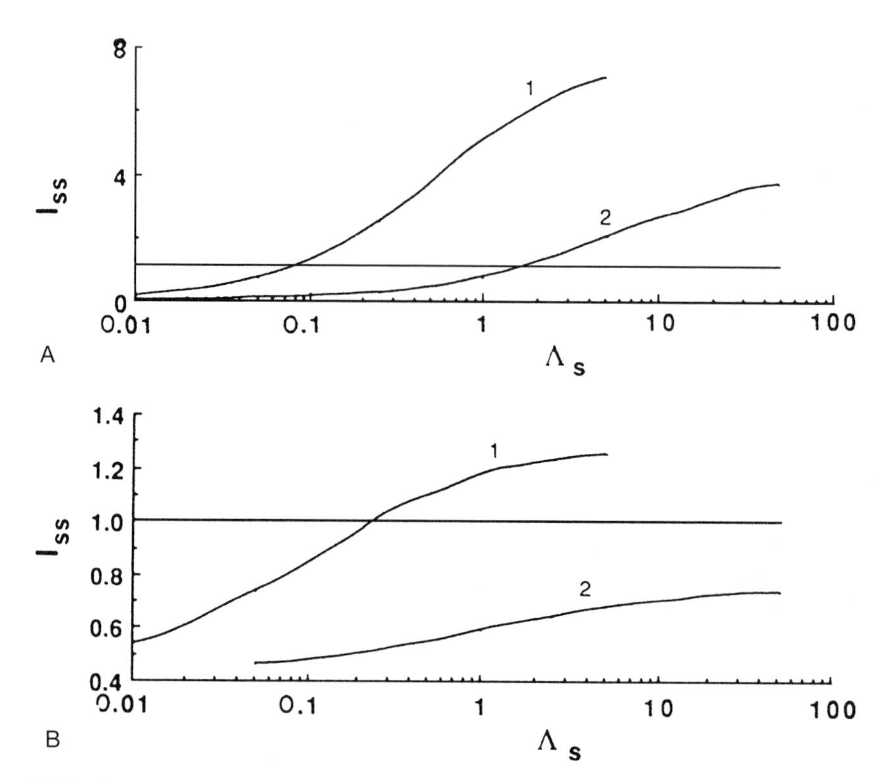

FIG. 8 Working curves of I_T vs. Λ_S, $\alpha = 0.5$, and $L = 0.1$ (A) and 1 (B). Curve 1, $nf(E_1 - E^{0'}) = 5.87$; curve 2, $nf(E_1 - E^{0'}) = 5.87$. The horizontal line corresponds to zero feedback current. (From Ref. 14. Copyright 1992 American Chemical Society.)

of finite substrate size in terms of a dimensionless radius $h = a_s/a$ (where a_s is the substrate radius) (14). When h is large the behavior approximates that for an infinite substrate, and as it gets smaller, the behavior tends towards an insulating substrate. Empirically, when $h = h^\infty$, given by

$$h^\infty = 1 + 1.5L \tag{7}$$

the substrate can be considered infinite. At small distances there is already a substantial difference between the steady-state feedback current for the inert substrate and that of a finite-sized active substrate with radius $h = 0.1$.

Since the lowest experimentally achievable distance is somewhat smaller than one tenth of a radius, the authors conclude that it should be possible to identify objects 10–20 times smaller than the tip.

3. *Diffusion-Controlled Processes*

At fast heterogeneous kinetics at both electrodes, the dimensionless tip current I_T follows an analytical expression that is accurate to 0.7% over a distance interval $0.05 \leq L \leq 20$ (15):

$$I_T^c(L) = \frac{i_T}{i_{T,\infty}} = 0.68 + 0.78377/L + 0.3315 \exp(-1.0672/L) \tag{8}$$

On the other hand, to describe infinitely slow sample kinetics and an *RG* of 10, the following equation has been derived, which is accurate to within 0.5% over the same interval:

$$I_T^{ins}(L) = \frac{i_T}{i_{T,\infty}} = 1/\{0.15 + 1.5385/L + 0.58 \exp(-1.14/L) + 0.0908 \exp[(L - 6.3)/1.017L]\} \tag{9}$$

The fitting of experimental current-distance curves to one of the above equations allows one to determine the position of the sample surface at $L = 0$ in the scanner's coordinate system. The current-distance curve with the negative feedback strongly depends on the shielding ratio *RG* of the electrode, but the positive feedback effect is not noticeably influenced by the *RG* value unless these values are considerably below 10 (see Chapter 5). In practice, the value of $i_{T,\infty}$ is determined experimentally to ensure the curves are, like Eqs. (8) and (9), free of fitting parameters like diffusion coefficients. Essentially, the quality of any extracted parameters obtained from fitting procedures relies on the correspondence of the real tip geometry and the modeled one. Especially, nm-size electrodes may deviate from the ideal symmetrical flat-ended geometry. The recession of the disk electrode from the apex of the insulator sheath, which may result from certain preparation procedures, has been introduced in a modified simulation (71).

From Eqs. (8) and (9) the shape of distant-dependent voltammograms under Nernstian conditions can be easily derived, as described extensively in Ref. 16, by dividing with the potential-depending factor θ defined above:

$$i(E, L) = \frac{i_T(L)}{\theta} \tag{10}$$

4. *Quasi-reversible Processes*

In general, finite kinetics at both tip and sample electrode can be determined using the SECM. However, it is generally more popular to investigate sample kinetics that limit the mediator turnover and to work with fast tip kinetics. The different basic concepts of studying tip and sample kinetics will be treated separately.

5. *Finite Heterogeneous Tip Kinetics*

Assuming a uniform accessibility of the tip surface, e.g., a uniform concentration of electroactive species, an analytical approximation of the tip feedback current can be derived (see Chapter 5). For convenience, we repeat the main equations here. Such a model represents a thin layer cell (TLC) with a diffusion-limiting current expressed by Eq. (8). The approximate equation for a quasi-reversible steady-state voltammogram is as follows (11):

$$I_T(L) = \frac{0.68 + 0.78377/L + 0.3315\exp(-1.0672/L)}{\theta + 1/\kappa} \tag{11}$$

with the kinetic parameter κ and the mass-transfer coefficient m_0:

$$\kappa = k^0 \exp - \frac{\alpha nf(E - E^{0'})}{m_0} \tag{12}$$

$$m_0 = 4D_O \frac{0.68 + 0.78377/L + 0.3315\exp(-1.0672/L)}{\pi a} = \frac{I_T(L)}{\pi a^2 nFc} \tag{13}$$

Equation (11) becomes exact for fast kinetics ($\kappa \to \infty$: Eq. (10) $\to$ Eq. (11)). Moreover, errors are negligible at $L \ll 1$, since this situation corresponds to the TLC. Based on the ideas of Zoski and Oldham (10), a more precise approximation was derived by Mirkin et al. (72) (for details see Chapter 5).

$$I_T(E, L) = \frac{0.78377}{L(\theta + 1/\kappa_{TLC})} + \frac{0.68 + 0.3315\exp(-1.0672/L)}{\theta\left(1 + \dfrac{\pi}{\kappa'\theta}\dfrac{2\kappa'\theta + 3\pi}{4\kappa'\theta + 3\pi^2}\right)} \tag{14}$$

where

$$\theta = 1 + \frac{D_O}{D_R}\exp\frac{nF}{RT}(E - E^{0'}) \tag{15}$$

$$\kappa_{TLC} = \frac{2}{\pi}\frac{d}{a}\kappa'\frac{D_O + D_R}{D_R} \tag{16}$$

$$\kappa' = k^0\frac{\pi a}{4D_O}\exp\frac{\alpha nF}{RT}(E - E^{0'}) \tag{17}$$

6. *Finite Heterogeneous Sample Kinetics*

The rate of an irreversible heterogeneous reaction at the sample can be found by fitting an experimental current-distance curve to Eqs. (18) and (19) (17):

$$I_T^{kin}(L) = I_S - 1 - \frac{I_T^{ins}}{I_T^c} + I_T^{ins} \tag{18}$$

$$I_S(L) = 0.78377/L(1 + 1/\Lambda) + \frac{[0.68 + 0.3315\exp(-1.0672/L)]}{1 + \frac{11/\Lambda + 7.3}{\Lambda(110 - 40L)}} \tag{19}$$

where $\Lambda = k_f d/D_R$. This equation is accurate to within 1–2% in the range of $0.1 \leq L \leq 1.5$ and $-2 \leq \log \kappa \leq 3$.

7. SECM of Partially Blocked Samples

The theory described above has been developed for sample surfaces that are uniform over areas of the tip size. However, some substrates, such as self-assembled monolayers (SAM), reveal even smaller details that cannot be resolved by SECM, but for which a mean effective rate constant k_{eff} can be determined. The theory has been developed for a blocking film with disk-shaped defects (18) for the kinetically-controlled regime (20) and irreversible kinetics (21) and is based on an effective medium approach of Szabo et al. (19):

$$k_{\text{eff}} = \frac{4(1 - \theta)D}{\pi R_d} \tag{20}$$

$$k_{\text{eff}} = \frac{4(1 - \theta)D}{\pi R_d\left(1 + \frac{\pi}{\kappa}\frac{2\kappa + 3\pi}{4\kappa + 2\pi^2}\right)} \tag{21}$$

So far, in reported SECM studies the rate constant and the transfer coefficient have been sufficient to describe the experimental results. However, the ohmic resistance of solids may lead to potential gradients inside bulk materials, which have not been dealt with up to now. This effect has to be taken into account, especially at poorly conducting semiconductors and thin films on insulators.

B. Reaction Rate Imaging

For surface analysis all SECM modes—the feedback, generation/collection, and direct modes—have been used to visualize lateral differences in heterogeneous electron transfer properties. At this point, we shall briefly review a few aspects of surface imaging concerning the acquisition and interpretation of data that are relevant for most experiments before presenting individual studies. Details of SECM imaging are discussed in Chapter 4.

In the feedback mode, tips with a large glass shielding are preferable as they heighten the contrast in current between active and inactive parts of the surface. On the other hand, there is a greater danger of mechanical contact with the surface, which may damage both the tip and the sample. The compromise between these two factors will depend on the size of the electrode and the sample roughness.

Moreover, the response may be complicated by the interplay of topographic and kinetic effects. Several approaches to solving this problem have been reported that use different experimental techniques to realize a constant distance mode (20–23). The vertical positions recorded during scanning yield a topographic image, whereas the current can be related to local kinetics. In the feedback mode with very small tip-sample separations ($d \ll a$), the lateral resolution corresponds to the tip diameter. With increasing distance, the apparent size of those objects, which reveal a positive feedback and have an electrochemically inactive vicinity, appears to grow in SECM images. The authors' group determined experimentally that, independent of the mediator's nature, the size s obtained is larger than the real size s_0 by a distance-dependent term (24).

$$s = s_0 + 1.64d \tag{22}$$

At small distances, the area of the sample that provides the current signal corresponds to that of the tip, and, therefore, the microelectrode characteristics of the tip are transferred to the fractions of the sample. At larger distances, the area is slightly enlarged because of the diffusion occurring parallel to the surface and may in practice be twice the tip area. The equation above is of the same kind as Eq. (7) but uses slightly different variables and is based on experimental data. A way to compensate for image distortions resulting from sample surface roughness has been reported by Ellis et al. (25).

Diffusion is also responsible for a second phenomenon. Even on samples that provide abrupt lateral transitions between different properties, the SECM response changes smoothly, producing apparently blurred images. This means that the current or the rate constant measured at a certain location is influenced by the surface properties in the vicinity. Several attempts were made to compensate for this undesired effect. Numerically, the diffusional broadening was successfully deconvoluted by deblurring the image using a Laplacian of a Gaussian (LOG) distribution (26). Since the choice of appropriate filter parameters could not be related to experimental conditions, this method has not become common practice. Alternatively, the diffusional broadening can be avoided in principle by introducing a redoxinactive scavenger (12,21) as shown in Figure 9. This technique, developed by the authors' group and called a chemical lens, has been successfully applied in surface modification (27) and analysis (28), with a resolution down to one third of the tip diameter. The interacting sample area has been sharply delimited. However, up to now there has not been any suitable theoretical

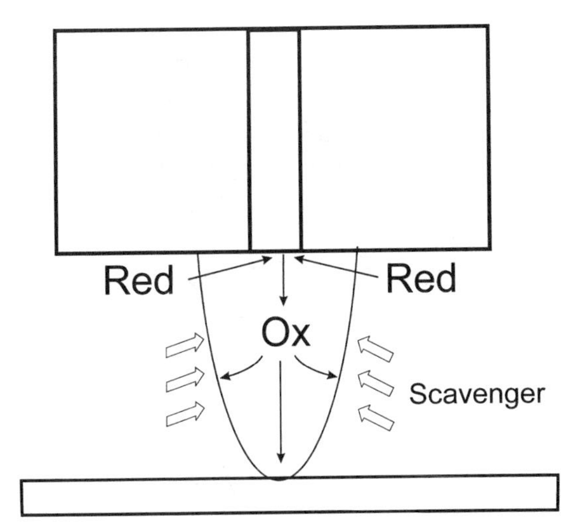

FIG. 9 Scheme of the chemical lens illustrating the focussing effect that is caused by the introduction of a redox inactive scavenger reacting with the tip-generated species Ox. The resolution achievable in practice is one third of the tip diameter.

simulation of the quantitative interpretation in terms of kinetic parameters. The concept of introducing a redox active scavenger instead (29) reduces diffusional broadening, but since the current is dominated by the reaction with the scavenger, this method will presumably be restricted to surface modification. At the beginning of SECM development, Engström et al. showed that lateral resolution can be improved by generating an unstable form of the mediator at the tip (30,31).

III. STUDIES OF HETEROGENEOUS ELECTRON TRANSFER

A. Heterogeneous Kinetics at the Tip and the Substrate

Historically, lateral variations in heterogeneous rate constants that are related to conductivity were the first properties to yield a contrast in SECM imaging. Due to their easy accessibility and defined shape, interdigitated structures (IDS), which consist of alternating stripes of an electrical conductor and an insulator, were frequently used as a model system for imaging studies, as shown in Figure 10 (13,30,32). Above conductors, higher currents are detected, whereas insulators yield lower currents. Using a chromium coating on the gold bars of an IDS, Bard et al. were able to switch from positive feedback above the oxide-free coating to negative above the oxide-covered

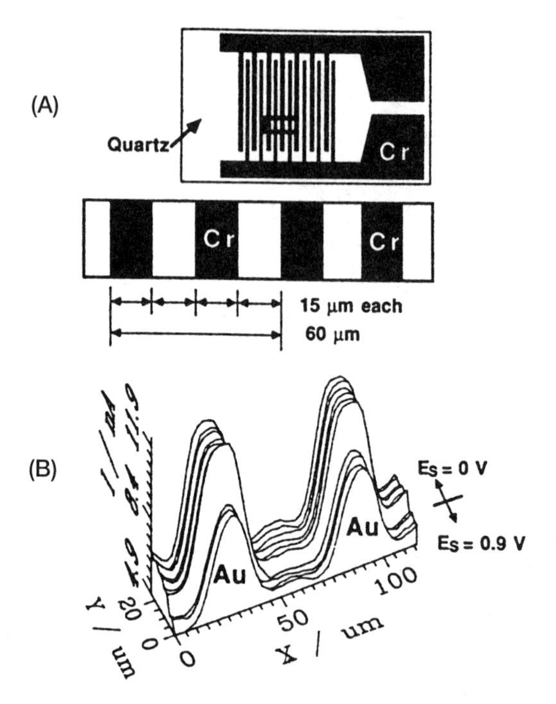

FIG. 10 (A) Schematic structure of an IDA (upper) and the enlargement of the boxed area (lower). (B) Scan of poly(vinylferrocene)-modified Au/Cr IDA electrodes (Au, 60 μm periodicity) in 8 mM $Ru(bipy)_3(ClO_4)_3$, 100 mM tetrabutylammonium phosphate/acetonitrile, $E_T = +1.5$ V vs. AgQRE. E_S was changed from +0.9 V (front) to 0 V (rear). (From Ref. 33. Copyright 1990 American Chemical Society.)

chromium (33). These findings demonstrate that the key parameter for feedback behavior is the rate of heterogeneous electron transfer; the use of electrical conductors is not sufficient to evoke a positive feedback.

When electron transfer reactions at both the tip and the sample electrode are fast, the overall rate is limited by diffusion only and in accordance to the simulation of the feedback behavior by Kwak and Bard (4). There are two approaches to measurements of heterogeneous electron transfer processes at electrodes by SECM. Processes can be measured at the tip electrode above a conductive substrate by taking advantage of the increased mass transfer to the tip. An example of such a process is the study of the oxidation of ferrocene in acetonitrile at a Pt electrode (72). Voltammograms taken at different values of d are shown in Figure 11. At a large d-value, curve 1, the process is undistinguishable from a nernstian reaction. At smaller d-values, a deviation from reversibility is observed and by fitting the voltam-

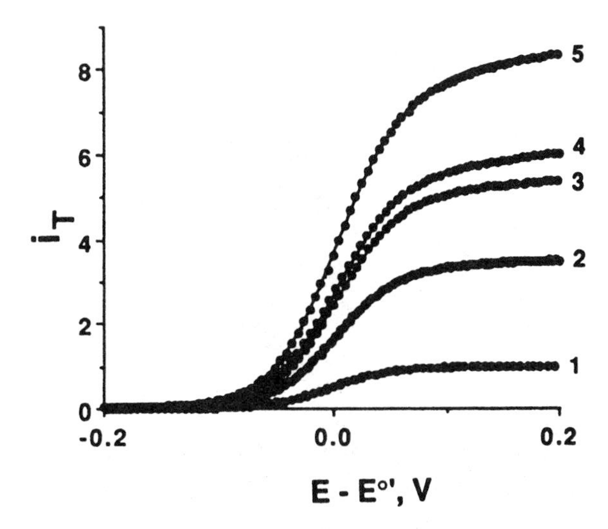

FIG. 11 Tip steady-state voltammograms for the oxidation of 5.8 mM ferrocene in 0.52 M $TBAPF_6$ in acetonitrile at a 2.16 μm diameter Pt tip. Solid lines calculated from Eq. (10). Tip-substrate separation decreases from line 1 (infinity) to line 5 (0.108 μm). (From Ref. 72. Copyright 1993 American Chemical Society.)

mograms, the values of α and k^0 can be obtained. This was carried out for curves 2–5, yielding $\alpha = 0.37 \pm 0.02$ and $k^0 = 3.7 \pm 0.6$ cm/s. SECM was shown by this experiment to allow measurement of a fast heterogeneous electron transfer at steady state and without problems of iR compensation.

A similar approach was used with a mercury pool substrate. In this case the small tip can be pushed into the mercury pool to trap a thin layer of solution between tip and Hg (Fig. 12). This allows very small d-values to be obtained and decreases the danger of breaking the tip with close approach to substrate (73). In this case the value of d can be obtained from the limiting i_T-value, compared to $i_{T,\infty}$. This approach was used to determine the rate constant for the reduction of C_{60} in 1,2-dichlorobenzene (ODCB) and in benzonitrile (PhCN) solution (74). Typical voltammograms are shown in Figure 13. Fitting these yielded k^0-values of 0.46 (ODCB) and 0.12 (PhCN) cm/s. Details about the procedure used for curve fitting by the voltammetric approach can be found in Refs. 11, 72, and 74.

The alternative approach to examine heterogeneous electron transfer kinetics on the substrate is to hold the tip at a potential where the reaction is mass transfer controlled and study the approach curve as a function of substrate potential. For example, one can generate iron(II) by reduction of iron(III) at the tip and study the oxidation of iron(II) at the substrate. One

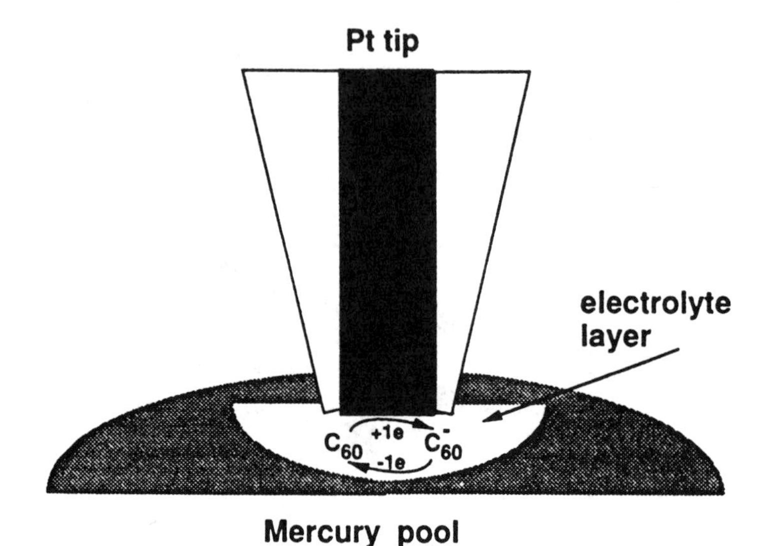

FIG. 12 Schematic representation of the TLC formed inside the Hg pool. C_{60} is reduced at the Pt tip to produce C_{60}^{-}, which is reoxidized at the Hg anode. The solution layer is shown greatly enlarged for clarity; the actual thickness is smaller than the electrode radius a. (From Ref. 74. Copyright 1993 American Chemical Society.)

can then obtain k from the fit of the experimental curve to the theoretical ones (e.g., Figs. 5 and 6 in Chapter 1) and then determine k^0 and α from a plot of log k vs. potential. This method was used in a study of iron(II) oxidation on carbon (45) and, more recently, in a study of hydrogen oxidation on platinum (75).

B. SECM Studies of Various Materials

1. Carbon Electrodes

An easily assessable and well-researched model system—the glassy carbon electrode—is frequently used for studies of heterogeneous electron transfer with quasi-reversible and irreversible kinetics (34). Moreover, carbon particles can be spray-coated on electrode surfaces to modify its properties, and carbon fibers have been used as microelectrodes of defined diameter.

Wipf and Bard used glassy carbon to study the redox behavior of $Fe^{2+/3+}$ in 1 M H_2SO_4. While the carbon fiber tip enabled a diffusion-limiting reduction of Fe^{3+}, the C substrate oxidized Fe^{2+} at a finite rate (35). This system provided a direct method of varying the driving force for the sample

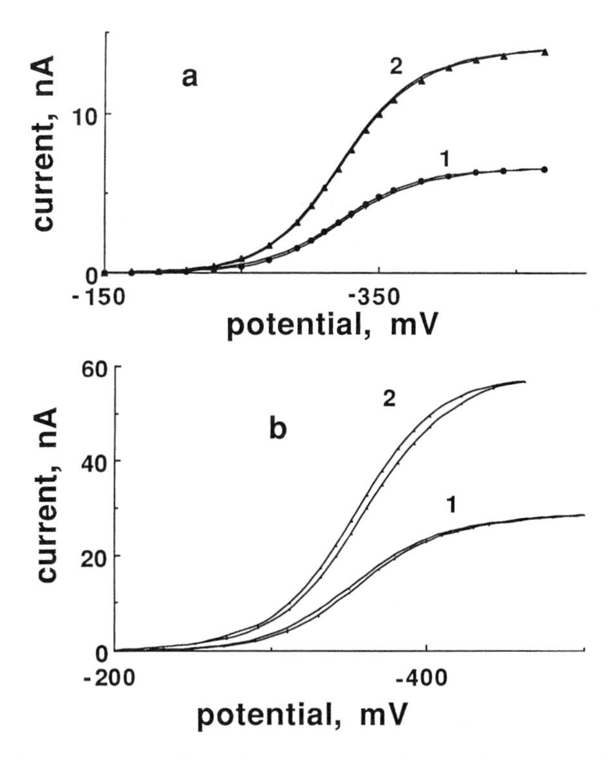

FIG. 13 Steady-state cyclic voltammograms of the solution containing 5 mM $Ru(NH_3)_6Cl_3$ in 0.2 M KNO_3 at a carbon (a) and platinum (b) UME: 1, with the tip far from the substrate; 2, voltammograms obtained with tip within Hg, i.e., in thin layer cell configuration. The potential scan rate was 5 mV/s (a) and 2 mV/s (b). Circles and triangles are calculated as explained in Ref. 73 with parameter values: $k^0 = 0.11$ cm/s, $\alpha = 0.40$, $E^{0'} = -319$ mV (circles); $k^0 = 0.15$ cm/s, $\alpha = 0.44$, $E^{0'} = -318$ mV (triangles). (From Ref. 73.)

reaction by altering the potential. Current-distance curves are given in Figure 14. The modeling of that experiment was reported in a later publication (14). In addition, ruthenium hexaamine has been used as a mediator providing quasi-reversible kinetics. By fitting a current-distance curve a rate k^0 of 0.076 cm/s was obtained under the assumption of $\alpha = 0.5$. The different electrochemical behavior of glassy carbon and gold with regard to $Fe^{2+/3+}$ resulted in different current responses above the respective areas. Therefore, Wipf and Bard could image the chemical composition on the surface of a composite electrode and visualized the active gold spots (36).

The SECM has been applied to image holes and humps obtained on a gold surface spray-coated with carbon (37,38). Compared to the gold, the

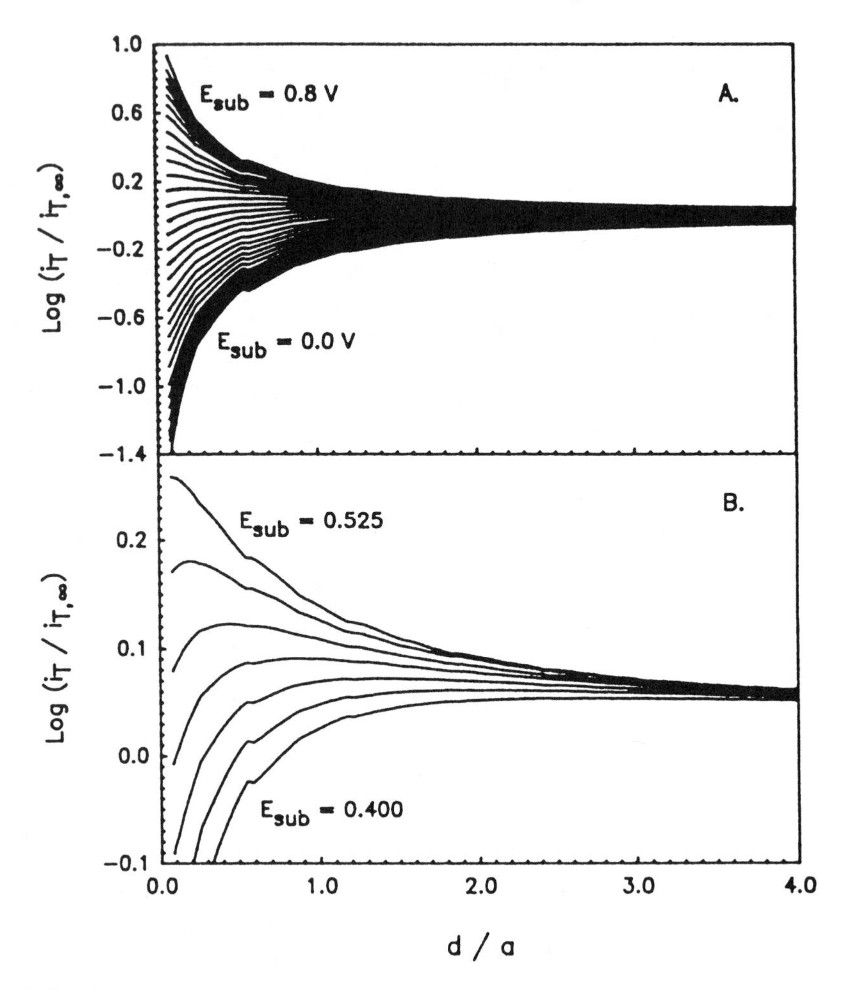

FIG. 14 Logarithmic current-distance curves for a 2.1 mM Fe(III) solution in aq. 1 M H_2SO_4. Tip electrode is a 11 μm carbon fiber at a potential of −0.6 V; substrate is glassy-carbon. (A) Curves for different substrate potentials at 25 mV increments between 0.8 V and 0 V. (B) Region of 0.525 V and 0.4 V. (From Ref. 35 by permission of The Electrochemical Society, Inc.)

ferrocenemonocarboxylic acid (FMCA) showed sluggish electron transfer on the carbon. Subsequent tempering of the structures increased the density of electrochemically reactive carbon particles, as shown in Figure 15. In an early work on distance-dependent resolution of imaging of mixed insulators

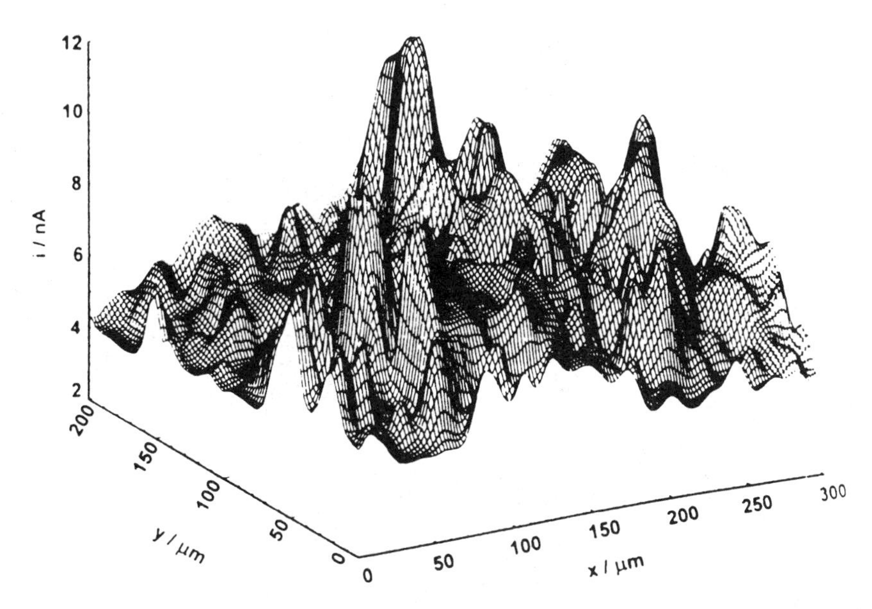

FIG. 15 Gold on silicon wafer completely coated with carbon spray. Tempered 6 hours at 453 K in air; SECM image using a 10 μm Pt tip at a speed of 7 μm/s in aq. 2 mM ferrocene monocarboxylic acid, 100 mM phosphate buffer at pH 7.2 and 100 mM KCl containing 10% ethanol. (From Ref. 37.)

and conductors, a 50 μm Pt wire was fixed on the surface of an insulating sample and imaged by a 5.5 μm carbon tip (39).

2. *Metal Oxides and Semiconductors*

Electron transfer to metal and semiconductor surfaces may in principle lead to chemical reactions such as dissolution. This chapter focuses on electron transfer reactions; aspects of surface modification are dealt with in Chapter 13.

Denuault et al. characterized pH-dependent and pH-modifying processes on polycrystalline platinum electrodes in aqueous sodium sulfate solutions at pH 4 and 7 to obtain a deeper insight into the mechanisms involved by introducing a pH sensor above the sample's surface (40,41). They monitored hydrogen evolution, Pt oxide formation, oxygen evolution, and hydrogen adsorption and desorption at different sample potentials. The pH was observed by measuring protons amperometrically as shown in Figure 16. In this way, a pH shift of 2.3 units was detected at the tip during desorption of hydrogen on the underlying platinum sample. The precision achieved allowed new conclusions on mechanistic details.

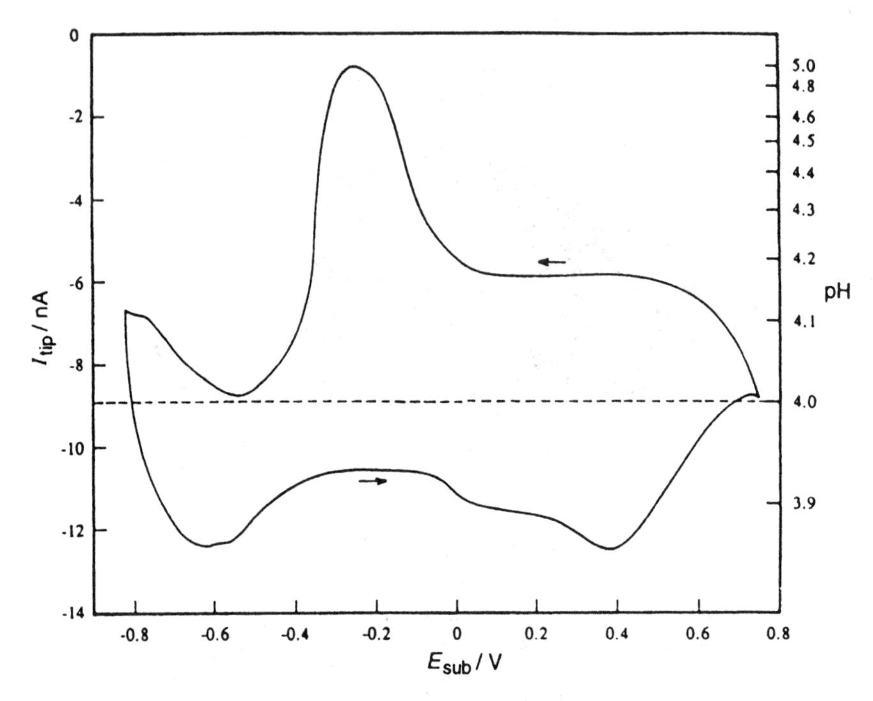

FIG. 16 Tip-substrate voltammogram in aq. 10^{-4} M H_2SO_4, 500 mM Na_2SO_4. Potential sweep rate v_S = 50 mV/s, 500 μm Pt sample and 25 μm Pt tip at E_T = -1.2 V, all potentials refer to Hg/Hg_2SO_4 in sat. K_2SO_4. (From Ref. 40. Reproduced by permission of The Royal Society of Chemistry.)

As part of an interdigitated array, Bard et al. switched layers of chromium between an oxide-covered surface revealing negative feedback and pure metal showing positive feedback (33). As noted above, these findings demonstrated that the key parameter for feedback behavior is the rate of heterogeneous electron transfer and not the electrical conductivity.

White et al. studied electron transfer reactions at a tantalum surface covered by 2.5 nm of native tantalum-(V)-oxide (42). The SECM detected microscopic electroactive sites with a diameter between 4 and 100 μm. Interestingly, some sites turned out to be active only for the reduction of $Ru(NH_3)_6^{3+}$, while others were also capable of oxidizing iodide. The authors studied the kinetics of mediator interaction under various sample potentials and locations in detail to determine parameters relevant for the growth of such tantalum oxide films. An example is given in Figure 17.

The heterogeneous ET reaction of outer-sphere redox couples at the semiconductor electrodes of WSe_2 and n-doped silicon has been reported by

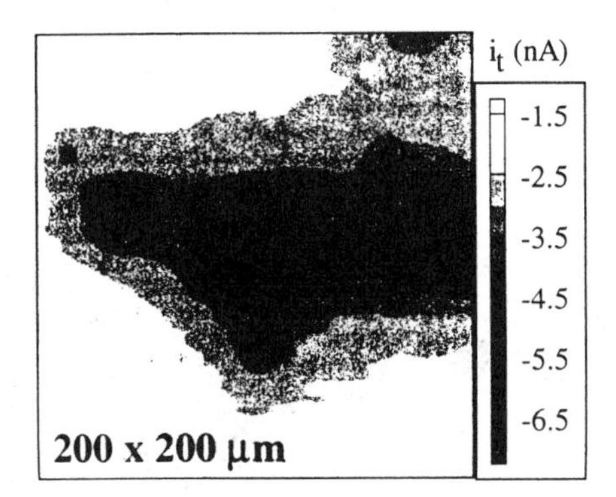

FIG. 17 Redox-active sites on a Ta/Ta_2O_5 electrode. Image recorded in a distance of about 4 μm in 10 mM Kl and 100 mM K_2SO_4. The sample was biased at +1.0 V vs. Ag/AgCl, the tip at 0.0 V. (From Ref. 42. Copyright 1999 American Chemical Society.)

Horrocks et al. (43). Compared to traditional techniques, the usage of the SECM benefited from the absence of ohmic resistance, charging currents, and relative insensitivity to parallel processes such as corrosion. The data obtained were interpreted quantitatively with regard to the transfer coefficient and heterogeneous rate constant. It turned out that reactivity also depends on local properties such as density of steps, cracks, and pits. Obviously, information that can be correlated with the nature of the location is of considerably more value than that from integral techniques averaged over large areas of samples.

3. *Thin Films*

SECM experiments in the field of thin films on electrode surfaces have attracted considerable attraction, since very high lateral resolution in the mid-nanometer scale can be achieved in combination with STM like conically shaped tips. The materials dealt with include self-assembled monolayers (SAM), fullerene derivatives, and inorganic salts.

Forouzan and Bard (18) studied the formation of hexadecanethiol SAMs on gold by SECM in combination with voltammetric techniques. With increasing duration of adsorption, the positive feedback initially observed turned negative. From the finite sample kinetics the authors were able to obtain quantitative data on the adsorption kinetics and fractional coverage. Wittstock et al. (44) locally desorbed a SAM consisting of dodecylthiolate on a gold surface for the generation of enzymatically active patterns using the direct mode. Eventually, as shown in Figure 18, they were able to image the generated pattern by exploiting the differences between heterogeneous

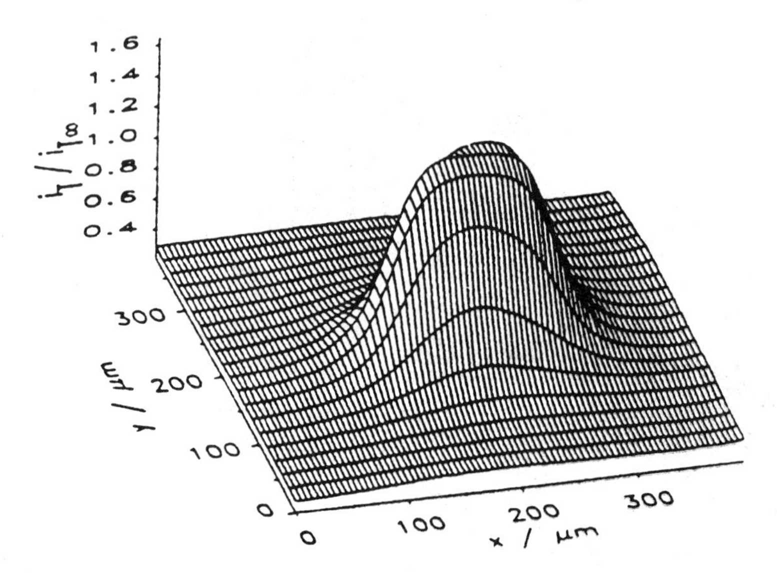

FIG. 18 Electrochemically desorbed spot of dedecylthiolate on gold imaged by a 10 μm Pt tip in 20 mM $K_4[Fe(CN)_6]$ and 100 mM phosphate buffer (pH 7.4) at 10 μm/s. (From Ref. 44.)

kinetics on bare gold surface and the parts, still covered by SAMs. The group has also worked on functionalized SAMs in biological applications.

Films of fullerenes (C_{60} and C_{70}) have shown interesting electrochemical behavior (45,46). Neither the C_{60} nor the completely reduced C_{60}^{-} film is a good electronic conductor, although a partly reduced film displayed enhanced conductivity (47). Recently, the same group reported investigations on a film of *tert*-butylcalix[8]arene-C_{60}. As a consequence of the reduction of the inner fullerene, the latter escaped from the calixarene basket and soluble C_{60}^{-} species entered the acetonitrile solution, as observed in an SECM collection experiment with simultaneous measurements with a quartz crystal microbalance.

The Masuhara group showed that even direct electron transfer at a solid/solid interface can be exploited for imaging by studying grains of Prussian blue and Prussian white in films with submicrometer resolution (48,49). The current of a sharpened tungsten tip penetrating the film was held at a constant value of 0.5 nA by a feedback loop, and the positions of the vertical scanner were recorded, resulting in an image (Fig. 19). This concept resembles that of the electrochemical STM, but the tip current detected does not result from a tunneling process; it is a faradaic current of the four-electron oxidation of Prussian white, $K_4Fe_4^{II}[Fe^{II}(CN)_6]_3$, to Prussian blue, $Fe_4^{III}[Fe^{II}(CN)_6]_3$ (or vice

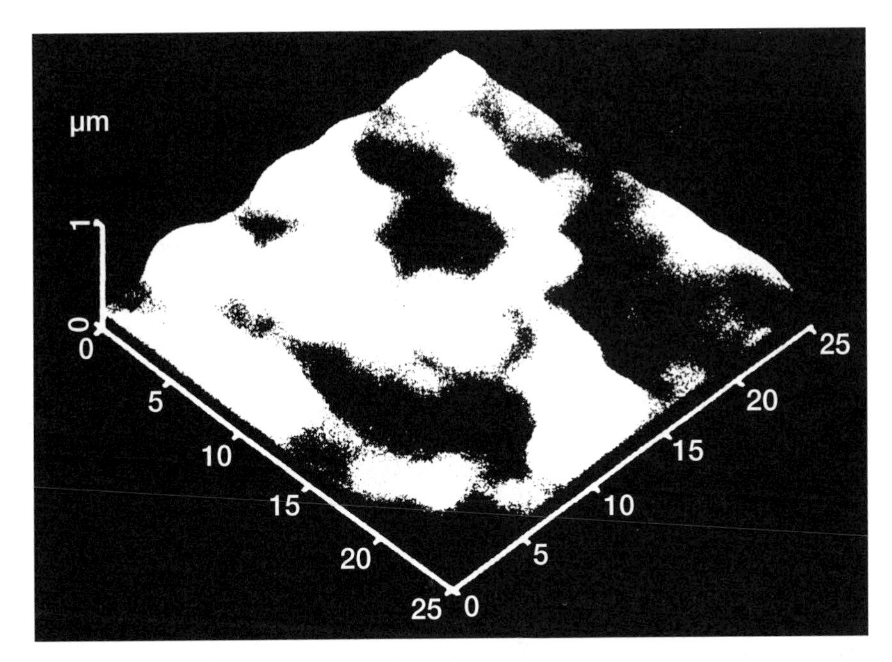

FIG. 19 Using direct electron transfer at the solid/solid interface, an image of the crystal grains in a Prussian white film in 20 mM KCl (pH 4) could be obtained with a penetrating tip. 25 μm $\times$ 25 μm, $E_S = -0.2$ V and $E_T = +0.3$ V, tip current 500 pA. Prussian white is continuously oxidized to Prussian blue. (From Ref. 48, with permission from Elsevier Science.)

versa), and was accompanied by apparent changes in the film composition. Similar experiments were even carried out successfully in air with dried samples at high voltages of +4 V and −8 V, respectively, and low currents of 0.1 nA. The anodic tip current induced the oxidation of the (III/II) to the (III/III) compound, while the reduction to the (II/II) compound occurred at the substrate. The authors state that the electrolysis of coordinated and interstitial water also contributed to the current.

4. Ion Conducting Polymers

Redox active ion-conducting polymers can be obtained by attaching redox centers either electrostatically or covalently to a functionalized polymer (50). The electron transfer is coupled to ingress of ions into and egress of ions from the film to maintain electroneutrality inside. Depending on the nature of the polymer, the charge induced by oxidation has to be compensated either by anions being incorporated or cations being expelled. Opposite reactions are necessary during the reduction process. The main goals concerning this class of material are the study of mechanisms of charge transport, homogeneous and heterogeneous electron transfer, counterion ejection, and thermodynamic parameters.

While moving a conically shaped STM-like tip towards a 200 nm thick Nafion® film loaded with $Os(bpy)_3^{2+}$, Mirkin et al. went through five different situations, nicely characterizing principal tip–sample interactions with ionically conducting polymers (as shown in Chapter 1, Fig. 6) (51). Above the film, redox active species were oxidized at the tip in negligible amounts. While penetrating the film, the current started to increase. Approaching further yielded a positive feedback with the underlying ITO substrate, until, finally, direct tunneling to the substrate set in. Using the results above and a recorded tip CV (Chapter 1, Fig. 7), the authors estimated properties inside the film, namely the diffusion coefficient $D = 1.2 \times 10^{-9}$ cm^2/s, $\infty = 0.52$, and $k^0 = 1.6 \times 10^{-4}$ cm/s, which were in agreement with the results of other experiments. Earlier, the Bard group investigated several different polymer films on electrodes: poly-(vinylferrocene), *N,N′*-bis[3-(trimethoxysilyl)propyl]-4,4′-bipyridinium dibromide, and the above-mentioned loaded Nafion film (33). The feedback behavior depended on the nature of the polymer and the mediator as well as tip and sample potentials. These results nicely illustrate the SECM's capability of distinguishing between chemically different sites. Protonated films of poly(4-vinylpyridine) loaded with highly charged redox active anions like $[Fe(CN)_6]^{3-/4-}$ were studied by Anson and coworkers. As shown in Figure 20, these films evoked a positive feedback at low concentrations of mediator, which gave way to negative feedback when the reactant concentration was increased to the point that electron propagation through the polymer became the current-limiting factor. Additional information was obtained as cathodic currents flowed through these films. Consequently, the anions were expelled and could be monitored by the tip facing the polymer. Such potential pulse experiments were carried out in the G/C mode and revealed a delay of the collection current that could be associated with the time required for the ions to traverse the coating before being ejected (52). Later, adding imaging techniques to voltammetry, this group observed a significant aging effect on the redox conductivity of the polymer loaded with $IrCl_6^{2-}$ and $[Fe(CN)_6]^{3-}$, respectively. Using the SECM, they explained this macroscopic phenomenon in terms of strong lateral inhomogeneities that developed over time and proposed possible structural changes (53).

More recently, Fan et al. worked on poly(vinylferrocene), which contains covalently bound redox centers. Again, they were able to characterize the five distance-dependent stages of tip polymer interaction (54). While scanning inside the film, these samples did not show a plateau, but rather a

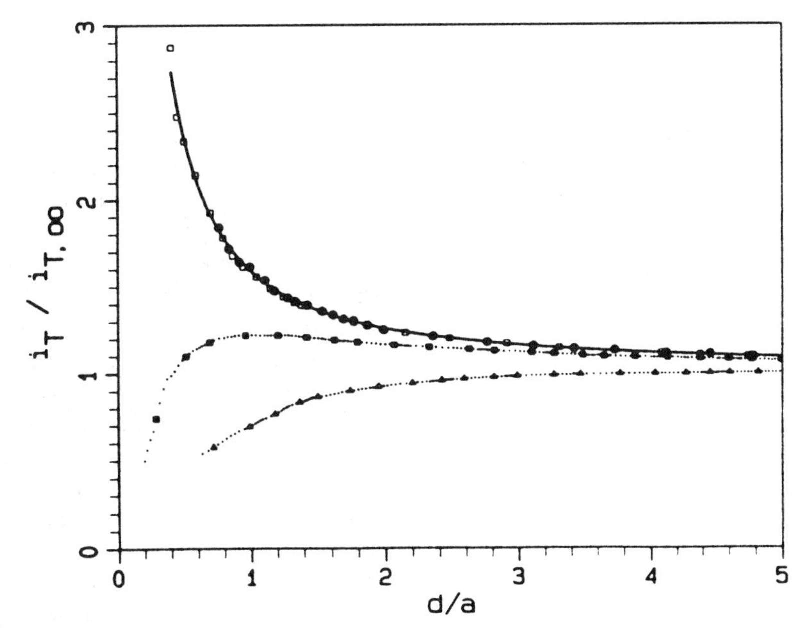

FIG. 20 Normalized steady-state tip current as a carbon tip is removed towards a glassy carbon electrode. (□) Bare substrate in 1 mM $K_3[Fe(CN)_6]$, 100 mM KCl, 5 mM HCl; (●) PVP-coated sample, 0.21 mM $K_3[Fe(CN)_6]$, 100 mM KCl, 10 mM HCl; (■) PVP-coated sample, 1 mM $K_3[Fe(CN)_6]$, 100 mM KCl, 5 mM HCl; (▲) PVP-coated sample, 5 mM $K_3[Fe(CN)_6]$, 100 mM KCl, 5 mM HCl. (From Ref. 52. Copyright 1992 American Chemical Society.)

tilted current curve, due to the high inherent ohmic resistance. The specific resistance of the polymer was found to be material could be calculated from the slope as 3×10^{-6} ohm-cm.

Fan studied the transport properties of three ferrocene derivatives carrying different types of charge inside two gels of polyacrylamide and polyacrylate (55). The latter gel formed a passivation layer by electrophoresis, which behaved as a cation-exchange membrane. Chronoamperometry was applied while penetrating the film and revealed diffusion coefficients slightly below those inside the solution. From this, the author concluded that the transport occurs via water-filled domains.

5. Conducting Polymers

The electronic properties of conducting polymers can be varied greatly by changing their degree of doping (56). Electrochemical or, alternatively, mediated chemical oxidation is used to increase polymer conductivity, which may be up to 10 orders of magnitude greater than in the uncharged polymer. Both aspects of the variation of conductivity and the ion movement related to doping/dedoping processes can be studied by SECM. Scientific interest is directed at potential-induced effects and the imaging of properties, studies with samples of polypyrrole and polyaniline, and derivatives of polythiophene.

Polypyrrole. Bard et al. analyzed the changes of conductivity at a polypyrrole (PPy) film on a platinum carrier as a function of its potential. Undoped films revealed sluggish heterogeneous kinetics in the feedback mode, whereas in the oxidized state fast kinetics were observed. T/S cyclic voltammetry has been developed to record the hysteresis and time-dependence of the effects, i.e., the tip current was measured as a function of the sample potential as shown in Figure 21 (57). Later the group studied the movement of redox active ions like bromide, ferrocyanide, and cationic metal complexes caused by the oxidation and reduction of a film (58). The PPy samples depended most significantly on the nature of the cation, indicating that the cationic conductivity resulted from the chosen preparation procedure. Tóth et al. incorporated micelles of a styrene-methacrylic acid block copolymer, which acted as polyanions, into PPy films (59). As illustrated in Figure 22, during charging/discharging experiments they monitored the flux of potassium potentiometrically and cadmium and ruthenium hexaamine amperometrically. The group of Heinze and Borgwarth demonstrated that the SECM can provide a contrast in images by exploiting differences in chemical properties. They scanned above a blend of polypropylene (PP) particles that had been coated with polypyrrole (24) and pressed into a smooth film. Due to the low bulk content of the conducting polymer of just 1.63%, the kinetics were very slow. To distinguish clearly between regimes of PP and PPy, the experiment shown in Figure 23 had to be carried out at low currents in the picoampere regime.

Polyaniline. The specific property of polyaniline (PAni) of oxidation in two waves followed by deprotonation and protonation in the two low oxidation states results in a complex square scheme of charging/discharging reactions involving five compounds. These processes are of considerable interest since they strongly affect the conductivity and color of PAni films. Denuault et al. studied the reaction pathways as a function of proton concentration and ionic strength by measuring the current for proton reduction, as shown in Figure 24 (60,61). In addition, they monitored the flux of the chloride anions in situ by potentiometry during cyclic voltammetric experiments (62). Compared to other techniques, these investigations using ultramicroelectrodes benefited from avoiding ohmic drops. The group of

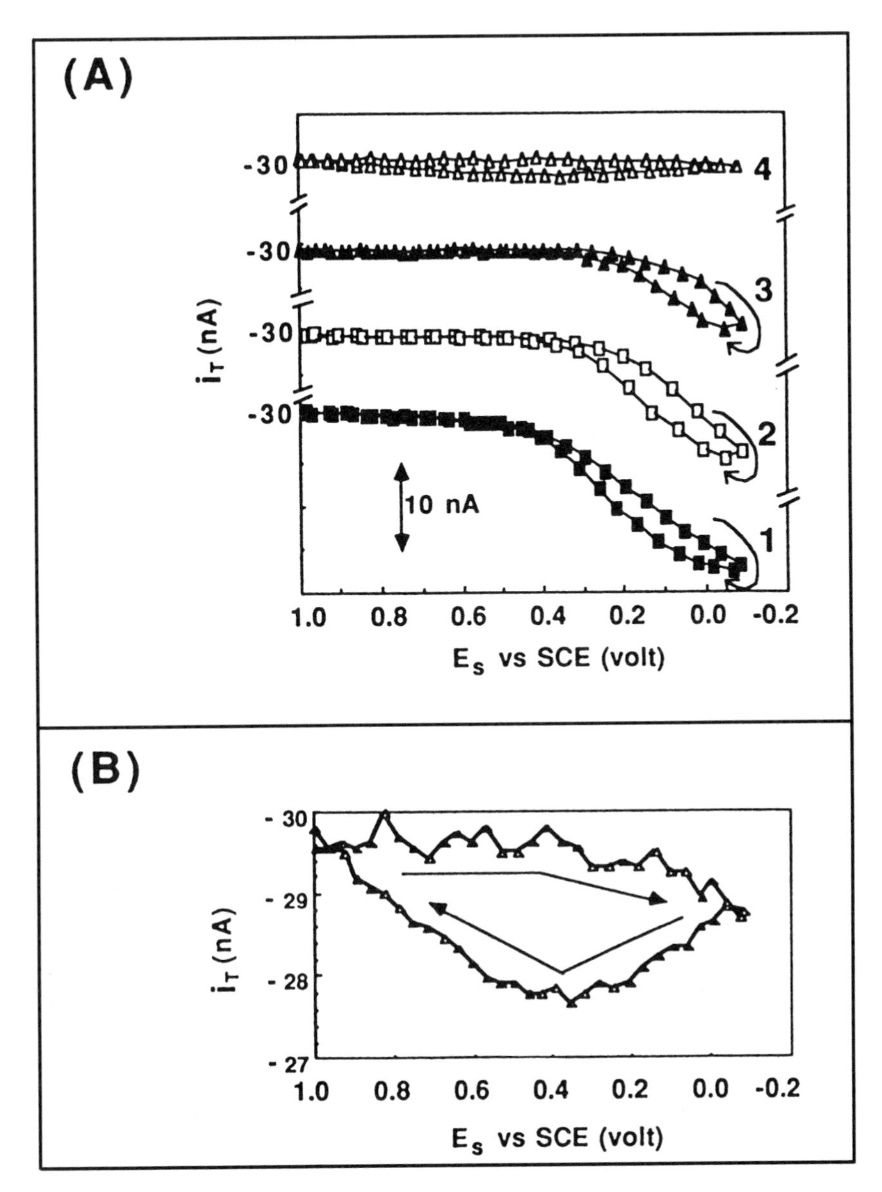

FIG. 21 T/S voltammogram using an 11 μm carbon tip biased at -0.415 V vs. SCE in 10 mM $Ru(NH_3)_6Cl_3$ and 100 mM K_2SO_4 above polypyrrole on Pt, recorded at scan rates of 1) 20 mV/s, 2) 50 mV/s, 3) 200 mV/s, 4) 10 V/s in a distance of about 6 μm. (From Ref. 57 by permission of The Electrochemical Society, Inc.)

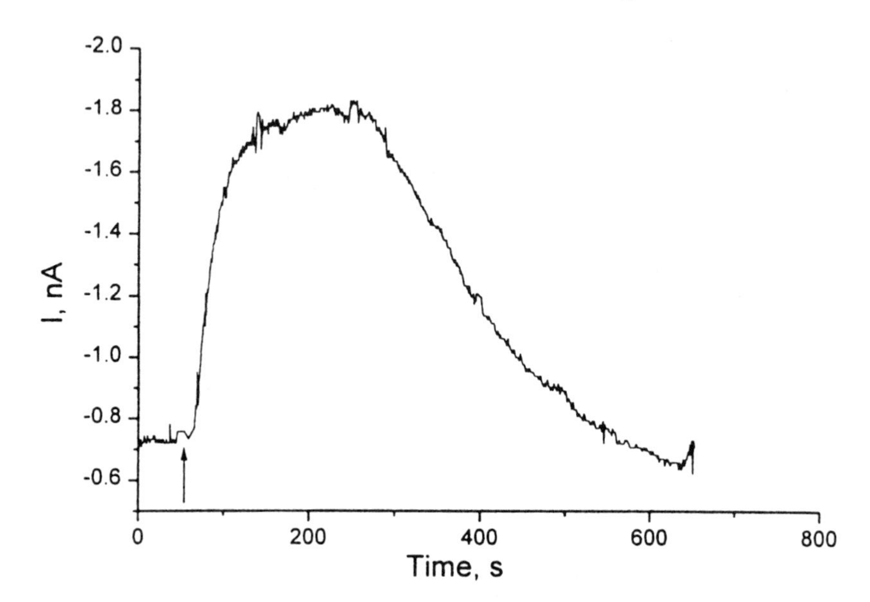

FIG. 22 Ejection of cadmium ions during oxidation of a polypyrrole film at +0.5 V vs. SCE. Monitored by an amalgamated 50 μm gold electrode biased at −0.8 V. Electrolyte: 100 mM KCl. (From Ref. 59. Copyright 1999 American Chemical Society.)

Heinze and Borgwarth were able to directly image the degree of doping as the key chemical property of conducting polymers (24). Moreover, they were able to dope stripes of a PAni film by locally generating an oxidant to yield a polymeric interdigitated array, which, when imaged, showed differences in conductivity (Fig. 25) (63). Mandler et al. polymerized a monolayer of anilinium that was electrostatically attached to a self-assembled monolayer

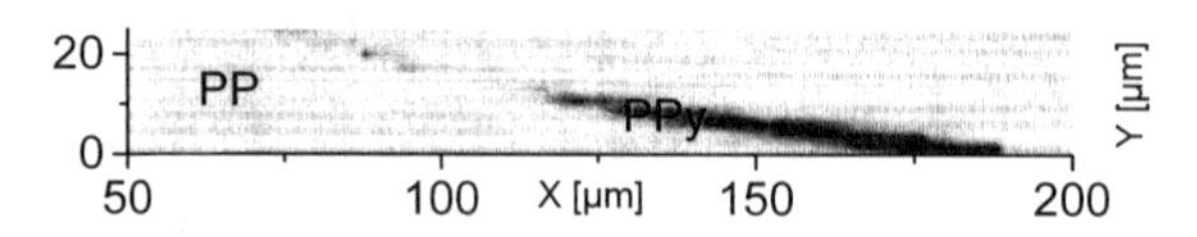

FIG. 23 Image of a blend containing 98.37% polypropylene (PP) and 1.63% polypyrrole (PPy). Taken by a 10 μm Pt tip at v = 0.6 μm/s at about d = 5 μm distance at E_T = +0.420 V and currents ranging from 170 to 270 pA. The electrolyte was aq. 5 mM $K_3[Fe(CN)_6]$, 1 M Na_2SO_4. (Adapted from Ref. 24.)

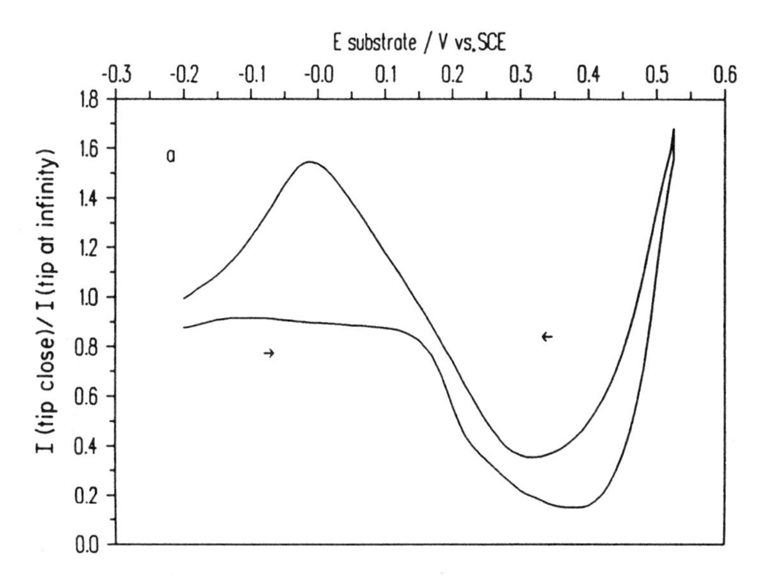

FIG. 24 Tip-substrate CV (above) and sample CV (below) of 3.1 μm polyaniline on platinum in 10 mM HCl; $I_{T,\infty}$ = 650 nA. Scan rate was 5 mV/s. (From Ref. 60, with permission from Elsevier Science.)

of sulfonate-terminated thiols. The presence of the PAni as a conducting layer was proved by the approach curves in the feedback mode and comparison with polymer-free samples given in Figure 26 (64).

Polythiophene Derivatives. While scanning above rough polymer samples, the current recorded in the feedback mode may change due to either morphological effects or lateral variations of polymer properties. To image morphology and conductivity separately, Heinze and Borgwarth verified the "picking mode" (21) at spots of poly-(4,4″-dithiomethyl-2,2′:5′,2″-terthiophene) coated on glass substrates to separate morphological effects from chemical properties during imaging (65); the results are shown in Figure 27. The groups of Bard and Decker characterized the electron transfer of the mediator methyl viologen and a poly-(3,3″-didodecyl-2,2′:5′,2″-terthiophene) covered electrode at different oxidation levels, as shown in Figure 28 (66). Adapting methods used for liquid/liquid interfaces, they established that the electron transfer occurred at the liquid/polymer interface and determined rate constants between 10^{-5} cm/s for the undoped film and 10^{-1} cm/s for the doped one.

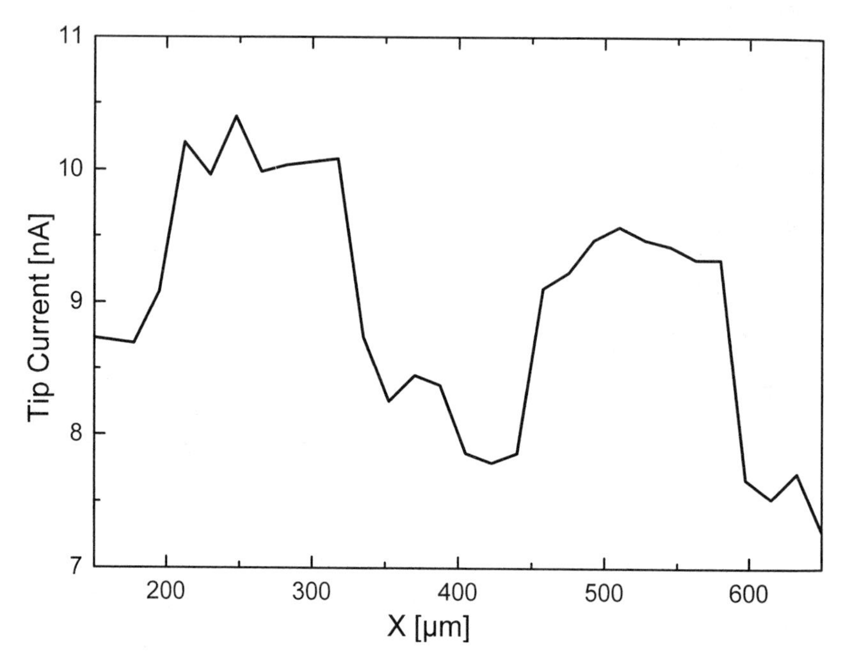

FIG. 25 Scanning above two lines of an interdigitated array generated previously by bromine-mediated oxidation of polyaniline in stripes. The 25 μm Au tip reduces iodine under diffusion control in 5 mM I_2, 20 mM KBr, 100 mM HCl at scan speed $v_X = 10$ μm/s. High currents indicate the doped and therefore conducting parts of PAni. (Adapted from Ref. 63.)

IV. APPLICATIONS

The experiments reported here illustrate the basic principles and SECM methods used to obtain data about heterogeneous electron transfer reactions. These have been adapted to numerous fields, including some discussed in other chapters of this book, such as phase transfer reactions, corrosion, liquid/liquid junctions, and biology. Due to the unique SECM concept of addressing chemical reactions locally by chemical means and the degree of maturity reached in recent years, SECM has potential in applied research. The authors regard areas of heterogeneous catalysis and the in situ study of sensors as the most promising fields. This view is supported by a few publications pointing in these directions. Mandler et al. used methylviologen as mediator with a redox potential significantly below that of protons at

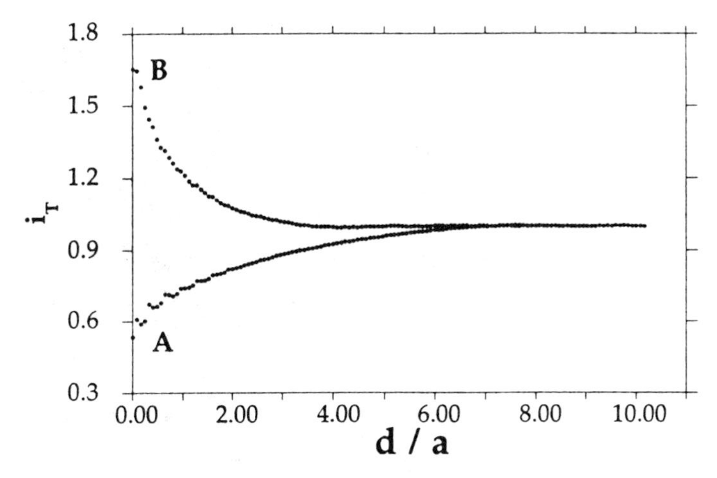

FIG. 26 Normalized feedback current-distance curves obtained with 25 μm Pt tip in 2 mM $Fe(phen)_3^{2+}$ and 100 mM HCl above a monolayer of polyaniline immobilized by a sulfonate terminated self-assembled monolayer on a gold substrate. (From Ref. 64. Copyright 1998 American Chemical Society.)

neutral pH. Using an amalgamated electrode, they generated the reduced form, which was not able to reduce protons directly itself but did at the surface of a platinum catalyst. The feedback behavior proved to depend on the pH but not on the size of the platinum electrodes. This clearly indicated that the reaction was chemical in nature and not an electrochemical response (67).

In the future, applied activities will become increasingly important in the area of sensors and biosensors. The key issue here is to locate the electrochemically active parts and to quantify the kinetics of transfer and the turnover. Schuhmann et al. used the SECM to study the cross-talk in a glucose microsensor array (68). They were able to monitor directly H_2O_2 diffusing between different microbands and causing an amperometric response in the vicinity of its generation. Our group studied in situ a conductometric sensor for chlorine based on a conducting polymer layer coated onto a membrane. From the SECM experiments it could be concluded that the charge propagation inside the polymer which limited the response time was strongly affected by the presence of randomly distributed larger polymer particles (69).

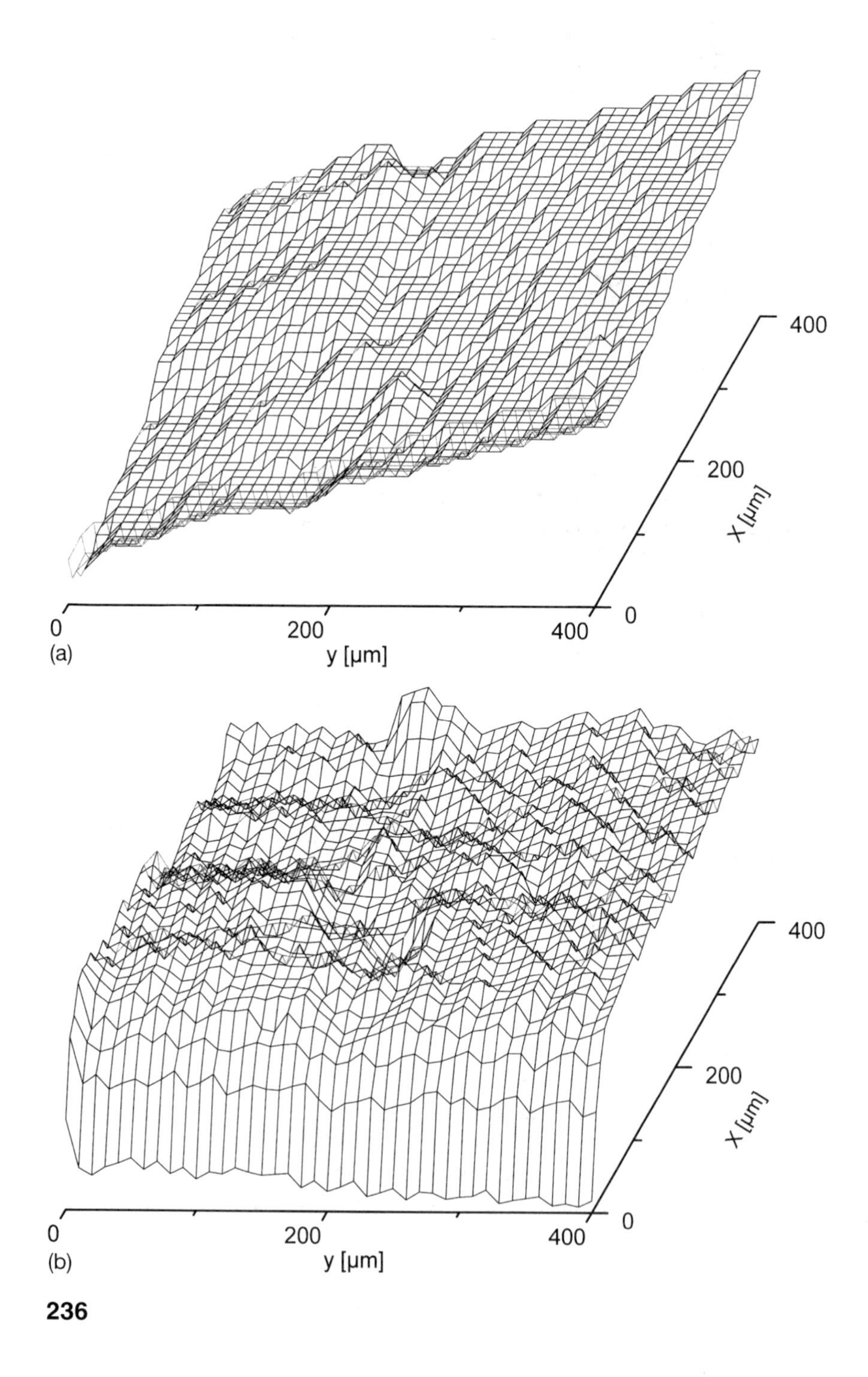
400
200
x [µm]
0
0
200
400
(a)
y [µm]
400
200
x [µm]
0
0
200
400
(b)
y [µm]

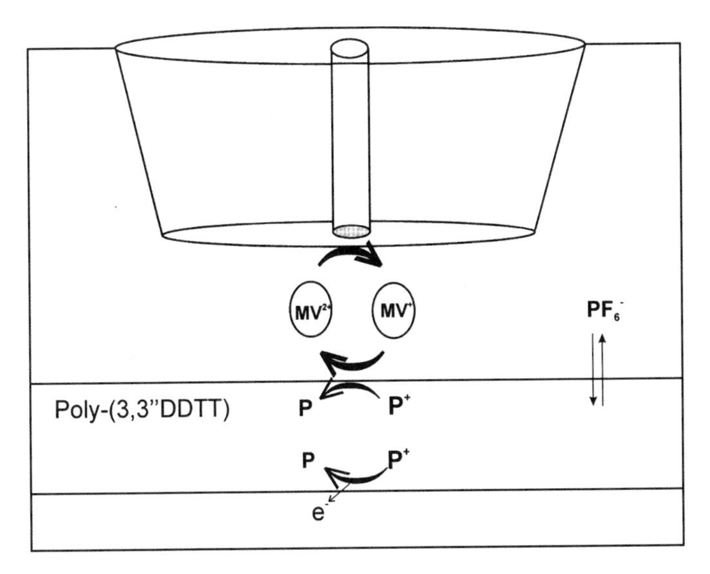

FIG. 28 Schematic view of the processes involved in a SECM experiment with a poly-(3,3″-DDTT)–coated sample electrode. (Adapted from Ref. 66.)

V. CONCLUSION AND OUTLOOK

The aims of current research are to extend the range of applications by, first, extending the methodology and, second, increasing the lateral resolution to lower nanodimensions. On the one hand, producing smaller electrodes that can be held in close proximity to the sample will give rise to large current densities. On the other hand, the development of more sensitive detection principles will enable the monitoring of apparently slow kinetics at the sample. The solution of these challenges opens new fields, especially in heterogeneous catalysis, but also in the material and life sciences (70).

FIG. 27 (a) Morphology of poly-(4,4″-dithiomethyl-2,2′:5′,2″-terthiophene) spots coated on glass substrates obtained in the "picking" mode. (b) Current induced by local conductivity measured in constant distance. The electrolyte was aq. 50 mM $K_3[Fe(CN)_6]$, 1 M Na_2SO_4. The approaches of the 25 μm Au tip occurred at 50 μm/s until the current doubled due to convective effects. (Adapted from Ref. 65.)

REFERENCES

1. MV Mirkin. Microchim Acta 130:127, 1999.
2. AT Hubbard, FC Anson. In: AJ Bard, ed. Electroanalytical Chemistry. Vol. 4. New York: Marcel Dekker, 1970, p 129.
3. J Ufheil, K Borgwarth, J Heinze. Unpublished work.
4. J Kwak, AJ Bard. Anal Chem 61:1221, 1989.
5. (a) R Guckenberger, M Heim, G Cevc, HF Knapp, W Wiegrabe, A Hillebrand. Science 266:1538, 1994. (b) F-RF Fan, AJ Bard. Science 270:1849, 1995.
6. (a) Y Shao, MV Mirkin. J Electroanal Chem 439:137, 1997. (b) Y Shao, MV Mirkin. Anal Chem 70:3155, 1998. (c) NJ Evans, M Gonsalves, NJ Gray, AL Barker, JV Macpherson, PR Unwin. Electrochem Comm 2:201–206, 2000.
7. J Heinze. Angew Chem Int Ed Engl 32:1268, 1993.
8. (a) M Fleischmann, S Pons, DR Rolison PP Schmidt, eds. Ultramicroelectrodes. Morganton: Datatech Systems Inc., 1987. (b) MI Montenegro, MA Querios, JL Daschbach, eds. Microelectrodes: Theory and Application. Dortrecht, Netherlands: Kluwer Academic Publishers, 1991. (c) RM Wightman, DO Wipf. In: AJ Bard, ed. Electroanalytical Chemistry. Vol. 15. New York: Marcel Dekker, 1988, p 267.
9. AJ Bard, LR Faulkner. Electrochemical Methods, Fundamentals and Applications. New York: Wiley, 1980.
10. KB Oldham, CG Zoski. J Electroanal Chem Interfacial Electrochem 256:11, 1988.
11. MV Mirkin, AJ Bard. Anal Chem 64:2293, 1992.
12. K Borgwarth. Oberflächenanalyse und Strukturierung mit Hilfe des Rasterelektrochemischen Mikroskops und des Rastertunnelmikroskops. Aachen, Germany: Shaker Verlag, 1996.
13. AJ Bard, F-RF Fan, J Kwak, O Lev. Anal Chem 61:132, 1989.
14. AJ Bard, MV Mirkin, PR Unwin, DO Wipf. J Phys Chem 96:1861, 1992.
15. MV Mirkin, F-RF Fan, AJ Bard. J Electroanal Chem 328:47, 1992.
16. X Liu, J Lu, C Cha. J Electroanal Chem 295:15, 1990.
17. C Wei, AJ Bard, MV Mirkin. J Phys Chem 99:16033, 1995.
18. F Forouzan, AJ Bard. Isr J Chem 37:155, 1997.
19. (a) R Zwanzig. Proc Natl Acad Sci USA 87:5856, 1990. (b) R Zwanzig, A Szabo. Natl Biophys J 60:671, 1991. (c) A Szabo, R Zwanzig. J Electroanal Chem 314:307, 1991.
20. DO Wipf, AJ Bard. Anal Chem 65:1373, 1993.
21. K Borgwarth, D Ebling, J Heinze. Ber Bunsenges Phys Chem 98:1317, 1994.
22. M Ludwig, C Kranz, W Schuhmann, HE Gaub. Rev Sci Instrum 66:2857, 1995.
23. PI James, LF Garfias-Mesias, PJ Moyer, WH Smyrl. J Electrochem Soc 145: L64, 1998.
24. K Borgwarth, C Ricken, DG Ebling, J Heinze. Fresenius Anal Chem 356:288, 1996.
25. KA Ellis, MD Pritzker, TZ Fahidy. Anal Chem 67:4500, 1995.
26. C Lee, DO Wipf, AJ Bard, K Bartels, AC Bovik. Anal Chem 63:2442, 1991.

27. (a) K Borgwarth, J Heinze. J Electrochem Soc 146:3285, 1999. (b) K Borgwarth, FM Boldt, C Hess, T Nann, J Heinze. In: First International Workshop on Scanning Electrochemical Microscopy. Freiburg, Germany: ISE, 1997, p L23.
28. FM Boldt, K Borgwarth, J Heinze. In preparation.
29. (a) Z Tian, Z Fen, X Zhuo, J Mu, C Li, H Lin, B Ren, Z Xie, W Hu. Faraday Discuss Chem Soc 94:37, 1992. (b) YB Zu, L Xie, BW Mao, Z Tian. Electrochim Acta 43:1683, 1998.
30. RC Engstrom, T Meany, R Tople, RM Wightman. Anal Chem 59:2005, 1987.
31. RC Engstrom, B Small, L Kattan. Anal Chem 64:241, 1992.
32. C Lee, CJ Miller, AJ Bard. Anal Chem 63:78, 1991.
33. C Lee, AJ Bard. Anal Chem 62:1906, 1990.
34. RL McCreery. In: AJ Bard, ed. Electroanalytical Chemistry. Vol. 17. New York: Marcel Dekker, 1991, p 221.
35. DO Wipf, AJ Bard. J Electrochem Soc 138:469, 1991.
36. DO Wipf, AJ Bard. J Electrochem Soc 138:L4, 1991.
37. G Wittstock, H Emons, M Kummer, JR Kirchhoff, WR Heineman. Fresenius J Anal Chem 348:712, 1994.
38. G Wittstock, H Emons, TH Ridgway, EA Blubaugh, WR Heineman. Anal Chim Acta 298:285, 1994.
39. J Kwak, AJ Bard. Anal Chem 61:1794, 1989.
40. Y-F Yang, G Denuault. J Chem Soc Faraday Trans 92:3791, 1996.
41. Y-F Yang, G Denuault. J Electroanal Chem 418:99, 1996.
42. SB Basame, HS White. Langmuir 15:819, 1999.
43. BR Horrocks, MV Mirkin, AJ Bard. J Phys Chem 98:9106, 1994.
44. G Wittstock, R Hesse, W Schuhmann. Electroanalysis 9:746, 1997.
45. C Jehoulet, YS Obeng, Y-T Kim, F Zhou, AJ Bard. J Am Chem Soc 114:4237, 1992.
46. J Chlistunoff, D Cliffel, AJ Bard. Thin Solid Films 257:166, 1995.
47. DE Cliffel, AJ Bard, S Shinkai. Anal Chem 70:4146, 1998.
48. H Sugimura, N Shimo, N Kitamura, H Masuhara, K Itaya. J Electroanal Chem 346:147, 1993.
49. H Sugimura, N Kitamura. In: H Masuhara, ed. Microchem Proc JRDC-KUL Jt Int Symp. Amsterdam: North-Holland, 1994, p 521.
50. (a) RW Murray. In: AJ Bard, ed. Electroanalytical Chemistry. Vol. 13. New York: Marcel Dekker, 1984, p 191. (b) AR Hillman. In: AR Hillman, ed. Electrochemical Science and Technology of Polymers. New York: Elsevier, 1987, p 103.
51. MV Mirkin, F-RF Fan, AJ Bard. Science 257:364, 1992.
52. J Kwak, FC Anson. Anal Chem 64:250, 1992.
53. IC Jeon, FC Anson. Anal Chem 64:2021, 1992.
54. F-RF Fan, MV Mirkin, AJ Bard. J Phys Chem 98:1475, 1994.
55. F-RF Fan. J Phys Chem B 102:9777, 1998.
56. (a) K Doblhofer, K Rajeshwar. In: TA Skotheim et al., eds. Handbook of Conducting Polymers, 2nd ed. New York: Marcel Dekker, 1998, pp 531–588. (b)

G Inzelt. In: AJ Bard, ed. Electroanalytical Chemistry. Vol. 18. New York: Marcel Dekker, 1994, p 89. (c) J Heinze. Top Curr Chem 152:1, 1990.
57. J Kwak, C Lee, AJ Bard. J Electrochem Soc 137:1481, 1990.
58. M Arca, MV Mirkin, AJ Bard. J Phys Chem 99:5040, 1995.
59. I Kapui, R Gyurcsányi, G Nagy, K Tóth, M Arca, E Arca. J Phys Chem B 102:9934, 1999.
60. MHT Frank, G Denuault. J Electroanal Chem 354:331, 1993.
61. MHT Frank, G Denuault. J Electroanal Chem 379:399, 1994.
62. G Denuault, MHT Frank, LM Peter. Faraday Discuss 94:23, 1992.
63. K Borgwarth, C Ricken, DG Ebling, J Heinze. Ber Bunsenges Phys Chem 99: 1421, 1995.
64. I Turyan, D Mandler. J Am Chem Soc 120:10733, 1998.
65. K Borgwarth, D Ebling, J Heinze. Electrochim Acta 40:1455, 1995.
66. M Tsinonsky, AJ Bard, D Dini, F Decker. Chem Mater 10:2120, 1998.
67. Y Selzer, I Turyan, D Mandler. J Phys Chem B 103:1509, 1999.
68. DJ Strike, A Hengstenberg, M Quinto, C Kurzawa, M Koudelka-Hep, W Schuhmann. Mikrochim Acta 131:47, 1999.
69. AW Synowczyk. Sensorik auf Basis leitfähiger Polymere. Aachen: Shaker Verlag, 1997, pp 153–159.
70. RA Clark, SE Zerby, AG Ewing. In: AJ Bard, I Rubenstein, eds. Electroanalytical Chemistry. Vol. 20. New York: Marcel Dekker, 1999, p 227.
71. F-RF Fan, J Kwak, AJ Bard. J Am Chem Soc 96:9669, 1996.
72. MV Mirkin, TC Richards, AJ Bard. J Phys Chem 97:7672, 1993.
73. MV Mirkin, AJ Bard. J Electrochem Soc 139:3535, 1992.
74. MV Mirkin, LOS Bulhoes, AJ Bard. J Am Chem Soc 115:201, 1993.
75. J Zhou, Y Zu, AJ Bard. J Electroanal Chem. In Press.

7

KINETICS OF HOMOGENEOUS REACTIONS COUPLED TO HETEROGENEOUS ELECTRON TRANSFER

Patrick R. Unwin

University of Warwick
Coventry, England

I. INTRODUCTION

Scanning electrochemical microscopy (SECM) has proven a powerful approach for measuring the kinetics of homogeneous reactions coupled to heterogeneous electron transfer. For these investigations, both the tip and the substrate are electrodes, and the tip/substrate configuration essentially functions as a variable-gap ultra-thin-layer cell. The classes of reaction investigated hitherto are primarily those involving irreversible chemical reactions of electrogenerated species (1–6) and chemical reactions sandwiched between successive electron transfers, as in the ECE/DISP scheme (7) (defined in Sec. IV).

The origins of SECM homogeneous kinetic measurements can be found in the earliest applications of ultramicroelectrodes (UMEs) to profile concentration gradients at macroscopic (millimeter-sized) electrodes (1,2). The field has since developed considerably, such that short-lived intermediates in electrode reactions can now readily be identified by SECM under steady-state conditions, which would be difficult to characterize by alternative transient UME methods, such as fast scan cyclic voltammetry (8).

The aim of this chapter is to assess the various SECM methods that are currently available for measuring homogeneous reactions and to identify areas in which further progress is possible. The approach will be to treat the methodologies within the context of various electrochemical mechanisms, beginning with simple irreversible first-order chemical reactions following electron transfer (i.e., the EC_i mechanism), progressing to second-order (irreversible) follow-up chemical reactions (i.e., EC_{2i} processes), and finally ECE/DISP processes. This is a logical progression in terms of increasing the complexity of the electrode processes considered. Moreover, it is useful to begin with the EC_i mechanism, as most of the SECM techniques covered in this chapter have been applied practically to this class of reaction, and it thus provides a useful point from which to judge the various methods available.

There are three methods of SECM that can be used to characterize and measure homogeneous reactions of the type identified above: the feedback mode, the tip generation/substrate collection (TG/SC) mode, and the substrate generation/tip collection (SG/TC) mode. For illustrative purposes, these modes are compared schematically in Figure 1 for a case where a species B produced by reduction of a species A at a generator electrode undergoes a first-order chemical reaction in solution (characterized by a rate constant, k_1).

In the closely related feedback and TG/SC modes (Fig. 1a), the heterogeneous electron transfer process is driven at a diffusion-controlled rate at a disc-shaped tip UME (radius a). A larger collector (substrate) electrode, positioned directly below the tip, typically at a distance $d \leq 2a$, is held at a potential where the oxidation of B back to A is diffusion-controlled. With this electrode configuration, a competition is established between (1) the diffusion of B from the tip to the substrate, with the consequent regeneration and feedback diffusion of A, and (2) the chemical decomposition of B to products (generally considered to be electroinactive at the potentials of interest). The kinetics of the chemical reaction are probed by varying the tip/substrate distance, and thus the diffusion time of B to the substrate, while measuring the tip current i_T (in the feedback mode) or i_T, and the substrate current i_S (in the TG/SC mode). When the tip/substrate diffusion time, of the order d^2/D_B (where D_B is the diffusion coefficient of B) is short compared to the lifetime of B, of the order $1/k_1$, most of B generated reaches the substrate for conversion to A, with the result that i_T and i_S approach values expected for positive feedback in the absence of homogeneous kinetics. In the fast kinetic limit ($d^2/D_B \gg 1/k_1$) the tip response is similar to that for negative feedback and i_S tends to zero, since little B survives the passage across the gap. Between these two limits, kinetic measurements are possible.

From a practical viewpoint, feedback measurements are easier to implement than TG/SC measurements in terms of both the electrochemical and positioning instrumentation required. In the feedback mode, it is often unnecessary to bias the substrate externally, allowing the use of a simple two- or three-electrode arrangement in which the tip UME is the sole working electrode. The rationale for making feedback measurements under these conditions was discussed in Chapter 1 and is considered in detail elsewhere (9). In brief, and with reference to the example in Figure 1, an externally unbiased substrate electrode will be bathed predominantly in A, the only electroactive species initially present in solution, provided that the substrate is sufficiently large and the A/B couple is sufficiently reversible. Under these

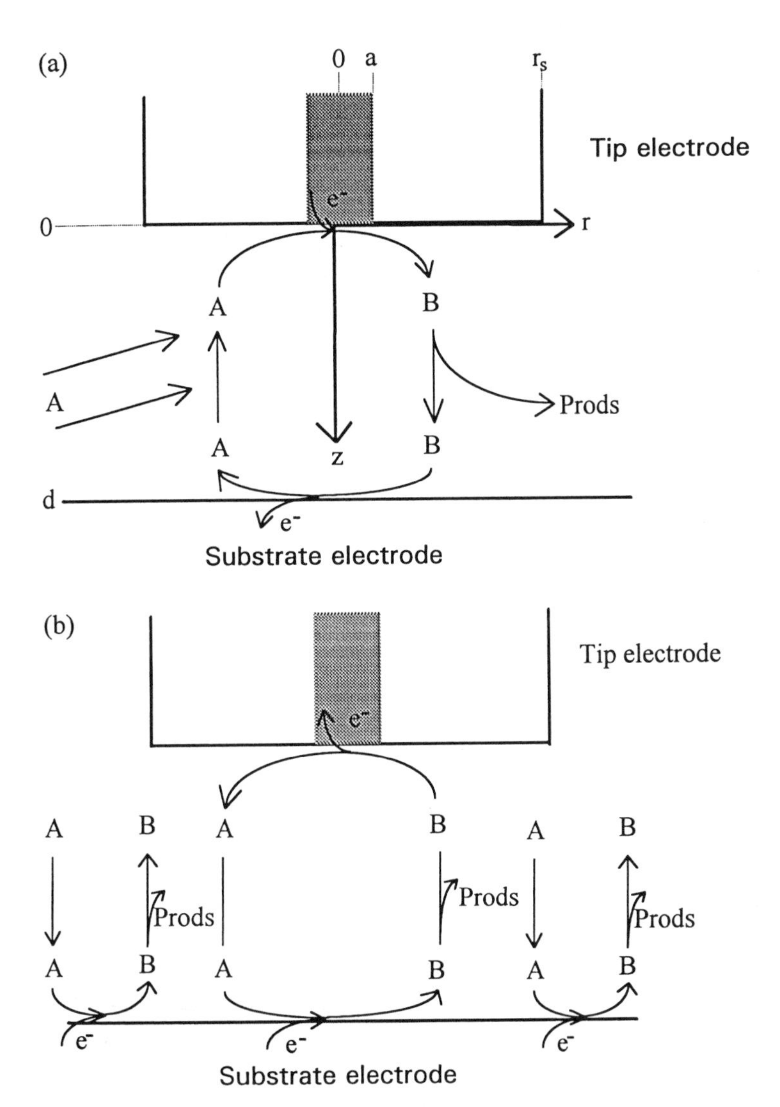

FIG. 1 Principles of (a) the feedback (TG/SC) modes and (b) the SG/TC mode for measuring follow-up chemical reactions of electrogenerated species. Species B, produced from a solution precursor species, A, at a generator electrode undergoes decomposition during transit to a collector electrode. Note that in the case of the feedback (a) and SG/TC (b) modes, only the tip current is generally measured, whereas for the TG/SC mode (a), the tip and substrate currents are of interest. The coordinate system and notation for the SECM geometry are shown. The schematic is not to scale; typically, $r_S \geq 10a$ and $d \leq a$.

conditions, the substrate potential is thus poised by the solution at a value where the oxidation of B to A is driven at a rate close to diffusion-control.

A bipotentiostat is required if substrate currents are also to be measured, as in the TG/SC mode. For this mode, the substrate electrode is shrunk to a size where background noise is minimized, but all B reaching the substrate is still collected. The factors governing the minimum substrate size for this type of measurement are outlined in Sec. II.A. For this case, it is necessary to align the tip and substrate UMEs with good precision, typically demanding the use of high resolution remote-controlled positioners on all three axes. For feedback mode experiments, it is only necessary to accurately control the position of the tip in the direction normal to a large substrate electrode, which greatly simplifies the instrumentation requirements and costs. However, as will become apparent from Sec. II.A, the TG/SC mode offers advantages over the feedback mode for the study of rapid follow-up chemical reactions.

In the SG/TC mode (Fig. 1b), species B is electrogenerated at the substrate electrode and the reverse process, in which B is oxidized to A, occurs at the tip. In experiments hitherto, the substrate electrode has generally (1,2,6,10), but not exclusively (4,11), been macroscopic (millimeter dimensions) with a disk of micrometer dimensions serving as the collector electrode. The substrate response in the SG/TC mode is thus typically transient (2,4,6). As with the feedback and TG/SC modes discussed above, the tip/substrate separation is a key experimental variable. Since the substrate electrode is generally large, high-resolution positioning is only required in the direction perpendicular to the surfaces of the tip and substrate.

II. EC_i PROCESSES

The EC_i process is the simplest mechanism involving homogeneous reactions coupled to heterogeneous electron transfer at an electrode (12). Historically, dual electrode methods have proven especially useful for studying the kinetics of the C_i step, including the rotating ring-disc electrode (RRDE) (13), twin electrodes in a thin layer cell (14), double channel or tubular electrodes (15), interdigitated channel array electrodes (16), and paired microband electrodes (17). Common to all of these methods is the use of a second (collector) electrode to measure the fraction of an electrogenerated species that survives the transit from a nearby generator electrode. Overall, the following processes occur:

Generator electrode: $$\mathrm{A} + \mathrm{n}e^- \rightarrow \mathrm{B} \tag{1}$$

Solution: $$\mathrm{B} \xrightarrow{k_1} \text{Products} \tag{2}$$

Collector electrode: $$\mathrm{B} - \mathrm{n}e^- \rightarrow \mathrm{A} \tag{3}$$

Although SECM operates on broadly similar principles to other double electrode methods, the SECM geometry offers several advantages, in terms of the range of kinetics (especially the upper limit) that can be studied and the precision with which measurements can be made, as discussed below.

A. Theory

A general overview of SECM theoretical methods was given in Chapter 5, and it is beyond the scope of this chapter to provide a detailed description of the mechanics of the numerical methods that have been used to treat SECM problems involving coupled homogeneous kinetics (3,4,6,7). Some discussion of the formulation of SECM models for homogeneous kinetics is warranted, however, since an understanding of the assumptions made in tackling the problems provides a basis for assessing the conditions under which the various treatments apply.

1. *Feedback and Tip Generation/Substrate Collection Modes*

For both the feedback and TG/SC modes, Eqs. (1)–(3) are relevant, with the tip UME as the generator electrode and the substrate as the collector electrode. As discussed above, the distinction between these two methods is that in the TG/SC mode, both the tip and substrate currents are measured, whereas in the feedback mode only the tip current is of interest.

Theoretical characteristics for the EC_i mechanism, in the feedback mode under chronoamperometric and steady-state conditions, were obtained initially by solving numerically the diffusion equations in the axisymmetric cylindrical geometry defining the tip and substrate, in the SECM configuration (Fig. 1). The results were later generalized and simplified equations presented for the collection efficiency and the relationship between the substrate and tip responses (TG/SC mode) under steady-state conditions (5).

To treat the tip and substrate responses, it is necessary to solve the diffusion equations for the species of interest. In the coordinate system of Figure 1, the equations for A and B are:

$$\frac{\partial c_A}{\partial t} = D_A \left(\frac{\partial^2 c_A}{\partial r^2} + \frac{1}{r}\frac{\partial c_A}{\partial r} + \frac{\partial^2 c_A}{\partial z^2} \right) \tag{4}$$

$$\frac{\partial c_B}{\partial t} = D_B \left(\frac{\partial^2 c_B}{\partial r^2} + \frac{1}{r}\frac{\partial c_B}{\partial r} + \frac{\partial^2 c_B}{\partial z^2} \right) - k_1 c_B \tag{5}$$

where r and z are, respectively, the coordinates in the directions radial and normal to the tip UME surface, starting at its center. D_i and c_i are the diffusion coefficient and concentration of species i (either A or B), and t is time.

Prior to a chronoamperometric feedback experiment, only A is present at a bulk concentration, c_A^*. The initial condition (over all space) is thus:

$$t = 0, \text{ all } r, \text{ all } z: \qquad c_A = c_A^*; \; c_B = 0 \tag{6}$$

The tip electrode potential is stepped to a value to effect the diffusion-controlled reduction of A [Eq. (1)], while the substrate potential is maintained at a value so that the reverse reaction [Eq. (3)] occurs at a diffusion-controlled rate. A and B are considered to be inert on the insulating sheath surrounding the tip microdisk electrode. Provided that $RG = r_S/a \geq 10$, where r_S is the radial distance from the center of the UME to the edge of the insulating sheath, defined in Figure 1, it is reasonable, given the time scale and nature of most SECM measurements, to assume that A and B attain their bulk solution values beyond the radial edge of the tip/substrate domain (3) (or to employ a zero-flux condition at this point on the radial boundary) (18,19). The axisymmetric cylindrical geometry also implies that there is zero radial flux of A and B at the central axis. The boundary conditions for the EC_i process are therefore:

$$z = 0, \; 0 \leq r \leq a \text{ (tip electrode)}: c_A = 0; \; D_A \frac{\partial c_A}{\partial z} = -D_B \frac{\partial c_B}{\partial z} \tag{7}$$

$$z = 0, \; a \leq r \leq r_s \text{ (insulating sheath)}: D_A \frac{\partial c_A}{\partial z} = D_B \frac{\partial c_B}{\partial z} = 0 \tag{8}$$

$$z = d, \; 0 \leq r \leq r_s \text{ (substrate electrode)}: c_B = 0; \; D_B \frac{\partial c_B}{\partial z} = -D_A \frac{\partial c_A}{\partial z} \tag{9}$$

$$r > r_s, \; 0 \leq z \leq d \text{ (radial edge of tip/substrate domain)}: \quad c_A = c_A^*; \; c_B = 0 \tag{10}$$

$$r = 0, \; 0 < z < d \text{ (axis of symmetry)}: D_A \frac{\partial c_A}{\partial r} = D_B \frac{\partial c_B}{\partial r} = 0 \tag{11}$$

General solutions to SECM problems can be obtained by introducing the following dimensionless variables:

$$R = r/a \tag{12}$$

$$Z = z/a \tag{13}$$

$$C_i = c_i/c_A^* \qquad (i = \text{A, B}) \tag{14}$$

$$\tau = tD_A/a^2 \tag{15}$$

$$K = k_1 a^2 / D_A \tag{16}$$

$$\beta = D_B / D_A \tag{17}$$

The tip and the substrate currents are governed by the fluxes of A and B over the electrode surfaces:

$$i_T = 2\pi n F D_A \int_0^a (\partial c_A / \partial z)_{z=0}\, r\, dr \tag{18}$$

$$i_S = 2\pi n F D_B \int_0^{r_s} (\partial c_B / \partial z)_{z=d}\, r\, dr \tag{19}$$

In these equations, F is Faraday's constant. These currents are conveniently normalized by the n-electron steady-state diffusion-limited current for the reduction of A at a tip positioned at an effectively infinite distance from the substrate (20):

$$i_{T,\infty} = 4nFaD_A c_A^* \tag{20}$$

so that the tip and substrate currents are given by:

$$i_T / i_{T,\infty} = (\pi/2) \int_0^1 (\partial C_A / \partial Z)_{Z=0}\, R\, dR \tag{21}$$

$$i_S / i_{T,\infty} = (\beta\pi/2) \int_0^{R_S} (\partial C_B / \partial Z)_{Z=L}\, R\, dR \tag{22}$$

where

$$R_s = r_s / a \tag{23}$$

$$L = d/a \tag{24}$$

All treatments of SECM problems involving homogeneous reactions have been for $\beta = 1$, i.e., with reference to the above example, A and B were considered to have a common diffusion coefficient. This is usually a good approximation, and all of the theoretical results presented in this chapter are for this particular case. It should be noted, however, that the simple positive feedback problem has been treated without this assumption, revealing that when $\beta \neq 1$, the form of the chronoamperometric characteristic is sensitive to the β-value, but the final steady-state tip current is β-independent (21).

In order to explain the effects of irreversible follow-up chemical reactions on the SECM response, it is first useful to briefly review the feedback and TG/SC characteristics under diffusion-controlled conditions, without the additional complications from homogeneous kinetics. Typical theoretical results for the chronoamperometric tip and substrate responses are shown in Figure 2 for a range of tip/substrate separations. These particular data were obtained numerically with the alternating direction implicit finite difference method (ADIFDM) (4,22).

The tip characteristics in Figure 2 are identical to an original treatment of the positive feedback response simulated using a Krylov integrator (19). At very short times (too short for the scale on Fig. 2), the normalized tip current is independent of d/a, since the diffusion field adjacent to the UME is small compared to the size of the gap and the tip shows the behavior predicted for a simple microdisk electrode under chronoamperometric control (23). With time, the diffusion field extends towards, and ultimately intercepts, the substrate causing a current flow at this detector electrode. Since the tip/substrate diffusion time is of the order d^2/D, the smaller the value of d/a, the sooner the substrate current begins to flow [in normalized time; Eq. (15)] and the more rapidly the currents at the tip and substrate attain steady-

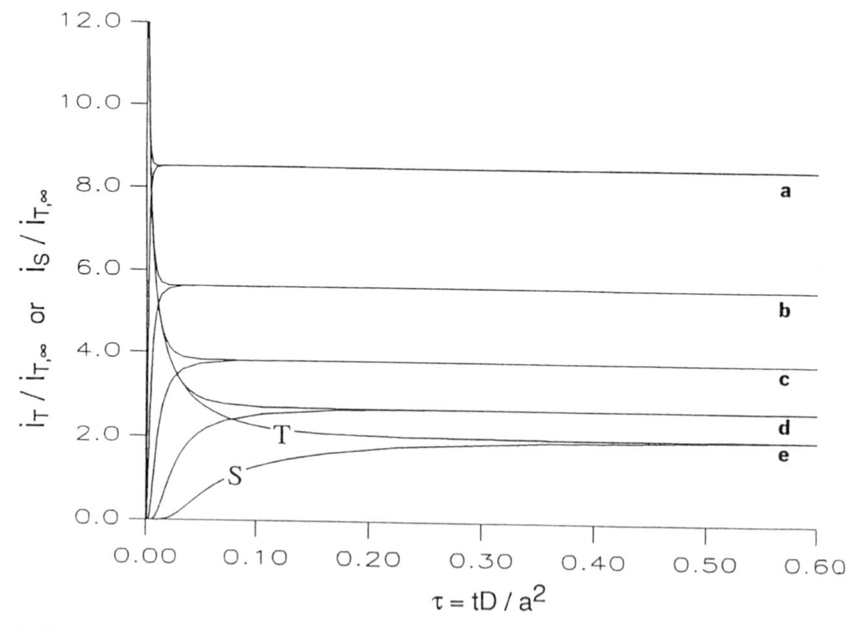

FIG. 2 Tip and substrate chronoamperometric characteristics (presented as normalized current magnitudes) for the TG/SC mode for diffusion-controlled positive feedback with no homogeneous kinetic complications. The data are for log d/a = (a) -1.0, (b) -0.8, (c) -0.6, (d) -0.4, and (e) -0.2. The tip and substrate currents are denoted by T and S, respectively, on the curves labeled e.

state values. Moreover, the closer the two electrodes are positioned together, the greater is the extent of positive feedback, reflected by increases in $i_T/i_{T,\infty}$ and $i_S/i_{T,\infty}$ as d/a is decreased (3,4).

For all of the tip/substrate separations shown in Figure 2, steady-state collection efficiencies

$$N_{SS} = (i_S/i_T)_{SS} \tag{25}$$

of essentially 100% are apparent. This behavior, which is characteristic of the SECM geometry, extends to large tip/substrate separations, as illustrated by the data in Table 1. For example, $N_{SS} > 0.99$ provided that $d/a < 2$. The very high collection efficiencies (with the tip and substrate currents enhanced by feedback) (24) that result when the normalized tip/substrate separation, $d/a \ll R_s$, make TG/SC measurements particularly sensitive for kinetic measurements, compared to many of the earlier dual electrode configurations mentioned in Sec. I.

For conditions where N_{SS} approaches 100%, the portion of the substrate electrode that collects the intermediate species, B, is governed by the normalized tip/substrate separation, d/a. This point is illustrated in Figure 3, which shows the flux of tip-generated species, B, over the surface of the substrate electrode. As $d/a \to 0$, the domain of the substrate that collects the tip-generated species approaches the size of the tip but increases as the tip is moved away from the substrate. Empirically, the minimum dimen-

TABLE 1 Steady-State TG/SC Collection Efficiencies for Diffusion-Controlled Positive Feedback with a Stable Mediator, Calculated with $R_S = 10$

log ($L = d/a$)	$N_{SS} = i_T/i_S$
−1.0	1.00
−0.8	1.00
−0.6	1.00
−0.4	1.00
−0.2	1.00
0.0	0.998
0.3	0.992
0.4	0.978
0.5	0.973

Source: Ref. 4.

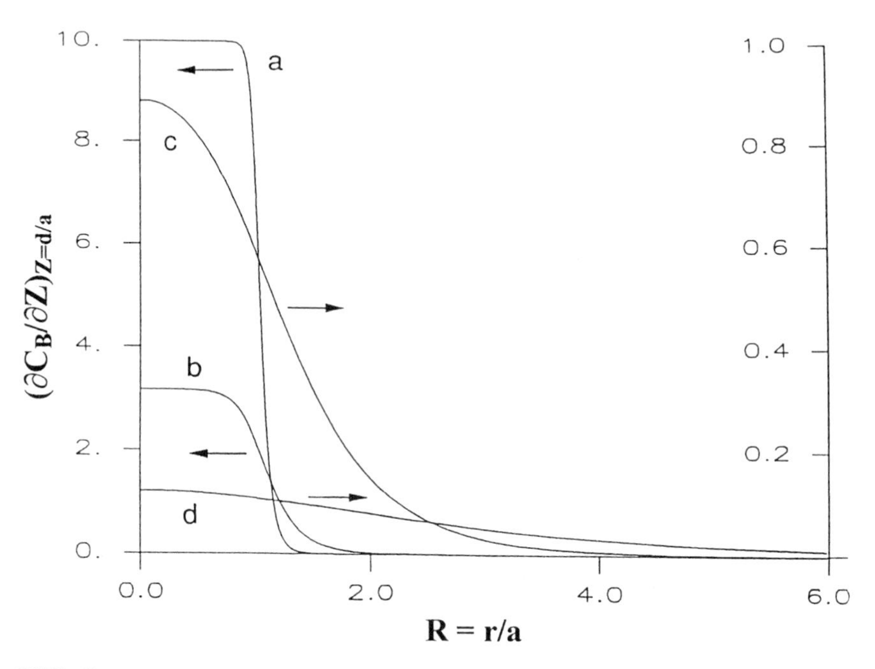

FIG. 3 Normalized flux of the tip-generated species, B, over the substrate (collector) electrode for diffusion-controlled kinetically uncomplicated conditions. The data are for log d/a = (a) −1.0, (b) −0.5, (c), 0.0, and (d) 0.5. The left-hand y-axis relates to a and b, while the right-hand axis is for c and d.

sionless substrate size, h/a, where h is the radius of the substrate required for a steady-state collection efficiency of effectively 100%, is given by (25):

$$h/a = 1 + 1.5(d/a) \tag{26}$$

Equation (26) has important experimental implications, since the noise, in the form of residual current, associated with a steady-state current signal is typically proportional to the area of the electrode. This equation thus serves as a useful guide of the minimum substrate size that can be employed experimentally to maximize the signal-to-noise ratio in TG/SC measurements.

To illustrate the effect of an irreversible follow-up chemical reaction on the tip feedback response, Figure 4 shows calculated chronoamperometric behavior for log d/a values of −1.0 and −0.4. To emphasize the short-time behavior, the data are presented as normalized current versus the inverse square root of normalized time. At the shortest times (largest $\tau^{-1/2}$), the i_T response is as for the simple electron transfer case discussed above, showing no dependence on K or d/a. On this time scale, the current is governed by

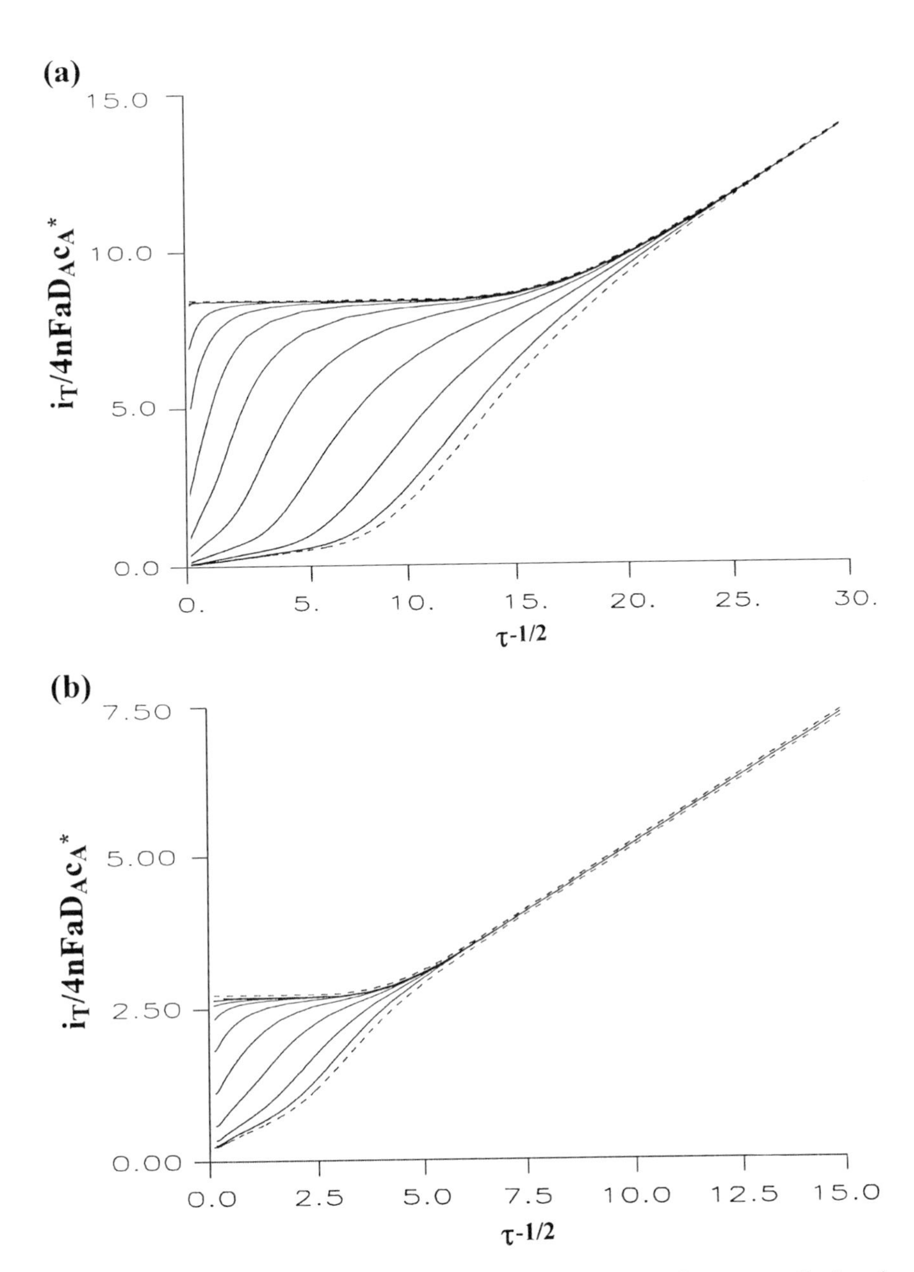

FIG. 4 Calculated tip feedback current transients for an EC_i process, displayed as a function of $\tau^{-1/2}$, for log $d/a = -1.0$ (a) and -0.4 (b). The upper and lower dashed curves in each diagram are, respectively, the behavior for diffusion-controlled positive and negative feedback. The curves between these extremes are for: (a) log $K = 3.0$ (lower solid curve), 2.5, 2.0, 1.5, 1.0, 0.5, 0.0, -0.5, -1.5, and -2.0 (upper solid curve); (b) log $K = 1.8$ (lower solid curve), 1.3, 0.8, 0.3, -0.2, -0.7, -1.2, -1.7, and -2.2 (upper solid curve).

the diffusion of species A to the electrode, which results in a Cotrellian response. Once the tip diffusion field senses the substrate, the magnitude of K has a significant effect on the subsequent current-time characteristics. As K increases, the behavior shifts from a limit where the response is essentially that for positive feedback to a situation where, as $K \rightarrow \infty$, the behavior tends to that for negative feedback with an inert substrate. In this latter case, the lifetime—and reaction layer thickness—of B approaches zero, so that the tip current is governed entirely by the hindered diffusion of A.

For intermediate values of K, the current-time behavior initially follows that for positive diffusion-controlled feedback, without kinetic complications, i.e., a quasi-steady-state current is rapidly attained, resulting in plateaux regions in the $i_T - \tau^{-1/2}$ plots. Under these conditions, the rates of diffusion are sufficiently rapid for the follow-up chemical reaction to be outrun. At times comparable to the lifetime of B, deviations from pure positive feedback behavior are observed, as the homogeneous chemical process begins to consume the tip-generated species. In general, the larger the K value, the earlier is the deviation from positive feedback, the more rapid is the current decay, and the smaller is the final steady-state current.

A discussion of the chronoamperometric feedback response is valuable, as potential step measurements have been used to obtain homogeneous kinetic parameters (3). Calculations of the chronoamperometric response also provide information required for the design of experiments to be made under steady-state conditions. Theoretical results have demonstrated that values of τ in excess of 100 may be necessary to achieve steady-state conditions (within a few percent) (3). For typical tip microdisk radii of 12.5 and 2.5 μm and a typical diffusion coefficient of 10^{-5} cm^2 s^{-1}, this corresponds to real times that are greater than 15 and 0.6 s, respectively. For the larger disk size, the time is longer than required for many other steady-state measurements with SECM (25).

Steady-state working curves of the tip feedback response as a function of K, for different tip/substrate separations are given in Ref. 3, and a set of curves for several distances is shown in Figure 5. As with many kinetic applications of SECM, this figure shows that the range of kinetics open to characterization can be expanded by decreasing the interelectrode separation. Moreover, at close tip/substrate separations, the feedback current varies most sharply with K, in principle allowing measurements to be made with the greatest precision.

To estimate the upper range of rate constants that can be accessed with SECM feedback, the closest normalized tip/substrate separation generally attainable can be assigned as $L \sim 0.1$. This puts an upper limit of approximately 100 on K for current measurements that can be distinguished reliably from the negative feedback case. For typical disk-shaped UMEs employed

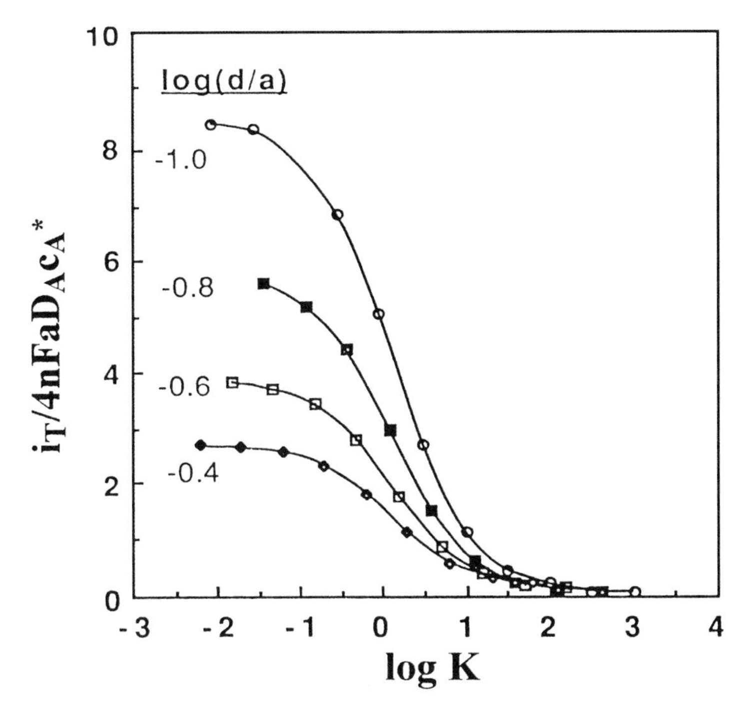

FIG. 5 Steady-state normalized tip feedback current—K working curves for several values of log d/a.

for kinetic measurements with SECM of a = 12.5 and 2.5 μm, together with $D = 10^{-5}$ cm^2 s^{-1}, first-order rate constants up to ~640 s^{-1} or 1.6 × 10^4 s^{-1}, respectively, should be accessible from SECM feedback measurements.

The time-dependent substrate collector response, with the tip under chronoamperometric control, has not been calculated for the EC_i mechanism. However, a formalism relating the steady-state substrate current to the steady-state tip generator current has been developed, resulting in (5):

$$i_S = i_T^{\text{con}} \left(\frac{i_T - i_T^{\text{ins}}}{i_T^{\text{con}} - i_T} \right) \tag{27}$$

where i_T^{con} and i_T^{ins} are the steady-state tip currents for the limits of pure diffusion-controlled positive ($K = 0$) and negative ($K \rightarrow \infty$) feedback, respectively. These currents are given by (5,26):

$$i_T^{con}/i_{T,\infty} = 0.78377/L + 0.3315\ \exp(-1.0672/L) + 0.68 \quad (28)$$

$$i_T^{ins}/i_{T,\infty} = (0.15 + 1.5358/L + 0.58\ \exp(-1.14/L) + 0.0908\ \exp[(L - 6.3)/(1.017L)])^{-1} \quad (29)$$

It has further been shown that N_{SS} depends on a single-valued dimensionless parameter (5):

$$\kappa = \frac{k_1 d^2}{D} \quad (30)$$

with the following empirical equation describing the relationship within about 1%

$$\kappa = 5.608 + 9.347\ \exp(-7.527N_{SS}) - 7.616\ \exp(-0.307/N_{SS}) \quad (31)$$

This shows the expected limiting behavior: as $\kappa \rightarrow 0$, $N_{SS} \rightarrow 1$; as $\kappa \rightarrow \infty$, $N_{SS} \rightarrow 0$.

The determination of kinetic parameters from collection efficiency measurements alone is most suitable when values of the latter lie between 0.1 and 0.8 (5), which corresponds to κ values in the range of ~0.5–10. Assuming d = 200 nm as the closest distance that is achievable reliably with disk-shaped UMEs of the type used for SECM kinetic measurements, this translates to an upper range for k_1 values of 1.25×10^4 to $2.5 \times 10^5\ s^{-1}$, for a typical D value of $10^{-5}\ cm^2\ s^{-1}$. Lower values of k_1 are, of course, accessible by employing larger d values.

2. *Substrate Generation/Tip Collection Mode*

In this mode, Eqs. (1)–(3) apply, but Eq. (1) is driven at a substrate electrode (larger than the tip), while the collection process occurs at the tip UME (see Fig 1b). In initial treatments of this mode (1,2), for various types of follow-up, and intermediate, chemical reactions, it was assumed that the tip UME was a passive sensor, acting as a local noninvasive probe of a time-dependent planar diffusion field at a larger substrate electrode. Although this approach gave interesting semi-quantitative insights into reaction kinetics and mechanisms, a more rigorous model, taking account of possible feedback effects between the substrate and tip, was later developed for the EC_i mechanism (6) and will be discussed here.

For the SG/TC mode, the diffusion equations outlined above [Eqs. (4) and (5)] still apply, together with the initial condition [Eq. (6)]. The boundary conditions for the substrate and the active portion of the tip are modified to reflect the fact that the substrate is now the generator and the tip is the collector. Additionally, the boundary condition for the radial edge of the tip/substrate domain assumes that there is planar diffusion of Red and Ox normal to the substrate electrode. This is handled by considering no radial flux of Red and Ox in this location, which is reasonable, provided that measurement times are not excessive (6). Consequently, Eqs. (7), (9), and (10) become:

$$z = 0,\ 0 \leq r \leq a\text{: } c_B = 0;\ D_A \frac{\partial c_A}{\partial z} = -D_B \frac{\partial c_B}{\partial z} \tag{32}$$

$$z = d,\ 0 \leq r \leq r_S\text{: } c_A = 0;\ D_B \frac{\partial c_B}{\partial z} = -D_A \frac{\partial c_A}{\partial z} \tag{33}$$

$$0 < z < d,\ r = r_S\text{: } D_A \frac{\partial c_A}{\partial r} = D_B \frac{\partial c_B}{\partial r} = 0 \tag{34}$$

Since for SG/TC measurements, the substrate is generally (1,2,6,10) [though not always (4,11)] several orders of magnitude larger than the tip, it can usually be assumed that the tip has no effect on the substrate characteristics, a fact that has been confirmed experimentally (4,6). In this situation, only the tip current is of interest for quantitative kinetic measurements of follow-up chemical reactions. After normalization, with respect to the steady-state n-electron current for the diffusion-controlled reduction of A, with the UME placed at a great distance from the substrate electrode [Eq. (20)], the collector tip current, $i_{T,c}$ is given by:

$$i_{T,c}/i_{T,\infty} = (\beta\pi/2) \int_0^1 (\partial C_B/\partial Z)_{Z=0}\ R\ dR \tag{35}$$

The normalized current-time response of the tip UME depends primarily on d/a, K, and β. As above, we consider here the case of $\beta = 1$. Tip chronoamperometric characteristics for a range of values of the normalized rate constant are shown in Figure 6 for log $d/a = -0.7$ and -0.2, respectively (typical of relatively small and large separations).

Considering first the $K = 0$ case, it can be seen that after an initial lag period, the tip current rises from zero to a steady-state value. The magnitude of the steady-state current and the rise time depend on the tip-substrate separation (6,10), such that the closer the tip and substrate, the more rapid the current rise, and the sooner a steady state is established.

In contrast, for finite values of K, the current does not reach a steady state, rather it rises to a peak value and then decreases at longer times. The values of both K and d/a have a marked effect on the overall shape of the tip current transient and the magnitude of the peak current. For a given tip/substrate separation, the larger the value of K, the smaller is the peak current, the shorter the time to reach the peak, and the steeper the decay curve following the peak. This can be explained by noting that by increasing the

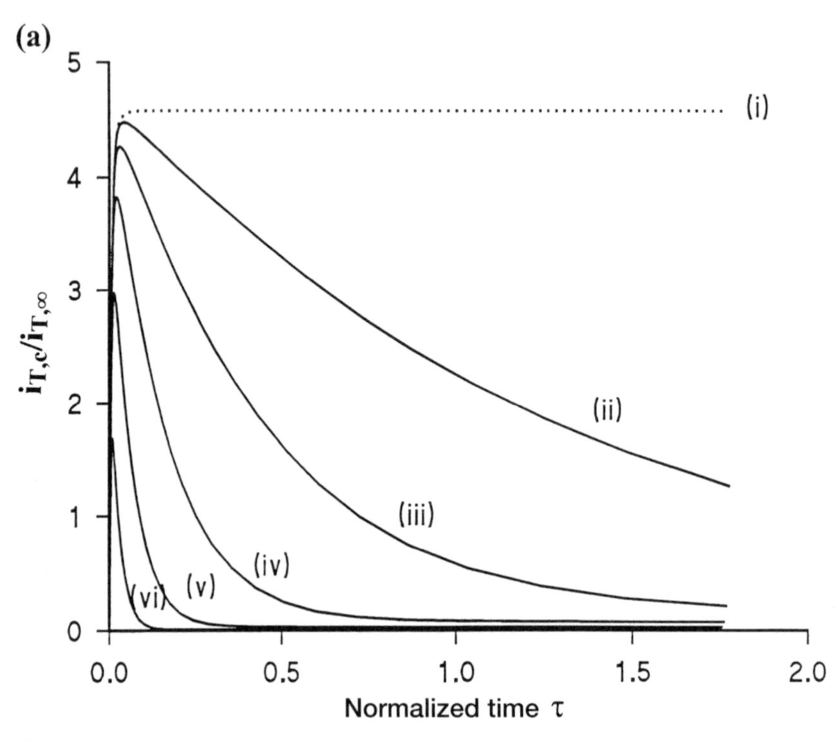
(a)
5
4
3
2
1
0
0.0
0.5
1.0
1.5
2.0
Normalized time τ
$i_{T,c}/i_{T,\infty}$
(i)
(ii)
(iii)
(iv)
(v)
(vi)

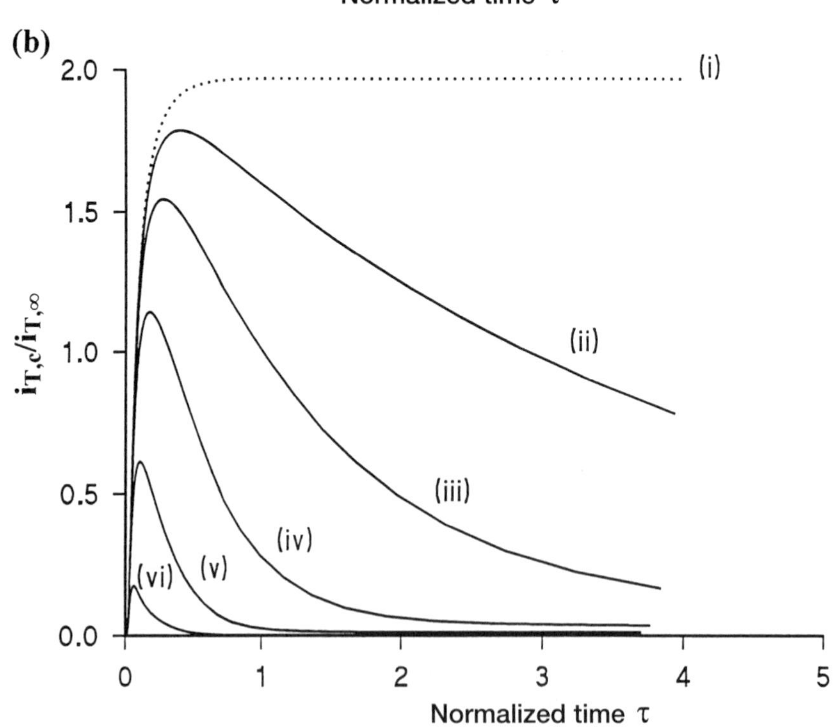
(b)
2.0
1.5
1.0
0.5
0.0
0
1
2
3
4
5
Normalized time τ
$i_{T,c}/i_{T,\infty}$
(i)
(ii)
(iii)
(iv)
(v)
(vi)

rate of the following chemical reaction, the fraction of substrate-generated B that reaches the tip in the first pass decreases, thereby diminishing the peak current. The significance of feedback between the substrate and the tip decreases with increasing rate constant, causing the rapid decay in the current and the appearance of the peak at shorter times.

As for feedback and TG/SC measurements (see Sec. II.A), it can be seen from Figure 6 that the tip current response becomes less sensitive to solution kinetics, the greater the spacing between the tip and substrate. For a given value of K, the peak current can be increased by decreasing the tip/substrate separation, but this also pushes the peak to shorter times. Thus, although close tip/substrate separations lead to high sensitivity in terms of tip current, the tip response has to be measured with the highest temporal resolution in this situation.

The characteristic features of the tip current transients in Figure 6 are the peak current, peak time, and post–half-peak time. These features show interesting dependences on the tip/substrate separation and K (Fig. 7). The magnitude of the normalized peak current (Fig. 7a) increases as the values of both K and d/a are decreased, whereas the time taken to reach the peak increases as the value of d/a is increased and K is decreased (Fig. 7b). This contrasting behavior allows both K and d/a to be determined from a single transient measurement with good precision (6). In particular, kinetics are determinable from SG/TC measurements with no prior knowledge of the tip/substrate separation (6). The post–half-peak time has a yet different dependence on d/a and K. This increases as K decreases but shows only a minimal dependence on the normalized interelectrode separation (Fig. 7c).

The range of k_1 values that are measurable with the SG/TC mode is largely governed by the time scale on which tip currents can be recorded. For example, towards the fast kinetic limit, a normalized peak current in excess of unity results for log $K = 2.5$, with tip/substrate separations closer than log $L < -0.8$ (Fig. 7). The normalized time at which the peak occurs is, however, smaller than 0.1. For a tip UME with a radius of 5 μm and a typical diffusion coefficient of 10^{-5} cm^2 s^{-1}, these figures relate to a rate constant in excess of 1.2×10^4 s^{-1}, but the corresponding peak time is less than 250 μs, placing some demands on the type of bipotentiostat required for practical measurements.

FIG. 6 SG/TC tip chronoamperometric characteristics for an EC_i process at normalized tip/substrate separations of log $d/a = -0.7$ (a) and -0.2 (b). The dotted lines (i) show the behavior for $K = 0$, while the solid lines are for log K values of: (ii) -0.5, (iii) 0.0, (iv) 0.5, (v) 1.0, and (vi) 1.5.

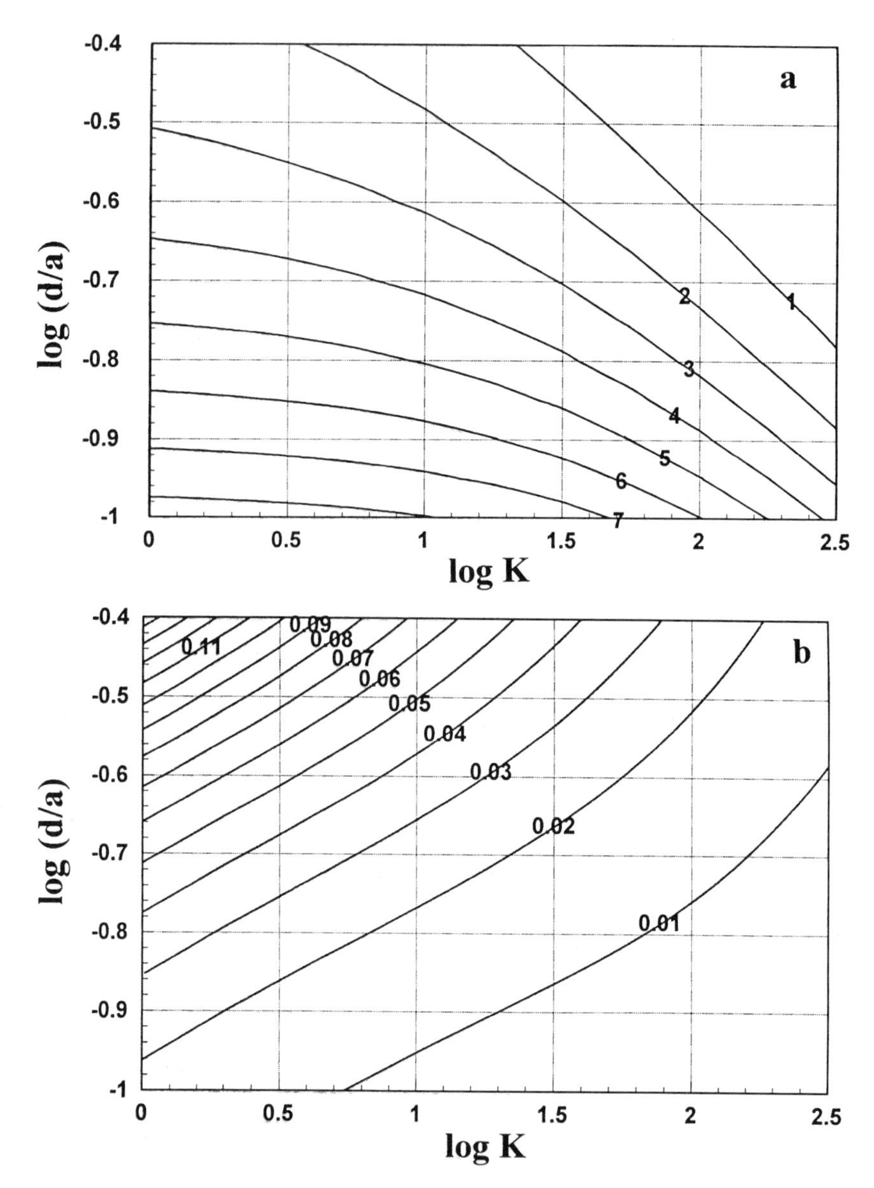

FIG. 7 Theoretical contour plots showing the variation of (a) peak current, (b) peak time, and (c) post–half-peak time with tip/substrate separation and normalized rate constant. The labels on the contours are values of the normalized current ratio ($i_{T,C}/i_{T,\infty}$) in (a) and normalized time (τ) in (b) and (c).

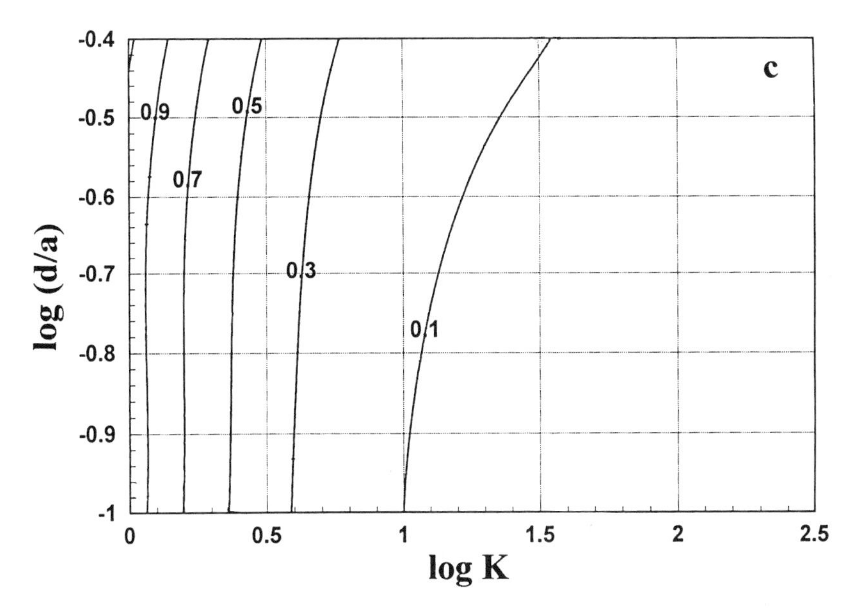

FIG. 7 Continued

B. Applications

1. Oxidation of N,N,*-Dimethyl-*p*-phenylenediamine (DMPPD) in Basic Aqueous Solution*

The oxidation of DMPPD is a model EC_i process, which has been used to characterize the capabilities of several electrochemical techniques to measure follow-up chemical kinetics (3,6,27–29). A fast two-electron oxidation, with the loss of a proton, constitutes the E step [Eq. (36)], which may be followed by a pH-dependent deamination (C) step [Eq. (37)]:

$$\underset{\mathbf{I}}{H_2N{-}C_6H_4{-}N(CH_3)_2} \underset{-H^+}{\overset{-2e^-}{\rightleftharpoons}} \underset{\mathbf{II}}{HN{=}C_6H_4{=}N^+(CH_3)_2} \tag{36}$$

$$\mathbf{III} \xrightarrow[k_1]{+OH^-} \text{(NH=}C_6H_4\text{=O)} + H_3C\text{–NH–}CH_3 \qquad (37)$$

The first SECM studies of this scheme employed chronoamperometric and steady-state feedback measurements with a 25 μm-diameter Pt UME tip and an unbiased macroscopic Pt disk substrate (12.5 mm diameter) (3).

In general, steady-state feedback measurements involve measuring $i_T - d$ characteristics (approach curves), typically by translating the tip towards or away from a substrate, in the perpendicular direction, with the UME held at a potential to effect the diffusion-limited electrolysis of the species of interest. The tip scan rate has to be sufficiently slow to ensure that a true steady-state is established, as discussed in Sec. II.A. Given the relatively large size of the probe employed for the DMPPD system, measurements were made pointwise, with the tip moved a fixed distance, after which it was stopped and the steady-sate current recorded (3).

Typical feedback approach curves for DMPPD oxidation in solutions of pH 10.20, 10.78, and 11.24 are shown in Figure 8, along with the behavior for diffusion-controlled positive and negative feedback. The tip electrode process was defined by Eq. (36), while the unbiased macroscopic substrate was poised at a potential where the reverse reaction occurred at a diffusion-controlled rate:

$$\mathbf{II} + 2e^- + H^+ \rightarrow \mathbf{I} \qquad (38)$$

It is clear from Figure 8 that, as the pH is raised, the current at any particular distance shifts away from the positive feedback response towards the behavior predicted for negative feedback due to the consumption of II in solution, according to Eq. (37).

Experimental data were found to analyze well in terms of the EC_i theory (3). Using $i_T - K$ working curves for each of the distances in Figure 8, the tip currents were converted to values of K and then κ. It follows from the definition of the latter kinetic parameter [Eq. (30)] that, for a given pH, a plot of κ values versus d^2 should be linear and intercept the origin. Figure 9 shows that this is the case; k_1 of 19 s^{-1} (pH 11.24), 5.8 s^{-1} (pH 10.78), and 1.4 s^{-1} (pH 10.20) were obtained from the slopes of the lines (3). The corresponding second-order rate constants (approximating the measured activity of OH^- by concentration), $k_2 = k_1/[OH^-]$, were found to be reasonably consistent: 1.1×10^4, 9.5×10^3, and 9.0×10^3 mol^{-1} dm^3 s^{-1}, respectively.

Chronoamperometric measurements of the oxidation of DMPPD were made over a range of pH from 7.80 to 12.42, with a fixed tip/substrate

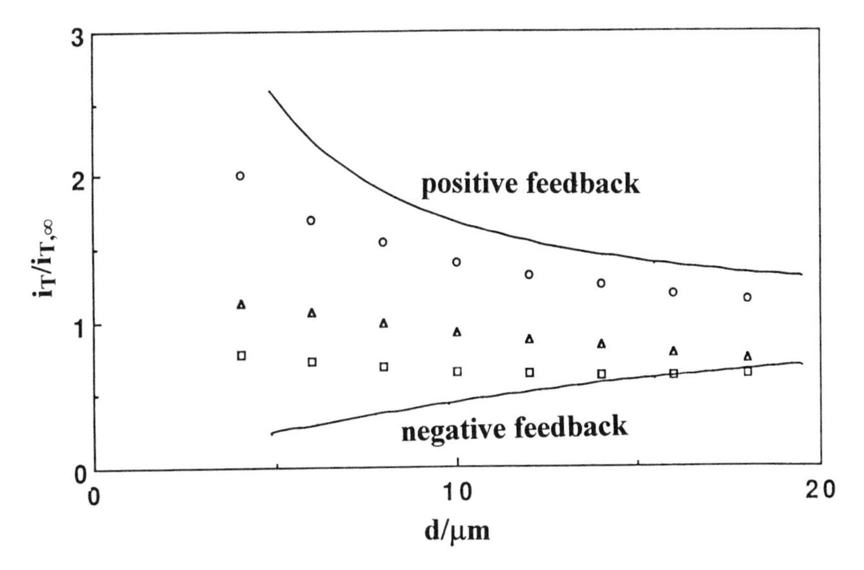

FIG. 8 Steady-state feedback $i_T - d$ characteristics for the oxidation of 5×10^{-4} M DMPPD at pH 10.20 (○), 10.78 (△), and 11.24 (□). The solid lines represent the theoretical behavior for diffusion-controlled positive feedback ($K = 0$) and negative feedback ($K \rightarrow \infty$).

separation of 6.5 μm (3). Typical experimental data are shown in Figure 10, alongside theoretical transients based on the mean rate constant derived from the steady-state measurements described above. For pH 7.80, the deamination process was assumed to proceed to a negligible extent, while for the other pH, the rate constants used to fit the data were $k_1 = 7.8$ s^{-1} (pH 10.90), 24 s^{-1} (pH 11.38), 42 s^{-1} (pH 11.63), and 260 s^{-1} (pH 12.42). Even with the large-diameter UME and tip/substrate separations employed for this study, the ability of transient chronoamperometric measurements to probe follow-up chemical reactions over a wide dynamic range, and relatively fast kinetics, is clearly evident from Figure 10.

As discussed in Sec. II.A, the analysis of SECM chemical kinetic data assumes that the electroactive precursor (A in the terminology of our example case) and the electrogenerated species (B) have the same diffusion coefficient, i.e., $\beta = 1$. A simple approach for confirming whether this holds is to measure the tip currents at a fixed (close) tip/substrate separation under positive feedback and then SG/TC control. For a chemically stable redox couple, the ratio of the tip feedback and collector currents under steady-state conditions reveals β directly (1):

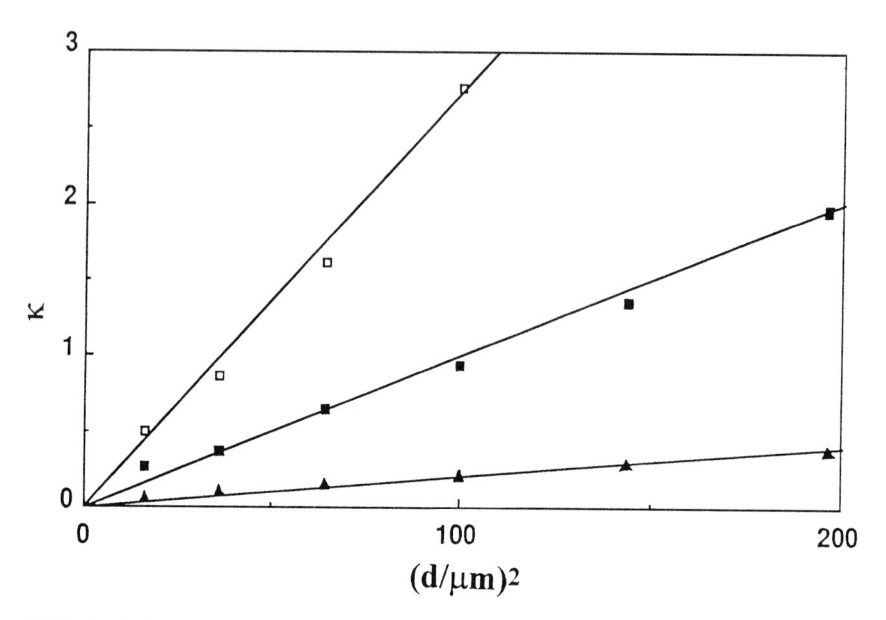

FIG. 9 Kinetic analysis of the data in Figure 8 [pH 10.20 (△), pH 10.78 (■), and pH 11.24 (□)].

$$\beta = \frac{i_{T,c}}{i_T^{con}} \tag{39}$$

without any prior knowledge of the tip/substrate separation, the electrode sizes, or the concentration of the species of interest. Figure 11 shows typical tip transients in the feedback and SG/TC modes for the oxidation of DMPPD at pH 7.65, where the follow-up chemical reaction [Eq. (37)] is negligible. The tip/substrate separation was ~5 μm, as inferred from the steady-state feedback current, which is β-independent (10) (although this information is not needed for the analysis). For the I/II couple, it was found that $\beta = 0.98 \pm 0.02$, indicating that the diffusion coefficients of I and II were effectively the same (10).

The kinetics of the deamination step [Eq. (37)] of oxidized DMPPD have also been investigated with the SG/TC mode, with a 2 mm square substrate generator electrode and a 25 μm diameter Pt disk UME collector electrode. Figure 12 shows a selection of typical tip current responses for the oxidation of I at the substrate electrode and the collection of II at the tip, with the solution at three pH values (10.94, 11.36, and 11.54) sufficient to promote the deamination process. In making these measurements, the substrate potential was stepped from 0.0 V (vs. a silver quasi reference

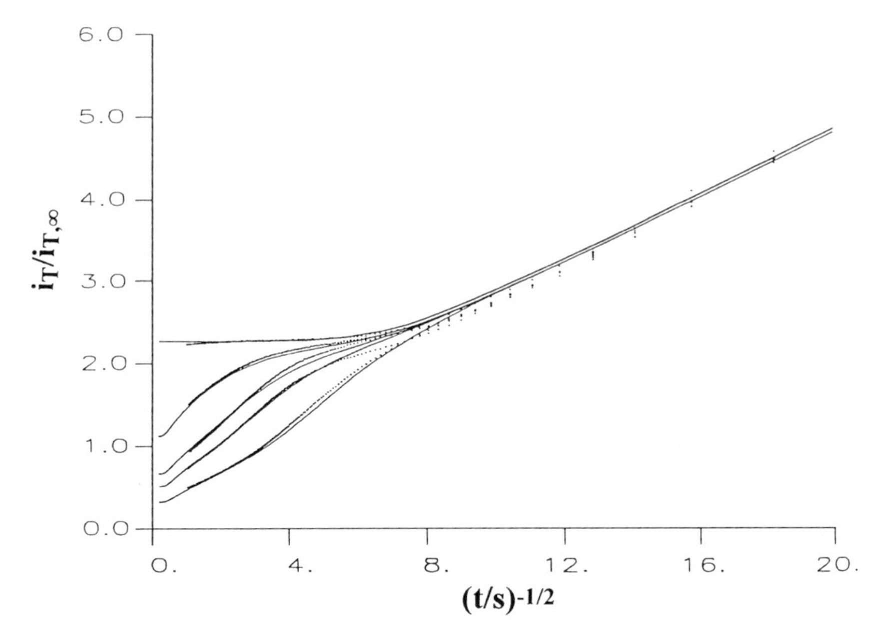

FIG. 10 Potential step chronoamperometric feedback characteristics (· · ·) with a 25 μm diameter Pt tip at a tip/substrate separation of 6.5 μm, for the oxidation of DMPPD at pH 7.80 (upper curve), pH 10.90, pH 11.38, pH 11.63, and pH 12.42 (lower curve). The solid lines show the corresponding theoretical behavior for each pH using the rate constants defined in the text.

electrode, AgQRE), where there were no Faradaic processes, to 0.5 V, where the oxidation of DMPPD was driven at a diffusion-controlled rate. The tip UME potential was held at 0.0 V to drive reaction (38) at a diffusion-controlled rate.

The shapes of the experimental tip transients were found to be consistent with the theoretical predictions for an EC_i process (Sec. II.A), with the collector current initially rising to a peak value and then decaying at longer times. In each case, the interelectrode separation and diffusion coefficient of DMPPD were known, and thus the rate constant was the only variable in the fitting procedure. The transients at different pH yield a consistent rate constant for the deamination step, with values for k_1 of 8.8 s^{-1} [(a), pH 10.94], 23 s^{-1} [(b), pH 11.36], and 33 s^{-1} [(c) pH 11.54]. For comparative purposes, these values yielded a second-order rate constant, $k_2 = k_1/[OH^-]$ $= 1.0\ (\pm 0.1) \times 10^4\ mol^{-1}\ dm^3\ s^{-1}$, in good agreement with the feedback measurements discussed above.

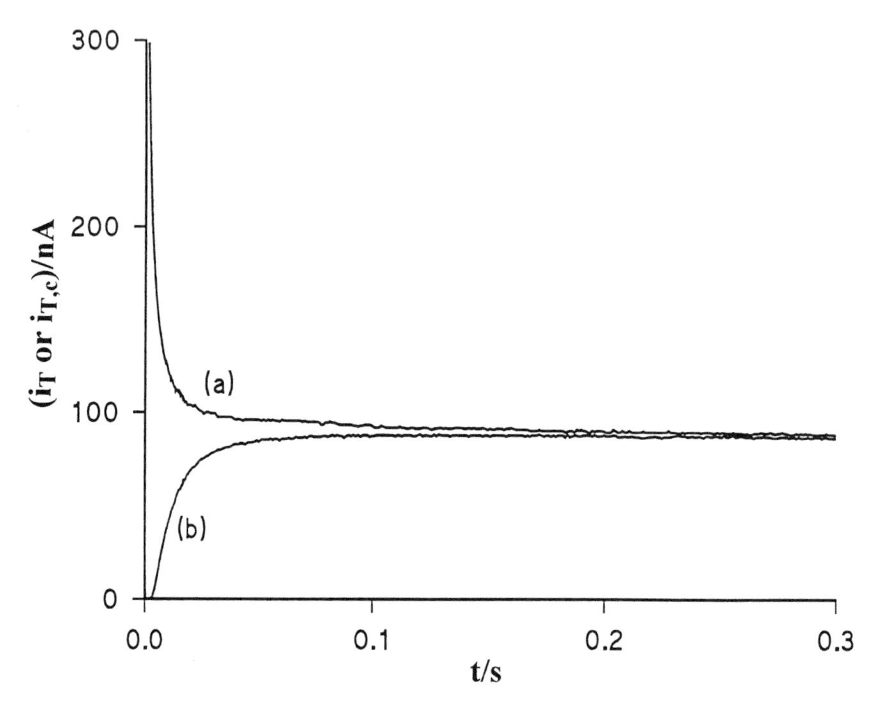

FIG. 11 (a) Tip feedback and (b) SG/TC tip collector chronoamperometric characteristics for the oxidation of 4 mM DMPPD at pH 7.65, with a fixed tip/substrate separation of 5.0 μm. The transients reach similar steady-state values indicating that the ratio of diffusion coefficients of the electrode reactant and product (β) is close to unity.

Although the interelectrode separations were known with good precision in this set of experiments (6), the extent to which SG/TC transients could determine both the rate constant and interelectrode separation was also examined. Figure 13 shows the sets of interelectrode spacings and rate constants that are consistent with the peak currents and peak times derived from the transients in Figure 12. The intersection of the two contours on each

→

FIG. 12 Tip collector chronoamperometric responses for the oxidation of 4 mM DMPPD in the SG/TC mode, under the following conditions: (a) pH 10.54, $d = 2.75$ μm; (b) pH 11.36, $d = 2.0$ μm; (c) pH 11.54, $d = 2.5$ μm. The experimental data (——) have been fitted to theoretical responses using the values of k_1 cited in the text.

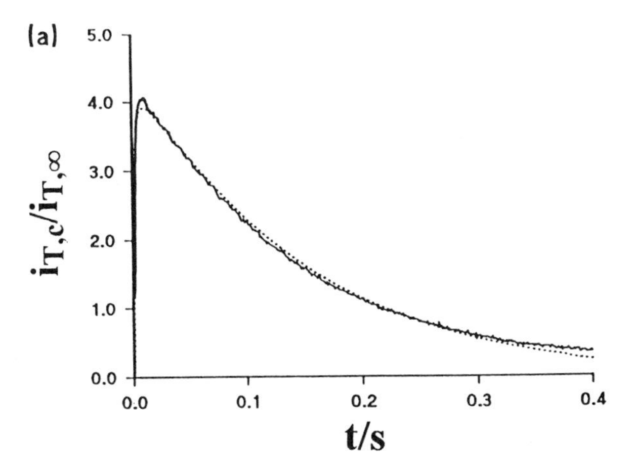
(a)
5.0
4.0
3.0
2.0
1.0
0.0
$i_{T,c}/i_{T,\infty}$
0.0
0.1
0.2
0.3
0.4
t/s

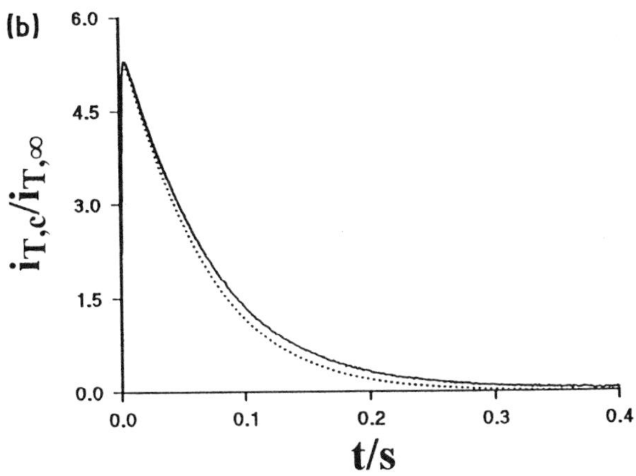
(b)
6.0
4.5
3.0
1.5
0.0
$i_{T,c}/i_{T,\infty}$
0.0
0.1
0.2
0.3
0.4
t/s

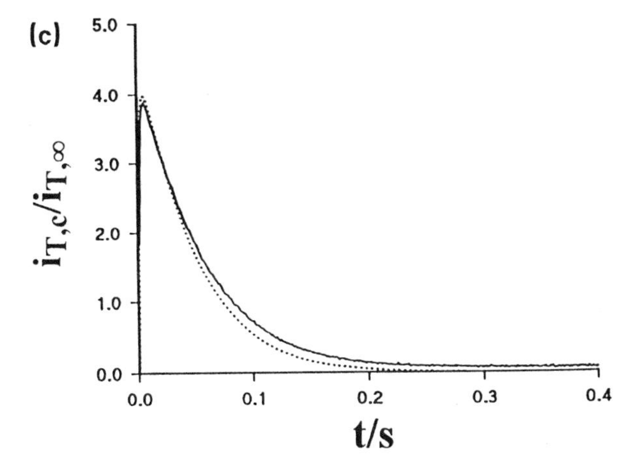
(c)
5.0
4.0
3.0
2.0
1.0
0.0
$i_{T,c}/i_{T,\infty}$
0.0
0.1
0.2
0.3
0.4
t/s

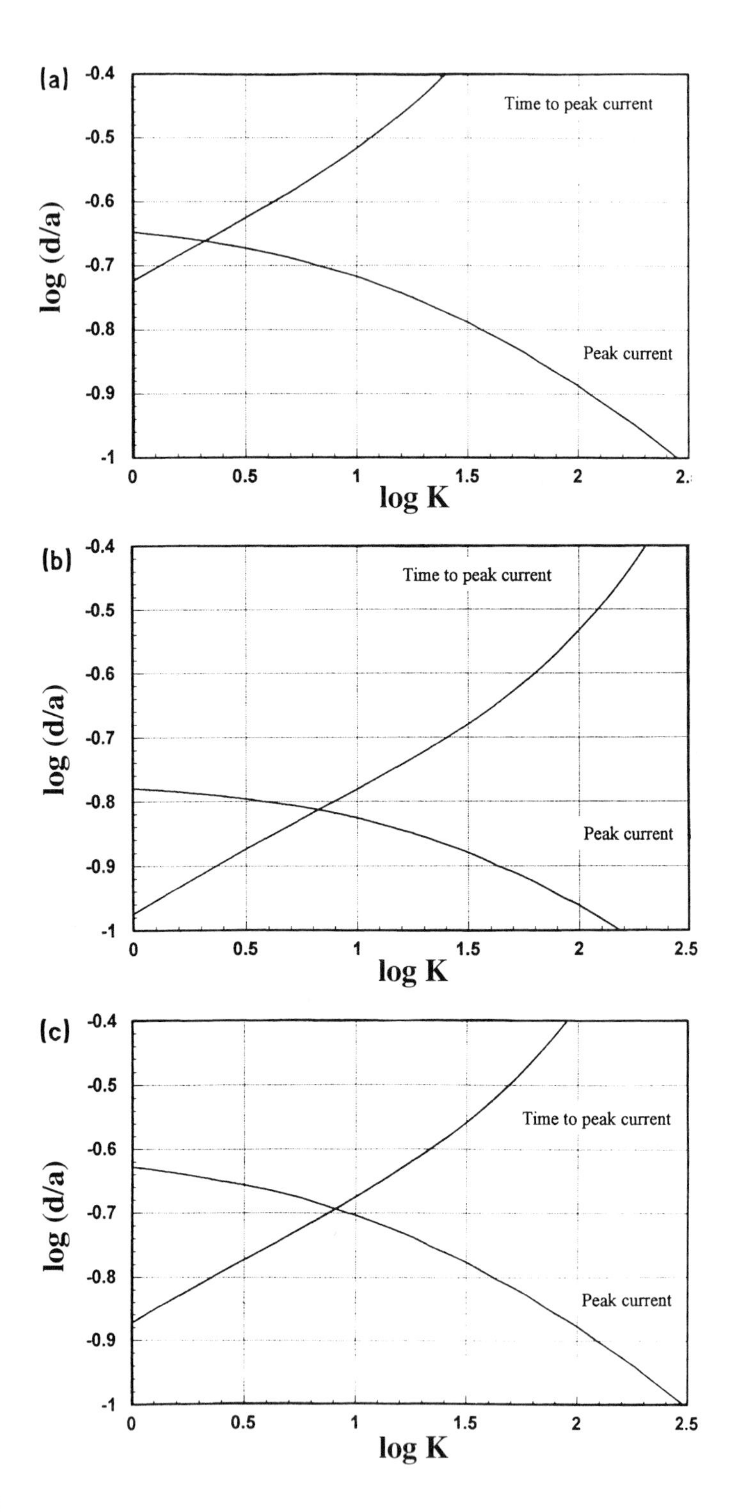
(a)
Time to peak current
Peak current
log (d/a)
log K
-0.4
-0.5
-0.6
-0.7
-0.8
-0.9
-1
0
0.5
1
1.5
2
(b)
Time to peak current
Peak current
log (d/a)
log K
2.5
(c)
Time to peak current
Peak current
log (d/a)
log K

plot yields the following values of d and k_1, respectively: (a) 2.7 μm and 9.2 s^{-1} (pH 10.94); (b) 1.9 μm and 30 s^{-1} (pH 11.36 and; (c) 2.6 μm and 37 s^{-1} (pH 11.54). The values of interelectrode separation are close to those assumed in fitting the entire transients above, while the rate constants are in good agreement with those obtained by fitting the full tip transients, yielding $k_2 = 1.15\ (\pm 0.15) \times 10^4$ mol^{-1} dm^3 s^{-1}. These results clearly illustrate that following chemical reaction rate constants and interelectrode separations can be determined with good precision and accuracy from the measurement of the peak current and peak time from tip collector transient measurements in the SG/TC mode.

2. *Reduction of [Cp*Re(CO)$_2$(p-N$_2$C$_6$H$_4$OMe)] [BF$_4$] in Acetonitrile*

SECM feedback measurements have been applied, together with cyclic voltammetry and controlled potential electrolysis (30), to the reduction of the rhenium aryldiazenido complex [Cp*Re(CO)$_2$(p-N$_2$C$_6$H$_4$OMe)] [BF$_4$] (III(BF$_4$); Cp* = η-C$_5$Me$_5$). A one-electron reduction results in the 19-electron complex, IV [Eq. (40)], which then decomposes to give products [Eq. (41)].

$$\underset{\textbf{III}}{[\mathrm{Cp^*Re(CO)_2}(p\text{-}\mathrm{N_2C_6H_4OMe})]^+} + e^- \rightarrow \underset{\textbf{IV}}{\mathrm{Cp^*Re(CO)_2}(p\text{-}\mathrm{N_2C_6H_4OMe})} \tag{40}$$

$$\textbf{IV} \xrightarrow{k_1} \mathrm{Cp^*Re(CO)_2(N_2)} \tag{41}$$

The normalized steady-state approach curve of tip current versus tip/substrate separation for the reduction of III(BF$_4$) is shown in Figure 14. This was obtained with a 25 μm diameter Pt UME, biased at −0.8 V versus AgQRE to affect the diffusion-controlled reduction of III, as the probe was translated towards a 1 mm diameter Pt disk substrate, biased at 0.0 V versus AgQRE to promote the diffusion-controlled oxidation of IV. When analyzed in terms of EC$_i$ theory, the approach curve yielded a value $k_1 = 145$ s^{-1}, which was in excellent agreement with that determined by cyclic voltammetry at sweep rates between 10 and 50 V s^{-1}.

3. *Identification of Intermediates in the Oxidation of Borohydride*

The oxidation of borohydride at gold electrodes in aqueous alkaline solution is a highly complex multistage process (11,31,32). On relatively long vol-

FIG. 13 Contours showing the sets of values of log (d/a) and log K derived from the measurements of the peak current and peak time from the transients in Figure 12. Data are for pH: (a) 10.94, (b) 11.36, and (c) 11.54.

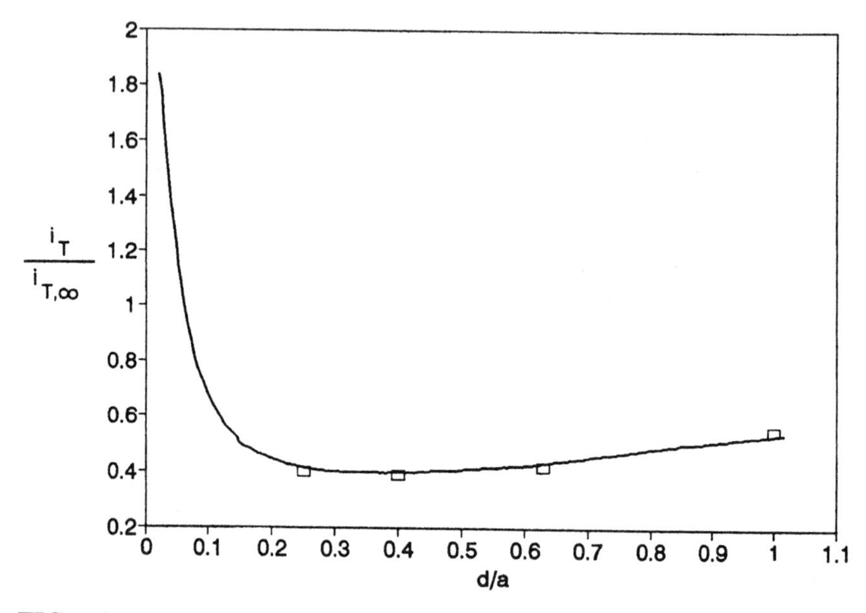

FIG. 14 Tip feedback approach curve (solid line) for the reduction of 1.0 mM [Cp*Re(CO)$_2$(p-$N_2C_6H_4$OMe)] [BF_4] in acetonitrile at a 25 μm diameter Pt UME translated towards a Pt substrate at 0.05 μm s^{-1}. The remaining experimental details are given in the text. The squares are theoretical values for $k_1 = 145\ s^{-1}$.

tammetric time scales—either slow scan speed (100 mV s^{-1}) experiments at large electrodes, or steady-state measurements at moderate-sized microdisc electrodes ($a = 25\ \mu$m)—voltammograms were found to be irreversible, with close to eight electrons involved in the oxidation process. This was attributable to the following overall reaction:

$$BH_4^- + 8\ OH^- \rightarrow BO_2^- + 6\ H_2O + 8e^- \tag{42}$$

With faster scan cyclic voltammetry, a new two-electron anodic peak was detected, at more negative potentials, for the first stage of the oxidation process, with an accompanying cathodic peak on the reverse scan (11). The ratio of the forward to the reverse peak currents increased towards unity as the scan rate was raised to ~200 V s^{-1} (Fig. 15). This behavior was attributed to the initial two-electron process being accompanied by a fairly rapid follow-up chemical reaction and was successfully analyzed in terms of an E_qC_i process (quasi-reversible electron transfer followed by a first-order irreversible chemical process), with a rate constant for the chemical step, $k_1 = 250\ s^{-1}$.

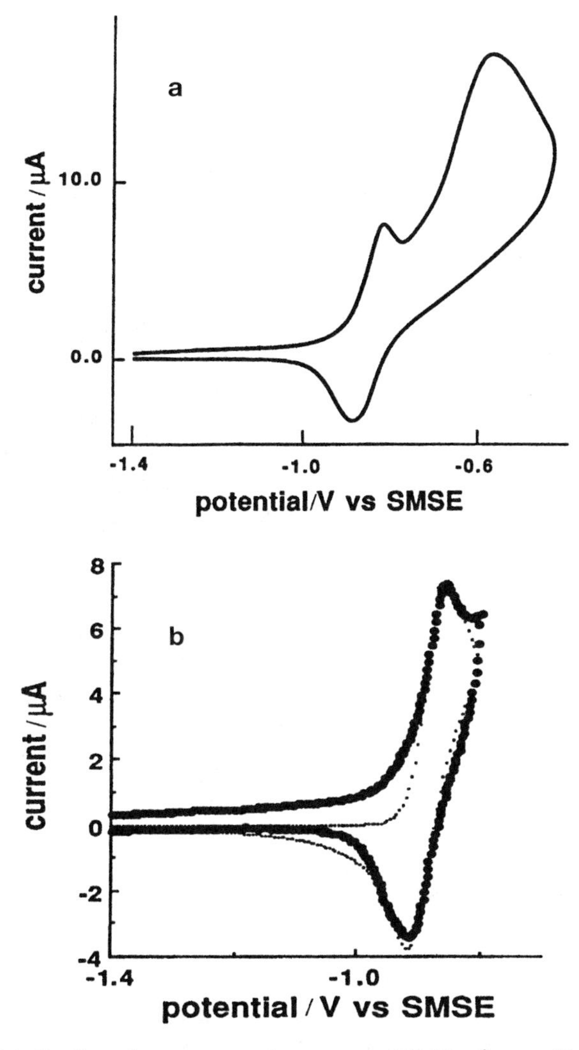

FIG. 15 (a) Cyclic voltammogram (sweep rate 200 V s^{-1}) at a 50 μm diameter Au microdisk for the oxidation of 10 mM sodium borohydride in 1 M aqueous NaOH. (b) As (a), but with a reversal potential of -0.8 V. The fine points are the experimental data, and the bold points represent the behavior computed for an E_qC_i mechanism for a two-electron process with $D = 1.6 \times 10^{-5}$ cm^2 s^{-1}, a standard rate constant of 0.3 cm s^{-1}, and a transfer coefficient of 0.5.

SECM SG/TC experiments were carried out to prove that the product of the initial two-electron oxidation process diffused into the solution, where it would react homogeneously and irreversibly. For these measurements, a 10 μm diameter Au tip UME was stationed ~1 μm above a 100 μm diameter Au substrate electrode. With the tip held at a potential of -1.3 V versus saturated mercurous sulfate electrode (SMSE), to collect substrate-generated species by reduction, the substrate electrode was scanned through the range of potentials to effect the oxidation of borohydride. The substrate and tip electrode responses for this experiment are shown in Figure 16. The fact that a cathodic current flowed at the tip, when the substrate was at a potential where borohydride oxidation occurred, proved that the intermediate formed in the initial two-electron transfer process (presumed to be monoborane), diffused into the solution. An upper limit of 500 s^{-1} was estimated for the rate constant describing the reaction of this species (with water or OH^-), based on the diffusion time in the experimental configuration. This was consistent with the results of the cyclic voltammetry experiments (11).

III. EC_{2i} PROCESSES

In the EC_{2i} process, an initial electron transfer step is followed by a second-order irreversible chemical reaction (typically a dimerization process, as considered in the practical examples in Sec. III.B). The use of SECM to characterize the kinetics of the second-order chemical reactions is based on the same principles as for the EC_i case, discussed in Sec. II, with a generator electrode employed to electrogenerate the species of interest [B, see Eq. (1)], which is collected at a second electrode. The second-order process involving the consumption of B to form electroinactive products occurs in the gap between the two electrodes:

$$\text{Solution:} \qquad 2\,\mathrm{B} \xrightarrow{k_2} \text{Products} \tag{43}$$

Rigorous theoretical treatments for this mechanism have been developed for the feedback and TG/SC modes (4,5), which are reviewed in Sec. III.A. These modes (4,5,8) along with the SG/TC mode (1), have found applications in several areas, as discussed in Sec. III.B.

A. Theory

The diffusion equation for A in the SECM configuration is Eq. (4). That for B now reflects the second-order decomposition kinetics:

$$\frac{\partial c_B}{\partial t} = D_B\left(\frac{\partial^2 c_B}{\partial r^2} + \frac{1}{r}\frac{\partial c_B}{\partial r} + \frac{\partial^2 c_B}{\partial z^2}\right) - k_2 c_B \tag{44}$$

For the feedback and TG/SC modes, all of the remaining equations, defining

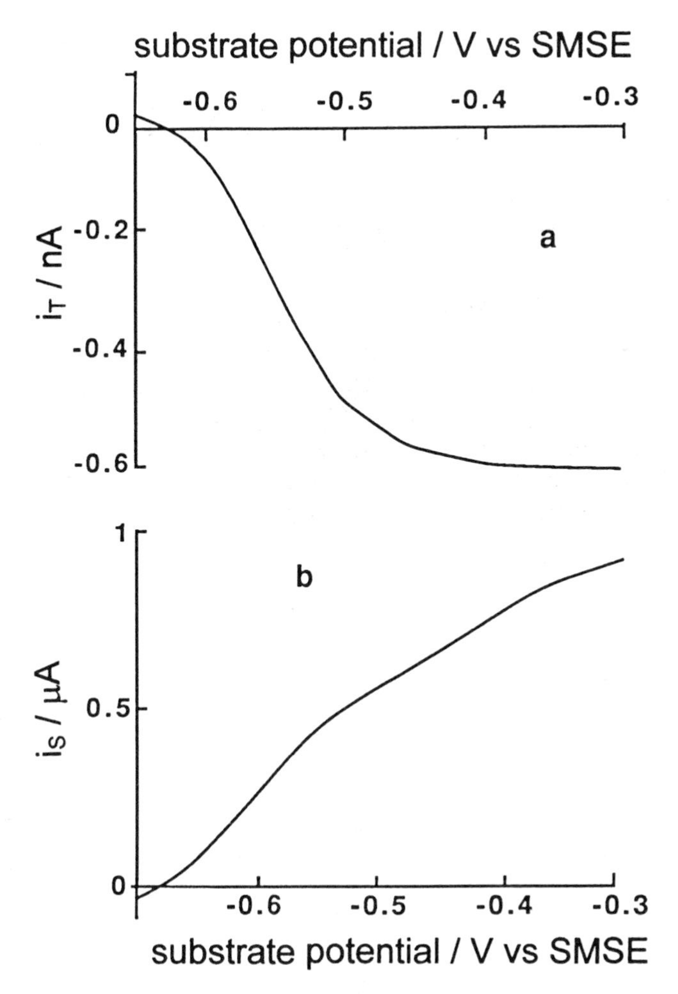

FIG. 16 Steady-state tip (a) and substrate (b) currents as a function of substrate potential for SG/TC measurements. A 10 μm diameter Au tip, biased at −1.3 V versus SMSE, was positioned ~1 μm above a 100 μm diameter substrate electrode, which was swept through the range of potentials indicated.

the initial and boundary conditions, and the tip and substrate currents, are as given in Eqs. (6)–(22), except that the normalized homogeneous rate constant is given by:

$$K_2 = k_2 a^2 c_A^* / D_A \tag{45}$$

Theoretical results have been presented for both chronoamperometric and steady-state TG/SC conditions (4). Typical tip and substrate chronoamperometric characteristics for log $d/a = -0.4$ and $K_2 = 1$, 10, and 100 are shown in Figure 17, along with the corresponding kinetically uncomplicated diffusion-controlled behavior, which has already been discussed (Sec. II.A). As expected, the tip (feedback) behavior, with follow-up kinetics, is qualitatively similar to the EC_i case, considered in Sec. II.A, with an increase in K_2 serving to decrease the tip current compared to the pure positive feedback case.

The general substrate (collector) chronoamperometric characteristics for the TG/SC mode, with kinetics, deserve greater attention, as this mode was not considered in the initial treatment of the EC_i mechanism (3), but was treated for the EC_{2i} process (4). At the shortest times, the substrate current, with the range of rate constants considered, follows the behavior for the kinetically uncomplicated case (discussed in detail in Sec. II.A). For longer times, the following chemical reaction competes with the diffusion of B from the tip to the substrate, decreasing the flux, and hence current, at that electrode. This leads to a peak in the $i_S - \tau$ characteristic, as shown in Figure

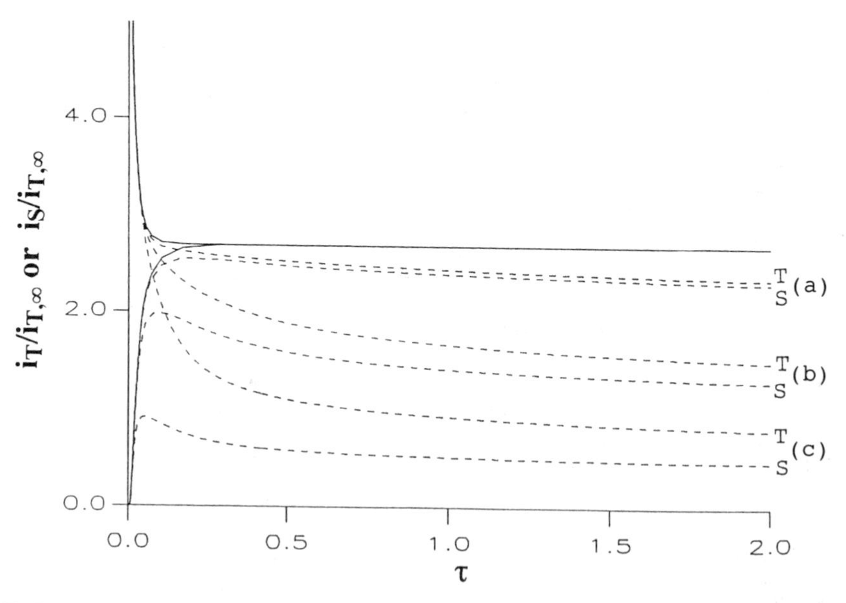

FIG. 17 TG/SC tip and substrate characteristics for the EC_{2i} mechanism at log $(d/a) = -0.4$. The solid line is the behavior in the absence of following homogeneous reactions, while the labeled dashed lines are for K_2 = (a) 1, (b) 10, and (c) 100. The labels T and S denote tip and substrate, respectively.

17. The magnitude and position of the peak varies with K_2 in a similar manner to the collector peak for the SG/TC mode with kinetics (Sec. II.A), i.e., the peak decreases in size and moves to shorter times as K_2 increases. In contrast to the SG/TC mode, however, the collector response for the TG/SC mode attains a steady-state value at long times, due to the hindered diffusion of A to the tip, which ensures the generation of a continuous supply of B for collection at the substrate.

Hitherto, the feedback and TG/SC modes have found exclusive application in the quantitative study of dimerization kinetics under steady-state conditions (4,5,8). Table 2 provides an extensive list of normalized steady-state tip and substrate currents, as a function of K_2 and d/a, which can be used for the analysis of experimental data. The characteristics in Table 2 display the general trends already identified for follow-up chemical reactions (Sec. II.A). (1) For a given d/a value, the tip current varies from a limit corresponding to pure positive feedback (as $K_2 \rightarrow 0$) to one for negative feedback (as $K_2 \rightarrow \infty$), while the collection efficiency varies from unity to zero as K_2 increases towards infinity. (2) The feedback and collection currents become most sensitive to kinetics, the closer the tip/substrate separation. Additionally, increasingly fast kinetics become accessible as the tip/substrate separation is minimized.

The following empirical relationship between the steady-state collection efficiency and a generalized parameter, κ_2, allows the measurement of rate constants with an error of no greater than 5–10%, provided $0.3 \leq N_{SS} \leq 0.9$ (5):

$$\kappa_2 = 104.87 - 9.948N_{SS} - 185.89/N_{SS}^{1/2} + 90.199/N_{SS} + 9.389/N_{SS}^2 \tag{46}$$

where

$$\kappa_2 = k_2 d^3 c_A^* / Da \tag{47}$$

As pointed out in Sec. II.A, in terms of determining rapid kinetics, SG/TC measurements are of greatest interest. From Eq. (46), $N_{SS} \sim 0.3$ corresponds to $\kappa_2 \sim 70$. Assuming $d = 200$ nm is the closest attainable tip/substrate separation and a typical value of $D = 10^{-5}$ cm^2 s^{-1}, the upper limit on the range of measurable rate constants can be estimated as $k_2 \leq (8.75 \times 10^{10}$ cm^{-1} s^{-1}) a/c_A^*. The absolute value of k_2 then depends on the value of a and c_A^*, with a large value of the former parameter and small value of the latter, favoring the measurement of rapid kinetics. Even for moderate values of $a = 2.5$ μm and $c_A^* = 10^{-3}$ mol dm^{-3}, diffusion-controlled homogeneous reactions are potentially accessible with this technique.

TABLE 2 Theoretical Normalized Tip and Substrate Steady-State Currents for the E_rC_{2i} Mechanism

$\log_{10}$	$K_2 = 0.5$		$K_2 = 1$		$K_2 = 2$		$K_2 = 3$		$K_2 = 5$		$K_2 = 10$	
$(L = d/a)$	$i_T/i_{T,\infty}$	$i_S/i_{T,\infty}$	$i_T/i_{T,\infty}$	$i_S/i_{T,\infty}$	$i_T/i_{T,\infty}$	$i_S/i_{T,\infty}$	$i_T/i_{T,\infty}$	$i_S/i_{T,\infty}$	$i_T/i_{T,\infty}$	$i_S/i_{T,\infty}$	$i_T/i_{T,\infty}$	$i_S/i_{T,\infty}$
−0.8	4.759	4.759	4.255	4.236	3.630	3.630	3.052	3.047	2.778	2.756	2.186	2.144
−0.7	3.907	3.893	3.490	3.479	2.981	2.976	2.671	2.658	2.289	2.256	1.813	1.756
−0.6	3.226	3.221	2.889	2.873	2.480	2.450	2.226	2.187	1.909	1.860	1.517	1.448
−0.5	2.694	2.684	2.418	2.392	2.082	2.037	1.875	1.818	1.616	1.544	1.296	1.200
−0.4	2.278	2.260	2.050	2.011	1.775	1.710	1.604	1.523	1.392	1.291	1.131	1.001
−0.3	1.955	1.924	1.766	1.708	1.529	1.448	1.399	1.288	1.225	1.089	1.012	0.840
−0.2	1.705	1.658	1.549	1.468	1.362	1.239	1.247	1.099	1.105	0.925	0.931	0.708
−0.1	1.514	1.448	1.386	1.276	1.232	1.071	1.138	0.946	1.023	0.791	0.882	0.600
0.0	1.371	1.282	1.265	1.123	1.125	0.934	1.065	0.820	0.972	0.680	0.860	0.508
0.1	1.265	1.148	1.178	0.997	1.079	0.821	1.019	0.714	0.946	0.586	0.859	0.429
0.2	1.188	1.039	1.123	0.892	1.041	0.723	0.995	0.623	0.939	0.503	0.874	0.361

$\log_{10}$ $(L = d/a)$	$K_2 = 15$		$K_2 = 20$		$K_2 = 25$		$K_2 = 30$		$K_2 = 40$		$K_2 = 50$	
	$i_T/i_{T,\infty}$	$i_S/i_{T,\infty}$	$i_T/i_{T,\infty}$	$i_S/i_{T,\infty}$	$i_T/i_{T,\infty}$	$i_S/i_{T,\infty}$	$i_T/i_{T,\infty}$	$i_S/i_{T,\infty}$	$i_T/i_{T,\infty}$	$i_S/i_{T,\infty}$	$i_T/i_{T,\infty}$	$i_S/i_{T,\infty}$
−0.8	1.871	1.831	1.672	1.630	1.538	1.499	1.431	1.377	1.265	1.210	1.166	1.099
−0.7	1.568	1.503	1.396	1.336	1.288	1.216	1.192	1.126	1.068	0.993	0.977	0.898
−0.6	1.310	1.236	1.180	1.098	1.084	1.000	1.012	0.925	0.911	0.814	0.834	0.734
−0.5	1.124	1.023	1.021	0.908	0.943	0.826	0.884	0.763	0.805	0.671	0.726	0.601
−0.4	0.993	0.851	0.906	0.754	0.842	0.685	0.792	0.633	0.722	0.556	0.673	0.501
−0.3	0.899	0.712	0.828	0.629	0.774	0.570	0.735	0.525	0.678	0.460	0.638	0.413
−0.2	0.839	0.597	0.779	0.526	0.738	0.475	0.706	0.436	0.660	0.379	0.628	0.339
−0.1	0.807	0.502	0.760	0.439	0.727	0.395	0.701	0.361	0.665	0.311	0.639	0.277
0.0	0.801	0.421	0.764	0.365	0.738	0.326	0.718	0.296	0.690	0.254	0.670	0.224
0.1	0.814	0.351	0.786	0.302	0.766	0.268	0.751	0.241	0.730	0.204	0.715	0.179
0.2	0.842	0.291	0.821	0.247	0.806	0.211	0.795	0.194	0.780	0.163	0.770	0.141

TABLE 2 Continued

$\log_{10}$	$K_2 = 100$		$K_2 = 200$		$K_2 = 300$		$K_2 = 500$		$K_2 = 1000$	
$(L = d/a)$	$i_T/i_{T,\infty}$	$i_S/i_{T,\infty}$	$i_T/i_{T,\infty}$	$i_S/i_{T,\infty}$	$i_T/i_{T,\infty}$	$i_S/i_{T,\infty}$	$i_T/i_{T,\infty}$	$i_S/i_{T,\infty}$	$i_T/i_{T,\infty}$	$i_S/i_{T,\infty}$
−0.8	0.864	0.797	0.629	0.577	0.532	0.472	0.433	0.365	0.335	0.254
−0.7	0.728	0.655	0.552	0.470	0.470	0.384	0.392	0.296	0.311	0.205
−0.6	0.639	0.536	0.498	0.382	0.432	0.312	0.369	0.239	0.303	0.164
−0.5	0.581	0.439	0.464	0.311	0.413	0.252	0.361	0.192	0.307	0.131
−0.4	0.546	0.359	0.451	0.252	0.410	0.203	0.367	0.153	0.324	0.102
−0.3	0.535	0.292	0.458	0.202	0.424	0.161	0.389	0.119	0.356	0.078
−0.2	0.545	0.236	0.484	0.160	0.456	0.126	0.430	0.091	0.403	0.058
−0.1	0.574	0.189	0.527	0.125	0.506	0.096	0.458	0.069	0.466	0.042
0.0	0.620	0.148	0.585	0.095	0.570	0.070	0.556	0.050	0.542	0.030
0.1	0.680	0.115	0.655	0.072	0.644	0.053	0.635	0.036	0.626	0.021
0.2	0.746	0.088	0.729	0.053	0.723	0.039	0.701	0.026	0.711	0.015

Source: Ref. 4.

B. Applications

1. Reductive Dimerization of Dimethyl Fumarate and Fumaronitrile in N,N-*Dimethylformamide*

The reductive hydrodimerization of activated olefins was considered to be a good test system for initial SECM studies of EC_{2i} processes due to the fact that the dimerization rate constant could be tuned via the substituent activating the olefinic double bond (33–35). For the relatively slow dimerization of the dimethyl fumarate (DF) radical anion, a large UME (25 μm diameter Pt electrode) was employed so that the value of K_2 [Eq. (45)] was in the range where the SECM response was sensitive to the kinetics. Experiments were carried out with solutions of 5.15 and 11.5 mM DF in *N,N*-dimethylformamide (DMF), containing 0.1 M tetrabutylammonium tetrafluoroborate as a supporting electrolyte. The oxidation of *N,N,N′,N′*-tetramethyl-1,4-phenylenediamine (TMPD), included in solutions at a concentration of ~4 mM, was used as a calibrant of the tip/substrate separation, through positive feedback measurements.

Normalized steady-state feedback current-distance approach curves for the diffusion-controlled reduction of DF and the one-electron oxidation of TMPD are shown in Figure 18. The experimental approach curves for the reduction of DF lie just below the curve for the oxidation of TMPD, diagnostic of a follow-up chemical reaction in the reduction of DF, albeit rather slow on the SECM time scale. The reaction is clearly not first-order, as the deviation from positive feedback increases as the concentration of DF is increased. Analysis of the data in terms of EC_{2i} theory yielded values of $K_2 = 0.14$ (5.15 mM) and 0.27 (11.5 mM), and thus fairly consistent k_2 values of 180 M^{-1} s^{-1} and 160 M^{-1} s^{-1}, respectively. Due to the relatively slow follow-up chemical reaction, steady-state TG/SC measurements carried out under these conditions yielded collection efficiencies close to unity over the range of tip-substrate separations investigated ($-0.5 \leq \log d/a \leq 0.0$) (4).

Typical steady-state tip and substrate electrode responses for TG/SC voltammetric measurements, on the reduction of 28.2 mM fumaronitrile (FN), with a tip ($a = 5$ μm) − substrate ($r_S = 50$ μm) separation of 1.8 μm are shown in Figure 19. These were obtained by scanning the tip (generator) potential at 100 mV/s through the FN/$FN^{\bullet -}$ wave while holding the substrate (collector) electrode at a potential to detect $FN^{\bullet -}$ by diffusion-controlled oxidation to FN. The voltammograms provide a clear illustration of the ability of the SECM TG/SC mode to pick up relatively unstable intermediates with good sensitivity.

Tip and substrate steady-state approach curves for TG/SC measurements with four different FN concentrations are shown in Figure 20, along with

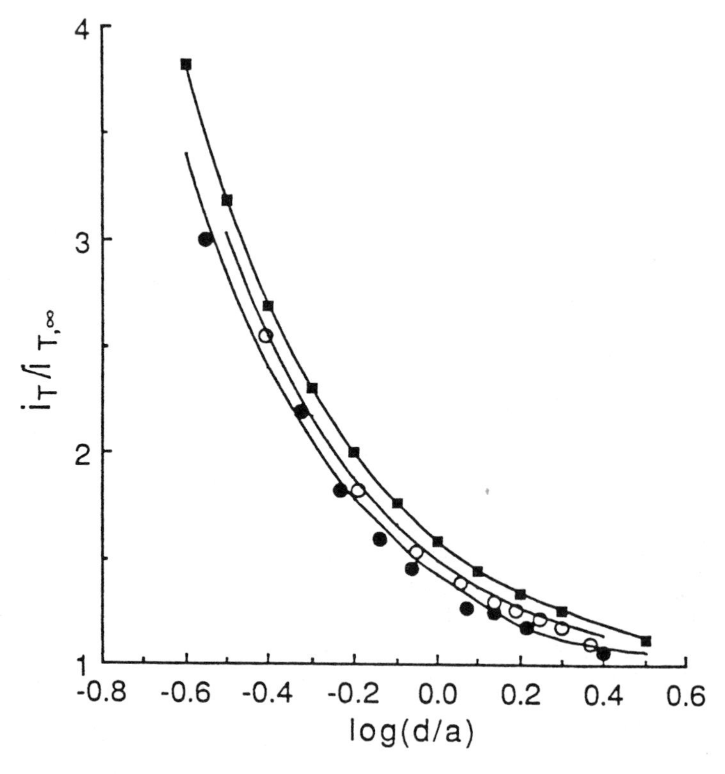

FIG. 18 Normalized steady-state tip feedback current-distance behavior for the reduction of DF at concentrations of 5.15 mM (○) and 11.5 mM (●), along with the best theoretical fits (solid lines) for $K_2 = 0.14$ and 0.27, respectively. The behavior obtained for simple diffusion-controlled positive feedback measurements on the oxidation of TMPD is also shown (■).

theoretical analyses obtained with $K_2 = 10$ (1.5 mM), 24 (4.1 mM), 200 (28.2 mM), and 1000 (121 mM). Together, these measuremnts—over almost two orders of magnitude in [FN]—yielded a consistent dimerization rate constant of 2.0 (±0.4) × 10^5 M^{-1} s^{-1}. This was in reasonable agreement with the rate constant of 6 (±3) × 10^5 M^{-1} s^{-1} estimated from earlier simultaneous electrochemical electron spin resonance studies (35), but considerably lower than that determined by RRDE measurements (35). The SECM data were considered to be more reliable than those obtained in the RRDE configuration, due to the higher sensitivity of collection efficiency measurements in the SECM geometry (4).

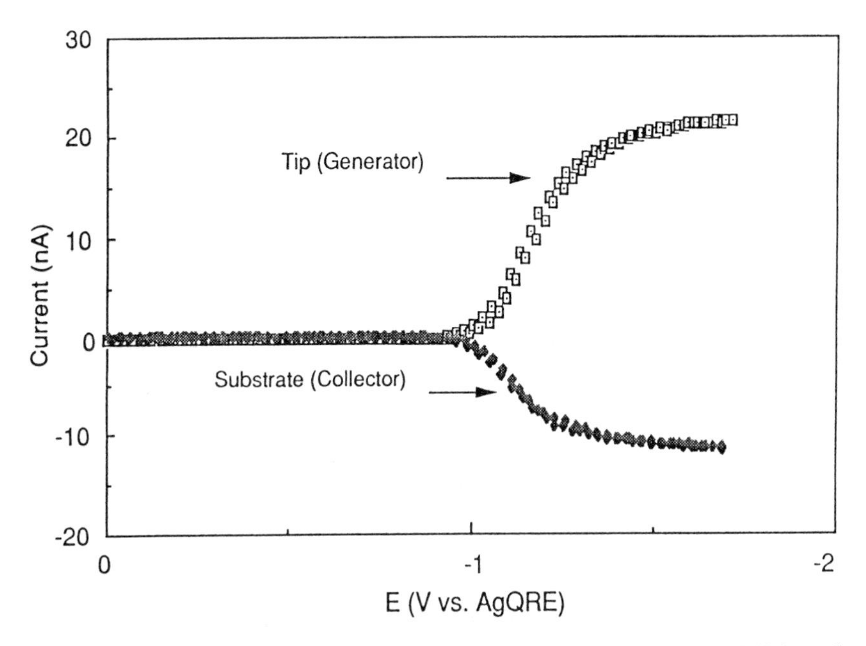

FIG. 19 Voltammograms for the reduction of 28.2 mM FN in the TG/SC mode at a tip/substrate separation, $d = 1.8\ \mu m$. The tip was scanned at 100 mV s^{-1} over the potential range shown, while the substrate potential was maintained at 0.0 V vs. AgQRE.

2. *Detection of Acrylonitrile Radical Anion*

The most powerful illustration to date of the ability of steady-state TG/SC measurements to identify short-lived intermediates in electrode processes comes from studies in which the radical anion of acrylonitrile (AN) was detected in the electrohydrodimerization reaction (8). Prior to these investigations, the mechanism of the AN electrohydrodimerization process was unclear. On the basis of SECM measurements, it was shown that the reaction proceeded analogously to other activated olefins, including those highlighted in Sec. III.B. Electrogeneration of the radical anion, $R^{\cdot-}$, is followed by dimerization and then protonation to yield the hydrodimer:

$$R + e^{-} \rightleftharpoons R^{\cdot-} \tag{48}$$

$$2\,R^{\cdot-} \xrightarrow{k_2} R_2^{2-} \tag{49}$$

$$R_2^{2-} \xrightarrow{H^+} R_2H^{-} \xrightarrow{H^+} R_2H_2 \tag{50}$$

SECM measurements were carried out in dry DMF at low AN concentration to minimize possible complications from side reactions. A typical

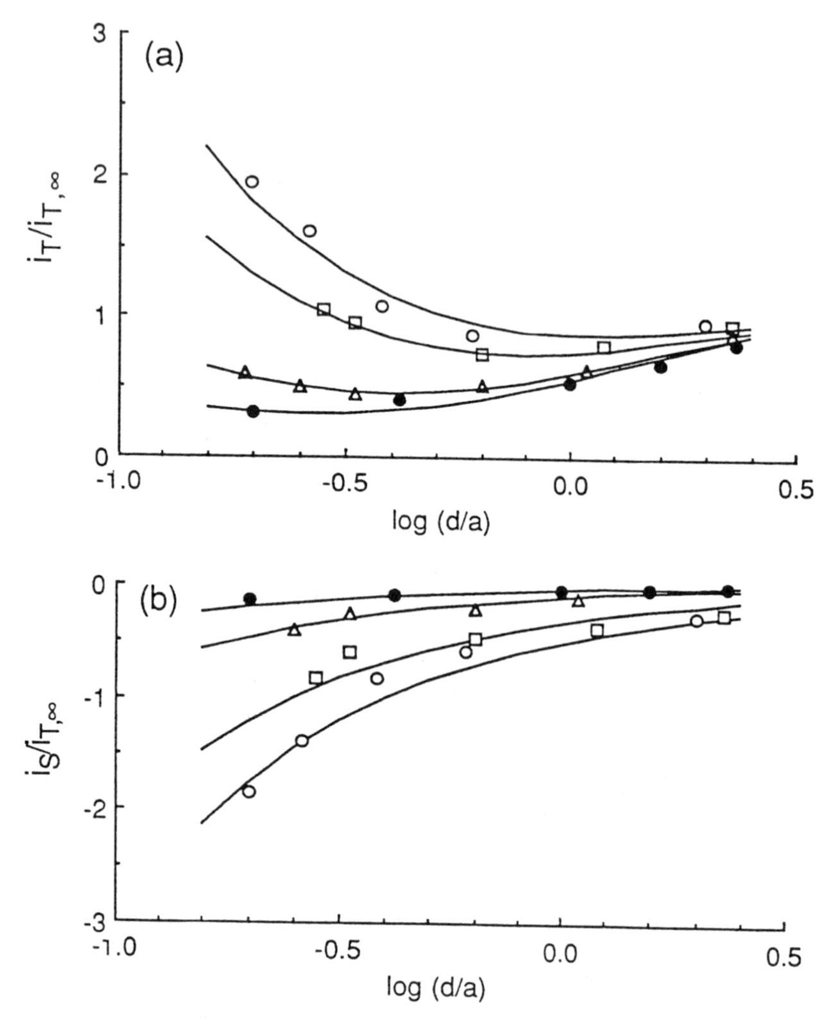

FIG. 20 Tip (a) and substrate (b) current-distance behavior for the reduction of FN at the tip and collection of $FN^{\cdot -}$ at the substrate (TG/SC mode). The data relate to [FN] = 1.5 mM (○), 4.12 mM (□), 28.2 mM (△), and 121 mM (●). The solid lines are the best theoretical fits with the values of K_2 cited in the text.

TG/SC voltammogram obtained with a 5 μm diameter Au tip and a 60 μm diameter Au substrate is shown in Figure 21. The effective interelectrode spacing, as deduced from positive feedback measurements on the oxidation of ferrocene, was 1.36 μm. The tip was scanned through the range of potentials corresponding to the reduction of AN, while the substrate was held

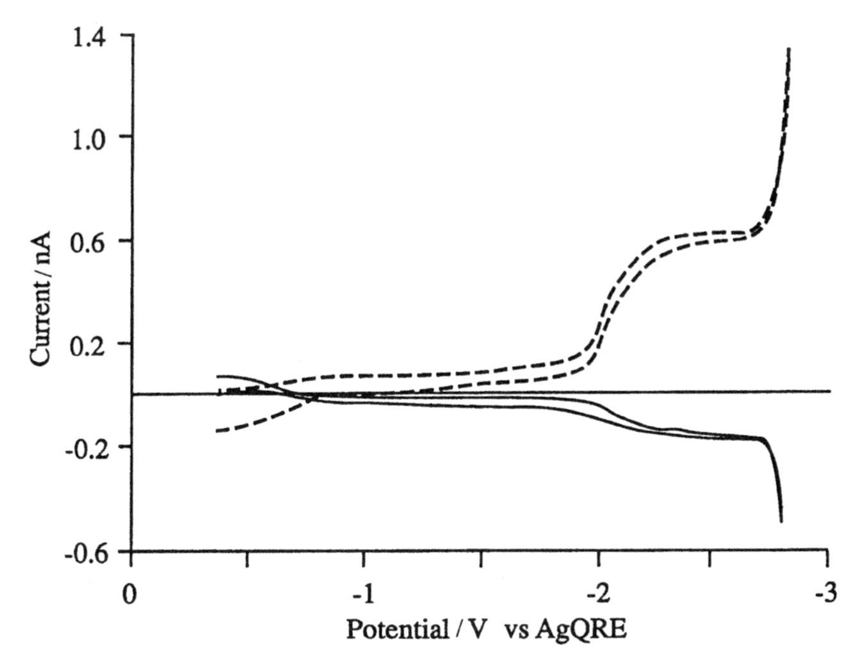

FIG. 21 TG/SC voltammograms for the reduction of 1.50 mM AN. The tip potential was scanned at a rate of 100 mV s^{-1} from -0.4 to -2.8 V vs. AgQRE, while the substrate potential was maintained at a potential of -1.75 V. The tip current is indicated by the dashed line, while the substrate current is denoted by the solid line. The small wave that appears at the tip and substrate at -0.6 V on the reverse scan was attributed to the presence of an intermediate (R_2H^-).

at a potential where any $AN^{\bullet -}$ reaching its surface would be detected by diffusion-controlled oxidation to AN. An anodic substrate current was observed to flow when the tip was in the potential region for the reduction of AN, consistent with the collection of $AN^{\bullet -}$. The small size of the collection efficiency, $N_{SS} \sim 0.29$, in this particular case indicated that an appreciable fraction of $AN^{\bullet -}$ was lost by reaction in solution during transport from the tip to the substrate.

The kinetics of the homogeneous reaction of $AN^{\bullet -}$ were determined by measuring collection efficiencies as a function of interelectrode separation for three different AN concentrations (Fig. 22). There was reasonable agreement between experiment and EC_{2i} theory in all three cases, with the closest results obtained with the lowest AN concentration, where contributions from competing polymerization side reactions were less important. Taking account

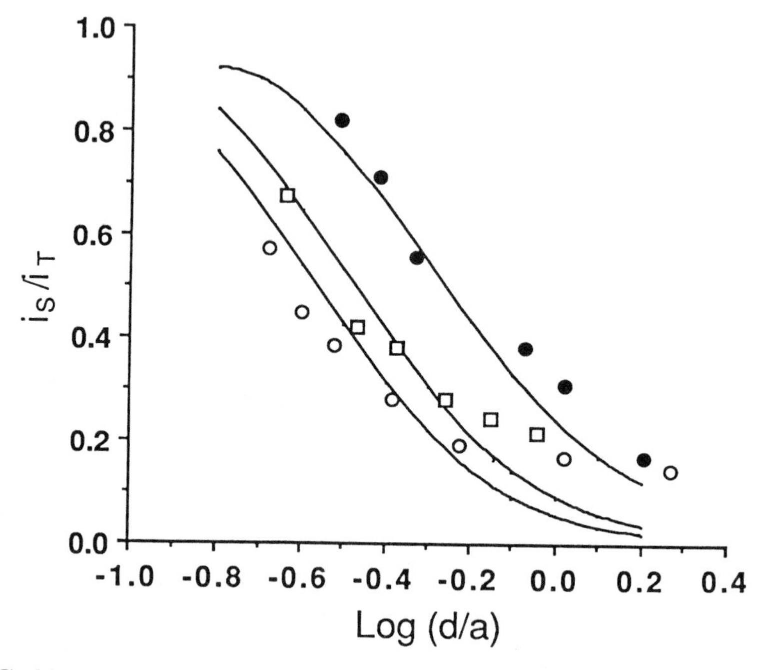

FIG. 22 Steady-state TG/SC collection efficiencies as a function of tip/substrate separation for AN reduction at the tip for [AN] = 0.58 mM (●), 1.50 mM (□), and 1.91 mM (○). The solid lines through each set of experimental data are the best theoretical fits obtained with K_2 = 100 (0.58 mM), 500 (1.50 mM) and 1000 (1.91 mM).

of all data sets, a dimerization rate constant, $k_2 = (6 \pm 3) \times 10^7\ M^{-1}\ s^{-1}$, was deduced (8).

3. Oxidative Dimerization of 4-Nitrophenolate in Acetonitrile

The rate constant for the rapid oxidative dimerization of 4-nitrophenolate (ArO^-) in acetonitrile solutions has been measured using the TG/SC mode (5), in which the following processes occurred:

Tip: $$ArO^- - e^- \rightarrow ArO^\bullet \tag{51}$$

Gap: $$2\ ArO^\bullet \rightarrow (ArO)_2 \tag{52}$$

Substrate: $$ArO^\bullet + e^- \rightarrow ArO^- \tag{53}$$

Tip/substrate separations were established with positive feedback measurements on the one-electron reduction of benzoquinone and the one-electron oxidation of ferrocene. The former mediator was found to be preferable in that it did not appear to interfere with the reaction of interest. In contrast, ferrocene oxidation occurred prior to the wave for the oxidation of ArO^-, and a side reaction involving the reaction of $ArO^{\bullet}$ with Fc was found to compromise the electrochemical response for the $ArO^-/ArO^{\bullet}$ couple (5).

Typical steady-state tip and substrate current approach curves for the oxidation of different concentrations of ArO^- are shown in Figure 23. A general observation is that as the concentration of ArO^- increases, the tip and substrate currents—at a particular distance—decrease, due to the second-order nature of the follow-up chemical reaction. The experimental approach curves are shown alongside theoretically derived curves for a spread of normalized rate constants, K_2, from which it can be seen that there is reasonable agreement between the observed and predicted trends. From measurements of both feedback currents, for all three ArO^- concentrations investigated, and collection efficiencies, for the lowest two concentrations, a radical dimerization rate constant of 1.2 ($\pm$0.3) $\times$ 10^8 M^{-1} s^{-1} was determined (5), which was in reasonable agreement with that determined earlier using fast scan cyclic voltammetry (36).

4. *Detection of Nicotinamide Adenine Dinucleotide Radical*

The initial stage in the reduction of nicotinamide adenine dinucleotide cation (NAD^+) in aqueous solution results in the formation of a radical ($NAD^{\bullet}$), which rapidly dimerizes (37). Through investigations with the SG/TC mode, Engstrom and coworkers (1) provided clear evidence that $NAD^{\bullet}$, electrogenerated at a glassy carbon substrate, could be detected by oxidation to NAD^+ at a 25 μm diameter Pt tip UME positioned as close to the substrate as possible. Although no quantitative formation was obtained on the rate constant for the dimerization process, this early work provided a good illustration of the ability of SECM to detect short-lived (submillisecond lifetime) intermediates in electrode reactions.

IV. ECE/DISP PROCESSES

The ECE/DISP framework covers many two-electron oxidation and reduction processes (38). Using a cathodic process as an example, the reduction of an initial species, A, producing B is followed by a first-order homogeneous reaction yielding an intermediate (C), which is often more easily reduced at the electrode (to give D) than the starting material. Under these conditions $E^o_{C/D} \ll E^o_{A/B}$, and the second electron transfer can occur in the solution, by disproportionation of B and C, so that overall the following steps have to be considered:

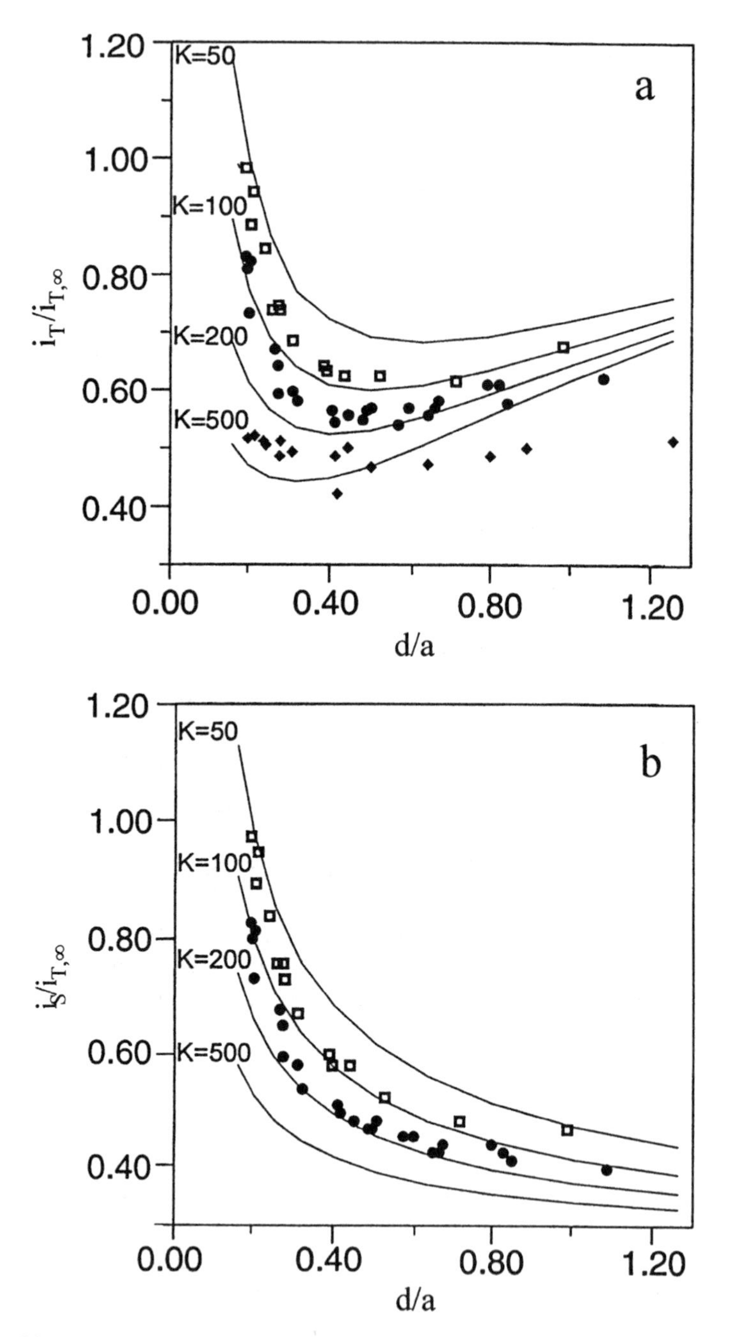

FIG. 23 (a) Normalized tip steady-state feedback current-distance behavior for 4-nitrophenolate oxidation at concentrations of 0.20 mM (□), 0.40 mM (●), and 0.71 mM (◆). Solid lines are the characteristics for normalized rate constants, K_2 of 50 (upper curve), 100, 200, and 500 (lower curve). (b) Collector current responses for two of the experiments in (a) with concentrations of 0.20 mM (□) and 0.40 mM (●).

Electrode:	$A + e^- \rightarrow B$	$(E^o_{A/B})$	(54)
Solution:	$B \xrightarrow{k_1} C$		(55)
Electrode:	$C + e^- \rightarrow D$	$(E^o_{C/D} \ll E^o_{A/B})$	(56)
Solution:	$B + C \xrightarrow{k_d} A + D$		(57)

The limiting ECE situation applies when Eqs. (54)–(56) are important, while the DISP scenario is where Eqs. (54), (55), and (57) define the overall process. The terminology DISP1 is used when the first-order decomposition of B to C is rate-limiting, while DISP2 defines the case where Eq. (57) is rate limiting. The difference in the reduction potentials for the A/B and C/D couples is often so great that the disproportionation reaction is driven at a diffusion-controlled rate, so that the problem reduces to considering the competition between ECE and DISP1 processes.

Distinguishing between ECE and DISP1 mechanisms by many conventional electrochemical methods is difficult. However, in common with generation-collection measurements at other double electrode geometries (39) and earlier applications of double potential step transient methods (40,41), generation-collection and feedback measurements with SECM have been shown to allow an unequivocal mechanistic assignment (7). Moreover, SECM allows the measurement of larger rate constants for Eq. (55) than the aforementioned techniques.

With reference to the above scheme, the study of ECE/DISP1 processes by the feedback and TG/SC modes involves reducing A at a diffusion-controlled rate at the tip UME, with the substrate held at a potential where the oxidation of B is diffusion-controlled. Provided that the substrate potential is not too positive of $E^o_{A/B}$, the ECE mechanism will involve the reduction of C at the substrate as well as the tip. The resulting diffusional and chemical processes within the tip/substrate domain for ECE and DISP1 pathways are shown schematically in Figure 24. As for the EC_i and EC_{2i} processes, considered in Secs. II and III, the route to measuring the rate constant for the conversion of B to C is to establish a competition between the diffusion of B to the substrate and the decomposition step. As discussed in Sec. IV.A, mechanistic resolution comes from measuring the substrate current simultaneously with the tip current, since the conversion of C to D at the substrate in the ECE reaction causes a (partial) cathodic current to flow, whereas the DISP1 reaction only involves an anodic current at the substrate for the oxidation of B to A.

A. Theory

We provide a brief outline of the formulation of the SECM problem for ECE/DISP1 processes, as a foundation for the theoretical results that follow.

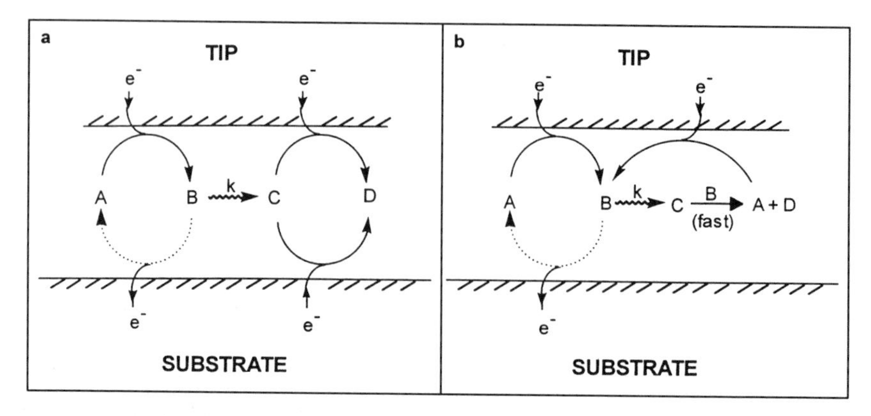

FIG. 24 Diffusional and chemical processes occurring within the tip-substrate domain for the TG/SC mode with (a) an ECE pathway and (b) a DISP1 pathway.

The feedback and TG/SC modes have been considered under steady-state conditions and with the tip under potential step control (7).

1. *DISP1 Mechanism*

The time-dependent diffusion equations for this case, appropriate to the axisymmetric cylindrical SECM geometry are (7):

$$\frac{\partial c_A}{\partial t} = D_A \left[\frac{\partial^2 c_A}{\partial r^2} + \frac{1}{r}\frac{\partial c_A}{\partial r} + \frac{\partial^2 c_A}{\partial z^2} \right] + k_1 c_B \tag{58}$$

$$\frac{\partial c_B}{\partial t} = D_B \left[\frac{\partial^2 c_B}{\partial r^2} + \frac{1}{r}\frac{\partial c_B}{\partial r} + \frac{\partial^2 c_B}{\partial z^2} \right] - 2k_1 c_B \tag{59}$$

The corresponding boundary conditions for the situation defined are identical to Eqs. (7)–(11). The initial condition (required for the simulation of tip potential step chronoamperometry), completing the definition of the problem, is Eq. (6). The tip and substrate currents are evaluated, respectively, from Eqs. (18) and (19), and the problem is readily cast into dimensionless form using the variables defined in Eqs. (12)–(17).

2. *ECE Mechanism*

The diffusion equations for species A, B, and C that have to be considered for this case are:

$$\frac{\partial c_A}{\partial t} = D_A \left[\frac{\partial^2 c_A}{\partial r^2} + \frac{1}{r}\frac{\partial c_A}{\partial r} + \frac{\partial^2 c_A}{\partial z^2} \right] \tag{60}$$

$$\frac{\partial c_B}{\partial t} = D_B \left[\frac{\partial^2 c_B}{\partial r^2} + \frac{1}{r}\frac{\partial c_B}{\partial r} + \frac{\partial^2 c_B}{\partial z^2} \right] - k_1 c_B \tag{61}$$

$$\frac{\partial c_C}{\partial t} = D_C \left[\frac{\partial^2 c_C}{\partial r^2} + \frac{1}{r}\frac{\partial c_C}{\partial r} + \frac{\partial^2 c_C}{\partial z^2} \right] + k_1 c_B \tag{62}$$

The boundary conditions for this three-species problem, with the electrode potentials as defined above, are:

$$z = 0,\ 0 \le r \le a: \qquad c_A = 0,\ D_A \frac{\partial c_A}{\partial z} = -D_B \frac{\partial c_B}{\partial z},\ c_C = 0 \tag{63}$$

$$z = 0,\ a \le r \le r_S: \qquad D_A \frac{\partial c_A}{\partial z} = D_B \frac{\partial c_B}{\partial z} = D_C \frac{\partial c_C}{\partial z} = 0 \tag{64}$$

$$z = d,\ 0 \le r \le r_S: \qquad c_B = 0,\ D_B \frac{\partial c_B}{\partial z} = -D_A \frac{\partial c_A}{\partial z},\ c_C = 0 \tag{65}$$

$$r > r_S,\ 0 \le z \le d: \qquad c_A = c_A^*,\ c_B = 0,\ c_C = 0 \tag{66}$$

$$r = 0,\ 0 < z < d: \qquad D_A \frac{\partial c_A}{\partial r} = D_B \frac{\partial c_B}{\partial r} = D_C \frac{\partial c_C}{\partial r} = 0 \tag{67}$$

The initial condition for potential step chronoamperometry is:

$$t = 0,\ \text{all } r,\ \text{all } z: \qquad c_A = c_A^*,\ c_B = 0,\ c_C = 0 \tag{68}$$

The tip and substrate currents are evaluated from:

$$i_T = 2\pi F \left[D_A \int_0^a (\partial c_A/\partial z)_{z=0}\, r\, \mathrm{d}r + D_C \int_0^a (\partial c_C/\partial z)_{z=0}\, r\, \mathrm{d}r \right] \tag{69}$$

$$i_S = 2\pi F \left[D_B \int_0^{r_S} (\partial c_B/\partial z)_{z=d}\, r\, \mathrm{d}r - D_C \int_0^{r_S} (\partial c_C/\partial z)_{z=d}\, r\, \mathrm{d}r \right] \tag{70}$$

The problem is reduced to dimensionless form in the same way as for the DISP1 case.

3. *Theoretical Characteristics*

For generality, all of the theoretical predictions have been derived assuming a common diffusion coefficient, D, for the species involved in the electrode process. Since the steady-state tip current in bulk solution depends on K (42), the tip and substrate currents are best normalized with respect to the tip current for the reduction of A, with $K = 0$. Denoted by $i_{T,1e,\infty}$, this has the form of Eq. (20), with $n = 1$.

Although theoretical results have been reported for both time-dependent and steady-state conditions, only the latter are discussed here, as these are most readily applied to the study of ECE/DISP1 processes (7). Typical steady-state tip and substrate current-distance characteristics for the DISP1 mechanism are shown in Figure 25 for a range of K values. It is first useful to identify the behavior in the limits of small and large K. As $K \rightarrow 0$, all of B generated at the tip reaches the substrate and the currents at the two electrodes are identical to the case of pure one-electron positive feedback. For $K \rightarrow \infty$, a two-electron current flows at the tip, but B does not reach the substrate. Consequently, as the interelectrode separation is decreased, no current flows at the substrate, while the tip current diminishes and follows the behavior for negative feedback, but with the transfer of two electrons. Between these two limits, kinetic measurements are possible. Qualitatively, the tip and substrate characteristics are similar to the EC_i and EC_{2i} cases, discussed in Secs. II.A and III.A. (1) For a given distance, an increase in K serves to decrease the tip and substrate currents. (2) It becomes increasingly easier to resolve different kinetic cases and to measure fast kinetics by minimizing the tip-substrate separation.

Steady-state tip and substrate current-distance characteristics for the ECE mechanism, with a range of rate constants, are shown in Figure 26. Although the shapes of the tip approach curves are qualitatively similar to the DISP1 case (Fig. 25), it is interesting to note that the tip current, for a given value of K, is generally lower (see, e.g., Fig. 27). The reason for this is clear from the schematic in Figure 24, where it can be seen that in the ECE mechanism C is able to pick up an electron from the substrate as well as the tip. A further interesting consequence of the reduction of C at the substrate for the ECE mechanism is that the substrate current changes sign, from cathodic at the largest tip/substrate separations considered, where the reduction of C dominates, to anodic at closer separations, where the oxidation of B becomes increasingly important. This variation in the substrate current with distance (Figs. 26 and 27) is a ready diagnostic of an ECE mechanism.

From the approach curves in Figures 25 and 26, it follows that normalized rate constants up to approximately $K = 50$ are measurable for both mechanisms, provided that tip/substrate separations as small as $d/a = 0.1$ are attainable. The absolute range of k_1 determinable then depends on a [Eq. (16)]. Assuming $D = 10^{-5}$ cm^2 s^{-1} and a 1 μm diameter tip, it should be possible to measure rate constants in excess of 10^5 s^{-1}, similar to the EC_i case considerd earlier. If, however, the aim is simply to discriminate between the ECE and DISP1 pathways, this can be achieved at much larger K, since in the ECE case the substrate is able to sense C at large tip/substrate separations, even when K is large. This is clear from the inset to Figure 26, where even for $K = 100$, the substrate current is greater than zero for $0.2 < d/a \leq 1.6$.

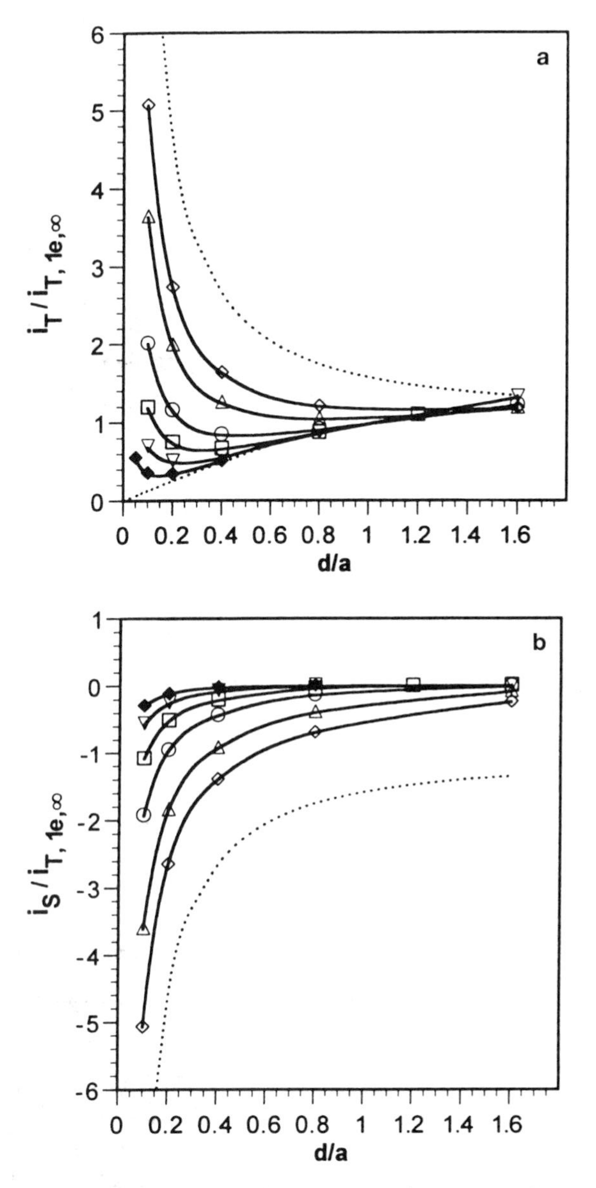

FIG. 25 Variation of the normalized steady-state tip (a) and substrate (b) currents with tip/substrate separation for a DISP1 pathway, in the TG/SC mode, with $K = 1$ (◇), 2 (△), 5 (○), 10 (□), 20 (▽), and 50 (◆). In (a) the upper dotted line represents one-electron pure positive feedback ($K = 0$) and the lower dotted line is for two-electron pure negative feedback ($K \rightarrow \infty$). In (b) the dotted line is the substrate current for pure positive feedback.

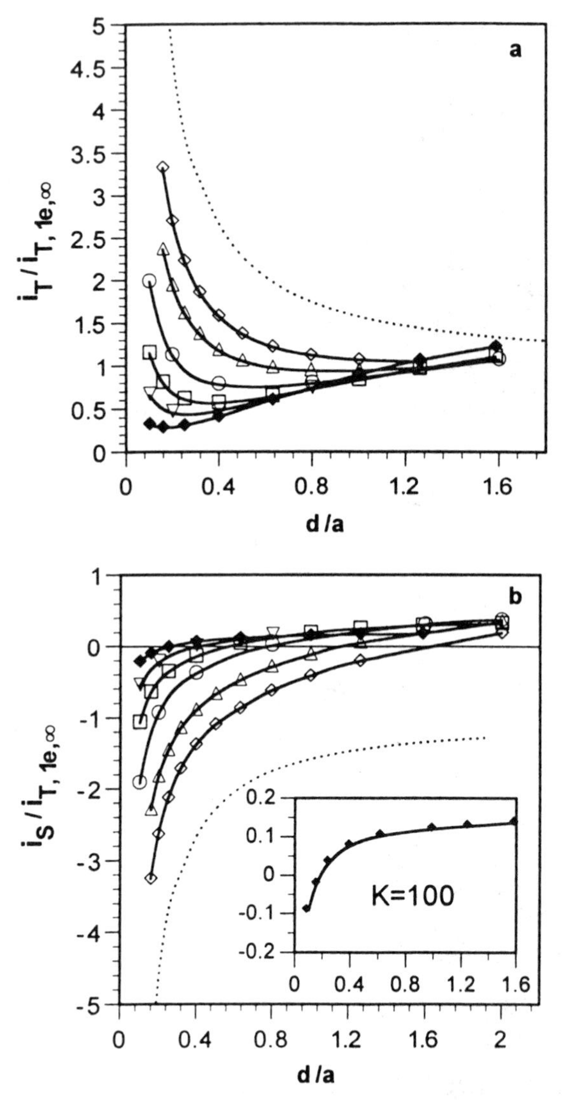

FIG. 26 Variation of the normalized steady-state tip (a) and substrate (b) currents with tip/substrate separation for an ECE pathway, in the TG/SC mode, with $K = 1$ (◇), 2 (△), 5 (○), 10 (□), 20 (▽), and 50 (◆). In (a) the upper dotted line represents one-electron pure positive feedback ($K = 0$). In (b) the dotted line is the substrate current for pure positive feedback. The inset shows the substrate current for $K = 100$.

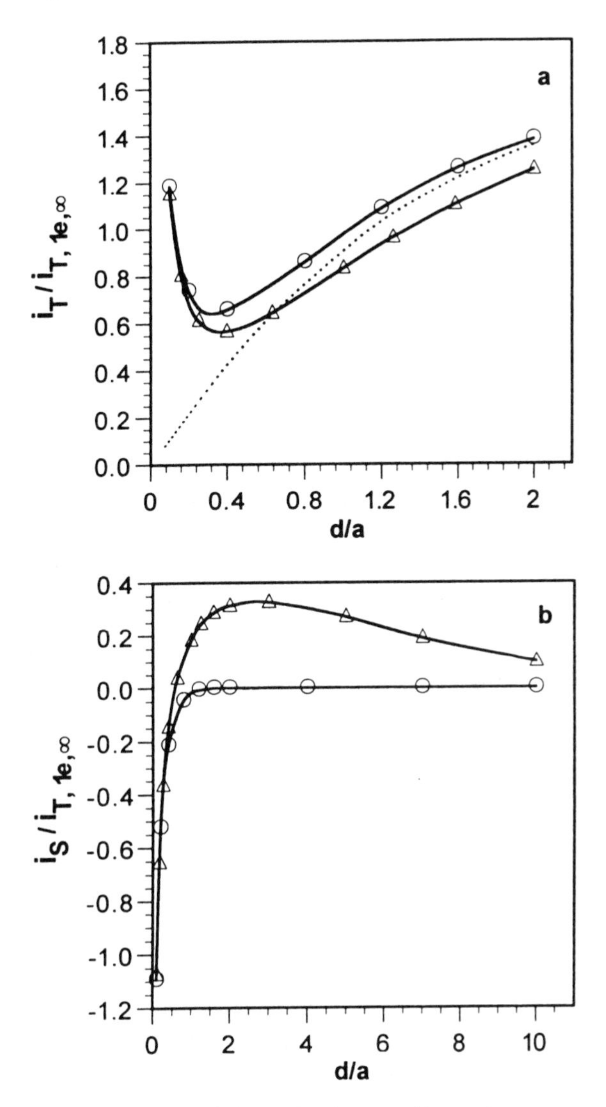

FIG. 27 Comparison of the tip (a) and substrate (b) steady-state approach curves for the DISP1 (○) and ECE (△) cases with $K = 10$, under TG/SC control. The dotted line in (a) is for two-electron negative feedback ($K \rightarrow \infty$).

B. Applications

1. *Reduction of Anthracene in DMF*

To demonstrate the utility of SECM measurements, the reduction of anthracene (AC) in DMF in the presence of phenol (PhOH) was studied, as an example of a DISP1 process, yielding 9,10-dihydroanthracene (ACH_2) (40,41):

$$AC + e^- \rightleftharpoons AC^{\bullet -} \tag{71}$$

$$AC^{\bullet -} + PhOH \xrightarrow{k_1} ACH^{\bullet} + PhO^- \tag{72}$$

$$ACH^{\bullet} + AC^{\bullet -} \xrightarrow{k_d} ACH^- + AC \tag{73}$$

$$ACH^- + PhOH \xrightarrow{\text{fast}} ACH_2 + PhO^- \tag{74}$$

Equation (72) has been shown to be rate-limiting, and this can be considered as pseudo first-order in the presence of excess PhOH.

Experiments were carried out in DMF with 4 mM AC, and PhOH at concentrations in the range 0.1–0.43 M, with tetrabutylammonium tetrafluoroborate as a supporting electrolyte. Distances were established using the positive feedback response for the oxidation of decamethylferrocene, which was added to solutions at mM levels. A 7 μm diameter C fiber UME was employed as the tip, while the substrate was a 60 μm diameter Au disk electrode. Typical steady-state tip and substrate approach curves for the diffusion-limited reduction of AC at the tip and the oxidation of $AC^{\bullet -}$ at the substrate are shown in Fig. 28. At a qualitative level, the tip current shows the features predicted for ECE/DISP1 processes, identified in Sec. IV.A, while the substrate current—always of opposite sign to the tip current—points unambiguously to a DISP1 pathway.

A series of tip feedback current-distance curves for a range of PhOH concentrations is shown in Figure 29. This shows that a PhOH increases, the current decreases due to the consumption of $AC^{\bullet -}$ in the gap, causing a diminution in the extent of positive feedback for the $AC/AC^{\bullet -}$ couple. Analysis of all data sets allowed values for the effective first-order rate constants to be determined, from which the second-order rate constant, $k_2 = k_1/[PhOH] = (4.4 \pm 0.4) \times 10^3\ M^{-1}\ s^{-1}$, was deduced. This value was found to be in good agreement with that measured earlier by double potential step chronoamperometry (40), confirming the applicability of SECM in the study of ECE/DISP1 processes.

2. *Oxidation of Epinephrine in Aqueous Solution*

The oxidation of epinephrine proceeds via an initial rapid two electron–two proton process to yield the corresponding quinone (Q). Considering this as an E step, Q undergoes deprotonation and cyclization (C step), producing

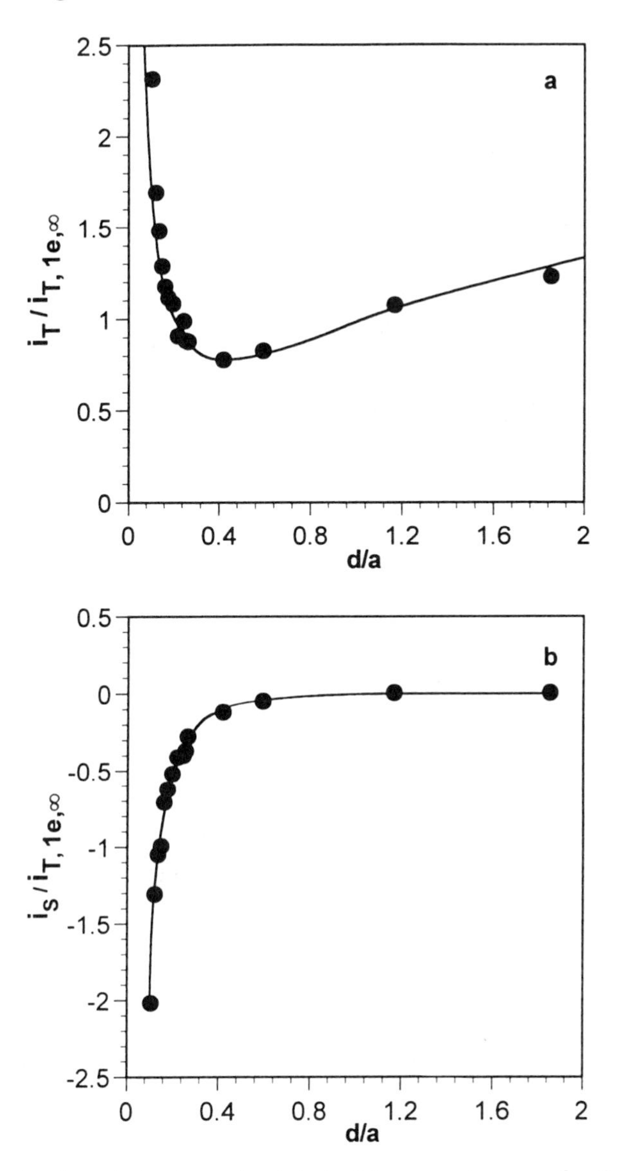

FIG. 28 Steady-state tip (a) and substrate (b) approach curves for the reduction of 4.1 mM anthracene (AC) at the tip, in the presence of PhOH, and the collection of $AC^{\bullet -}$ at the substrate.

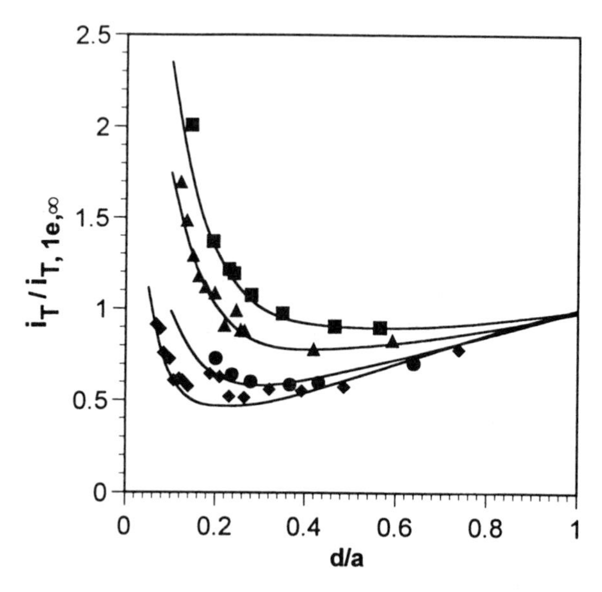

FIG. 29 Normalized steady-state tip approach curves for the reduction of 4 mM AC, with PhOH present at concentrations of 0.1 (■), 0.154 (△), 0.23 (●), and 0.43 (♦) M. The solid lines through each data set correspond to $K = 4$, 6, 12.5, and 22 (from top to bottom).

leucoadrenochrome (L), which is easily oxidized at the electrode in a two electron–two proton process, forming adrenochrome (A) (43). This second group of electron transfers, which can be considered as an E step, can also occur in solution by disproportionation of Q and L. Although this latter mechanistic pathway was termed ECC (2), it is, in fact, a DISP process. By identifying the intermediate Q and product A, using the SG/TC mode, a second-order DISP2 process was considered to provide the best fit to tip current-distance curves. The possibility of feedback processes (in the detection of Q), modifying the apparent concentration profiles was not, however, considered in this early work (2).

V. CONCLUSIONS AND FUTURE PERSPECTIVES

The aim of this chapter has been to show how SECM, operating as an ultra-thin-layer cell, can be used to quantitatively study electrode processes involving coupled chemical reactions. The advantages of SECM are that a wide range of calculable mass transport rates can be achieved with high reproducibility simply by changing the interelectrode separation and the tip size. Crucially, high mass transfer rates can be obtained through the employment of small electrodes at close tip/substrate separations, facilitating the study of rapid reactions.

Of the three SECM modes that can be used to study electrode reaction mechanisms—the TG/SC, feedback, and SG/TC modes—the former is the most powerful for measuring rapid kinetics. With this approach, fast follow-up and sandwiched chemical reactions can be characterized under steady-state conditions, which are difficult to study even with rapid transient techniques such as fast scan cyclic voltammetry or double potential step chronoamperometry, where extensive corrections for background currents are often mandatory (44). At present, first- and second-order rate constants up to $\sim 10^5$ s^{-1} and 10^{10} M^{-1} s^{-1}, respectively, should be measurable with SECM. The development of smaller tip and substrate electrodes that can be placed closer together should facilitate the detection and characterization of electrogenerated species with submirosecond lifetimes. In this context, the introduction of a fabrication procedure for spherical UMEs with diameters

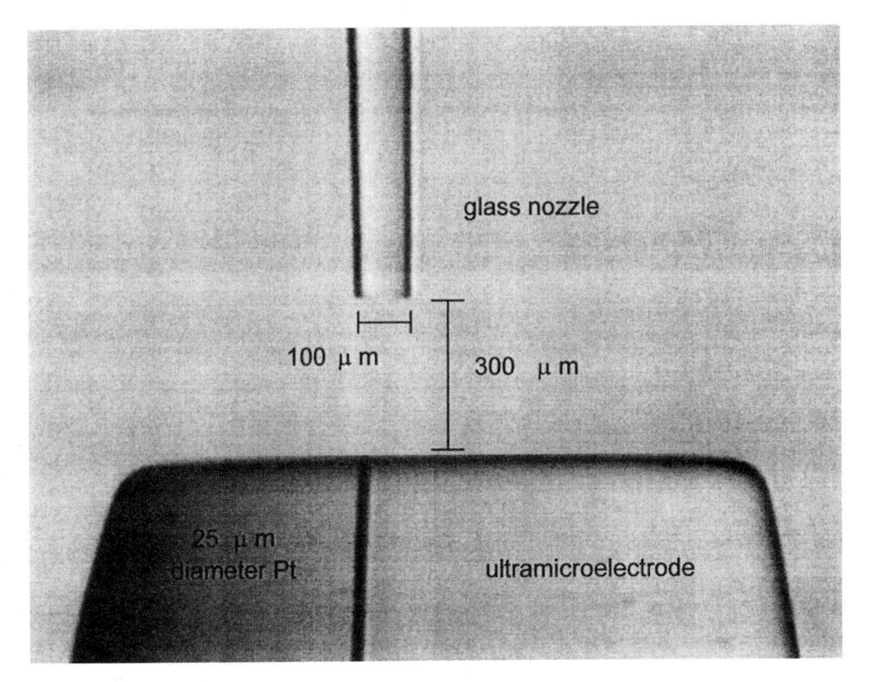

FIG. 30 Micrograph of a microjet electrode in which solution impinges at high velocities from a micropipette onto a disc UME. In this particular case the UME and nozzle have diameters of 25 μm and 100 μm (internal), respectively.

down to 1 μm, through the self-assembly of gold nanoparticles at the end of a pulled glass micropipette, using 1,9-nonanedithiol as a molecular glue, is an interesting development (45). It should be possible to position this type of probe very close to a substrate electrode, facilitating the detection of short-lived intermediates in the TG/SC mode in particular.

There is some scope for diversifying the range of mechanisms to which SECM can be applied. In particular, all three modes considered in this chapter could prove useful in the study of the catalytic EC′ mechanism and its variants (38), which describe a diversity of significant electrode processes. It may also be possible to study preceding chemical reactions with SECM, using the interelectrode gap to vary mass transport to the tip via hindered diffusion. The range of accessible rate constants and sensitivity with which measurements could be made would, however, be unlikely to compete with alternative methods.

SECM technology has had spin-offs in the development of several hydrodynamic UMEs, which employ high-resolution micropositioners to assemble microelectrochemical flow cells. When these devices are considered

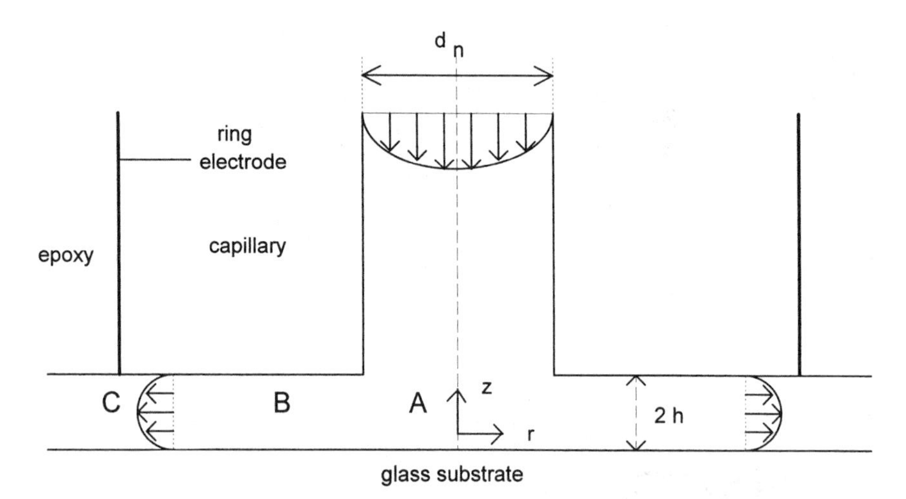

FIG. 31 Schematic of the radial flow microring electrode (not to scale), with the ring electrode deposited on the outer wall of a pulled capillary. Solution impinges from a nozzle at region A and undergoes a transition (in region B) to fully developed parabolic flow in region C. The axi-symmetric cylindrical geometry of the device is characterized by the coordinates r and z. The inner diameter of the nozzle, d_n, is typically 100 μm, and the diameter of the thin electrode is 140–180 μm. The ring electrode itself is 0.1–0.5 μm thick, while the nozzle/substrate spacing, $2h$, is variable in the range 5–40 μm.

under the general SECM umbrella, the range of reaction mechanisms and rates that can be investigated widens considerably. For example, the microjet electrode (micro-wall tube electrode) (46–48), in which a UME is aligned underneath a micropipette (Fig. 30) through which solution flows with mean solution velocities in the tens of meters per second region, enables the study of CE processes with higher rate constants than any other UME technique (49). Likewise, the radial flow microring electrode (50) (Fig. 31), which has proven a promising device for examining fast heterogeneous electron transfer processes (51), could find application in the study of CE, ECE/DISP, and EC′ mechanisms. There is also the possibility of developing dual-series electrode configurations based on these geometries, which would further diversify the types of processes that could be studied.

ACKNOWLEDGMENTS

Various aspects of our SECM program at Warwick University have been supported by the EPSRC (UK). I am grateful to Dr. Julie Macpherson for useful comments on earlier drafts of this chapter. The contributions of Dr. Rachel Martin to parts of the work described in this chapter are much appreciated.

REFERENCES

1. RC Engstrom, M Weber, DJ Wunder, R Burgess, S Winquist. Anal Chem 58: 844, 1986.
2. RC Engstrom, T Meaney, R Tople, RM Wightman. Anal Chem 59:2005, 1987.
3. PR Unwin, AJ Bard. J Phys Chem 95:7814, 1991.
4. F. Zhou, PR Unwin, AJ Bard. J Phys Chem 96:4917, 1992.
5. DA Treichel, MV Mirkin, AJ Bard. J Phys Chem 98:5751, 1994.
6. RD Martin, PR Unwin. J Chem Soc Faraday Trans 94:753, 1998.
7. C Demaille, PR Unwin, AJ Bard. J Phys Chem 100:14137, 1996.
8. F. Zhou, AJ Bard. J Am Chem Soc 116:393, 1994.
9. DO Wipf, AJ Bard. J Electrochem Soc 138:469, 1991.
10. RD Martin, PR Unwin. Anal Chem 70:276, 1998.
11. MV Mirkin, H Yang, AJ Bard. J Electrochem Soc 139:2212, 1992.
12. CP Andrieux, P Hapiot, JM Savéant. Chem Rev 90:723, 1990.
13. WJ Albery, ML Hitchman. Ring Disc Electrodes. Clarendon Press, Oxford, UK, 1971.
14. LB Anderson, CN Reilly. J Electroanal Chem 10:538, 1965.
15. PR Unwin, RG Compton. Comp Chem Kinet 29:173, 1989.
16. TY Ou, S Moldoveanu, JL Anderson. J Electroanal Chem 247:1, 1988.
17. T Varco Shea, AJ Bard. Anal Chem 59:2101, 1987.
18. J Kwak, AJ Bard. Anal Chem 61:1221, 1989.

19. AJ Bard, G Denuault, RA Freisner, BC Dornblaser, LS Tuckerman. Anal Chem 63:1282, 1991.
20. Y Saito. Rev Polarogr Jpn 15:177, 1968.
21. RD Martin, PR Uwin. J Electroanal Chem 439:123, 1997.
22. DW Peaceman, HH Rachford. J Soc Ind Appl Math 3:28, 1955.
23. D Shoup, A Szabo. J Electroanal Chem 140:237, 1982.
24. C Lee, J Kwak, FC ANson. Anal Chem 63:1501, 1991.
25. AJ Bard, MV Mirkin, PR Unwin, DO Wipf. J Phys Chem 96:1861, 1992.
26. MV Mirkin, F-RF Fan, AJ Bard. J Electroanal Chem 328:47, 1992.
27. LKJ Tong. J Phys Chem 58:1090, 1954.
28. LKJ Tong, K Liang, WR Ruby. J Electroanal Chem 13:245, 1967.
29. K Aoki, H Matsuda. J Electroanal Chem 94:157, 1978.
30. TC Richards, AJ Bard, A Cusanelli, D Sutton. Organometallics 13:757, 1994.
31. JH Morris, HJ Gysling, D Reed. Chem Rev 85:51, 1985.
32. MV Mirkin, AJ Bard. Anal Chem 63:532, 1991.
33. MM Baizer. Tetrahedron Lett 973, 1963.
34. L-SR Yeh, AJ Bard. J Electrochem Soc 124:355, 1977.
35. IB Goldberg, D Boyd, R Hirasama, AJ Bard. J Phys Chem 78:295, 1974.
36. P Hapiot, J Pinson, N Yousfi. New J Chem 16:877, 1992.
37. CO Schmakel, KSV Santhanam, PJ Elving. J Am Chem Soc 75:5083, 1975.
38. AJ Bard, LR Faulkner. Electrochemical Methods. Wiley. New York. Ch. 11, 1980.
39. PR Unwin. J Electroanal Chem 297:103, 1991.
40. C Amatore, JM Savéant. J Electroanal Chem 107:353, 1980.
41. C Amatore, M Gareil, JM Savéant. J Electroanal Chem 147:1, 1983.
42. M Fleischmann, F Lasserre, J Robinson. J Electroanal Chem 177:115, 1984.
43. MD Hawley, SV Tatwawadi, S Piekarski, RN Adams. J Am Chem Soc 89:447, 1967.
44. RM Wightman, DO Wipf. Electroanalytical Chemistry (AJ Bard, Ed). Marcel Dekker, New York, Vol. 15, p 267, 1989.
45. C Demaille, M Brust, M Tsionsky, AJ Bard. Anal Chem 69:2323, 1997.
46. JV Macpherson, S Marcar, PR Unwin. Anal Chem 66:2175, 1994.
47. JV Macpherson, MA Beeston, PR Unwin. J Chem Soc Faraday Trans 91:899, 1995.
48. JV Macpherson, PR Unwin. Anal Chem 69:5045, 1997.
49. RD Martin, PR Unwin. J Electroanal Chem 397:325, 1997.
50. JV Macpherson, PR Unwin. Anal Chem 70:2914, 1998.
51. JV Macpherson, CE Jones, PR Unwin. J Phys Chem 102:9891, 1998.

8

CHARGE-TRANSFER AT THE LIQUID/LIQUID INTERFACE

Michael V. Mirkin

Queens College–City University of New York
Flushing, New York

Michael Tsionsky

Gaithersburg, Maryland

I. INTRODUCTION

The structure and dynamics of the liquid/liquid interface have been the focus of considerable research activity (for review of the electrochemistry of ITIES, see Ref. 1). In addition to fundamental interest, the charge-transfer (CT) reactions occurring at the liquid/liquid interface, i.e., electron transfer (ET), simple ion transfer (IT), and facilitated IT, are relevant to important technological systems from chemical sensors to drug delivery in pharmacology to solvent extraction in hydrometallurgy (1,2). At the interface between two immiscible electrolyte solutions (ITIES), one can conduct reactions that cannot occur at solid electrodes, e.g., reactions involving species of very different polarities. These reactions find applications in electrochemical catalyses and synthesis (for examples, see Ref. 3). The liquid/liquid interface has been suggested as a simple model for biological and artificial membranes (4).

The state of knowledge in the field of CT at the liquid/liquid interface is different from homogeneous ET and ET at the metal/solution interface, where the well-established theory has been validated by a number of experiments (5). Several conflicting models have been proposed for charge-transfer processes at the ITIES, and experimental data necessary for testing these models have yet to be obtained. For example, Marcus suggested that classical ET theory should be applicable to the ITIES and proposed expressions for reorganization energy and rate constant (6). Other authors argued that Marcus model may not correctly describe ET at the ITIES (7). Schiffrin et al. (8) studied interfacial ET implicitly assuming that the potential dependence of the rate constant obeys the Butler-Volmer equation. At the same time, the apparent potential dependence of the ET rate can be attributed to the change in concentrations of the reactants at the interface rather than to activation control (1b,9). Similarly, the kinetic data reported for IT reactions at the ITIES do not allow one to choose between the activation mechanism (10) and slow diffusion through the boundary layer proposed as the rate-determining step (11,12).

The information available about the structure of the ITIES is also sparse. In spite of recent progress in spectroscopic studies (13) and molecular dynamics simulations (1e,14), questions remain about the nature and properties of the boundary (mixed-solvent) layer, which is supposed to separate two liquid phases (1). Since the thickness of the interface and physical localization of the ET reaction are largely unknown, the microscopic description of this process remains problematic.

There are two fundamentally different ways of studying CT reactions at the ITIES by electrochemical methods. A more conventional approach is based on application of external voltage across a polarizable interface, i.e., the ITIES formed by two liquid phases containing no common ion. The applied voltage provides the driving force for an ET or IT reaction. The thermodynamics and kinetics of such a reaction can be evaluated from the current versus voltage curves (e.g., cyclic voltammograms) as one does at a solid electrode. The applicability of this approach is limited by severe experimental problems caused by uncompensated *i*R-drop in highly resistive nonaqueous solvents and double-layer charging current. Unlike the metal/liquid interface, faradaic and nonfaradaic processes occurring at the ITIES are less directly accessible by electrochemical measurements, and the extraction of mechanistic information from the measured current is not straightforward (1). In most cases, the discrimination between ET and IT is difficult.

An alternative approach is to probe CT reactions at a nonpolarizable ITIES where two immiscible liquids contain a common ion. The interfacial potential drop ($\Delta_w^o\varphi$) in this case is governed by the ratio of common ion concentration in water and in organic phase. By changing local chemical composition in the interfacial region, one can selectively induce the desired ET or IT reaction without applying any external voltage across the ITIES. This approach is easy to implement using scanning electrochemical microscopy (SECM), which is capable of probing CT reactions at the ITIES in the same way as ET at a solid substrate. The interfaces between water and various organic solvents were found to be mechanically stable and sharp on a submicrometer scale (15). With SECM one can quantitatively separate interfacial ET from IT processes and minimize resistive potential drop and double-layer charging effects. In this chapter we discuss SECM studies of ET, electrochemical catalysis, and IT reactions occurring at the liquid/liquid interface.

II. ELECTRON TRANSFER

A. Electron Transfer at the Clear ITIES

1. Principles of SECM/ITIES Measurements

Conventional electrochemical experiments at the ITIES are based on a four-electrode configuration with a reference and an auxiliary electrodes positioned in each liquid phase (16,17). A four-electrode potentiostat is used to apply a voltage between the reference electrodes and to measure the current flowing between auxiliary electrodes. The interfacial charge transfer is assumed to be rate limiting, and the whole potential drop occurs mostly within a thin interfacial layer. However, the ITIES in such an experiment is not microscopically probed directly, and the nature of CT reaction (i.e., ET vs. IT) typically remains uncertain.

Slow ET between redox species confined to two immiscible solvents was first observed by Guainazzi et al. (18). Several different theoretical and experimental studies of ET between redox species at the ITIES have been reported in the last several years (6–8,15,19–21). Severe experimental problems complicate extraction of the kinetic parameters from conventional electrochemical measurements at the ITIES (e.g., by cyclic voltammetry). Besides the difficulty of discrimination between ET and IT, there are also distortions from the double-layer charging current and *i*R-drop in the highly resistive nonaqueous solvents, and the limited potential window for studying ET in the absence of currents controlled by IT (1,16).

Unlike conventional electrochemical techniques, in SECM measurements the ITIES is poised by concentrations of the potential-determining ions, providing a constant driving force for the ET process. This eliminates some of the problems mentioned above. In a typical SECM/ITIES experiment, a tip ultramicroelectrode (UME) with a radius a is placed in the upper liquid phase containing reduced form of redox species, R_1. The tip is held at a positive potential, and R_1 reacts at the tip surface to produce the oxidized form of the species, O_1. When the tip approaches the ITIES, the mediator can be regenerated at the interface via bimolecular redox reactions between O_1 in the upper phase and R_2 in the bottom phase:

$$R_1 - e \rightarrow O_1 \qquad \text{(tip)} \tag{1}$$

$$O_1 + R_2 \rightarrow R_1 + O_2 \qquad \text{(liquid/liquid interface)} \tag{2}$$

During these reactions electrons transfer from the bottom phase into the upper phase. To maintain electroneutrality in both phases, the transfer of common ions (either anions or cations or both) occurs simultaneously with ET reaction.

Thus, the overall charge transfer reaction at the ITIES consists of several steps shown in Figure 1:

Diffusion of organic mediator species in the tip UME/ITIES gap
Diffusion of aqueous mediator to and from the interface
ET between organic and aqueous mediators
Transfer of the common ion (negative charge in Figure 1)

In principle, any of these steps can determine the rate of the overall reaction occurring at the ITIES. Unlike previous measurements at the polarized ITIES, the SECM/ITIES technique allows one to distinguish between slow IT, ET, and diffusion-controlled processes.

The electric current across the ITIES (i_s) caused by a multistage serial process can be expressed as (22):

$$1/i_s = 1/i_T^c + 1/i_{ET} + 1/i_d + 1/i_{IT} \tag{3}$$

where i_T^c, i_{ET}, i_d, and i_{IT} are the characteristic limiting currents for the above four stages. When IT and mediator diffusion in the bottom phase are not

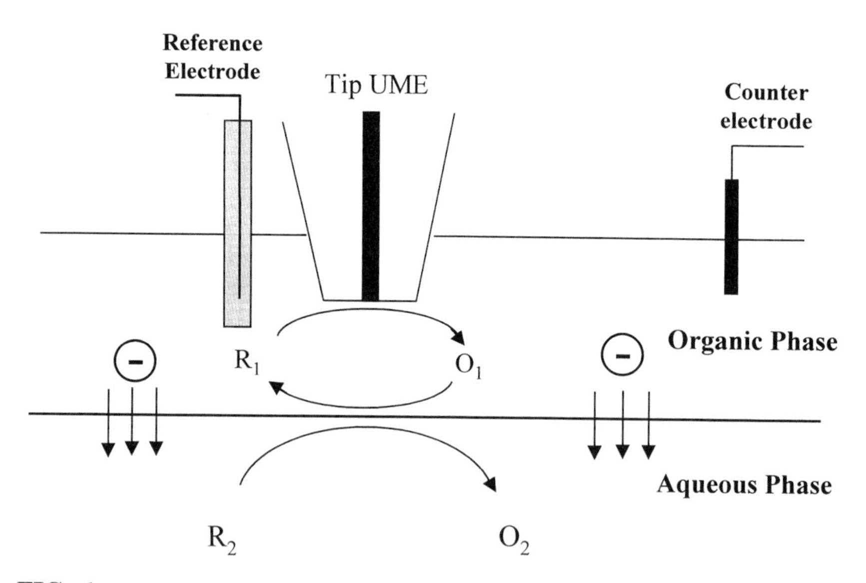

FIG. 1 Schematic diagram of the feedback mode SECM measurements of the kinetics of ET between organic and aqueous redox species. Electroneutrality was maintained by transfer of common ions shown as negative charges.

rate limiting,* the following equations can be used to extract the first-order effective heterogeneous ET rate constant for approach curves (15):

$$I_T^k = I_S^k(1 - I_T^{ins}/I_T^c) + I_T^{ins} \tag{4}$$

$$I_S^k = 0.78377/L(1 + 1/\Lambda) + [0.68 + 0.3315 \exp(-1.0672/L)] /[1 + F(L, \Lambda)] \tag{5}$$

where I_T^c, I_T^k, and I_T^{ins} represent the normalized currents for diffusion-controlled regeneration of redox mediator, finite substrate kinetics, and insulating substrate (i.e., no mediator regeneration), respectively, at a normalized tip-substrate separation, $L = d/a$, I_s^k is the kinetically controlled substrate current; $\Lambda = k_f d/D_R$, where k_f is the apparent heterogeneous rate constant (cm/s) and D_R is the diffusion coefficient of the reduced mediator in the top phase; and $F(L, \Lambda) = (11 + 2.3\Lambda)/[\Lambda(110 - 40L)]$. These currents are normalized by the tip current at an infinite tip-substrate separation, $i_{T,\infty} = 4nFaD_R c_R$. The analytical approximations for I_T^c and I_T^{ins} are:

$$I_T^c = 0.78377/L + 0.3315 \exp(-1.0672/L) + 0.68 \tag{6}$$

$$I_T^{ins} = 1/(0.15 + 1.5358/L + 0.58 \exp(-1.14/L) + 0.0908 \exp[(L - 6.3)/(1.017L)]) \tag{7}$$

2. *SECM Apparatus and Procedure*

The basic apparatus used for the SECM/ITIES experiments is essentially the same as for any other SECM measurements (see Chapter 2). Before SECM measurements, the tip UME is positioned in the top phase and biased at a potential where the tip process is diffusion-controlled. Typically, both liquid phases are placed into a small (2 mL) beaker. In this arrangement the less dense phase forms the top layer. If one wants the less dense phase to be the bottom phase, a small (10–50 μL) volume of it can be stabilized at the end of glass capillary inside the other phase (23). A list of liquid/liquid systems used for ET studies by SECM is shown in Table 1.

The mechanical stability of the liquid/liquid boundary, accuracy, and reproducibility of the measurements were demonstrated in earlier SECM/ITIES experiments. The SECM approach curve (i_T vs. d) in Figure 2 was obtained with a 5-μm-radius tip immersed in the aqueous solution (top phase) containing 5 mM $FcCOO^-$ (ferrocene carboxylic acid) and 0.1 M KCl, and nitrobenzene (NB) (bottom phase) containing no electroactive species (15). In this configuration, the NB layer acts as an electrical insulator,

---

*A kinetic study based on a model that accounts for the depletion of the reactant in the bottom phase was reported recently (Barker AL, Unwin PR, Bard AJ. J. Phys. Chem. B 103:7260, 1999).

TABLE 1 Redox Systems Studied by SECM/ITIES Technique

Interface	Redox system	Transferred ion	Ref.
Water/Nitrobenzene	$Ru(bpy)_3^{3+}/Fc$	Fc^+	15
	FcCOO/Fc	TEA^+, ClO_4^-, Fc^+	15
	$IrCl_6^{2-}/Fc$	FC^+	24
	$Fe(phen)_3^{3+}/Fc$	Fc^+	24
	$Mo(CN)_8^{3-}/Fc$	Fc^+	24
	$Fe(CN)_6^{3-}/Fc$	Fc^+	24
Benzene/Water	$ZnPor^+/Ru(CN)_6^{4-}$	ClO_4^-	25
	$ZnPor^+/Mo(CN)_8^{4-}$	ClO_4^-	26
	$ZnPor^+/Fe(CN)_6^{4-}$	ClO_4^-	26,27
	$ZnPor^+/Fe^{2+}$	ClO_4^-	26
	$ZnPor^+/V^{2+}$	ClO_4^-	26
	$ZnPor^+$/Co(II) Sepulchrate	ClO_4^-	26
Water/Dichloroethane	$Fe(CN)_6^{4-}$/TCNQ	$TPAs^+$	28
Water/Benzonitrile	B_{12}/Dibromocyclohexane	ClO_4^-	23

and the UME current shows negative feedback and in a good agreement with the related theory (29). The tip current for micrometer-sized SECM tips as well as for a 25-nm Pt UME (15) was oscillation-free. It was concluded that the interface is sharp on the nanometer scale and z-coordinate corresponding to the zero-distance point can be easily determined. This is essential for any quantitative SECM measurements. In later SECM/ITIES publications the shape of the approach curve to insulating ITIES served as a test for a proper tip geometry and stability of the liquid/liquid boundary.

The measurements are reliable only if the experimental data fit the theory at small tip/ITIES distances (i.e., $d/a \leq 0.1$). The reliability of measurements can also be verified by fitting experimental i_T versus d curves to the theory for conductive substrate. The maximum normalized feedback current for such a process should be at least 6.

3. *Separation of ET and IT Processes*

In most conventional electrochemical studies carried out at either macro- or micro-ITIES, the same interface area was available for both ET and IT processes. In contrast, the ET process in SECM experiments occurs at a micrometer-size area of the ITIES confronting the tip, while charge-compensating IT can occur at any point on the large (on the order of cm^2) phase boundary. Thus, the interfacial ET can be probed without complications caused by IT. This assumption was checked for the ET between Ru(III)-2,2′-bipyridine ($Ru(bpy)_3^{3+}$) and ferrocene (Fc) in NB:

$$Fc^{\circ}_{(NB)} + Ru(bpy)^{3+}_{3(w)} \rightarrow Fc^{+}_{(NB)} + Ru(bpy)^{2+}_{3(w)} \tag{8}$$

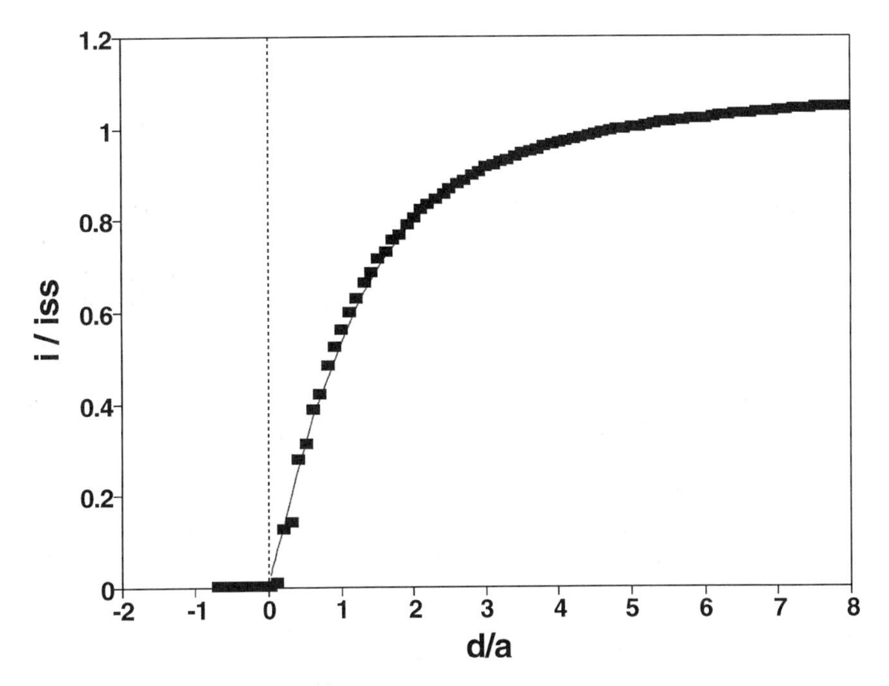

FIG. 2 SECM approach curve for a 5-μm-radius Pt tip UME approaching NB from an aqueous solution containing 5 mM FcCOONa and 0.1 M NaCl. Positive distances correspond to the tip in water; negative distances correspond to tip penetration into NB. The tip potential was held at +400 mV vs. Ag/AgCl, sufficiently positive that the oxidation of $FcCOO^-$ was diffusion-controlled. The NB contained no electroactive species, so the interface behaved as an insulator. Solid line represents SECM theory for an insulating substrate [Eq. (7)]. (From Ref. 15.)

A very large difference of standard potentials of these redox couples (~1 V) suggested that the interface process should be rapid. However, in the absence of supporting electrolyte in NB the overall process rate was limited by charge compensating IT. Using the slower ET reaction between FcCOOH and Fc:

$$Fc^{\circ}_{(NB)} + FcCOO_{(w)} \rightarrow Fc^{+}_{(NB)} + FcCOO^{-}_{(w)} \tag{9}$$

the transition from slow IT to slow ET was observed when concentration of supporting electrolyte in NB (tetraethylammonium perchlorate) increased from 0 to 10 mM. It was found that IT limitations can be avoided by keeping the concentration of common ion in both phases at least 10 times higher than redox mediator concentration in the top phase.

A shallow minimum is typically present in the approach curves when interfacial process is partially limited by common ion transport. This characteristic feature was never observed for ET-controlled process and can be used to detect IT limitations.

4. Potential Drop Across a Nonpolarized ITIES

ET reaction is among the most fundamental and thoroughly studied chemical processes. The predictions of Marcus theory have been tested and verified for a number of chemical and biological systems from bimolecular reactions in solutions (30) to intramolecular ET in proteins (31) to electrochemical reactions at metal electrodes modified by self-assembled organic monolayers (32) and at semiconductors (33). The ET process at the ITIES is a very interesting intermediate case linking homogeneous and heterogeneous ET.

Experimental studies of ET at the ITIES by conventional electrochemical methods are scarce (8,34–36) and the data are often complicated by coupling of the interfacial ET and IT reactions and experimental artifacts (1,16).

The interpretation of the experimental results depends on the adopted model of the interface. One of the questions is about potential profile in the interfacial region. Only the potential drop that exists between the two reacting molecules can contribute to the driving force for ET reaction. But it is not clear what fraction of the total Galvani potential difference ($\Delta_w^o\varphi$) at the liquid junction actually drops between the reacting molecules. In most experimental studies it was implicitly assumed that the potential dependence of the rate constant obeys the Butler-Volmer equation. This may be the case if the reactants are separated by the sharp phase boundary (6a) (Fig. 3A) or a thin ion-free layer (Fig. 3B). If the reactants can penetrate the phase boundary (6b,c), or if ET reaction occurs within a fairly thick mixed solvent layer (Fig. 3C), the fraction of the potential drop contributing to the ET driving force should be negligible or at least much smaller than the total $\Delta_w^o\varphi$ value.

Girault (1b) pointed out that the apparent potential dependence of the ET rate may be attributed to the change in concentration of the reactants near the interface rather than to activation control. This model, further developed by Schmickler (9), postulates that the rate constant is essentially potential-independent because the potential drop across the compact part of the double-layer at the ITIES is small. In this model, the ET rate dependence on the interfacial potential drop is only due to the diffuse layer effect similar to Frumkin effect at metal electrodes.

Some of these questions were clarified by studying the potential dependence of the ET rate at the ITIES using SECM. The potential drop across the ITIES ($\Delta_w^o\varphi$) can be quantitatively controlled and varied by changing the ratio of concentrations of the potential-determining ions in the two liquid phases. If only one type of ion (e.g., ClO_4^- anion) partitions between two solvents:

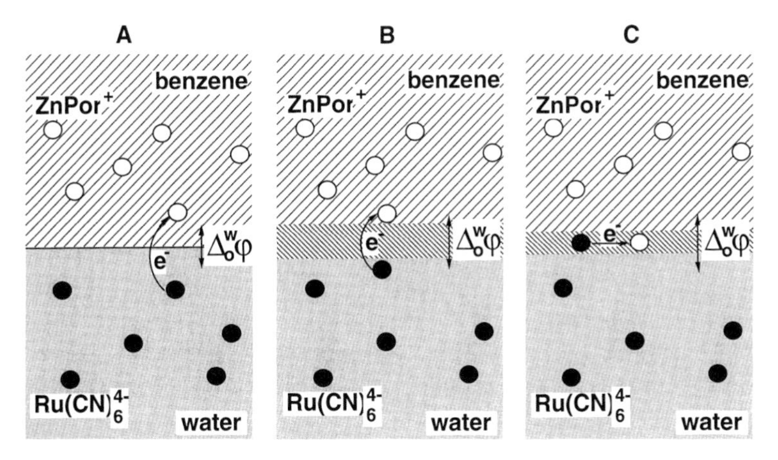

FIG. 3 Different models of interfacial ET. (A) Aqueous and organic redox species are separated by the sharp interfacial boundary. (B) Interfacial potential drop across a thin ion-free layer between redox reactants. (C) ET reaction occurs within a nm-thick mixed solvent layer. No potential drops between reactant molecules.

$$\Delta_w^o\varphi = \Delta_w^o\varphi^o_{ClO_4^-} - 0.059\log([ClO_4^-]_w/[ClO_4^-]_o) \tag{10}$$

This approach was used to study the potential dependence of the rate constant for ET at the ITIES by SECM with no external potential applied (25). The heterogeneous rate constant of ET between $ZnPor^+$ in benzene and aqueous $Ru(CN)_6^{4-}$ were measured with NaCl and $NaClO_4$ dissolved in water and tetrahexylammonium perchlorate ($THAClO_4$) in benzene. Perchlorate was the only ion common in both phases. The transfer of this ion between the two phases maintained electroneutrality during ET process.

Since perchlorate concentration in organic phase was kept constant ($[ClO_4^-]$ = 0.25 mM) in all experiments in (25,26), $\Delta_w^o\varphi$ was a linear function of $[ClO_4^-]_w$ with the slope close to 60 mV per decade:

$$\Delta_w^o\varphi = \text{const} - 0.06\log[ClO_4^-]_w \tag{11}$$

Equation (11) was verified experimentally by measuring steady-state voltammograms of different oxidation reactions in the following cell:

$$\begin{aligned}&Ag/AgCl/H_2O,\ KCl(\text{sat'd})/H_2O,\ 0.1\ M\ NaCl,\ 0.01-2\ M\ NaClO_4\\ &\quad //\text{benzene},\ 0.25\ M\ THAClO_4,\ 5\ mM\ ZnPor\ (\text{or } 5\ mM\ Fc)/Pt\end{aligned} \tag{12}$$

Reversible half-wave potentials ($E_{1/2}$) of Fc an ZnPor oxidation in benzene at a 25-μm electrode were measured with respect to the aqueous Ag/AgCl reference. The dependencies of $E_{1/2}$ versus $\log[ClO_4^-]_w$ were linear with a ~60 mV/decade slope in agreement with Eq. (11) (25).

Equation (11) gives the relative values of the potential drop across the ITIES, which is one of two components of the total driving force for ET reaction. The driving force for interfacial ET is given by (26):

$$\Delta G^\circ = -F(\Delta E^\circ + \Delta_w^o\varphi) \tag{13}$$

where ΔE° is the difference of formal potentials of the organic and the aqueous redox couples (26) (see also Sec. II.A). Since the diffusion coefficients of oxidized and reduced forms are usually similar, one can assume that the formal potential is approximately equal to the reversible half-wave potential. As the $E_{1/2}$ values for both aqueous and organic redox species were measured with respect to the same aqueous reference electrode, the difference of these values is

$$\Delta E_{1/2} = \Delta E^\circ + \Delta_w^o\varphi \tag{14}$$

Although the absolute value of $\Delta_w^o\varphi$ cannot be found without an extrathermodynamic assumption (1), $\Delta E_{1/2}$ gives the absolute value of the driving force.

5. *Potential Dependence of Heterogeneous Rate Constant of ET at the ITIES*

The scheme of SECM measurements of the rate of the ET between $ZnPor^+$ in benzene and aqueous $Ru(CN)_6^{4-}$ (25) was analogous to the one in Figure 1. The tip electrode generates $ZnPor^+$ ions by oxidation of ZnPor. $ZnPor^+$ diffuses to the ITIES, where it is reduced back to ZnPor by $Ru(CN)_6^{4-}$:

$$ZnPor - e \rightarrow ZnPor^+ \qquad \text{(tip)} \tag{15}$$

$$ZnPor^+ + Ru(CN)_6^{4-} \rightarrow ZnPor + Ru(CN)_6^{3-} \qquad \text{(ITIES)} \tag{16}$$

The rate of the mediator regeneration via reaction (16) was evaluated from the tip current. When no oxidizable species was present in the aqueous phase, i_T versus d dependence followed the SECM theory for an insulator (curve 5 in Fig. 4). In the presence of $Ru(CN)_6^{4-}$, the tip current increased with the decrease in d; at higher $[Ru(CN)_6^{4-}]$, the i_T versus d dependencies approached the diffusion limit given by curve 1 in Figure 4.

The values of the effective heterogeneous rate constant for reaction (16) were obtained by fitting the families of approach curves obtained at different concentrations of $Ru(CN)_6^{4-}$ and $NaClO_4$ in water phase to Eqs. (4) and (5) (Fig. 4). At low concentrations of $Ru(CN)_6^{4-}$, k_f was proportional to $[Ru(CN)_6^{4-}]$ for any given values of $[ClO_4^-]_w$ (Fig. 5). As expected for reaction (16), which is the first order in $Ru(CN)_6^{4-}$,

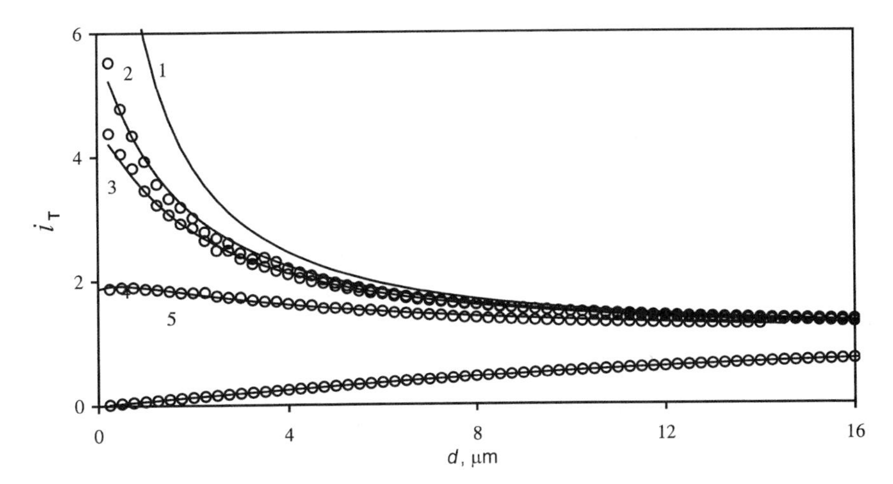

FIG. 4 SECM current-distance curves for a 12.5-μm-radius Pt tip UME in benzene solution approaching the water/benzene interface. Benzene contained 0.5 mM ZnPor and 0.25 M $THAClO_4$. The aqueous phase contained 0.1 M NaCl, 0.01 M $NaClO_4$ and (curve 2) 50, (curve 3) 5, (curve 4) 0.5, or (curve 5) 0 mM $Ru(CN)_6^{4-}$. Curve 1 is the theoretical curve for a diffusion-controlled process obtained using Eq. (6). The tip potential was held at 0.95 V vs. Ag/AgCl, corresponding to the plateau current of first oxidation of ZnPor. The tip was scanned at 0.5 μm/s. Circles are experimental data and lines represent theoretical fit obtained from Eqs. (4) and (5). (From Ref. 25.)

$$k_f = \text{const}[Ru(CN)_6^{4-}]\exp(-\Delta G^{\#}/RT) \tag{17}$$

where $\Delta G^{\#}$ is the free energy barrier. For lower overvoltages, the Butler-Volmer type approximation can be used:

$$\Delta G^{\#} = \alpha\Delta G^{\circ} = -\alpha F(\Delta E^{\circ} + \Delta_w^o\varphi) \tag{18}$$

where ΔE° is the difference of standard potentials of two redox couples and α is the transfer coefficient. At a given concentration of $Ru(CN)_6^{4-}$, the apparent rate constant depends on the potential drop across the ITIES, which increases with a decrease in $NaClO_4$ concentration:

$$k_f = \text{const}[Ru(CN)_6^{4-}]\exp(-0.06\alpha F\log[ClO_4^-]_w/RT) \tag{19a}$$

or

$$\log k_f = \text{const} + \log[Ru(CN)_6^{4-}] - \alpha\log[ClO_4^-]_w \tag{19b}$$

In agreement with Eq. (19b), the log k_f versus $\log[ClO_4^-]_w$ dependencies at different concentrations of $Ru(CN)_6^{4-}$ were linear (Fig. 6a). At lower $[ClO_4^-]_w$,

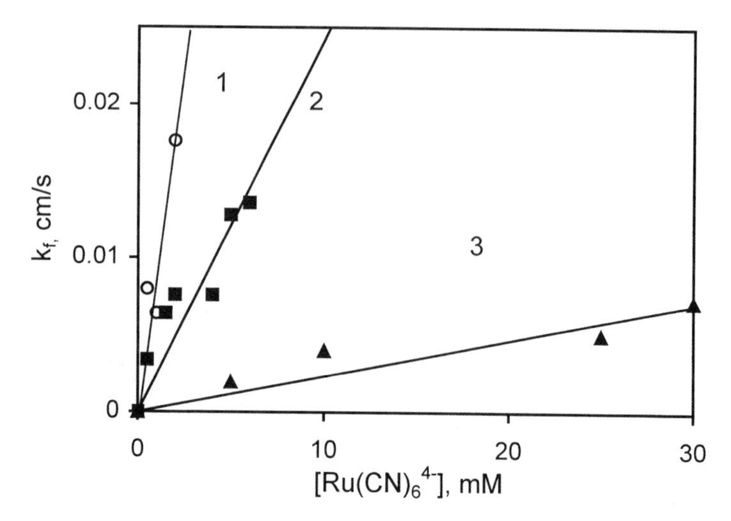

FIG. 5 Dependence of the effective heterogeneous rate constant on $[Ru(CN)_6^{4-}]$ at different concentrations of $NaClO_4$ in water. $[ClO_4^-]_w$ was 0.01 (1), 0.1 (2), and 1 M (3). (From Ref. 25.)

corresponding to a more positive $\Delta_w^o\varphi$, the ET rate approached the diffusion limit and log k_f versus $\log[ClO_4^-]_w$ curves leveled off. Clearly, this effect is more significant at higher concentration of $Ru(CN)_6^{4-}$ (curve 1 in Fig. 6A), for which the k_f values (at the same $\Delta_w^o\varphi$) are higher. Two similar transfer coefficient values were found from the linear portions of the Tafel plots, i.e., $\alpha = 0.49 \pm 0.1$ for 50 mM $Ru(CN)_6^{4-}$ and $\alpha = 0.56 \pm 0.05$ for 5 mM $Ru(CN)_6^{4-}$. The potential dependence of bimolecular rate constant ($k = k_f/[Ru(CN)_6^{4-}]$) presented in Figure 6B shows similar slope of Tafel dependence over more than 1.5 order of magnitude in k.

The following conclusions about the mechanism of ET process and the ITIES structure were made from the obtained potential dependencies of the ET rate constant:

The linear Tafel plots and α values close to 0.5 indicate that conventional ET theory, e.g., Buttler-Volmer model, is applicable to heterogeneous reactions at the ITIES.*

The participants of the ET reaction are separated by a thin interfacial boundary, which they do not significantly penetrate (Fig. 3A). Otherwise, the potential drop between two redox molecules would be

*More recent data (Liu B, Mirkin MV. J Am. Chem. Soc. 121:8352, 1999) indicate that the rate of ET involving neutral organic species can be independent of $\Delta_w^o\varphi$.

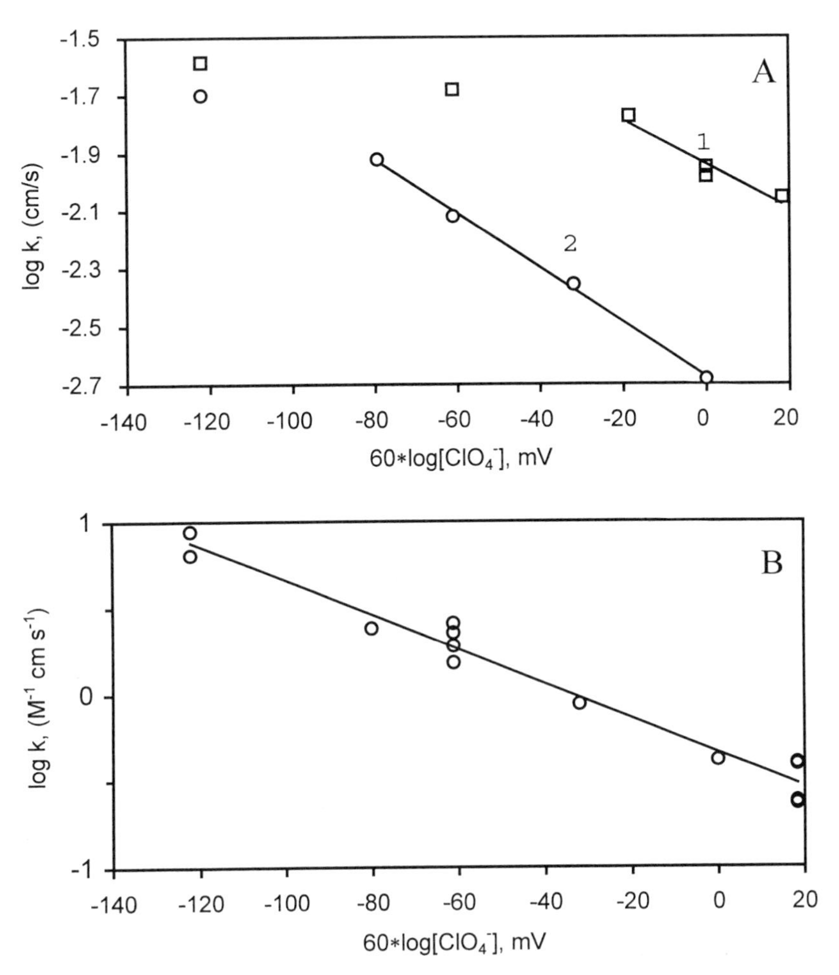

FIG. 6 (A) Dependencies of the effective heterogeneous rate constant on potential drop across the ITIES at 50 (1) and 5 mM (2) concentrations of $Ru(CN)_6^{4-}$. For other experimental parameters see Figure 4. $\Delta_w^o\varphi$ is expressed in terms of $[ClO_4^-]_w$ according to Eq. (11). (B) Potential dependence of the effective bimolecular rate constant. (From Ref. 25.)

much smaller than the total $\Delta_w^o\varphi$, and α would be much smaller than 0.5. Similarly, the experimental results do not support a picture of ET occurring within a fairly thick mixed-solvent layer (Fig. 3C).

A sharp-boundary model (6a) rather than a model assuming a significant penetration of species into a mixed-solvent layer (6b, c) is applicable to the ET at the interface between two very low miscibility solvents, like water and benzene. This does not exclude the possibility of a thin ion-free layer at the interface separating participants of the redox reaction (Fig. 3B). Such a layer would result in the smaller ET rate constant but would not affect the α value.

6. *Driving Force Dependence of the ET Rate*

The driving force for an ET reaction at the ITIES consists of two components, i.e., the difference of standard potentials of redox mediators and the interfacial potential drop [Eq. (14)]. The dependence of k_f on ΔE° was studied for reactions between $ZnPor^+$ in benzene and the series of similar cyanide complexes in water (26):

$$ZnPor^+ + R_w \rightarrow ZnPor + O_w \tag{20}$$

where $R_w = Ru(CN)_6^{4-}$, $Mo(CN)_8^{4-}$, or $Fe(CN)_6^{4-}$.

Assuming that standard rate constants (i.e., k_f at $\Delta G^\circ = 0$) are similar, the differences in ET rates for these reactants may only be due to the different ΔE° values. The formal potentials of $Ru(CN)_6^{4-}$, $Mo(CN)_8^{4-}$, and $Fe(CN)_6^{4-}$ couples measured by cyclic voltammetry are 750, 590, and 235 mV versus Ag/AgCl, respectively. Accordingly, the feedback current obtained with $Mo(CN)_8^{4-}$ in water was higher than with $Ru(CN)_6^{4-}$. With $Fe(CN)_6^{4-}$, the ET rate is much higher, and the overall process was diffusion-controlled at any $[ClO_4^-]_w$.

More quantitatively, one can compare potential dependencies of k_f (Tafel plots) obtained for the different redox species (Fig. 7). For $Ru(CN)_6^{4-}$ and $Mo(CN)_8^{4-}$, the Tafel plots in Figure 7 are linear, and the transfer coefficients are very close to 0.5. The horizontal shift between these two straight lines corresponds to the ~150 mV difference between the formal potential of the $Ru(CN)_6^{4-}$ and $Mo(CN)_8^{4-}$ redox couples. For $Fe(CN)_6^{4-}$ and two more negative aqueous species (V^{2+} and Co(II) Sepulchrate) the ET rate is too high to measure, and the SECM response was diffusion-controlled at any $\Delta_w^o\varphi$ value (Fig. 7).

B. Electron Transfer Across a Molecular Monolayer

To check the predictions of the Marcus theory of ET (6), one has to carry out measurements at higher overpotentials. This can be done by separating the electron donor and acceptor by a well-defined spacer (32), which decreases the ET rate. Molecular monolayers of long-chain saturated phospholipids (Fig. 8) have been used as a spacer in ITIES studies (26,27).

1. *Adsorption of Phospholipids at the Interface*

Amphiphilic lipid molecules dissolved in an organic solvent spontaneously form a monolayer film at the water/organic interface. The orientation of the

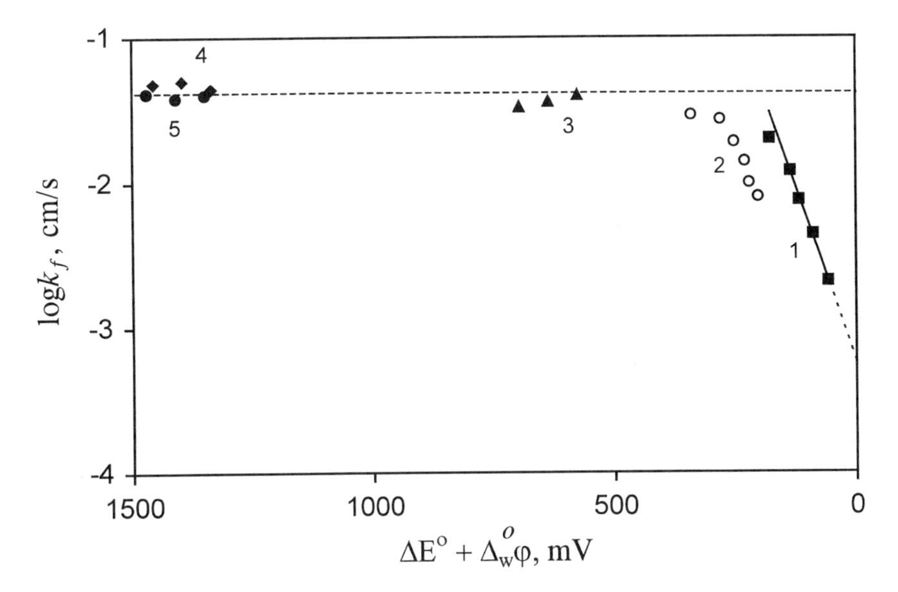

FIG. 7 Potential dependencies of the ET rate between $ZnPor^+$ and various aqueous redox species. The abscissa plots the driving force for ET given by Eq. (14). The aqueous phase contained 7 mM of $Ru(CN)_6^{4-}$ (1), $Mo(CN)_8^{4-}$ (2), $Fe(CN)_6^{4-}$ (3), Co(II) Sepulchrate (4), and V^{2+} (5). The aqueous supporting electrolyte was 0.1 M NaCl, 0.01–2 M $NaClO_4$ (1–4) or 0.5 M H_2SO_4 (5). For other parameters see Figure 4. Horizontal dash line shows the diffusion limit for the ET rate measurements by SECM with 25-μm tip (~0.03 cm/s). (From Ref. 26.)

monolayer is such that the hydrophilic head group is immersed in water, while the hydrophobic tail remains in the organic phase. The first attempt to probe ET across a phospholipid monolayer adsorbed at the ITIES was reported by Cheng and Schiffrin, who found that a monolayer makes the ET rate immeasurably slow (37).

SECM measurements of the interfacial ET between $ZnPor^+$ in benzene and $Fe(CN)_6^{4-}$ in water also showed a strong blocking effect of phospholipids adsorbed at the ITIES (26,27). In the presence of the monolayer of 1,2-diacyl-*sn*-glycero-3-phosphocholine with 10 methylene groups in a hydrocarbon chain (C-10; the abbreviations C-10, C-12, C-14, etc. will be used in this section to designate different 1,2-diacyl-*sn*-glycero-3-phosphocholines with the corresponding numbers of methylene groups in the hydrocarbon chain), at the ITIES the k_f value, which was immeasurably high at a clear ITIES, decreased and became accessible by SECM measurements.

The blocking effect of the lipid film can be attributed to the decrease in either the ET or IT rate. Kakiuchi and coworkers (39) and Schiffrin and

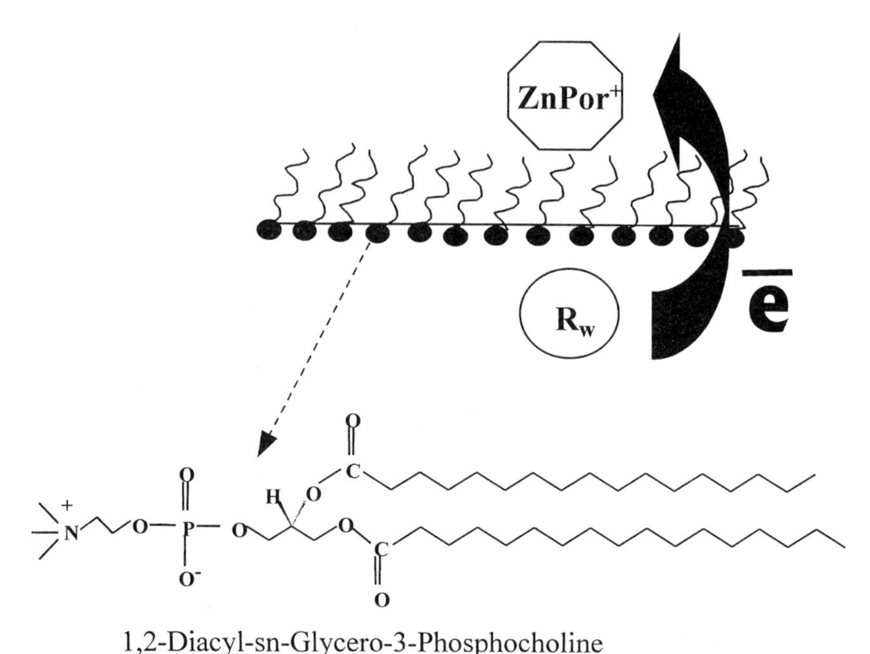

FIG. 8 The ITIES modified with a monolayer of phospholipid. The insert shows the structure of synthetic saturated phosphatidyl choline lipid. (From Ref. 26.)

coworkers (38) extensively studied the effect of different lipid monolayers on transfer rates of various ions, including ClO_4^-. The largest decrease in IT rate caused by the presence of a compact phospholipid monolayer at the ITIES was by a factor of 5. The effects observed in Ref. 26 were orders of magnitude larger. This observation and the characteristic shape of approach curves discussed in Sec. II.A allowed the authors to rule out the possibility of the slow IT effect.

A typical set of SECM current-distance curves for different concentrations of C-10 under equilibrium conditions is shown in Figure 9 for $Ru(CN)_6^{4-}$ species in the aqueous solution and ZnPor in benzene. The ET rate decreases monotonically as the lipid concentration increases from 0 (curve 2) to 48 μM (curve 6). These curves show the effect of increasing coverage of the interface with lipid film. The ET at both clear ITIES and covered with lipid contributed to the obtained rate constant values.

Adsorption isotherms (dependencies of k_f vs. lipid concentration) were obtained for two different aqueous redox species, $Ru(CN)_6^{4-}$ and $Fe(CN)_6^{4-}$, and several different lipids (Fig. 10). The decrease in k_f indicates that the

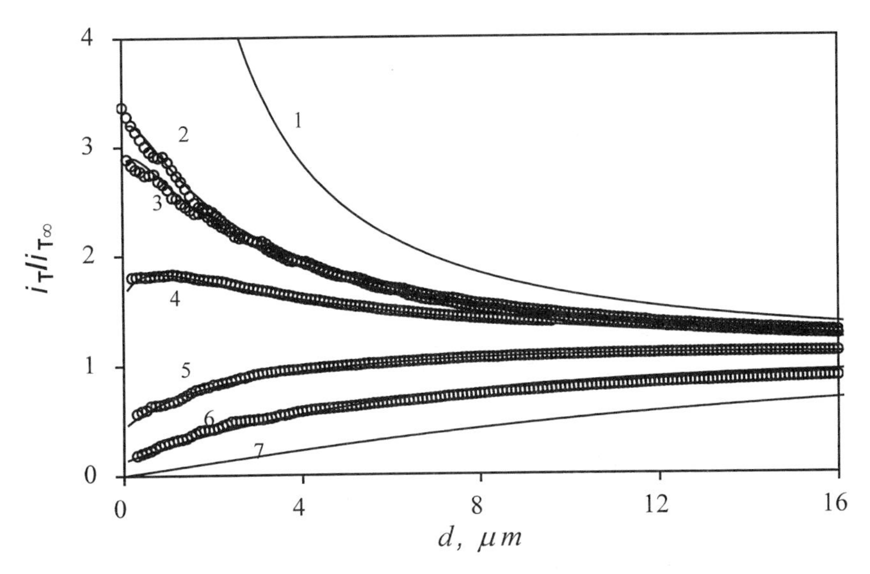

FIG. 9 Effect of lipid concentration on the shape of the SECM current-distance curves. The aqueous solution was 0.1 M NaCl, 0.1 M $NaClO_4$, and 7 mM $Na_4Ru(CN)_6$. The concentration of C-10 lipid in benzene was 0 (2), 1.12 (3), 3.54 (4), 12.4 (5), and 47.8 μM (6). Circles are experimental data. Solid lines represent the theory for the following k_f values (cm/s): (1) ∞ (diffusion-controlled process), (2) 0.0145, (3) 0.013, (4) 0.0075, (5) 0.002, (6) 0.0006, and (7) 0 (pure negative feedback). The tip was scanned at 1 μm/s. (From Ref. 26.)

fraction of the interfacial area covered with lipid increases with increasing lipid concentration in benzene and results in a lower ET rate. At high concentration of any lipid (i.e., >50 μM) the ET rate between $Ru(CN)_6^{4-}$ and $ZnPor^+$ becomes immeasurably slow.

The k_f versus [lipid] dependencies for $Fe(CN)_6^{4-}$ are different. Although the ET rate decreases markedly with increasing concentration of lipid, it does not approach zero at higher concentrations, but reaches the limiting value at about 50 μM. This saturation suggests the formation of a complete phospholipid monolayer at the ITIES. The formation of compact phospholipid monolayers at similar lipid concentrations was observed at water/nitrobenzene (40a) and water/dichloroethane (40b) interfaces.

The presence of defects and pinholes is always a concern in experimental work with monolayers (32b). The following findings indicated that the defect density in the lipid film is too low to affect ITIES/SECM measurements (26):

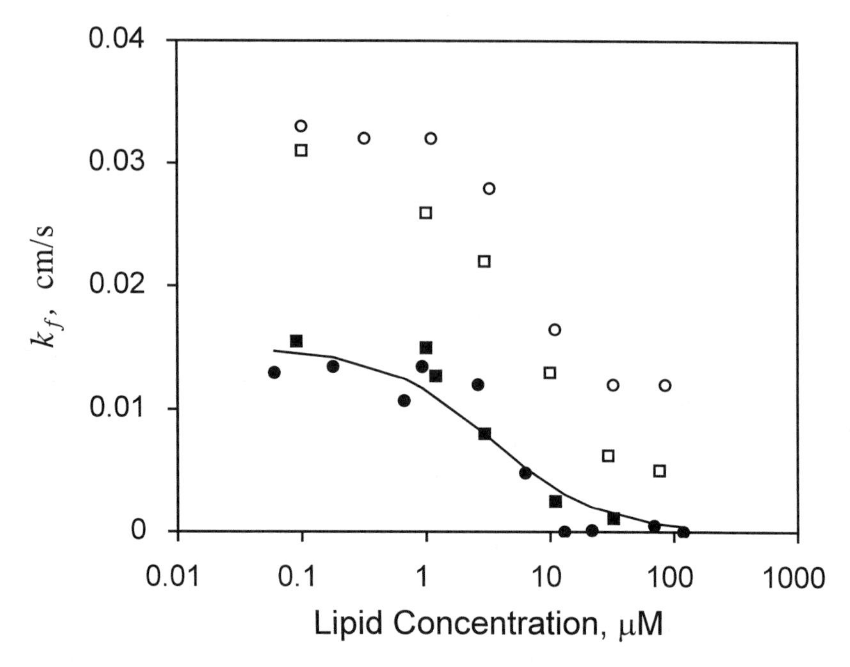

FIG. 10 Dependence of the rate constant of ET between $ZnPor^+$ and $Ru(CN)_6^{4-}$ (■ and ●) or $Fe(CN)_6^{4-}$ (□ and ○) on lipid concentration in benzene. The number of methylene groups in the lipid hydrocarbon chain was: 10 (■ and ○), 12 (□), and 20 (●). The organic phase contained 0.25 M $THAClO_4$, 0.5 mM ZnPor, and lipid. The water phase was 0.1 M NaCl, 0.1 M $NaClO_4$, and 7 mM $Na_4Ru(CN)_6$ (■ and ●) or $Na_4Fe(CN)_6$ (□ and ○). ET rate constants were obtained by fitting experimental approach curves (Fig. 9) to the theory Eq. (4), (5). The solid line represents Frumkin isotherm [Eq. (21)] with $B = 2 \times 10^5\ M^{-1}$ and $a = 0.25$. (From Ref. 26.)

The extent of the blocking effect apparently depends on the driving force for ET. When the driving force is small (e.g., $\Delta E^o \sim 100$ mV for $Ru(CN)_6^{4-}$ and $ZnPor^+$), the formation of any long-chain phospholipid monolayer (i.e, from C-10 to C-20) results in a ET rate below the lower limit for SECM measurements. This shows that the density of the possible defects in the monolayer is too low to produce detectable feedback current under experimental condition of Ref. 26.

The ET between $Fe(CN)_6^{4-}$ and $ZnPor^+$, for which ΔE^o is about 0.5 V larger, occurs at a measurable rate via tunneling through the monolayer. If the electrons were transferred through pinholes rather than across the lipid monolayer, one would expect the ET rate to be independent of the spacer length. In contrast, from Figure 10 one can see that the longer lipid, C-12 (open squares), inhibits ET more strongly than C-10 at the same concentration (open circles).

Time dependencies measured in Ref. 26 showed that the adsorption of lipid at the ITIES is fully reversible. The lipid adsorption on and desorption from the ITIES occur at about the same time scale. The monolayer self-assembled at the ITIES is in dynamic equilibrium with dissolved lipid molecules that allows any pinhole defects to be healed. In contrast, self-assembly on the solid/liquid interface often involves irreversible adsorption and requires high purity of chemicals, complex pretreatment, and preparation protocols to obtain a low-defect-density molecular monolayer (32).

The equilibrium adsorption of phospholipids at the ITIES has previously been studied by several groups (38–43) and shown to follow Frumkin isotherm:

$$B[\text{lipid}] = \theta \exp(-2a\theta)/(1 - \theta) \quad (21)$$

where a is a small lateral interaction parameter, θ is the coverage of the ITIES with lipid, and the standard Gibbs energy of adsorption, $\Delta G^{o}_{ads} = -RT\ln B$, is about 35–40 kJ/mole (41,43c). The best fit to experimental data (solid line in Fig. 10) was obtained with $a = 0.25$ and $B = 2 \cdot 10^{5}$.

2. *Driving Force Dependence of the ET Rate Across a Lipid Monolayer*

The potential dependence of the ET rate across a lipid monolayer was measured using the methodology described above for a clear ITIES (25,26). The driving force for ET reaction was evaluated as a difference between the two half-wave potentials measured with respect to the same aqueous Ag/AgCl reference. The decrease in $[ClO_4^-]_w$ resulted in an increase in driving force for ET [reaction (20)]. Figure 11 illustrates the driving force dependence of the ET rate across a C-10 lipid monolayer adsorbed on the water/benzene interface for $ZnPor^+$ and three different aqueous redox species. For the $Fe(CN)_6^{4-}/ZnPor^+$ reaction, the $(\Delta E^{o} + \Delta_w^o\varphi)$ values were within the range of 570–700 mV. At lower overpotentials $(\Delta E^{o} + \Delta_w^o\varphi \leq 630$ mV), the Tafel plot was linear with a slope corresponding to $\alpha = 0.59$. At larger overpotentials, this curve levels off as predicted by Marcus theory (5a):

$$\Delta G^{\neq} = (\lambda_o/4)(1 + \Delta G^{o}/\lambda_o)^2 \quad (22)$$

where λ_o is the reorganization energy and ΔG^{o} is given by Eq. (18).

At higher overpotentials (i.e., $\Delta G^{o} > \lambda_o$), Eq. 22 predicts the existence of the inverted region, in which the increasing driving force results in a decrease in the ET rate (5,32c). This effect cannot be observed at metal

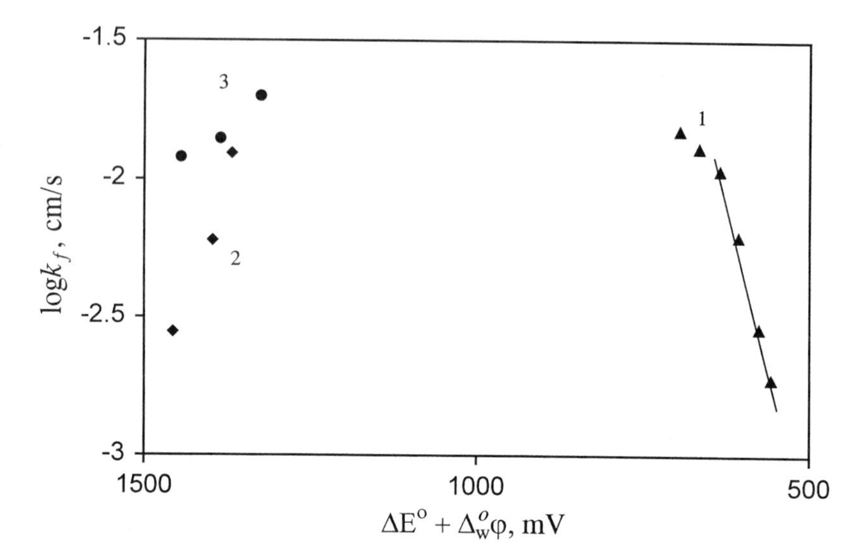

FIG. 11 Driving force dependence of the ET rate between $ZnPor^+$ in benzene and various aqueous redox species across a monolayer of C-10 lipid. The organic phase contained 0.25 M $THAClO_4$, 0.5 mM ZnPor, and 100 μM C-10. The aqueous phase contained 7 mM of $Fe(CN)_6^{4-}$ (1), Co(II) Sepulchrate (2), and V^{2+} (3). For other parameters, see Figure 7. (From Ref. 26.)

electrodes, where a continuum of electronic states exists in the metal below Fermi level. In contrast, the concentrations of redox species in both liquid phases are finite, and one can expect to see the inverted behavior when ΔG^o is sufficiently large. Accordingly, the rate constant of the $ZnPor^+/V^{2+}$ reaction decreased significantly when the driving force increased by 120 mV (filled circles in Fig. 11). These data were obtained with 0.5 M H_2SO_4 used as a supporting electrolyte in the aqueous phase. The protonation of the lipid layer at low pH might have affected its barrier properties.

Additional evidence of the inverted behavior was obtained using another redox couple with a large negative standard potential, $CoSep^{3+/2+}$. A large decrease in the ET rate was observed for the $ZnPor^+/CoSep^{2+}$ system with increasing ΔG^o (diamonds in Fig. 11). One might try to explain the observed behavior by a diffuse layer effect. Unlike $Ru(CN)_6^{4-}$, $Mo(CN)_8^{4-}$, and $Fe(CN)_6^{4-}$ species, both V^{2+} and $CoSep^{2+}$ are cationic. The decrease in interfacial concentrations of such species may result in the lower ET rate. However, according to the theory in Ref. 9, the diffuse layer effect should be significant mostly for species residing in organic phase. Besides, the rate of ET between cationic (but significantly more positive) Fe^{2+} species and

$ZnPor^+$ increased with the increase in driving force (26). No inverted behavior was observed for this reaction because of a much lower driving force (i.e., 270–400 mV). Additional experimental data are needed to estimate the reorganization energy from the potential dependence of the rate constant.

3. *Distance Dependence of ET*

The results in Figure 10 suggest that the longer lipid (C-12) produces a stronger blocking effect on ET between $Fe(CN)_6^{4-}$ and $ZnPor^+$ than the shorter one (C-10). To probe the distance dependence of ET, the k_f was measured for the reaction between $Fe(CN)_6^{4-}$ and $ZnPor^+$ at the ITIES modified with monolayers of different saturated phospholipids (C-10, C-12, C-14, and C-16) (26,27). The feedback current decreases markedly with an increasing number of methylene groups (n) in a hydrocarbon chain at any concentration of ClO_4^- in the aqueous phase (Fig. 12). With the C-16 lipid, the results were less reproducible, and the k_f versus n dependence tends to level off. This may be due to the partial penetration of $ZnPor^+$ into the lipid monolayer. The unknown depth of $ZnPor^+$ penetration impairs quantitative analysis of k_f versus n results in terms of the distance dependence of ET. Even for C-10 to C-14 lipids fitting the experimental points to Eq. (23):

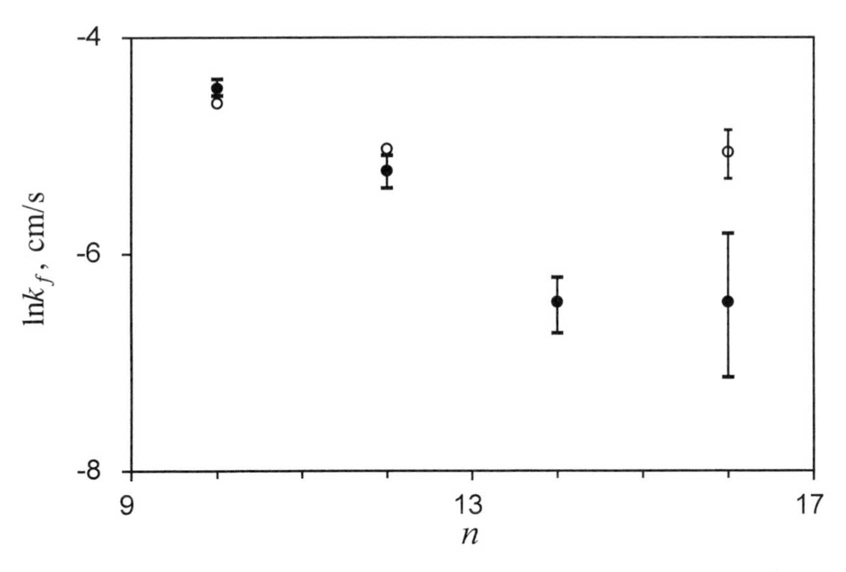

FIG. 12 Distance dependence of the ET rate between $ZnPor^+$ and $Fe(CN)_6^{4-}$. The abscissa plots the number of the methylene groups in the lipid hydrocarbon chain. The concentration of $NaClO_4$ was 0.1 M, corresponding to a 635 mV driving force for the ET reaction. For other parameters, see Figure 7. (From Ref. 26.)

$$k_f(n) = k_f(0)\exp(-\beta n) \tag{23}$$

yielded a β value about one half of that usually reported for saturated hydrocarbon spacers (31,32,44). However, the penetration depth must be significantly less than the length of the hydrocarbon tail in C-10. Otherwise, the ET rate would not decrease by several orders of magnitude in the presence of C-10 monolayer, and the lengthening of the hydrocarbon chain from C-10 to C-14 would not cause further decrease in k_f.

4. Effect of the Lipid Monolayer Structure on the ET Rate

The studies at solid electrodes showed that the introduction of delocalized conjugated units and aromatic groups in a molecular monolayer affects the ET rate (32b). The SECM has been used to compare the ET rate across C-16 and similarly sized dipalmytoyl phosphocholine (DPHC) (Fig. 13). The latter phospholipid has a *trans,trans,trans*-1,6-diphenyl-1,3,5-hexatriene chain. This prototype of polyene systems has been previously used as a fluorescent probe (45). Its nonpolar hydrocarbon structure is compatible with membranes and can serve as an electron relay in a blocking monolayer film. In addition to its conjugated chain, DPHC also contains one saturated C-16 alkyl chain, which is easily incorporated into the C-16 monolayer.

Similarly to saturated lipids, increasing concentration of DPHC in benzene resulted in the formation of a compact film and a stronger blocking of the interfacial ET. The limiting ET rate across the complete monolayer of DPHC was 2.4 times higher than that obtained with C-16 (Fig. 14).

One possible interpretation of the results shown in Figure 14 is that ET occurs more readily through the delocalizated conjugated chain of the DPHC. If so, this pathway should be quite efficient given that only half of the chains contain more conductive conjugated assembly. Alternatively, the structure of the DPHC layer might be less compact and allow greater penetration of $ZnPor^+$ into lipid film. The ET rate constant across DPHC layer is similar to the k_f for a much shorter saturated lipid (C-10) (Fig. 12).

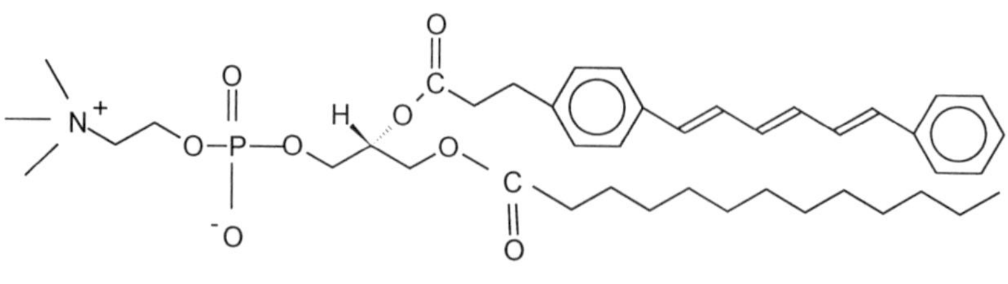

FIG. 13 Structure of the synthetic conjugated phosphatidyl choline lipid, 2-(3-(diphenylhexatrienyl)propanoyl)-1-hexadecanoyl-glycero-3-phosphocholine (DPHC).

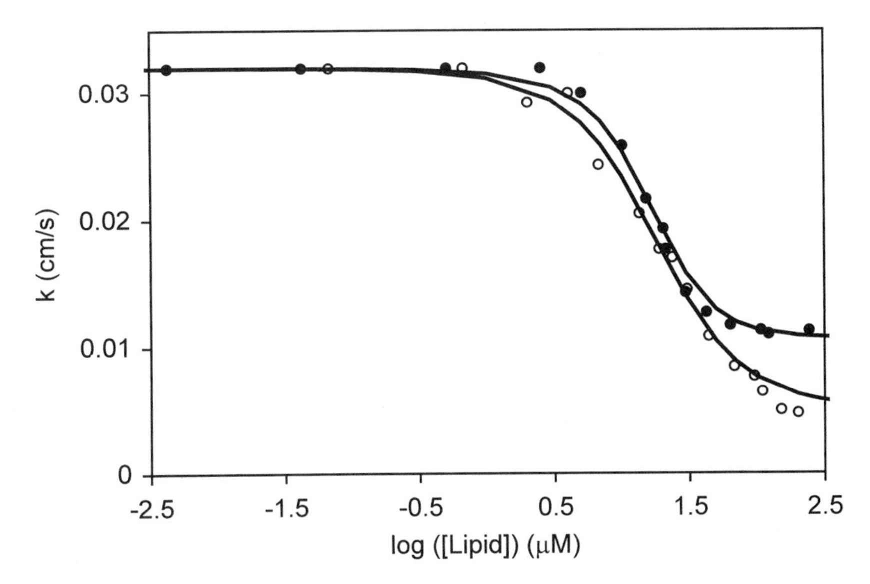

FIG. 14 Heterogeneous ET rate constant, k_f, as a function of the DPHC (●) and C-16 (○) concentrations in the organic phase. The organic phase contained 0.25 M $THAClO_4$, 0.5 mM ZnPor, and lipid. The aqueous phase was 0.1 M NaCl, 0.1 M $NaClO_4$, and 7 mM $Na_4Fe(CN)_6$. 25-μm Pt tip was scanned at 1 μm/s. (From Ref. 27.)

The authors of Ref. 27 used the above approach to probe the distribution of molecules in mixed phospholipid monolayers. It was shown that the lipid can embed itself into the preexisting monolayer of another lipid. First, a complete monolayer of C-16 was formed at the benzene/water interface. The ET at this interface was slow ($k_f = 0.005$ cm/s). The addition of DPHC to the organic phase caused an increase in the apparent rate constant (Fig. 15, curve b). The increase in k value was observed only with mole fraction of DPHC (x_{DPHC}) in solution ≥ 0.3. The k_f reached a constant value and ceased to increase at $x_{\mathrm{DPHC}} \sim 0.55$. The opposite effect was induced by increasing the mole fraction of C-16 ($x_{\mathrm{C\text{-}16}}$) in organic phase in the presence of monolayer of DPHC (Fig. 15, curve a), i.e., the sharp decrease in k_f was observed at $x_{\mathrm{C\text{-}16}} > 0.4$.

While the approach curves for monolayers of pure C-16 and DPHC lipids were reproducible, the k_f obtained at mixed monolayers of lipids showed larger variations. Some values of k_f obtained from different approach curves were very close to those for pure C-16 and DPHC monolayers. This was explained by segregation of lipids in the monolayer into separate do-

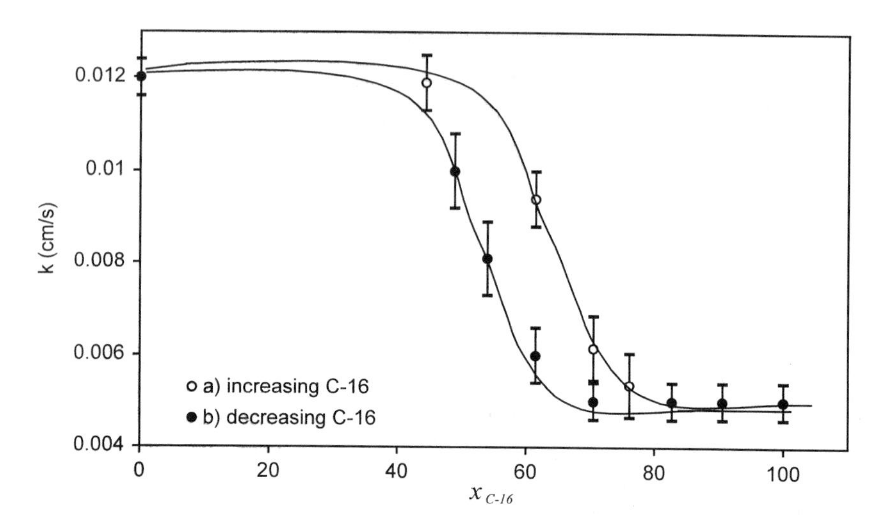

FIG. 15 Variation of the rate constant of ET across mixed lipid monolayer adsorbed at the ITIES. (a) Incorporation of C-16 into preexisting monolayer of DPHC. (b) Incorporation of DPHC into preexisting monolayer of C-16. For other experimental details, see Figure 15. (From Ref. 27.)

mains composed mostly of pure C-16 or DPHC. The apparent k_f value depends on the nature of the domain facing the UME tip.

To prove the existence of separate domains in the mixed monolayer, the tip was scanned laterally above the interface. At a clear ITIES or the interface covered with a monolayer of a single phospholipid (either DPHC or C-16), the current fluctuations during lateral scan did not exceed 5–10% (curves 1 and 2 in Fig. 16A). Much stronger fluctuations of the tip current were observed at mixed lipid films (curve 3, Figure 16A). The lateral movement of the UME along the interface revealed the existence of 10–30 μm domains composed of different lipids. These numbers are in a good agreement with domain sizes found from fluorescent probe microscopy of DPHC adsorbed at the air/water interface (46). As expected, the forward and reverse lateral scans of the same portion of the interface separated by 2-minute intervals showed similar current responses (Fig. 16B). It was concluded that only small changes in domain structure of the monolayer occur on this time scale.

5. *Temperature Dependence of ET Rate Across Lipid Monolayer Adsorbed at the ITIES*

It is well known that lipid monolayers adsorbed at the air/water interface can undergo phase transitions at certain values of pressure and temperature

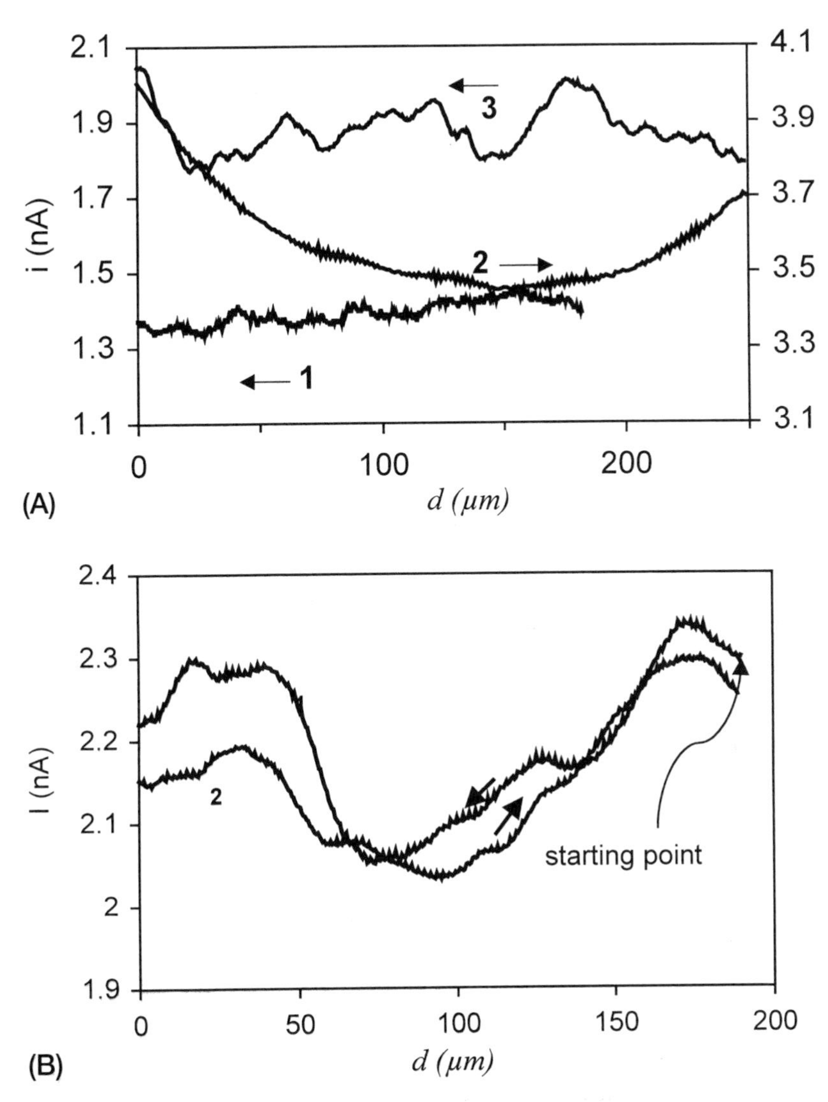

FIG. 16 (A) Images of the ITIES obtained by lateral scanning of the tip UME: (1) the tip is far (d ≈ 250 μm) from the interface; (2) a clear ITIES without adsorbed lipid (d ≈ 3–5 μm); (3) ITIES covered with mixed lipid monolayer (d ≈ 3–5 μm). The mole fractions of DPHC and C-16 in organic phase were 0.7 and 0.3, respectively. The total lipid concentration in the organic phase was 62 μM. (B) Forward (1) and backward (2) lateral scans of the tip UME above the mixed lipid monolayer adsorbed at the ITIES. Curves 1 and 2 were recorded within a 2-minute time interval. The composition of the monolayer was the same as for curve 3 in (A). (From Ref. 27.)

(39c,40,42,47). Different phases detected in lipid monolayer, i.e., gas, liquid-flexible, liquid-extended, solid-tilted, and solid-vertical (42), have been characterized with respect to typical values of area per lipid molecule, conformation, and tilt angle of hydrocarbon chain. Such transitions may change the separation distance between aqueous and organic redox species and hence the ET rate.

The SECM has been used to measure temperature dependence of ET rate and detect phase transitions occurring in phospholipid monolayers adsorbed on the ITIES (27). The temperature dependencies of the ET rate between $ZnPor^+$ and $Fe(CN)_6^{4-}$ across monolayers of C-10 and C-16 are shown in Figure 17 for the temperature range from 5 to 25°C. The ET rate across a C-10 monolayer slightly decreased when the temperature was changed from 25 to 15°C (Fig. 17, curve a). However, there was a sharp drop in k_f for the temperature decrease between 15 and 10°C. The further decrease in the temperature had a small effect on k_f value, and the slope of the $\ln k_f - 1/T$ dependence was about the same as that for the 15–25°C range. The sharp decrease in the ET rate between 15 and 10°C may be caused by a phase transition in the C-10 monolayer adsorbed at the ITIES. The 10–15°C temperature range agrees well with the results of phase tran-

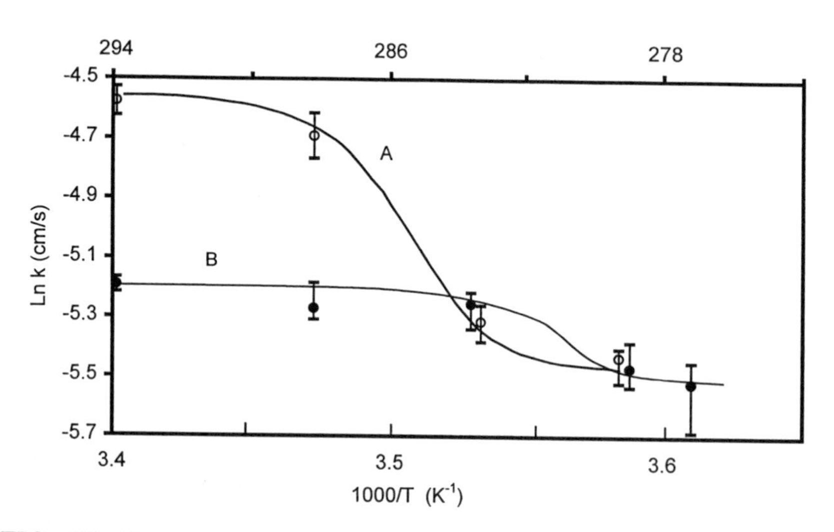

FIG. 17 Temperature dependence of ET rate between $ZnPor^+$ and $Fe(CN)_6^{4-}$ across the ITIES with adsorbed monolayer of C-10 (A) and C-16 (B). The organic phase contained 0.25 M $THAClO_4$, 0.5 mM ZnPor, and 60 μM C-10 or 110 μM C-16. The water phase was 0.1 M NaCl, 0.1 M $NaClO_4$, and 7 mM $Fe(CN)_6^{4-}$. The temperature was controlled within ±0.01°C. The standard deviations were calculated using at least four different data points. (From Ref. 27.)

sition study by impedance technique (39c). In comparison, no sharp decrease in k_f was observed for a monolayer of C-16 (Fig. 17, curve b) that does not undergo phase transitions in the temperature range 5–25°C (37).

III. ION TRANSFER AT THE ITIES

Although most amperometric SECM experiments involved ET reactions at the tip and/or substrate, interfacial IT processes can also be probed. Historically, the first IT reactions studied by SECM were ion-exchange processes at ionically and electronically conductive polymer films (48). The ions of interest were electrochemically active (e.g., $Fe(CN)_6^{3-}$ or Br^-) to enable amperometric detection at the tip. It was shown more recently that the tip process can be an IT reaction rather than an ET process if a micropipet electrode is used as an amperometric probe (49). In this section we consider two different types of IT reactions employed in SECM studies, i.e., facilitated IT and simple IT.

A. Facilitated Ion Transfer

In a typical facilitated IT reaction an ion (most often, a cation, M^+) is transferred from aqueous solution into the organic phase. A complex species formed by this ion and a ligand (L, initially present in organic phase) is easier to transfer than M^+ itself. Reactions of this type are widely used in chemical sensors [ion-selective electrodes, liquid ion-exchangers (50)]. For SECM experiments, an aqueous solution of M^+ is placed inside a micropipet, which serves as a tip electrode. The facilitated IT reaction at the micropipet tip is

$$M^+_{(w)} + L_{(o)} \rightarrow ML^+_{(o)} \qquad \text{(at the pipet tip)} \tag{24}$$

With the tip biased at a sufficiently positive potential and the concentration of M^+ inside a pipet much higher than concentration of L in the outer solvent, the tip current is limited by diffusion of L to the pipet orifice. When the tip approaches the bottom (aqueous) layer, M^+ is released from the complex and transferred to the aqueous solution, and L is regenerated (Fig. 18A):

$$ML^+_{(o)} \rightarrow M^+_{(w)} + L_{(o)} \qquad \text{(at the ITIES)} \tag{25}$$

The ITIES formed at the pipet tip is polarizable, and the voltage applied between the micropipet and the reference electrode in organic phase provides the driving force for facilitated IT reaction. The interface between organic (top) and water (bottom) layers is nonpolarizable, and the potential drop, $\Delta_w^o\varphi$, is governed by the ratio of concentrations of the common ion (e.g.,

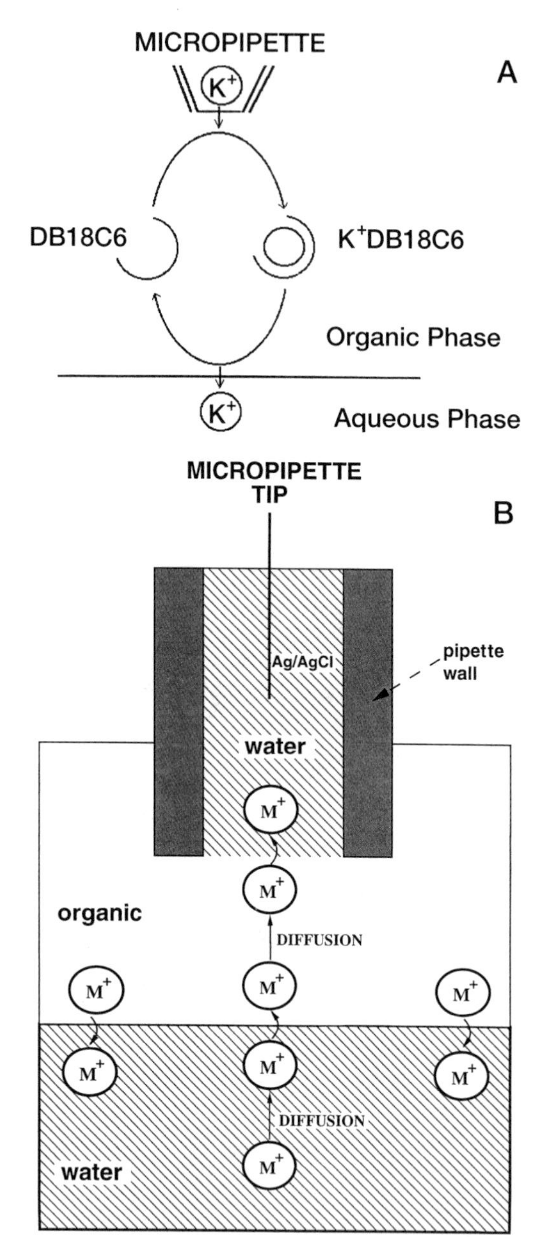

FIG. 18 Schematic representation of the SECM operating in the facilitated (A) and simple (B) IT feedback mode. (A) Potassium ions are transferred from the pipet into DCE by interfacial complexation with DB18C6 [Eq. (26)] and from DCE to

ClO_4^-) in two phases. The $\Delta_w^o\varphi$ value determines the rate of reaction (25). The species L serves as a mediator in reactions (24) and (25), and its consumption at the tip and regeneration via a heterogeneous IT at the ITIES are mathematically equivalent to the conventional (i.e., ET-based) SECM feedback scheme. The tip current enhancement (positive feedback) is observed when reaction (25) is rapid, and the negative feedback occurs if L is not regenerated at the ITIES.

Six stages of the overall process can affect the tip current, i.e.: (1) diffusion of M^+ inside the pipet, (2) facilitated IT at the pipet tip, (3) diffusion of L in organic solvent, (4) IT at the water/organic interface, (5) diffusion of M^+ in the bottom aqueous phase, and (6) charge compensation by transfer of supporting electrolyte between the water and organic phase. In principle, any of these stages can be rate-limiting, and a careful choice of experimental conditions is essential for meaningful kinetic analysis. Typically, step 5 is fast and the charge compensating IT (stage 6) does not affect the overall process rate because both concentrations of the common ion in organic and aqueous layers are much higher than concentration of L. The concentration of M^+ should be sufficiently high to exclude the possibility of diffusion limitations inside the pipet (stage 1). The rate of the IT reaction at the pipet/solution interface can be made rapid by biasing the tip at a sufficiently positive potential. Under these conditions, only diffusion of the L in organic phase or/and interfacial dissociation reaction may control the overall process rate. If reaction (25) is rapid, the SECM theory for a conductive substrate is applicable; otherwise finite IT kinetics should be observed.

To our knowledge, only two facilitated IT processes have been used in SECM experiments. The transfer of K^+ from the aqueous solution inside the pipet into 1,2-dichloroethane (DCE) assisted by dibenzo-18-crown-6 (DB18C6) (49)

$$K^+_{(w)} + DB18C6_{(DCE)} \rightarrow [K^+DB18C6]_{(DCE)} \qquad \text{(at the pipet tip)} \qquad (26)$$

has been extensively studied and shown to be a mechanistically simple one-step interfacial complexation reaction (1b). With the concentration of K^+ inside a pipet at least 50 times higher than the concentration of DB18C6 in

---

the bottom aqueous layer by interfacial dissociation mechanism [Eq. (27)]. Electroneutrality in the aqueous layer is maintained by transfer of TBA^+ across the interface. (From Ref. 49.) (B) Positive feedback is due to IT from the bottom (aqueous) layer into the organic phase. Electroneutrality in the bottom layer is maintained by reverse transfer of the common ion across the ITIES beyond the close proximity of the pipet where its concentration is depleted.

DCE, the tip current was limited by diffusion of DB18C6 to the pipet orifice. When the tip approached the bottom aqueous layer, the regeneration of DB18C6 occurred via interfacial dissociation:

$$[K^{+}DB18C6]_{(DCE)} \rightarrow K^{+}_{(w)} + DB18C6_{(DCE)} \qquad \text{(at the ITIES)} \qquad (27)$$

Thus, a positive feedback current was observed when the tip approached the water/DCE interface. The negative feedback was observed when the tip approached glass insulator.

Since the SECM response depends on rates of both heterogeneous reactions (26) and (27), their kinetics can be measured using methodologies developed previously for ET processes. The mass-transfer rate for IT measurements by SECM is similar to that for heterogeneous ET measurements, and the standard rate constants of the order of 1 cm/s should be measurable. With relatively large pipet probes used in Ref. 49 (i.e., $a \geq 10$ μm), the mass-transfer coefficient was <0.01 cm/s, i.e., too low to measure the standard rate constant of potassium transfer [1.3 cm/s (51)]. The SECM response was diffusion controlled. In contrast, a kinetically controlled regeneration of the mediator was observed for a slower reaction of proton transfer assisted by 1,10-phenanthroline.

Conventional SECM theory is not applicable to micropipet tips because the ratio of the glass radius to the aperture radius (RG) is typically much less than 10 [the typical RG value is ~1.1 (52)]. An approach curve for facilitated transfer of potassium could only be fit to the theory for a diffusion-controlled positive feedback assuming a near-hemispherical shape of the meniscus (49). But the later video-microscopic study showed that the ITIES formed at the micropipet tip is flat (52). Neither was it possible to fit an i_T − d curve obtained when a micropipet tip approached an insulator (49). Both conductive and insulting curves can be fit to the theory developed recently for small RG (53) (see Chapter 5). The theory accounting for finite kinetics of facilitated IT at the ITIES has yet to be developed.

B. Simple Ion Transfer

Simple IT reactions are numerous and of great importance for biological systems (40). Unlike both ET and facilitated IT processes, no mediator species is involved in a simple IT reaction. It can be induced in two different ways: either by application of external voltage across the interface or by depleting the concentration of the common ion in one of two liquid phases near the ITIES. White and coworkers used the first approach to probe localized iontophoretic fluxes of electroactive species (e.g., Fe^{2+}) through artificial membranes and hairless mouse skin (54). The tip scanned over the membrane surface detected the species driven through pores by applied voltage. In addition to flux, profile maps such measurements allow quantitative determination of the rates of mass transport in membrane pores. Matsue et al. (55a) probed the transfer of ions through alamethicin channels in a bilayer lipid membrane (BLM) induced by voltage applied across the membrane. The flux of electroactive ions (e.g, I^-) was monitored with a Pt UME positioned near BLM. Different permselectivity values were measured for several cations and anions. Unlike typical SECM experiments, CT in that work occurred at a large interface. Such measurements are likely to suffer from resistive and capacitive effects.

The second approach is based on using a microprobe to perturb equilibrium in one of the phases near the interfacial boundary. The transfer of electrochemically active ions and neutral molecules across the liquid/liquid interface can be studied with a metal tip positioned near the phase boundary (56). The interfacial flux was induced by using a disk-shaped UME to deplete the concentration of transferred species in one of two phases. The transfer processes at an air/water and hydrogel/solution interfaces can be studied similarly (56).

A rather general theory for such processes developed in Refs. 56b and 56c is applicable to both steady-state conditions and transient experiments. The model accounts for reversibility of the transfer reaction and allows for diffusion limitations in both liquid phases. The possibility of different diffusion coefficients in two phases is also included. The steady-state situation is defined by three dimensionless parameters, i.e., $K_e = c_o/c_w$ (the ratio of bulk concentrations in organic and aqueous phases), $\gamma = D_o/D_w$ (the ratio of diffusion coefficients), and $K = k_o a/D_o$ (normalized rate constant for the transfer from organic to water). The effects of these parameters on the shape of current-distance curves are shown in Figure 19. The tip current (at a given distance) increases strongly with both K and γK_e values but is largely independent of individual values of K_e and γ as long as their product is constant. The diffusion in the bottom phase significantly affects the SECM response at $\gamma K_e \lesssim 10$. While this theory provides a very general description of transfer of neutral species, in the case of IT some modifications may be necessary to include potential dependence of the rate constant and ensure electroneutrality in the bottom phase.

Interestingly, this approach allows the determination of the diffusion coefficient and concentration of species in the bottom phase with the tip positioned in the top phase. The microprobe does not have to enter or contact the second phase, which may allow measurements in the media unsuitable for electrochemical experiments (e.g., very nonpolar solvents) (56c).

An important advantage of the replacement of the metal tip with a micropipet in IT measurements is the possibility of probing numerous processes involving ionic reactants but no redox species. One should also notice that

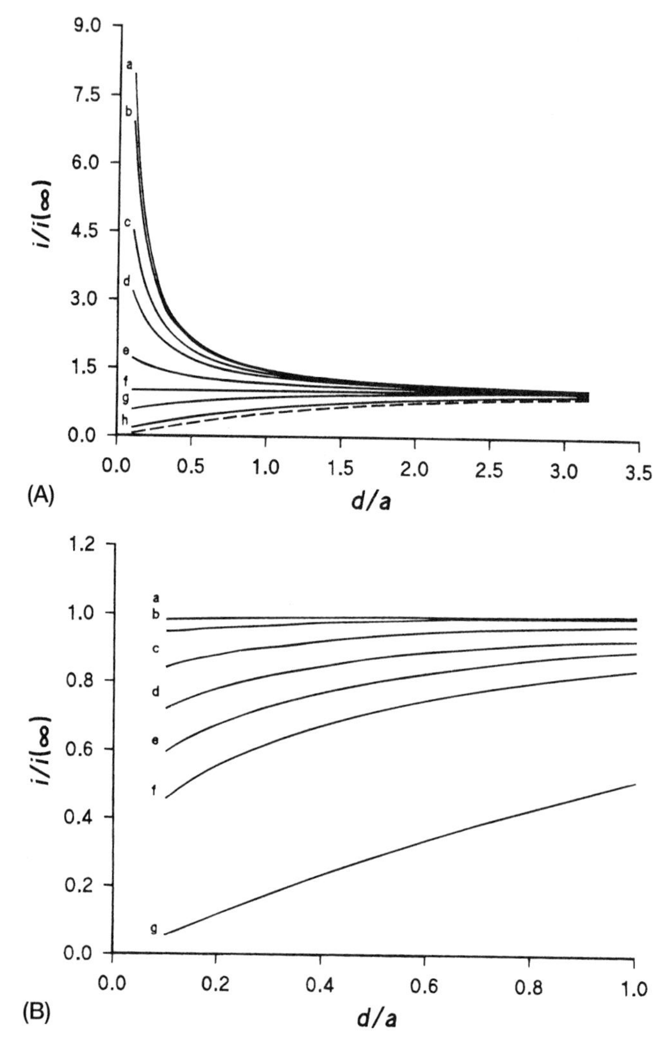

FIG. 19 Simulated normalized steady-state current as a function of normalized tip-interface distance for (A) $K = 10^5$, $\gamma = 1$, and K_e = (a) 1000, (b) 50, (c) 10, (d) 5, (e) 2, (f) 1, (g) 0.5, and (h) 0.1. The dashed line is the theory for the blocking interface. (B) $K_e = 1$, $\gamma = 1$, and K = (a) 1000, (b) 20, (c) 5, (d) 2, (e) 1, (f) 0.5, (g) 0.0. (From Ref. 56c.)

a mechanistic study of interfacial transfer of reducible (or oxidizable) ions is problematic because it is hard to distinguish between the IT and ET pathways (see, e.g., Ref. 55b). A scheme of the simple IT-based SECM process is shown in Figure 18B. A cation (M^+) is transferred from organic phase into the aqueous filling solution inside a pipet:

$$M^+_{(o)} \rightarrow M^+_{(w)} \qquad \text{(at the pipet tip)} \tag{28}$$

In this case, the top and the bottom liquid phases contain the same ion at equilibrium. A micropipet tip is used to deplete concentration of this ion in the top solvent near the ITIES. This depletion results in the ion transfer across the ITIES:

$$M^+_{(w)} \rightarrow M^+_{(o)} \qquad \text{(at the ITIES)} \tag{29}$$

Reaction (28) can produce positive feedback if concentration of M^+ in the bottom phase is sufficiently high. Any solid surface (or a liquid phase containing no specific ion) acts as an insulator in this experiment. The interface between the top and the bottom layers is nonpolarizable, and the potential drop is determined by the ratio of concentrations of the common ion (i.e., M^+) in two phases.

Two different arrangements were used to probe the transfer of tetraethylammonium cation (TEA^+) between water and DCE (53). Cell 1 includes a pipet filled with an aqueous solution containing supporting electrolyte (LiCl) and immersed in a DCE solution containing TEA^+:

Ag/AgTPB/0.4 mM TEATPBCl + 0.01 M BTPPATBP//10 mM LiCl/AgCl/Ag
Outer DCE solution pipet
(Cell 1)

where BTPPATPB = bis(triphenylphosphoranylidene)ammonium tetraphenylborate and TEATPBCl = tetraethylammonium tetrakis[4-chlorophenyl]-borate. When a pipet is biased at a sufficiently positive potential, the steady-state diffusion-limiting current is observed due to essentially spherical diffusion of TEA^+ to the pipet orifice. A family of approach curves in Figure 20 was obtained by moving such a pipet towards the aqueous bottom layer containing different concentrations of TEACl. With no TEA^+ in the bottom layer (curve 1) the ITIES blocks the diffusion to the pipet and a pure negative feedback is observed. The experimental approach curve fits well the theory for an insulating substrate with RG = 1.1.

The current-distance curve obtained with equal concentrations of TEA^+ in the top and bottom phases (curve 2) is essentially flat, i.e., $I_T \approx 1$ at any L. This can be expected (56c) because the diffusion coefficients of TEA^+ in water and DCE are similar and interfacial IT is quite fast. The overall process rate is limited by diffusion of TEA^+ in the bottom aqueous layer. When

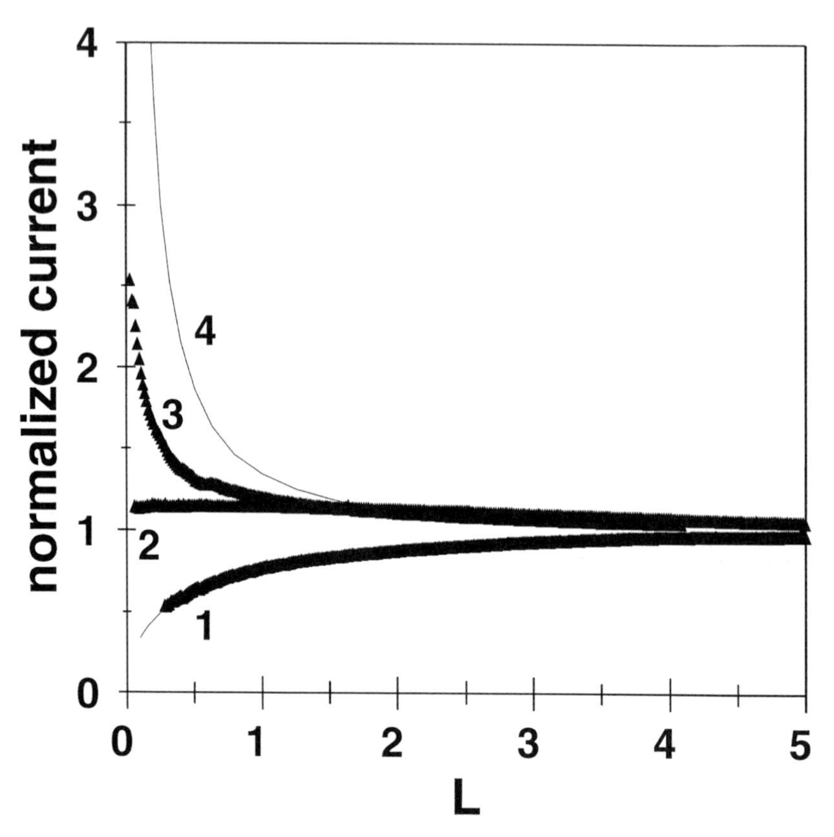

FIG. 20 Current-distance curves obtained with different concentrations of TEACl in the bottom aqueous layer and constant composition of the organic phase (0.4 mM TEATPBCl + 10 mM BTPPATPB in DCE). The filling aqueous solution contained 10 mM LiCl. Concentration of TEACl in the bottom aqueous phase was (1) 0, (2) 0.4, and (3) 10 mM. Aqueous phase also contained 10 mM LiCl. Solid lines represent SECM theory for an insulating substrate (curve 1) and pure positive feedback (curve 4). The radius of the pipet orifice was (1) 18, (2) 10, and (3) 12 μm. The tip was scanned at 1 μm/s. (From Ref. 53.)

$c_{TEA^+,w}$ is higher than $c_{TEA^+,o}$, an increase in the I_T at closer distances is observed (curve 3). The feedback current increases with $c_{TEA^+,w}$ due to the faster diffusion in the bottom layer.

At $c_{TEA^+,w} \gtrsim 15c_{TEA^+,o}$ diffusion in the bottom layer cannot be rate-limiting, and either a pure positive feedback or finite heterogeneous kinetics should be observed (25). But the measured $I_T - \mathrm{L}$ curve remained significantly lower than the theoretical curve for a pure positive feedback (curve

4) even at high $c_{TEA^+,w}/c_{TEA^+,o}$ ratios (e.g., $c_{TEA^+,w}/c_{TEA^+,o} = 25$ in curve 3). Under these conditions the overall process rate is most likely limited by the interfacial IT reaction (i.e., $TEA^+_{(w)} \rightarrow TEA^+_{(o)}$). This is somewhat surprising because of a large standard rate constant [$k^o \geq 0.1$ cm/s (57)] measured for this reaction at a polarized ITIES. One of the possible reasons for slow kinetics is that the common ion concentration in the bottom layer affects the heterogeneous IT rate constant, k_f. The k_f was shown to follow the Butler-Volmer equation (1):

$$k_f = k^o \exp\left(-\frac{\alpha F}{RT}(\Delta_w^o\varphi - \Delta_w^o\varphi^o)\right) \tag{30}$$

where F is the Faraday constant, $\alpha \approx 0.5$ is the transfer coefficient, and $\Delta_w^o\varphi^o$ is the standard potential of TEA^+ transfer. The interfacial potential drop, $\Delta_w^o\varphi$, is governed by the ratio of TEA^+ concentration in water and the organic phase (1a):

$$\Delta_w^o\varphi = \Delta_w^o\varphi^o + (RT/F)\ln(c_{TEA^+,w}/c_{TEA^+,o}) \tag{31}$$

The combination of Eqs. (30) and (31) gives

$$k_f = k^o(c_{TEA^+,w}/c_{TEA^+,o})^{-\alpha} \tag{32}$$

The decrease in the rate constant with increasing $c_{TEA^+,w}/c_{TEA^+,o}$ ratio may allow probing faster IT reactions with no complications associated with slow diffusion in the bottom phase. One should also notice that, unlike previously studied ET processes at the ITIES, the rate of the reverse reaction cannot be neglected. The difference is that in the former experiments no ET equilibrium existed at the interface because only one (reduced) form of redox species was initially present in each liquid phase (15,25). In contrast, reaction (29) is initially at equilibrium and has to be treated as a quasi-reversible process (56c). Probing kinetics of IT reactions at a nonpolarizable ITIES under steady-state conditions should be as advantageous as analogous ET measurements (25). The theory required for probing simple IT reactions with the pipet tips has not been published to date.*

It was shown recently (52) that pipets filled with an organic solvent are suitable for quantitative voltammetric measurements in aqueous media (see also Sec. III.C). Using such a pipet as an SECM tip one can probe IT from organic phase to water and perform imaging in aqueous solutions. The $I_T - L$ curves obtained with organic-filled pipets immersed in an aqueous solution and approaching the ITIES were similar to those in Figure 20 (53). The

*This approach was further developed by Selzer Y, Mandler D. J. Phys. Chem. B 104:4903, 2000.

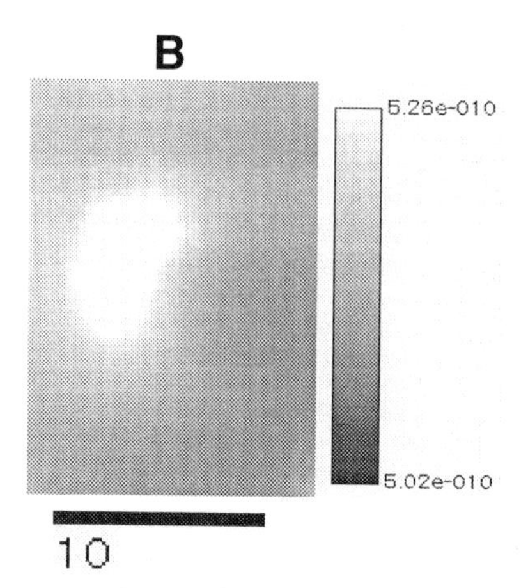

FIG. 21 Constant height mode gray scale image of a 5-μm pore in a polycarbonate membrane obtained with a 3-μm pipet tip. The filling DCE solution contained 10 mM TBATPBCl. The aqueous phase contained 0.4 mM TEACl + 10 mM LiCl. The scale bar corresponds to 10 μm. The tip scan speed was 10 μm s^{-1}. (From Ref. 53.)

change from negative to positive feedback occurs with increasing concentration of TEA^+ in the bottom (organic) phase, but the pure positive feedback could not be observed, as discussed above.

The ion transfer feedback mode of the SECM can also be used for imaging of solid/liquid and liquid/liquid interfaces. The topography of a solid substrate can be imaged in the same way as insulators are imaged in conventional SECM experiments. For membrane/solution and liquid/liquid interfaces, the maps of IT reactivity of the interfacial boundary can be obtained. A high stability of the pipet response is required for SECM imaging. The diffusion-limiting current at a pipet should remain practically constant for 10 minutes (the approximate time required to scan one SECM image). The response instabilities are usually caused by transfer of supporting electrolyte into the pipet, which results in precipitation and clogging of the pipet tip. It can be minimized by proper choice of both aqueous and organic supporting electrolytes and by keeping the pipet potential within the potential window limits.

Figure 21 shows a constant height mode gray-scale image of a 5-μm pore in the 10-μm-thick polycarbonate membrane obtained with a 3-μm-diameter pipet. The fluxes of electroactive species through porous membranes have previously been imaged using generation/collection mode of the SECM operation (54). In contrast, the feedback mode image in Figure 21 was obtained without any current flowing across the interface and is free from diffusion broadening—hence a significantly higher lateral resolution. Such an image represents distribution of IT reactivity on the interfacial boundary. Filling a pipet with an organic solvent was essential for experiments with polycarbonate membranes, which are soluble in organic media. The same setup can be employed to map ion transfers occurring at a polymer/solution interface and in living cells. A water-filled micropipet should be used as a SECM tip for imaging in organic media.

The spatial resolution of the feedback mode SECM imaging is governed by the size of the microprobe (58) (see also Chapter 4). Although the reported SECM experiments employed relatively large pipets (i.e., 3–50 μm in diameter), the nanopipet electrodes as small as 4-nm-radius have been fabricated and can potentially be used as SECM tips (51). Another advantage of this approach is the possibility to work without redox mediator in solution. This technique should be most useful for probing IT rates in biological systems.

C. Micropipet Probes

Important advantages of micropipets for electrochemical studies of the ITIES have been demonstrated by Girault's group (59). Unlike metal UMEs, small pipets are easy to make using a pipet puller. Pipets with radii as small as a few nm have recently been prepared and used for voltammetric and SECM experiments (51,60). However, a careful surface pretreatment is necessary to ensure reproducibility of the response and avoid major deviations from the theory (52). When a glass pipet is immersed in an organic solvent, a small amount of the filling aqueous solution flows out of the pipet and forms a submicrometer-thick layer on the hydrophilic outer wall. This makes the effective area of the ITIES much larger than the geometrical area of the pipet orifice. The diffusion current to such a pipet is significantly (more than two times) higher than the theoretical prediction based on the orifice radius (52). This effect is equally important for any charge-transfer process at a micropipet (i.e., simple and facilitated IT and ET) as long as its rate is controlled by diffusion of species in the external solution to the pipet orifice.

The aqueous layer can be eliminated by silanizing the outer pipet wall to render it hydrophobic. To silanize only the outer surface of the pipet, the flow of argon should be passed through the pipet from the back while its tip is immersed in a solution of trimethylchlorosilane. It is crucial to avoid silanization of the inner wall if the pipet is going to be filled with an aqueous solution. Otherwise the outer organic solvent gets drawn inside a pipet whose inner surface is hydrophobic. This procedure allows one to achieve quantitative agreement between experimental and theoretical values of diffusion current to the pipet (the theory modifications required to account for a small RG are discussed in previous sections and in Chapter 5).

A pipet with a silanized inner wall can be filled with an organic solvent and used for measurements in aqueous solutions. The inner wall can be silanized by putting the back of the pipet in a solution of trimethylchlorosilane and then pushing the solution towards the tip by a syringe from the back. The solution is removed from the pipet in about 30 minutes with a syringe, and the silanized pipet should be allowed to dry in the air overnight. In this procedure only a minor portion of the outer pipet wall is silanized by the vapor emerging from the tip. In contrast, when silanization is done by dipping the pipet tip into trimethylchlorosilane, both the inner and the out walls become silanized. However, voltammetric responses of organic-filled pipets silanized by these two approaches are practically identical. Silanized pipets retain their hydrophobic properties for about one week after preparation.

A considerable ohmic resistance of a pipet may impair kinetic measurements. When the filling solution is aqueous the resistive potential drop in it is typically negligible even for nanopipets (51). For organic-filled pipets the resistive effect is more important. For example, the resistance of a typical 10-μm-radius pipet filled with DCE and containing 0.01 M of supporting electrolyte was found to be about 10 mΩ, and it is approximately inversely proportional to the pipet radius. This corresponds to $\sim$15 mV iR-drop when 0.5 mM TEA^+ is transferred from the outer aqueous phase into a 10-μm-radius pipet. The resistive potential drop can be minimized by using higher concentrations of supporting electrolyte inside a pipet and by improving pulling programs to produce shorter (patch-type) pipets. Fortunately, the ohmic resistance of a pipet is less important for SECM setup than for voltammetric measurements because the pipet tip is biased at a potential corresponding to the diffusion current plateau.

IV. PROCESSES WITH COUPLED HOMOGENEOUS REACTIONS

Many interesting processes occurring at the liquid/liquid interface involve coupled homogeneous chemical reactions. In principle, electrochemical methods used for probing complicated mechanisms at metal electrodes (61) can be employed at the ITIES. However, many of these techniques (e.g., rotating ring-disk electrode or fast-scan cyclic voltammetry) are hard to adapt to liquid/liquid measurements. Because of technical problems, few studies of multistep processes at the ITIES have been reported to date (1,62).

The SECM/ITIES setup offers a number of ways to study complicated mechanisms that cannot be realized with a metal electrode. The ITIES can serve as a controllable source (or sink) of reagents and charges. Utilizing preferential solubilities of different species in either aqueous or organic medium, one may be able to simplify the analysis of different mechanisms by probing successive steps of a complex process independently. For example, consider a hypothetical ECE mechanism involving water-soluble species A and X, and species B confined to organic phase. Under SECM conditions the following reactions occur:

$$A + le \rightarrow A^- \quad \text{(tip)} \tag{33a}$$

$$A^- + B \rightarrow X \quad \text{(ITIES)} \tag{33b}$$

$$X + le \rightarrow X^- \quad \text{(tip)} \tag{33c}$$

As a sufficiently negative tip potential both ET reactions are diffusion controlled, and the rate of the overall process is limited by heterogeneous reaction (33b). Its rate constant can be determined from the current-distance curves as discussed in Sec. II. The kinetic analysis of a more complicated ECE mechanism can be reduced to the measurement of an effective heterogeneous rate constant at the ITIES.

A recent SECM study of electrochemical catalysis at the ITIES was based on a similar concept (23). The ITIES was used as a model system to study catalytic electrochemical reactions in microemulsions. Microemulsions, i.e., microheterogeneous mixtures of oil, water, and surfactant, appear attractive for electrochemical synthesis and other applications (63). The ITIES with a monolayer of adsorbed surfactant is of the same nature as the boundary between microphases in a microemulsion. The latter interface is not, however, directly accessible to electrochemical measurements. While interfacial area in a microemulsion can be uncertain, the ITIES is well defined. A better control of the ITIES was achieved by using the SECM to study kinetics of electrochemical catalytic reduction of *trans*-1,2-dibromocyclohexane (DBCH) by Co(I)L (the Co(I) form of vitamin B_{12}):

$$[\text{Co(II)L} \sim \text{H}]^+ + e \rightarrow \text{Co(I)L} \sim \text{H} \quad \text{(tip)} \tag{34}$$

$$2\text{Co(I)L} \sim \text{H} + \text{RBr}_2 \rightarrow 2[\text{Co(II)L} \sim \text{H}]^+ + 2\text{Br}^- + \text{olefin} \quad \text{(ITIES)} \tag{35}$$

Reactions (34) and (35) together represent electrochemical catalysis of the reduction of DBCH. From earlier voltammetric studies, the elementary rate-determining step was thought to be (64)

$$\text{Co(I)L} \sim \text{H} + \text{RBr}_2 \rightarrow [\text{Co(II)L} \sim \text{H}]^+ + \text{RBr}^{\cdot} + \text{Br}^- \tag{36}$$

Under SECM conditions, reaction (35) is mathematically equivalent to kinetically controlled regeneration of the mediator at a solid substrate and can be characterized by the value of the effective heterogeneous rate constant. This greatly simplifies the analysis of the SECM current-distance curves, which can be fit to Eqs. (4) and (5). The dependencies of the rate of reaction (35) on various experimental parameters, i.e., reactant concentrations, interfacial potential drop, and the presence of adsorbed surfactant, were investigated. Unlike uncomplicated ET discussed in the previous section, the overall rate of reaction (35) was independent of the interfacial potential drop, and the apparent second-order rate constant depended on the concentration of vitamin B_{12}. The rate of DBCH reduction was only slightly affected by adsorbed long-chain surfactants. These results suggest that bimolecular ET reaction (36) is not the rate-limiting step of the process, and the interfacial reaction is not fully described by a simple second-order rate law, as it was found previously. Thus, the use of SECM/ITIES technique

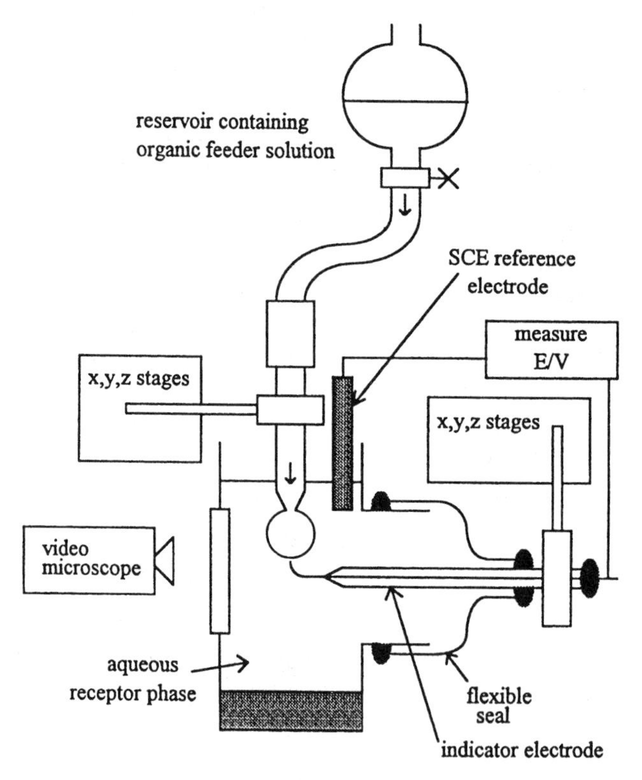

FIG. 22 Schematic diagram of the SECM apparatus employed for probing interfacial reactions at the ITIES. An UME tip is used to measure the local concentration of a reactant or product in the near-interface region of an expanding droplet. (From Ref. 65.)

allowed uncovering details of interfacial complexity in a reaction that may not be obvious from conventional voltammetric studies in microemulsions.

A novel approach to studying chemical reactions at the ITIES was described by Slevin and Unwin (65). They combined the SECM operating in the potentiometric collection mode with a previously developed dropping electrolyte electrode (66). The Ag/AgCl microelectrode tip was used to measure concentration profile of Cl^- produced by hydrolysis of triphenylmethyl chloride (TPMCl):

$$TPMCl_{(org)} + H_2O \rightarrow TPMOH_{(org)} + Cl^-_{(aq)} + H^+_{(aq)} \tag{37}$$

which occurred at the interface between the growing DCE droplet and water (Fig. 22). The rate constant for this reaction was obtained by fitting experimental concentration profiles to numerical solution of convective diffusion equation. The main advantage of this technique over a conventional SECM setup is in a predictable flux of a reactant delivered to the continuously renewable interface. In contrast with feedback and GC modes of the SECM operation, this approach is useful when all reactants are initially present in the system rather that being generated at a tip or a substrate electrode. The precise tip/drop alignment and accuracy in distance determination are critical for quantitative applications of this method.

ACKNOWLEDGMENTS

MVM gratefully acknowledges the support by grants from the Petroleum Research Fund administrated by the American Chemical Society and PSC-CUNY Research Award Program.

REFERENCES

1. (a) Girault HH, Schiffrin DJ. In: AJ Bard, ed. Electroanalytical Chemistry. Marcel Dekker: New York, 1989, p 1. (b) Girault HH. In: JO Bockris, BE Conway, RE White, eds. Modern Aspects of Electrochemistry. Vol. 25. New York: Plenum Press, 1993, p 1. (c) Samec Z, Kakiuchi T. In: H Gerischer, CW Tobias, eds. Advances in Electrochemical Science and Electrochemical Engineering. Vol. 4. New York: VCH, 1995, p 297. (d) Volkov AG, Deamer DW, eds. Liquid-Liquid Interfaces. Theory and Methods. Boca Raton: CRC Press, 1996. (e) Benjamin I. Chem Rev 96:1449, 1996.
2. (a) Arai K, Ohasawa M, Kusu F, Takamura K. Bioelectrochem Bioenerg 31: 65, 1993. (b) Lo TC, Baird MHI, Hanson C, eds. Handbook of Solvent Extraction. New York: John Wiley & Sons, 1994. (c) Vanysek P. TRAC—Trends Anal Chem 12:357, 1993.

3. (a) Rusling JF. Acc Chem Res 24:75, 1991. (b) Cunnane VJ, Murtomaki L. In: Volkov AG, Deamer DW, eds. Liquid-Liquid Interfaces. Theory and Methods. Boca Raton, FL: CRC Press, 1996, p 401.
4. (a) Girault HH, Schiffrin DJ. In: Allen MJ, Usherwood PNR, eds. Charge and Field Effects in Biosystems. Turnbridge Wells, England: Abacus Press, 1984, p 171. (b) Ohkouchi T, Kakutani T, Senda M. Bioelectrochem Bioenerg 25: 71:81, 1991. (c) Gennis RB. Biomembranes. New York: Springer, 1995.
5. (a) Marcus RA, Sutin N. Biochim Biophys Acta 811:265, 1985. (b) Barbara PF, Meyer TJ, Ratner MA. J Phys Chem 100:13148, 1996.
6. (a) Marcus RA. J Phys Chem 94:1050, 1990. (b) Marcus RA. J Phys Chem 94:4152, 1990; addendum, J Phys Chem 94:7742, 1990. (c) Marcus RA. J Phys Chem 95:2010, 1991; addendum, J Phys Chem 99:5742, 1995.
7. Smith BB, Halley JW, Nozik AJ. Chem Phys 205:245, 1996.
8. (a) Geblewicz G, Schiffrin DJ. J Electroanal Chem 244:27, 1988. (b) Cheng Y, Schiffrin DJ. J Electroanal Chem 314:153, 1991.
9. Schmickler WJ. Electroanal Chem 428:123, 1997.
10. (a) Gurevich YY, Kharkats YI. J Electroanal Chem 200:3, 1986. (b) Shao Y, Girault HH. J Electroanal Chem 282:59, 1990.
11. Kakiuchi T. J Electroanal Chem 322:55, 1992.
12. Kontturi K, Manzanares JA, Murtomaki L, Schiffrin DJ. J Phys Chem 101: 10801, 1997.
13. (a) Eisenthal KB. Chem Rev 96:1343, 1996. (b) Naujok RR, Higgins DA, Corn RM. J Chem Soc Faraday Trans 91:2114, 1995. (c) Conboy JC, Messmer MC, Richmond GL. J Phys Chem 100:7617, 1996.
14. (a) Benjamin I. Acc Chem Res 28:233, 1995. (b) Benjamin I. Science 261: 1559, 1993. (c) Schweighofer KJ, Benjamin I. J Electroanal Chem 391:1, 1995. (d) Wilson MA, Pohorille A. J Am Chem Soc 118:6580, 1996.
15. Wei C, Bard AJ, Mirkin MV. J Phys Chem 99:16033, 1995.
16. Senda M, Kakiuchi T, Osakai T. Electrochim Acta 36:253, 1991.
17. Samec Z, Marecek V. J Electroanal Chem 200:17, 1986.
18. Guainazzi M, Silvestry G, Survalle G. J Chem Soc Chem Commun 200, 1975.
19. (a) Kharkats YI, Volkov AG. J Electroanal Chem 184:435, 1985. (b) Kharkats YI, Ulstrup J. J Electroanal Chem 308:17, 1991.
20. Girault HH. J Electroanal Chem 388:93, 1995.
21. (a) Senda M. Anal Sci 10:649, 1994. (b) Senda M. Electrochim Acta 40:2993, 1995. (c) Katano H, Maeda K, Senda M. J Electroanal Chem 396:391, 1995.
22. Andrieux CP, Saveant J-M. In: Murray RW, ed. Molecular Design of Surfaces. New York: John Wiley & Sons, 1992, p 267.
23. Shao Y, Mirkin MV, Rusling JF. J Phys Chem B 101:3202, 1997.
24. Selzer Y, Mandler D. J Electroanal Chem 409:15, 1996.
25. Tsionsky M, Bard AJ, Mirkin MV. J Phys Chem 100:17881, 1996.
26. Tsionsky M, Bard AJ, Mirkin MV. J Am Chem Soc 119:10785, 1997.
27. Delville M-H, Tsionsky M, Bard AJ. Langmuir 14:2774, 1998.
28. Solomon T, Bard AJ. J Phys Chem 99:17487, 1995.
29. Kwak J, Bard AJ, Anal Chem 61:1221, 1989.
30. Miller JR, Calcaterra LT, Closs GL. J Am Chem Soc 106:3047, 1984.

31. (a) Winkler JR, Gray HB. Chem Rev 92:369, 1992. (b) McLendon G, Hake R. Chem Rev 92:481, 1992.
32. (a) Chidsey CED. Science 251:919, 1991. (b) Finklea HO. In: Bard AJ, ed. Electroanalytical Chemistry. Vol. 19. New York: Marcel Dekker, 1996, p 109. (c) Miller CJ. In: Rubinstein I, ed. Physical Electrochemistry: Principles, Methods, and Applications. New York: Marcel Dekker, 1995, p 27.
33. Lui H, Prieskorn JN, Hupp JT. J Am Chem Soc 115:4027, 1993.
34. Samec Z, Marecek V, Weber J, Homolka D. J Electroanal Chem 126:105, 1981.
35. Cunnane VJ, Schiffrin DJ, Beltran C, Geblewicz G, Solomon T. J Electroanal Chem 247:203, 1988.
36. Nakatani K, Chikama K, Kitamura N. Chem Phys Lett 237:133, 1995.
37. Cheng Y, Schiffrin DJ. J Chem Soc Faraday Trans 90:2157, 1994.
38. Cunnane VJ, Schiffrin DJ, Fleischman M, Geblewicz, G, Williams D. J Electroanal Chem 243:455, 1988.
39. (a) Kakiuchi T, Kondo T, Senda M, Bull Chem Soc Jpn 63:3270, 1990. (b) Kakiuchi T, Kondo T, Kotani M, Senda M. Langmuir 8:169, 1992. (c) Kakiuchi T, Kotani M, Noguchi J, Nakanishi M, Senda M. J Colloid Interface Sci 149: 279, 1992.
40. (a) Kakiuchi T, Yamane M, Osakai T, Senda M. Bull Chem Soc Jpn 60:4223, 1987. (b) Girault HH, Schiffrin DJ. J Electroanal Chem 179:277, 1984.
41. Kakiuchi T, Nakanishi M, Senda M. Bull Chem Soc Jpn 61:1845, 1988.
42. Larsson K, Quinn PJ. In: Gunstone FD, Harwood JL, Padley FB, eds. The Lipid Handbook. London: Chapman & Hall, 1994, p 401.
43. (a) Girault HH, Schiffrin DJ. Biochim Biophys Acta 857:251, 1986. (b) Wandlowski T, Marecek V, Samec Z. J Electroanal Chem 242:277, 1988. (c) Kakiuchi T, Nakanishi M, Senda M. Bull Chem Soc Jpn 62:403, 1989.
44. (a) Becka AN, Miller CJ. J Phys Chem 96:2657, 1992. (b) Smalley JF, Feldberg SW, Chidsey CED, Linford MR, Newton MD, Liu Y-P. J Phys Chem 99:13141, 1995.
45. (a) Piknova B, Marsh D, Thompson TE. Biophys J 71:892, 1996. (b) Wu JR, Lentz BR. J Flouresc 4:153, 1994. (c) Allen MT, Miola L, Whitten DG. J Am Chem Soc 110:3198, 1998.
46. (a) Walker RA, Conboy JC, Richmond GL. Langmuir 13:3070, 1997. (b) Mingins J, Taylor JAG, Pethica BA, Jackson CM, Yue BYT. J Chem Soc Faraday Trans 1 78:323, 1982. (c) Taylor JAG, Mingins J, Pethica BA. J Chem Soc. Faraday Trans 1 72:2694, 1976. (d) Yue BYT, Jackson CM, Taylor JAG, Mingins J, Pethica BA. J Chem Soc Faraday Trans 1 72:2685, 1976.
47. Miller A, Mohwald H. Phys Rev Lett 56:2633, 1986.
48. (a) Kwak J, Anson FC. Anal Chem 64:250, 1992. (b) Lee C, Anson FC. Anal Chem 64:528, 1992. (c) Denuault G, Troise Frank MH, Peter LM. Faraday Discuss 94:23, 1992. (d) Troise Frank MH, Denuault G. J Electroanal Chem 379:405, 1994. (e) Arca M, Mirkin MV, Bard AJ. J Phys Chem 99:5040, 1995.
49. Shao Y, Mirkin MV. J Electroanal Chem J Electroanal Chem 439:137, 1997.
50. (a) Buhlmann P, Pretch E, Bakker E. Chem Rev. 98:1593, 1988. (b) Bradshaw JS, Izatt RM. Acc Chem Res 30:338, 1997.

51. Shao Y, Mirkin MV. J Am Chem Soc 199:8103, 1997.
52. Shao Y, Mirkin MV. Anal Chem 70:3155, 1998.
53. Shao Y, Mirkin MV. J Phys Chem B 102:9915, 1998.
54. (a) Scott ER, White HS, Phipps JB. J Membr Sci 58:71, 1991. (b) Scott ER, White HS, Phipps JB. Anal Chem 65:1537, 1993. (c) Bath BD, Lee RD, White HS, Scott ER. Anal Chem 70:1047, 1998.
55. (a) Matsue T, Shiku H, Yamada H, Uchida I. J Phys Chem 98:11001, 1994. (b) Yamada H, Akiyama S, Inoue T, Koike T, Matsue T, Uchida I. Chem Lett 147, 1998.
56. (a) Slevin CJ, Umbers JA, Atherton JH, Unwin PR. J Chem Soc Faraday Trans 92:5177, 1996. (b) Slevin CJ, Macpherson JV, Unwin PR. J Phys Chem 101: 10851, 1997. (c) Barker AL, Macpherson JV, Slevin CJ, Unwin PR. J Phys Chem 102:1586, 1998.
57. (a) Marecek V, Lhotsky A, Racinsky S. Electrochim Acta 40:2905, 1995. (b) Beattie PH, Delay A, Girault HH. Electrochim Acta 40:2961, 1995.
58. (b) Borgwarth K, Ricken C, Ebling DG, Heinze J. Fresenius J Anal Chem 356: 288, 1996.
59. Taylor G, Girault HH. J Electroanal Chem 208:179, 1986. (b) Campbell JA, Girault HH. J Electroanal Chem 266:465, 1989. (c) Campbell JA, Stewart AA, Girault HH. J Chem Soc Faraday Trans 85:843, 1989. (d) Stewart AA, Shao Y, Pereira CM, Girault HH. J Electroanal Chem 305:135, 1991.
60. Wei C, Bard AJ, Feldberg SW. Anal Chem 119:8103, 1997.
61. Bard AJ, Faulkner LR. Electrochemical Methods. New York: John Wiley & Sons, 1980, pp. 428–487.
62. Kong Y-T, Imabayashi S-I, Kakiuchi T. Anal Sci 14:121, 1998.
63. (a) Rusling JF. In: Bard AJ, ed. Electroanalytical Chemistry. Vol. 18. New York: Marcel Dekker, 1994, p 1. (b) Rusling JF. In: Bockris JO, Conway BE, White RE, eds. Modern Aspects of Electrochemistry. Vol. 26. New York: Plenum Press, 1994, p 44.
64. Connors TF, Arena JV, Rusling JF. J Phys Chem 92:2810, 1988.
65. Slevin CJ, Unwin PR. Langmuir 13:4799, 1997.
66. Koryta J, Vanysek P, Brezina M. J Electroanal Chem 67:263, 1976.

9

IMAGING MOLECULAR TRANSPORT ACROSS MEMBRANES

Bradley D. Bath

ALZA Corporation
Mountain View, California

Henry S. White

University of Utah
Salt Lake City, Utah

Erik R. Scott

Medtronic Corporation
Minneapolis, Minnesota

I. INTRODUCTION

Traditional models of molecular transport in porous membranes are based on measurements of spatially averaged fluxes, incorporating details of the membrane structure obtained from independent experimentation and theory (1). While such transport models are successful in predicting the overall flux across a membrane, the microscopic features incorporated within the models are often left untested by direct measurements. In this chapter, we describe spatially resolved measurements of the transport of molecular species across porous membranes using scanning electrochemical microscopy (SECM) (2–6). Absolute values of the rate of transport of a chemical species through a *single* pore can be directly measured using SECM, and "images of the molecular flux" across a porous membrane can be obtained with submicrometer resolution. SECM provides a means to identify microscopic structures in the membrane that are associated with molecular transport paths and mechanisms.

The capabilities of SECM make it a nearly ideal method for studying molecular transport across porous membrane samples. In a typical application, the SECM tip is rastered at constant height across the surface of a membrane mounted in a diffusion cell (Fig. 1). An electroactive species is

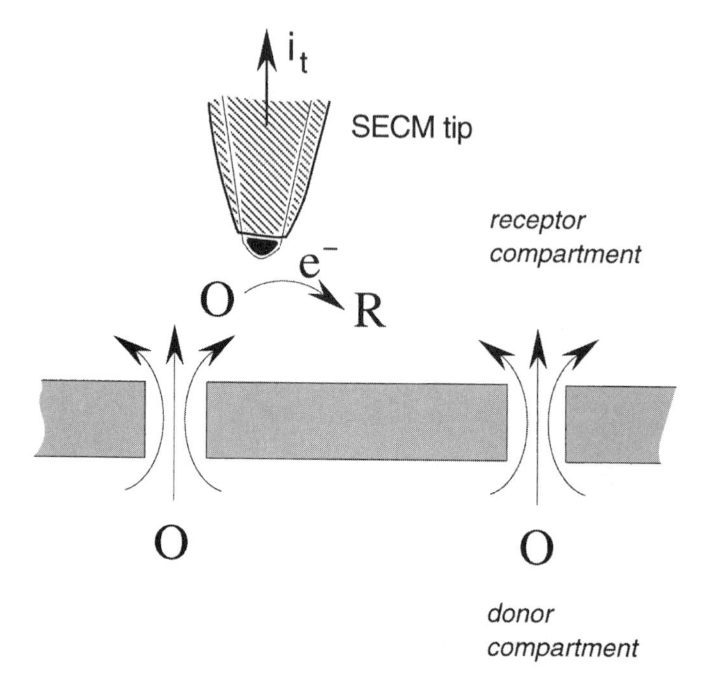

FIG. 1 SECM measurement of localized molecular flux through a porous membrane. The SECM tip current, i_t, resulting from the redox reaction, $O + e^- \rightarrow R$, is largest when the tip is directly above a pore.

dissolved in the lower (donor) compartment and is either allowed to freely diffuse across the membrane or is driven across the membrane by an applied force. The SECM tip in the upper (receptor) compartment is poised at a potential such that the electroactive species is oxidized or reduced at the mass transport-limited rate as it emerges from the pores in the membrane. The magnitude of the Faradaic current, i_t (measured as the SECM tip is rastered across the membrane surface), is proportional to the local concentration of the electroactive species. Since the concentration above the membrane surface is greater at regions where the transmembrane flux is large, the resulting SECM image directly reflects the local transport pathways in the membrane. For example, Figure 2 shows an image of the iontophoretic transport of $Fe(CN)_6^{-4}$ through a 1 mm^2 section of mouse skin (7). The bright region in the SECM image corresponds to a localized region where the flux of $Fe(CN)_6^{-4}$ through the skin sample is very high; this flux is associated with a hair follicle, which acts as a low-resistance path for molecular transport. The contrast in this SECM image reflects differences in the local rate of mass transport across the membrane, rather than topographical features on the surface.

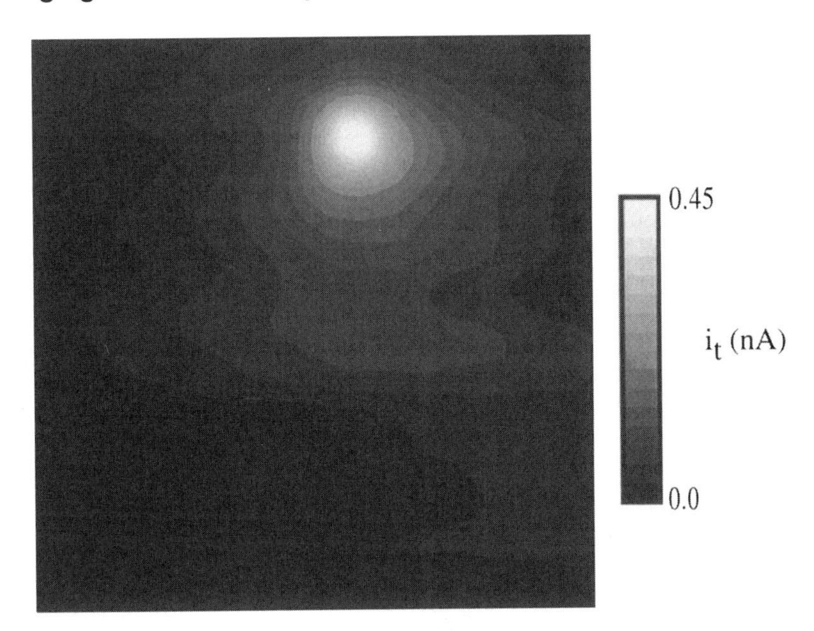

FIG. 2 SECM image of $Fe(CN)_6^{-4}$ transport across a 1 mm^2 region of hairless mouse skin. Donor solution: 0.1 M $K_4Fe(CN)_6$ in 0.1 M NaCl. Receptor solution: 0.1 M NaCl. (From Ref. 7.)

SECM has been used to investigate transport across both synthetic porous membranes (7–10) and biological membranes (7,9,11–16). The flux across the membrane usually results from a concentration gradient that is established by a difference in the concentration of electroactive species in the receptor and donor compartments of the diffusion cell. The flux may also be a function of an external force applied across the membrane. For example, SECM has been employed to study electroosmotic transport in permselective Nafion membranes (10) and pressure-driven fluid flow within dentine (11,12). The general goal of these studies is to determine the paths and mechanisms of transport *within* the membrane. However, once the molecule emerges from the membrane, it is usually transported solely by diffusion into the bulk solution of the receptor compartment. As discussed in detail below, this greatly simplifies the analysis of SECM data, since relatively straightforward analytical expressions based on Fick's laws of diffusion can be used to relate the measured tip current to the local flux of molecules within the membrane. This capability is key to characterizing transport across heterogeneous biological membranes, where the molecular flux may vary significantly as a function of spatial position.

Membrane imaging techniques based on the measurement of the local solution conductivity using scanning ion conductance microscopes have also been reported by Burnette and Ongipipattanakul (17) and by Hansma and coworkers (18). A unique advantage of SECM relative to ion conductance microscopy is that the individual fluxes of different chemical species in a multicomponent solution may be determined by adjusting the electrochemical potential of the SECM tip to a value where the electroactive species of interest is reduced or oxidized. Since different chemical species have different electrochemical reduction potentials, it is possible to analyze for the individual membrane fluxes of several electroactive species in the same experiment (14).

SECM has also been used to measure transport dynamics in membranes (10). In this application the current at the tip is recorded as a function of time following a chemical or physical perturbation that alters the rate of transport across the membrane (e.g., a change in the electric field across the membrane). No other analytical technique or microscopy provides a comparable level of direct information about the pathways and dynamics of molecular transport in membranes.

The basic principles, methodology, and theory employed in SECM imaging of porous membranes are presented in this chapter. Many of the experimental procedures (e.g., tip preparation), as well as the basic SECM instrumentation, employed in imaging porous membranes are similar to that described in Chapters 2 and 3. The interested reader should consult these chapters or the original literature for complete details.

II. PRINCIPLES OF IMAGING POROUS MEMBRANES

SECM images of porous membranes represent contour maps of the concentration of a redox species *after* it has emerged from the membrane pore and is freely diffusing in the receptor solution. Thus, the interpretation of SECM images requires careful consideration of the flux of the species away from the pore. This is a relatively straightforward problem for transport from a single pore; however, the situation becomes very complex in the case of a membrane in which the spacing between pores is small and the diffusional fields of the pores begin to overlap. The following sections outline the basic mathematical description of the shape of the SECM image of a single pore and the procedure for determining the absolute value of the flux through the pore, as well as the pore radius. The single pore model is very useful in the analysis of SECM images of real membranes.

A. Concentration Distribution above a Pore Opening

Transport within a pore can result from diffusion, migration, convection, or some combination of these mechanisms. For a relatively large (>1 μm radius) unobstructed pore, the composition of the solution within the pore is essentially identical to that of the donor/receptor solutions that are in contact with the pore; the mechanism of transport inside the pore is generally well understood in this case. For instance, in the absence of an applied field or a pressure gradient, transport of a molecule in such a pore occurs by diffusion, with the molecular diffusivity inside the pore being equal to the value measured in the bulk solution outside the pore. For complex biological samples, such as skin, or for membranes with very small pores (i.e., radius < 10 nm), the composition of the solution inside the pores may be significantly different than that of the donor/receptor solutions, preventing, a priori, knowledge of the transport mechanism within the pore. Electrosmotic flow may also occur in pores that are permselective towards either anions or cations (10,19–23). Fortunately, in the following theoretical description of the SECM experiment, it is only necessary to consider the flux *after* the molecules have emerged from the pore (Fig. 1). In essence, the pore opening is treated as a source of the molecular species, without regard to the transport processes that occur within the pore. Information concerning transport phenomena occurring inside the membrane pores (e.g., electroosmotic velocity) is extracted from the SECM data once the flux in the pore has been determined. The methods for doing this depend on the specific details of membrane structure as well as the mechanism of molecular transport (see Sec. III).

In the following discussion, we consider an ideal situation where the flux of a molecular species is localized to a single pore; the membrane is otherwise impermeable to the molecule. Although this model is only an approximation of real samples, the resulting theory remains quite useful in the quantitative analysis of porous membranes, provided that the pores are not too closely spaced. The membrane separates donor and receptor solutions; the donor solution contains an electroactive molecule that is transported across the membrane and detected by the SECM tip on the receptor side of the membrane (Fig. 1).

In the presence of an excess concentration of supporting electrolyte in the receptor compartment (which is the case in all experiments reported to date), and in the absence of convective flow, the flux of the species away from the opening of a small pore is dominated by radial diffusion. Assuming, for simplicity, that the pore opening is hemispherical with radius r_0, the rate of transport from the pore opening, Ω(mol/s), can be readily obtained from the solution of the continuity equation, $\nabla^2 C(r) = 0$, where $C(r)$ is the concentration at radial distance, r, measured from the center of the pore opening. Using Fick's law and employing the appropriate boundary conditions for the experiment [$C(r) = 0$ as $r \rightarrow \infty$ and $C(r) = C_S$ at $r = r_O$] yields (1,24,25):

$$\Omega = 2\pi D C_S r_O \tag{1}$$

where D is the diffusion coefficient for the electroactive species *after* it has entered the receptor compartment and C_s is the concentration of the molecule at the surface of the pore opening. The concentration profile above the pore opening, C(r), is given by the following:

$$C(r) = (r_O/r)C_S \tag{2}$$

Figure 3 shows several isoconcentration contours (solid curves) surrounding a hemispherical-shaped pore opening.

Pore openings on real membrane surfaces rarely have a hemispherical shape. A disk-shaped opening or a flat, irregular opening are more typical descriptions. However, the flux of molecules from a microscopic pore diverges very rapidly in a radial pattern, resulting in approximately hemispherical-shaped isoconcentration contours. To illustrate this, consider a disk-shaped pore opening of radius a. The rate of mass transport and concentration distribution associated with a disk-shaped pore are given by the following (24,26):

$$\Omega = 4DC_S a \tag{3}$$

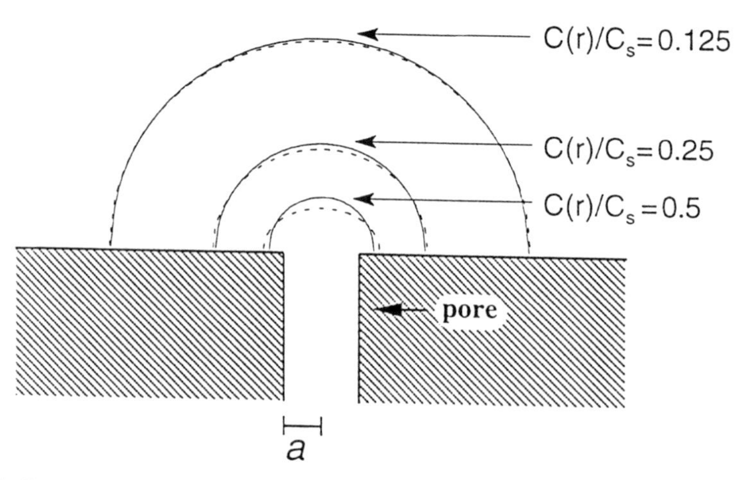

FIG. 3 Steady-state isoconcentration contours projected in the x-z plane corresponding to semi-infinite diffusion from hemispherical (solid lines) and disk-shaped (dashed lines) pore openings. Contours are plotted for $C_s/2$, $C_s/4$, and $C_s/8$, where C_s is the concentration at the surface of the pore. The disk-shape pore drawn in the figure has a radius a; the radius of the corresponding hemispherical pore opening (not shown), r_0, is equal to $2a/\pi$.

and

$$C(r, z) = \frac{2C_s}{\pi} \tan^{-1} \frac{a\sqrt{2}}{((r^2 + z^2 - a^2) + ((r^2 + z^2 - a^2)^2 + 4z^2a^2)^{1/2})^{1/2}} \tag{4}$$

Here r and z are the radial and axial coordinates, respectively, in the two-dimensional cylindrical coordinate system. Comparison of Eqs. (1) and (3) indicates that the rate of mass transport from a disk is equivalent to that of a hemisphere with an effective radius, r_{eff}, given by

$$r_{eff} = \frac{2a}{\pi} \tag{5}$$

Theoretical isoconcentration contours corresponding to a disk-shaped pore opening, Eq. (4), are plotted in Figure 3 (dashed lines) for comparison with contour lines corresponding to a hemispherical pore opening. Whereas the two-dimensional projections of the contours for the hemispherical pore are perfect semicircles, the contours resulting from diffusion from a disk-shaped source are flattened in the center because of planar diffusion. This is most evident in the contour for $C(r)/C_S = 0.5$. However, at larger distances, e.g., $5a$, the contours for the disk and hemispherical geometry are indistinguishable.

The isoconcentration contours above a pore can be experimentally evaluated by SECM (7). In this experiment, an electroactive molecule is placed in the donor solution and allowed to freely diffuse across the membrane. The concentration of a redox molecule, after it emerges into the receptor solution, is measured throughout the solution volume adjacent to the pore opening. Concentration is determined by measuring the voltammetric limiting current, i_t, at the SECM tip that corresponds to the oxidation or reduction of the redox molecule. Assuming a hemispherical tip geometry, the relationship between i_t and C(r) is given by (25)

$$i_t = 2\pi nFDC(r)r_t \tag{6}$$

where r_t is the SECM tip radius, F is Faraday's constant, and n is the number of electrons transferred per molecule. Once C(r) is determined in the volume above the pore, isoconcentration contours (either a hemispherical surface or a semicircular line) above the opening are constructed by choosing an arbitrary value of C(r) and finding the locus of points above the pore opening that corresponds to that value. The radius of this hemispherical surface or semicircular line can be determined by fitting the resulting isoconcentration contour to the equation of a hemisphere or semicircle.

Figure 4a shows isoconcentration contours around a single 7-μm-radius track-etched pore in mica, measured during the steady-state diffusional trans-

A)

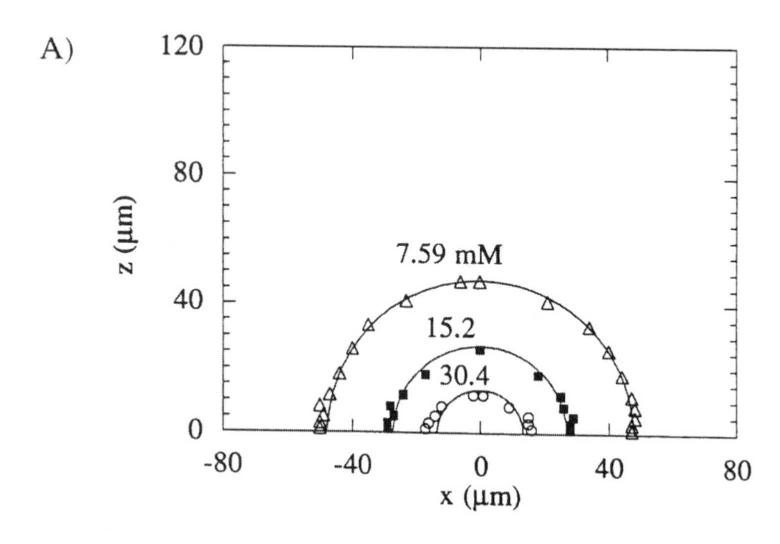

B)

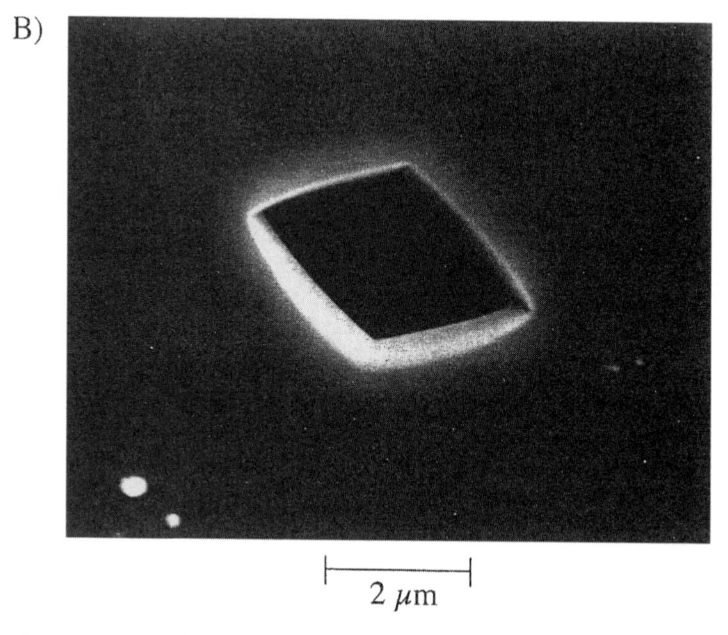

FIG. 4 (A) $Fe(CN)_6^{-3}$ isoconcentration contours above a 7 ± 1-μm-radius pore in a 10-μm-thick mica membrane. Concentrations corresponding to each contour are shown on the figure. Donor solution: 0.1 M $K_3Fe(CN)_6$ in 10 mM NaCl; receptor solution: 10 mM NaCl. (From Ref. 7.) (B) Scanning electron micrograph of a typical pore in the mica membrane. Although the pore opening has a rhomboidal-shaped opening, the isoconcentration contours are hemispherical, a result of radial diffusion away from the pore. (From Ref. 8.)

port of $Fe(CN)_6^{3-}$ through the pore (7). Best fit semicircular isocontour lines [based on the assumption of a hemispherical pore opening, Eq. (2)] are plotted for comparison. The isoconcentration contours are nearly perfectly semicircular in shape, except at distances close to the electrode surface where planar diffusion becomes significant. These results demonstrate that the radial divergence of the diffusive flux from a microscopic pore, regardless of its real shape, results in the pore appearing as if it were hemispherical in shape. The track-etched pore in mica, in fact, has an opening that is neither hemispherical nor disk shaped—instead the pore has a rhomboidal cross-sectional shape (Fig. 4b) a result of the chemical etching process used to create the pore (8). To measure the true shape of the pore would require scanning the tip very close to the surface at distances much smaller than the dimensions of the pore.

As a side note, the above procedure for determining the isoconcentration contours also provides sufficient information to determine the steady-state rate of mass transfer through the pore, Ω. Since mass continuity requires that $rC(r) = r_O C_S$ [see Eq. (2)], the measurement of r and C(r) for any arbitrary isoconcentration contour is equivalent to determining the value of the product $r_O C_S$. This latter value, in turn, can be substituted into Eq. (1) to obtain Ω. Similarly, for a disk-shaped opening, $rC(r) = r_{eff} C_s = (2a/\pi) C_s$; the product aC_s is then substituted into Eq. (3) to calculate Ω. The flux of molecules ($mol/cm^2 \cdot s$) within the pore is obtained by dividing Ω by the cross-sectional area of the pore. Although relatively precise and accurate values of Ω can be determined by this method (7), the sampling of C(r) in the solution volume above the pore is cumbersome and time consuming. A more rapid method for measuring Ω is described in Sec. II.C.

B. Relationship Between Tip Current and Pore Transport Rate

It is straightforward to show that the current measured at the SECM tip, when the tip is positioned above the pore opening, is proportional to the rate of transport of the redox molecule inside the pore. This is an important result that is implicitly applied in the qualitative interpretation of SECM images of porous membranes.

Consider a disk-shaped pore opening. Transport of the molecule from the pore opening into the bulk solution occurs by diffusion and is given by Eq. (3). With the tip is positioned directly at the surface of the pore opening (z = 0), the tip current, $i_t(z = 0)$ is proportional to the concentration at the surface of the pore opening, C_s, and Eq. (6) is rewritten as

$$i_t(z = 0) = 2\pi nFDC_s r_t \tag{7}$$

Combining Eqs. (3) and (7) yields the relationship between the SECM tip current and the transport rate within the pore (10)

$$i_t(z = 0) = (\pi n F r_t / 2a)\Omega \tag{8}$$

Equation (8) indicates that the current measured at the SECM tip, $i_t(z = 0)$, is directly proportional to Ω. At steady state, mass continuity requires that the rate of transport from the pore into the receptor compartment be equal to the rate of transport within the pore. Thus, $i_t(z = 0)$ is also proportional to the rate of transport at any point within the pore. The molecular flux in the pore, N, is obtained by simply dividing Ω by the cross-sectional area of the pore. Note that in deriving Eq. (8), no restrictions have been placed on the mechanism of transport within the pore. Thus, $i_t(z = 0)$ is proportional to the flux in the pore, independent of whether the flux is due to diffusion, migration, or convection. It can be shown that the tip current at any arbitrary separation distance, z, is also proportional to the flux in the pore.

In principle, Eq. (8) can be used to compute Ω in a single pore if the pore radius, a, is known. Implicit in Eq. (8) is the assumption that the tip does not interfere with the molecular diffusion from the pore. This is a reasonable approximation when the tip radius is very small compared with the pore radius, i.e., $r_t \ll a$. However, in many experiments, r_t is comparable to or even larger than a, and the tip can not be treated as a noninterfering probe when it is placed directly at the pore opening. Thus, Eq. (8) is useful only as an approximate relationship between the tip current and Ω. Furthermore, the radius of the pore, a, may not be known a priori in real samples. A quantitative method of analysis that circumvents both of these problems is presented in Sec. II.D. Regardless of tip interference effects, the key result here is that the magnitude of the SECM tip current is proportional to the molecular flux in the pore.

C. Image Shape

It is relatively straightforward to compute the shape of the SECM image that corresponds to the transport of a molecule from an individual pore. Consider a typical experiment in which the SECM tip is rastered in the x, y plane across a flat membrane at a constant height, z_t. A schematic of this experiment with the coordinate system is shown in Figure 5. The current measured at the SECM tip is proportional to the concentration of the electroactive species at tip position x, y, z; thus, computing the shape of the image requires knowledge of how the concentration varies as the tip is moved in either the x or y direction (while keeping z at z_t). We first recall that the concentration distribution above a pore with hemispherical opening

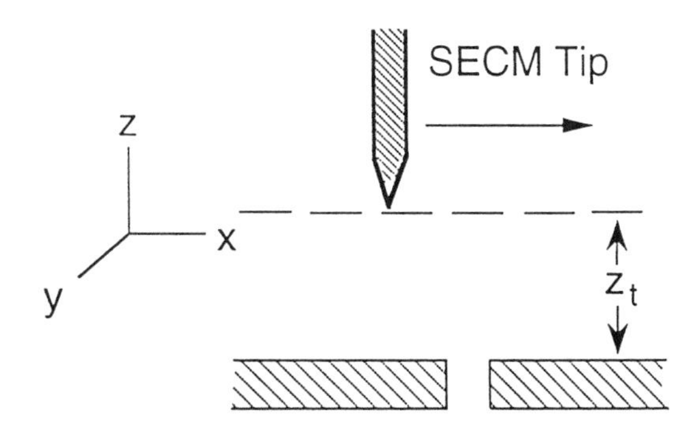

FIG. 5 Schematic diagram showing the SECM tip scanned in the x, y plane at constant membrane-to-tip separation, z_t.

is given by Eq. (2). Mapping the radial coordinate r into the Cartesian coordinate system yields

$$C(x, y, z) = \frac{C_s r_o}{(x^2 + y^2 + z^2)^{1/2}} \tag{9}$$

Since the SECM tip current is proportional to C(x, y, z), a plot of Eq. (9) with $z = z_t$ is equivalent to the shape of a SECM image in which the tip current is plotted as a function of x, y. Figure 6 shows several examples of theoretical C(x, y, z) curves (normalized to C_s) for different tip-to-substrate distances, z_t. The curves are obtained by setting $y = 0$ and allowing x to vary, corresponding to a scan across the middle of the pore at height z_t. The SECM image has a Lorentzian line shape, reaching a maximum directly above the pore opening.

A similar result is obtained for a pore that has a disk-shaped opening. In this case, Eq. (4) describes the two-dimensional concentration distribution above the pore. Converting Eq. (4) to the Cartesian coordinates using $r = (x^2 + y^2)^{1/2}$ allows the image of $C(x, y, z)/C_s$ to be readily computed for the constant height experiment. The results are also plotted in Figure 6 for comparison with a hemispherical pore opening. In calculating $C(x, y, z)/C_s$, we have used Eq. (5) to scale the radius of the disk-shaped pore opening to that of a hemispherical opening so that the rate of mass transport is identical in both cases. The curves in Figure 6 demonstrate that the SECM images of disk and hemispherical openings are indistinguishable when the tip is scanned at a distance greater than a few pore radii from the surface. A significant difference in the concentration profiles can be resolved only at very small separations.

Experimental confirmation of the Lorentzian shape of the SECM image is shown in Figure 7 for experiments involving the diffusive transport of

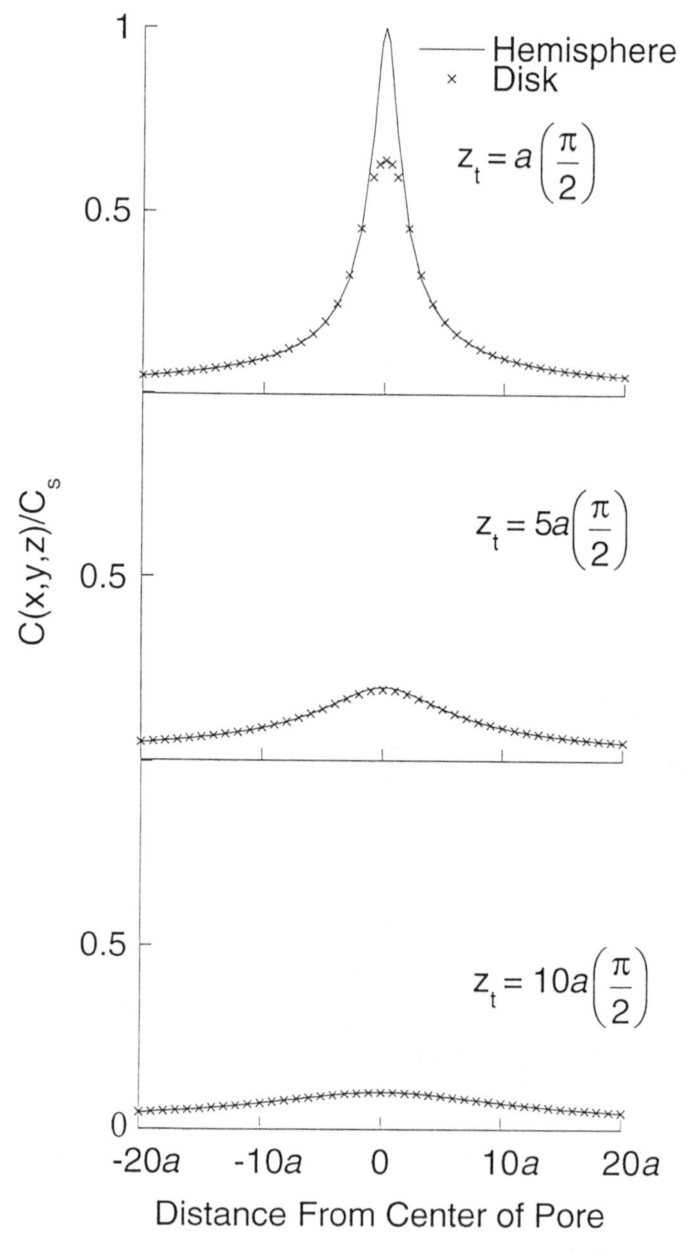

FIG. 6 Normalized concentration above a pore at constant membrane-to-tip separations, z_t. Values of z_t (in units of pore radius, *a*) are indicated on the figure. Solid lines correspond to a hemispherical pore opening; crosses (x) correspond to a disk-shape pore opening.

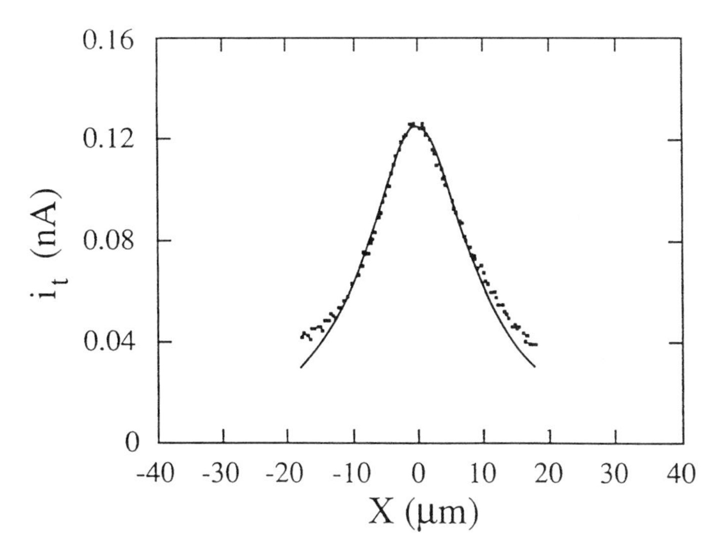

FIG. 7 Plot of experimental SECM-tip current (points) measured above a track-etched pore in mica. The solid line corresponds to Eq. (9) with a $z_t = 9\ \mu m$, after converting C(x, y, z) to tip current, i_t, using Eq. (6). The image corresponds to the diffusion of $Fe(CN)_6^{-4}$ through the membrane. (From Ref. 8.)

$Fe(CN)_6^{-4}$ through a small pore in mica (8) (e.g., see Fig. 4). The figure shows a comparison of the experimental i_t profile (points) with the profile (solid line) calculated from Eq. (9). The tip-to-substrate separation, z_t, in this experiment is $\sim 9.0\ \mu m$. It is clear from these results that the theoretical expression and experimental data are in good agreement within $\pm 10\ \mu m$ of the center of the pore. At greater lateral distances, the theory underestimates the experimental values. This is probably due to a slight overlap of diffusion fields from neighboring pores and the noncircular shape of the pore opening (see Fig. 4b). However, the agreement between theory and experiment is very good, validating the assumption of diffusional transport away from the pore opening.

The advantages of placing the SECM tip as close as possible to the membrane surface (i.e., small z_t) are readily apparent from inspection of Figure 6 and Eq. (9). First, for a scan across the center of a hemispherical pore opening, the SECM tip current is proportional to $(z_t^2 + x^2)^{-1/2}$. Thus, the lower limit of detection of species emerging from the pore increases as z_t decreases. Second, Eq. (9) and the results presented in Figure 6 indicate that the width of the SECM peak above a pore decreases with decreasing z_t—the peak width at half of the maximum current can be shown to be equal to $z_t\sqrt{12}$. Clearly, it is advantageous to scan the SECM tip as close to the surface as possible to optimize both resolution and sensitivity.

D. Quantitative Measurement of Pore Flux and Radius

In addition to providing images of localized transport, SECM measurements provide sufficient information to quantify the absolute flux of a molecular permeant through an individual pore, a particularly important parameter in characterizing transport across membranes. As discussed in Sec. II.A, one method of doing this is based on mapping the isoconcentration contours to obtain the product of the pore radius and surface concentration, r_oC_s or aC_s. A considerably simpler and faster method is described here. We limit the discussion to a pore with a disk-shaped opening.

Diffusive transport of a species from a disk-shaped pore into the receptor solution yields a concentration profile given by Eq. (4). The distribution profile is uniquely determined by two parameters: the surface concentration, C_s, and the pore radius, a. These parameters are also the two unknowns in Eq. (3) that are required to compute the rate of mass transport through the pore.

To evaluate C_s and a, the SECM tip current can be measured as a function of r and z and used to establish the concentration profile C(r,z) near the pore opening. The analysis is greatly simplified by measuring only the concentration profile along the hypothetical centerline axis of the pore that extends away from the pore surface into the receptor compartment (10). In the cylindrical coordinate system, this axis corresponds to $r = 0$. From Eq. (4), the concentration profile along this axis is given by the following equation:

$$C(z) = \frac{2C_s}{\pi} \tan^{-1}\left(\frac{a}{z}\right) \tag{10}$$

Equation (10) indicates that C(z) is also uniquely determined by C_s and a. Values of these parameters are obtained by fitting Eq. (10) to the experimental C(z) profiles. Once C_s and a are determined, they are used in Eq. (3) to compute Ω.

Figure 8 shows theoretical profiles of normalized concentration, C(z)/C_s, computed from Eq. (10) for pore radii, a, ranging from 1 to 200 μm. This plot is instructive in considering the sensitivity of experimental data to the value of a, one of the two unknown parameters used in the curve-fitting analyses. The curves in Figure 8 demonstrate that the C(z) profile is sensitive to a for values less than ~200 μm. For values of a greater than ~200 μm, the concentration profiles approach the expected profile for diffusion from

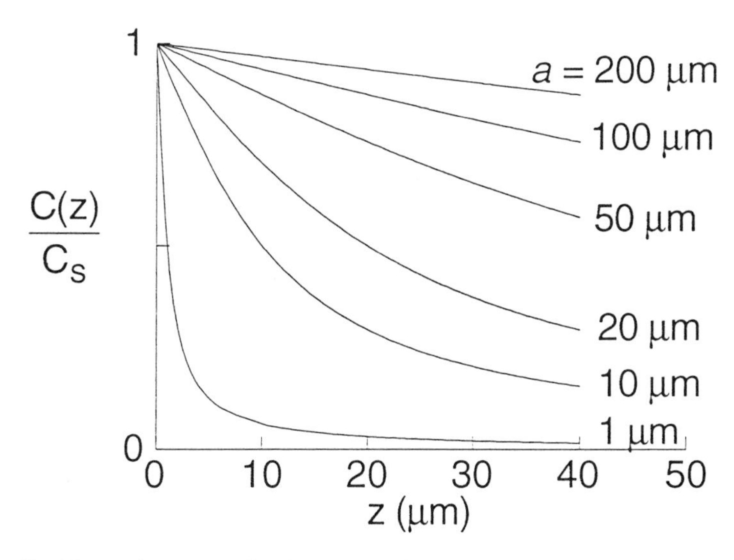

FIG. 8 Theoretical normalized concentration profiles above a disk-shaped pore opening as a function of the pore radius [Eq. (10)]. C_s is the concentration at the pore surface.

a planar surface. In this limiting case, C(z) is independent of *a*, and the analysis is no longer useful.

The above method of determining Ω has proved very successful in analyzing transport through both synthetic and biological membranes. As one example, Figure 9 shows a plot of C(z) vs. z and the statistical least-squares best fit of Eq. (10) to the data obtained by varying the parameters C_s and *a* (27). The data correspond to the steady-state iontophoretic transport of hydroquinone through a single hair follicle in mouse skin, a topic presented in more detail in Sec. III.B. The curve fitting procedure yields C_s = 0.89 mM and a = 18 μm—these values are substituted into Eq. (3) to obtain a mass transport rate of 5.9×10^{-14} mole/sec through the hair follicle.

E. Resolution Issues

The analysis presented in the preceding sections is based on the model of a single pore in an otherwise impermeable membrane. Of course, real membranes deviate in one or more ways from this highly idealized model. It is worthwhile to briefly consider the limitations in applying the single-pore model to the SECM analysis of real membranes.

Figure 10 shows isoconcentration contours above the surfaces of three membranes, each having slightly different structural or transport character-

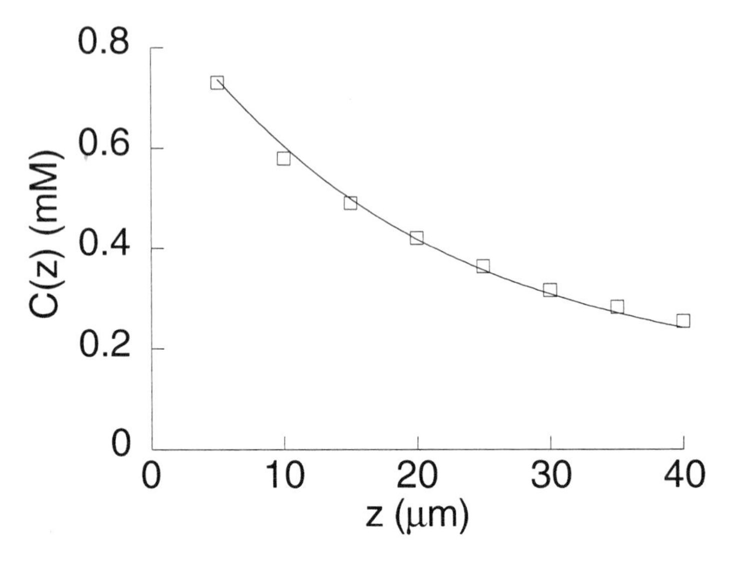

FIG. 9 Experimental concentration profile of hydroquinone (HQ) above a hair follicle in hairless mouse skin. The experiment corresponds to diffusion of HQ through the hair follicle. (Donor solution: 0.2 M HQ and 0.2 M NaCl; receptor solution: 0.2 M NaCl.) The solid line corresponds to the best fit of Eq. (10) to the data. (From Ref. 27.)

istics. Membrane I contains two pores in an otherwise impermeable phase. The spacing between the two pores is large compared with the pore radii, resulting in little overlap between the concentration profiles. In this case, the pores can be analyzed separately using the analysis presented above for a single ideal pore. Widely spaced track-etched pores in mica are representative of this limiting case.

Membrane II depicts a membrane that contains many closely spaced pores, such that overlap occurs between the diffusional fields of each pore. Far away from the pore openings, the isoconcentration contours no longer have a hemispherical shape, because of overlapping fluxes from neighboring pores. This case is frequently encountered in real membranes, including examples discussed later in Sec. III. Clearly, the ability to resolve individual pores depends on how close the tip is positioned above the surface and the size of the tip. If the SECM tip radius, r_t, is significantly smaller than the spacing between two pores and is positioned at a tip-to-membrane distance smaller than the pore radius ($z_t \ll a$), then the individual pores in the membrane can be readily resolved. These conditions are often difficult to achieve, especially on rough surfaces where z_t is necessarily large in order to prevent

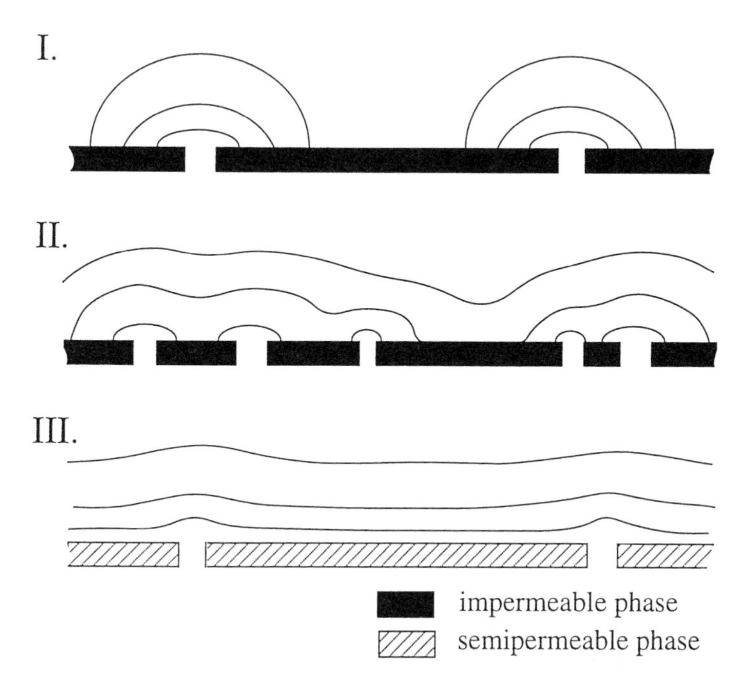

FIG. 10 Isoconcentration contours above porous membranes. Transport in membranes I and II occurs exclusively through pores in an otherwise impermeable phase. In membrane III, transport occurs through both pores and the semipermeable phase.

the tip from crashing into the surface. In such instances, the flux of molecules from two adjacent pores cannot be completely resolved.

Quantitative measurement of the rate of mass transfer, Ω, from a pore whose diffusional field overlaps that of a neighboring pore is generally not as difficult as one might first anticipate. The reason for this is that the tip can be positioned much closer to the pore opening when it is not being rastered across the surface. When the tip is very close to the pore opening, the current is dominated by the flux of molecules from that pore, with very little interference from neighboring pores. Analysis of i_t as a function of z_t, as described in Sec. II.D, yields the rate of mass transport. Values of i_t at larger z_t are more affected by overlapping profiles; thus, the range of z_t for measuring i_t may be significantly reduced.

Membrane III in Figure 10 depicts another nonideal behavior encountered in imaging membranes. Here, discrete pores are surrounded by a homogeneous phase that is moderately permeable to the species of interest. For instance, the homogeneous phase may contain many pores that are too small to be resolved by SECM. Transport across the homogeneous phase presents several difficulties in the SECM analysis. First, it is more difficult to resolve the discrete pores in the membrane. There are two reasons for this: (1) the flux in the pore is reduced because of competitive transport through the neighboring homogeneous regions, and (2) the isoconcentration contours on the receptor side are a combination of radial diffusion from the pores and planar diffusion from the homogeneous region; the increase in planar diffusion reduces the ability to resolve the flux from a pore. Second, it is difficult to quantify the flux across the homogeneous region by SECM, since the flux in this region is frequently very small and thus i_t is also small. Compounding the problem is the fact that the rate of mass transport, Ω, from the semipermeable region is often large since this region occupies the largest fraction of exposed surface area (frequently >95% of the surface). Thus, it is difficult to quantify the overall mass transport rate for such a membrane using SECM alone. In this situation, one must measure the total transmembrane flux by an independent method. This usually involves more classical transport measurements, such as sampling the bulk concentration of the molecular permeant in the receptor compartment as a function of time. The mass transport rate through the pores can still be measured as before using SECM and subtracted from the total rate to estimate the flux through the homogeneous region. This procedure has been reported for analysis of transport across skin, where approximately 75% of the flux is associated through a handful of discrete pores, and the remainder is associated with the more homogeneous tissues of skin (15).

F. Imaging Strategies

Several strategies recently developed for imaging membrane structure and quantifying molecular fluxes are depicted in Figure 11. Experimentally, the

FIG. 11 Imaging modes for SECM analysis of porous membranes. The SECM tip is rastered at constant height in each of the following modes. (I) *Tip collection mode*. The electroactive species is present in high concentration in the donor solution and absent in the receptor solution. The tip senses changes in concentration resulting solely from transport through the pores. (II) *Negative feedback mode*. The donor and receptor solutions contain equal concentrations of the electroactive species so there is no driving force for diffusional transport across the pore. Variation in negative feedback due to surface topography and pore openings gives rise to the contrast in the SECM image. (III) *Combined tip collection and negative feedback mode*. The concentration in the donor solution is larger than that in the receptor solution. Characteristics of the SECM image are due to both negative feedback and a net diffusional transport of the electroactive species across the pore. Convective transport

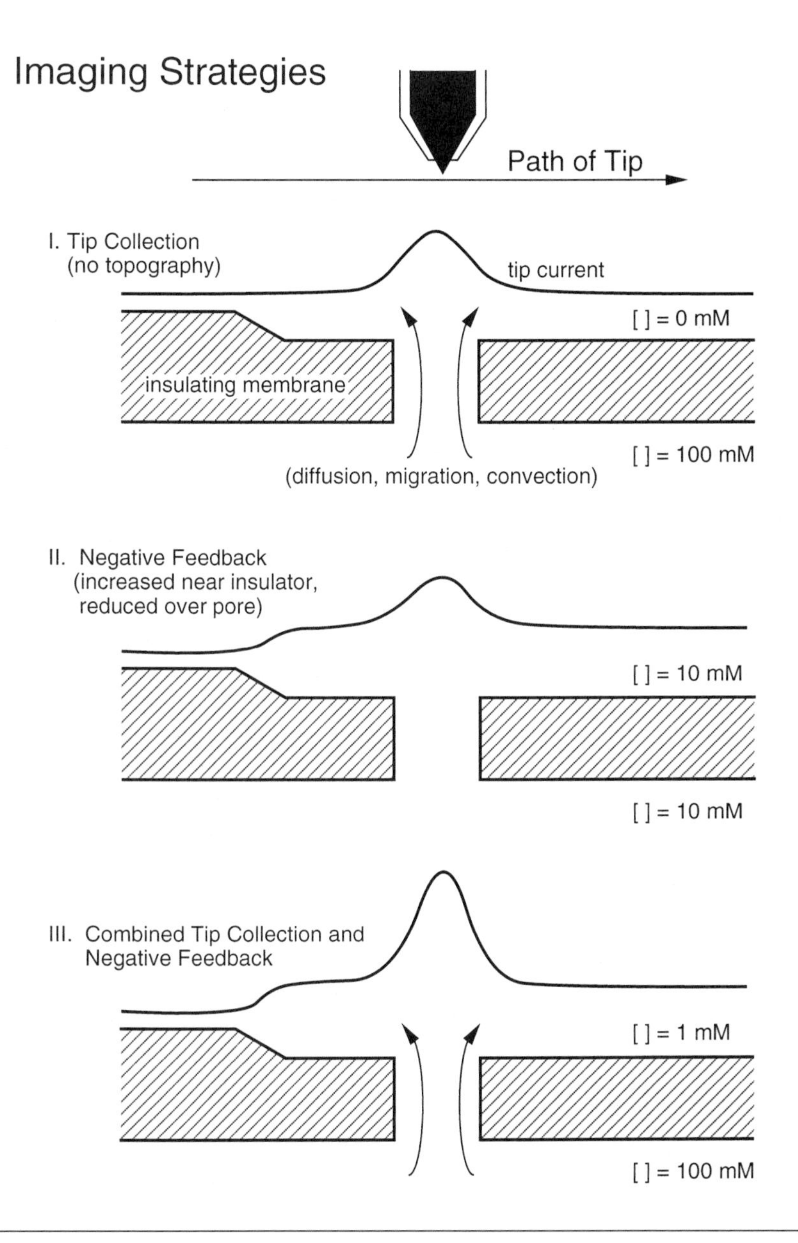

and migration can occur in any of these modes if a pressure or electric field gradient is applied across the membrane. In these cases, the tip current directly over the pore may be significantly enhanced because of the increase in mass transport of the molecular permeant through the pore.

primary difference between these methods lies in the relative concentrations of the redox species in the donor and receptor solutions. The choice of imaging method depends on the properties of the membrane as well as the specific goals of the study. In the following descriptions, the SECM tip is assumed to be rastered at a constant absolute height across the membrane surface. However, the distance between the tip and membrane, z_t, may vary due to topographical features on the surface of the membrane.

1. Tip Collection Mode

Tip collection mode is the simplest SECM mode for imaging membranes and the first to be employed for this purpose (8,14). The membrane separates a donor solution, which contains a redox species, from a receptor solution in which the redox species is absent. The SECM tip current arises solely from the transport of redox species across the membrane—a large tip current indicates a localized region in which the spatial flux in the membrane is large, often due to a discrete pore. No information regarding the membrane topography is obtained in this mode. However, as discussed in the preceding sections, the theoretical relationship between the tip current and the flux in the pore is relatively straightforward in this case.

Tip collection mode has been used for imaging both synthetic and biological membranes. In the absence of an applied field, the redox species is transported across the membrane by diffusion. The flux across the membrane may be relatively small in this case, resulting in difficulties in obtaining an SECM image. For ionic species, this problem is circumvented by passing a current across the membrane to induce migration of the species. The applied current can be increased to achieve a molecular flux that is sufficiently large for SECM imaging. If the membrane is permselective to ions, an applied current will cause electroosmotic flow, which can be used to enhance the flux of neutral molecules. Similarly, hydrostatic or osmotic pressure can be used to enhance the flux of neutral and ionic molecules across the membrane.

Because the theory is relatively simple and well developed, the tip collection mode has been employed extensively for quantitative measurements of pore flux and radius.

2. Negative Feedback Mode

In the negative feedback mode, the concentration of redox species is the same on both sides of the membrane; no concentration gradient exists between the donor and receptor compartments to drive diffusive transport across the pore (11–13). This mode of imaging can be used to obtain topographical maps of the surface where the variation in the SECM tip current arises from differences in the tip-to-sample separation (28). When the tip is far from the sample surface, a steady-state current is measured at the tip,

$i_{t\infty}$, that is proportional to the concentration of redox species in the receptor compartment, Eq. (6). When the tip is brought close to the sample surface, diffusion of the redox species to the tip is hindered by the presence of the sample surface. This results in a decrease in tip current relative to $i_{t\infty}$. A three-dimensional image of the surface is obtained by rastering the tip, at a fixed height, across the sample surface. The topography of the surface is then obtained from the normalized tip current ($i_t/i_{t,\infty}$) using the expressions for negative feedback described in Chapter 5. Qualitatively, a decrease in $i_t/i_{t,\infty}$ corresponds to a decrease in the tip-to-sample separation and thus a raised surface. An increase in $i_t/i_{t,\infty}$ indicates that either (1) the tip-to-membrane surface separation is large or (2) the tip is above a pore. In the latter case, the increase in i_t results from transport of the molecule from the donor-side solution to the tip, via the pore. If an external driving force is used to enhance transmembrane transport, modulation of the force during imaging allows one to decouple the topographic and pore flux components of the SECM tip current (11–13).

3. Combined Tip Collection and Negative Feedback Mode

As is implied by the name, this imaging strategy is a combination of the two previous modes. Here, a concentration gradient exists across the sample. However, unlike the pure tip collection mode, the concentration of redox species in the receptor compartment has a nonzero value. Thus, the tip current, $i_{t\infty}$, has a nonzero value at large tip-to-sample separations. This tip current is reduced when the tip is brought close to an insulating portion of the sample because of hindered diffusion, similar to that described above for the *negative feedback mode*. Because there is a concentration gradient present between the donor and receptor solutions, the redox species diffuses through the pores in the sample. When the tip is scanned over these pores, the tip current increases. This increase in tip current is similar to the tip response in *tip collection mode* (the difference is that $i_{t\infty} = 0$ in tip collection mode). Information on both the surface topography of the membrane and local flux of the redox species across the membrane is obtained in this imaging mode (9).

G. Tip Perturbations

Qualitative interpretation of SECM images of a membrane is generally possible without detailed consideration of how the SECM tip influences the diffusion field around a pore opening. However, more quantitative analyses of pore transport generally require that the SECM tip behave as a noninteracting probe. This means that the rate of consumption of electroactive molecules at the SECM tip must be sufficiently small so that the diffusion field around the pore is not significantly altered. A detailed analysis of this problem has not yet been presented; however, as described below, Nugues and Denuault have reported numerical simulations demonstrating how the diffusion field above a membrane can be altered by the SECM tip (9).

The perturbation of the diffusion field can be minimized by using the smallest SECM tip as possible. The reason for this is twofold. First, the amount of electroactive material consumed by the tip decreases in proportion to the tip radius, r_t, Eq. (6). Second, the thickness of the concentration profile around the tip, which results from electrochemical reaction at the tip, is proportional to r_t. Thus, a larger tip alters the distribution of the electroactive species over a larger volume of space.

In general, the consumption of the electroactive molecules at the tip results in a skewing of the concentration profile above the pore opening. If this interaction is sufficiently large, then the concentration profile around the pore opening will no longer have a hemispherical shape (Sec. II.A) and the theoretical treatment for determining the rate of mass transport (Sec. II.D) will no longer be valid. In addition, because the tip reaction consumes electroactive material, the gradient across the membrane is increased, resulting in a local increase in the rate of mass transport across the membrane.

In addition to using small tips, one can also reduce the tip interaction by increasing the tip-to-surface distance, z_t, since the gradient of the diffusion layer around the pore decreases rapidly (as the inverse square of the distance) as one moves away from the pore opening. Thus, at large distances, consumption of the electroactive molecule at the tip has a relatively smaller effect on the profile. Of course, the drawbacks to imaging at large z_t are a reduction in the signal-to-noise ratio and lower image resolution. While there is no exact treatment for this problem, quantitative agreement between SECM-measured concentration profiles and theory, such as that shown in Figures 4, 7, and 9, empirically demonstrates that the perturbation of the concentration profile by the tip is negligibly small when z_t is greater than about one tip radius from the surface.

Nugues and Denuault have simulated the molecular flux across a porous membrane, computing the concentration profiles of a SECM tip placed in very close proximity to the membrane surface (9). Figure 12a shows the simulated profiles when the tip is positioned directly above a pore opening. As noted above, the consumption of redox species at the tip results in an increased concentration gradient in the pore. As a result, the diffusive flux of redox species through the pore is increased. An interesting situation arises when the sample thickness is smaller than the tip radius. The tip (positioned at the surface of a pore) can theoretically scavenge molecules beyond the sample, probing the concentration of redox species in the donor compartment (9).

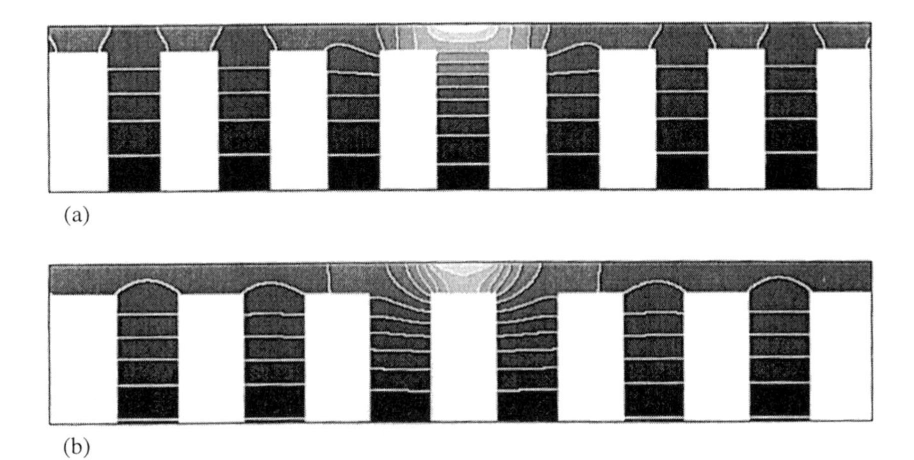

FIG. 12 Simulated isoconcentration contours in a membrane containing closely spaced pores. The simulation shows a perturbation of the contours inside a pore resulting from depletion of the electroactive species by the SECM tip. (a) Tip positioned above a pore; (b) tip positioned between two pores. (From Ref. 9.)

Simulated profiles are shown in Figure 12b for the case where the tip is positioned between two closely spaced pores. Here the tip and its insulating sheath partially block the pore, effectively reducing the diffusive flux of the electroactive molecule through the pore. The simulated profiles illustrate the complex nature of probing mass transport across membranes at small tip-to-sample separations.

III. APPLICATIONS

Synthetic membranes, e.g., track-etched polycarbonate, have been frequently employed as idealized membrane models in the development of SECM-based methodologies. In these studies, the pore dimensions and spacing, as well as the transport mechanism in the pore, are typically well defined a priori, allowing a rigorous testing of the SECM theory. On the other hand, biological membranes, e.g., skin, have a very complex structure with the possibility of parallel transport paths and mechanisms. The analysis of the SECM images and data in these cases is generally less direct. However, the initial studies in this area indicate that SECM is a very powerful technique for quantifying localized transport across biological membranes.

In the following sections, examples of SECM investigations of transmembrane transport are arranged in terms of the structural complexity of the membrane, beginning with the simplest membranes. Molecular transport in these examples occurs by a variety of mechanisms including simple diffusion, diffusion with migration, pressurized convective flow, and electroosmotic flow. The diversity of membrane structures and transport mechanisms that have been investigated by SECM indicates a general utility of this scanned-probe microscopy for studying membrane transport.

A. Synthetic Membranes

1. *Diffusion and Migration across Porous Mica and Polycarbonate Membranes*

Porous Mica. The initial use of SECM for studying transport across a membrane involved porous mica membranes (8). The pores in this membrane are produced by a track-etch process in which defect tracks are produced in a thin sheet of mica by exposure to a collimated beam of ^{238}U fission fragments. These tracks are then etched in hydrofluoric acid for a specified time to produce pores. The pores have a rhomboidal cross section with 60° and 120° included angles and with edge lengths ranging from 1.0 to 3.0 μm; an electron micrograph of the opening of one such pore is shown in Figure 4b.

The porous mica membrane represents the simplest type of membrane studied by SECM. The pores are essentially straight with a uniform cross section throughout the membrane. The pores are sufficiently large and unobstructed so that the solution composition inside the pore is approximately the same as that in the bulk solution contacting the membrane. Thus, the diffusivity of a molecular permeant is not expected to change when it enters into the pore. For these reasons, one can compute, with a fairly high level of precision, the expected diffusive flux through a single pore and compare it with the flux measured by SECM.

Figure 13 shows a schematic diagram of the apparatus used to image porous mica. The mica membrane is positioned in a diffusion cell, such that it separates a *donor* solution containing both 0.1 M $K_4Fe(CN)_6$ and 0.1 M NaCl from a *receptor* solution containing only 0.1 M NaCl. $Fe(CN)_6^{-4}$ is transported across the membrane and is detected by the SECM tip as it emerges from the pores into the receptor solution. To detect $Fe(CN)_6^{-4}$ exiting from the pores, the potential of the SECM tip is held at 0.5 V versus the saturated calomel electrode (SCE), sufficiently positive of the standard electrode potential of the $Fe(CN)_6^{-4/-3}$ redox couple ($E^{O\prime}$ = 0.185 V vs. SCE) to cause the mass-transfer limited oxidation of $Fe(CN)_6^{-4}$. This experiment corresponds to the "tip collection" mode described in Sec. II.F. The flux of $Fe(CN)_6^{-4}$ through the pores can be enhanced by driving a current, i_{app}, across the membrane, dramatically improving the signal-to-noise ratio in the SECM images. Experimentally, this is readily accomplished by using the commer-

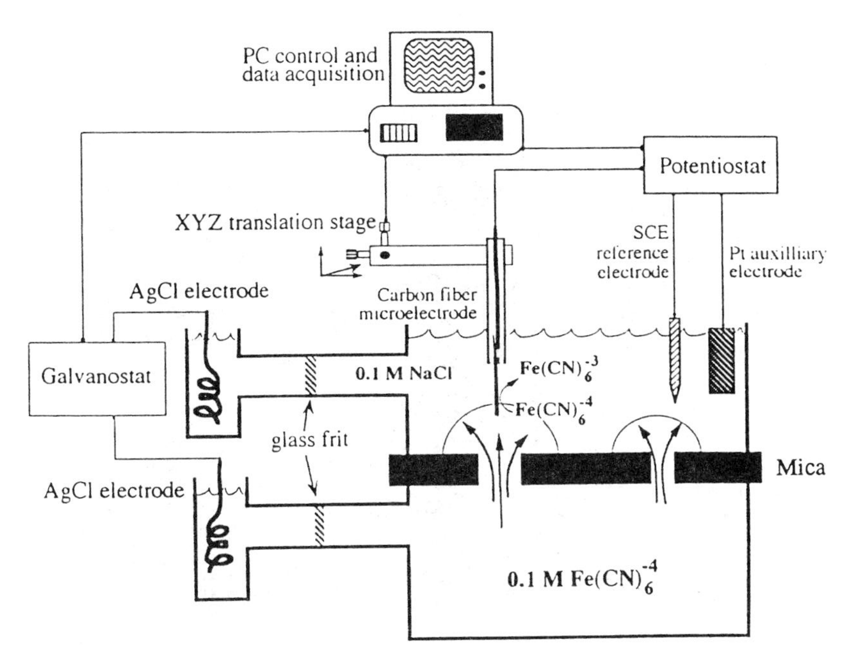

FIG. 13 Schematic drawing of the scanning electrochemical microscope used for imaging molecular transport across a membrane. The membrane separates the donor and receptor solutions. The electroactive species in the donor solution diffuses across the membrane or may be driven across the membrane by an electrical current supplied by the galvanostat. The SECM tip detects the electroactive species as it emerges from the membrane. (From Ref. 16.)

cial galvanosat and two large Ag/AgCl electrodes as depicted in Figure 13. When the cathode (negative terminal) is in the donor compartment, the applied current results in an upward migration of $Fe(CN)_6^{-4}$. In this situation, the net flux of $Fe(CN)_6^{-4}$ is due to combined diffusion and migration and is described by the Nernst-Planck equation.

Figure 14 shows an SECM image of a 250 μm^2 region of the mica membrane for $i_{app} = 20\ \mu A$. The gray-scale contrast in the image indicates the magnitude of the current measured at the SECM tip, i_t, as a function of the lateral position of the tip (lighter regions indicate higher currents). The four circular bright regions observed in the image are due to an increased concentration of $Fe(CN)_6^{-4}$ above pore openings on the membrane surface. As described in Sec. II.B, the tip current is directly proportional to the flux of the redox-active permeant inside the pore, Eq. (8). Thus, the regions of high tip current in these images are associated with the localized flux of

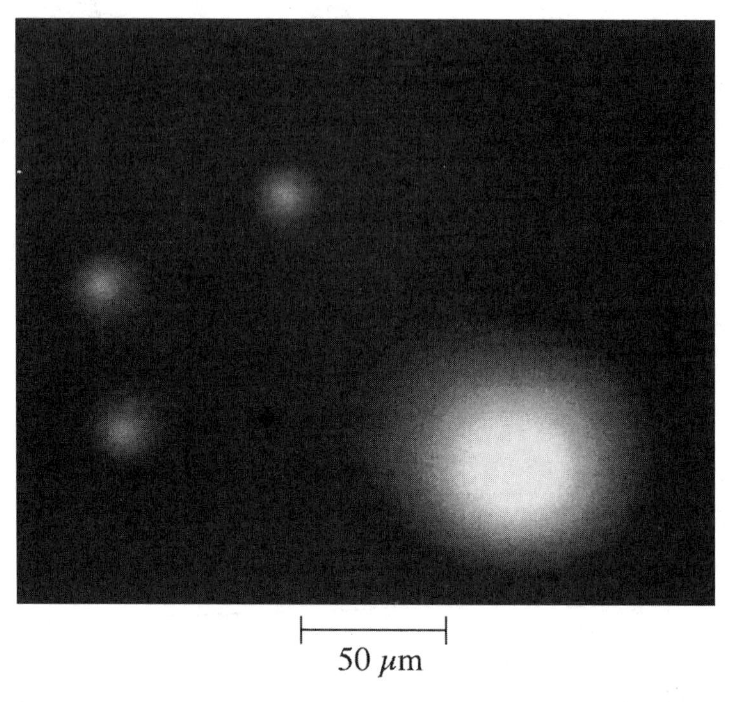

FIG. 14 SECM image of a porous mica membrane obtained while applying a 20 μA current across the membrane. The bright spots in the image correspond to the diffusion and migration of $Fe(CN)_6^{-4}$ across the membrane. Donor solution: 0.1 M $K_4Fe(CN)_6$ and 0.1 M KCl. Receptor solution: 0.1 M KCl. (From Ref. 8.)

$Fe(CN)_6^{-4}$ across the membrane pores. In essence, the SECM images thus provide a means to *visualize* the flux of $Fe(CN)_6^{-4}$ across the membrane, a general concept that is employed in all applications of SECM to study membrane transport.

Three of the SECM peaks in Figure 14 have apparent widths of ~15 μm. This value is ~10 times greater than the average pore dimensions, a result of the SECM tip intercepting the quasi-radial flux of $Fe(CN)_6^{-4}$ from the pores at a nonzero distance from the membrane surface. Figure 14 shows that the flux through one of the pores is significantly larger than the others, indicating a larger pore opening. The SECM-measured density of pores, based on multiple images of the porous mica sample, is 3×10^3 cm^{-2}, in reasonable agreement with the density of pores measured by SEM, 4.5×10^3 cm^{-2}. The density of pores is sufficiently low that the diffusional fields of individual pores do not overlap to any significant extent, corresponding to the ideal type I membrane depicted in Figure 10.

The beneficial effect of inducing a migrational flux of the redox species through the pores in mica is presented in Figure 15, which shows a series of SECM images obtained as a function of applied constant current, i_{app}. In the absence of an applied current, the diffusional flux of $Fe(CN)_6^{-4}$ from the pores in mica is just barely discernible in the SECM image (top image). The tip current corresponding to the $Fe(CN)_6^{-4}$ flux through two pores is significantly enhanced by increasing i_{app}, a finding that reflects the fact that the highly charged anion carries a significant fraction of the current across the membrane. A smaller enhancement in the flux is expected when the transference number of the redox permeant is lowered (for instance, by decreasing its concentration in the donor solution relative to the concentration of the background electrolyte). The experimental results shown in Figure 15 were obtained with equal concentrations of $Fe(CN)_6^{-4}$ and KCl in the donor compartment, solution conditions that ensure a fairly large transference number for $Fe(CN)_6^{-4}$, and thus a significant increase in the flux of $Fe(CN)_6^{-4}$ as i_{app} is increased. The use of an applied current to drive the molecular permeant across the membrane at enhanced rates is completely general and can be used to enhance the quality of SECM images in many instances. Additional examples are given throughout the remainder of this chapter.

Porous Polycarbonate. Commercially available porous polycarbonate membranes have a relatively simple structure and have been used by several groups to test SECM methodology. Figure 16a shows an image obtained by Nugues and Denuault of a Nucleopore membrane (6 μm thickness, 5-μm-radius pores, 1.6×10^4 pores/cm^2) (9). This image is obtained in the combined tip collection/negative feedback mode (see Fig. 11) with a solution of 80 mM $Fe(CN)_6^{-4}$/0.1 M KCl in the donor compartment and a solution of 0.80 mM $Fe(CN)_6^{-4}$/0.1 M KCl in the receptor compartment. The difference in concentration drives the diffusional transport of $Fe(CN)_6^{-4}$ across the membrane. This image has a particularly large signal-to-noise ratio, with 16 active pores identified in the image. Similar to the image of porous mica (Fig. 14), the flux through some pores is clearly grater than others.

A much more difficult test of the resolution achievable by SECM is shown in Figure 16b, which presents an image of a Poretics polycarbonate membrane (15 μm thickness, 45-nm-radius pores, $\sim 10^5$ pores/cm^2) (7). Here the pore size is very small and the pore density is large, making it difficult to resolve the molecular flux from individual pores. In particular, the high pore density causes the diffusional fields of pores to overlap to a large extent (Type II membrane, Fig. 10). Nevertheless, the image in Figure 16b shows that it is possible to visualize molecular fluxes through individual 45-nm-radius pores. The large background current is due to overlap of the diffusional fields of neighboring pores. In principle, this background current can

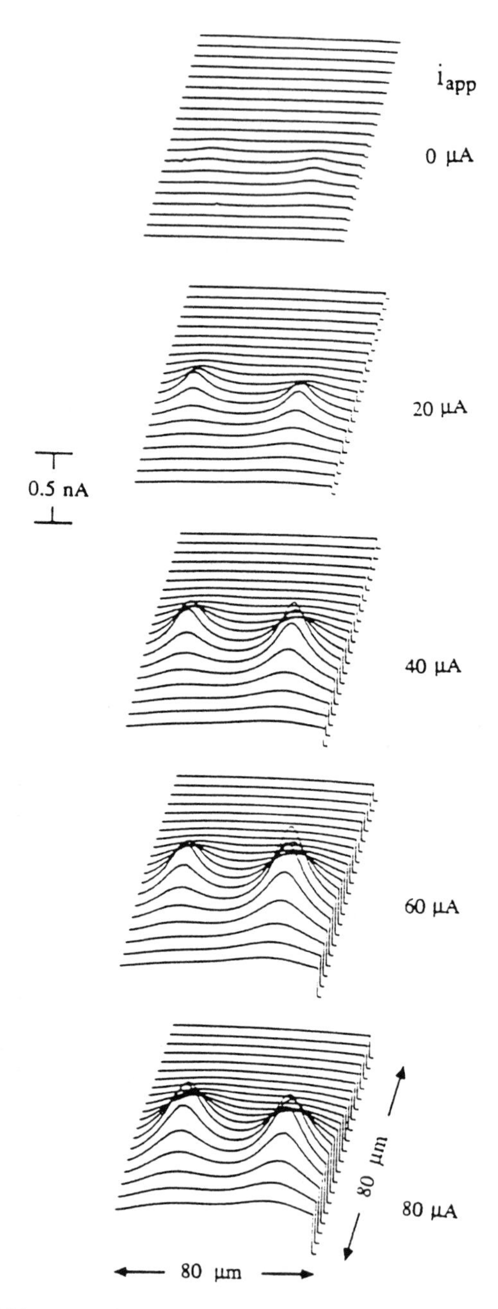

FIG. 15 SECM images of a porous mica membrane as a function of the applied current, i_{app}. The peaks in the images correspond to the diffusion and migration of $Fe(CN)_6^{-4}$ across the membrane. The donor solution contained 0.1 M $K_4Fe(CN)_6$ and 0.1 M KCl. The receptor solution contained 0.1 M KCl. (From Ref. 8.)

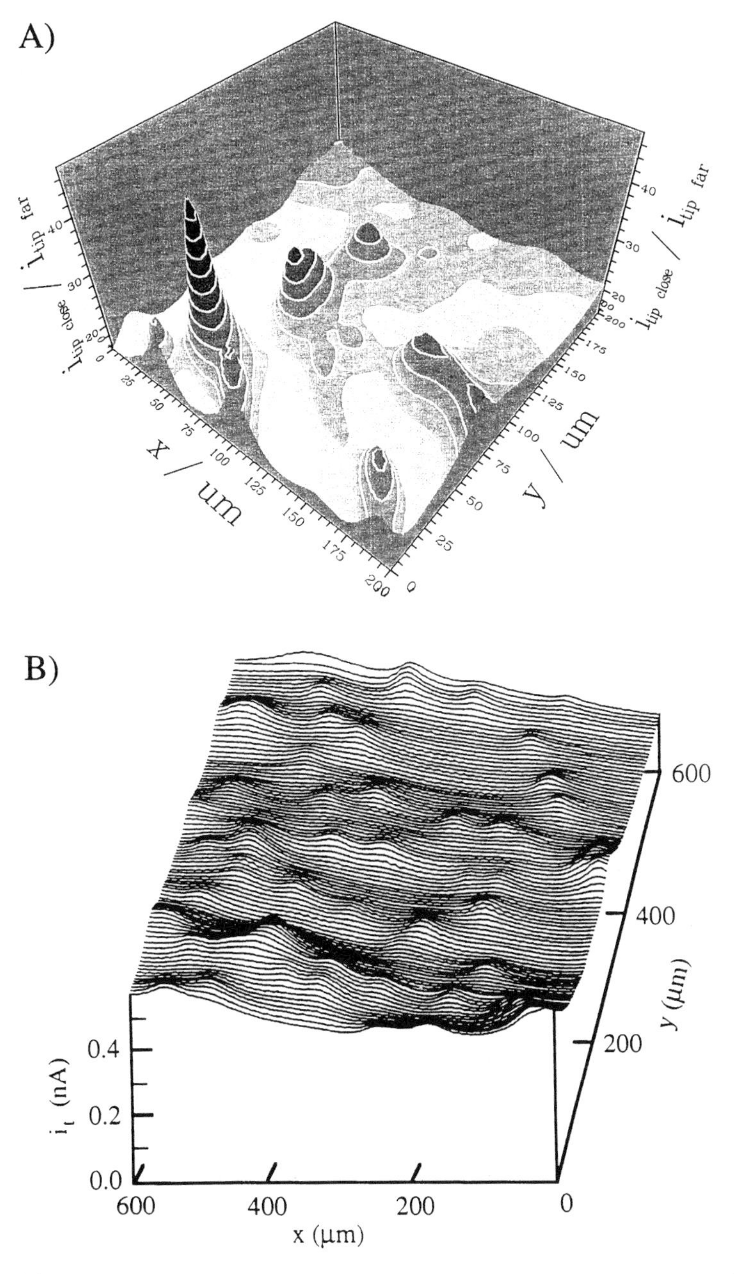

FIG. 16 SECM images of (A) Nucleopore polycarbonate membrane containing 5-μm-radius pores at a density of 1.6×10^4 pore/cm^2. (From Ref. 9.) (B) Poretics polycarbonate membrane containing 45-nm-radius pores at a density of $\sim 10^5$ pores/cm^2. (From Ref. 7.) The images show the transport of $Fe(CN)_6^{-4}$ across the membrane.

be reduced relative to the peak current directly over a pore by scanning closer to the surface with smaller tips.

2. *Electroosmosis across Nafion Membranes*

SECM has been recently used by Bath et al. in the investigation of electroosmotic convective flow (10). Electroosmosis occurs in porous membranes employed in fuel cells, representing an important practical issue in the parasitic crossover of fuel (e.g., methanol) and in the flooding/drying of the fuel cell electrodes (29,30). Electroosmotic flow across skin, a naturally occurring ion-selective membrane, is also of interest in transdermal drug delivery, as discussed in Sec. III.B.2.

SECM studies of electroosmosis in the persulfonated polymer Nafion have been investigated using the single-pore membrane shown in Figure 17. The membrane consists of a ~200 μm layer of mica in which one large (~50 μm radius) pore has been introduced by pushing a sharpened metal wire through the mica. The pore in the mica layer is filled with the cation-selective polymer Nafion to impart charge and size selectivity to the pore. Cation transport numbers in Nafion approach unity (31–34), and electroosmotic flow of water readily occurs in this material. The Nafion-filled, single pore membrane is referred to hereafter as the "mica/Nafion" membrane.

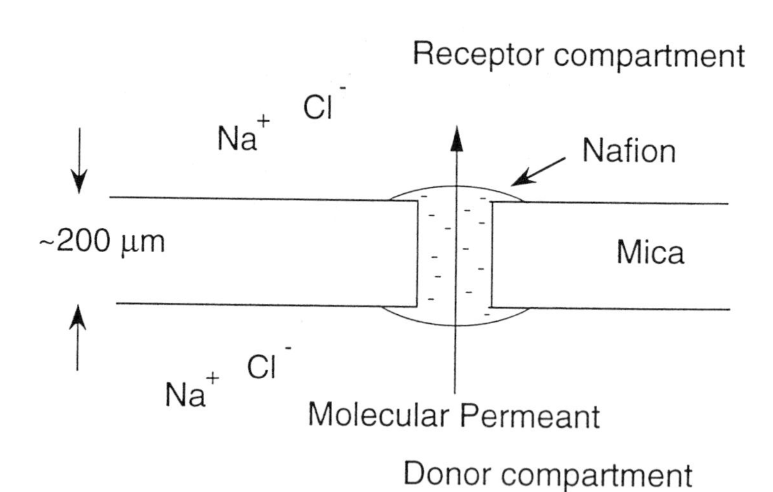

FIG. 17 Schematic diagram of a Nafion-filled pore in mica. Transport of a redox-active permeant through the pore occurs by diffusion, electroosmotic flow, and migration. The membrane separates donor (lower) and receptor (upper) compartments containing equal concentrations of NaCl. The donor solution contains the permeant. (From Ref. 10.)

When a current is applied across the membrane, transport of a species may occur by diffusion and electroosmotic flow if the species is electrically neutral and by diffusion, electroosmotic flow, and migration if the molecule is charged. The generic term *iontophoresis* describes the application of an applied current to control the overall transport rate of a species across a porous membrane. The experimental arrangement for studying iontophoretic transport is similar to that described in the previous section for the imaging of porous mica and polycarbonate membranes. The mica/Nafion membrane separates a donor solution containing a redox-active molecular permeant and electrolyte (e.g., NaCl) from a receptor solution containing only the electrolyte (tip collection mode) (Fig. 13). The SECM tip is used to detect the permeant as it emerges from the pore. In the experiments described below, the tip is held directly above a pore. The tip current, i_t, is then measured as a function of the current applied across the membrane, i_{app}.

Figure 18 shows voltammograms recorded at the SECM tip, positioned directly above the Nafion-filled pore, for three redox-active permeants having different electrical charges (manovalent cation, neutral and monovalent anino) (35). The magnitude of the voltammetric limiting current at the tip is directly proportional to the permeant flux in the pore [Eq. (8)]; thus, a comparison of the voltammetric response of the SECM tip provides a quick semiquantitative understanding of the influences of i_{app} and molecular charge on transport behavior in Nafion.

The middle panel in Figure 18 shows voltammograms recorded at the SECM tip during the steady-state transport of the neutral molecule, acetaminophen, through the Nafion-filled pore at i_{app} = 10, 0, and -10 μA. The response at i_{app} = 0 corresponds to diffusion of acetaminophen across the membrane. The flux of acetaminophen is enhanced by a factor of $\sim$2 when a positive current, i_{app} = 10 μA, is applied across the membrane. By definition, a positive current is carried across the membrane by the migration of electrolyte cations (Na^+) from the donor compartment to the receptor compartment (the transference number for the electrolyte anions is essentially zero). Because acetaminophen is electrically neutral, the observed flux enhancement must solely be due to transport of acetaminophen by electroosmotic convective flow, a result of the upward flow of Na^+ in the Nafion-filled pore. When the direction of the current is reversed, i.e., i_{app} = -10 μA, the flux of acetaminophen is reduced to a negligible level, a consequence of electroosmotic flow occurring from receptor to donor compartment, opposing the diffusional flux of acetaminophen through the Nafion-filled pore.

The SECM data for transport of a monovalent anion (ascorbate) and a monovalent cation ([(trimethylammonio)methyl]ferrocene, $FcTMA^+$) are consistent with the cation-selective properties of Nafion. Figure 18 (left

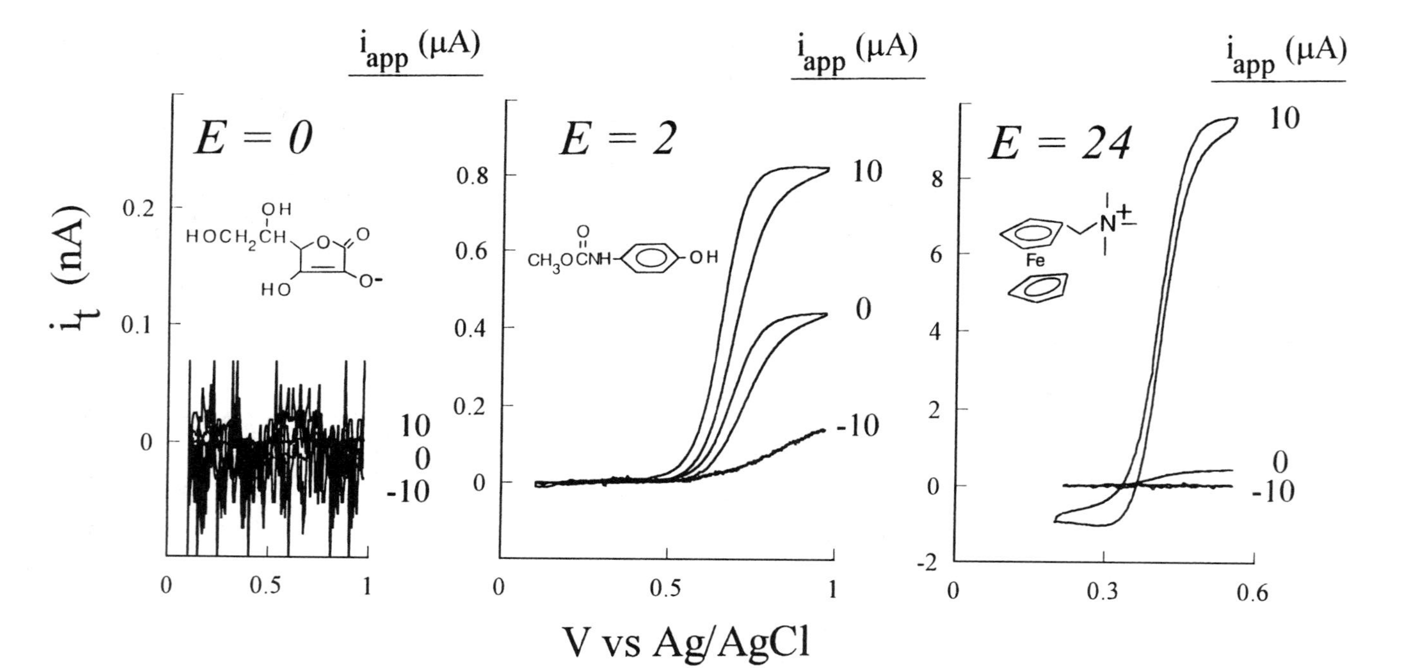

FIG. 18 Voltammetric response at the SECM tip (i_t) above the Nafion-filled pore as a function of the molecular permeant charge and applied current (i_{app}). The magnitude of the SECM tip current is proportional to the rate of transport of the molecules shown on the figure. The voltammetric response at $i_{app} = 0$ corresponds to the diffusional flux through the pore. The voltammograms recorded at $i_{app} = 10\ \mu A$ correspond to iontophoretic transport from the donor solution to the receptor solution. The voltammograms recorded at $i_{app} = -10\ \mu A$ correspond to iontophoretic transport from the receptor solution to the donor solution. E is the enhancement factor due to electroosmosis in the forward direction (equal to the ratio of the flux at $i_{app} = 10\ \mu A$ to that at $i_{app} = 0\ \mu A$). (From Ref. 35.)

panel) shows that ascorbate is not transported through the pore, regardless of the applied current; the ascorbate anion is electrostatically repelled from the negative fixed charges incorporated in the Nafion polymer. Figure 18 (right panel) shows that the diffusional flux at $i_{app} = 0$ for the monovalent cation, $FcTMA^+$, is approximately equal to that of acetaminophen (note the change in y-axis scale). However, the flux of $FcTMA^+$ is enhanced by a factor of 24 at $i_{app} = 10\ \mu A$, in comparison with an enhancement of ~2 for acetaminophen under the same conditions. The difference is due to the dominant contribution of electrical migration to the net flux of $FcTMA^+$, which is absent for the neutral molecule, acetaminophen.

A number of fundamental parameters that describe iontophoretic transport in Nafion (e.g., convective flow velocities, solute drag coefficient) can be evaluated from the SECM experiment. Consider, the overall iontrophoretic flux of a neutral molecular permeant:

$$N_{iont} = -D_p \nabla C_p(z) + C_p(z) v_{eo} \tag{11}$$

where v_{eo} is the effective velocity associated with electroosmotic convective transport of the solute molecule. The value of v_{eo} depends on the applied current, i_{app}, and ion-selective properties of the membrane. Substitution of the flux N_{iont} [Eq. (11)] into the mass conservation law ($\nabla \cdot N_{iont} = 0$) and integration over the length of the pore yields the steady-state flux between the donor and receptor compartments (23):

$$N_{iont} = \frac{v_{eo} \kappa C_s^{D*}}{\left(1 - \exp\left(\frac{-v_{eo} \ell}{D_p}\right)\right)} \tag{12}$$

where κ is the equilibrium coefficient describing the partitioning of the redox molecule between the Nafion and aqueous solutions.

Equation (12) relates the pore flux, N, to the convective velocity, v_{eo}. It also represents the relationship between i_t and i_{app}, since the SECM tip current is proportional to the flux inside the pore ($i_t \propto N$), and since v_{eo} may be assumed as being proportional to i_{app}. Thus, the dependence of i_t on i_{app} for a neutral molecule (e.g., acetaminophen; Fig. 18, middle panel) can be qualitatively understood by examining the behavior of N in the limits of large positive and negative flow velocities.

1. $N_{iont} = v_{eo} \kappa C_s^{D*}$ in the limit $v_{eo} \to +\infty$.

In this limit, the flux of redox-active solute molecule from the donor compartment to the receptor compartment is due entirely to convection. Thus, N_{iont} is proportional to the convective velocity, v_{eo}, and one expects that i_t should increase linearly with increasing i_{app}.

2. $N_{iont} = 0$ in the limit $v_{eo} \rightarrow -\infty$.

In this limit, convective flow from the receptor to the donor compartment opposes diffusion, driving the net flux of solute molecules to zero. One expects i_t to approach zero as i_{app} is made increasingly more negative.

The above limiting behaviors can be compared with the results presented in Figure 19, which shows the SECM tip current, i_t, plotted against the applied iontophoretic current, i_{app}. These data correspond to the electroosmotic transport of the neutral molecule, hydroquinone (HQ), across the mica/Nafion membrane (10). As noted above, the flux of a molecular permeant through the pore, N_{iont}, is proportional to i_t. Similarly, the electroosmotic flow velocity, v_{eo}, is proportional to i_{app}. Thus, Figure 19 is qualitatively equivalent to a plot of N_{iont} versus v_{eo}. The anticipated limiting behaviors are immediately apparent in Figure 19: i_t increases in proportion to i_{app} for large positive values of i_{app} and asymptotically approaches zero as i_{app} is made increasingly more negative.

Although it is beyond the scope of this chapter, a relatively straightforward analysis of the SECM data shown in Figure 19 allows one to compute the convective velocity of the solute, v_{eo}, and the electroosmotic drag coef-

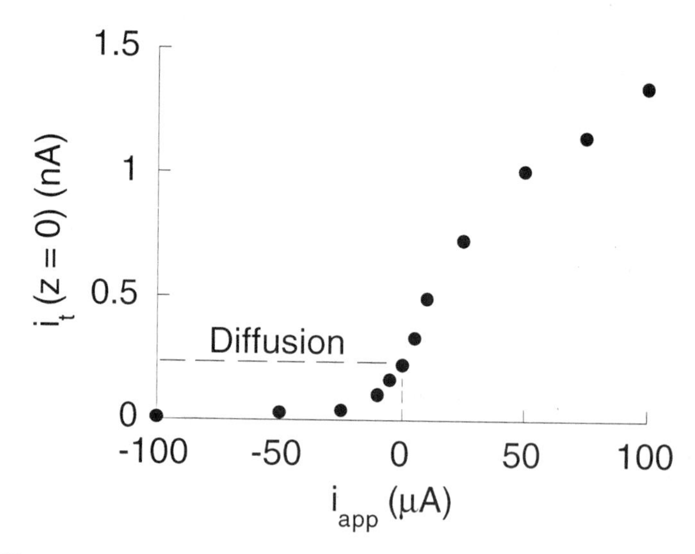

FIG. 19 SECM-tip current measured directly above a pore opening (i_t) as a function of the applied iontophoretic current (i_{app}). The current corresponds to the oxidation of hydroquinone (HQ). The donor compartment contains 0.2 M HQ and 0.2 M NaCl. The receptor compartment contains 0.2 M NaCl. Tip radius = 0.5 μm. (From Ref. 10.)

ficient at any specified applied current. For instance, for i_{app} = 50 μA, the convective flow velocity of HQ in the Nafion-filled pore is about 1 μm/s and approximately 1 molecule of HQ is transported through the pore per 2000 Na^+ (10).

As discussed in Sec. II.D, the absolute values of pore flux and pore radius can be determined by measuring the concentration profile along the centerline axis of the pore that extends away from the pore opening into the receptor solution. Figure 20 shows representative examples of experimental HQ concentration profiles, corresponding to diffusive (i_{app} = 0 μA, squares) and iontophoretic transport of HQ (i_{app} = 50 μA, triangles) in solutions containing 0.1, 0.2, and 2.0 M NaCl. The solid lines correspond to theoretical profiles [Eq. (10)] and are based on best fits of the data obtained by varying C_s and a. The excellent fits of Eq. (10) to the data indicate that diffusion is the predominant mode of transport away from the pore openings, that the pore opening can be approximated as being disk shaped, and that the SECM tip does not significantly alter the concentration profiles.

A set of typical results of the curve-fitting analysis (C_s and a) are summarized in Table 1 for diffusive and iontophoretic transport through individual pores. For comparison, values of the pore radii estimated by optical microscopy are also included. The latter values were obtained with significant difficulty because of the low optical contrast between the Nafion-filled pores and surrounding regions of mica. Two conclusions concerning SECM measurements of pore radii can be drawn from the data in Table 1. First, values of a that are determined during diffusive and iontrophoretic transport agree, on average, to within 10%. This consistency demonstrates the reproducibility of the SECM measurement, the soundness of the assumption of pure diffusional transport outside the pore, and the sensitivity of the experimental data to the pore dimension. Second, SECM-determined pore radii are in reasonable agreement (within 20%) with the corresponding values determined by optical microscopy, suggesting that the SECM analysis provides a relatively accurate estimation of the pore dimensions. Because of the difficulty in measuring pore radii by optical microscopy, the SECM-determined values are probably more reliable. A more rigorous test using better defined pores is necessary to fully address the issue of measurement accuracy.

Table 1 also summarizes SECM-measured values of N for diffusion (i_{app} = 0 μA) and iontophoretic transport (i_{app} = 50 μA), using the symbols N_{diff} and N_{iont}, respectively, to indicate these values. Values of N were computed from Eq. (3) using a and C_s obtained by analyses of the HQ concentration profiles. Bath et al. found $N_{diff} \approx 7 \times 10^{-10}$ mol/(cm^2·s) for pure diffusion (i_{app} = 0 μA), increasing by approximately a factor of five under applied iontophoretic conditions, i_{app} = 50 μA (10).

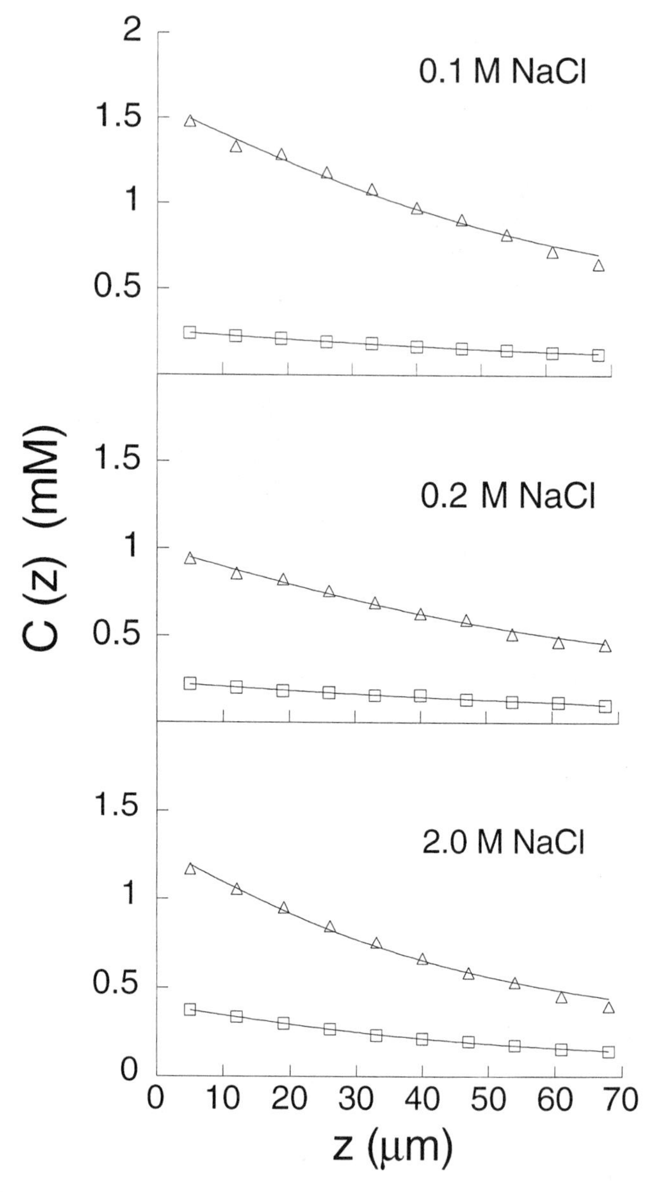

FIG. 20 Concentration profiles of HQ above Nafion-filled pores in mica membranes. The data correspond to passive diffusion ($i_{app} = 0$ μA, squares) and iontophoretic transport ($i_{app} = 50$ μA, triangles). The solid lines represent best fits of Eq. (10) to the data, obtained by varying the parameters C_s and a. The concentrations of NaCl in the donor/receptor compartments are indicated on the figure. The concentration of HQ in the donor compartment is 0.2 M in all experiments. (From Ref. 10.)

TABLE 1 Diffusional and Iontophoretic Transport Parameters for Hydroquinone in Nafion

Pore radius[a] (μm)	Diffusion, i_{app} = 0 μA C_s^R (z = 0) (mM)	a (μm)	$N_{diff} \times 10^{10}$ (mol/cm$^2\cdot$s)	Iontophoresis, i_{app} = 50 μA C_s^R (z = 0) (mM)	a (μm)	$N_{iont} \times 10^{10}$ (mol/cm$^2\cdot$s)	E
58	0.50	58.4	10.0	2.9	56.8	59.8	5.8
53	0.22	51.8	5.0	1.73	42.3	48.0	7.9
59	0.11	60.0	2.1	0.62	55.1	13.2	5.6
62	0.26	67.4	4.4	1.59	56.8	32.7	6.1

[a]Determined by optical microscopy.
Source: Ref. 10.

SECM can also be used to measure electroosmotic transport dynamics (10). For instance, Figure 21 shows the transient response of the SECM-tip current during HQ transport following changes in the applied iontophoretic current. As in the steady-state measurements, the SECM tip is positioned directly above a pore to measure the flux of HQ from the donor compartment

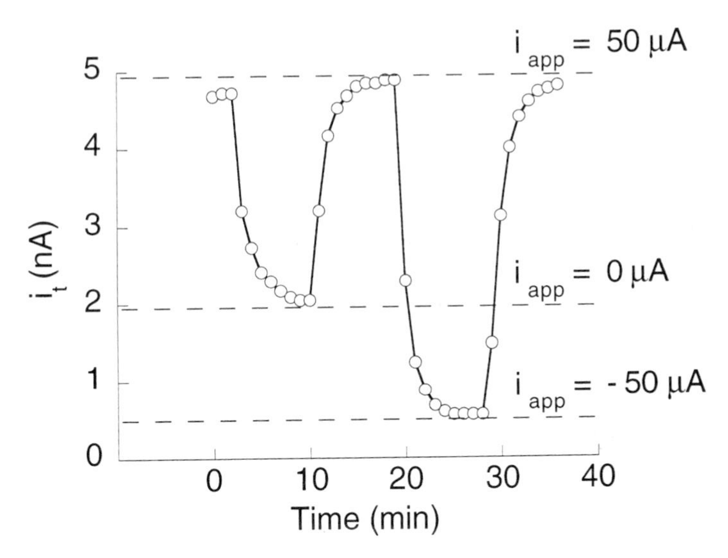

FIG. 21 Transient SECM-tip current (i_t(z = 0)) measured while switching the iontophoretic current, i_{app}, between −50, 0, and 50 μA. E_t = 1.0 V vs. Ag/AgCl. The donor compartment contains 0.1 M NaCl and 0.1 M hydroquinone. The receptor compartment contains 0.1 M NaCl. (From Ref. 10.)

to the receptor compartment. $i_t(z = 0)$ is measured as the iontophoretic current i_{app} is varied between 50, 0, and -50 μA. The time required to obtain a new steady-state i_t is on the order of 10 minutes. Because i_{app} is controlled by the galvanostatic circuitry, the iontophoretic current in the pore and, thus, the steady-state electroosmotic flow are established very rapidly following any change in i_{app}. It follows that the slow transient response of the SECM tip reflects the establishment of new steady-state HQ concentration profiles in the Nafion-filled pore following the change in current. The time necessary to establish the HQ concentration profiles in a pore of length ℓ can be estimated from the root-mean-square diffusional displacement of HQ, which is given by $\langle z^2 \rangle^{1/2} = (2D_p t)^{1/2}$. Replacing $\langle z^2 \rangle^{1/2}$ with ℓ, and using $\ell \sim 200$ μm and the reported diffusivity of hydroquinone in Nafion, $D_p = 3.7 \times 10^{-7}$ cm^2/s, yields a value of $t \sim 9.0$ minutes, in good agreement with Figure 21.

B. Biomaterials

1. *Diffusion and Convection in Dentine and Laryngeal Cartilage*

Recent studies have employed SECM to investigate molecular transport in human dentine and laryngeal cartilage. Convective and diffusive transport in these porous biomaterials is of interest in the biomedical and biophysical communities. To study these materials, thin slices (typically ~100 μm) of dentine and cartilage are prepared and mounted in a diffusion cell suitable for SECM imaging. Thus, although dentine and cartilage are bulk biomaterials (rather than membranes), the general experimental procedures are basically the same as that described in the previous sections.

Dentine. Dentine is a calcareous material located between the pulp and enamel in the tooth. Convective fluid flow within the 1–2 μm diameter tubules of this material is associated with dentinal hypersensitivity and the treatment of this condition. Unwin and coworkers have investigated convective flow in individual tubules in SECM experiments in which a fluid is forced through the tubules by pressurized flow (11,12). Images are obtained in negative feedback mode employing a solution of 10 mM $K_4Fe(CN)_6$ and 0.5 M KCl on both sides of the membrane. A hydrostatic pressure gradient across the slice of dentine is introduced by raising the level of solution in the donor compartment relative to the solution in the receptor compartment. Transport of $Fe(CN)_6^{-4}$ results from the pressurized fluid flow; the flux of $Fe(CN)_6^{-4}$ is then used to determine the fluid velocity in the tubule.

Figure 22 shows 30 μm^2 SECM images of dentine recorded using a 1-μm-radius tip. SECM images were obtained (1) with a pressure of 2 kPa across the dentine sample and (2) in the absence of a pressure difference. The peak in the SECM image in Figure 22a is interpreted as being due to

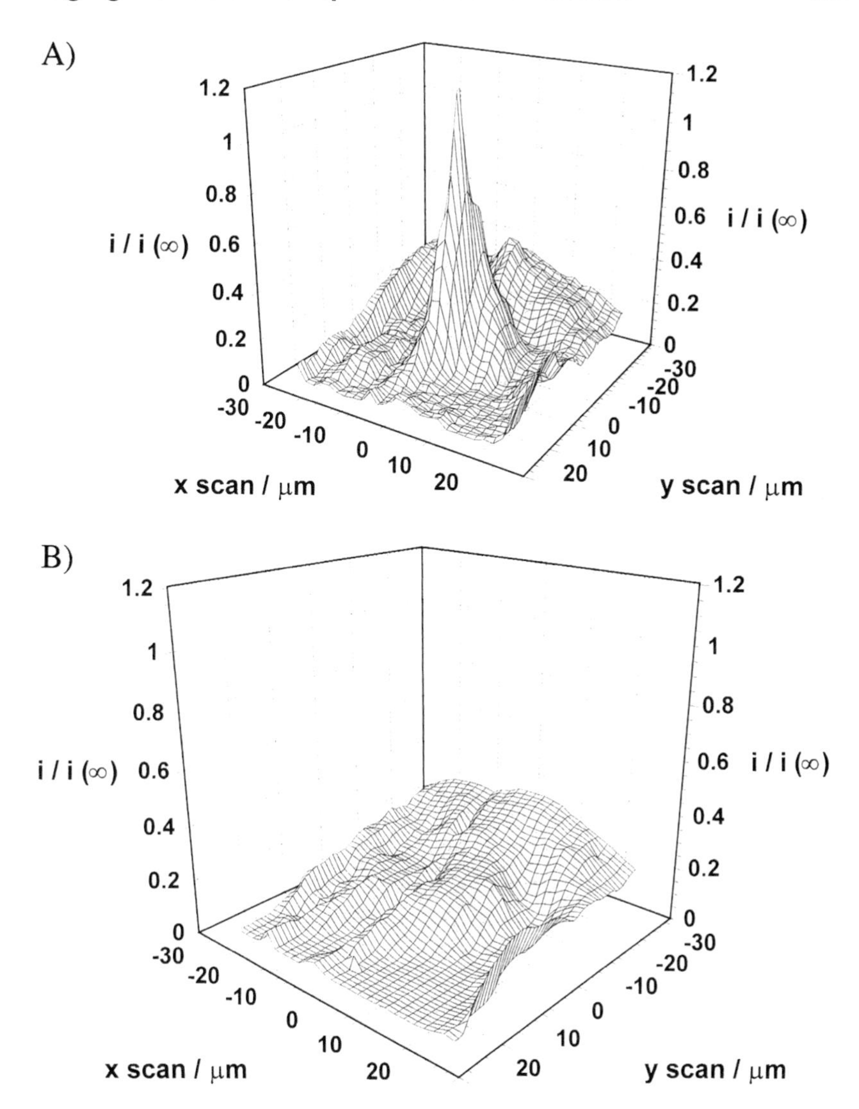

FIG. 22 SECM images of a dentine surface. The images correspond to the transport of $Fe(CN)_6^{-4}$ through a single pore in dentine and the topography of the dentine surface. The data are obtained (A) with a pressure head of 20 cm aqueous solution across the dentine slice and (B) with no applied pressure. The pressure gradient in (A) results in convective transport of $Fe(CN)_6^{-4}$ through the pore. The donor and receptor solutions each contains 10 mM $K_4Fe(CN)_6$ and 0.5 M KCl. (From Ref. 11.)

the convective transport of $Fe(CN)_6^{-4}$ through a *single* pore. The SECM tip current, i_t, corresponding to flow from a single tubule is given by (11)

$$i_t/i_{t\infty} = 0.94U^{1/2}aD^{-1/3}\upsilon^{-1/6}b^{-1/2} \times [2.464 - 1.063(d/b) + 1.5794(d/b)^2] \quad (13)$$

where $i_{t\infty}$ is the current produced at the tip when the tip is far from the sample surface, υ is the kinematic viscosity of the solution, b is an empirical constant (~2 μm), d is the tip to sample separation, D is the diffusion coefficient, a is the radius of the pore, and U is the average velocity of solution through the pore. Using Eq. (13) and the data from Figure 22a, the velocity of solution through the single pore in dentine is computed to be 0.39 cm/s. The SECM data demonstrate that localized flow rates in single tubules are significantly different than the mean flow rate obtained from the flow measurements on a bulk sample.

Using a similar experimental procedure, Unwin and coworkers investigated the effect of chemical blocking agents on the molecular transport through the tubules in dentine (12). In these experiments, the ability to resolve individual pores was not a concern and a 25-μm-radius tip was used to image relatively large areas (400 μm^2) of the surface. Figure 23a shows the normalized tip current, $i_t/i_{t\infty}$, over the dentine surface without an applied pressure. The variation in the SECM tip current is simply a result of hindered diffusion because of the topographical structure of the dentine surface (28). Figure 23b shows the same area when 2 kPa of pressure is applied across the sample. Localized regions of increased $Fe(CN)_6^{-4}$ flux are observed because of convective solution flow through pores from donor to receptor compartments.

SECM images of the same area were then recorded after removing the pressure gradient and introducing a chemical reagent (calcium oxalate) into the receptor solution that deposits onto the surface and blocks the pores. Figure 24a shows an image of the dentine surface after treatment with the blocking agent. Comparison of this SECM image with the image in Figure 23a suggests that the flux of $Fe(CN)_6^{-4}$ to the tip has been decreased across the entire sample, i.e., the calcium oxalate has formed a thin film on the dentine sample that effectively reduces the tip to sample separation and thus hinders $Fe(CN)_6^{-4}$ flux to the tip. In Figure 24b, the 2 kPa of pressure is reapplied across the sample and the SECM image is recorded. This SECM image is essentially identical to that obtained in the absence of applied pressure (Fig. 24a), demonstrating that the deposited calcium oxalate layer inhibits convective flow in the dentine sample.

Nugues and Denuault have also used SECM to study *diffusive* transport in the pores of dentine (9). In these experiments, the concentration of $Fe(CN)_6^{-4}$ was 1 mM in the receptor compartment and 100 mM in the donor

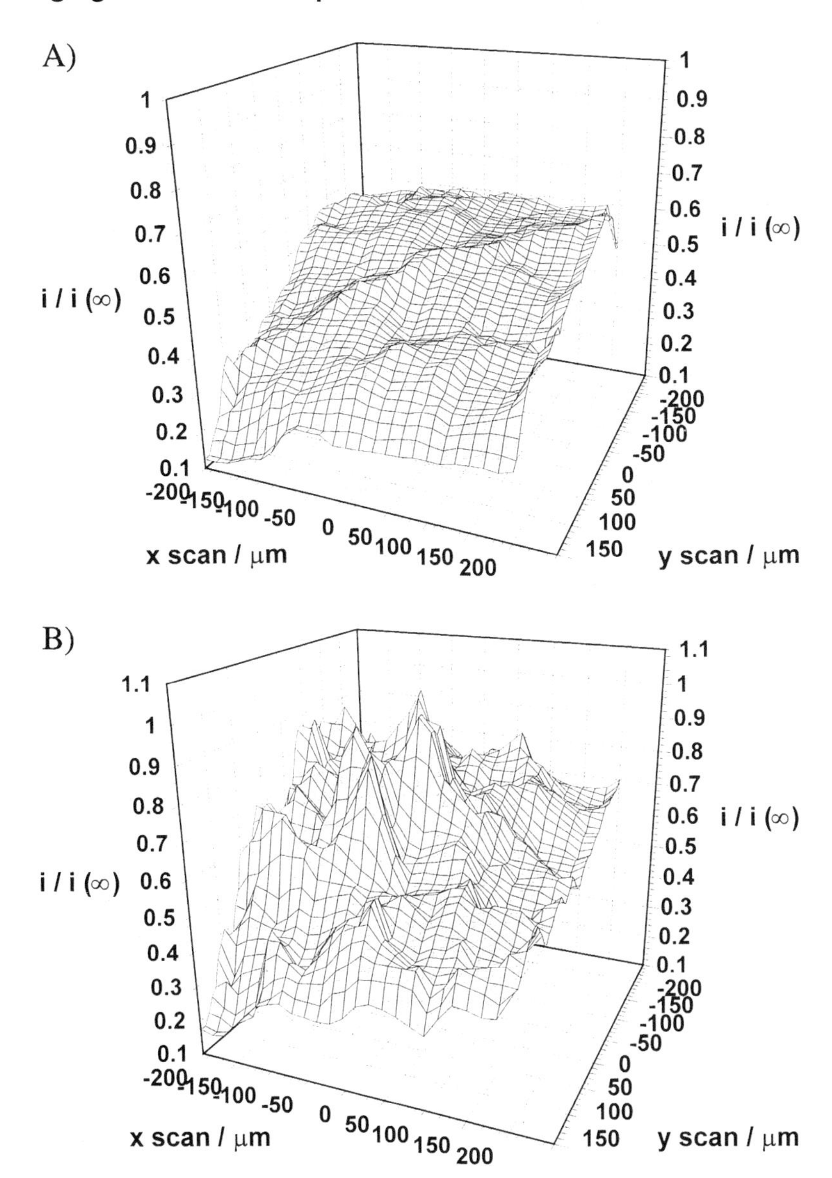

FIG. 23 SECM images recorded over a slice of dentine, resulting from oxidation of $Fe(CN)_6^{-4}$. Images were recorded (A) with no pressure applied across the dentine slice and (B) with a fluid pressure of 2 kPa across the slice. The donor and receptor solutions each contains 10 mM $K_4Fe(CN)_6$ and 0.5 M KCl. (From Ref. 12.)

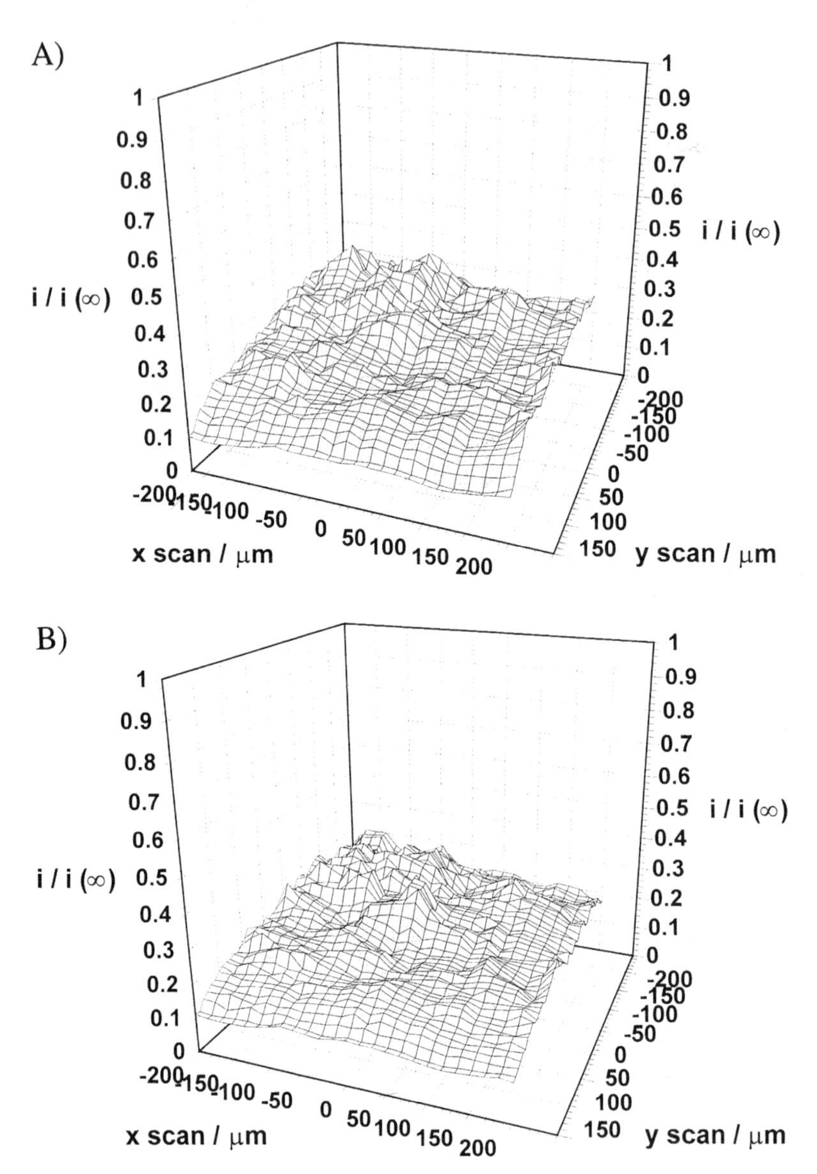

FIG. 24 SECM images recorded over a slice of dentine after treatment of the surface with calcium oxalate. Images were recorded (A) with no pressure applied across the dentine slice and (B) with a fluid pressure of 2 kPa across the slice. The donor and receptor solutions each contains 10 mM $K_4Fe(CN)_6$ and 0.5 M KCl. (From Ref. 12.)

compartment, corresponding to the combined tip collection/negative feedback imaging mode. With these solution conditions, the SECM image represents a convolution of the surface topography with the high localized diffusive $Fe(CN)_6^{-4}$ flux associated with pores in the dentine. Figure 25a shows a scan over a 30 μm^2 region of a dentine sample. The light regions are associated with localized high concentrations of $Fe(CN)_6^{-4}$ resulting from diffusive flux of $Fe(CN)_6^{-4}$ through individual pores. Conversely, the darker regions correspond to regions of low tip current. These regions correspond to pores that are inactive, possibly because of occlusions inside the tubules. Nugues and Denuault used numerical simulation to study the effect of a blocked pore on the diffusion of $Fe(CN)_6^{-4}$ through neighboring pores. The simulation of the concentration profile above a blocked pore (Fig. 25b) suggests that $Fe(CN)_6^{-4}$ that diffuses across the membrane through a neighboring unblocked tubule can be transported into the blocked pore region, i.e., the blocked pore acts as a sink of $Fe(CN)_6^{-4}$. A consequence of this effect is that the tip current is greatly reduced in the surrounding regions of the SECM image where a blocked pore exists. An example of this occurs in the right side of the image shown in Figure 25a.

Laryngeal Cartilage. Macpherson et al. (13) have used SECM to study molecular transport in laryngeal cartilage. Cartilage acts as a load-bearing material in the human body. Understanding transport mechanisms in this material is of interest for several reasons. First, the rate at which water is redistributed under applied loads determines the viscoelastic properties of the material. Second, the physiological functioning of the material is determined by the diffusive-convective transport of nutrients and metabolites (13).

Figure 26a shows a large-area SECM image of a thin slice of cartilage obtained with the sample mounted in a diffusion cell. The donor and receptor compartments contain identical solutions of the redox mediator $Ru(NH_3)_6^{3+}$. The variation in tip current reflects the hindered diffusion of $Ru(NH_3)_6^{3+}$. Thus, the image corresponds to the surface topography of the cartilage sample (negative feedback mode). This tip current image is transformed into a topographical image (Fig. 26b) using the empirical expression:

$$d/a = 0.984[i_t]^2 + 0.7876[i_t] + 0.0532 \qquad (14)$$

where d is the tip-to-sample separation and a is the radius of a pore. Equation 14 is an extension of the idea that the current at the SECM tip in *negative feedback mode* is purely a function of the tip-to-sample separation (28). The topographical SECM image reveals a highly heterogeneous cartilage surface.

After the image in Figure 26a was obtained, an osmotic pressure difference equivalent to 0.75 atm was applied across the sample by dissolving polyethylene glycol in the receptor compartment. This pressure difference

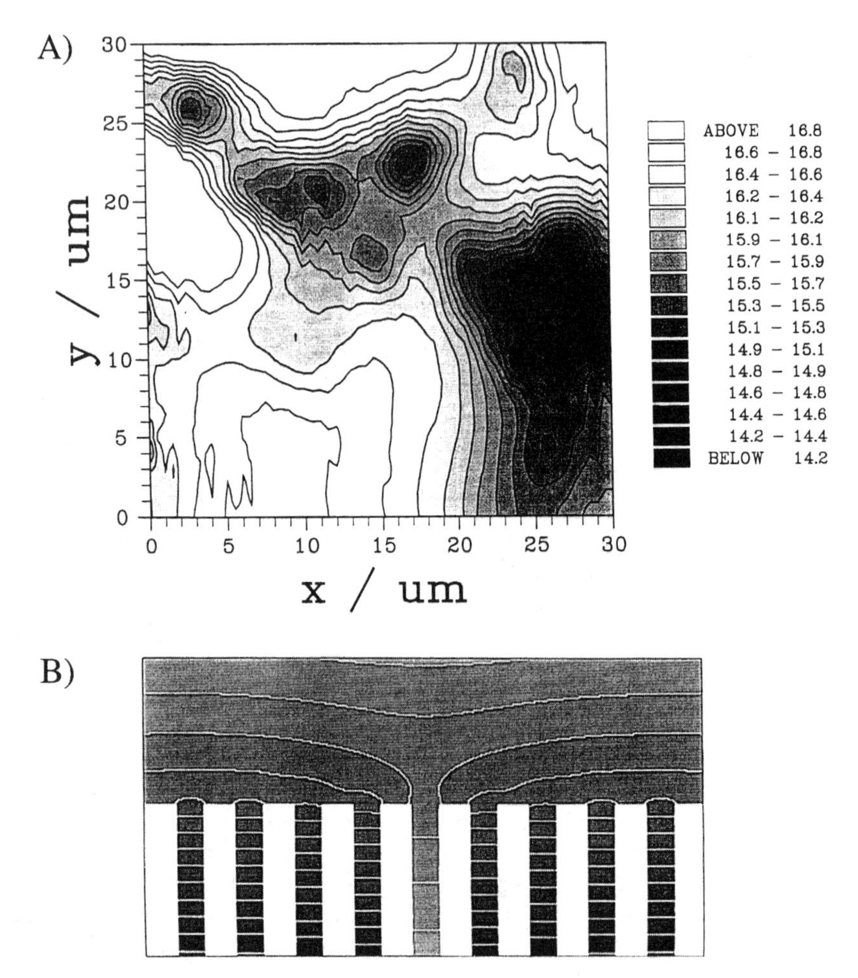

FIG. 25 (A) SECM image of a dentine surface. The image corresponds to the diffusion of $Fe(CN)_6^{-4}$ through the pores in dentine. The dark regions correspond to low normalized tip current, indicating blocked pores, whereas the bright regions correspond to high normalized tip current, indicating open pores. Donor solution: 100 mM $K_4Fe(CN)_6$. Receptor solution: 1 mM $K_4Fe(CN)_6$. (B) Simulation of the isoconcentration profiles for a membrane containing one blocked pore. The simulation shows the sink effect of the blocked pore. (From Ref. 9.)

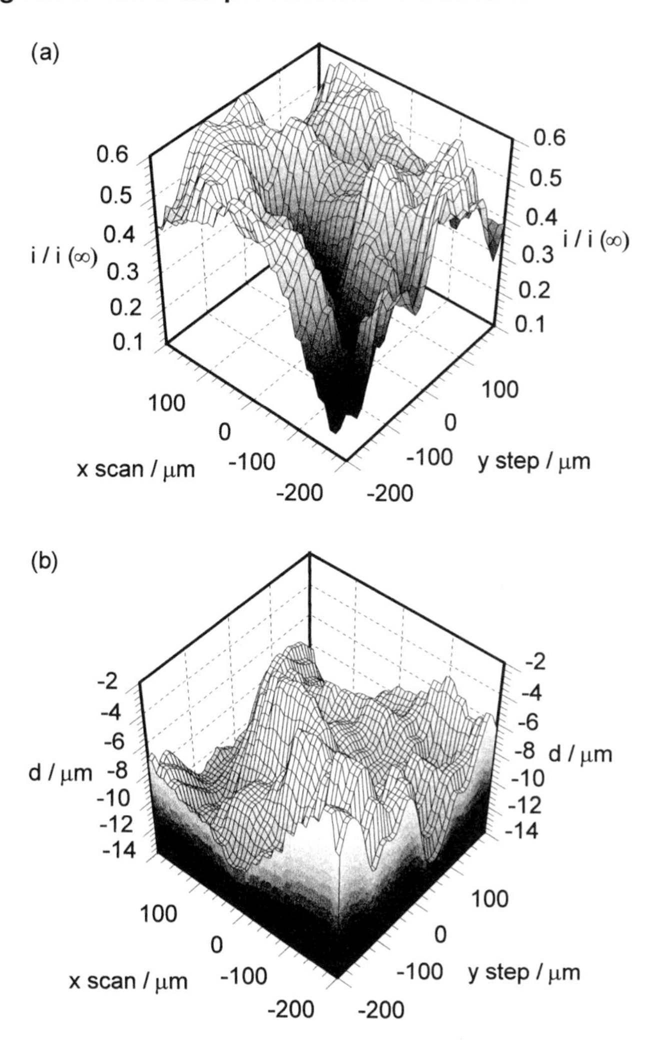

FIG. 26 (a) SECM image of a sample of cartilage. The image corresponds to the reduction of $Ru(NH_3)_6^{+3}$ at the tip above the cartilage surface. The donor and receptor solutions each contain 10 mM $Ru(NH_3)_6Cl_3$ and 0.2 M KNO_3. (b) Corresponding topographical map of the surface after transforming the data using Eq. (14) in the text. (From Ref. 13.)

induced fluid flow across the sample, and SECM images thus obtained reflect both surface topography and the flux of $Ru(NH_3)_6^{3+}$ across the sample. Macpherson et al. (13) demonstrate that this image can be corrected for the topography of the cartilage sample by simply subtracting the current response in the absence of osmotic pressure. The resulting image, plotted as $\Delta i_t/i_{t\infty}$, where Δi_t is the difference in currents in the presence and absence of the pressure gradient, corresponds to the flux of $Ru(NH_3)_6^{3+}$ across the sample.

Figure 27a shows the SECM image obtained at 0.75 atm after subtracting the image at 0 atm. This image is a visualization of the transport of $Ru(NH_3)_6^{3+}$ away from the cartilage surface due to convective fluid flow. The

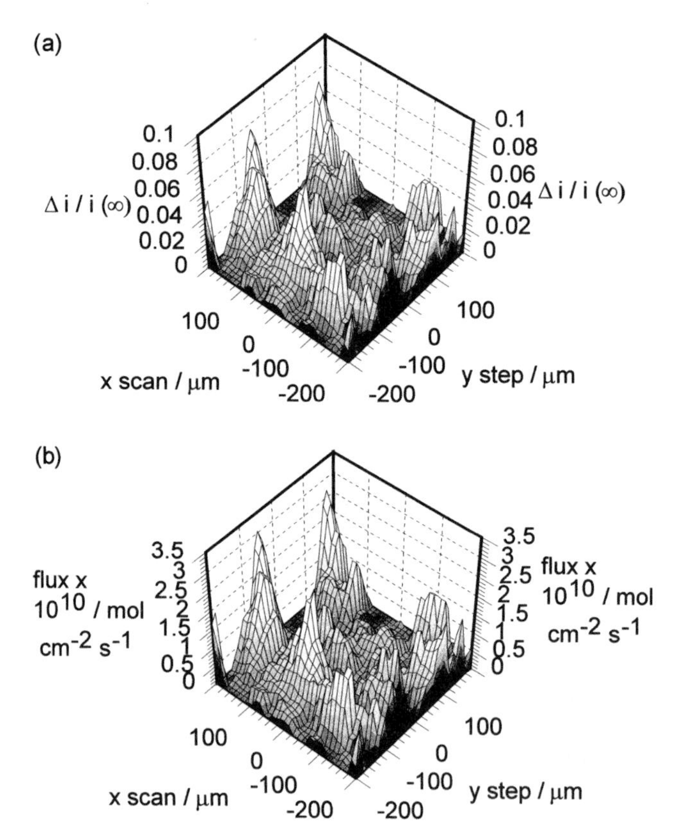

FIG. 27 (a) SECM image of a sample of cartilage. The flux is due solely to an applied osmotic pressure across the sample (P = 0.75 atm). The donor and receptor solutions each contain 10 mM $Ru(NH_3)_6Cl_3$ and 0.2 M KNO_3. (b) Corresponding interfacial flux image. (From Ref. 13.)

flux of $Ru(NH_3)_6^{3+}$ at the cartilage/aqueous interface (N) is thus described by (13)

$$N = -\nu_z K_p C^* \tag{15}$$

where ν_z is the fluid velocity in the cartilage, K_p is the partition coefficient for $Ru(NH_3)_6^{3+}$ between the aqueous and cartilage phases, and C^* is the concentration of $Ru(NH_3)_6^{3+}$ in the donor/receptor solutions. Experimental values of $\Delta i_t/i_{t\infty}$ are converted to the flux, N, using a model developed by the same authors. A plot of N at the cartilage sample is shown in Figure 27b. When the flux image (Fig. 27b) is compared with the topographical image (Fig. 26b), the active transport areas appear to be correlated with recessed areas in the cartilage sample surface. The results suggest that fluid flow follows paths that exist around the cells at the surface of the cartilage sample.

2. *Iontophoresis Across Skin*

The ability to deliver drugs through the skin at well-defined controlled rates has been a goal in the pharmaceutical community for many years. A promising technique for reaching this goal is iontophoresis. In iontophoresis, a potential difference is applied between two electrodes that are placed in contact with the skin. The drug molecule, which can be either an ion or a neutral molecule, is contained in a solution beneath one of the electrodes. An electrical current is driven across the skin tissue, resulting in the transdermal transport of the drug. Ionic molecules are transported across the skin by diffusive-migrational transport. The transport of neutral molecules can also be enhanced by the electrical current. The reason for this is that skin tissue has a net negative charge at physiological pH (36) and behaves as an ion-selective membrane. As with synthetic ion-selective membranes, e.g., Nafion, electroosmotic flow occurs in skin when a current is passed through it. However, skin is a very complex membrane, and molecular transport across it is localized to a small number of active sites. This section briefly describes the use of SECM to determine the pathways and mechanisms of molecular transport across skin tissues during iontophoresis. The SECM images and data reported here were obtained in the *tip collection mode*.

The experimental setup is essentially identical to that described in Sec. III.A and Figure 13. The synthetic membrane in the diffusion cell is replaced with a section of excised skin tissue. The donor compartment contains a redox-active species that is transported across the tissue due to the concentration gradient or an applied current. The two large Ag/AgCl electrodes and a galvanostat are used to drive the iontophoretic current across the skin.

Investigations of molecular transport across skin using SECM were first performed by Scott et al. to identify regions of localized transport (7). Figure

2 shows a SECM image of a 1 mm^2 region of hairless mouse skin (HMS). For this experiment, 0.1 M $Fe(CN)_6^{-4}$ is placed in the donor compartment, and an applied iontophoretic current of 40 $\mu A/cm^2$ is driven across the HMS sample. Both donor and receptor compartments contain 0.1 M NaCl as supporting electrolyte. The tip current corresponds to the oxidation of $Fe(CN)_6^{-4}$ above the skin surface in the receptor compartment. The bright region in Figure 2 corresponds to a large SECM tip current and thus to a region where the molecular flux of $Fe(CN)_6^{-4}$ is largest. It is clear from this and other images that iontophoretic transport of $Fe(CN)_6^{-4}$ is highly localized, occurring through discrete pores in skin. These pores have been shown to correspond to individual hair follicles, as determined by optical microscopy (37). The corresponding isoconcentration contours for $Fe(CN)_6^{-4}$ above this pore are shown in Figure 28. Using this data set and the method of analysis discussed in Sec. II.A, a pore radius of 13 μm and a mass transport rate for $Fe(CN)_6^{-4}$ of 1.6×10^{-13} mol/s were determined for this pore (7).

Scott et al. also performed experiments examining *competitive* molecular transport through a hair follicle in HMS (14). In these experiments, both Fe^{2+} and Fe^{3+} are present in the donor compartment and are simultaneously

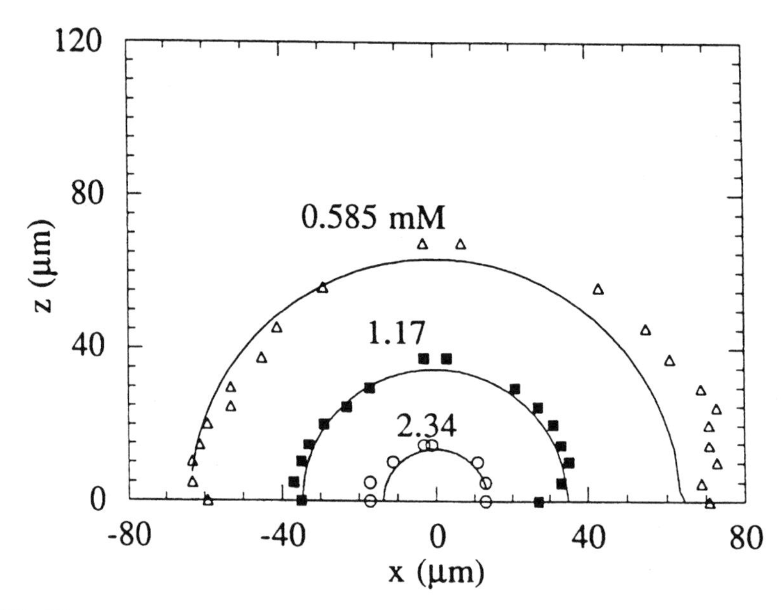

FIG. 28 SECM isoconcentration lines above a pore in skin. The concentration of $Fe(CN)_6^{-4}$ corresponding to each isoconcentration line is given on the figure. The least-squares fits to the data are shown as solid lines. i_{app} = 40 $\mu A/cm^2$. (From Ref. 16.)

transported across the skin sample. Figure 29a shows a series of one-dimensional SECM line scans (length = 500 μm) across the center of a hair follicle. The line scans are offset on the x-axis by the potential at which the SECM tip is held while recording the scan. Fe^{3+} is reduced at the tip at the diffusion-controlled rate when the SECM tip is biased to a value negative of ~0.5 V vs. Ag/AgCl. Thus, the peak heights in this potential region are directly proportional to the flux of Fe^{3+} through the hair follicle. Increasing

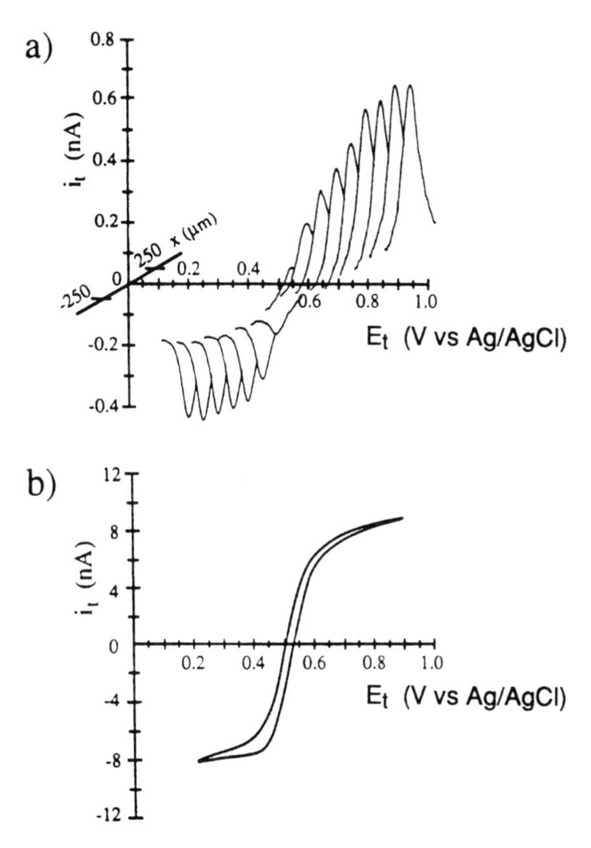

FIG. 29 (a) SECM line profiles above a pore in skin, measured as a function of the tip potential, E_t. Each of the 16 Gaussian-shape segments is a plot of i_t as a function of the tip position (the x-coordinate). The donor solution contains 0.05 M $FeCl_3$ and 0.05 M $FeCl_2$. The anodic currents at $E_t > 0.5$ V vs Ag/AgCl correspond to detection of Fe^{2+} as it emerges from the pore. The cathodic currents at $E_t < 0.5$ V correspond to detection of Fe^{3+}. The envelope of curves in (a) resembles the conventional voltammetric response (b) of a microelectrode immersed in a bulk solution of 0.01 M $FeCl_3$ and 0.01 M $FeCl_2$. (From Ref. 14.)

the tip potential to values positive of 0.5 V results in diffusion-controlled oxidation of Fe^{2+}. The peak heights in this potential region are proportional to the flux of Fe^{2+} in the hair follicle. The magnitudes of the peak heights for Fe^{3+} and Fe^{2+} suggest that the rate of transport of Fe^{2+} through the hair follicle is about 50% greater than that of Fe^{3+}. For comparison, Figure 29b shows the voltammetric response of the SECM tip immersed in a solution containing 0.01 M $FeCl_3$, 0.01 M $FeCl_2$, and 0.08 M NaCl, demonstrating that the rate of transport for these two species is the same in a homogeneous solution. Thus, the difference in the transport rates of these ions in skin suggest that Fe^{2+} and Fe^{3+} interact to different degrees with the pore structures comprising skin. In particular, the SECM data suggest that the ionic mobility of Fe^{3+} in skin may be significantly smaller than that of Fe^{2+}.

Recently, Bath et al. have investigated molecular transport through hair follicles to define the mechanisms by which the molecules traverse skin (38). Measurements across bulk skin tissues have shown a preferential transport of cations across the skin. These studies suggest that skin is net negatively charged and should exhibit transport behavior analogous to that of Nafion membranes (Sec. III.A.2). The SECM images in Figure 30 clearly illustrate the phenomenon of electroosmotic flow during iontophoretic transport of hydroquinone, HQ, through an individual hair follicle. The images are obtained by rastering the SECM tip over a 300 μm^2 area surrounding the hair follicle (holding the tip potential, E_t, at +0.9 V vs. Ag/AgCl to ensure diffusion controlled oxidation of HQ at the tip). The middle image corresponds to the diffusion of HQ across the skin, a consequence of the concentration gradient between the donor and receptor compartments. During cathodic iontophoresis ($i_{app} = -50\ \mu A$), preferential cation transport occurs from the receptor to the donor compartment, inducing a net solvent flow in the same direction. Electroosmotic flow inhibits the diffusion of the neutral HQ species through the hair follicle and reduces its molecular transport to nearly background levels, as indicated by the small currents produced at the SECM tip recorded under these conditions (top image). During anodic iontophoresis ($i_{app} = 50\ \mu A$), preferential cation transport occurs from the donor to the receptor compartment, and electroosmotic flow is in the same direction as diffusion. The transport of HQ is significantly increased in this case, as indicated by the much larger currents measured at the SECM tip (lower image).

IV. FUTURE DIRECTIONS

SECM-based methods provide unique experimental capabilities for imaging and quantifying molecular transport in porous membranes. No other experimental method or technique currently exists that provides a means to visu-

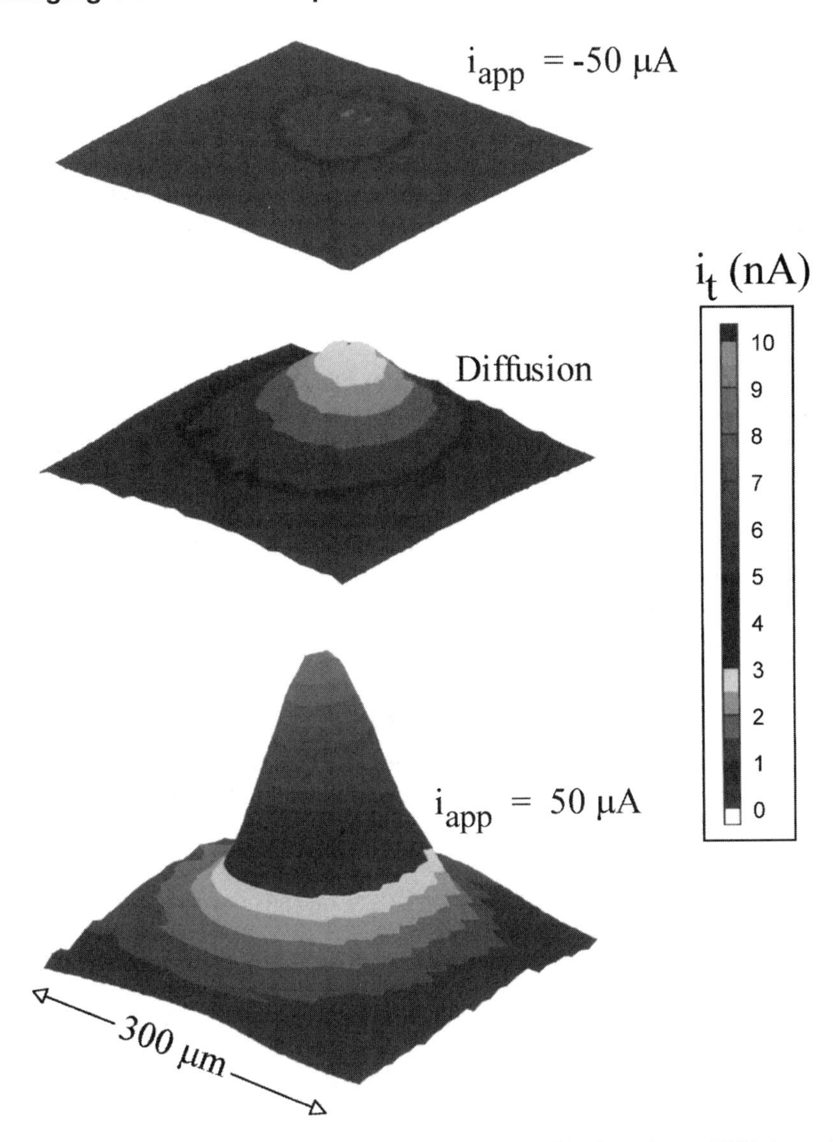

FIG. 30 SECM images of electroosmotic transport of hydroquinone (HQ) through a single hair follicle in hairless mouse skin. The middle image corresponds to the diffusional flux of HQ in the absence of an applied current ($i_{app} = 0$). The top image ($i_{app} = -50\ \mu A$) corresponds to electroosmotic flow from the receptor solution to the donor solution, opposing the diffusional flux of HQ. The bottom image ($i_{app} = 50\ \mu A$) corresponds to electroosmotic flow from the donor solution to the receptor solution, enhancing the flux of HQ in the hair follicle. (From Ref. 38.)

alize localized molecular flux across a membrane or the ability to quantify transport rates at the single pore level. The transport of molecules from individual pores as small as tens of nanometers has been imaged. Further improvements in resolution are anticipated with the development of procedures that allow the routine fabrication of tips of submicrometer dimensions. The detection of very small numbers of molecules, perhaps even single molecules, as they are transported across membranes may eventually be achieved by combining the concepts outlined in this chapter with scanned probe methods that employ fluorescence-based detection instead of amperometric detection.

The future of applications of SECM in investigations of transmembrane transport appears very promising. The first images of molecular transport across biological materials are just appearing, and the results are particular encouraging. With the development of higher resolution microscopes, it is possible to imagine the imaging of ion fluxes across ion channels in cell membranes. The real-time monitoring capabilities of SECM also provide an unprecedented ability to monitor membrane transport dynamics, an area important in designing membrane-based devices for energy conversion (e.g., fuel cells) as well as in biological systems.

ACKNOWLEDGMENTS

We gratefully acknowledge the contributions and insights of Dr. J. Bradley Phipps (ALZA Corp) to the development of the SECM methodology for imaging transport across porous membranes. Research on SECM imaging of porous membranes in the authors' laboratory has been supported by ALZA Corp and the Office of Naval Research.

REFERENCES

1. EL Cussler. Diffusion: Mass Transfer in Fluid Systems. New York: Cambridge University Press (1997).
2. RC Engstrom, M Weber, DJ Wunder, R Burgess, S Wenquist. Anal Chem 58: 844, 1986.
3. RC Engstrom, T Meaney, R Rople, RM Wightman. Anal Chem 59:2005, 1989.
4. AJ Bard, F-RF Fan, J Kwak, O Lev. Anal Chem 61:132, 1989.
5. AJ Bard, F-RF Fan, VV Mirkin. In: AJ Bard, ed. Electroanalytical Chemistry, Vol. 18. New York: Marcel Dekker, 1994, p 243.
6. FR Fan, AJ Bard. Science 267:871, 1995.
7. ER Scott, HS White, JB Phipps. Anal Chem 65:1537, 1993.
8. ER Scott, HS White, JB Phipps. J Membrane Sci 58:71, 1991.
9. S Nugues, G Denuault. J Electroanal Chem 408:125, 1996.
10. BD Bath, RD Lee, HS White, ER Scott. Anal Chem 70:1047, 1998.

11. JV Macpherson, MA Beeston, PR Unwin, NP Hughes, D Littlewood. J Chem Soc Faraday Trans 91:1407 (1995).
12. JV Macpherson, MA Beeston, PR Unwin, NP Hughes, D Littlewood. Langmuir 11:3959, 1995.
13. JV Macpherson, D O'Hare, PR Unwin, CP Winlove. Biophys J 73:2771, 1997.
14. ER Scott, HS White, JB Phipps. Solid State Ionics 53–56:176, 1992.
15. ER Scott, AI Laplaza, HS White, JB Phipps. Pharm Res 10:1699, 1993.
16. ER Scott, JB Phipps, HS White. J Invest Derm 104:142, 1995.
17. RR Burnett, B Ongipipattanakul. J Pharm Sci 77:132, 1988.
18. PK Hansma, B Drake, O Marti, SAC Gould, CB Prater. Science 243:641, 1989.
19. BR Breslau, IF Miller. Ind Eng Chem Fundam 10:555, 1971.
20. RF Probstein. Physiochemical Hydrodynamics. New York: Butterworth, 1989.
21. SM Sims, WI Higuchi, K Peck. J Colloid Interface Sci 155:210, 1993.
22. MJ Pikal, Adv Drug Deliv Rev 9:201, 1992.
23. V Srinivansan, WI Higuchi. J Pharm Sci 6:133, 1990.
24. J Crank. The Mathematics of Diffusion. New York: Oxford University Press Inc., 1956.
25. AJ Bard, LR Faulkner. Electrochemical Methods. New York: John Wiley and Sons, 1980.
26. Y Saito. Rev Polarogr 15:177, 1968.
27. BD Bath. University of Utah, unpublished results.
28. J Kwak, AJ Bard. Anal Chem 61:1221, 1989.
29. TE Springer, TA Zawodzinski, S Gottesfeld. J Electrochem Soc 138:2334, 1991.
30. X Ren, W Henderson, S Gottesfeld. J Electrochem Soc 144:L267, 1997.
31. PN Pintauro, D Bennion. Ind Eng Chem Fundam 23:234, 1984.
32. MW Verbrugge. J Electrochem Soc 136:417, 1989.
33. MW Verbrugge. J Electrochem Soc 137:886, 1990.
34. N Lakshminarayaniah. Chem Rev (Washington, D.C.) 65:492, 1965.
35. BD Bath, HS White, ER Scott. Anal Chem 72:433, 2000.
36. RR Burnette, D Marrero. J Pharm Sci 75:738, 1986.
37. RD Lee, HS White, ER Scott. J Pharm Sci 85:1186, 1996.
38. BD Bath, HS White, ER Scott. Pharm Res. 17:471, 2000.

10

POTENTIOMETRIC PROBES

Guy Denuault

University of Southampton, Southampton, England

Géza Nagy

Janus Pannonius University, Pécs, Hungary

Klára Tóth

Technical University of Budapest, Budapest, Hungary

I. INTRODUCTION

Most scanning electrochemical microscopy (SECM) experiments are conducted in the amperometric mode, yet microelectrodes have for many years been used as potentiometric devices. Not surprisingly, several SECM articles have described how the tip operated in the potentiometric mode. In this chapter we aim to present the background necessary to understand the differences between amperometric and potentiometric SECM applications. Since many aspects of SECM are covered elsewhere in this monograph, we have focused on the progress made in the field of potentiometric microelectrodes and presented it in the context of SECM experiments. Starting with an historical perspective, the key discoveries that facilitated the development and applications of micro potentiometric probes are highlighted. Fabrication techniques and recipes are reviewed. Basic theoretical principles are covered as well as properties and technical operational details. In the second half of the chapter, SECM potentiometric applications are discussed. There the differences between the conventional amperometric mode are developed and emphasized.

Potentiometric probes are the oldest forms of electrochemical sensors. They can conveniently be used for studying many interesting chemical systems not accessible to voltammetric techniques. In particular, alkali and alkaline earth metal ion concentrations, of importance in biological systems, can be monitored with potentiometric ion-selective electrodes but not with voltammetric techniques because these species are not electroactive in aqueous media. This is the main advantage of ion-selective tips beside the high selectivity of their potentiometric signal.

Ion-selective electrodes (ISE) are membrane-based devices with internal filling solution and internal reference electrode or with internal solid contact. Ion-selective electrodes may be classified according to the nature of the ion-selective membrane or the shape and size of the electrode arrangement. The main types of membrane electrodes are:

Glass membrane electrodes, e.g., H^+, Na^+, K^+, Ag^+ selective glass electrodes.

Crystalline (or solid-state) membrane electrodes, e.g., F^-, CN^-, Cl^-, Pb^{2+}, Cu^{2+}, Cd^{2+} selective electrodes.

Liquid membrane electrodes: (1) classical ion-exchangers (with mobile positively and negatively charged sites as hydrophobic cations or hydrophobic anions), for example K^+, Cl^- selective electrodes, (2) liquid ion-exchanger based electrodes (with positively or negatively charged carriers, ionophores), for example Ca^{2+}, NO_3^- selective electrodes, (3) neutral ionophore based liquid membrane electrodes (with electrically neutral carriers, ionophores), for example Na^+, K^+, NH_4^+, Ca^{2+}, Cl^- selective electrodes.

A. A Brief History of Ion-Selective Electrodes

Since the historic discovery of Cremer (1), selective potentiometric probes (transducers) offer a good example of the ideal analytical tool that gives quantitative chemical information about the composition of complex sample solutions. The pH-sensitive glass membrane electrode was introduced soon after and contributed considerably to the development and to the acceptance of instrumental methods of chemical analysis. The investigation (2) of the reproducible alkaline error of glass electrodes led to the development of glass electrodes sensitive to sodium and potassium ions. Eisenman et al. (3) reported first the preparation of a well-behaved sodium ion–selective glass electrode. Glass membrane–based electrodes are still popular sodium ion potentiometric sensors. A long time passed between the early glass electrodes and the appearance of useful new ion-selective electrodes.

In 1936 Tendeloo (4) investigated the dependence of the potential response of fluorite, CaF_2 membranes, on calcium ion. A linear relationship was found between the negative logarithm of the calcium ion activity and the emf of the cell. However, the results could not be easily reproduced (5).

Kolthoff and Sanders (6) almost made the first anion-selective electrode with their study of silver halide disks. Close to Nernstian response was measured for chloride, bromide, and iodide using the corresponding silver halide membranes. They even noticed that the presence of chloride or bromide ions or of a strong oxidizing agent ($KMnO_4$) does not influence the iodide response of the silver iodide disk.

In the 1950s several research groups studied different membrane electrodes prepared from ion exchanger materials (7–10). Their potentiometric response was not found to be selective.

Pungor was luckier in this respect, since he succeeded in preparing the first reliable ion-selective electrode. With his coworkers he embedded different sparingly soluble inorganic precipitates in hydrophobic membrane materials and managed to develop selective membrane electrodes with near Nernstian response to different ions. The first report (11) of a paraffin membrane loaded with silver iodide appeared in 1961. Later, Pungor-type heterogeneous membrane electrodes were made with cold vulcanized silicon rubber matrix material (12). Pungor realized the practical importance of his discovery and initiated the fabrication and commercialization of heterogeneous ion-selective electrodes on an industrial scale. Frant and Ross, the founders of the Orion Company, followed a similar path with their lanthanum fluoride crystal–based fluoride-selective and liquid ion-exchanger–based calcium-selective electrodes (13,14).

Ruzicka and Tjell proposed a universal, porous graphite rod–based ion-selective electrode (15). In the early 1970s Freiser developed a simple way of preparing ion-selective electrodes by dip-coating the stripped end of a wire with a thin ion-selective membrane. Using Aliquat 336S (methyltricaprylammonium chloride; General Mill) ion exchanger as active membrane material dissolved in 1-decanol, a high number of different anion-sensitive so-called coated wire electrodes were prepared and described (16). This kind of electrode became quite popular despite the heavily disputed stability of the measuring membrane/metal interfacial potential.

Simon's group, ETH Zurich, showed first that neutral organic molecules such as the antibiotics nonactin and valinomycin could be used effectively as selective complex forming ligands, ionophores, in ion-selective membrane electrode preparations (17–19). This was an important step forward since, in general, classical ion-exchanger–based electrodes exhibit moderate selectivity. This is due to a relatively poor selectivity of the electrostatic interactions at the heart of the ion-exchange process. The neutral ionophore is a selective complexing agent that excludes every ionic species except the one fitting its host-guest chemistry. In the early days the ionophores were dissolved in diphenyl ether and used mostly in liquid ion-selective films. Later they were incorporated in plasticized PVC membranes.

Shatkay introduced PVC as a membrane matrix into ion-selective electrode research (20). But it was Thomas's brilliant description of an amazingly simple preparation procedure that helped researchers to make thin plasticized PVC membranes (21). For more than two decades electrically charged and neutral ligands (ionophores) containing membranes have been most often made following that recipe.

The first neutral ionophore molecules, valinomycin and nonactin, were borrowed from nature, but soon thereafter good success was also achieved with synthetic ionophores. Simon's results initiated intensive research aiming to design, prepare, and test different types of organic molecules in ion-selective potentiometric sensor preparations. This work is still in progress today.

B. Review of Micro ISE in Clinical Chemistry and Physiology

While electroanalytical chemists were busy researching new selective ionophores and membrane compositions, life scientists and companies aiming for clinical applications made steps towards the miniaturization of ion-selective devices. In clinical chemistry the small sample volumes make measurements difficult. Figure 1 illustrates a range of specially shaped miniaturized ion-selective electrodes developed for these conditions. As for amperometric tips, there are two dimensions to consider: the size of the sensing element and the size of the surrounding insulating material. The latter will limit the range of applications. Currently the smallest sensor tip diameters are in the range of 20–50 nm.

1. *Integrated Circuit Technology*

The technology developed for semiconductor electronic industries found applications in miniature ion-selective sensors. In the early 1970s Bergveld (22) was the first to notice that metal oxide field effect transistors (MOSFET) could be sensitized with ion-selective materials (the device was named ISFET) to measure different ionic species. The original idea was to mass-produce with the efficiency of the high-tech production lines cheap, uniform, miniaturized sensing elements capable of performing different analytical tasks. The same element could contain several sensing transistors, measuring electronic circuits, operational amplifiers, etc. The incompatibility of materials used in electronic device fabrication with the conditions in wet chemical analysis, however, restricted the development of commercial chemically sensitive field effect transistor devices. Mostly pH-sensitive versions are used today.

Photolytographic technology with electrochemical finish can produce microelectrode arrays of the same potentiometric sensor or arrays of different sensors in the same electrode body. Clinical analyzers routinely use this kind

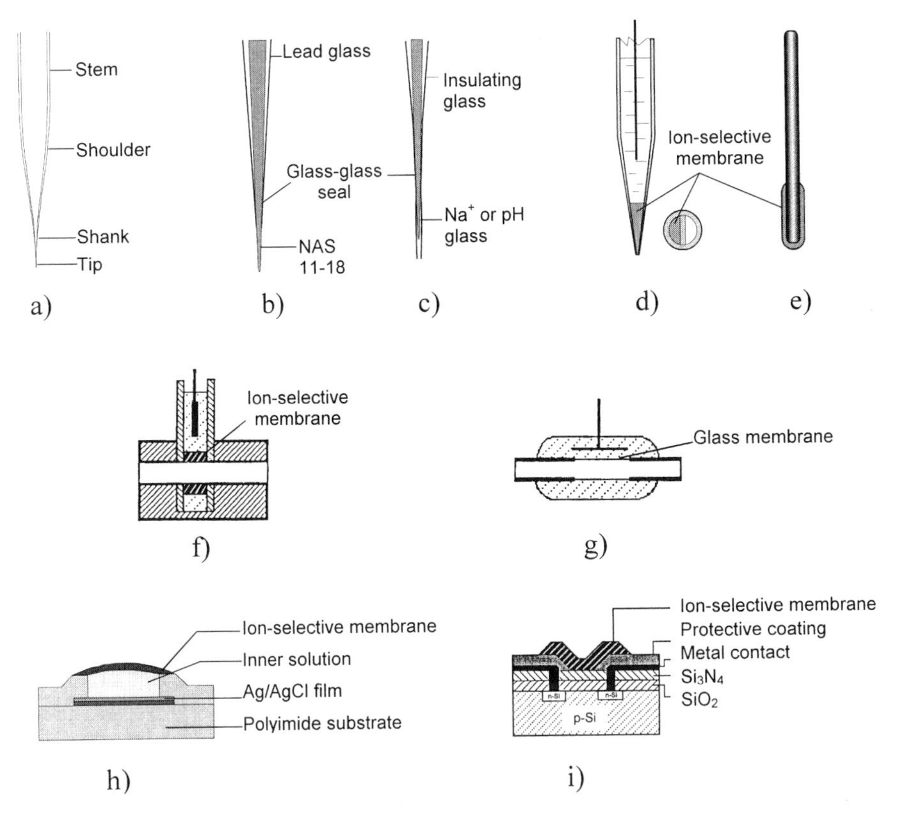

FIG. 1 Schematic cross sections of selected microelectrode configurations. (a) Nomenclature for parts of microelectrode, (b) Na^+-sensitive microelectrode (22), (c) recessed-tip Na^+-sensitive microelectrode (27), (d) liquid ion-exchanger micropipette electrode (38), (e) coated wire electrode (16), (f) flow-through ISE (e.g., NOVA 6, Boehringer ISE 2020), (g) micro-capillary glass electrode of tubular shape (e.g., Radelkis OP-266), (h) planar sensor fabricated by microelectronic technology (93), (i) ISFET sensor (94).

of sensor cartridges. Devices prepared with semiconductor technology have a sensing element size in the range of 5–200 μm; the body holding the elements has, of course, bigger dimensions (23,24). This type of planar electrode has been employed to image in real time the release of protons from a cation-exchange resin (25).

2. *Glass Microelectrodes*

In micro-capillary electrodes, the active measuring membrane is the section of the internal wall of a micro-capillary. A few microliters of sample solution are introduced inside the measuring capillary. To improve the reproducibility of the measurements, the whole electrode can be easily incorporated in a thermostated electrically shielded metal block.

Life scientists aiming for intracellular applications needed much smaller devices. Small-diameter glass capillaries or glass micropipettes often used in biological micromanipulation provided an excellent solution to the preparation of ultra-micro ISE for intracellular applications. Caldwell (26) miniaturized the pH-sensitive glass electrode and successfully measured the intracellular pH of the large size crab muscle fiber. He made the electrode stem from the same glass as the active part but coated the stem with shellac. In all other aspects, Caldwell's microelectrode had the same structure as the conventional glass electrode, namely a miniature bulb, microcapillary stem, internal filling buffer, and silver/silver chloride internal reference electrode. Hinke (27) prepared needle-type microelectrodes from sodium- or potassium-sensitive glasses and used them for intracellular measurements of these ions. The sealed end of the micropipette was made out of ion-sensitive glass and served as the measuring part. To prepare a microelectrode a closed ion-sensitive glass micropipette was placed into an open-ended micropipette made of insulating glass. The two glasses were sealed together along the conical sides leaving a 100 μm long protruding measuring glass tip. This configuration cannot give good spatial resolution; for intracellular measurements the entire protruding tip had to be inside the cell. Therefore, ordinary cells could not be studied with such microelectrodes. Lev and Buzhinsky (28) managed to decrease the length of the protruding tip to a few micrometers by employing a sealing wax between the two concentric glass cones. This was a considerable improvement, but the sampled area was still bigger than the lumen of an average size cell. Thomas (29,30) tried to solve this problem and produced so-called recessed tip glass pipette electrodes. In this case the pipette made of insulating glass is longer than the active sealed-end measuring pipette. The measuring pipette is sealed inside the lumen of the insulating one. With this the penetration into biological structures does not need to be so deep. Keeping in mind SECM applications, the use of recessed tips does not bring considerable improvement.

Pucacco and Carter (31) described a method to coat the open end of an insulating glass micropipette with a thin glass film to form a flat micro pH-sensitive electrode. They blew a very thin-walled bubble from uranium-containing glass. Under careful heating the insulating pipette was pushed into the bubble, which sealed the open end with a thin glass film. After cooling the excess glass was broken off from the sides. Levy and Coles (32), using a similar procedure, made a 4 μm double-barrel version with appropriate suction inside the pH barrel. These electrodes look more promising, although the preparation procedure is very difficult.

Yamaguchi and Stephens (33) proposed to make recessed pH-measuring glass microelectrodes by placing a short capillary made of pH-sensitive glass inside another glass capillary. In the first step of preparation they suggest heating the end of this concentric tube assembly to recess the outer glass to the level of the inner glass while decreasing the diameter of the inner capillary. The next step consists of pulling the assembly with a conventional pipette puller to produce a microelectrode with the end of the outer capillary open. The pH-sensitive pipette is usually sealed, recessed by about 2 μm, and functions well. First methanol, then distilled water, and finally a hydrochloric acid or sodium chloride internal filling solution is introduced inside the lumen of the inner glass capillary.

NAS 11-18 glass is usually chosen to make the measuring tip of ion-selective sodium microelectrodes. The insulating body of the electrodes is made of aluminosilicate glass. In case of protruding measuring tips, an open-ended pipette is first made with the insulating glass. Then another pipette is pulled from the active measuring glass capillary. Its end is sealed with a micro forge and a short section is broken off by scratching. This closed-end conical section is dropped, tip down, into the end of the electrode body pipette. It falls down inside and stops with the measuring tip protruding. The two pipettes are sealed together above the tip. Microelectrodes with a recessed measuring tip can be made in a similar manner.

When short response times and short sampling depths are needed, reversed tip glass microelectrodes can be made by introducing the pointed end of an active sealed-end measuring tip into the narrow open end of the electrode body pipette. The pointed end of the active tip ends up inside the body and the narrow tip of the body is sealed together with the wide part of the measuring tip.

Potassium-selective glass microelectrodes were also in use in the early days of potentiometric physiological studies (34,35). They were made with NAS 27-8 or NAS 27-4 glass. The sealed-end active measuring glass tip was glued inside the glass micropipette body by wax. However, the glass-based ion-selective potassium electrodes do not possess the necessary selectivity and are nowadays very seldom used.

3. *Electrodes of the Second Kind*

In principle these can also be used for ion activity measurement. Life scientists first tried to take advantage of electrodes of the second kind as early as the 1950s. Mauro (36) measured chloride ion concentrations inside the squid giant axon with a micro chloride electrode that he prepared by cementing a 15 μm diameter silver chloride–coated silver wire into the end of a glass capillary.

A much smaller micropipette silver electrode was prepared for chloride measurements by Kerkut and Meech (37). They introduced a concentrated ammoniac silver nitrate solution into borosilicate glass micropipettes and kept the micropipettes for 10 hours in 20% formaldehyde solution. The formaldehyde diffusing into the narrow opening reduced the silver and formed a metallic silver plug inside the tip of the micropipette. Silver chloride–coated silver wire was used as inner reference electrode in the silver nitrate internal filling solution.

Saunders and Brown (38) reported a method to prepare recessed tip silver/silver chloride microelectrodes. In the first step of this procedure they electrochemically sharpened the end of a silver wire (diameter 125 μm) by dipping it into basic sodium cyanide solution and anodizing it under voltage control. This was followed by an electrochemical chloride-coating step in 100 mM HCl solution.

4. *Micropipette Liquid Ion-Selective Microelectrodes*

Conventional size ion-selective liquid membrane electrodes were well known in the early 1970s (39). The measuring membrane was first made of a thin porous membrane (Millipore) soaked with an ion-selective liquid consisting of an organic electrolyte or of a neutral ionophore dissolved in a water-immiscible, high molecular weight organic liquid. These sensors were often commercialized in the form of an electrode preparation kit. The kit contained an electrode body in pieces with washers, O-rings, a membrane-holding screw cap, small tools to facilitate assembling, a few Millipore membranes, and an instruction booklet. The internal filling solution and the organic liquid was also supplied.

Liquid membrane electrodes are three (or four) component membrane based devices with positively charged, negatively charged, or electrically neutral ionophores. The main component of the liquid membrane is the ionophore, or mobile carrier—a compound with binding properties to different ions and responsible for the ion selectivity. The carriers are incorporated into membrane matrices, most commonly into plasticized poly(vinyl chloride). Liquid membranes often contain lipophilic salt additives also to enhance the permselectivity of hte membrane.

Since the glass micropipettes had already been used extensively in micro-manipulations and in field potential measurements the introduction of the ionophore-containing liquid membrane into the tip of a glass micropipette was an obvious step in ISE miniaturization. Walker (40) pioneered the development of micropipette liquid ion-exchanger–based ion-selective electrodes for the measurement of potassium and chloride ions. As expected, the organic ion-selective liquid inside the tip formed a measuring membrane separating the internal filling solution and the sample. Sub-micrometer size electrodes could be made in this way, but the preparation procedure, although universal, requires skills (see Sec. I.C). Stability problems affected the first experiments (41). The glass is hydrophilic, and water rapidly displaces the water-immiscible liquid away from the glass wall. An electrolyte bridge slowly forms between the internal and external solutions on either side of the membrane, and the electrode ceases to function. Surface treatment procedures giving a hydrophobic character to the internal glass wall were needed for the ion-selective glass micropipette electrodes to become practically acceptable. Walker treated the ready-made micropipettes with organic silicon compounds, and the lifetime of the microelectrodes increased dramatically. Different successful procedures have since been proposed for the preparation of these electrodes (see Sec. I.C).

Natural neutral carriers such as valinomycin and nonactin were first employed in liquid membranes for the selective detection of potassium and ammonium ions. This discovery in Simon's group initiated the design of several synthetic acyclic and macrocyclic compounds such as crown ethers and more recently calixarene derivatives. Their use in liquid membrane technology led to the elaboration of highly selective membrane electrodes. These ionophores have in common a stable conformation with a polar binding "cavity" and lipophilic shell. The first ion-selective microelectrodes based on neutral ionophores were introduced in the mid-1970s (42,43) and used for physiological measurement of ion activities in living cells.

The invention of liquid membrane microelectrodes and the progress in the design and synthesis of neutral ligands opened up new perspectives for the development of ion-selective microelectrodes with excellent selectivity.

The great advantage of the micropipette electrode configuration is that the same procedure can be used to prepare electrodes sensitive to different species, simply by changing the liquid ion-selective cocktails. Different ion-selective cocktails and ion-selective compounds (ionophores) have been elaborated (44). Several are available from Fluka AG, Corning, Orion, etc. as organic solutions delivered in a vial ready for electrode preparation. A typical cocktail is made of an organic solvent of low viscosity, such as 2-nitrophyl octyl ether or bis(ethylhexyl)sebacate, in which a few weight percent (5–10 wt%) of the ionophore is dissolved. Most cation-selective ionophore-based cocktails contain lipophilic anionic sites, e.g., tetraalkylammonium tetraaryl borate, potassium tetrakis (4-chlorophenyl)borate, or tetradodecylammonium tetrakis(4-chlorophenyl)borate. The role of such membrane additives is to enhance ion selectivity, to improve response time, and to reduce membrane resistance (45).

The major problem in developing neutral carrier–based membrane solution (cocktail) is to produce ion-selective microelectrodes of relatively low resistance. In order to keep the microelectrode resistance low, the membrane composition is optimized by altering the membrane solvent, the neutral carrier concentration, the percentage and the kind of lipophilic salt additives. The amount of salt additive is carefully adjusted to preserve ion selectivity.

A survey of ionophores available for ion-selective microelectrodes and membrane solution compositions (ion-selective cocktails) is shown in Table 1.

In the early days of ion-selective micropipette measurements, the cation-exchanger, potassium tetra (*p*-chloro) phenylborate based organic liquid commercialized by Corning (K^+ exchanger, type 477317), was a very popular potassium-measuring cocktail. Its limited selectivity only allowed certain applications, but this electrically charged ion-sensing material advantageously provided low electrode resistance. Simon's group (42,43) produced the neutral ionophore, valinomycin-based potassium-measuring cocktail (Fluka No 60031), which showed much better selectivity at the expense of higher electrode resistance. Bis-crown ether–type neutral ionophore (BME 44), developed by Pungor's group, were found to work well in potassium ion–measuring micropipettes (46).

Another Simon's ionophore (ETH 227) (Fluka 71176) is now the choice for making Na^+-measuring cocktail. Before this, monensine antibiotic–based sodium-sensitive pipettes were used (47) but suffered from high K^+ interference.

Different kinds of ion exchangers were designed for calcium ion measurement. The earlier versions (Orion or Corning calcium cocktails) used charged active measuring materials such as the calcium salt of dialkyl phosphoric acid mixed with an organic solvent, e.g., di-(octylphenyl)-phosphonate (DOPP) (48,49). Brown and coworkers (50) designed a micropipette calcium electrode using di-*p*-(1,1,3,3-tetramethylbutyl) phenyl phosphate calcium salt mixed with di-(*n*-octyl) phenylphosphonate. They used a PVC matrix to prepare the measuring membrane. To improve the selectivity of the calcium measurements, neutral ionophores were synthesized in Simon's group. The *N,N′*-di[-(11-ethoxycarbonyl) undecyl]*N, N′*-tetramethyl 3, 6-dioxa-octane diamide (ETH 1001) ionophore was found to work well in calcium micropipette electrodes (51).

Nonactin-based membrane solutions (52) can be used for the measurement of ammonium ion activity. A great deal of interest exists in designing lithium-selective electrodes because Li^+ is used clinically to control manic depression. As a result, micropipette Li^+ electrodes have also been reported (43).

With regard to selectivity, lifetime, and dynamic range, glass electrodes are superior to any other tools used for pH measurement, but the preparation of micrometer-size glass electrodes is not an easy task. To be able to use the universal and simple electrode preparation procedures for pH microelectrodes, liquid pH cocktails were also worked out. These cocktails usually use neutral tridodecylamine as active ionophore (53,54). Microelectrodes prepared with this cocktail can be used to monitor pH changes in the physiologically important pH range.

Classical (charged) ion exchangers are typically employed as active ingredients in potentiometric anion-selective electrodes. Therefore, their insufficient selectivity precludes their use in complex matrices containing high concentrations of other anions. The micropipette version of the different anion electrodes can be easily prepared. The commercial Orion 92-07-02 nitrate-measuring cocktail contains a tris(substituted 2-phenanthroline) nickel(II) exchanger, while the Corning 477316 cocktail uses tridodecylhexadecylammonium nitrate. Chlorate, perchlorate and iodide would interfere with the nitrate measurement if present.

The Corning 477313 cocktail works well for chloride micropipettes (55,56). These pipettes were used to measure the concentration of chloride, the major anionic component in biological media.

Methyltricaprylammonium chloride, Aliquat 336S, has often been used in anion-selective electrodes as active ion exchanger. Different Aliquat 336S–based micropipette electrodes were prepared and used in life science experiments (57). Using hydrophobic solvents the discrimination against chloride ions can be enhanced, leading to an improvement in selectivity. Crystal violet dye was also successfully used as ion exchanger in micropipette anion-measuring electrodes (58).

C. Fabrication of Micropipette Ion-Selective Microelectrodes

1. Early Recipes

According to Walker's method (40) micropipettes with 0.5–1 μm tips are made from borosilicate glass tubing previously cleaned in hot ethanol vapor. After pulling, the pipette tip is dipped in a 1% fresh solution of Siliclad (siloxane polymer, Clay Adams) in 1-chloronaphthalene. The solution is allowed to rise about 200 μm inside the shank. The pipette, held vertically with the tip up, is baked in an oven for an hour at 250°C. It can then be stored for a few days in dry air. Before use the pipettes are filled with the liquid ion exchanger cocktail (see Table 1) and with the internal filling solution. The ion exchanger cocktail is usually introduced first through the tip with capillary action and light suction. The internal filling solution (e.g., 0.1 M KCl) is often introduced through the back end of the pipette. The smaller the tip, the more difficult the filling-up procedure. The internal reference electrode most often consists of a 100 μm diameter chloride-coated silver wire. The back end of the electrode is normally sealed with silicon rubber or closed in some other way.

TABLE 1 Ion-Selective Cocktails for Micropipette Electrode Preparation

Ion	Ionophore	Ion-selective liquid (cocktail)	Reference
Cl^-	$C_{44}H_{28}ClMnN_4$	Cocktail: 5.00 wt% Chloride Ionophore 4.00 wt% 1-Decanol 1.00 wt% Tetradodecylammonium tetrakis(4-chlorophenyl)borate 90.00 wt% 2-Nitrophenyl octyl ether	M. Hauser, W. E. Morf, K. Fluri, K. Seiler, P. Schulthess, W. Simon, Helv. Chim. Acta *73*, 1481 (1990)
		Liquid ion-exchanger microelectrode cocktail: 10.00 wt% Trioctylpropylammonium chloride 90.00 wt% 1,2-Dimethyl-3-nitrobenzene	R. N. Khuri, S. K. Agulian, K. Bogharian, Am. J. Physiol. *227*, 1352 (1974) C. M. Baumgarten, Am. J. Physiol. *241*, C258 (1981)
H^+	$[CH_3(CH_2)_{11}]_3N$	Cocktail A: 10.00 wt% Hydrogen Ionophore 89.30 wt% 2-Nitrophenyl octyl ether 0.70 wt% Sodium tetraphenylborate Cocktail B: 10.00 wt% Hydrogen Ionophore 89.30 wt% 2-Nitrophenyl octyl ether 0.70 wt% Potassium tetrakis(4-chlorophenyl)borate	P. Schulthess, Y. Shijo, H. V. Pham, E. Pretsch, D. Amman, W. Simon, Anal. Chim. Acta *131*, 111 (1981) D. Amman, F. Lanter, R. A. Steiner, P. Schulthess, Y. Shijo, W. Simon, Anal. Chem. *53*, 2267 (1981)
Li^+	H_3C, $N(CH_2)_6CH_3$, O, O, H_3C, H_3C, O, O, $N(CH_2)_6CH_3$, H_3C **ETH 149**	Cocktail: 9.70 wt% Lithium Ionophore 4.80 wt% Sodium tetraphenylborate 85.50 wt% Tris(2-ethylhexyl) phosphate	R. C. Thomas, W. Simon, M. Oehme, Nature *258*, 754 (1975)

K^+	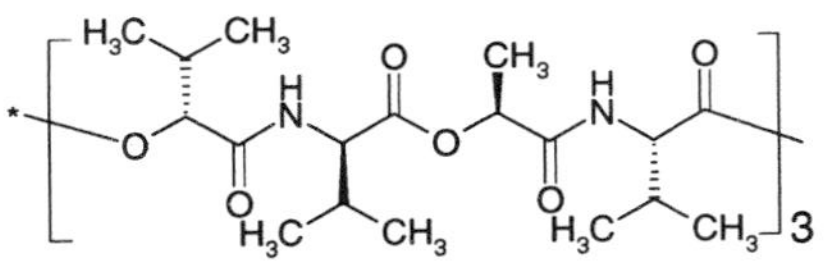 **VALINOMYCIN**	<u>Cocktail A:</u> 5.00 wt% Potassium Ionophore (Valinomycin) 25.00 wt% 1,2-Dimethyl-3-nitrobenzene 68.00 wt% Dibutyl sebacate 2.00 wt% Potassium tetrakis(4-chlorophenyl)borate	M. Oehme, W. Simon, Anal. Chim. Acata *86*, 21 (1976) P. Wuhrmann, H. Ineichen, U. Riesen-Willi, M. Lezzi, Proc. Natl. Acad. Sci. *76*, 806 (1979)
		<u>Cocktail B:</u> 5.00 wt% Potassium Ionophore (Valinomycin) 93.00 wt% 1,2-Dimethyl-3-nitrobenzene 2.00 wt% Potassium tetrakis(4-chlorophenyl)borate	D. Amman, P. Chao, W. Simon, Neurosci. Let. *74*, 221 (1987).
K^+	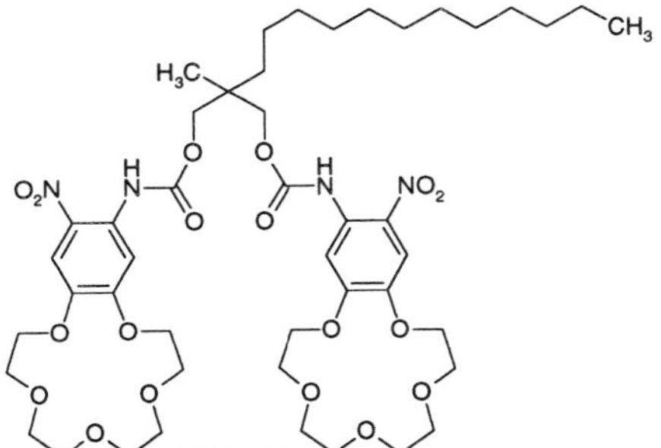 **BME 44**	<u>Cocktail A:</u> 5.00 wt% Potassium Ionophore (BME 44) 80.60 wt% 2-Nitrophenyl octyl ether 13.00 wt% Poly(vinyl chloride) high molecular weight 1.40 wt% Sodium tetraphenylborate	J. Tarcali, G. Nagy, K. Tóth, E. Pungor, Anal. Chim. Acta *178*, 231 (1985)
		<u>Cocktail B:</u> 5.00 wt% Potassium Ionophore (BME 44) 83.20 wt% 2-Nitrophenyl octyl ether 10.00 wt% Poly(vinyl chloride) high molecular weight 1.80 wt% Potassium tetrakis(4-chlorophenyl)borate	

TABLE 1 Continued

Ion	Ionophore	Ion-selective liquid (cocktail)	Reference
Na^{+}	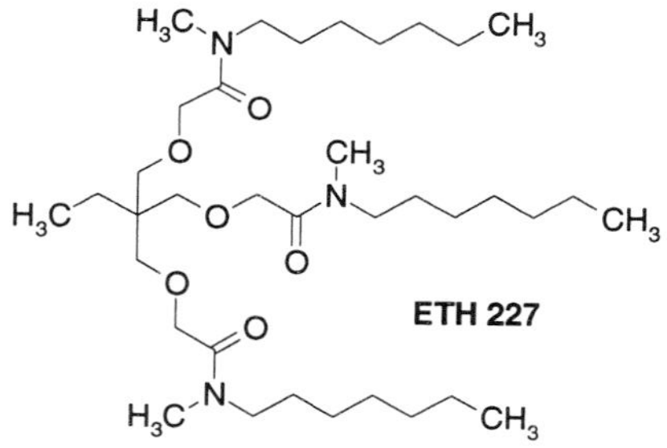 ETH 227	Cocktail A: 10.00 wt% Sodium Ionophore (ETH 227) 89.50 wt% 2-Nitrophenyl octyl ether 0.50 wt% Sodium tetraphenylborate	R. A. Steiner, M. Oehme, D. Amman, W. Simon, Anal. Chem. *51*, 351 (1979)
		Liquid ion-exchanger microelectrode cocktail A: 2.10 wt% Potassium tetrakis(4-chlorophenyl)borate 12.10 wt% Trioctylphosphine oxide 85.80 wt% Tris (2-ethylhexyl) phosphate	C. J. Cohen, H. A. Fozzard, S. S. Cheu, Circ. Res. *50*, 651 (1982) L. G. Palmer, M. M. Civan, J. Membrane Biol. *33*, 41 (1977)
		Liquid ion-exchanger microelectrode cocktail B: 1.50 wt% Potassium tetrakis(4-chlorophenyl)borate 98.50 wt% Tris (2-ethylhexyl) phosphate	T. J. Century, M. M. Civan, J. Membrane Biol. *40*, 24 (1978)
		Liquid ion-exchanger microelectrode cocktail C: 2.70 wt% Potassium tetrakis(4-chlorophenyl)borate 10.00 wt% Monensin 85.80 wt% 1,2-Dimethyl-3-nitrobenzene	P. P. Craig, C. Nicholson, Science *194*, 725 (1976) K. Kotera, N. Satake, M. Honda, M. Fujimoto, Membr. Biochem. *2*, 323 (1979) M. Fujimoto, M. Honda, Jap. J. Physiol. *30*, 859 (1980) S. Fabczak, Acta Protozool. *26*, 129 (1987)
Ca^{2+}	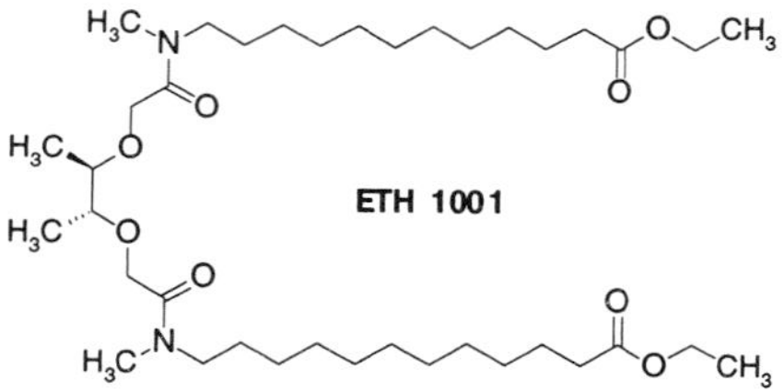 ETH 1001	Cocktail A: 10.00 wt% Calcium Ionophore (ETH 1001) 89.00 wt% 2-Nitrophenyl octyl ether 1.00 wt% Sodium tetraphenylborate	F. Lanter, R. A. Steiner, D. Amman, W. Simon, Anal. Chim. Acta *135*, 51 (1982) R. Y. Tsien, T. J. Rink, J. Neurosci. Methods *4*, 73 (1981)
		Cocktail B: 21.50 wt% Cocktail A 3.50 wt% Poly(vinyl chloride) high molecular weight 75.00 wt% Tetrahydrofuran	E. Ujec, E. E. O. Keller, N. Kùriùz, V. Pavlik, J. Machek, Bioelectrochem. Bioenerg. *7*, 363 (1980)

Ion / Ionophore	Cocktail	References
Ca^{2+} 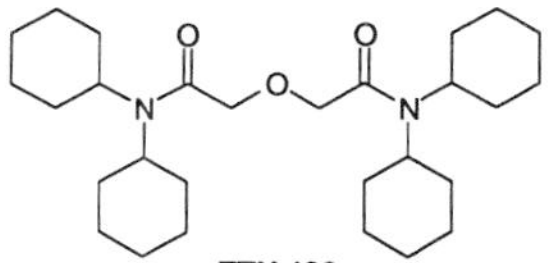 ETH 129	Cocktail A: 5.00 wt% Calcium Ionophore (ETH 129) 94.00 wt% 2-Nitrophenyl octyl ether 1.00 wt% Sodium tetraphenylborate Cocktail B: 21.50 wt% Cocktail A 3.50 wt% Poly(vinyl chloride) high molecular weight 75.00 wt% Tetrahydrofuran	D. Amman, T. Bührer, U. Schefer, M. Müller, W. Simon, Pflügers Arch. *409*, 223 (1987) H. Mack Brown, S. K. Marron, Anal. Chem. *62*, 2153 (1990) D. Amman, P. Caroni, Methods in Enzymol. *172*, 136 (1989)
Mg^{2+} 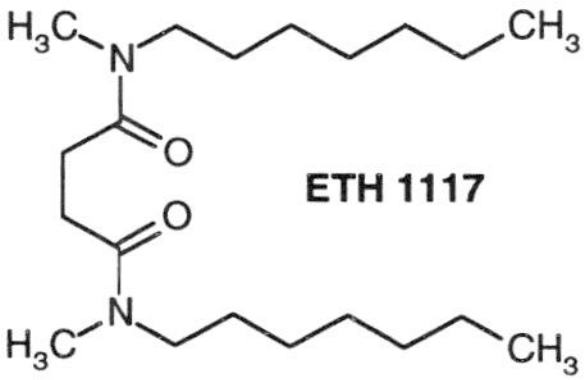 ETH 1117	Cocktail: 20.00 wt% Magnesium Ionophore I (ETH 1117) 79.00 wt% Propylene carbonate 1.00 wt% Sodium tetraphenylborate	F. Lanter, D. Erne, D. Amman, W. Simon, Anal. Chem. *52*, 2400 (1980)
Mg^{2+} ETH 5214	Cocktail A: 10.00 wt% Magnesium Ionophore (ETH 5214) 87.00 wt% 2-Nitrophenyl octyl ether 3.00 wt% Potassium tetrakis(4-chlorophenyl)borate	Z. Hu, T. Bührer, M. Müller, B. Rusterholz, M. Rouilly, W. Simon, Anal. Chem. *61*, 574 (1989) U. E. Spichiger, Electroanalysis *5*, 739 (1993) D. Amman, P. Caroni, Methods in Enzymol. *172*, 136 (1989)
	Cocktail B: 10.00 wt% Magnesium Ionophore (ETH 5214) 62.00 wt% 2-Nitrophenyl octyl ether 25.00 wt% Chloroparaffin (60% chlorine) 3.00 wt% Potassium tetrakis(4-chlorophenyl)borate	Z. Hu, F. Ye, P. Zhao, D. Qi, Huaxue Xuebao *49*, 1483 (1991) D. Amman, P. Caroni, Methods in Enzymol. *172*, 136 (1989)

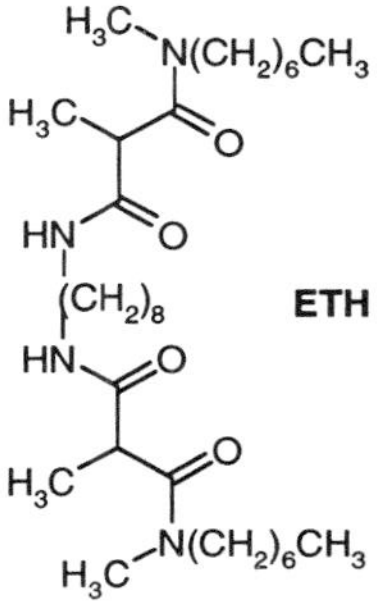

TABLE 1 Continued

Ion	Ionophore	Ion-selective liquid (cocktail)	Reference
Mg^{2+}	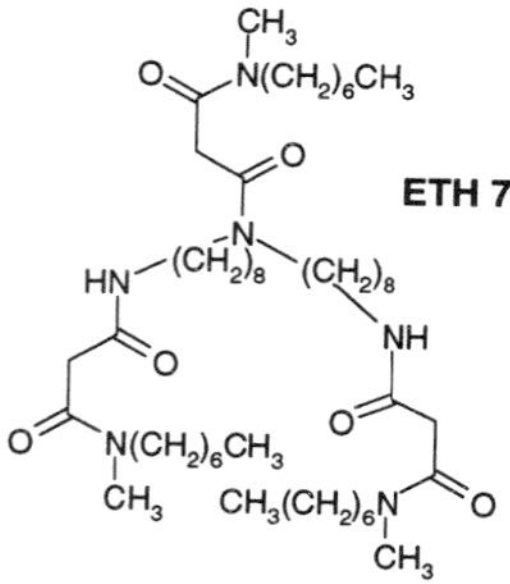 ETH 7075	Cocktail: 2.20 wt% Magnesium Ionophore (ETH 7075) 18.00 wt% 2-Nitrophenyl octyl ether 1.10 wt% Tetradodeylammonium tetrakis(4-chlorophenyl)borate 0.75 wt% Potassium tetrakis(4-chlorophenyl)borate 3.00 wt% Poly(vinyl chloride) high molecular weight 75.00 wt% Cyclohexanone	U. Schaller, U. E. Spichiger, W. Simon, Pflügers Arch. *423*, 338 (1993)
NH_4^+	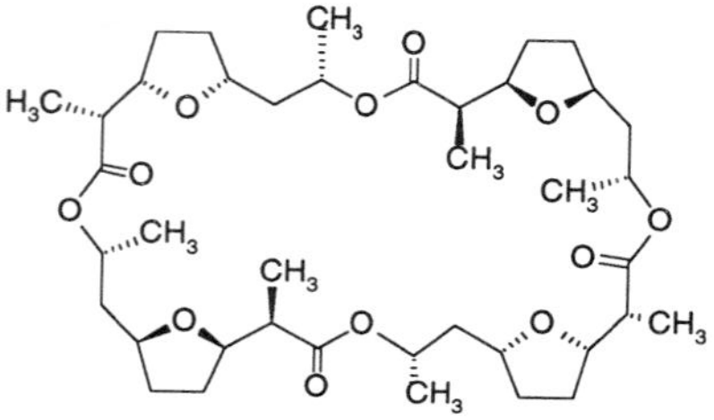 NONACTIN	Cocktail A: 6.90 wt% Ammonium Ionophore (Nonactin) 0.70 wt% Potassium tetrakis(4 chlorophenyl)borate 92.40 wt% 2-Nitrophenyl octyl ether Cocktail B: 3.50 wt% Ammonium Ionophore (Nonactin) 0.35 wt% Potassium tetrakis(4 chlorophenyl)borate 0.95 wt% Poly(vinyl chloride) high molecular weight 32.95 wt% Dibutyl sebacate 62.50 wt% Tetrahydrofuran	T. Bührer, H. Peter, W. Simon, Pflügers Arch. *412*, 359 (1988) D. De Beer, J. C. van den Heuvel, Talanta *35*, 728 (1988) F. Fresser, H. Moser, M. Nair, J. Exp. Boil. *147*, 227 (1991) G. H. Henrriksen, A. J. Bloom, R. M. Spanswick, Plant. Physiol. *93*, 271 (1990)

Ion	Ionophore	Membrane composition	Reference
Zn^{2+}	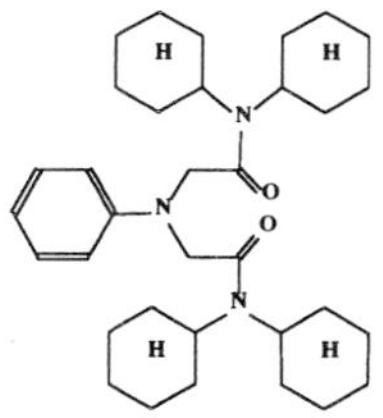 H H N N O O N H H	<u>Cocktail A</u>: 5.00 wt% Zinc Ionophore 3.20 wt% Potassium tetrakis(4 chlorophenyl)borate 91.80 wt% 2-Nitrophenyl octyl ether <u>Cocktail B</u>: 5.00 wt% Zinc Ionophore 3.20 wt% Potassium tetrakis(4 chlorophenyl)borate 81.80 wt% 2-Nitrophenyl octyl ether 10.00 wt% Poly(vinyl chloride) high molecular weight	C. Wei, A. Bard, G. Nagy, K. Tóth, Anal. Chem. *67*, 1346 (1995)

Different laboratories used and suggested modified versions of the original procedure. Different glass-cleaning procedures were found simpler, easier, or more efficient with less failure rate. Some of the large-scale procedures used for micropipette electrode preparation start with very effective cleaning of the glass. The glass capillaries to be used for microelectrode preparation are soaked for several hours in a 1:1 mixture of concentrated sulfuric acid and 30% hydrogen peroxide (59). After intensive washing and drying the capillaries can be stored before pipette pulling. Neher and Lux (60) advise fire polishing the pipette, a procedure in which the broken tips of the freshly made micropipettes are slightly melted, then dipped for 1–5 seconds in 4% dichlorodimethylsilane in carbon tetrachloride for silanization. The pipettes are then, without drying or baking, immersed in boiling methanol under slight vacuum. When the methanol replaces the air in the lumen of the tubing, the pipette is transferred into distilled water and finally into 0.1 M KCl internal filling solution. The liquid ion-exchanger solution is then introduced through the tip by capillary action.

2. *Typical Fabrication Procedures*

Micropipettes for liquid membrane microelectrodes can be fabricated from different glass tubing (borosilicate, aluminosilicate, etc.) with a typical diameter of about 2 mm. Commercially available glass tubes come in single, double, or multibarrel type. Prior to use the glass tubes should be soaked in a 1:1 (v/v) mixture of concentrated sulfuric acid and 30% hydrogen peroxide for 24 hours, then washed thoroughly with doubly distilled water and dried at 120°C for 30 minutes. A laser puller– or a heating coil–based puller (e.g., Stoetling) is then used to produce the capillary end. The tip diameter should be within the range of 1–20 μm and the ratio of tip length to stem radius smaller than 5. For silanization the single-barrel micropipettes are first filled with dimethyldichlorosilane or trimethylsilyldimethylamine vapor by syringe injection and then placed in a covered Petri dish, which also contains a few drops of the silanizing reagent. The silanization is completed by allowing the glass surface of the micropipette to react with the silanizing reagent vapour for 2 hours at 120°C in an oven. Any silanizing reagent remaining in the micropipette is removed by connecting it from the back to a water pump. The internal filling solution containing the ion of interest and potassium chloride is backfilled by applying a slight overpressure with a syringe. The tips are then frontfilled with the corresponding ion-selective cocktail by capillary action and slight suction. Finally, a Ag/AgCl wire is introduced to complete the ion-selective microelectrode (ISME).

Double-barrel capillaries are silanized by introducing the silanizing vapor into one channel while continuously flushing the other channel with nitrogen. The channel not silanized is later used for the outer reference electrode. The silanized channel is filled with the ion-selective solution (cocktail), as described previously.

Air bubbles entrapped between the organic and the water phases often break the electrical path. Small tools such as glass filaments or cats and rabbits whiskers are combined with skill and great patience to remove the insulating bubble. With liquid phase silanization techniques submicrometer size tips are easily blocked by insoluble plugs. Thus, very small pipette tips and the ion-selective channel of double barrel electrodes are usually vapor silanized (61). In preparing double-barrel electrodes the reference barrel must remain open. Therefore, it must not be silanized, otherwise the ion-selective membrane cocktail easily plugs both barrels. Double and multi-barrel electrode preparation necessitates an overpressure of nitrogen gas in the reference barrel to prevent silanizing vapors from entering into it. The tip diameter of ion-selective micropipette electrodes can be as small as 20–50 nm. However, the preparation and application of such a delicate device is not routine. On the other hand, an experienced researcher can easily prepare electrodes with 0.5–3 μm diameter tips with a success rate of 50–80%.

II. BASIC THEORY

Potentiometric ion-selective electrodes are passive probes, which in contrast to voltammetric sensors do not convert the analyte in the sample. The response of an ISE depends linearly on the logarithm of the activity (concentration) of a potential determining ion (primary ion) in the presence of other ions. The schematic layout of a complete potentiometric cell including an ion-selective electrode is shown in Figure 2. The electrochemical notation of the cell assembly is given as:

$$\underbrace{\text{Ag} \mid \text{AgCl, KCl}}_{\text{Reference electrode}} \parallel \text{salt bridge} \parallel \underbrace{\text{sample} \mid \text{membrane} \mid \text{internal solution, AgCl} \mid \text{Ag}}_{\text{Ion-selective electrode}}$$

where vertical lines indicate phase boundaries while double lines mark a liquid/liquid interface.

The cell potential is the sum of a number of local potential differences generated at the solid/solid, solid/liquid, and liquid/liquid interfaces within the cell and measured between the two reference electrodes. At zero current, the cell potential, E_{cell}, is given by

$$E_{cell} = E_{ISE} - E_{ref} + E_{jnc} \tag{1}$$

where E_{ISE} is the potential of the ion-selective electrode, E_{ref} the potential of the reference electrode, and E_{jnc} the liquid junction potential arising at the interface of the sample solution (or test solution) and the internal electrolyte

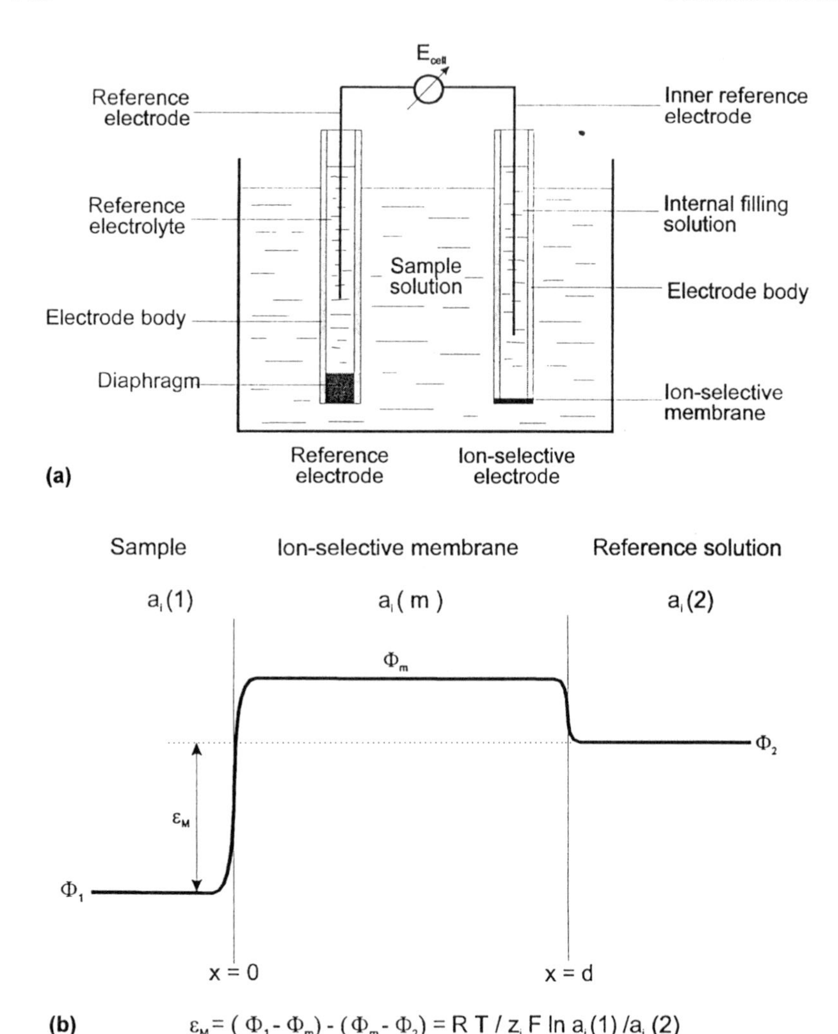

FIG. 2 (a) Schematic diagram of a potentiometric measuring circuit. (b) Potential profile across an ion-selective membrane of fixed sites, symmetrically bathed with the sample (1) and the reference solution (2).

solution of the reference electrode. In analytical applications one chooses electrolyte combinations to eliminate liquid junction potentials.

In ion-selective electrode potentiometry, the cell potential reflects the dependence of the membrane potential on the primary ion activity (concentration). According to the Teorell-Meyer-Sievers (TMS) theory, the sum of three different potential contributions, the phase boundary potentials generated by ion-exchange processes at both interfaces, $(\phi_1 - \phi_{m,1})$ and $(\phi_{m,2} - \phi_2)$, and the intermembrane diffusion potential, $(\phi_{m,1} - \phi_{m,2})$, constitutes the membrane potential. If the membrane composition is constant and there are no concentration gradients within the membrane, then the membrane diffusion potential is zero and the membrane potential can be described by phase boundary potentials (see Fig. 2b). This approach is also used to treat the response of ISE made with a range of membranes.

For an ideally selective electrode, the measured cell potential is described by the Nernst equation:

$$E_{\text{cell}} = \text{constant} \pm 2.303 \frac{RT}{z_i F} \log(a_i) \tag{2}$$

where a_i refers to the activity of the primary ion i in the sample solution, z_i is the charge of the primary ion i, R the gas constant, T the absolute temperature, and F the faraday constant. $2.303RT/z_iF$ is the Nernstian slope, which is equal to $59.16/z_i$ mV (25°C). The constant term includes all primary ion i activity independent terms.

In practice, however, deviation from the ideal electrode behavior is common, and an additional contribution to the total measured ion activity has to be considered due to the presence of the interfering ion j in the sample solution. Under these conditions the ion-selective electrode potential can be approximated by the Nikolsky-Eisenman equation:

$$E_{\text{cell}} = \text{constant} \pm 2.303 \frac{RT}{z_i F} \log\left(a_i + \sum_{j \neq i} K_{ij}^{pot} (a_j)^{z_i/z_j} \right) \tag{3}$$

where a_j and z_j are the activity and charge of the interfering ion j. The weighting factor K_{ij}^{pot} in Eq. (3) is the selectivity coefficient. Naturally, if the value of K_{ij}^{pot} is small, the contribution of the interfering ion to the total cell potential is negligible. The selectivity coefficients can be determined potentiometrically and are tabulated. Their values are used as a guideline when designing ISE-based potentiometric experiments.

III. PROPERTIES AND BEHAVIOR OF ION-SELECTIVE PROBES

A. Electrode Function and Detection Limit

The electrode function gives the dependence between E_{cell} of an ion-selective electrode cell assembly and the logarithm of the potential determining ion (primary ion) activity in the sample solution. Typically, ISEs exhibit Nernstian electrode function in the concentration range of 10^{-1} and 10^{-5} M. De-

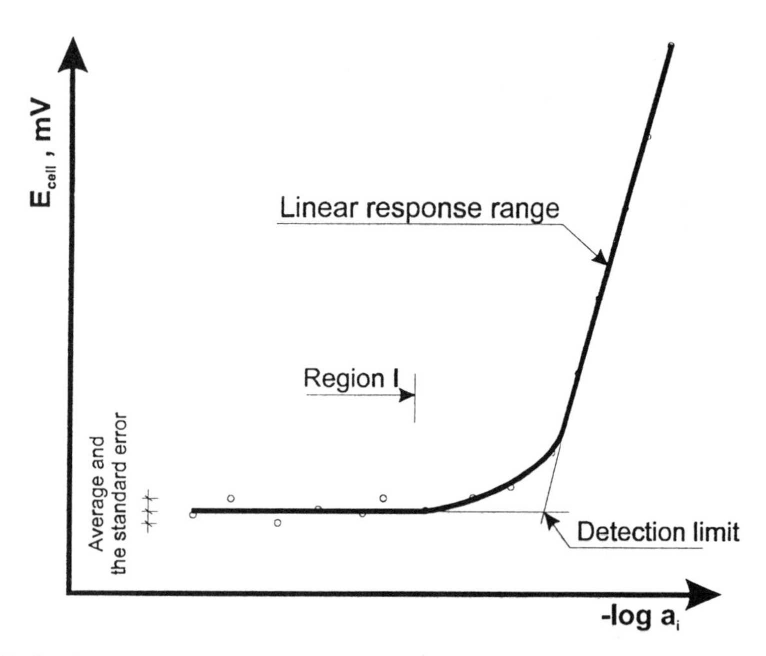

FIG. 3 Schematic diagram of the response function of an ion-selective electrode (a_i) denotes the activity of the primary ion, i. Region I is the primary ion activity independent activity range, in which the cell potential deviates from the average value by a multiple of the standard error. (From Ref. 95.)

viation of linearity occurs at low activities, below 10^{-5} M. The detection limit usually varies with the nature of the ion-selective membrane material, and for ionophore-based electrodes it is theoretically determined by the stability of the ionophore cation complex. The detection limit is typically around 10^{-6} M and can be determined as shown in Figure 3. It is important to note that micropipette electrodes are usually calibrated before and after the experiment. If the two calibrations show significant differences, the experiment has to be discarded and repeated.

B. Selectivity Coefficient

The selectivity factor, K_{ij}^{pot}, is a measure of the preference of the ion-selective sensor for the interfering ion *j* relative to the primary ion *i* to be measured. A selectivity factor below 1 indicates that the preference is for the primary ion *i*. Two methodologies are used to measure the selectivity factor experimentally:

1. The fixed interference method (FIM). The potential of an ion-selective electrode is measured in solutions of constant activity of interfering ion (a_j) and varying activity of the primary ion (a_i). The selectivity coefficient, K_{ij}^{pot}, is calculated from the relevant calibration graph plotted for the ion of interest, i. The intersection of the extrapolated linear portions of the response curve indicates the value of a_i^*, which is used to calculate K_{ij}^{pot} from the Nikolsky-Eisenman equation:

$$K_{ij}^{pot} = \frac{a_i^*}{a_j^{z_i/z_j}} \tag{4}$$

   With the knowledge of the selectivity factor the experimental error due to poor ion-selectivity can be calculated.

2. The separate solution method (SSM). The potential of a cell comprising an ISE and a reference electrode is measured in two separate solutions: one containing only the primary ion i (E_i), the other containing the interfering ion j (E_j), at the same activity ($a_i = a_j$). The value of the selectivity coefficient can be calculated on the basis of the Nikolsky-Eisenman equation:

$$\log K_{ij}^{pot} = \frac{E_j - E_i}{2.303RT/z_iF} + (1 - z_i/z_j)\log a_i \tag{5}$$

   The method is relatively simple and especially suitable to compare the selectivity of novel ion-selective electrodes prepared with different membranes. The data thus calculated may not be representative for mixed sample solutions. For further details, see Ref. 62.

C. Stability and Reproducibility of E_{cell}

In ion-selective potentiometry the stability and reproducibility of E_{cell} depends on the type of ISE used. Carefully optimized liquid membrane electrodes in flow-through arrangements for clinical applications yield standard deviations in E_{cell} smaller than 0.1 mV. This corresponds to less than $z_i \times$ 0.4% error in the determination of activities. Ion-selective microelectrodes have poorer characteristics due to technical (e.g., silanization) and electrical (high resistance; see below) reasons.

D. Response Time

The response time is one of the most critical characteristics of the potentiometric probe in continuous analyzers and in in situ monitoring. Most often the activity step method is used for its determination (Fig. 4). The response time is affected by several factors such as the properties of the ion-selective

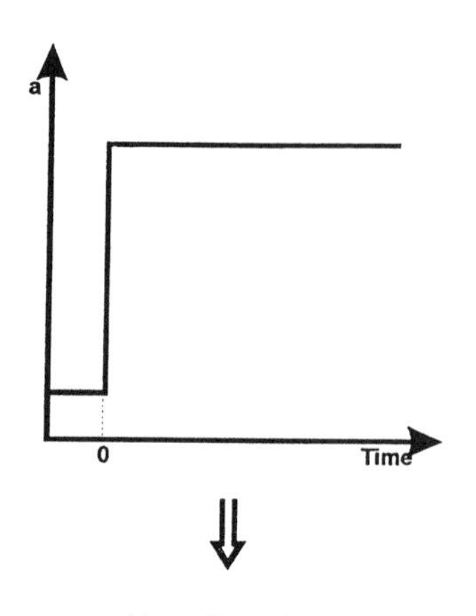

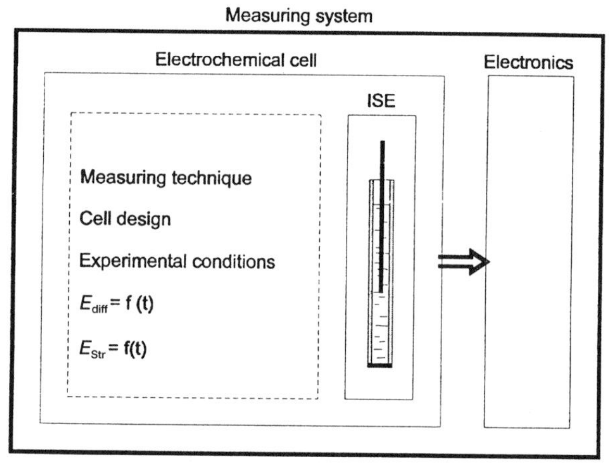

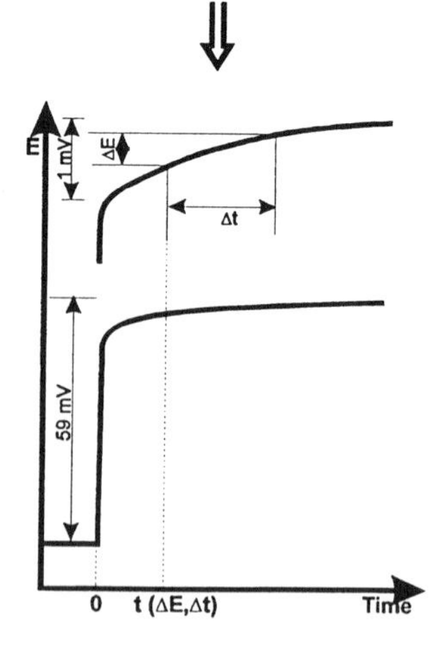

FIG. 4 Schematic diagram of the measuring system used with the activity step method.

electrode, the measuring technique used to perform the activity step, the experimental conditions (e.g., cell design and flow rate, the time dependence of other sources of potential in the cell including diffusion and streaming potentials), and finally the electronics used to record the transient signal. For solid-state and most liquid membrane electrodes, the rate-limiting step is the rate of ion transport in the stagnant solution layer adherent to the sensor. However, the response time of microelectrodes with high internal resistance never reflects transport processes towards or within the membrane. Rather it is determined by the *RC* time constant of the measuring circuit or the ion-selective microelectrode itself (63):

$$E_{cell}(t) = E_{cell}(\infty) + [E_{cell}(\infty) - E_{cell}(0)]e^{-t/RC} \tag{6}$$

where $E_{cell}(0)$ is the cell potential prior to the change in activity, $E_{cell}(\infty)$ the cell potential long after the change in activity (i.e., once the potentiometric response has stabilized), R the internal resistance of the electrochemical cell and C the capacitance of the amplifier input.

The response time, t_α, is defined as the time taken by the measuring assembly to reach a given percentage of the final cell voltage following a given change in activity (where α indicates the given percentage) or by the slope $\Delta E/\Delta t$ of the dynamic response curve (64).

E. Handling and Instrumentation

Most ion-selective microelectrodes are very delicate tools, and working with them requires training and patience. Compared to conventional size electrodes, their only advantage is that they can operate in small volume samples and thus can provide very fine spatial resolution.

The microelectrodes are sensitive to mechanical stress; their tip can easily break off if they touch any hard surface. Some are also sensitive to static electrical shocks, which are often generated during handling. Ion-selective micropipettes and glass microelectrodes have very high resistance. Typically a submicrometer micropipette prepared with a neutral ionophore-containing membrane has an impedance as high as 10^{11} Ω, while micropipettes made with charged ion exchangers or glass membrane microelectrodes usually have a lower resistance, $\sim 10^8$–10^9 Ω. This high resistance has serious effects on the electrode performance. The glass body of the electrode needs to be electrically shielded, otherwise the electrode will detect the potential difference across the glass instead of across the membrane. Special instrumentation, arrangement, and shielding are required for the measurements.

The input impedance of the measuring apparatus has to be a few orders of magnitude higher than the electrode impedance, otherwise the input bias current may produce a significant voltage error. This is important because the impedance of the electrode can change considerably over time. The voltage error, i.e., the maximum bias current passing through the electrode multiplied by the largest electrode resistance, should not exceed 0.5 mV. Because of the logarithmic relationship, a 0.5 mV error already corresponds to a 2% (assuming $z_i = 1$, 4% assuming $z_i = 2$) error on the activity measured.

FET operational amplifiers with high input impedance and low input current available today make it easy to construct an instrument suitable for high-resistance microelectrodes. Aiming to gain applications in life sciences, a variety of potentiometric instruments designed for microelectrode measurements are on the market. These commercial high impedance electrometers usually have several measuring channels. Their reference and ion-selective inputs have similar input impedance. The inputs are connected to appropriate voltage followers, themselves connected to a differential amplifier stage that measures the voltage difference between the ISE and reference electrodes, amplifies it if necessary, and conveys it for display, recording, or further processing. It is advantageous to position the impedance matching, i.e., the voltage follower, very close to the electrodes. The shorter the connecting leads, the less electrical noise they will pick up. In SECM applications the voltage follower should be located inside a small shielded box attached to the electrode holder. In addition the whole experimental set-up should be placed inside a Faraday cage. The use of battery-powered electrometers inside the cage is highly recommended.

In microelectrode potentiometric measurements, a very noisy voltage output can be expected without appropriate shielding. One can check the electrode condition by hand-waving close to the cell. An absence of noise spikes indicates an electrode failure.

When properly shielded and connected to the appropriate apparatus, ion-selective microelectrodes should show the response and selectivity similar to conventional size electrodes. They should be calibrated in a solution of similar composition to that of the sample solution to be analyzed. The response time of the electrode itself should be comparable to that of conventional electrodes. However, with these electrodes the diffusion of the ions through the stagnant layer near the ion-selective membrane has a smaller effect on the transient response than the time constant of the measuring circuitry. Measuring instruments often contain a variable capacitor to adjust the delay caused by the capacitance of the input stage.

IV. POTENTIOMETRIC MEASUREMENTS IN SCANNING PROBE MICROSCOPIES OTHER THAN SECM

The earliest forms of scanning electrochemical microscopic experiments were carried out in the potentiometric mode. Evans (65) reported the use of a traveling reference electrode to map equipotential surfaces in solution and calculate corrosion rates on a water pipeline. Isaacs (66,67) revived the idea in 1972 and coined the acronym SRET for scanning reference electrode technique. For a long time the instrument was home-made, often with a primitive scanning device such as an arm of an X-Y recorder (66,68). The spatial resolution was at best submillimeter. A computer-controlled instrument is now manufactured by Uniscan Instruments Ltd and commercialized by EG&G. The spatial resolution is much improved—around 20 μm (69). The scanning reference electrode has now been superseded by the scanning vibrating electrode technique (70), known as SVET, which achieves higher spatial resolution (71) and improved sensitivity. SVET instruments are also manufactured by Uniscan Instruments Ltd and marketed by EG&G.

Both SRET and SVET perform potentiometric measurements with micrometer-sized quasi-reference electrodes (typically platinum micro-cones electrochemically coated with platinum black) held precisely above the sample surface. So far these commercial instruments have not been designed to accommodate ion-selective microelectrodes. However, some groups have reported the use of home-made instruments to probe the concentration of specific ions (see Sec. V.B).

V. POTENTIOMETRIC MEASUREMENTS IN SECM

A. Asserting the Absolute Tip-Substrate Distance

In potentiometric scanning electrochemical microscopy, it is necessary to determine the absolute distance between the microelectrode and the target under investigation. There are basically two reasons for this. On the one hand, it is important to restrict the travel of the probe to avoid touching the sample because damage could occur to both surfaces; on the other hand, quantitative information regarding concentration profiles can only be extracted with the knowledge of the tip-substrate distance. The latter will be developed in subsequent sections. Experimentally the tip position is precisely known but is only relative to a reference location in the solution. Thus, it is common to exploit the dependence of the tip response on tip-substrate distance to estimate the absolute distance between the tip and the sample surface. In practice one fits an experimental approach curve to the theoretical dependence and adjusts an offset until the two curves coincide. In the amperometric mode the tip interferes with the concentration profiles of redox species in the tip-substrate gap, and this yields a unique tip current-distance dependence. In the potentiometric mode, however, the tip is passive and does not interfere with the concentration profiles, although it shields the sample surface from the bulk solution. Several procedures have been reported to assess the tip-substrate distance in these conditions. The simplest consists in approaching the tip under visual observation with a microscope until it is seen to barely touch the substrate surface. This position is then defined as the zero on the tip-substrate distance scale. The method is adequate when the user does not require an accurate estimate of the tip-substrate distance. More sophisticated procedures have been reported and are described in the following sections.

1. *Switching from Amperometric to Potentiometric Response*

Metal oxide based ion-sensitive devices have an advantage in that they can sustain Faradaic reactions and therefore be operated in the conventional amperometric mode. Because of the geometry of the electrode and of the tip-substrate gap, the rate of the Faradaic process is diffusion controlled and the tip current provides a measure of the tip-substrate distance (see Chapter 5). Subsequently, the tip is electrochemically reconditioned to perform potentiometric measurements. The whole procedure is carried out in situ to retain precise control of the tip position. Horrocks et al. (72) and Tóth et.al. (73) recorded the steady-state current for oxygen reduction on an antimony microdisk and showed that the current decreases when moving the electrode close to an inert substrate. This is in agreement with the principle of hindered diffusion, and the current-distance curve was exploited to determine the absolute tip-substrate distance. The experiment was conducted as follows. The tip potential was held in the plateau region for oxygen reduction. The approach curve was then recorded from the bulk to the substrate to avoid crashing the tip. After returning to the bulk the tip potential was held well positive of the oxygen reduction wave to regenerate the antimony oxide and reestablish the pH function. After a few seconds the tip was switched to the potentiometric mode and used to probe pH profiles and acquire pH maps over the surface.

This approach is, of course, only possible with solid-state ISE and only with those rapidly electrochemically regenerated. Unless removed by purging, oxygen is always present in the solution and alleviates the need for addition of a redox mediator that could alter the rest potential of the antimony electrode and interfere with the pH. Moreover, the normalization of the approach curve with the limiting current in the bulk removes the need for the knowledge of the bulk concentration and diffusion coefficient of oxygen.

2. *The Conductimetric Approach*

In the absence of redox mediator, the method described previously is not applicable. However, an alternative approach based on conductimetry is possible when the tip does not inhibit the passage of current. An ac impedance technique was exploited to determine the absolute tip-substrate distance between an enzyme-coated microelectrode tip and a sample (74). In the ac mode a sinusoidal voltage (respectively current) is applied between the tip and the counterelectrode normally located in the bulk of the solution. The sinusoidal current (respectively voltage) response is sent to a lock-in amplifier or a frequency response analyzer, and the real component of the impedance is recorded as the tip moves closer to the substrate. This component is in fact a measure of the solution resistance between the microelectrode tip and the counterelectrode. Several parameters determine the magnitude of this resistance: the conductivity of the solution and of the substrate, the tip-substrate distance, and the geometry of the tip. An equivalent circuit is derived to calculate the dependence of the resistance on the tip-substrate distance and produce a theoretical approach curve. The dc mode offers a somewhat simpler alternative to the ac mode. The instrumentation only consists of a voltage source and a current follower, while the procedure is simply to apply a fixed potential between the probe and the counter/reference electrode located in the bulk and measure a small current between the two. The technique, basically analogous to scanning ion-conductance microscopy (75,76), was demonstrated with a Ag/AgCl tip and a Ag/AgCl auxiliary electrode (77) (see Figs. 5 and 6). A constant dc potential difference applied between the two electrodes produces a steady current from two reversible reactions: formation of silver chloride on the tip and dissolution of silver chloride on the auxiliary electrode. As the experiment is conducted in the dc mode, the double-layer capacitance of the electrodes is ignored and the tip-solution-auxiliary electrode system behaves as a purely resistive equivalent circuit. Similarly, the charge transfer resistance at the counterelectrode/solution interface is neglected because of the large surface area and the rapidity of the reaction. The treatment of the tip impedance depends largely on its geometry. Two kinds of tip are available: a solid microdisk and a micropipette. If a microdisk is very small, then its charge transfer resistance may determine the overall resistance and the dependence on tip-substrate distance is lost. This is not a problem with the formation and dissolution of silver chloride, since the charge transfer is very fast. The overall resistance is therefore made of the charge transfer resistance at the tip in series with the solution resistance. Only the latter varies with the tip-substrate distance. The micropipette, on the other hand, has a large internal electrode whose charge transfer resistance can be neglected. However, the open micropipette has an internal solution resistance determined by the conductivity of the filling solution, the same electrolyte as in the SECM cell to avoid a liquid junction potential. The internal solution resistance also depends on the pipette geometry, diameter, and shank but does not depend on the tip-substrate distance. The total resistance therefore consists of the internal tip solution

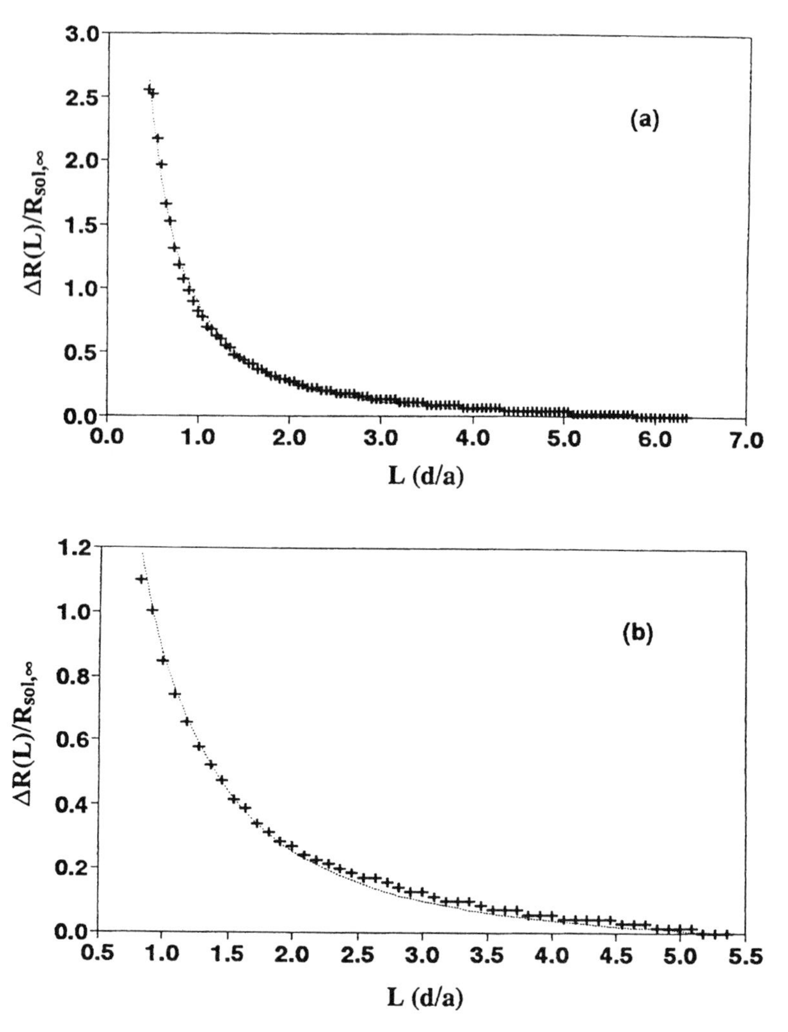

FIG. 5 Solution resistance vs distance curves for Ag/AgCl micropipette electrode over (a) Pt and (b) Teflon targets. Solution, 0.1 M KCl; applied voltage 50 mV between the two Ag/AgCl electrodes; micropipette diameter, 20 μm (RG = 10). (+++) Experimental data; (●●●) theoretical curves. (From Ref. 77.)

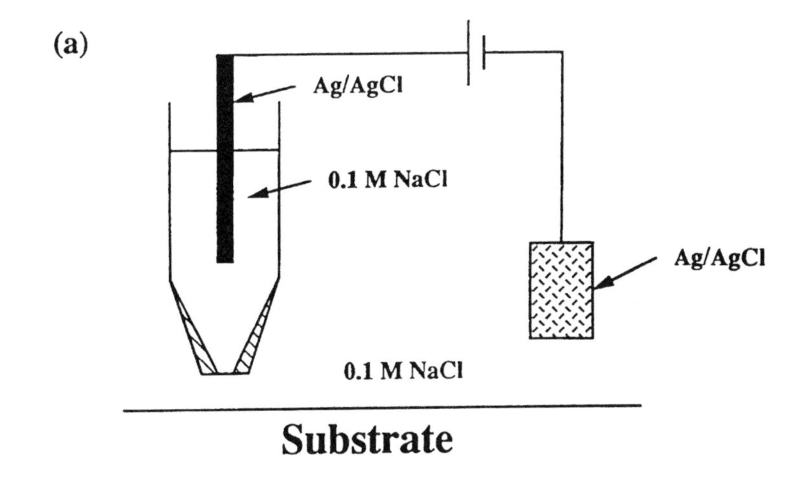

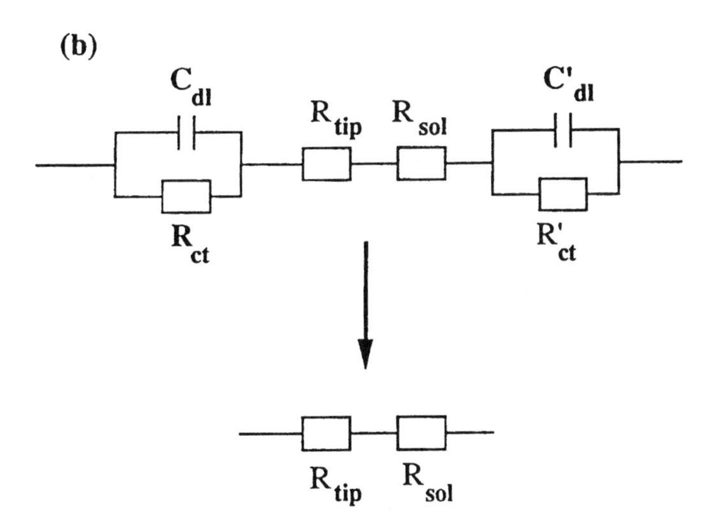

FIG. 6 (a) Setup for distance measurements using Ag/AgCl micropipette electrode. (b) Equivalent circuits corresponding to the electrode configuration shown in (a). (From Ref. 77.)

resistance in series with the external solution resistance; again, only the latter depends on the absolute tip-substrate distance. In both cases the tip-substrate distance is determined by fitting the experimental resistance profile to its theoretical equivalent. In fact, the equation for the potential distribution is analogous to the steady-state diffusion equation, and the theoretical dependence of the resistance is very easily derived from the amperometric feedback approach curve to an insulating substrate. The following equations illustrate the analogy and data treatment. In the steady state the diffusion equation is written as

$$\nabla^2 c = 0 \tag{7}$$

and the current density j is given by

$$j = nFD\nabla c \tag{8}$$

where n, F, D, and c have their usual meaning. Far away in the bulk the steady state tip current to a microdisk, $i_{T,\infty}$ is

$$i_{T,\infty} = 4nFDc^b a \tag{9}$$

where c^b is the bulk concentration and a the tip radius. Similarly, the potential distribution is written as

$$\nabla^2 \phi = 0 \tag{10}$$

and the current density, j is given by

$$j = \kappa \nabla \phi \tag{11}$$

where ϕ is the potential in solution and κ the conductivity of the solution. The analogy between Eqs. (7) and (8) and (9) and (10) leads to the following relationships:

$$R_{sol}(d)/R_{sol,\infty} = i_{T,\infty}/i_T(d) \tag{12}$$

where $R_{sol}(d)$ and $i_T(d)$ are respectively the solution resistance and tip current at a given tip substrate distance d. The solution resistance between the microdisk and the counterelectrode at infinity, $R_{sol,\infty}$ is given by (78)

$$R_{sol,\infty} = 1/4\kappa a \tag{13}$$

The total resistance between the probe and the secondary electrode, $R_{total}(d)$, is calculated from the applied voltage and the measured current. As discussed previously, this resistance is the sum of two terms

$$R_{total}(d) = R_{tip} + R_{sol}(d) \tag{14}$$

where R_{tip} is a constant equal to the charge transfer resistance for a microdisk or to the internal solution resistance for a micropipette. When the tip is far away in the bulk, the substrate does not interfere with the ionic pathway between the two electrodes and $R_{sol}(\infty)$ is a constant. However, when the tip approaches the substrate, the latter dramatically blocks the ionic pathway and $R_{sol}(d)$ increases. R_{tip}, on the other hand, is not a function of d and it cancels out when the left-hand side of Eq. (12) is rewritten as follows:

$$\frac{R_{\text{sol}}(d)}{R_{\text{sol},\infty}} = \frac{R_{\text{total}}(d) - R_{\text{total}}(\infty)}{R_{\text{sol},\infty}} + 1 \qquad (15)$$

Combining Eqs. (12) and (15) leads to

$$\frac{R_{\text{total}}(d) - R_{\text{total}}(\infty)}{R_{\text{sol},\infty}} = \frac{i_{T,\infty}}{i_T(d)} - 1 \qquad (16)$$

where $R_{\text{total}}(d)$ and $R_{\text{total}}(\infty)$ are measured, $R_{\text{sol},\infty}$ is calculated with Eq. (13), and $i_T(d)/i_{T,\infty}$ is the familiar amperometric approach curve for hindered diffusion described by the approximate analytical equations discussed in Chapter 5. Thus, the experimental resistance profile can be compared with the theoretical profile and the absolute tip-substrate distance derived.

It is important to stress one major difference between ac and dc modes. In the ac mode a conducting substrate participates in the conduction process and reduces the solution resistance observed. On the contrary, an insulating substrate blocks the ionic pathway and increases the solution resistance observed. In the dc mode both conducting (provided the applied potential is sufficiently small that no Faradaic process occurs on the substrate) and insulating substrates block the conduction pathway and increase the resistance between the tip and the auxiliary electrode (see Fig. 5).

3. *Fit to Theoretical Concentration Profile*

When the electrode resistance is very large, e.g., with an ion-selective microelectrode, the method described above does not work. An alternative approach consists of fitting the experimental concentration profile calculated from the tip potential to the theoretical profile predicted for the substrate geometry and activity. When the system investigated is at steady state, there is a unique relationship between tip potential and tip-substrate distance and an absolute distance scale can be determined. However, the procedure fails when the system is not at steady state because most ISE are unable to follow rapid changes in concentration.

The experiment is carried out under diffusion control. Theoretical concentration profiles are calculated by solving Fick's second law of diffusion in the steady state with boundary conditions appropriate to the solution domain and to the substrate, taking into account its geometry and the type of reaction occurring on it. Assumption is made that the redox species are stable and not involved in a homogeneous reaction in solution. Two geometries known to produce steady-state concentration profiles have been considered (72,77): the hemisphere and the microdisk. The former only requires a radial dimension, and the diffusion equation can be solved analytically. The latter, on the other hand, necessitates cylindrical coordinates and the solution becomes much more complex. With the latter a closed form analytical expression is available for the concentration profile above the center of the disk. The tip potential is determined by the concentration at the tip coordinates when the tip is small relative to the substrate. However, the situation is more complicated when the tip is comparable to or larger than the substrate. In this case the tip intersects a range of concentration profiles and the tip potential has to be calculated from the average concentration over the tip surface. There again the expression for the average concentration depends on the geometry of the substrate.

Since the concentration profile reflects the concentration in the bulk solution and at the substrate surface (or the flux at the substrate), the nature of the fitting depends on whether the substrate produces or consumes ISE detectable species. When the substrate acts as a source the concentration depends on the distance away from the substrate and on the flux or concentration at the substrate surface. Thus the fitting involves two unknowns: the surface flux or surface concentration (assuming that the substrate radius is known) and the offset needed to turn the relative distance scale into an absolute scale. When the substrate acts as a sink, the bulk concentration is known or measured. In the limiting diffusion-controlled condition where the surface concentration is zero, the offset is the only unknown.

While the theoretical treatment appears simpler than that of the feedback amperometric mode, in practice quantitative agreement between theoretical and experimental concentration profiles is more difficult because the tip travels through the diffusion layer due to the substrate activity. Thus, the tip introduces convection within the substrate diffusion layer while shielding the substrate from the bulk. In the amperometric mode diffusion is restricted to the tip-substrate gap and thus shielding and convection are not a problem. On the whole, the fit between experimental and theoretical concentration profiles works well when the tip is at least one tip radius away from the surface. Shielding dominates at closer tip-substrate distances where the tip significantly affects the concentration profile and the fit is very poor.

As stated earlier, the theoretical concentration profile is calculated on the assumption that the species of interest is not involved in a homogeneous reaction. With a pH probe such as the antimony-antimony oxide tip, this assumption is not valid because the buffer capacity of the solution significantly affects the concentration profile. For example, protons generated at the substrate react with the base form of the buffer and the concentration profile is no longer solely diffusion controlled. Similarly hydroxide anions produced by the substrate titrate the acid form of the buffer. A high-capacity buffer reestablishes the bulk conditions very close to the substrate, while a low-capacity buffer allows the concentration profile to extend far away from the substrate before returning to the bulk concentration. These effects were quantitatively predicted and theoretical pH profiles were found to agree with experimental profiles (72). The agreement worsens when the tip-substrate distance is less than three times the target radius. The buffering capacity can be observed on images of the antimony tip potential recorded over a Pt target undergoing water reduction. The pH variations and the region where pH changes are larger in a low buffer concentration than in a high buffer concentration. This approach was successfully used to map, on the micrometer scale, pH changes due to electrochemical reactions, corrosion processes, biocatalytic reactions, and metabolic processes of microorganisms.

4. *Double Tip Techniques*

Several of the procedures described in the previous sections can be advantageously carried out with double barrel tips. Such a probe consists of two capillaries (see Sec. V.B), one of which acts as the potentiometric sensor, while the other is used to determine the tip-substrate distance. For example (79), a gallium microdisk was combined with an ion-selective (K^+) potentiometric probe to image K^+ activity near the aperture of a capillary (see Fig. 7). Similarly (77), a double barrel tip with one channel as an open Ag/AgCl micropipette for solution resistance measurement and the other channel as an ion-selective neutral carrier–based microelectrode for potentiometric measurements was successfully used to image concentration distributions for NH_4^+ (Fig. 8) and Zn^{2+} (Fig. 9). While dual-channel tips facilitate the approach of the substrate and permit a direct determination of the absolute tip-substrate distance, their difficult fabrication severely limits their use. Reference 80 compares the above methods.

B. Considerations Specific to the Potentiometric Mode

1. *Ohmic Drop Effects*

When performing potentiometric measurements with a traveling probe, one needs to take into account the effect of ohmic drop. Whereas this effect is the basis of SRET measurements, it becomes a nuisance in potentiometric SECM applications. If the substrate is an electrode involved in a Faradaic process, the current flowing between the substrate and the counterelectrode leads to potential gradients in solution. The tip will be sensitive to the potential distribution, and this may overcome the signal due to the concentration change for the ion of interest. This is particularly pronounced if the reference electrode associated to the tip is located far away in the bulk and of course if the solution conductivity is low. To remedy this situation some researchers have used double barrel electrodes where one channel acts as the ion-sensitive element and the other acts as a reference electrode (81,82). In the life sciences intracellular measurements are usually carried out in this way. Alternatively, it is possible to subtract the ohmic drop from the tip

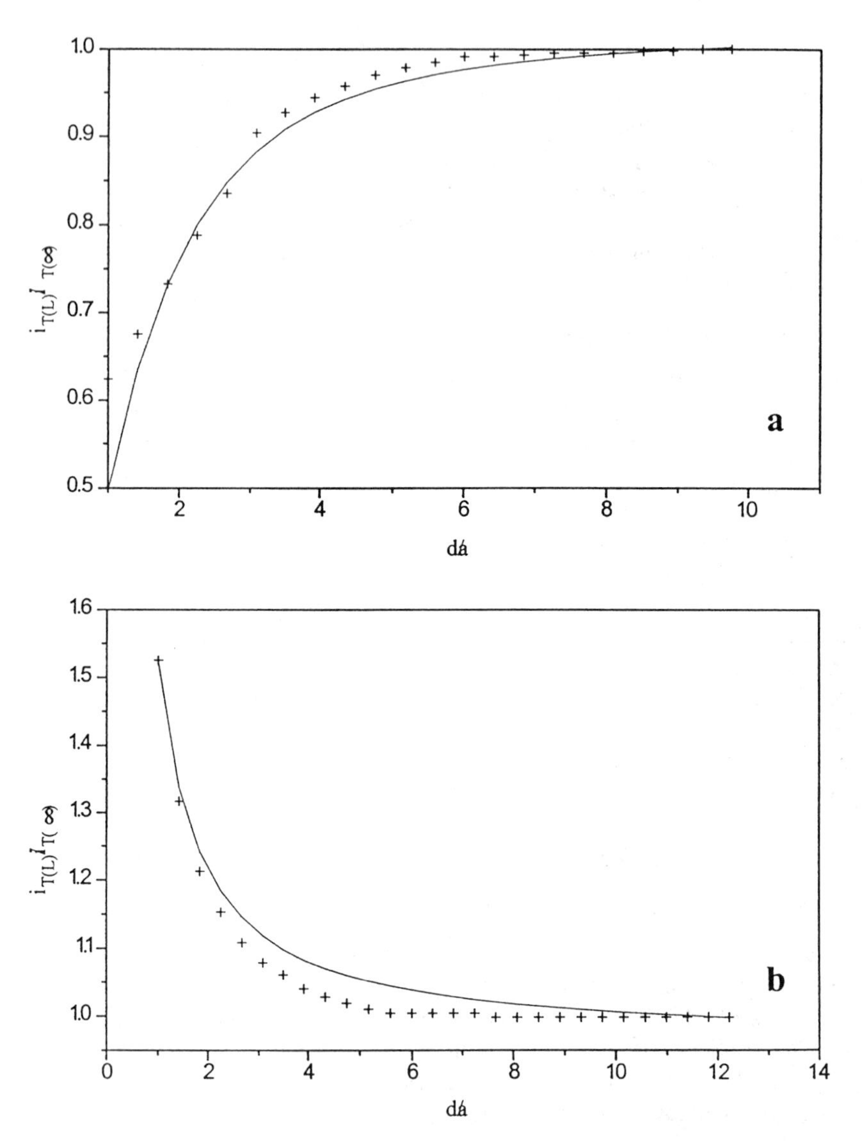

FIG. 7 Feedback current-distance curves for a gallium SECM tip of 22-μm diameter operating in amperometric mode in 0.1 M KCl and 5 mM $Ru(NH_3)_6Cl_3$: (a) Pt surface; (b) Teflon surface. Pluses represent experimental data, while the solid lines are theoretical data. The tip potential was held at -0.6 V vs. Ag/AgCl reference electrode. (From Ref. 79.)

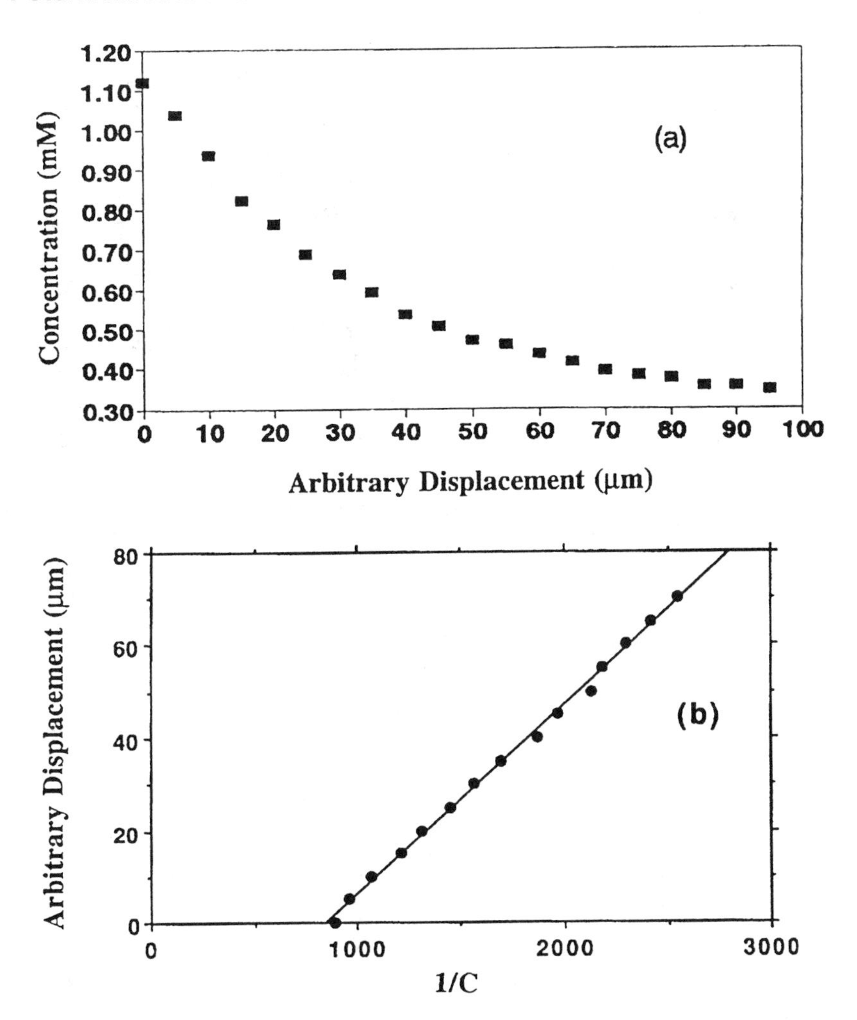

FIG. 8 (a) NH_4^+ concentration profile above a 20-μm-diameter urease gel target (in the z-direction) over the center of the target. Tip diameter: 18 μm; bulk concentration, 0.1 M urea and 0.1 M Tris, pH 7. (b) Plot of arbitrary distance vs 1/c using the data in part (a). (From Ref. 77.)

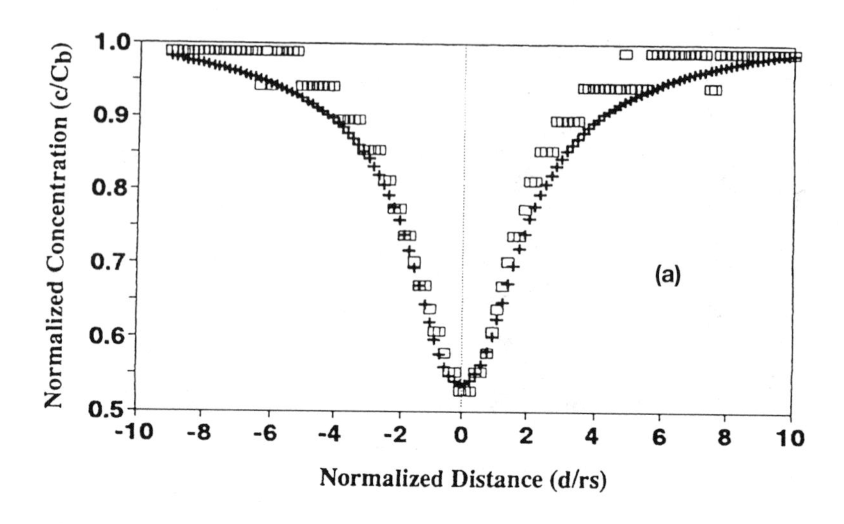

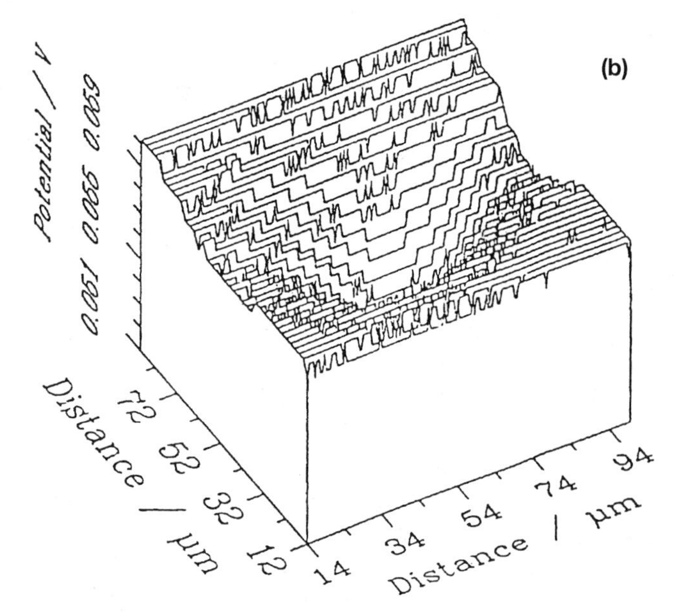

FIG. 9 Image of the Zn^{2+} concentration profile around a 25-μm-diameter Hg-coated Au electrode in 0.84 mM $Zn(NO_3)_2$ + 0.1 M NH_4Cl + 50 mM pH 7 Tris-HCl buffer. (a) Radial direction across the center of the electrode at a constant height of 11 μm from the surface. Electrode potential held at -1.15 V; tip diameter 10 μm; tip scan rate 2 μm s^{-1}. C_b bulk concentration, r_s target radius. Empty squares represent experimental data, while pluses are theoretical data. (b) Corresponding surface plot. (From Ref. 80.)

response. This requires an estimate of the solution resistance between the tip and its reference electrode and a measure of the substrate current (83).

2. *Mapping and Resolution*

The principles of potentiometric imaging are far simpler than those of amperometric imaging. In the amperometric mode the tip directly interacts with the concentration profiles of the redox mediator by consuming one redox form and producing the other. Hence the measurement severely alters the concentration profiles. In the potentiometric mode the tip is passive. Its response depends solely on the concentration profiles it encounters. It shields the substrate from the bulk solution but does not consume or produce a species. Unlike the amperometric mode there is no edge effect and the insulator around the tip plays no role. The tip potential reflects the concentration of the ion of interest, averaged over the sensing region. Thus the tip is only sensitive to the solution in contact with its sensing element. The geometry and dimensions of the sensing element therefore determine the spatial resolution (84). To improve the spatial resolution one tries to miniaturize the tip and the insulator around it. The smaller the sensing element, the better the spatial resolution; the smaller the overall tip end, the less shielding there is.

In the potentiometric mode the single sensor tip is only able to image the distribution of a specific ion. A complex multibarrel tip is needed to follow different species simultaneously. This is in contrast to the amperometric mode, where a redox mediator is selected by adjusting the tip potential. Thus, by having more than one redox mediator in solution, it is possible to produce several amperometric images simply by changing the tip potential each time. To our knowledge no one has made use of this idea, but in principle it is feasible.

C. Applications

In the following sections we consider several potentiometric applications. Many articles do not refer directly to scanning electrochemical microscopy, but all are closely related to the SECM principles and were therefore included in the present chapter. We have not included a section on applications related to potentiometric probing of biological substrates since they are covered in Chapter 11 of this volume.

1. *pH Probing*

Of all the applications, the use of potentiometric tips to follow pH profiles near electrodes, corroding surfaces, or membranes appears the most common. A short selection of these studies is included below.

Klusmann and Schultze (84) made a comparison between theoretical and experimental pH profiles near planar and microdisk electrodes for constant flux and constant concentration sources. They also investigated the spatial resolution for pH mapping. Horrocks et al. (72) reported the results of SECM experiments where potentiometric pH-selective tips were used to image local pH changes in several model chemical systems. These included platinum microelectrodes during water reduction (Fig. 10), a corroding disk of silver iodide in aqueous potassium cyanide, a disk of immobilized urease hydrolyzing urea, and a disk of immobilized yeast cells in glucose solution (Fig. 11). A detailed investigation of proton concentration profiles close to silver iodide–based ion-selective membranes was reported by the same group (73). pH measurements were carried out in the vicinity of a semipermeable membrane operating in a chlor-alkali cell (82). The tip consisted of two channels: one filled with antimony to form a pH sensing disk, the

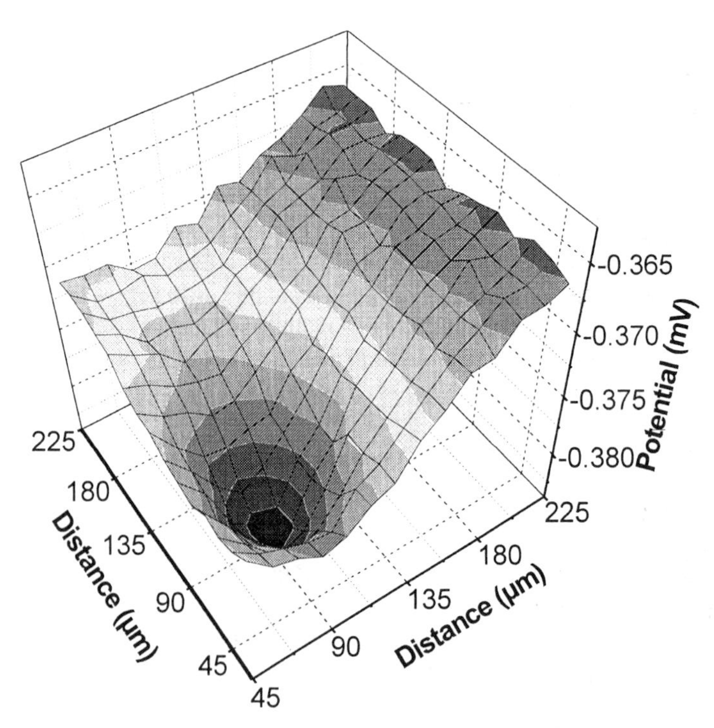

FIG. 10 Image of the pH profile around a 10 μm Pt target reducing water. The potential of the target was -2 V and the current was 0.5 μA. The diameter of the Sb tip was 15 μm, the tip-surface distance was 33 μm, and the tip scan rate was 10 μm/s. The solution was 1 mM phosphate buffer (pH 7).

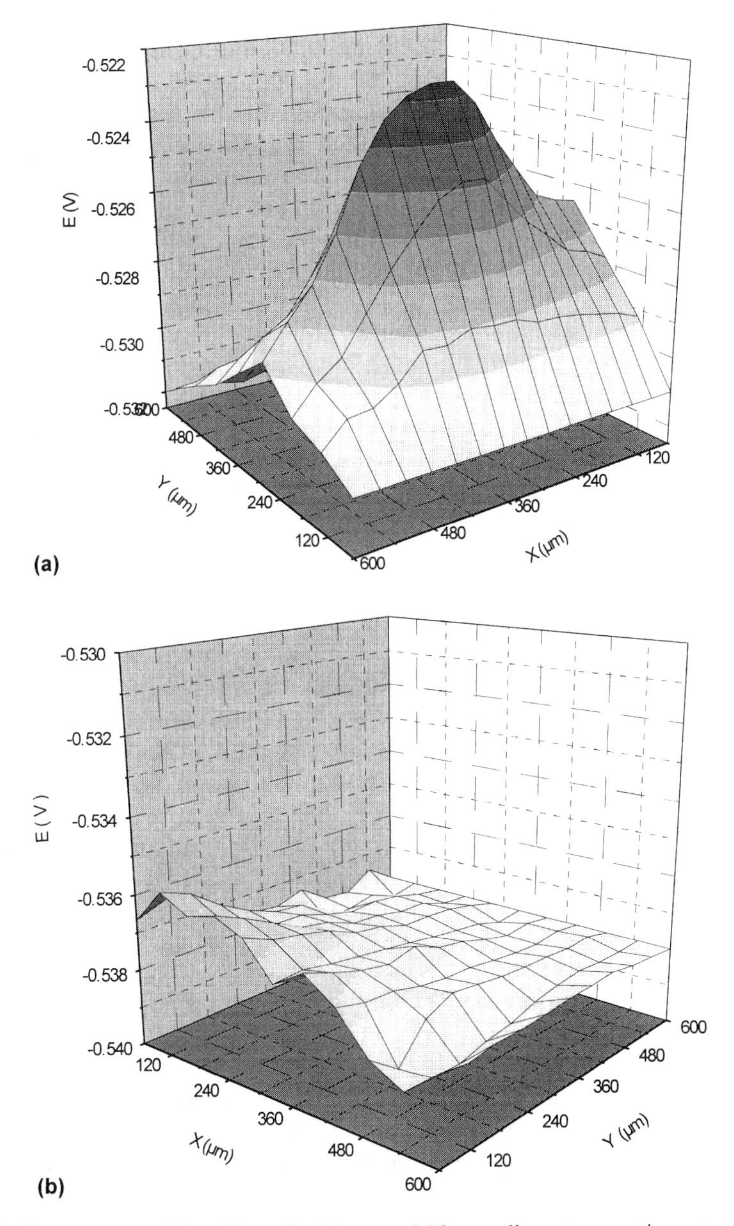

FIG. 11 Images of the pH profile (a) around 80-μm-diameter yeast/agarose target. The tip diameter was 15 μm, the tip scan rate was 10 μm/s, and the solution contained 5 mM glucose in 1 mM phosphate buffer. The tip-target distance was 20 μm; (b) after treatment of the yeast culture with 1 g/dm^2 $NaHSO_3$.

other left open to act as a Luggin capillary and eliminate the effect of iR drop. This study showed the extent of penetration and leakage of hydroxides through the membrane. Similarly, pH profiles were recorded near bilayer lipid membranes with an antimony microelectrode (85). pH measurements were also performed in the vicinity of a cathode evolving hydrogen gas using an antimony microelectrode (86). The effects of a buffering agent, glycine, were investigated. Tanabe et al. (87) used potentiometric and amperometric SECM techniques to investigate the ionic environment above pits on stainless steel surfaces. They report the potentiometric detection of Cl^- with Ag/AgCl tips and of protons with 7 μm diameter carbon fibers; the principle of potentiometric detection of protons with a carbon fiber is not clear. Park et al. (81) fabricated double barrel tips (one filled with a proton-sensitive liquid exchange resin, the other acting as a local reference electrode) and probed the OH^- build-up generated by cathodic reactions on Al_3Fe in contact with Al 6061 in a galvanic cell. Luo and coworkers (88) also reported the fabrication of potentiometric pH tips to investigate localized corrosion.

2. *Close-Proximity Mode Experiments*

As in amperometric applications, many experiments can be conducted in the close proximity mode where the tip is moved very close to the substrate surface and a perturbation is applied to the sample. This perturbation may take several forms, typically potentiostatic or galvanostatic excursions if the sample is acting as an electrode, but also optical illumination with a laser beam, change of solution, etc. The tip response is then recorded as a function of time following the application of the perturbation. In these conditions potentiometric detection offers two advantages over amperometric detection: (1) the range of ions detectable is extended to nonelectroactive species such as alkali metals, and (2) the tip response is selective. There are, however, some drawbacks. Because of the high impedance of the electrometer, the response time is worse in potentiometric applications where the t_{90} is rarely below 30 s. This must be compared to the millisecond time scale available with amperometric responses (89). Ohmic drop may also affect the tip potential.

Troise Frank et al. (83) reported the use of potentiometric tips to probe ion fluxes near surfaces. Ten and 50 μm diameter silver microdisk electrodes were used to probe the concentration profile due to the diffusion of Ag^+ to and from a planar Ag electrode. A similar experiment was carried out with a 50 μm diameter micro-cylinder substrate electrode. The tip response was recorded at various locations above the substrate while simultaneously applying a potential square wave to the substrate to induce the dissolution and plating of silver. These experiments demonstrated the principle of the technique and confirmed the possibility of obtaining spatially resolved information. Moreover, these experiments clearly showed that the tips response times were in fact very good ($t_{90} < 10$ ms) (Fig. 12). Undoubtedly this was due to their very low impedance.

In a separate experiment, 10 and 50 μm diameter Ag/AgCl microdisks were used to monitor the flux of Cl^- consumed and generated by a film of polyaniline undergoing redox cycles. The tip potential was recorded while cycling the potential of the polyaniline film. The results obtained (Fig. 13) provide direct evidence for the ingress and egress of Cl^- ions and support the mechanism previously proposed for the oxidation of polyaniline. Subsequent experiments carried out in the amperometric mode for the reduction of protons were found to be in complete agreement with the potentiometric data (90,91).

Horrocks et al. (74) performed similar tip-substrate voltammetric experiments where the potentiometric pH tip response was recorded while scanning the potential of a substrate electrode. Significant pH variations in the vicinity of a gold electrode reducing oxygen were observed with antimony tips.

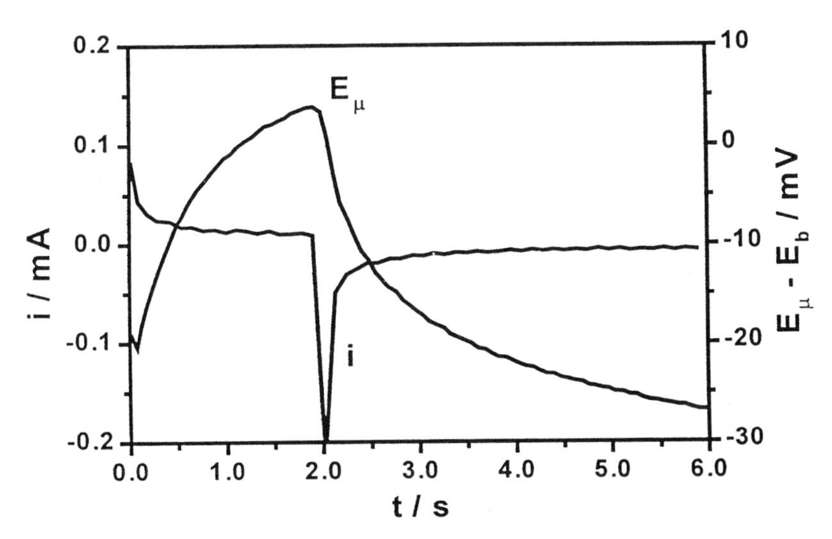

FIG. 12 Ag tip potential with respect to the bulk electrode potential as a function of time while pulsing the substrate electrode from 0 to 50 mV vs. Ag/Ag^+ for 2 s then from -50 mV for 4 s. The tip-substrate distance is <10 μm. The tip is a 50-μm-diameter silver disk, the substrate is a large planar silver electrode. The substrate current is shown on the left-hand scale. (From Ref. 83.)

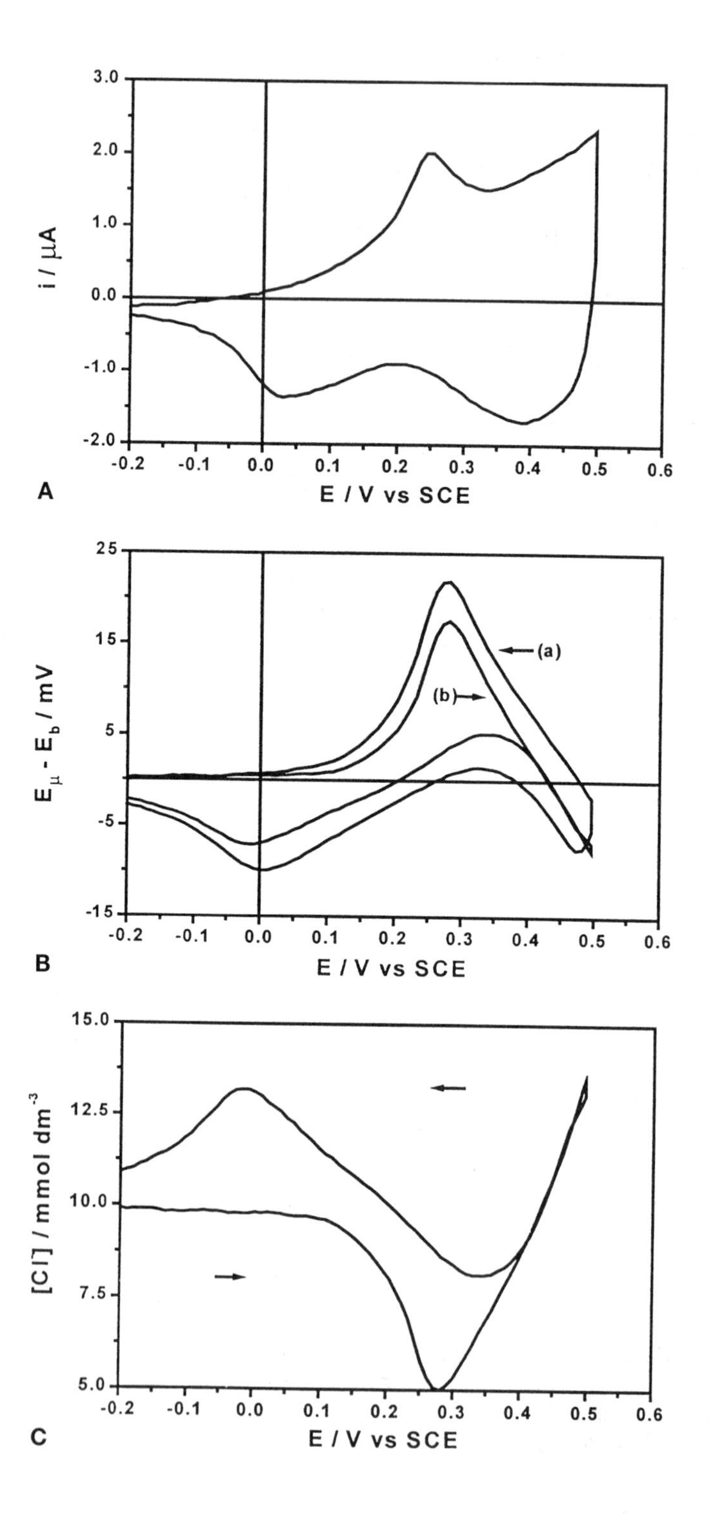

3.0
2.0
1.0
0.0
-1.0
-2.0
i / μA
-0.2
-0.1
0.0
0.1
0.2
0.3
0.4
0.5
0.6
A
E / V vs SCE
25
15
5
-5
-15
E_μ - E_b / mV
(a)
(b)
-0.2
-0.1
0.0
0.1
0.2
0.3
0.4
0.5
0.6
B
E / V vs SCE
15.0
12.5
10.0
7.5
5.0
[Cl⁻] / mmol dm⁻³
-0.2
-0.1
0.0
0.1
0.2
0.3
0.4
0.5
0.6
C
E / V vs SCE

Kemp et al. (92) operated in the close proximity mode to study the concentration of chloride ions generated near the surface of UV-irradiated titanium dioxide in the presence of aqueous 2,4-dichlorophenol. The Cl^- concentration was measured potentiometrically with a 50 μm diameter Ag/AgCl microdisk. Experiments were performed at several tip-substrate distances, and the concentration of chloride was recorded as a function of time after turning on and turning off the UV irradiation.

VI. CONCLUSION

One should not see the potentiometric mode as an alternative but as a complement to the amperometric mode. The potentiometric mode offers the possibility of performing measurements on non–redox active species, e.g., not detectable amperometrically in aqueous solutions. Another advantage is the increased selectivity of the potentiometric response compared to the Faradaic response. The logarithmic response is also useful when the species of interest is in low concentrations. Since the tip is passive, only substrate generation–tip detection experiments are possible. However, the tip does not alter the concentration profile of the chemically or electrochemically generated species. The potentiometric micropipette tips have very small dimensions, typically 1 μm or less, and the shielding is minimal.

However, one should not forget that the potentiometric mode has several drawbacks. The fabrication of most potentiometric tips is much more involved than that of the conventional amperometric tips. It is necessary to have the electrode made by a very skilled technician. Despite this, even when following a proven recipe the success rate is relatively low. The response of potentiometric tips is not always Nernstian and a calibration is required before and after performing the experiment. The behavior typically varies from one microelectrode to another. Potentiometric tips cannot rely on positive and negative feedback diffusion, thus it is difficult to assess the tip-substrate distance from the tip response. Several approaches are available, but most are cumbersome. In the potentiometric mode the response

FIG. 13 (A) Cyclic voltammogram of a polyaniline film on Pt (area 9.3×10^{-4} cm^2) recorded at 5 mV s^{-1} in 0.01 mol dm^{-3} HCl. (B) Micro-indicator (Ag/AgCl) response recorded as a function of the potential of the Pt electrode supporting the polyaniline film. The probe, a 10-μm-diameter Ag/AgCl microdisk with an RG circa 7, is located very close to the polymer film. (a) uncorrected, (b) iR corrected. (C) Calculated local Cl^- concentration as a function of the potential of the Pt electrode supporting the polyanailine. (From Ref. 83.)

time depends on several time constants, namely of the measuring circuit, of the tip itself, and of the other electrode. The transient response is typically slower than with amperometric tip, but this is not so much a problem for potentiometric probing of one- and two-dimensional concentration distributions because the tip moves slowly and steep concentration gradients are not common. Slow response times can be a limiting factor when performing close-proximity experiments.

REFERENCES

1. M Cremer. Z Biol 47:562, 1906.
2. B Lengyel, E Blum. Trans Faraday Soc 30:461, 1934.
3. G Eisenman, DO Rudin, JV Casby. Science 126:831, 1957.
4. HJC Tendeloo. J Biol Chem 113:333, 1936.
5. RS Anderson. J Biol Chem 115:323, 1936.
6. IM Kolthoff, HL Sanders. J Am Chem Soc 59:416, 1937.
7. SK Sinha. J Indian Chem Soc 30:529, 1953.
8. SK Sinha. J Indian Chem Soc 32:35, 1955.
9. D Woermann, KF Boahoeffer, F Helfferich. Z Phys Chem 8:265, 1956.
10. JS Parsons. Anal Chem 30:1262, 1958.
11. E Pungor, E Hollos-Rokosinyi. Acta Chim Acad Sci Hung 27:63, 1961.
12. E Pungor, J Havas, K Toth. Z Chem 5:9, 1965.
13. MS Frant, JW Ross Jr. Science 154:1553, 1966.
14. JW Ross Jr. Science 156:1378, 1967.
15. J Ruzicka, JC Tjell. Anal Chim Acta 49:346, 1970.
16. CJ Coetzee, H Freiser. Anal Chem 40:2071, 1968.
17. Z Stefanac, W Simon. Chimia 20:436, 1966.
18. Z Stefanac, W Simon. Microchem J 12:125, 1967.
19. LAR Pioda, V Stankova, W Simon. Anal Lett 2:665, 1969.
20. R Bloch, A Shatkay, HA Saroff. Biophys J 7:865, 1967.
21. A Craggs, GJ Moody, JDR Thomas. J Chem Educ 51:541, 1974.
22. P Bergveld. IEEE Trans Biomed Eng BME 17:70, 1970.
23. E Lindner, VV Cosofret, S Ufer, TA Johnson, RB Ash, HT Nagle, MR Mauman, RP Buck. Fresenius J Anal Chem 346:584, 1993.
24. MR Neuman, RP Buck, VV Cosofret, E Lindner, CC Liu. IEEE Eng Med Biol June/July:409, 1994.
25. S Nomura, M Nakao, T Nakanishi, S Takamatsu, K Tomota. Anal Chem 69: 977, 1997.
26. PCJ Caldwell. J Physiol 126:169, 1954.
27. JAM Hinke. Nature 184:1257, 1959.
28. AA Lev, EP Buzhinsky. Tsitolgiya 3:614, 1961.
29. RC Thomas. J Physiol 210:82P, 1970.
30. RC Thomas. J Physiol 238:159, 1974.
31. LR Pucacco, NW Carter. Anal Biochem 73:501, 1976.
32. S Levy, JA Coles. Experientia 33:553, 1977.

33. H Yamaguchi, NL Stephens. Fed Proc 36:499, 1977.
34. W McD Armstrong, CO Lee. Science 171:413, 1971.
35. CO Lee, W McD Armstrong. J Membrane Biol 15:331, 1974.
36. A Mauro. Fed Proc Fed Soc Exp Biol 13:96, 1954.
37. GA Kerkut, RW Meech. Comp Biochem Physiol 19:819, 1966.
38. JH Saunders, HM Brown. J Gen Physiol 70:507, 1977.
39. JW Ross. Solid-state and liquid membrane electrodes. In: RA Durst, ed. Ion-Selective Electrodes. Washington, DC: Nat. Bur. Stand., 1969, pp 57–88.
40. JL Walker. Anal Chem 43:89A, 1971.
41. FW Orme. In: M Lavallee, OF Schanne, NC Herbert, eds. Glass Microelectrodes. New York: Wiley, 1969, pp 376–395.
42. M Oehme, W Simon. Anal Chim Acta 86:21, 1976.
43. RC Thomas, W Simon, M Oehme. Nature 258:554, 1975.
44. D Ammann. Ion-Selective Microelectrodes: Principle, Design and Application. New York: Springer-Verlag, 1986.
45. D Ammann, E Pretsch, W Simon, E Lindner, A Bezegh, E Pungor. Anal Chim Acta 171:119, 1985.
46. J Tarcali, G Nagy, K Tóth, E Pungor, G Juhász, T Kukorelli. Anal Chim Acta 178:231, 1985.
47. RP Kraig, C Nicholson. Science 194:725, 1976.
48. A Cragg, GJ Moody, JDR Thomas. Analyst 103:68, 1978.
49. J Ruzicka, EH Hansen, JC Tjell. Anal Chim Acta 67:155, 1973.
50. HM Brown, JP Pemberton, JD Owen. Anal Chim Acta 85:261, 1976.
51. M Oehme, M Kessler, W Simon. Chimia 30:204, 1976.
52. M Meyerhoff. Anal Chem 52:1532, 1980.
53. P Schulthess, Y Shijo, HV Pham, E Pretsch, D Ammann, W Simon. Anal Chim Acta 131:111, 1981.
54. D Amman, F Lanter, RA Steiner, P Schulthess, Y Shijo, W Simon. Anal Chem 53:2267, 1981.
55. TB Bolton, RD Vaughan-Jones. J Physiol 270:801, 1977.
56. JH Saunders, HM Brown. J Gen Physiol 70:507, 1977.
57. ME Rice, GA Gerhardt, G Nagy, PM Hierl, RN Adams. Neuroscience 15:891, 1985.
58. C Nicholson, JM Phillips. J Physiol 321:225, 1981.
59. F Vyskocil, N Kriz. Pflug Arch Ges Physiol 337:265, 1972.
60. E Neher, HD Lux. J Gen Physiol 61:385, 1973.
61. JA Coles, M Tsacopoulos. J Physiol 270(1):12, 1977.
62. Y Umezawa, K Umezawa, H Sato. Pure Appl Chem 67:507, 1995.
63. E Lindner, K Tóth, E Pungor. Dynamic Characteristics of Ion-Selective Electrodes. Boca Raton, FL: CRC Press, 1988.
64. E Lindner, K Tóth, E Pungor. Pure Appl Chem 58:469, 1986.
65. UR Evans. Iron Steel Inst 141:219, 1940.
66. HS Isaacs, G Kissel. J Electrochem Soc 119:1628, 1972.
67. HS Isaacs, B Vyas. Scanning reference electrode technique in Localized Corrosion. In: F Mansfeld, U Bertocci, eds., Electrochemical Corrosion Testing. ASTM STP 727:3, 1981.

68. CDS Tuck. Corrosion Sci 23:379, 1983.
69. KR Threthewey, DA Sargeant, DJ Marsh, S Haines. In: KR Threthewey, PR Roberge, eds. Modelling Aqueous Corrosion, 1994, p 417.
70. HS Isaacs. Corrosion Sci 28:547, 1988.
71. LF Jaffe, R Nuccitelli. J Cell Biol 63:614, 1974.
72. BR Horrocks, MV Mirkin, DT Pierce, AJ Bard, G Nagy, K Tóth. Anal Chem 65:1213, 1993.
73. K Tóth, G Nagy, BR Horrocks, AJ Bard. Anal Chim Acta 282:239, 1993.
74. BR Horrocks, D Schmidtke, A Heller, AJA Bard. Anal Chem 65:3605, 1993.
75. PK Hansma, B Drake, O Marti, SAC Gould, CB Prater. Science 243:641, 1989.
76. CB Prater, PK Hansma, M Tortonese, CF Quate. Rev Sci Instrum 62:2634, 1991.
77. C Wei, AJ Bard, G Nagy, K Tóth. Anal Chem 67:1346, 1995.
78. J Newman. J Electrochem Soc 113:501, 1966.
79. C Wei, AJ Bard, I Kapui, G Nagy, K Tóth. Anal Chem 68:2651, 1996.
80. K Tóth, G Nagy, C Wei, AJ Bard. Electroanalysis 7:801, 1995.
81. JO Park, C-H Paik, RC Alkire. J Electrochem Soc 143:L174, 1996.
82. Y Ogata, S Uchiyama, M Hayashi, M Yasuda, F Hine. J Appl Electrochem 20: 555, 1990.
83. MH Troise-Frank, G Denuault, LM Peter. Faraday Discuss 94:23, 1992.
84. E Klusmann, JW Schultze. Electrochim Acta 42:3123, 1997.
85. YN Antonenko, AA Bulychev. Biochim Biophys Acta 1070:279, 1991.
86. T Honda, K Murase, T Hirato, Y Awakura. J Appl Electrochem 28:617, 1998.
87. H Tanabe, Y Yamamura, T Misawa. Materials Sci Forum 185:991, 1995.
88. JL Luo, YC Lu, MB Ives. J Electroanal Chem 326:51, 1992.
89. AJ Bard, G Denuault, BC Dornblaser, RA Friesner, LS Tuckerman. Anal Chem 63:1282, 1991.
90. MH Troise-Frank, G Denuault. J Electroanal Chem 354:331, 1993.
91. MH Troise-Frank, G Denuault. J Electroanal Chem 379:405, 1994.
92. TJ Kemp, PR Unwin, L Vincze. Faradaray Trans 91:3893, 1995.
93. MR Neuman, RP Buck, VV Cosofrei, E Lindner, CC Liu. IEEE Eng Med Biol June/July:409, 1994.
94. P Bergveld. IEEE Trans Biomed Eng 17:70, 1970.
95. J Inczédy, T Lengyel, AM Ure. IUPAC Compendium of Analytical Nomenclature. 3rd ed. Blackwell Science, 1998.

11

BIOLOGICAL SYSTEMS

Benjamin R. Horrocks

University of Newcastle upon Tyne
Newcastle upon Tyne, England

Gunther Wittstock

Wilhelm-Ostwald-Institute of Physical and Theoretical Chemistry
University of Leipzig
Leipzig, Germany

From the inception of scanning electrochemical microscopy (SECM), attempts to apply the technique to investigations of biological or biochemically relevant problems have been made (1). Measurements with the SECM may be carried out in buffered solutions, a preferred environment for most biological samples, with a scanning electrode that does not touch the specimen and that interferes with the sample much less than tips of alternative scanning probe techniques, e.g., scanning tunneling microscopy (STM) and atomic force microscopy (AFM). SECM with electrochemical detection of metabolites can be used to map biochemical activity, thereby complementing techniques that image only the topography of biological specimens or biomolecules.

This chapter aims to summarize the progress made so far in order to stimulate a broader application in the life sciences. In particular, we hope that groups who do not consider electrochemistry to be their scientific "center of gravity" will be encouraged to make use of SECM. This compilation of methodological approaches and selected applications is intended to assist the specialist to assess the viability of the SECM for the investigation of biological and biochemical problems.

I. APPROACHES TO IMAGING BIOLOGICAL AND BIOCHEMICAL SYSTEMS

Biological specimens can, in principle, be imaged the same way as any other sample (see Chapter 4). The focus of this section will, however, rest clearly on approaches that exploit specific properties of biological systems to generate an image.

A. Negative Feedback Imaging

One of the earliest images using negative feedback was obtained from upper and lower surfaces of leaves immersed in a 0.02 M $K_4[Fe(CN)_6]$ and 0.1 M KCl solution (2). The highly charged mediator, $K_4[Fe(CN)_6]$, does not easily penetrate the cells, and therefore interference with biochemical processes in the leaves was reduced. Where cells protruded from the surface, low currents were recorded. The biochemical function of the specimen was not involved in the image-generation mechanism. Therefore, the imaging process does not differ from that at a rough, insulating inorganic surface.

Nevertheless, this imaging mode has stimulated an ongoing scientific discussion about the imaging principle in experiments with insulating biomolecules like DNA immobilized on nonconductive mica supports using scanning tunneling microscope (STM)-like apparatus. Guckenberger et al. (3) were the first to demonstrate molecular resolution imaging of DNA immobilized on mica surfaces. A gold electrical contact to the mica was located several millimeters laterally from the scanning tip and image area. Such images can be obtained only if the current is of the order of 1 pA and a high bias voltage (3–10 V) is applied. Initially it was suggested that electron tunneling between the tip and the distant gold contact may be promoted by special tunneling pathways within the biomolecules and on the mica surface (4). Fan and Bard (5) have proposed an alternative explanation (Fig. 1). There is evidence that a thin water film is formed at high relative humidity on the hydrophilic surfaces of mica and around the immobilized biomolecules. According to Fan and Bard, the tip apex remains in contact with the thin water layer, thus forming an electrochemical cell with the gold contact. Electrochemical reactions including oxygen or proton reduction and oxygen evolution can take place at the tip and the gold. The current between the tip and the gold pad is conducted by ion migration in the surface water film. Consequently, the image would not be based on tunneling but on faradaic processes with the feedback system of the STM apparatus adjusting the tip height so that the electrolytic current remains constant. Although performed with an STM apparatus, the method should be regarded as an electrochemical technique based on recording faradaic currents. This explanation has not yet found general acceptance in the community working in this field (6). A report on protein imaging from independent workers (7) has added new experimental evidence for the model proposed by Fan and Bard. Patel et al. (7) were able to demonstrate the presence and critical importance of a water film for "STM-like" imaging of proteins in a sequence of images and ap-

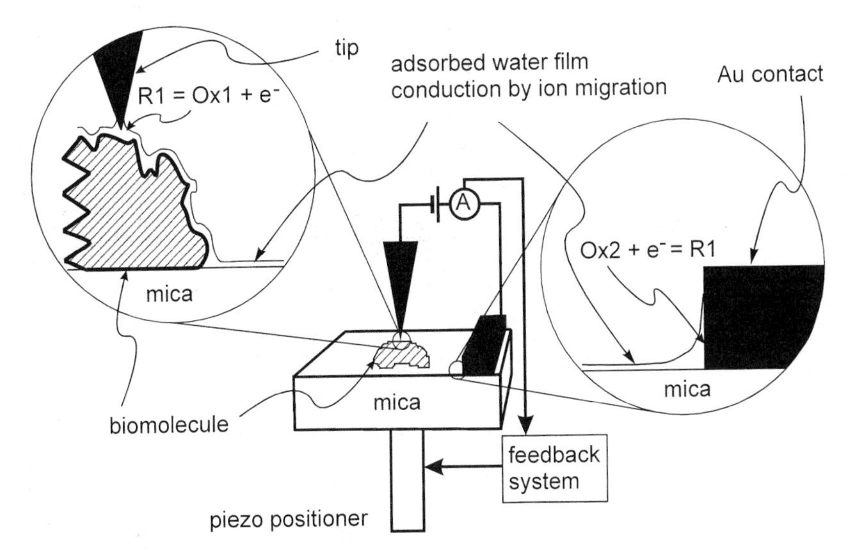

FIG. 1 Mechanism of image formation with an STM-like apparatus at biomolecules on insulating supports as proposed by Fan and Bard (5). At high relative humidity, both biomolecules and support are covered by a water film that permits ion migration and faradaic reactions at the tip apex and the distant gold contact. (Schematic prepared by chapter authors.)

proach curves in which the sample was exposed to a hydrated, dehydrated, and then rehydrated environment. Evidence for a water film capable of supporting faradaic processes has also been obtained from experiments in which silver clusters have been electrodeposited on the mica surface (8).

While the experiments in these studies may approach the ultimate resolution of SECM imaging, the main emphasis in this chapter is put on the utilization of biochemical reactions for imaging formation. Relevant applications of the negative feedback imaging mode are therefore only briefly summarized in Sec. II.A.

B. Nonspecific Positive Feedback Imaging

Positive feedback is obtained above electrically conductive samples. If a layer of biomolecules is placed onto a conducting sample, electron transfer at the covered areas is inhibited and the positive feedback does not reach the value recorded above the uncovered surface. In the example shown in Figure 2, a film of ubiquinone-10, deposited at the location ($130 < x/\mu m < 220$) reduces the positive feedback compared to the bare glassy carbon electrode. The biochemical function of ubiquinone as a cofactor in the respira-

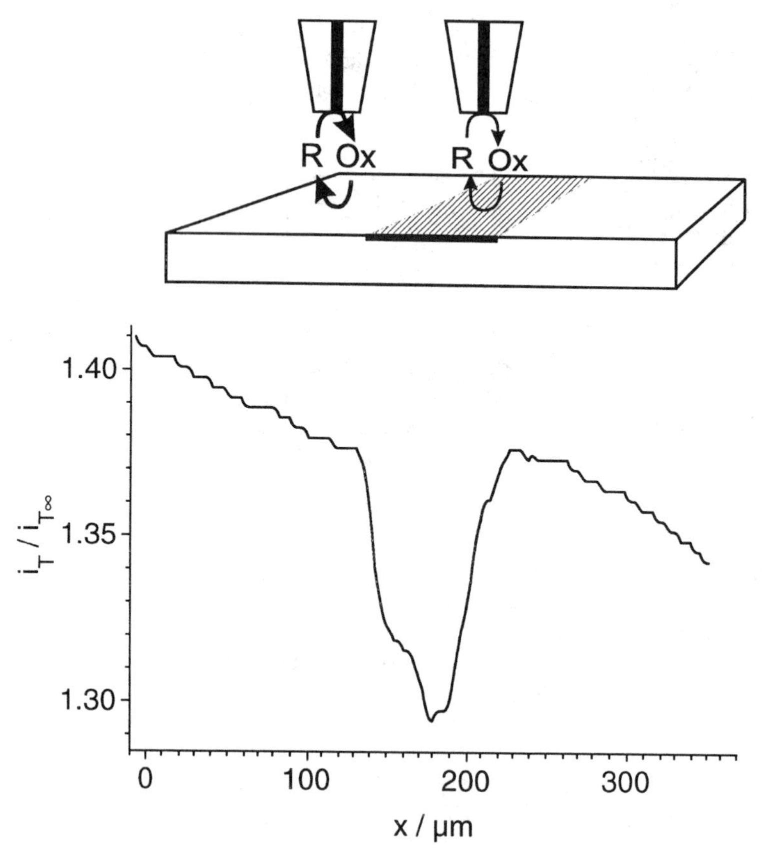

FIG. 2 Reduced positive feedback over a conducting support covered by a layer of biomolecules or other organic surfactants. The example shows a line scan above a glassy carbon electrode which was covered by a layer of adsorbed ubiquinone-10 at the location ($135 < x/\mu m < 210$). Experimental parameters; $E_T = +660$ mV (SCE), $E_S = -250$ mV (SCE), $a = 5$ μm, $v_t = 7$ μm s^{-1}, solution 2 mM ferrocene monocarboxylic acid, 0.1 M KCl, 0.1 M phosphate buffer, pH 7.4 (From G. Wittstock, unpublished, 1996.)

tory chain was not exploited for image generation. Consequently, this imaging mode, as the negative feedback imaging discussed above, does not provide any information about the biochemical function/activity. It may, however, help to locate immobilized biomolecules or other organic films on conducting samples. One should be aware that there is no simple mathematical relation between surface coverage of the biomolecule and the degree to which faradaic reactions are inhibited at the conducting support. The inhibition of heterogeneous electron transfer reactions by adsorbed thin films may depend on coverage, molecular structure of the adsorbate, supramolecular structure within the adsorbate layer, electrode potential, charge, size, and hydrophobicity of the mediator, and the nature and density of defects in the layer even for simple low molecular weight organic adsorbates (9).

C. Enzyme-Mediated Feedback Imaging

An imaging mode that contains more information related to the function of the biomolecule is illustrated in Figure 3. It may be termed enzyme-mediated positive feedback mode. The SECM mediator is chosen such that one of its redox forms can act as an artificial electron acceptor. The enzyme-mediated positive feedback imaging of glucose oxidase was reported by Pierce et al. (10). In this work glucose oxidase (GOx) was covalently attached to a nylon® surface. The working solution contained glucose and ferrocene monocarboxylic acid (Fc). Through deaeration of the electrolyte, the native electron acceptor O_2 was removed from the solution and the enzymatic reaction was halted. At the scanning microelectrode Fc was oxidized to the ferrocenium form (Fc^+). Ferrocenium derivatives serve as effective artificial electron acceptors for GOx (11). If the microelectrode is located above the en-

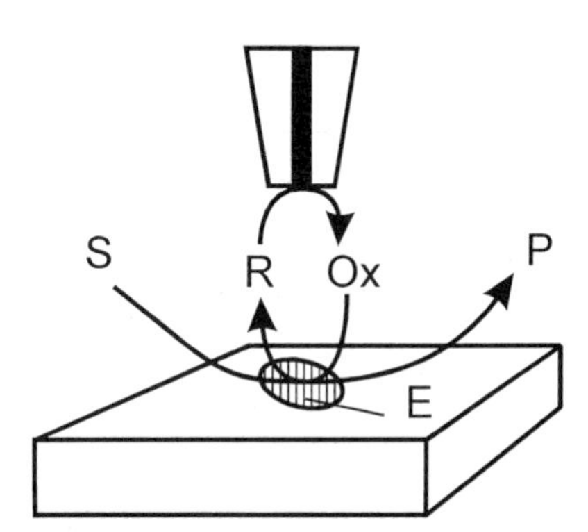

FIG. 3 Schematic of the enzyme-mediated positive feedback for an enzymatic oxidation of the substrate. E, Redox enzyme; S and P, substrate and product of the enzymatic reaction; R, SECM mediator added to the working solution, Ox, oxidized form of R generated by a faradaic reaction at the tip electrode. R has to be selected such that Ox is an active cofactor of the enzyme to be investigated. As discussed in the text, the following list of compounds provides a working example of this mode: E = glucose oxidase (GOx), S = glucose, P = gluconolactone, R = ferrocene monocarboxylic acid (Fc), and Ox = Fc^+. Of course the imaging principle can also be used for enzymatic reductions if a suitable mediator Ox is added to the working solution and is reduced at the tip electrode.

zyme-modified surface, the enzymatic reaction commences since Fc^+ formed at the tip electrode is available locally as an electron acceptor. During this reaction Fc is regenerated and becomes available again for oxidation at the microelectrode (positive feedback). At glucose concentrations much greater than the Michaelis constant, K_M, the magnitude of the effect depends on the enzyme turnover rate, the kinetics of the reaction between mediator and the reduced GOx, and the flux (mol cm^{-2} s^{-1}) of Fc^+ at the specimen surface (Fig. 4). As a result, the feedback depends on the bulk mediator concentration and the electrode radius. This constitutes an important difference to the diffusion-controlled positive feedback at conducting samples, where plots of normalized current $i_T/i_{T\infty}$ vs. normalized distance d/a superimpose for different mediator concentrations and electrode radii.

The following conditions must be met for successful imaging of enzymatic activity:

All solution constituents must be compatible with the diffusion-controlled reaction of the mediator at the tip (no inhibition of the tip electrolysis by the substrate or products of the enzymatic reaction).

There is no interference with the enzymatic reaction by dissolved species (no enzyme inhibitors, no alternative cofactors for the enzyme besides the oxidized mediator present).

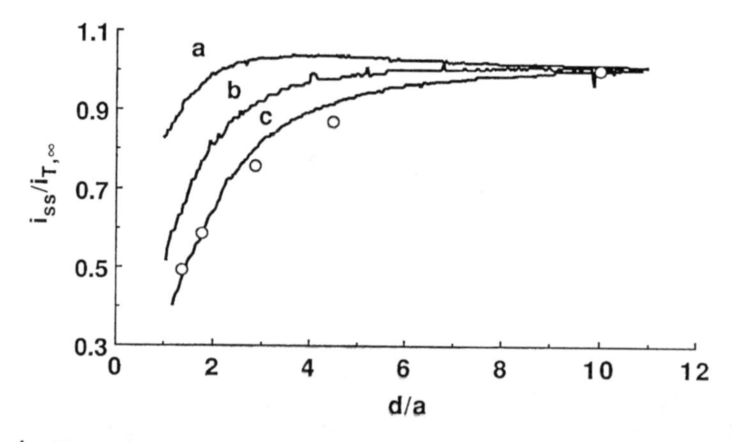

FIG. 4 Normalized current ($i_T/i_{T\infty}$, $i_{ss} = i_T$) vs. normalized distance d/a ($a = 4\ \mu m$, $RG = 20$) over modified nylon surfaces in solution of 0.5 mM ferrocene monocarboxylic acid, 0.1 M phosphate/perchlorate buffer pH 7.0. The surface was modified with (a) maximum glucose oxidase (GOx) coverage, (b) less than maximum GOx coverage, and (c) maximum GOx coverage. Curve c was recorded with no glucose in the working solution. For (a) and (b) 50 mM D-glucose was added to the working solution. (From Ref. 10. Copyright 1992 American Chemical Society.)

The enzymatic reaction has to produce a sufficient regeneration rate of the reduced mediator to compete with its mass transport from bulk solution to the tip electrode.

The third requirement has been expressed quantitatively using the results of digital simulation of enzyme-mediated positive feedback for the glucose oxidase–catalyzed reaction as (10)

$$k_{cat}\Gamma_{enz} \geq 10^{-3}(D_{Red}c_{Red})/a \qquad (1)$$

The left side of Eq. (1) summarizes the enzyme-dependent terms: the rate constant, k_{cat}, characterizes the catalytic reaction rate under conditions of substrate saturation (high glucose concentration in the example, reaction of mediator with reduced GOx is the rate-determining step). In general k_{cat} is characteristic of a particular combination of enzyme, mediator, and substrate. Γ_{enz} is the enzyme surface concentration or, in the case of a film of finite thickness, the product of enzyme concentration in the film and the film thickness. The right side of the expression summarizes the experimental terms. D_{Red} is the diffusion coefficient of the mediator. This criterion is most easily met at low mediator concentration, c_{red}, and large tip radius, a. On the other hand, decreasing c_{red} results in small currents, which require a more sensitive potentiostat. In work with simple inorganic samples this limitation is not present, and high mediator concentrations may be used to increase the tip current to a level that is convenient to measure.

The procedure described above can be refined by using two SECM mediators serving different purposes (10,12). One of the mediators should be a cofactor for the enzyme to be investigated (hydroquinone H_2Q). The other mediator is chosen so as not to interfere with the enzymatic reaction (methyl viologen, MV^{2+}). At a potential of -0.95 V (AgQRE) MV^{2+} is reduced at the tip, but not by the enzyme, and provides a negative feedback image mirroring the topography of the sample (Fig. 5a, low currents correspond to protruding regions). At the second working potential of $+0.82$ V (AgQRE), MV^{2+} does not react at the tip electrode. However, the electron acceptor for the enzyme is formed from the second mediator H_2Q. Therefore, feedback images recorded at that potential can be compared to the purely topographic image at -0.95 V to map the enzymatic activity (Fig. 5b, high currents over active enzyme).

While there is quite a variety of substances that can act as SECM mediators and that are artificial electron acceptors for GOx (Table 1), the extension to different redox enzymes is not straightforward, although examples have been reported.

Besides the enzymes listed in Table 1, imaging of NAD^+-dependent dehydrogenase activity should be feasible. This class of more than 300 redox

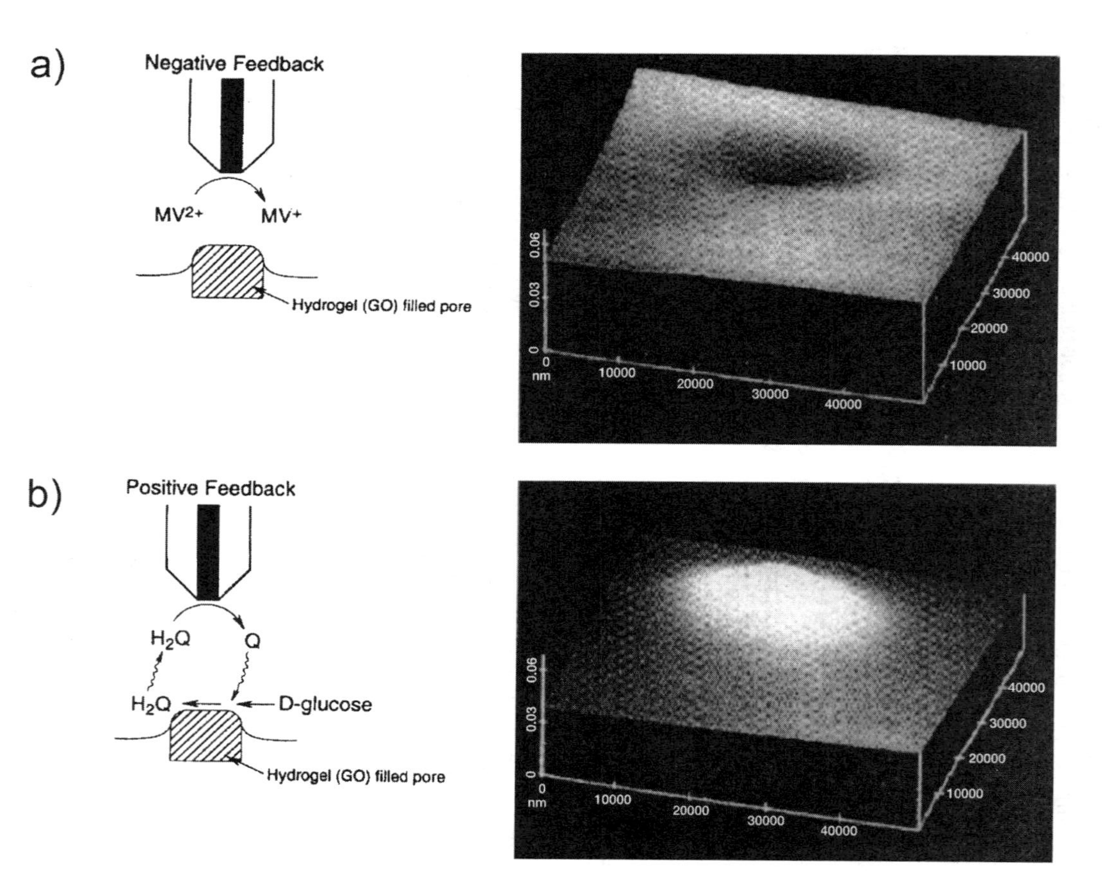
a)
Negative Feedback
MV2+
MV+
Hydrogel (GO) filled pore
b)
Positive Feedback
H2Q
Q
H2Q
D-glucose
Hydrogel (GO) filled pore
0.06
0.03
0
nm
10000
20000
30000
40000

enzymes uses nicotinamide adenine dinucleotide (NAD^+) as the electron-accepting cofactor, which is converted to its 1,4-dihydropyridines form (NADH). This reaction corresponds to the formal uptake of one hydride ion per NAD^+. The reoxidation of NADH proceeds only at high overpotentials at conventional carbon or noble metal electrodes and involves radical intermediates that lead to a rapid build-up of an inhibiting macromolecular organic layer on the electrode. Through the development of biosensors utilizing NAD^+-dependent dehydrogenases, a large body of literature has accumulated describing different strategies to enable stable oxidation of NADH at modified electrode surfaces (see Ref. 18 for a review). To date, there has been no extensive experimental comparison of the relative merit of the numerous approaches that might be adapted to the concept of enzyme-mediated SECM feedback imaging.

The use of other sensors, e.g., microenzyme electrodes, as scanning probes could become a very attractive opportunity and pave the way to activity mapping of an even wider range of redox enzymes. Tip positioning of such electrodes and the deconvolution of topographic feature and enzymatic effects should be more straightforward using current independent tip positioning (see Chapter 2).

D. Generation-Collection Mode

In the sense used in this chapter, generation-collection (GC) mode refers to experiments where the tip is used simply to sense redox active or electroactive species produced by the specimen under study. This usage is essentially the same as elsewhere in this volume except that the generator is a biochemical reaction or organism rather than another electrode.

Two cases may be distinguished: one where the tip radius is less than specimen dimensions and another in which the opposite holds. The latter type of experiment has hardly been explored, although the detection effi-

←

FIG. 5 SECM images ($50 \times 50\ \mu m^2$) of a glucose oxidase (GOx)–loaded hydrogel in the pore of a polycarbonate membrane together with schematics illustrating the imaging principle. (a) Negative feedback imaging at $E_T = -0.95$ V (AgQRE) using methyl viologen MV^{2+} as mediator and (b) enzyme-mediated feedback imaging of the same region at $E_T = +0.82$ V (AgQRE) using hydroquinone (H_2Q) as mediator which is converted to the quinone, an electron acceptor for GOx. Experimental conditions: $a = 4\ \mu m/RG = 10$; $v_t = 10\ \mu m\ s^{-1}$; in 100 mM D-glucose, 0.05 mM H_2Q, 0.1 mM $MVCl_2$, and 100 mM phosphate-perchlorate buffer at pH 7.0. Light shades depict higher normalized currents ($i_T/i_{T\infty}$). (From Ref. 12. Copyright 1993 American Chemical Society.)

TABLE 1 Direct Mapping of Enzymatic Activity with the Feedback Mode and Mediators Used

Enzyme imaged	Enzyme substrate	Mediator/Reaction type at the tip electrode, the reaction product formed at the tip interacts with the enzyme	Ref.
Glucose oxidase, EC 1.1.3.4	50–100 mM glucose	0.05–2 mM ferrocene monocarboxylic acid, dimethylaminomethyl ferrocene/oxidation 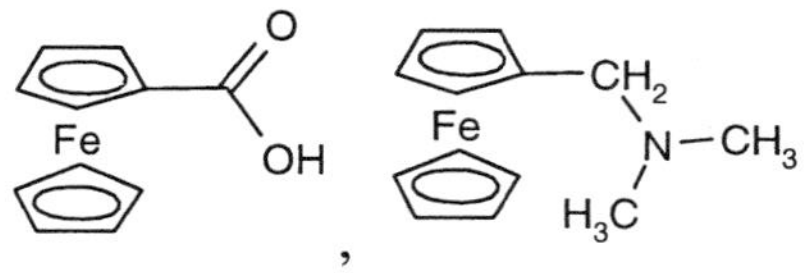	10, 13, 14
		0.05–2 mM $K_4[Fe(CN)_6]$/oxidation	10
		0.02–2 mM hydroquinone/oxidation 	10, 12
		0.5 mM [Os fpy $(bpy)_2$ Cl]Cl/oxidation fpy = 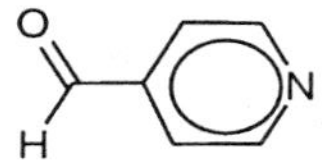 , bpy = 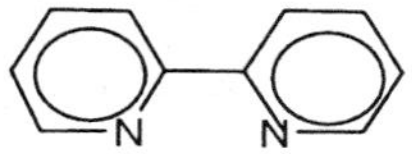	13

NADH-cytochrome *c* reductase, EC 1.6.99.3, within mitochondria	50 mM NADH	0.5 mM N,N,N′,N′-tetramethyl-*p*-phenylenediamine/oxidation 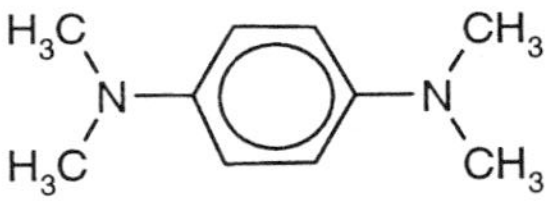	12
Diaphorase (NADH acceptor oxidoreductase), EC 1.6.99.	5.0 mM NADH	0.5 mM hydroxymethyl ferrocene/oxidation 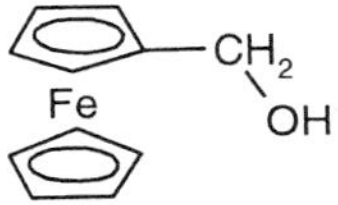	15
Horseradish peroxidase, EC 1.11.1.7	0.5 mM H_2O_2	1 mM hydroxymethyl ferrocenium/reduction 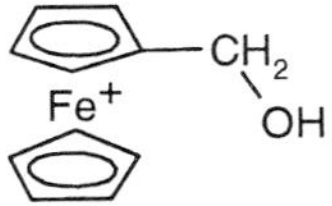	16
Nitrate reductase, EC 1.7.99.4	23–65 mM NO_3^-	0.25 mM methylviologen 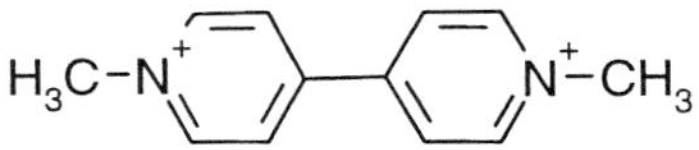	17

ciency of species generated at the specimen may be close to 100%, and such an experiment can be useful when the analyte is present at low concentration and spatially resolved data are not required (19). The more common situation is for a small tip to be rastered across the specimen surface to produce a map of the local concentration of a molecule generated or consumed at the specimen. GC mode experiments have some particular advantages and disadvantages compared to feedback mode experiments, and these will be discussed in this section with the aim of providing the reader with the information to decide if the technique is appropriate for a particular problem.

It should be noted that GC mode experiments with amperometric tips may contain a feedback component to the current if the electrochemical process at the tip is reversible and the tip-to-specimen distance is less than about $5a$. However, at greater distances or when employing a potentiometric tip, the tip acts approximately as a passive sensor, i.e., one that does not perturb the local concentration. This situation is quite distinct from feedback mode, where the product of the electrolysis at the tip is an essential reactant in the process at the specimen surface. This interdependence of tip and specimen reactions in feedback mode ensures that the biochemical process is confined to an area under the tip defined by the tip radius and diffusional spreading of the various reagents (20). In contrast, the biochemical process in GC mode is independent of the presence of the tip and may therefore occur simultaneously across the whole surface. In addition, the tip signal often does not directly provide information on the height of the tip above the surface; methods to overcome this limitation are described in Sec. I.D. Finally, since the tip process and the biochemical reaction at the specimen are independent, a wide range of microsensors may be employed as the tip, e.g., ion-selective microelectrodes, which are not applicable in feedback experiments.

1. *Immobilized Enzymes*

GC mode experiments have been applied to the study of immobilized oxidoreductases such as diaphorase and the glucose oxidase/glucose system (15,21), which has also been investigated by feedback with ferrocenyl mediators (10). In the feedback experiment, the catalytic cycle of the enzyme is dependent on ferrocenium species generated at the tip, as described in Sec. I.C. A generation-collection experiment utilizes a bulk solution concentration of the oxidized mediator, and the tip is poised at a potential sufficiently positive to detect ferrocene, or another reaction product such as H_2O_2, present near the interface (Fig. 6). The enzymatic reaction therefore occurs over the whole specimen, wherever there is active enzyme and a supply of substrate. In the feedback experiment, the tip current includes the flux of mediator from bulk solution as well as the flux of reduced mediator due to

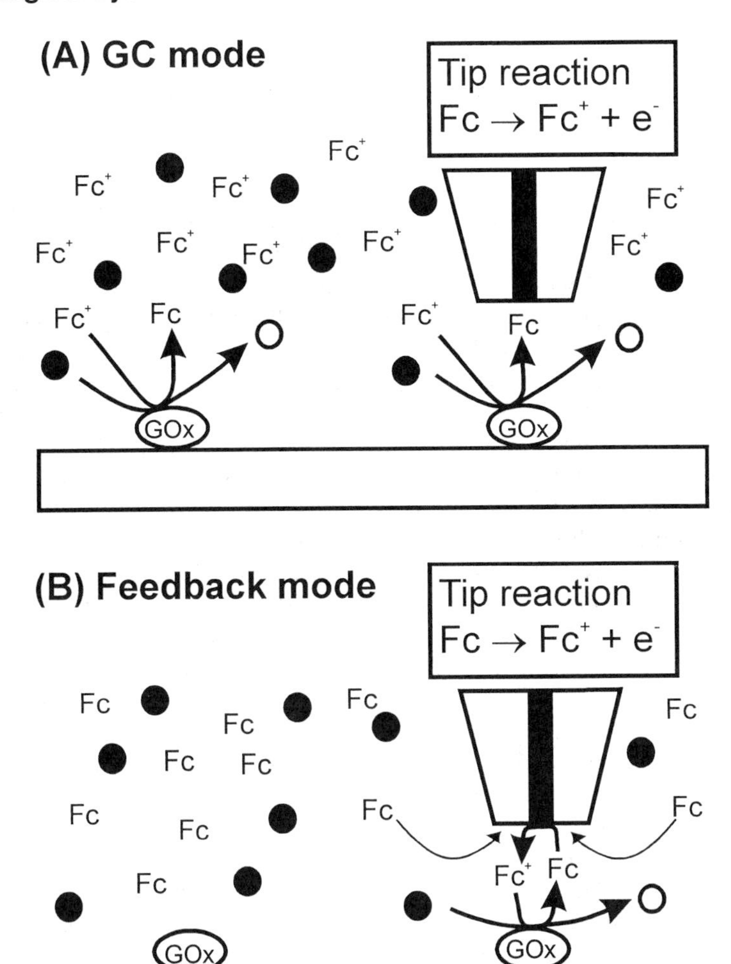

FIG. 6 Schematic illustrating the difference between (A) GC mode and (B) feedback mode detection of immobilized enzyme activity. The example enzyme is glucose oxidase (GOx) with a ferrocenyl redox mediator (Fc) detected at a Pt tip. (A) The bulk solution contains oxidized mediator (Fc^+) and the tip detects the reduced form amperometrically which is generated continually wherever active enzyme is present on the specimen. (B) The bulk solution contains reduced mediator, the tip generates ferrocenium locally and detects the enhanced flux due to feedback when there is active enzyme on the portion of the specimen beneath the tip. ●, Glucose; ○, gluconolactone.

the enzymatic reaction. In a GC mode experiment the tip current is entirely due to the enzymatic reaction and the background signal is effectively zero, which gives GC mode a sensitivity advantage over feedback mode (16). However, since ferrocene may be produced simultaneously across the entire surface, diffusional blurring will tend to result in a loss of image resolution. In the case where there are large areas of active enzyme, a defined steady-state concentration profile will not be established and quantitative measurements will not be possible. For these reasons imaging experiments in GC mode are only appropriate when there are small, well-separated spots of low enzyme activity on the specimen surface. The experimenter must therefore consider the relative importance of image resolution versus sensitivity in deciding between feedback and generation-collection experiments. One example where GC mode is appropriate owing to the low detection limit is in the development of ELISA sandwich assays using the SECM to address microspots of peroxidase-labeled antibody (16).

An additional difficulty in GC images with a macroscopic enzymatically active surface as generator consists in the shape of the approach curve (Fig. 7). The current initially rises on approaching the sample, but then reaches a maximum and falls sharply. The current decrease at small d is equivalent to the conventional negative feedback, except that here it is the sample reaction whose conversion is limited by hindered diffusion of the reagent from the bulk solution. It contrast to conventional negative feedback, where the approach curve depends only on the tip *RG*, the ratio between the reagent bulk concentration, and the enzymatic activity per unit area of sample is also important. The hindered diffusional transport will still provide enough reagent to saturate the enzyme kinetics (with respect to substrate) if the reagent concentration is high and the immobilized enzymatic activity is low. When looking for low quantities of enzymatic activity (as in immunoassays), this situation is likely to prevail if high reagent concentrations are practicable.

2. *Sensors and Tips*

GC mode SECM experiments have been used to detect localized activity of a variety of enzymes (see Table 2). Since microelectrodes are used in the life sciences to detect neurotransmitters (36), metal cations (37), and free radicals (38), a wide variety of probes may be combined with SECM positioning technology to obtain spatially resolved information. The GC mode experiments may be carried out with potentiometric or amperometric tips depending on the species to be sensed. The tip must be chosen to determine a product or reactant consumed by the enzyme reaction, e.g., H^+ for immobilized urease, H_2O_2 for immobilized glucose oxidase, or 4-aminophenolate for alkaline phosphatase. A list of enzymes and the tips employed is given in Table 2.

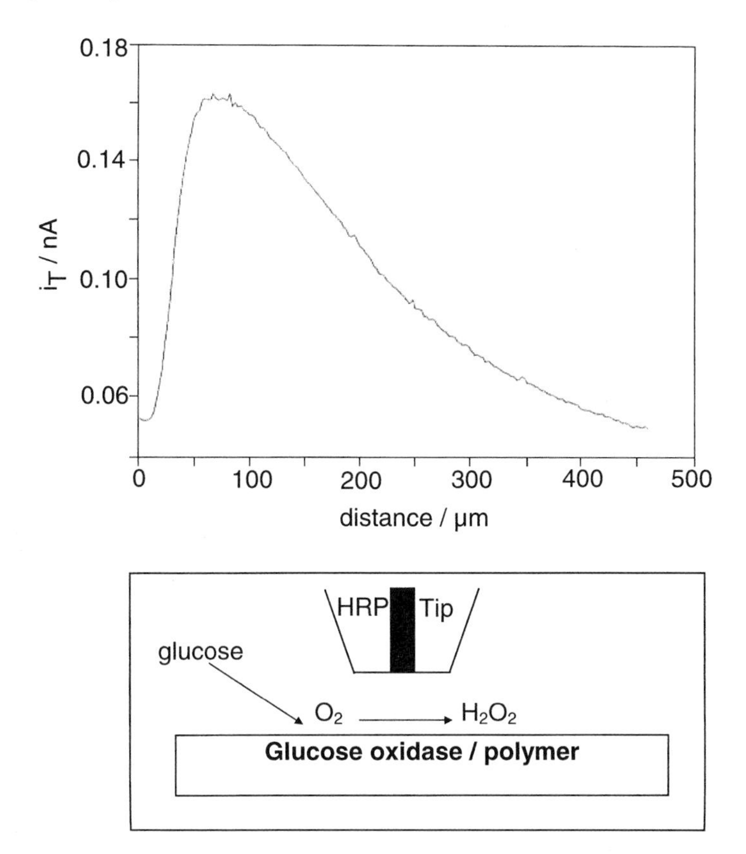

FIG. 7 Approach curve of a H_2O_2-selective amperometric tip electrode towards a polymer layer loaded with GOx. The tip was produced by immobilizing horseradish peroxidase onto the cross section of an insulated carbon fiber of 7 μm diameter. Solution composition: 1 mM glucose in air-saturated 0.2 M phosphate buffer, pH 7. (From Ref. 22. Copyright 1993 American Chemical Society.)

3. Distance Calibration and Control of Mass Transport

The major technical problems with the GC mode are the need for an independent distance measurement and the absence of a well-defined mass transport rate in some situations. These are linked in the sense that when mass transport is well defined, the distance dependence of the tip signal can be used to calibrate the tip-to-surface separation as well as to quantify the flux of analyte. One method of overcoming these problems is therefore the use of a microfabricated substrate where the enzymatic reaction is confined to a small disk-shaped region. The concentration profile of the products of the

TABLE 2 Examples of Generation-Collection Mode Experiments on Biochemical and Biological Systems Using a Variety of Tips

Specimen/Enzyme studied	Analyte	Tip	Ref.
Elodea leaf and *T. fluminensis* photosynthesis	O_2	Pt, amperometric	23, 24
Oxygen consumption/production by cultivated cells	O_2	Pt, amperometric	25–28
Glucose oxidase EC 1.1.3.4	H_2O_2	Pt, amperometric	21
Glucose oxidase EC 1.1.3.4 yeast	H_2O_2	C/wired enzyme amperometric	22
Urease EC 3.5.1.5 and brewers yeast	H^+	Sb, potentiometric	29
Urease EC 3.5.1.5 adsorbed on a gold electrode	NH_4^+	Liquid membrane ISE, potentiometric	30
NADPH-dependent oxidase in osteoclasts	$O_2^{\bullet -}$	Au/cytochrome *c* modified, amperometric	19
Single PC12 cell	Neurotransmitter	Carbon fiber	28
Horseradish peroxidase EC 1.11.1.7	Ferrocene mediators	Pt, amperometric	16, 31
Alkaline phosphatase EC 3.1.3.1	4-Aminophenolate	Pt, amperometric	32
NAD^+-dependent alcohol dehydrogenase EC 1.1.1.1	H^+	Sb, potentiometric	33
Ion diffusion in dentine	$Fe(CN)_6^{4-}$	Pt, amperometric	34
Dental materials (alloys)	O_2, metal ions	Pt, amperometric	35

enzymatic reaction may then be calculated from the diffusion equation. With such a defined steady-state concentration profile, the tip-substrate distance can be determined by fitting the approach curve of tip response distance to theory (39,40). This is described in more detail in Sec. II.B, where the use of the theory to obtain quantitative kinetic data is covered.

If the enzyme cannot be patterned in this way or the system under study does not consist of isolated spots of activity, then an independent distance measurement is necessary, even if only to avoid a tip crash. Observation of the tip via optical microscopy is sometimes possible, although it may be difficult and is not usually precise. Addition of a reversible redox couple to the system and monitoring the feedback current is possible (15). The redox couple should be chosen so as not to interfere with the system under study. This method is obviously applicable only to tips that may be used in a voltammetric mode, e.g., potentiometric antimony pH microelectrodes but not glass or liquid membrane microelectrodes. Irreversible redox processes, e.g., oxygen reduction, may also be used for distance calibration. The tip potential should be set sufficiently negative to drive the reduction of oxygen to water. This reaction is totally irreversible, and therefore negative feedback will be observed on approaching the surface, unless oxygen is being generated or consumed at the interface. The product, water, is not likely to interfere with the system under study. It should be noted that it is not essential to measure the oxygen concentration in the solution; all that is required is that oxygen reduction at the tip is diffusion limited, the tip radius is known, and the tip process does not adversely affect either the system under study or the tip response. It was observed that antimony electrodes operated at negative potentials regained their potentiometric response to pH if their potential was briefly returned to 0 V vs. SCE for a few seconds before switching to potentiometric measurements (29).

For high impedance electrodes, e.g., liquid membrane ion-selective tips (39), another method is essential. Using techniques borrowed from electrophysiology, it is possible to construct dual tips from theta glass (Chapter 3). One barrel of the double tip can be used for the liquid membrane, and the other may be filled with a low melting point metal such as gallium to make a voltammetric sensor (41).

Alternatively the second barrel can be filled with the electrolyte and used in the same manner as in ion-conductance microscopy (39). It is possible to relate the solution conductance between tip and counterelectrode to the normalized tip-to-substrate distance L. In fact, after normalizing the conductance $G(L)$ by the value with the tip far from the surface G_∞, the distance dependence of the conductance is identical to that for faradaic currents in feedback SECM with a redox mediator.

$$\frac{G(L)}{G_\infty} = \frac{i_\mathrm{T}(L)}{i_{\mathrm{T},\infty}} \tag{2}$$

This follows from the analogy between Fick's equation and Laplace's equation for the potential (22).

In the case of an electrolyte-filled glass micropipette, the filling solution contributes substantially to the total resistance. This component is distance independent and can be subtracted from the measured value (39). A similar technique can be applied with amperometric enzyme-based tips using a small amplitude AC potential perturbation. In this case, the contribution of the interfacial impedance is similarly constant and may be subtracted (22). Other distance calibrations have been proposed that do not depend on the measurement of tip currents. An example is the measurement of the damping of mechanical vibrations of the tip due to hydrodynamic forces that are distance-dependent (42,43). This may have wider application, though some additional equipment is required to produce and detect the oscillation of the tip. A method based on the use of a quartz tuning fork to sense the surface may also be of use in SECM experiments (44). In many experiments a distance measurement is only required before and after the GC mode is employed. However, the "current-independent" distance measurements may provide simultaneous topographic images and are likely to be essential for rough surfaces when a feedback loop can be employed to guide the tip over protrusions in the surface (43).

4. *Detection Limit*

By comparison with feedback methods, generation collection offers greater sensitivity to low activities of immobilized enzyme. An estimate of the minimum catalytic rate, k_cat, of the immobilized enzyme which can still be detected and quantified can be made on the basis of the analytical sensitivity of the tip collector. If c' is the detection limit of the tip, then enzyme kinetic data can be obtained if

$$k_\mathrm{cat}\Gamma_\mathrm{enz} \geq \frac{c'D}{r_\mathrm{S}} \tag{3}$$

where Γ_enz/mol cm^{-2} is the surface coverage of enzyme. From Eq. (3) it can be seen that sensitivity is increased for large specimen radius, r_S, low c', low D, and high surface coverage of enzyme. Equation (3) is similar in form to Eq. (1) for a feedback experiment with a redox enzyme. However, in the feedback case the contribution of the enzymatic flux to the tip signal must be distinguished from a significant background due to diffusion of mediator from the bulk. In the generation-collection experiment, this background signal is absent, and the sensitivity of the experiment to kinetics is directly proportional to the detection limit of the tip, c'. Reasonable values of $D = 5 \times 10^{-6}$ cm^2 s^{-1}, $r_S = 5 \times 10^{-3}$ cm, $\Gamma_{enz} = 10^{-12}$ mol cm^{-2}, and a rough estimate of $c' = 10^{-6}$ mol dm^{-3} indicates that enzymes with $k_{cat} > 1$ s^{-1} may be studied even at monolayer coverage. A wide range of nonredox enzyme reactions can be studied by generation-collection mode via detection of H^+. However, it should be noted that an exact theory is only available when the reaction product (H^+) does not undergo any solution phase reactions. If the solution is buffered to maintain pH close to the optimum for the enzyme, numerical treatments of the titration of the buffer components may be incorporated (29,40).

II. SELECTED APPLICATIONS

This section is a compilation of applications in which biological or biochemical samples have been investigated with the SECM. The examples are grouped according to the type of specimen investigated. Different methodological approaches to the study of similar samples are presented, wherever possible, to facilitate comparison of different experimental strategies.

A. Imaging Single Biomolecules

To date the high lateral resolution required for this task has only been achieved by "STM-like" negative topographic feedback imaging in which a tip maintains contact with an ultrathin water layer around the large biomolecule and the support (see Sec. I.A). When searching literature databases, the reader will find the original work indexed mainly under the keyword "STM." Indeed, for the vast majority of the studies commercial or modified STM apparatus and tips have been applied. Since the biological activity of the molecule is not sensed by this mode regardless of whether the current originates from faradaic reactions or from special tunneling mechanisms on these samples, those applications do not constitute the main focus of this chapter.

Some of the applications that fall clearly into this category are imaging of DNA on mica (3) and imaging of catalase (a globular protein) as well as fibrinogen (a filamentous protein) on a SAM-coated Au layer (7). Another review (45) is suggested to the interested reader for experimental details and for a more complete list of investigated biomolecules.

B. Imaging of Immobilized Enzymatically Active Layers

1. *Imaging Enzyme Activity on Insulating Supports and Within Cells and Organelles*

The imaging of enzyme activity at insulating supports can be achieved via feedback or generation-collection modes as described in Secs. I.C and I.D.

This is the simplest situation for mapping enzyme activity by the SECM, especially compared to imaging enzyme activity on the conductive surfaces as described in Sec. II.B. Attempts to obtain high-resolution images of enzyme activity therefore typically use an insulating support for the biological material. Although microscopic techniques, e.g., fluorescence microscopy, are commonly used to assay enzyme activity in even subcellular structures, the ability of the SECM to provide quantitative kinetic information may become useful in this context. We are aware of only one such study to date (12), in which feedback mode images of NADH–cytochrome *c* reductase system (EC 1.6.99.3) activity in the outer membrane of rat liver mitochondria were obtained. The redox mediator used was TMPD (tetramethyl-*p*-phenylene diamine) and the substrate NADH was added to the bulk solution at 50 mM. The mitochondria were fixed to an aminated glass slide with glutaraldehyde. The images showed a depression (negative feedback) in the absence of NADH, but positive feedback in the presence of substrate. Although this study was successful in detecting the enzyme activity and distinguishing individual mitochondria that were inactive (1–5% of population), the resolution was not high enough to map the distribution over the outer membrane surface of the mitochondria. The resolution limit was due to the rather large tip diameter, 8 μm; employed compared to the dimensions of the mitochondria, 3–5 μm. However, as noted by the authors, the limitation on tip size set by the detection relation [Eq. (1)] makes smaller tips much less sensitive to the enzymatic activity. For this reason, generation collection may be necessary to achieve the sensitivity needed for the application of smaller tips. The loss of lateral resolution typically encountered in GC experiments may be at least partially compensated by exploiting solution phase chemistry, as discussed below.

2. *Kinetics*

A feature of SECM is the quantitative theory available based on reaction-diffusion models (Chapter 5). The SECM may be used for kinetic studies in either the feedback or generation-collection modes. These two possibilities are described below from the point of view of studying immobilized enzyme kinetics.

Feedback. When an oxidoreductase enzyme is immobilized at the specimen surface, a redox mediator present in solution may be recycled by the diffusion-limited electrochemical process at the tip and electron exchange with the enzyme active site as described in Sec. I.C. The mass transport rate is defined by the tip radius and height of the tip above the specimen. The tip current depends on the mass transport rate and the enzyme kinetics. Kinetic information may therefore be obtained from the dependence of tip current on height, i.e., an approach curve. When the mediator is fed back from the specimen at a diffusion-controlled rate, the approach curve will be identical to that above a metallic conductor. In the opposite situation when the flux of mediator fed back from the specimen is much less than the flux of mediator to the tip from bulk solution, the approach curve will correspond to that above an insulating surface, i.e., pure negative feedback. In between these two limits, the approach curve will contain information on the steady-state rate of the enzymatic reaction and the shape of the approach curve as a function of substrate concentration may be used to investigate the reaction order (Fig. 4). A detailed study of the system glucose oxidase/glucose with several redox mediators and immobilization techniques has been reported (10). The specimen was either glucose oxidase covalently attached to nylon, immobilized in an albumin/glutaraldehyde hydrogel or in the form of a compact Langmuir-Blodgett (LB) film transferred to a glass slide. Under the assumptions that the enzyme layer was thin and highly permeable, the concentrations of glucose and mediator are uniform throughout the enzyme layer. The enzyme reaction kinetics can then be modeled with a simple heterogeneous rate law. Working curves for zero- and first-order enzyme kinetics (in redox mediator concentration) were computed by a finite difference technique for the case where the glucose concentration was sufficiently large to saturate the enzyme. The data were found to be best fit by zero-order kinetics for several choices of redox mediator, and this was interpreted as saturation of the enzyme kinetics. The maximal rate deduced from the quantity of immobilized enzyme ($0.15\ s^{-1}$) was low compared to the known bulk solution kinetics. In the case of the LB films of glucose oxidase, no significant enzyme-mediated feedback could be detected, and this was attributed to the relatively low surface coverage (a few monolayers) compared to the micrometer-thick hydrogel films. Matsue et al. (15,46) have reported a kinetic study of diaphorase immobilized on glass slides via treatment with 3-aminopropylsilane and glutaraldehyde. The redox mediator was ferrocenemethanol, which was oxidized at the tip and reduced at the glass slide by diaphorase in the presence of excess NADH. A slightly different approach to modeling the diffusion problem was taken; although the enzyme kinetics were also assumed to be saturated with respect to substrate, the full Michaelis-Menten expression for the dependence of the rate on the mediator was retained. Working curves of tip current versus active enzyme surface coverage at a given tip-surface distance and mediator concentration were computed using previously obtained values of the kinetic parameters of the immobilized enzyme, $K_M = 1 \times 10^{-6}$ M and $k_{cat} = 110\ s^{-1}$ (47). At the mediator concentrations employed, 0.5 mM, the kinetics were essentially zero order in this case also. For this system, surface coverages of 2×10^{-12} mol cm^{-2} diaphorase were calculated from the tip current. Since the current enhancement over negative feedback due to the enzymatic process is quite small (~10–20%) imaging experiments with enzyme-patterned substrates were useful to discriminate between pure negative feedback at the bare regions of the substrate and the enzyme-coated regions (Fig. 8).

Although these studies have shown that the feedback method is capable of quantifying kinetics of immobilized oxidoreductases, the theory has been developed entirely using numerical simulations and there is to date no convenient approximate analytical expression that can be used to fit the experimental data. The interested reader should therefore consult the original papers for details of the numerical methods and working curves (10,15). An

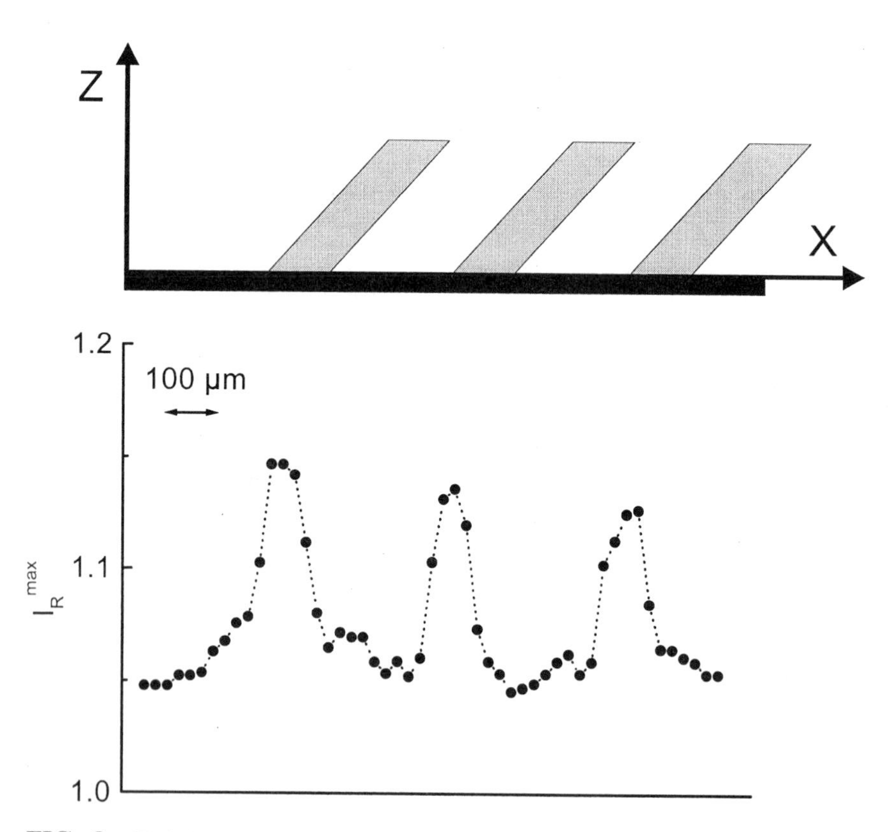

FIG. 8 Variation of tip current across a diaphorase patterned surface. Top, an illustration of the micropatterned surface. Enzyme-immobilized bandwidth, 100 μm; gap width, 200 μm. Bottom, profile of the normalized tip current along the diaphorase-micropatterned surface. The solution contained 0.5 mM ferrocenemethanol, 5.0 mM NADH, 0.1 M KCl, and 0.1 M phosphate buffer (pH 7.5). (From Ref. 15. Copyright 1995 American Chemical Society.)

approximate analytical treatment that is valid for general types of first-order kinetic processes, i.e., electrochemical, chemical, or enzymatic, at the specimen surface has been developed (48). However, as described above, zero-order kinetics is the typical case for immobilized enzymes unless the mediator concentration is low and the enzyme activity is high or the tip is far from the surface.

Generation Collection. The tip signal in a generation-collection experiment with an enzymatic reaction occurring at the specimen surface depends on the enzyme reaction kinetics and the rate of mass transport of the products to the tip. Quantitative kinetic investigations of immobilized enzymes can be made by generation-collection experiments when this mass transport is well defined. In principle a macroscopic specimen uniformly coated with enzyme could be used if the enzymatic process were switched on abruptly. This would be analogous to a chronoamperometric generation-collection experiment, however, such a mechanism is not usually available. In practice measurements have been made where the biochemical reaction is localized at micrometer-scale immobilized enzyme spots of defined geometry so that a steady-state diffusion field is produced (29,30,39). Technically the simplest experimental realization is a microdisk of immobilized enzyme. The concentration profile $c(L)$ of enzymatic reaction products above the center of a disk-shaped spot of enzyme is related to the flux (f/mol cm^{-2} s^{-1}) by established theory (29,40):

$$c(L) = \text{const.} \int_0^1 \frac{uf(u)}{\sqrt{L^2 + u^2}}\, du \tag{4}$$

where $c(L)$ is the concentration of reaction product monitored by the tip at a normalized height $L = d/r_S$ above the center of the specimen disk (r_S, radius of specimen). The flux at the specimen surface, $f(u)$, depends in general on the distance $u = r/r_S$ from the center of the disk.

Two limiting cases can be distinguished according to the distribution of the flux across the specimen surface. If the enzymatic process is limited by the diffusion of the substrate to the disk surface then the flux will be nonuniform across the disk in a manner analogous to the current density distribution at a microdisk electrode operating at the diffusion-limited current. This case can be distinguished experimentally if the bulk concentration of substrate and the disk radius are known by comparing the measured flux to the value for the diffusion-limited transport of substrate to the disk. However, this situation does not allow information to be gained on the enzyme kinetics, and a more interesting case occurs when the diffusion limited flux of substrate to the enzyme is greater than the flux of the enzyme-catalyzed reaction. This can often be achieved by increasing the substrate concentration above the K_M when the enzyme kinetics are saturated. In this case the flux distribution is determined by the distribution of enzyme and may be taken as approximately uniform across the specimen surface. Under these conditions the concentration profile above the specimen center simplifies to

$$\frac{Dc(L)}{fr_S} = \sqrt{L^2 + 1} - L \tag{5}$$

The kinetic flux, f/mol cm^{-2} s^{-1}, of the enzyme reaction is simply obtained by fitting the experimental approach curve $c(L)$ vs. L to Eq. (5). This analysis requires an independent measurement of the diffusion coefficient of the species detected at the tip in order to derive a value of the flux.

Experimentally the tip may be located over the center of the disk using either an optical microscope or by scanning the tip laterally at a constant height till the position of maximum tip response is obtained. As the maximum of the concentration profile is broad in the radial direction, a combination of these approaches is often best (29,39,40). Once the tip is centered over the specimen, an approach curve can be recorded using a slow tip speed (typically 1 μm s^{-1}) to minimize convective disturbance of the diffusion field. Further, since the theoretical treatment ignores entirely the presence of the tip, a small ratio of tip to specimen radii is preferable. Conventional wire-in-glass microelectrodes with $RG \sim 10$ are not expected to give accurate results for this reason and GC mode kinetic experiments have often employed tips based on finely drawn-out glass capillaries such as liquid membrane ion selective microelectrodes (39,41).

In principle the zero tip-to-specimen distance point can be included as a free parameter in the fitting process (30,39), but an independent measurement of the absolute tip-to-surface distance is preferable. This has been accomplished via several methods described in Sec. I.D.

Example Applications. Previous work has mostly been concerned with testing the theoretical models and obtaining proof of concept (29,39). Antimony pH tips were used to image the activity of urease immobilized in a disk of glutaraldehyde/BSA gel and to quantify the total flux of H^+ in the presence of a saturating concentration of urea (29). Wei et al. (39) showed that urease kinetics can also be quantified using an ammonium-selective neutral carrier–based tip to determine the concentration profile of NH_4^+ produced by the hydrolytic reaction

$$(NH_2)_2CO + 3H_2O \rightarrow 2NH_4^+ + HCO_3^- + OH^- \tag{6}$$

Since NH_4^+ is stable at pH 7, the solution could be strongly buffered during this experiment to control the pH at the specimen surface. More recently GC mode was used to study the effects of adsorption at a metal surface and the applied potential on enzyme activity (30). Jack Bean urease adsorbs strongly to soft metals, e.g., gold, through interaction with the disulfide links present on the outer surface of the enzyme. The adsorbed urease retained activity for the catalysis of urea when immobilized in this way on a 50 μm radius gold disk microelectrode and provided a useful model system to study the stability of enzymes on metal electrodes (Fig. 9). A steady-state diffusion field was observed corresponding to a flux of 1.2×10^{-12} mol s^{-1} of NH_4^+ at a concentration of urea (50 mM) where the enzyme kinetics were saturated. Application of a positive potential bias to the gold surface reduced the enzyme activity; this decrease occurred on a time scale comparable to r_S^2/D and the tip response was therefore diffusion limited. However, the recovery of the activity on returning the potential to 0 V was much slower and the kinetics of this process were not masked by diffusion. Double potential step experiments (Fig. 10) showed that the reduction in activity was reversible and not associated with any faradaic process, ruling out desorption or oxidation of the enzyme as the cause. The phenomenon was ascribed to an effect of the metal/electrolyte double layer environment on the enzyme. The kinetics of the modulation of enzyme activity during the potential step

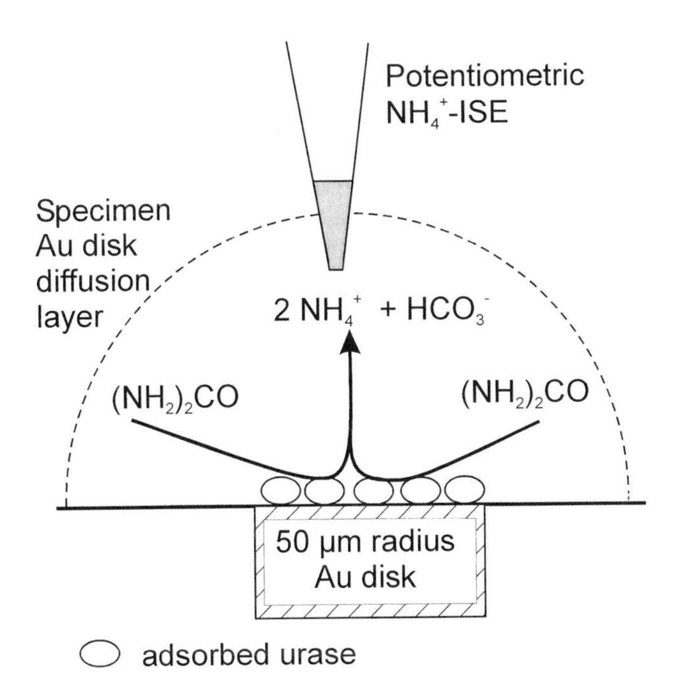

FIG. 9 Schematic diagram of the measurement of the activity of urease adsorbed on a gold microdisk using an ammonium-selective potentiometric SECM tip. (From Ref. 30. Copyright, the Royal Society of Chemistry, 1998.)

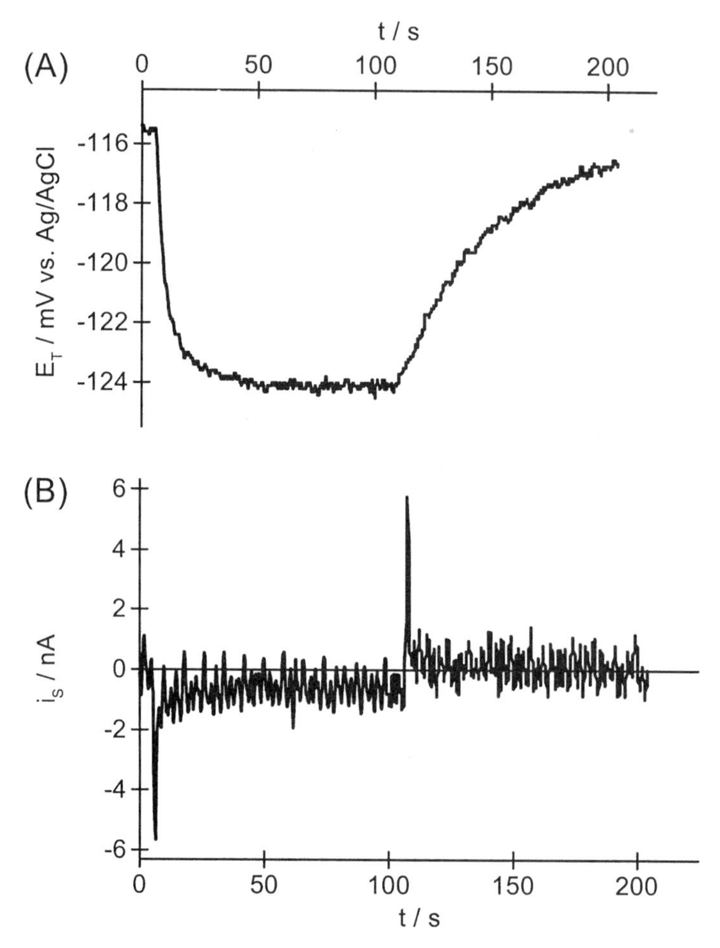

FIG. 10 Kinetic data obtained in the experiment illustrated in Figure 9 (30). (A) Tip potential and (B) gold electrode current during a potential step of the gold microdisk from 0 V to 0.4 V at $t = 0$ and back at $t = 100$ s. The solution contained 0.1 M Tris-HCl buffer adjusted to pH 7.0 and 50 mM urea. The tip potential and current data have been subjected to a 10-point moving average smoothing algorithm. (Copyright, the Royal Society of Chemistry, 1998.)

experiment were analyzed to show that the recovery of enzyme activity obeyed first-order kinetics with an observed rate constant of 0.022 s^{-1} (30). Kinetic studies of enzymatic activity at the solid/liquid interface with the SECM may provide complementary information to structural and spectroscopic investigations by techniques such as AFM (49) and ATR-IR spectroscopy (50).

3. *Enzymes on Conducting Surfaces*

The development of biosensors is an area where SECM may have an important role. Among the variety of biosensor design principles, amperometric enzyme electrodes enjoy a particular popularity. Probably the most successful commercial application is in decentralized blood glucose monitoring used in the treatment of diabetes mellitus. Other areas of application include fermentation control in food processing and biotechnological production of pharmaceuticals. Such sensing devices consist of a biorecognition layer (e.g., an enzyme layer) and a transducer (e.g., an amperometric electrode) (51). The amperometric glucose sensor of Clark and Lyons (52) introduced as early as 1962 was the first of its kind to unite the biorecognition layer and the transducer in a single device. Strong efforts are currently devoted towards the development of miniaturized biosensors. Overcoming size constraints would be a prerequisite for applications such as in vivo monitoring. Off-line analysis requires a reconsideration of sample volume as regulatory agencies in some countries, e.g., the United States, demand increasing efforts on the part of clinical labs to minimize the amount of blood drawn not only from neonates and geriatric populations, but also from critically injured or ill patients in general. Another important area for miniaturized sensors could be high-throughput testing procedures similar to those used in combinatorial chemistry where the sheer number of reaction vessels to be tested and manipulated calls for miniaturization. Simultaneous testing of the interaction between a library of substances against a range of target molecules with a microstructured multianalyte assay or sensing device would also fit with the philosophy of that approach.

Besides the size constraint, imposed either by the working environment of the sensor or the availability of the sample itself, experience with unmodified microelectrodes suggests that advantageous characteristics such as short response times and low convection dependence may be obtained. The miniaturization of enzyme electrodes will also allow the construction of multianalyte enzyme electrode arrays. SECM has already been used as a tool to gain a deeper understanding of the crosstalk, i.e., the response of electrodes to the analyte converted at the neighboring electrode or the signal modulation at the sensing electrode by the potential of the adjacent electrode. This is caused by common reactants and products, e.g., H_2O_2, diffusing from one electrode to the other. Active shielding was demonstrated by regions consuming the diffusing species by electrochemical or enzymatic reactions (53). Microcompartmentalization represents another new concept, which seems feasible with recent advances in the controlled design of solid/liquid interfaces (54,55). In such a device, different components of a sensor are immobilized in separate microscopic domains on one surface, where the environment for each component is optimized, and which are spaced so closely that efficient communication between different domains occurs through diffusing reactants (56). Nowall et al. (57) obtained microcompartmentalized electrodes by using a laser interference pattern and photobiotin that was incubated with fluorophor-tagged avidin. Segregation as seen by fluorescence microscopy corroborated the distribution of the heterogeneous electron transfer rate at the bare and derivatized regions as analyzed by SECM and imaging of electrogenerated chemiluminescence. Besides using SECM as a tool to assess heterogeneous redox kinetics, the potential for imaging biochemical reactivity on the micrometer scale should be very helpful for all of these developments. However, investigations of amperometric biosensors face an additional challenge compared to the examples described so far: Above a conducting sensor surface, the SECM feedback signal may be modulated by (1) surface roughness (58), (2) partial blocking of the specimen surface by the biorecognition layer or other sensor components [e.g., Fig. 2; (59)], and (3) lateral variation of the enzymatic activity [e.g., Fig. 12; (13)]. Typically at least two of these influences are effective simultaneously in a particular application.

The following part illustrates experimental strategies to cope with this situation. The preparation of the microstructured enzymatic layers itself can be carried out by SECM, too. This topic will be outlined in Sec. II.F very briefly, since a more detailed discussion will be provided in Chapter 13 of this volume.

Activity Mapping of Glucose Oxidase Bound to N-*Substituted Polypyrrole.* In a series of papers Kranz et al. (43,60,61) developed a technique to electropolymerize pyrrole in two-dimensional and three-dimensional structures by using the SECM tip as the auxiliary electrode. Through using an *N*-(ω-aminoalkyl) pyrrole as monomer, a derivatized polypyrrole (PPy) was deposited providing amino groups attached to the polymer backbone via a flexible spacer. Periodate-oxidized glucose oxidase was bound covalently to the amino functions at the outer polymer surface.

Using the enzyme-mediated feedback mode with an artificial electron acceptor for GOx formed at the tip as shown in Figure 3 would not give a clear result because of the overlapping effects illustrated in Figure 11. The overall feedback effect above the GOx-modified conducting polymer has three contributions: (1) mediator regeneration by an electron transfer reaction at the Au/polymer interface, (2) mediator regeneration by an electron transfer reaction at the polymer/solution interface, and (3) mediator regeneration by the enzymatic reaction at the outer polymer surface. The sum of these contributions has to be compared with the positive feedback above the uncovered Au electrode.

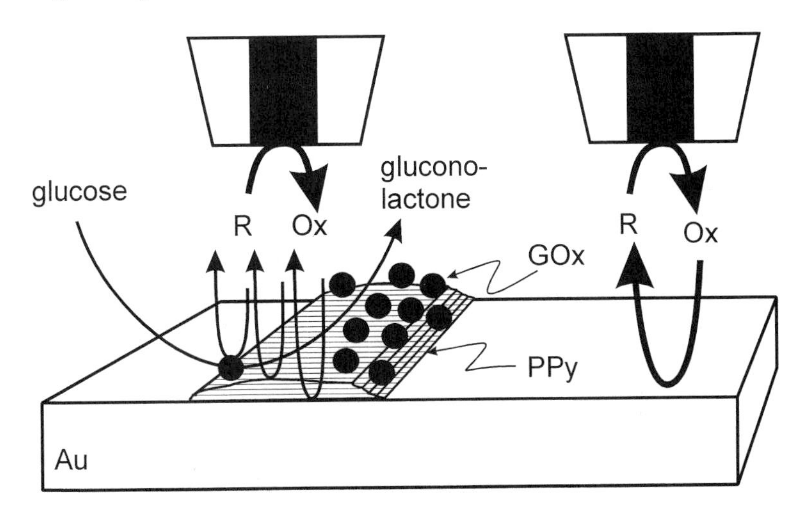

FIG. 11 Illustration of superposition of different contributions to the overall feedback effect above an GOx-modified polymer deposited on a conducting surface as investigated by Kranz et al. (13) (schematic prepared by chapter authors). The oxidized form of the mediator, Ox, acts as artificial electron acceptor for GOx; PPy, polypyrrole.

The conductivity of the *N*-substituted polypyrrole is considerably lower than that of unmodified PPy (62), and the regeneration rate of the mediator at the polymer/solution interface is slow. Since the hydrophobic polymer represents an additional diffusion barrier for highly charged mediator ions, the positive feedback current due to regeneration at the Au/polymer interface is reduced compared to the uncovered Au surface. If a mediator, such as $[Ru(NH_3)_6]^{3+}$, is chosen that does not react with GOx, reduced tip currents indicate the position of the polymer deposit (Fig. 12a). A qualitative similar image is obtained if an artificial electron acceptor for GOx is used, but the enzyme substrate β-D-glucose is not provided in solution (Fig. 12b) since the enzyme catalytic cycle requires both glucose and an electron acceptor. After addition of glucose to the solution, higher currents are recorded (Fig. 12c). The combined contributions of the electrochemical positive feedback at the Au/polymer interface and the enzyme-mediated positive feedback at the outer polymer surface almost equal the diffusion-controlled feedback above the uncovered Au electrode. The contributions of the enzymatic reaction can then be estimated as the difference between the images recorded in the presence of glucose (enzyme-mediated feedback plus conventional feedback) and in the absence of glucose (conventional feedback only).

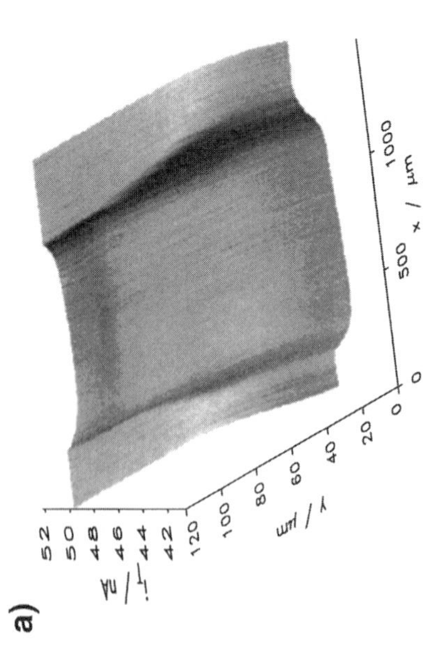
a)
i_T / nA
52
50
48
46
44
42
120
100
80
60
40
20
0
y / µm
0
500
1000
x / µm

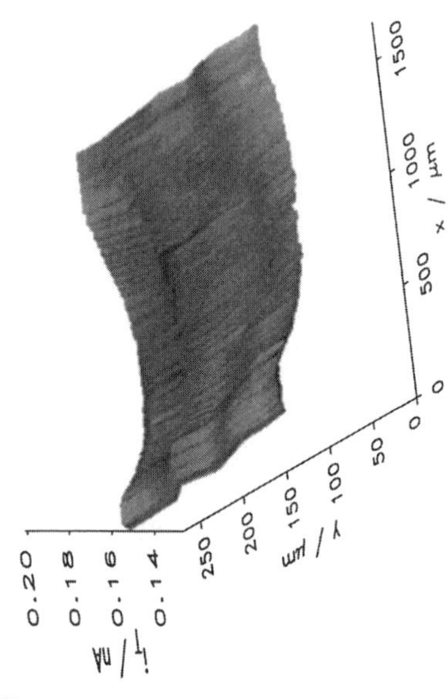
b)
i_T / nA
0.20
0.18
0.16
0.14
250
200
150
100
50
0
y / µm
0
500
1000
1500
x / µm

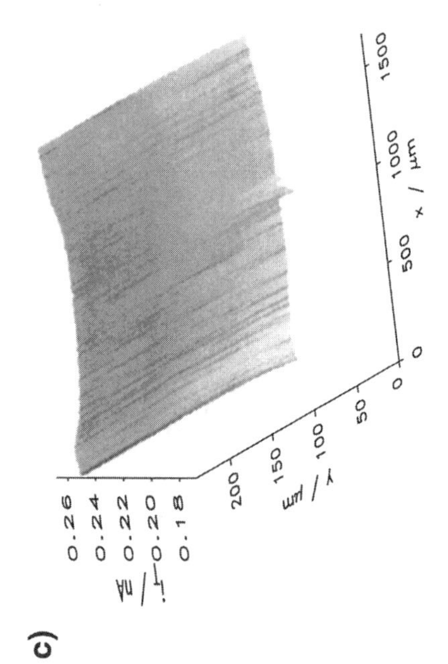
c)
i_T / nA
0.26
0.24
0.22
0.20
0.18
200
150
100
50
0
y / µm
0
500
1000
1500
x / µm

Investigation of Patterned Enzymatically Active Monolayers on Au. Another approach to microstructured enzymatically active regions makes use of the strong chemisorption of sulfur-containing organic molecules at Au and the tendency of long-chain alkanethiolate adsorbates on Au surfaces to arrange spontaneously into a densely packed self-assembled monolayer (SAM). Although such layers provide only monomolecular coverage of the Au surface, electrochemical reactions of solution constituents at the metallic electrode are almost completely inhibited if the alkyl chains contain more than eight carbon atoms. Among other techniques (see Ref. 21), SECM can be used for microstructuring of these layers (see Sec. II.F). The renewed Au surface is utilized to chemisorb a second amino-functionalized thiol or disulfide for covalent attachment of periodate-oxidized GOx (Fig. 26A).

To image the localized enzymatic activity in the feedback mode would be even more difficult than in the case of the enzyme-modified PPy lines (vide supra). There is negative feedback above the unaffected long-chain SAM and positive feedback above areas from which the long-chain alkanethiolate had been desorbed (see Fig. 2 for the principle). The cystaminium-bound enzyme layer is less effective for inhibition of the reaction of ferrocinium derivatives (64) at the Au surface than the densely packed long-chain alkanethiolate layer. Therefore, positive feedback would be observed above the enzyme-modified region due to the electrochemical feedback at the Au surface, although the positive feedback may be reduced to some extent compared to a bare Au surface.

When using an artificial electron acceptor for GOx and adding glucose to the solution, the additional contribution due to the enzymatic regeneration of the mediator would be small and difficult to separate from the electrochemical feedback. This is particularly problematic when the enzymatic activity decreases with time. Furthermore, as enzyme coverage varies the effects of increased electrochemical feedback (due to decreased blocking effects) and diminished enzyme-mediated feedback (due to a decreased amount of bound enzyme) would partly counteract each other.

FIG. 12 (a) SECM image of an *N*-substituted polypyrrole line with 10 mM $[Ru(NH_3)_6]^{3+}$ as mediator, $E_T = -350$ mV (Ag/AgCl). (b) SECM image of an *N*-substituted polypyrrole line after covalent attachment of GOx in the absence of glucose with 0.5 mM $[Os(bpy)_2fpyCl]^+$ as mediator, $E_T = +550$ mV (Ag/AgCl). (c) Same as (b) in the presence of 100 mM glucose. For all images $a = 12.5$ μm, $v_t = 5$ μm s^{-1}. (Reprinted from C Kranz, Wittstock, H Wohlschläger, W Schuhmann. Imaging of microstructured biochemically active surfaces by means of scanning electrochemical microscopy. Electrochim Acta 42:3105–3111. Copyright (1977), with permission from Elsevier Science.)

Under these circumstances the generation-collection mode represents the better choice (21) (Fig. 13). Experiments were carried out in an air-saturated solution. In the presence of glucose, the enzymatic reaction commences and gluconolactone and H_2O_2 are formed. The latter is amperometrically detected at the SECM microelectrode. High currents are measured above regions of H_2O_2 production and hence provide a map of GOx activity (Fig. 14A). Repetition of the experiment in the absence of glucose results in a featureless image of low currents (Fig. 14B) proving that the signals in Figure 14A are due to the oxidation of enzyme-generated H_2O_2.

During the experiments the Au electrode is at the open circuit potential which is established at a macroscopic blank area of Au located several millimeters away from the enzyme-modified regions. From thermodynamic considerations one might expect H_2O_2 to be oxidized at the Au surface. However, this reaction occurs only at high overpotentials if no catalytic sites are present on the electrode surface. The electrocatalytic noble metal surfaces are deactivated rapidly by the chemisorption of thiolates or other organic molecules. Therefore, H_2O_2 diffuses into the solution and can be detected by the microelectrode. In this study a Pt microelectrode was used. Alternatively, an enzyme-modified microelectrode could be applied as the tip (Table 2).

The tip radius was 25 μm for both electrochemical desorption and mapping of GOx activity. The area from which the alkanethiolates were desorbed

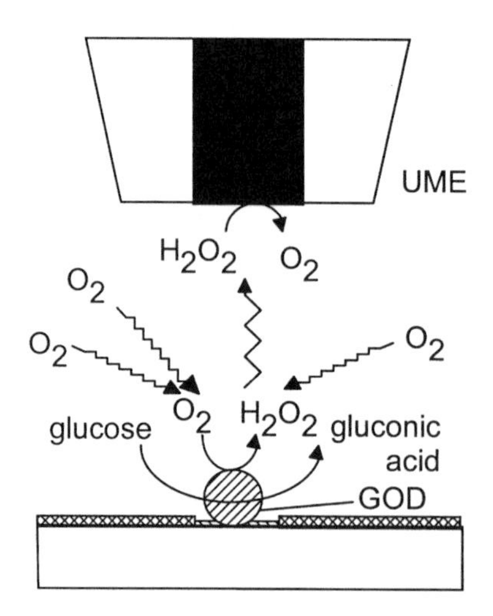

FIG. 13 Detection of GOx activity immobilized on a conductive support in the generation-collection mode; GOx, glucose oxidase; UME, microelectrode tip.

(A)

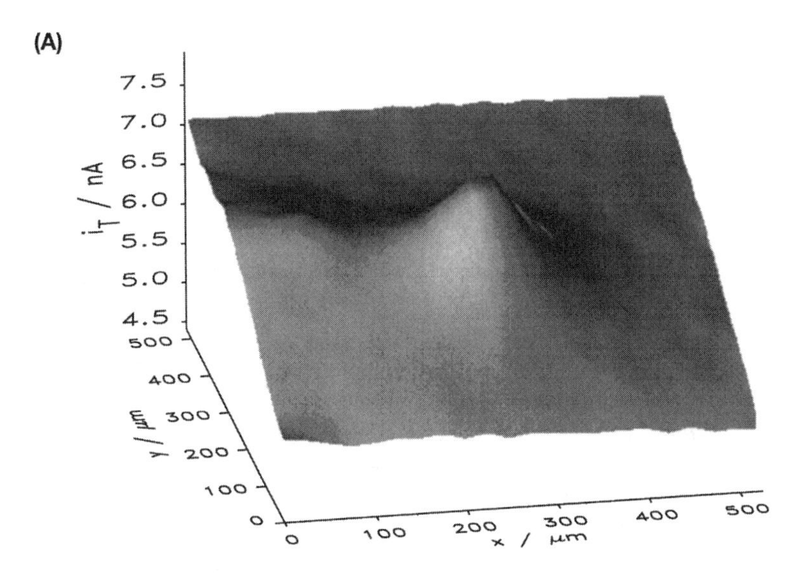

(B)

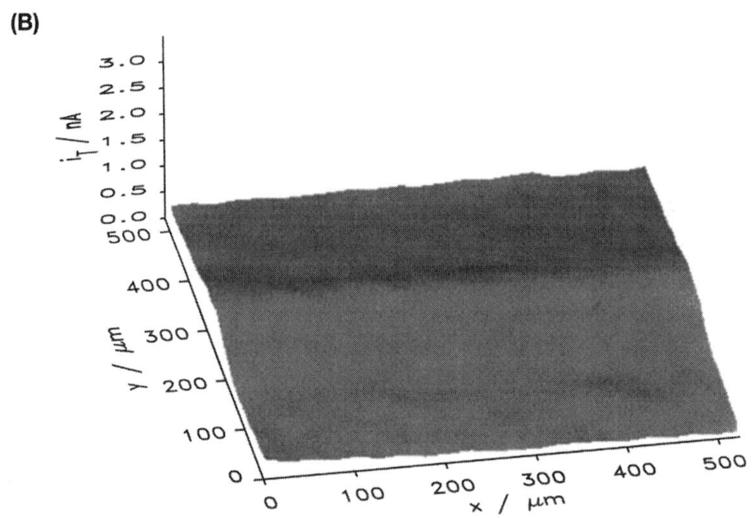

FIG. 14 Generation-collection mode images of GOx bound to a patterned SAM in the presence (A) and absence (B) of 50 mM glucose in air-saturated 0.1 M phosphate buffer (pH 7.4); $E_T = +400$ mV, $v_t = 15\ \mu m\ s^{-1}$). (From Ref. 21. Copyright 1997 American Chemical Society.)

gave feedback images with a full width at half maximum (FWHM) of about three times the diameter of the microauxiliary electrode (tip) used in the desorption step (63). One would therefore expect that the enzyme-modified surface would have similar dimensions. The image of the enzyme activity is, however, further blurred by the fast diffusion of H_2O_2. As discussed in Sec. I.D, this is a trade-off necessarily connected with the use of the generation-collection mode on large specimens. The image does not represent a steady-state situation as the enzymatic reaction produces H_2O_2 continuously leading to a slow increase of the bulk H_2O_2 concentration. This fact is obvious from the background currents in Fig. 14A, which rise steadily from the start of the image at the front left corner to the last recorded point at the far right corner during an image acquisition time of about 40 minutes.

Improving the Resolution of the Generation-Collection Mode in Biological Applications. The example presented above illustrates a dilemma when investigating enzymatically active layers on conducting supports: application of the feedback mode is generally problematic and the generation-collection mode is connected with a loss in lateral resolution and difficulties for kinetic analysis owing to the non–steady-state concentration profiles. Even over insulating supports the feedback mode may not be applicable when no redox active cofactor is involved in the enzymatic reaction. In other cases its use may be difficult because the enzymatic activity is below the limit of detection [Eq. (1)]. As discussed in Secs. I.C and I.D, the GC mode is more sensitive.

Introduction of a homogeneous chemical reaction which competes with the detection reaction at the tip for the species formed at the specimen represents a general approach for resolving these difficulties. Practical realizations of this concept have appeared occasionally in the SECM literature. Because of their significance for biological SECM application, a systematic discussion is provided here, even if the original application did not involve a biochemically active specimen (65).

The first systematic study on resolution enhancement for generation-collection mode imaging of biochemically interesting compounds was presented by Engstrom et al. as early as 1992 (66). A large Pt disk electrode embedded in insulating material served as model specimen for the experiments discussed here. The solution contained either $[Fe(CN)_6]^{4-}$ or epinephrine as oxidizable species. The specimen acted as a generator electrode to which a sweep was applied from potentials where no oxidation occurred to a potential causing a diffusion-controlled oxidation of $[Fe(CN)_6]^{4-}$ or epinephrine. The amperometric microelectrode tip was placed within 2 μm of the specimen and poised at a potential E_T to detect the local concentration of species electrogenerated at the specimen as a function of the specimen potential E_S (Fig. 15A). After movement of the electrode parallel to the specimen surface, a new potential sweep at the specimen was performed.

$[Fe(CN)_6]^{4-}$ undergoes a quasi-reversible reaction at the specimen electrode. The oxidized form $[Fe(CN)_6]^{3-}$ is produced on sweeping to positive potentials. After completing the sweep the specimen electrode returns to a potential where the generated $[Fe(CN)_6]^{3-}$ is rereduced. By increasing the sweep rate at the specimen electrode, the time available for $[Fe(CN)_6]^{3-}$ generation and diffusion away from the specimen electrode is diminished. Therefore, the distance at which the microelectrode tip can sense the source of $[Fe(CN)_6]^{3-}$ shrinks, leading to improved lateral resolution (Fig. 15A).

However, Engstrom et al. (66) did not find this approach very fruitful because kinetic limitations of the quasi-reversible redox system cause the interfacial concentrations to deviate considerably from Nernstian behavior at high scan rates. However, it is worthwhile to evaluate some possible adaptations of the concept to biological systems. For instance, a discontinuous generation of a redox-active species may be caused by chopped illumination. Light may be the energy source for light-dependent enzymes, enzyme complexes, cells, or organs. Alternatively, light may induce the formation of unstable compounds in solution which are substrates of fast

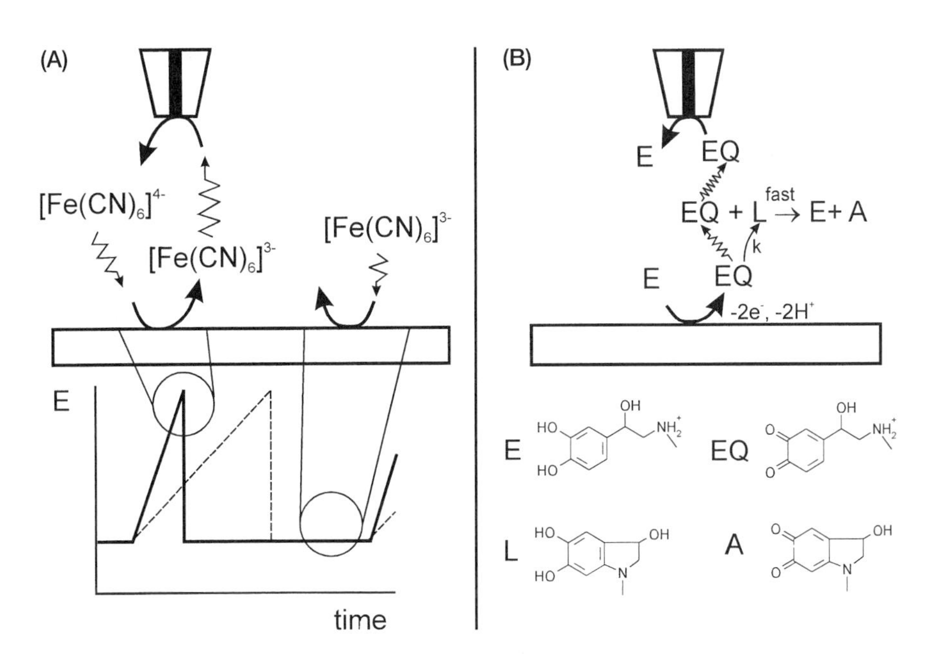

FIG. 15 Image improvement in generation-collection mode imaging as introduced by Engstrom et al. (66) (schematics prepared by the chapter authors).

enzymatic reactions. Such reactions are important as natural deactivation pathways for highly toxic radicals like the superoxide anion $O_2^{\cdot(-)}$ or those formed as a result of environmental stress (ozone or UV exposure). The time between light irradiation and the ejection of the product to be detected at the tip may be quite long because of diffusion barriers (membranes) or complicated multienzyme reaction cascades. Tsionsky et al. (23) monitored the oxygen release of single guard cells of an intact plant and found that the time between switching on illumination and reaching a constant oxygen reduction current to be of the order of 6 minutes. Since a complete on/off cycle of the enzyme activity needs to be performed at each microelectrode tip position, practical constraints for imaging are anticipated for a significant number of biological systems. Periodic addition of the enzyme substrate could also be envisaged in a stopped-flow setup. It remains to be demonstrated that such an arrangement would allow fast and complete solution exchange under the geometric conditions of a SECM cell.

Many enzymes exhibit a pH optimum of their activity. Working in a solution outside that optimum, an electrolysis of solution components involving H^+ or OH^- at the microelectrode can induce a local pH shift into the pH optimum of the enzyme activity, thereby "switching on" enzyme activity locally. The product of the enzymatic reaction may be collected at the microelectrode tip after applying the detection potential to it. This event would mark the end of the electrolysis causing the local pH shift. Because of the high diffusion coefficients of H^+ and OH^-, the locally changed pH will dissipate rapidly and "switch off" the enzyme activity. This principle has been demonstrated with fluorescence spectroscopic observation of the enzyme activity (67) and may soon find other applications as microelectrode-induced localized pH shifts have also been utilized in SECM microfabrication techniques (68,69) (see Chapter 13). It remains to be seen whether interesting enzyme systems exist that can be manipulated by rather small pH changes. The trade-in for all these discontinuous variations of reaction conditions at the specimen would be increased imaging times.

Engstrom et al. (66) demonstrated the use of a chemically irreversible redox reaction at the specimen as a way to improve resolution in the generation-collection mode (Fig. 15B). Epinephrine E was oxidized to its quinone EQ at the specimen electrode. EQ is unstable and decomposes to produce adrenochrome A via an intermediate leucoadenochrome L (71,72). EQ is reduced at the tip at potentials negative of +200 mV (SCE), whereas both A and EQ are reduced at potentials negative of −100 mV (SCE). Poising the microelectrode at +100 mV (SCE) allows the detection of EQ molecules that have not yet decomposed to A decreasing the distance at which the microelectrode tip can sense the source of EQ. The transformation rate EQ $\rightarrow$ A is pH-dependent, providing a tuning mechanism for the lifetime of EQ.

Engstrom et al. (66) used a potential sweep for these experiments as they were primarily interested in local concentrations. Using a potential sweep limits the extent of the diffusion layer above the specimen electrode and allows reestablishment of the bulk concentration of oxidizable compounds before initiating a new potential cycle at the next tip position. This scheme is associated with increased imaging time (vide supra). Indeed, when adapting this idea to biological systems, one will in many cases work under conditions of substrate saturation where the enzyme kinetics rather than the mass transport limits the conversion rate at the specimen, thus avoiding the need to reestablish the substrate bulk concentration close to the specimen surface. If the product to be detected at the tip undergoes a homogeneous reaction to a redox-inactive product, an image can be recorded in the usual way, i.e., without changing periodically the conditions in the cell at each tip position. On the other hand, tuning the decomposition rates by varying the solution pH may not always be feasible because of the limited pH ranges in which many enzymes are stable and active. The applicability of this mode still depends on finding an enzyme substrate that may be converted to a redox active product, which undergoes a following chemical reaction to a redox inactive compound within the time scale of the diffusion across the specimen-tip gap. These prerequisites greatly limit the range of biological applications for this imaging scheme.

Instead of a creating an unstable redox-active product in the enzymatic reaction, one may use a second enzymatic reaction to decrease the lifetime of the substances generated at the specimen. Since enzymes are available for many metabolites, this approach should be of rather general utility for biological samples and has already been demonstrated (21) for glucose oxidase immobilized on a specimen and catalase dissolved in solution (Fig. 16a). Under conditions of high glucose concentration H_2O_2 is continuously formed at the sample. It leads to increasing background currents during image acquisition as discussed in Sec. II.B and evident from Figure 16b and 16e, profile 1. Adding minute amounts of catalase to the solution phase eliminates the rising background and increases the lateral resolution because the diffusion length of H_2O_2 is limited (Fig. 16c and 16e, profile 2). The peak-to-baseline signal height is still 85% of that in the absence of catalase. One should note that the FWHM is an easily measurable quantity, which can be used to follow trends in resolution for similar images of the <u>same</u> object. Since resolution quantifies the ability of a method to resolve <u>different</u> closely spaced objects, an exact quantity describing resolution can only be derived from the gradient of the signal above a well-defined transition between surface regions of different properties (70)—a condition not fulfilled here. Therefore, it is not possible to differentiate how much the size of the enzyme spot, the tip diameter, and the diffusion of H_2O_2 contribute to the

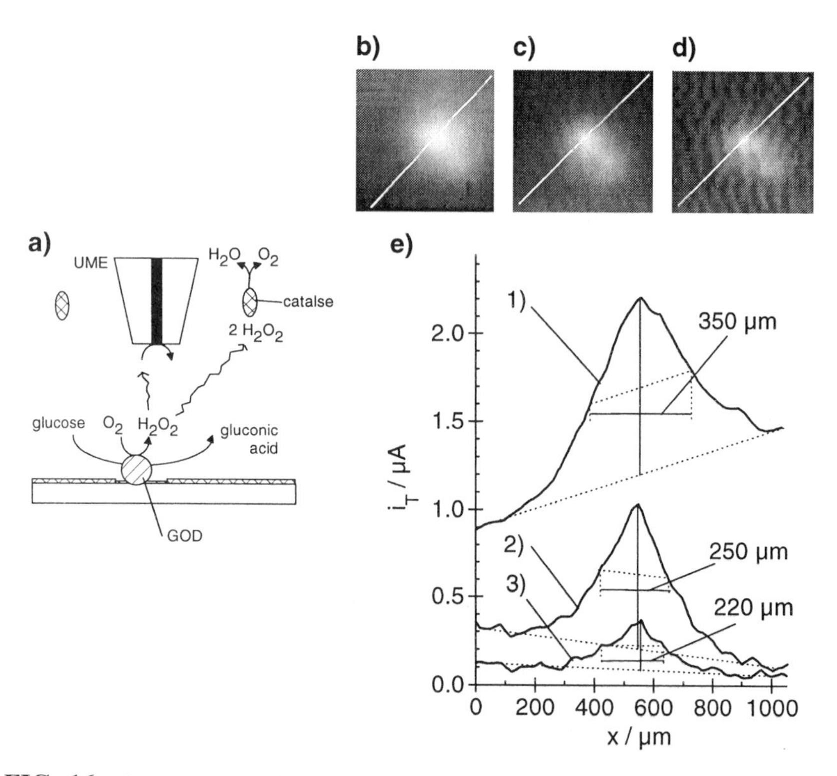

FIG. 16 Improvement of image resolution in the generation-collection mode by providing a catalyst for the solution phase follow-up reaction of the species to be detected at the tip electrode. (a) Schematic; (b)–(d) gray-shaded current images with location of cuts after addition of catalase in microliter suspension (20 mg mL^{-1} protein) per milliliter working solution—(b) none, (c) 0.005, (d) 0.03 (e) current profiles (1) for (b), (2) for (c), and (3) for (c). The FWHM is given for each curve; working solution, air-saturated 50 mM glucose + 0.1 M phosphate buffer (pH 7.4); sample was prepared as shown in Figure 26A. (Parts b–e from Ref. 21. Copyright 1997 American Chemical Society.)

signal width. To obtain optimum conditions for imaging, the catalase concentration must be adjusted to make the lifetime of H_2O_2 sufficiently long to diffuse across the specimen-tip gap but too short to accumulate in the solution phase. Unfortunately, the reaction rate does not follow a first-order rate law in this concentration range. Therefore, adjustment needs to be done empirically. Increasing the catalase concentration leads to a further small improvement of resolution and a considerable decrease of signal (Fig. 16d and 16e, profile 3). An even further addition of catalase to 0.15 μL suspension per mL working solution quenches the signal for H_2O_2 production (21) (not shown). This is a valuable observation because the high specificity of enzymes shows that the species detected at the microelectrode tip is in fact H_2O_2 and not another redox-active byproduct or solution impurity. Using an enzyme microelectrode as tip electrode as discussed in Sec. I.D would be another way to accomplish this task. Such methods of image verification will gain importance if more complicated biological samples are investigated and the mixture of redox active compounds released from the specimen is less well understood than in the model system investigated in Figure 16.

C. Extension of the Imaging Concept to Other Biomolecules

So far SECM applications have been considered where enzymes immobilized at a surface catalyze redox reactions of low molecular weight compounds. The reaction products are detected at the ultramicroelectrode tip under diffusion-controlled conditions. This approach requires that the biochemically active layer continuously generate or consume redox active (for amperometric detection) or charged (for potentiometric detection) species. Since the tip signal depends on the diffusion coefficients and/or convective effects as well as the local concentration, it is possible to image localized mass transport phenomena instead of localized chemical fluxes (Chapter 9). In a general sense it is a process that is recorded with lateral resolution.

Imaging the activity of other important biomolecules (e.g., antibodies, a large variety of different receptor molecules, DNA and DNA-binding proteins, which are capable of selective interaction with other chemical species) is challenging because only a single or a few binding events occur at a given binding site and there is no continuous transformation of a reagent. Therefore no continuous flux of species occurs above surfaces modified with such biomolecules. In the absence of a chemical flux, the electrode/electrolyte interface of the microelectrode separated from the specimen surface by a solution layer of finite thickness will be incapable of sensing the modified state of the surface. This limitation of SECM can be overcome by appropriate labeling techniques—a general approach well established in the life sciences. For SECM one chooses a reagent that (1) is able to bind selectively to the immobilized biomolecule of interest and (2) has an enzyme label that allows detection with one of the schemes discussed in Sec. I.

With these general constraints in mind, how well does SECM imaging compare to other microscopies using more conventional labels such as fluorescence (either direct fluorescence labeling or enzymatic generation of a fluorescing compound) and heavy-metal staining in scanning electron microscopy (SEM)? Owing to their optimized protocols, the accumulated experience and expertise, the commercial availability, and the speed of the image acquisition, the latter methods are likely to remain the first choice for procedures where they are already well established. SECM imaging, however, offers favorable possibilities for novel types of investigation. It can be done in solution and, in contrast to SEM, leaves the biomolecules in their native environment. It offers a very sensitive detection scheme, in particular the combination of amperometric detection and enzyme labeling. Unlike fluorescence microscopy it does not depend on the optical properties of the specimen surface (transparent or reflective), nor are there any disturbing effects of light absorption by solution components or signal losses at the solution/air interface. In addition, the imaging mechanism can often be modeled numerically and used to extract quantitative kinetic information (see Sec. II.B). The low throughput, currently the major drawback of all scanning probe techniques, may be increased when SECM detection principles are combined with microstructured specimens and with absolute positioning of the tip with respect to particular surface features. This would enable rapid and precise vertical and lateral positioning of the SECM tip without the need to search large areas of the sample before taking a reading.

1. Imaging Antibodies

Immobilized antibody (Ab) layers were first imaged using the SECM generation-collection mode by Wittstock et al. (32). This study provided a new tool for optimization of immobilization protocols in immunoassay development. The approach was demonstrated for an immunoassay of digoxin, which is the antigen (Ag). A periodate-oxidized (anti-digoxin) Ab was immobilized covalently at spots of 1600 μm diameter onto an amino-functionalized glass surface. During incubation of the layer with an excess of digoxin that was labeled with the enzyme alkaline phosphatase (Ag-E), all accessible and active binding sites formed an Ab:Ag-E complex. The density of the active sites could be imaged by measuring the generation of 4-aminophenolate (PAP) from 4-aminophenyl phosphate (PAPP) catalyzed by alkaline phosphatase. The detection was performed amperometrically in the collection mode at a potential where PAP exhibits redox activity but PAPP does not (Fig. 17). Generation-collection mode is the only option for the enzyme alkaline phosphatase because it is not a redox enzyme. Through applying Ab solutions of different concentrations to distinct surface regions of one specimen using an appropriate masking procedure, a convenient comparison was possible. Figure 18 shows line scans above five different spots with covalently immobilized Ab. Curve 1 is a result from a control experiment where no Ab had been incubated on the glass surface. The absence of any signal indicates that the Ag-E is not retained on the aminated glass surface by adsorption or covalent bonding to the NH_2-groups. Curves 2–5 correspond to increasing concentrations of the applied Ab solution. The current

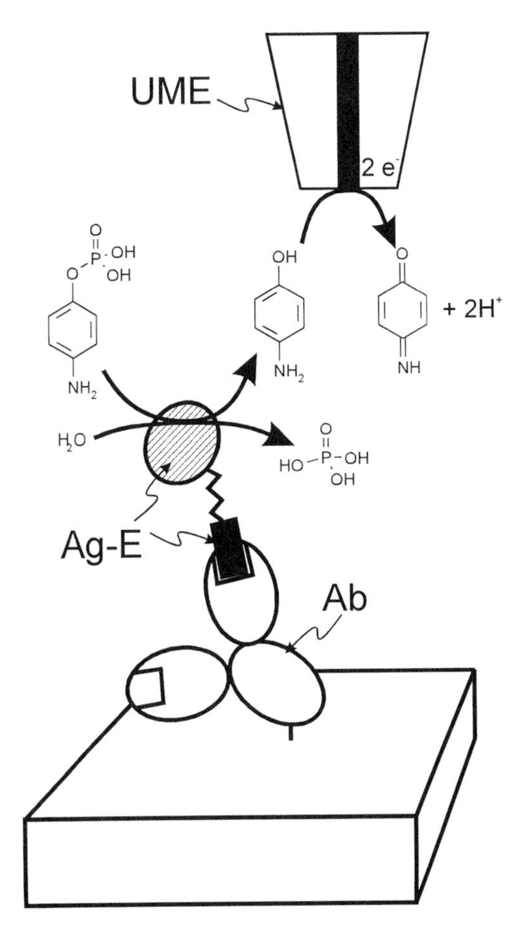

FIG. 17 Schematic of the detection of localized Ab layers via an alkaline phosphatase–labeled Ag and amperometric detection of 4-aminophenol (PAP) at the tip electrode. UME, Ultramicroelectrode tip.

rises only where Ag-E is bound by immobilized Ab as can be seen from the match between the signal width at half height and the diameter of the modified regions (1600 μm). The slight increase in the background currents results from the enrichment of the bulk solution with PAP during the experiment. Inhomogeneities of the modified regions were resolved only for the highest concentration Ab solution (curve 5).

Since nonspecific adsorption is regarded as a major source of systematic error in heterogeneous immunoassays, the adsorptive immobilization of different Ab solutions was studied on clean, nonaminated glass surfaces (Fig.

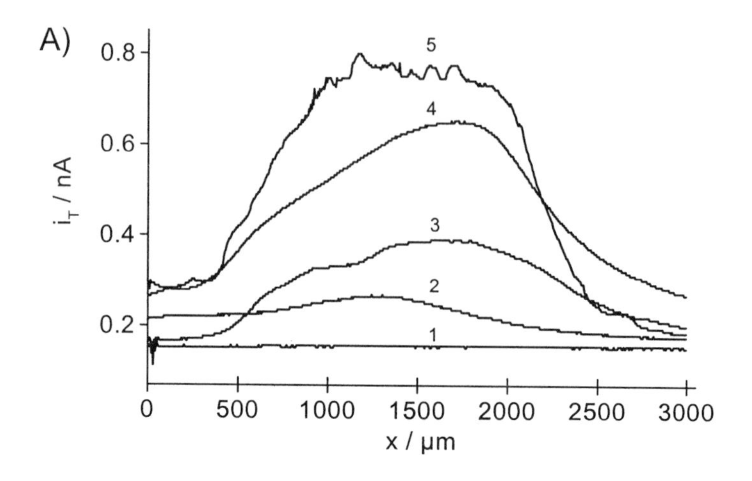

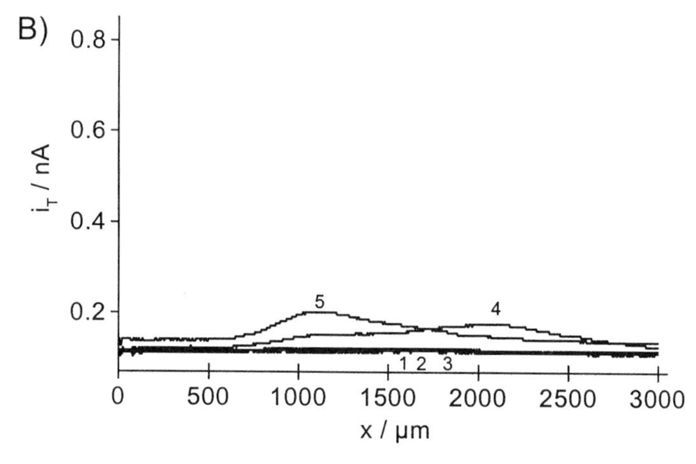

FIG. 18 Horizontal line scans across spots with immobilized monoclonal (anti-digoxin) Ab after reaction with enzyme-labeled digoxin. (A) covalent immobilization with the Ab solutions: (1) no Ab solution applied, (2) 1:8, (3) 1:4, (4) 1:2, (5) undiluted Ab stock solution 0.141 mg mL^{-1} protein. (B) Adsorptive immobilization with the Ab solutions: (1) no Ab solution applied, (2) 1:8, (3) 1:4, (4) 1:2, (5) undiluted Ab stock solution 0.326 mg mL^{-1} protein. Measurement solution 4 mM PAPP, 1 mM $MgCl_2$, 0.1 M KCl, 0.02% NaN_3, 0.1 M Tris-HCl, pH 9.0. v_t = 7.9 $\mu m\ s^{-1}$; d = 10–20 μm; a = 15 μm; E_T = +280 mV (Ag/AgCl/3 M NaCl). (From Ref. 32. Copyright 1995 American Chemical Society.)

18B). Curve 1 is a control experiment. The absence of a signal demonstrates that Ag-E is not adsorbed at the glass itself. Curves 2–5 were measured above areas treated with increasing concentrations of Ab. No significant differences to the control experiment were observed when the Ab concentration was kept low (curves 2, 3). However, detectable amounts of Ab are adsorbed if more concentrated Ab solutions are applied to the glass surfaces. The adsorbed Ab that are able to bind Ag-E molecules give a signal corresponding to 10% of the signal obtained from covalent attachment of the most concentrated Ab solution in Figure 18, i.e., curve 5.

2. *Small Spot Immunoassay*

In a series of papers Shiku et al. (16,31,73) developed a strategy for small spot heterogeneous enzyme immunoassays using an SECM to read out the signal, thus allowing microscopic sensing regions for the analysis of multiple antigens to be placed in close proximity on one chip. The assays are carried out in the sandwich format (Fig. 19). An antibody Ab against the analyte is immobilized onto glass. After application of the antigen Ag (i.e., the analyte) onto the surface, it is allowed to form an Ab:Ag complex. Subsequently, the surface is exposed to a solution of a second Ab labeled with the enzyme horseradish peroxidase (HRP). Microspots are obtained either by micropipetting the Ag (16) or the Ab (31) onto the surface. The continuous reaction used in the SECM detection is the same in each case:

$$2\text{FMA} + \text{H}_2\text{O}_2 + 2\text{H}^+ \xrightarrow{[\text{HRP}]} 2\text{FMA}^+ + 2\text{H}_2\text{O}$$

where FMA is hydroxymethyl ferrocene and FMA^+ is the corresponding ferrocenium derivative. Detection can conveniently be made in the generation-collection mode. In this case the working solution contains FMA. The current at the tip microelectrode results from the reduction of FMA^+ formed in the HRP-catalyzed reaction at the specimen (Fig. 19A). One may also work in a solution containing FMA^+ in the bulk phase using the feedback mode. In this mode FMA^+ is also reduced at the tip microelectrode, which provides FMA for the enzymatic reaction to occur and maintains the feedback (Fig. 19B). The comparison of both detection schemes (16) supports some general recommendations made in Sec. I. The generation-collection mode tends to achieve higher sensitivities because the background currents are lower and the current variations within an image frame are larger. On the other hand, the feedback image provides better lateral resolution of fine variations in enzyme activity. Using a species of limited lifetime (chemically unstable), as discussed in Sec. II.B, may represent a flexible strategy for combining the sensitivity and flexibility of the generation-collection experiments with lateral resolution close to that obtained from feedback experiments of similar systems.

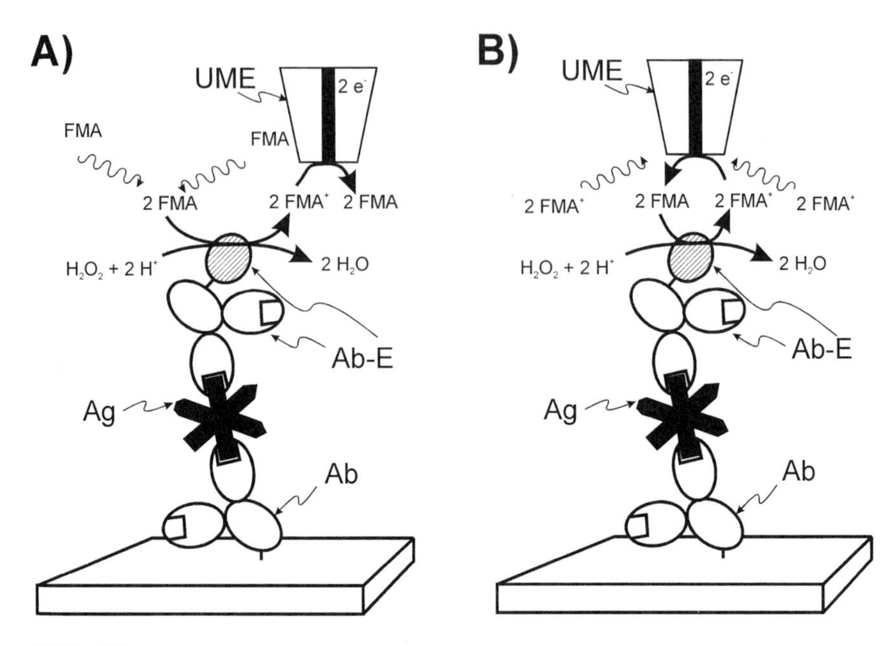

FIG. 19 Schematic representation of the detection scheme in the small spot immunoassays used by Shiku et al. (16) (schematic prepared by the chapter authors). (A) generation-collection mode; (B) feedback mode. UME, ultramicroelectrode tip; FMA/FMA$^+$, ferrocene/ferrocenium forms of hydroxymethyl ferrocene; Ab-E; antibody labeled with horseradish peroxidase; wavy lines indicate diffusion, solid lines are reactions.

Immunoassays have also been developed for carcinoembryonic antigen (CEA) (16), human chorionic gonadotropin (HCG) (31), and human placental lactogen (HPL) (31). CEA is a tumor marker with a molecular weight MW of 180,000–200,000. HCG (MW = 38,000) and HPL (MW = 22,000) are peptide hormones that are pregnancy indicators. Shiku et al. (31) describe a miniaturized, dual immunoassay for HCG and HPL with SECM as the detection method and a microstructured glass surface as the solid support. The glass surface was etched to create a plane of $300 \times 150\ \mu m^2$, which is 0.7 μm below the original glass surface. Within that "trough," two elevated, square-shaped regions of $50 \times 50\ \mu m^2$ served as sensing areas where (anti-HCG)-Ab and (anti-HPL)-Ab were immobilized by micropipetting. After exposure of the modified surface to a solution containing HCG and/or HPL, the surface was treated with the HRP-labeled antibodies (anti-HCG)-Ab-HRP and (anti-HPL)-Ab-HRP. Prior to the detection step, the elevated regions were efficiently relocated by topographic negative feedback imaging in a solution of FMA, in which diffusion-controlled oxidation of FMA was performed at the tip at +0.40 V (Ag/AgCl). After adding H_2O_2 to the working solution and switching the working potential of the tip microelectrode to 0.05 V, a value where FMA^+ can be reduced, the activity and location of the enzyme label HRP bound in the Ab:Ag:Ab-HRP complex could be detected (Fig. 20). If the analyte solution contained either HCG or HPL, HRP activity was found only above the active region with the corresponding antibody. When both analytes were present, corresponding levels of activity

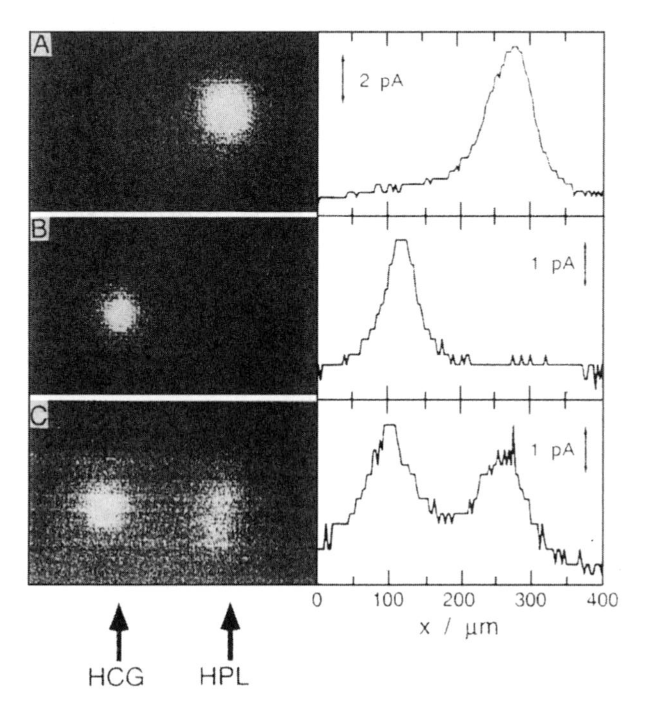

FIG. 20 SECM images and cross-sectional profiles of microstructured glass supports with (anti-HCG)-Ab and (anti-HPL)-Ab immobilized at two distinct regions, treated with (A) 56 ng mL^{-1} HPL; (B) 2.0 IU mL^{-1} HCG; (C) a mixture containing 31 ng mL^{-1} HPL + 0.63 IU mL^{-1} HCG. After rinsing, the glass support was dipped into a solution of 20 $\mu g\ mL^{-1}$ (anti-HCG)-Ab-HRP + 7 $\mu g\ mL^{-1}$ (anti-HPL)-Ab-HPL. Working solution, 1.0 mM FMA + 0.5 mM H_2O_2 + 0.1 M KCl + 0.1 M phosphate buffer, pH 7.0; v_t = 9.8 $\mu m\ s^{-1}$; E_T = +0.05 V (Ag/AgCl). (From H Shiku, Y Hara, T Matsue, I Uchida, T Yamauchi. Dual immunoassay of human chorionic gonadotropin and human placental lactogen at microfabricated substrate by scanning electrochemical microscopy. J Electroanal Chem 438:187–190. Copyright 1997, with permission from Elsevier Science.)

could be detected over both sensing areas. Electrochemical immunoassays for mouse immunoglobulin G as a model analyte were demonstrated using agglomerates of surface-modified microbeads (Sec. II.F) and quantitation by the GC mode of the SECM with alkaline phosphatase as the enzyme label (74). The close proximity of the amperometric tip and the antibody-modified area allows readings to be made without any incubation time to accumulate products of the enzymatic reactions. This feature represents an advantage over conventional enzyme-labeled immunosorbent assays (ELISA), which usually require an incubation step prior to detection.

D. Applications to Cells and Whole Organisms

In this section selected application of the SECM to problems of physiology and cell biology are discussed. The examples are chosen from widely differing biological systems; photosynthesis in the leaves of *Tradescantia fluminensis*, respiratory activity of cultivated cells, and the resorption of bone matrix in mammals. Although there are to date comparatively few examples of the use of SECM to study living organisms, electrochemical methods such as ion-selective microelectrodes have been of longstanding importance in the life sciences. The flexibility of the SECM approach should enable many new kinds of experiments and information to be obtained. With the commercialization of the technology, a significant growth in the application of SECM in the life sciences is possible.

1. Photosynthetic and Respiratory Activity in Single Guard Cells of an Intact Plant and in Single Cultivated Cells

Tsionsky et al. (23) recently reported an SECM study of the leaves of an intact plant, *T. fluminensis*. The SECM was used both to map the topography of a leaf and the rate of photosynthetic oxygen production in vivo. A leaf that was alive and still attached to the plant was held against the base of the electrochemical cell using mounting tape or soft O-rings. The electrochemical cell, plant, and xyz positioning stage were placed inside a glove bag to allow control of the gas composition (Fig. 21). Illumination of the leaf was achieved using a 1 mm diameter optical fiber to direct the output of a xenon lamp onto the leaf through the base of the electrochemical cell. Since conventional electrochemical redox mediators are often toxic to living organisms, the choice of mediator was restricted. Previous work had utilized ferrocyanide as the redox mediator (24), but for the study of the living plant oxygen was chosen (23). Although the electrochemical behavior of oxygen is not simple, it is a logical choice for the study of living systems, and as long as the tip potential is poised at a sufficiently negative value to ensure a diffusion-controlled tip current, standard SECM theory for negative feed-

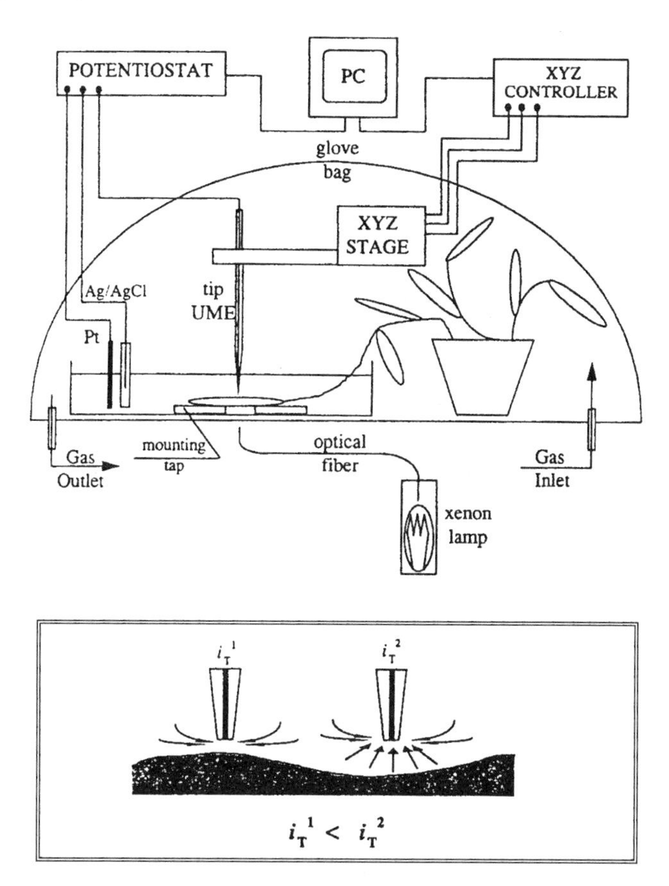

FIG. 21 Experimental apparatus for the study of leaf surfaces by Tsionsky et al. (23). (Copyright American Society of Plant Physiologists.)

back applies. Since oxygen reduction involves the uptake of protons, the solution was buffered to ensure a stable local pH at the leaf surface. Topographic images based on the negative feedback mode were obtained using the reduction of oxygen in an air-saturated solution in the dark so that the contribution of oxygen fluxes due to photosynthesis was negligible. Individual stomatal complexes were clearly resolved with the guard cells appearing as depressions in the tip current image and the stomatal pore in the center of the complex as a current peak since the height and tip current are inverted in a negative feedback experiment. Illumination of the leaf caused the production of an oxygen flux due to photosynthesis, and the approach curve for oxygen reduction showed a different behavior with the tip current rising above the value observed in the bulk solution. *T. fluminensis* possesses white regions in which chloroplasts are only present in the guard cells around the stomata. The stomata were sufficiently widely separated that the oxygen diffusion layers did not overlap and generation-collection mode imaging was possible. The images recorded under illumination showed a peak current roughly a factor of two above the oxygen background over the stomata. Use of the SECM to image the leaf topography, position the tip over a single stomatal complex, and observe directly the generation of oxygen in response to a step change in illumination was demonstrated. In fact, Tsionsky and coworkers were able to image the photosynthetic oxygen production in a single stomatal complex of a white leaf region and demonstrate directly for the first time the photosynthetic electron transport in single guard cells of an intact living plant (Fig. 22). This work illustrates the flexibility of the SECM technique; the experimenter can image topography and biochemical reactivity and then choose a particular location on the sample to make kinetic studies of the response of a specific portion of the system to an external stimulus in essentially one experiment.

Such an approach was also selected to study the reaction of cultivated cancer cells of the cell line SW-480 on addition of cyanide to the culture medium (25). In cyanide-free solution the concentration of oxygen is reduced around the cells. The reduced oxygen reduction current above the cells was used to locate them. The difference in the oxygen reduction current directly above a cell and far away from it formed a quantitative indicator for cellular activity. After addition of cyanide, the cellular activity could be followed with a temporal resolution on the minute time scale. Cyanide blocks the electron transport of the respiratory chain in mitochondria leading to cell death. However, Yasukawa et al. found that cellular activity first remained constant for 500 seconds, then rose until 800 seconds before decreasing until cell death. The critical concentration for cell death is 50 μM for this cell line (25). Because the CN^- ion has to pass the cell and mitochondria membranes, the local concentration inside the mitochondria lags behind the CN^- extracellular concentration. Modeling of the mass transport and comparison with the observed lag time suggests that the permeability of the membranes for CN^- in vivo is about $(1-3) \times 10^{-7}$ cm s^{-1} (25). Detection of molecular oxygen was exploited by the same authors (26) to characterize the activity of protoplasts in dark and under illumination. Changes of photosynthetic activity after injection of the inhibitor 3-(3,4-dichlorophenyl)-1,1-dimethylurea into an individual cell were studied (26).

2. *Calcium and Superoxide Measurement at Osteoclasts*

A recent application of SECM ideas has been in the study of the resorption of bone by specialized cells known as osteoclasts (19,75). Bone is a complex

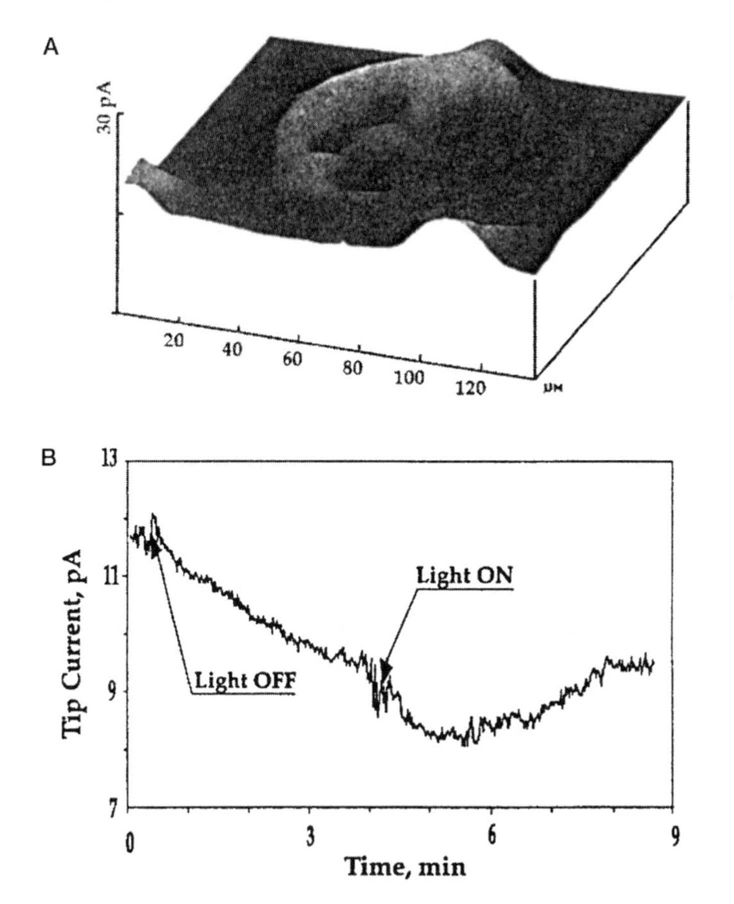

FIG. 22 (A) A single, illuminated stomatal complex in a white leaf region of *T. fluminensis* (800 μE m^{-2} s^{-1}). (B) The response of i_T to step changes in light availability when held over a stomate in a white leaf region. Measured current decreases after the leaf is placed in the dark and increases after reillumination (800 μE m^{-2} s^{-1}). Measurements for (A) and (B) were made with a 7-μm-diameter carbon microelectrode. (From Ref. 23. Copyright American Society of Plant Physiologists.)

material consisting of calcium phosphate–based minerals (hydroxyapatite), organic material (mostly collagen), and various cells. The osteoclast carries out the resorption of bone by the secretion of protons and hydrolytic enzymes which break down the minerals and proteins, releasing, among other species, Ca^{2+} (Fig. 23). In vivo, bone is a dynamic system with a balance between the resorption and formation of bone. An imbalance in these processes is implicated in many diseases, such as osteoporosis in which there

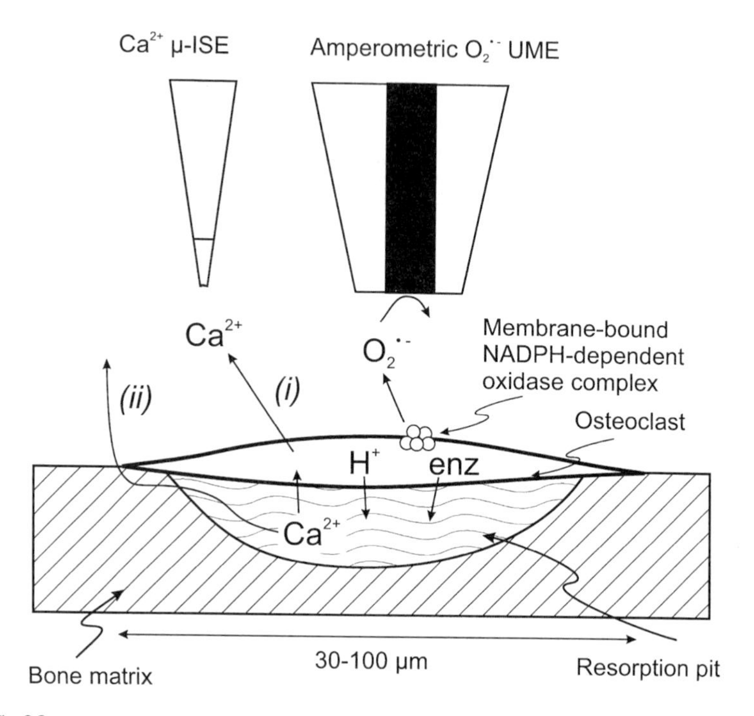

FIG. 23 A schematic of the processes involved in osteoclastic resorption of bone. Two SECM tips, a potentiometric Ca^{2+} sensor, and an amperometric superoxide anion sensor are shown. The osteoclast resorbs bone through secretion of protons and hydrolytic enzymes, enz, which break down the bone matrix. The disposal of Ca^{2+} released in this process may, in principle, occur via intracellular (i) or extracellular (ii) pathways.

is a loss of bone mineral. Assays for the resorptive activity of the osteoclast are usually carried out by incubating the cells on bone slices for 18–20 hours and then examining the bone under the electron microscope. The area and number of resorption pits can be determined from the micrographs. This assay is slow and imprecise. The rate of resorption has been probed more directly using a simplified SECM apparatus and liquid membrane calcium-selective microelectrodes to determine the ionized calcium released from the hydroxyapatite into the culture medium during resorption (Fig. 23). A significant background Ca^{2+} concentration (0.21 ± 0.08 mM) always exists in the culture medium near the bone surface due to dissolution of the bone by a purely chemical mechanism. The increase in Ca^{2+} due to the resorptive activity of the cells was detectable within approximately 10 minutes of incubation. The effect of fluid treatment (soaking in 10 mM NH_4F for 24 hours) of the bone slice was studied, and it was found that the background levels of calcium released by noncellular processes were reduced by three orders of magnitude ($\leq$19 μM). However, the osteoclasts were able to resorb the fluoride-treated bone albeit at a reduced rate. The standard SEM assay showed that there was no significant difference in the number of resorption pits on fluoride-treated and untreated bone, but the area of the individual pits on the fluoride-treated bone was 60% lower. This suggested that the cells were not inhibited from attaching to the treated bone or by any fluoride released into the medium on resorption, rather the resorption process is slowed simply by the greater resistance of fluorapatite to acidic dissolution.

In addition to the possibility of using the SECM to assay the activity of these cells, several open questions concerning the physiology of the osteoclast can now be addressed using the SECM to position a sensor over a single cell. The transport of products of bone resorption from the resorption pit (hemivacuole) to the external medium may occur via intracellular pathways (Fig. 23, i) or extracellular pathways such as diffusion around the cell periphery (Fig. 23, ii). Evidence for the transport of degraded collagen by the former process has recently been obtained by confocal microscopy (76). The pathways of Ca^{2+} transport are not known, but this and similar questions may in future be answered by straightforward SECM imaging experiments.

Free radicals, especially superoxide anion, are known to be important in the regulation of osteoclast activity and can be detected using an amperometric sensor consisting of cytochrome *c* covalently attached to a gold electrode (38,77). This method was used to detect the acute response of the cells to stimulation by parathyroid hormone (PTH) as a burst of superoxide anion generation (78). Other work has concentrated on the regulation of the membrane bound NADPH-dependent oxidase responsible for superoxide generation (19). A simplified SECM in which a manual or stepper motor driven stage was used to position the superoxide detector electrode near the osteoclasts on a bovine cortical bone slice (Fig. 23). The effects of inhibitors of protein kinases, membrane-permeable analogs of cAMP, and cholera toxin on the stimulation of superoxide anion production by PTH, pertussis toxin, and ionomycin were studied. cAMP-dependent inhibition was found to be dominant in controlling the superoxide anion production. For this type of collection mode experiment the SECM apparatus can provide real-time information and could be used in combination with an optical microscope to demonstrate that the time scale of the superoxide anion burst following stimulation by PTH was consistent with superoxide diffusion from the osteoclast to the tip (Fig. 24). Although the stimulation of the osteoclast by PTH has been generally accepted to be mediated indirectly by other cells of osteoblastic lineage, receptors for PTH in the osteoclast have been demonstrated (79). In combination with the short timescale of the response and the effects

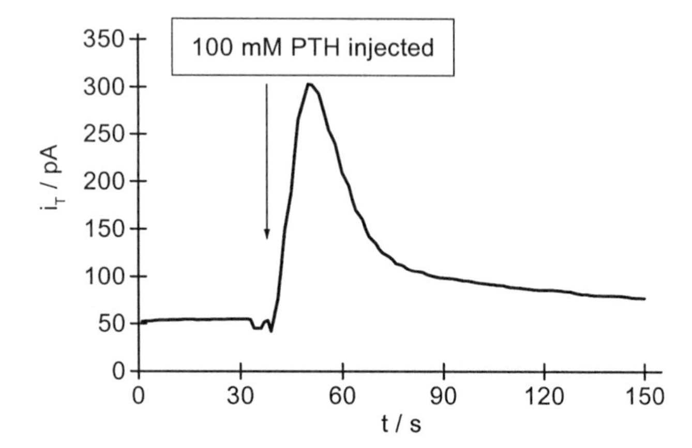

FIG. 24 Tip current response due to superoxide anion production on stimulation of osteoclasts cultured on bone by parathyroid hormone (PTH). The Au tip diameter was 125 μm and was modified with cytochrome *c* according to the procedures of Ref. 38. The tip-specimen distance was approximately 100 μm. (Reproduced by permission of the Society for Endocrinology from Ref. 19.)

of cholera and pertussis toxin, this suggests that the superoxide anion burst results from a direct action of PTH on the osteoclast via a G-protein coupled receptor.

E. Imaging Mass Transport in Biological Systems

The tip response in SECM is strongly dependent on local mass transport, and this may in fact be utilized to image local mass transport. Examples include the transport of oxygen and electroactive ions in cartilage (80), convective (81) and diffusional (34) transport in dentinal tubules, and ionic fluxes through skin (82), which are described in Chapters 9 and 12. In this section we discuss briefly experiments by Pohl, Antonenko and coworkers which, though not SECM, employ microelectrode techniques in a similar manner.

The systems studied by these authors are chemical fluxes at the bilayer lipid membrane, which are of importance in many cellular processes (33). The experimental system for the study of the reduction of acetaldehyde by alcohol dehydrogenase is shown in Figure 25. The pH microelectrodes employed were based on antimony-filled glass capillaries, which were pulled to a tip diameter of 5 μm. The pH gradient on the *trans* side due to the enzymatic reaction

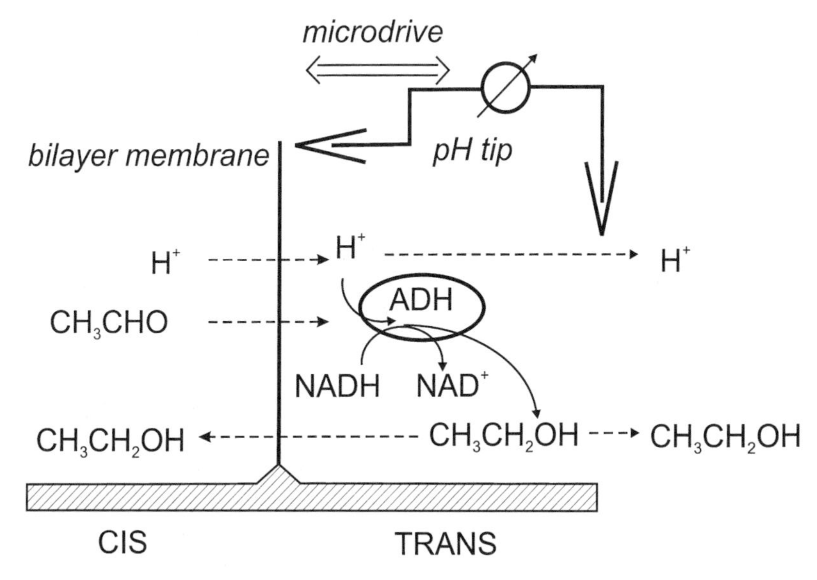

FIG. 25 The reduction of acetaldehyde catalyzed by alcohol dehydrogenase (ADH) in the presence of NADH proceeds at the *trans* side. The enzyme/coenzyme and substrate are added to opposite sides of the membrane. After the addition of acetate the membrane is highly permeable not only to acetaldehyde and alcohol, but to protons as well. pH profiles within the unstirred layer near the planar bilayer lipid membrane are measured with the help of a microelectrode driven by a hydraulic microdrive. (Schematic redrawn from Ref. 33 by the chapter authors. Copyright Academic Press, 1996.)

$$CH_3CHO + NADH + H^+ \rightarrow CH_3CH_2OH + NAD^+ \tag{7}$$

was measured by moving the pH microelectrode normal to the membrane at a speed of 2 μm s^{-1}. The zero distance point was determined by touching the membrane with the tip. The pH profile is determined by the kinetics of the enzymatic reaction and the thickness of the diffusion layer in the solution. The combined effects of transmembrane substrate diffusion, enzymatic reaction and product diffusion were incorporated in a quantitative model based on the Nernst diffusion layer approximation. These workers emphasized that presence of the diffusion layer has a significant effect on the apparent kinetics of important biological phenomena such as membrane permeation, carrier-mediated transport, and the reactions of enzymes immobilized on the membrane surface. Using similar technology it was possible to observe the dynamics of formation and dissipation of pH gradients in response to a voltage step (83), transmembrane osmotic fluxes (84), and ammonium ion permeation with a micro-ISE (85).

F. Micropatterning of Biologically Active Structures with the SECM

The formation of patterns of biomolecules on solid surfaces is an area of current interest in analytical chemistry and biochemistry. The interest in patterning layers of biomolecules on surfaces stems from the need for sensor arrays for multisample assays or for analysis of samples containing multiple analytes. Inorganic structured surfaces are often fabricated via photolithographic methods. Biomolecular layers, however, require structuring techniques that preserve the activity of the biomolecules. Therefore, many established processes involving either the exposure of the specimen to vacuum, elevated temperature, or the use of etchants are not applicable. STM and AFM have been considered for the formation of submicrometer scale features. The SECM has also been employed both as a fabrication tool and as a means to image or address the localized biochemical reactivity of the surface. Although scanning probe techniques are generally not competitive with photolithography or micro-contact printing in terms of speed, the ability of the SECM to accomplish both fabrication and chemically selective imaging seems attractive for prototyping of new devices containing microstructured biochemically active layers. In this section we discuss only such examples, a general discussion of microfabrication by the SECM is given in Chapter 13 of this volume.

1. Localized Deposition of Polymers as Matrix for Enzymes

Polypyrrole and similar conducting polymers belong to the most intensively studied matrices for the immobilization of biorecognition elements in biosensors (see Ref. 86 for a review). Electropolymerization of *N*-substituted or 3-substituted pyrrole monomers leads to a polypyrrole backbone to which biomolecules can be attached depending on the nature of the substituent. In a series of papers Kranz et al. (43,60,61) developed a technique to electropolymerize pyrrole in two-dimensional and three-dimensional structures by using the SECM tip as the auxiliary electrode and performing a galvanostatic pulse program (87) at the macroscopic specimen. Through using an *N*-(ω-aminoalkyl) pyrrole as monomer, a derivatized polypyrrole was deposited that contained amino groups attached to the polymer backbone via a flexible spacer (13). Glucose oxidase partially oxidized by periodate (to form aldehyde groups in the carbohydrate part of the enzyme) was bound covalently via Schiff base chemistry to the amino functions at the outer polymer surface. This approach is currently being further explored by the use of a variety of 3-substituted monomers (K. Habermüller and W. Schuhmann, personal communication, 1998). The resulting substituted polypyrroles possess a higher electrical conductivity and will allow wiring of enzymes enclosed or bound to three-dimensional polymer towers. This strategy aims to immobilize a larger quantity of biomolecules on a microscopic region in order to increase the currents that can be measured at amperometric enzyme microelectrodes.

2. *Patterning of SAMS*

Wittstock et al. (63) have demonstrated a route to the formation of patterns of enzymatic activity via the desorption of patches of alkanethiol self-assembled monolayers (SAMs) on gold surfaces. In this approach the tip is used as the auxiliary electrode in the potentiostatic circuit to localize the current flow. On application of a negative potential to the gold, the reductive desorption of alkanethiolate, known from conventional electrodes (88,89), is restricted to the region located directly below the tip. Well-defined areas of bare gold resulted and could be imaged by positive feedback experiments if the desorption solution contained 0.2 M KOH. Following desorption, the bare gold spots could be used for self-assembly of an ω-functionalized thiol (cystaminium dihydrochloride) and subsequent attachment of osmium complexes whose spatial distribution was mapped by x-ray photoelectron imaging. The pattern of osmium centers coincided with that observed for the desorbed patches of alkanethiolate imaged by the SECM. The formation of patterns of glucose oxidase was achieved by attachment of periodate-oxidized GOx to the ω-amino function of cystamine (Fig. 26A). The SECM was also used to confirm the pattern of enzyme activity by GC mode imaging of the distribution of H_2O_2 above the surface in a glucose-containing medium with an amperometric Pt tip (Fig. 14; Sec. II.B for a detailed discussion). A potential advantage of this method is the possibility to pattern the surface with many different enzymes by sequential application of desorption and enzyme attachment steps (21). Pattern size can be reduced and edge definition is enhanced significantly if an alternating current with a frequency in the kHz range is passed through the two-electrode cell formed by the microelectrode and the thiolate-covered gold electrode (90).

A related technique was demonstrated by Shiku et al. (91), who formed alkylsilane monolayers at glass substrates and used electrochemical generation of $OH^{\cdot}$ radicals at the tip via the Fenton reaction to locally destroy the SAM. Diaphorase could then be patterned on the surface by physical adsorption to the undamaged hydrophobic areas or via covalent linkages to the radical-attacked areas. The former process was imaged in feedback mode with a ferrocenyl mediator and showed a decreased current over the disk of destroyed SAM and a constant background of enzyme activity over the rest

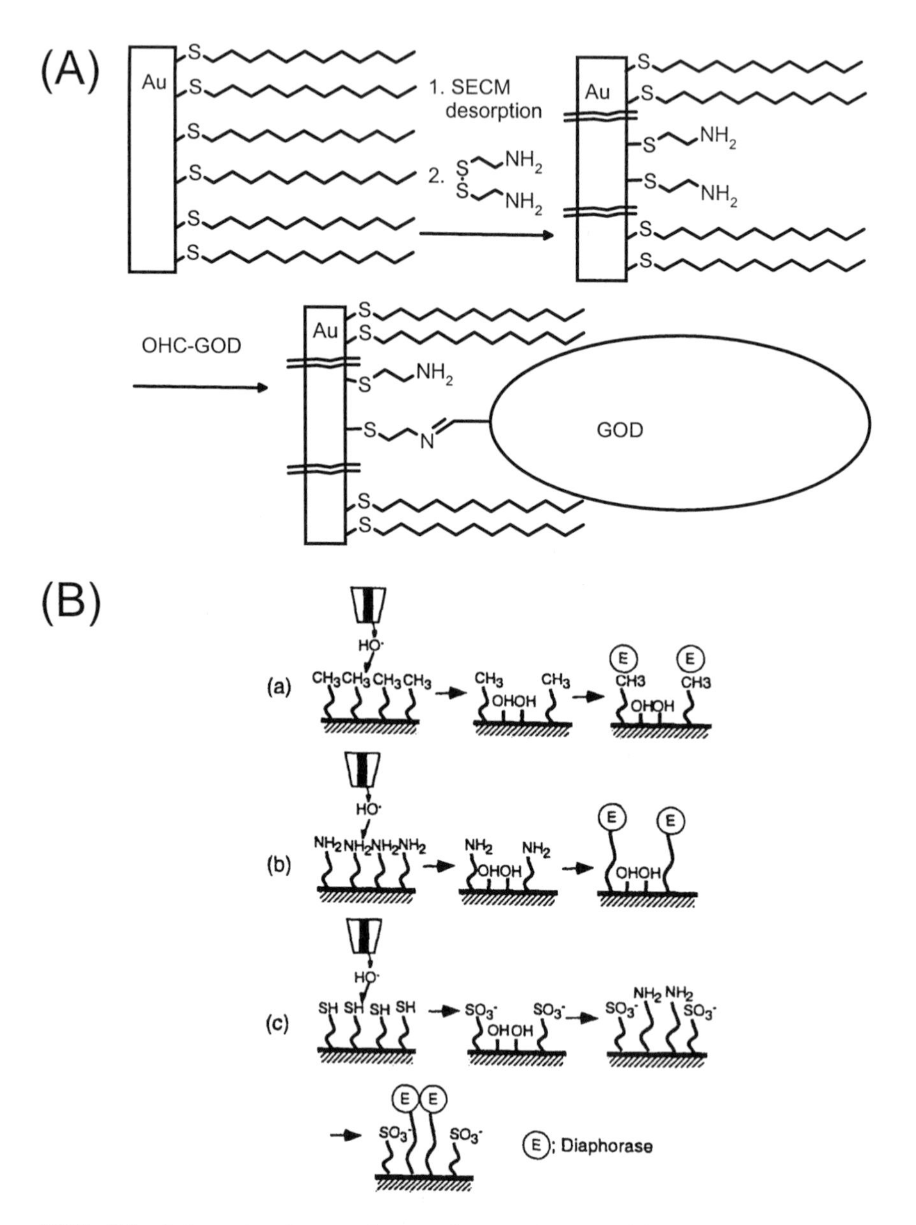

FIG. 26 Schematics for creating a microscopic spot of immobilized enzymatic activity by localized derivatization of self-assembled monolayers according to (A) GOx on an patterned gold alkanethiolate layer and (B) diaphorase on alkylsilanized glass. (Part A from Ref. 21. Copyright 1997 American Chemical Society. Part B from Ref. 91. Copyright 1997 American Chemical Society.)

of the specimen surface. Treatment of the damaged SAM with aminopropyltriethoxysilane produced disk-shaped patches of amino-terminated monolayer at these areas and subsequent attachment of diaphorase produced a surface with the opposite contrast in a feedback mode image (Fig. 26B). Wilbur et al. (92) used local reagent generation by the SECM tip in conjunction with biotin/avidin chemistry for microderivatization of carbon surfaces. Local derivatization was achieved by activating biotin hydrazine by an unknown electrochemical reaction. Activated biotin bound to the carbon surface and formed a microspot. Alternatively, bound biotin could be removed locally by generation of hydroxide ions at the tip to form a microscopic underivatized region surrounded by the biotin-derivatized surface.

3. Agglomerates of Functionalized Microbeads

Wijayawardhana et al. (14) used commercial paramagnetic surface-modified beads (2.8 μm and 1 μm mean diameter) to assemble biochemically active microspots. The use of the magnetic beads unites the advantages of homogeneous and heterogeneous assays. The bead suspension can be mixed with reagents and dosed as simply as a liquid. Separation of the sensing surface, i.e., the bead surface, from excess reagent, rinsing solution, and sample matrix can be made using a magnet to retain the beads against the test tube wall when pouring off the supernatant liquid. The microspots were created from a 1 μL drop extruded from a micropipette mounted on the SECM positioning system. A magnet was placed beneath the receiving, hydrophobic surface. As the pipette is brought slowly towards the support surface, the beads collect at the bottom of the drop under the influence of the magnet. After contact of the drop with the surface, the micropipette, with the drop still attached, was retracted, leaving the beads behind. The beads assemble in mounds whose size is determined by the number of beads but not by the area wetted by the drop (Fig. 27). If antibodies or other functionalities (e.g., streptavidin) have been immobilized on the beads, the microspots represent a platform on which assays can be performed. Biochemical activity of the assembled bead structures was imaged in GC mode using alkaline phosphatase as the label and in the feedback mode using glucose oxidase as the label.

4. Local Modulation of Enzyme Activity

An alternative approach to the patterning of enzymatic activity is to start with a substrate uniformly coated with enzyme and to locally deactivate the enzyme in order to generate the pattern. This can be achieved by the generation of denaturing agents at the tip, e.g., Br_2/HOBr (15). Although the mechanism of denaturation is not well understood, geometrically well-defined disks and lines of deactivated diaphorase could be produced and im-

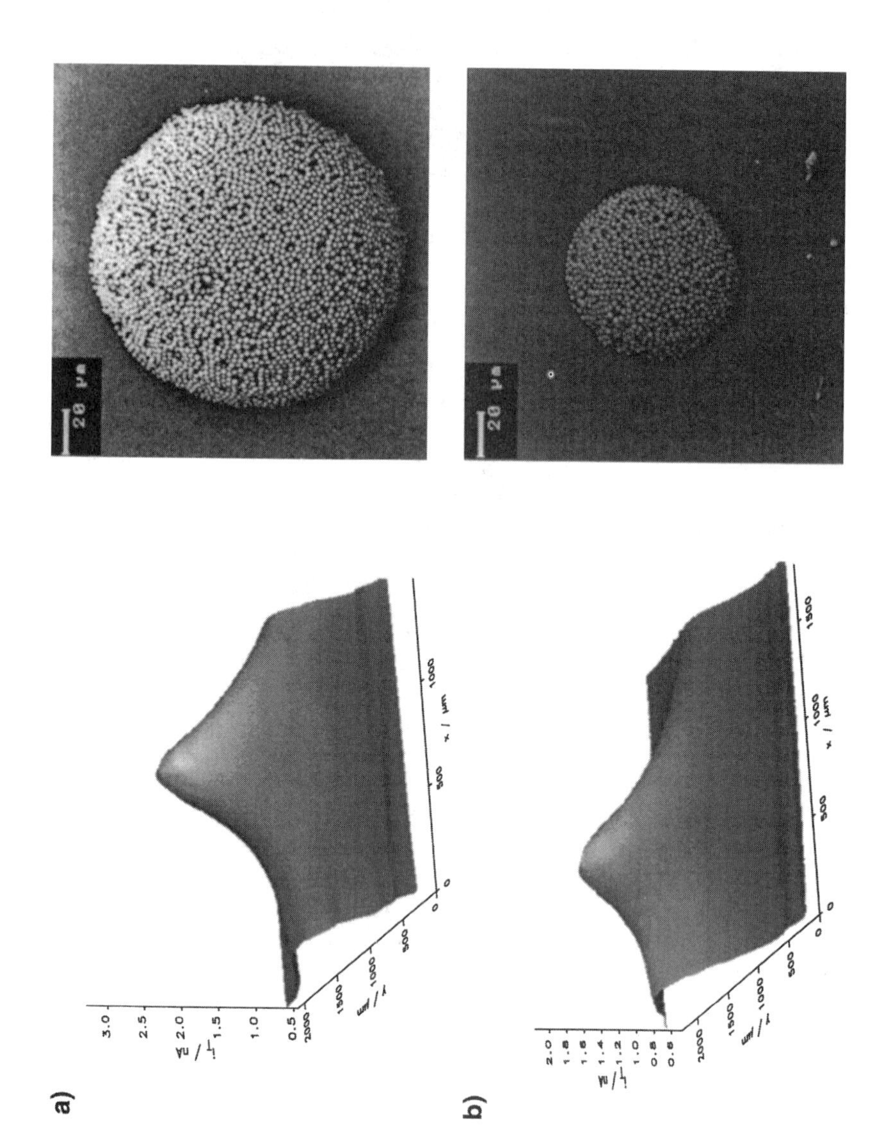
a)
b)
20 µm
20 µm
i_T / nA
x / µm
y / µm
3.0
2.5
2.0
1.5
1.0
0.5
2000
1500
1000
500
0
0
500
1000
2.0
1.8
1.6
1.4
1.2
1.0
0.8
0.6
1500

c)

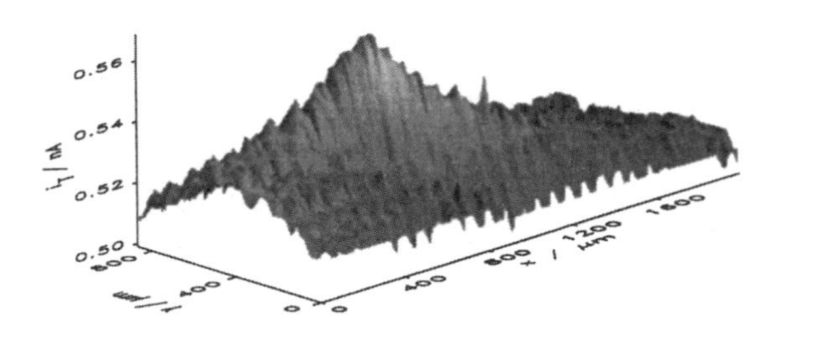

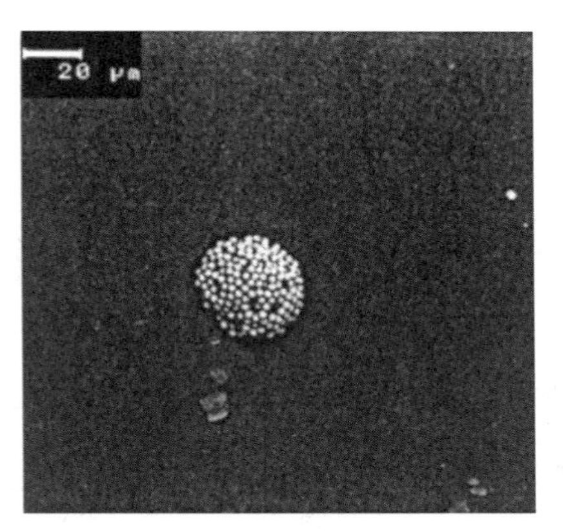

FIG. 27 SECM (left) and SEM (right) images of microbead agglomerates. The beads were coated with anti-mouse Ab saturated with alkaline phosphatase. SECM imaging in GC mode is based on oxidation of enzyme-generated 4-aminophenol. SEM images of representative bead agglomerates obtained after sputter-coating with gold illustrate the size of the agglomerate. Agglomerates were formed from 1 μL of bead suspension having the following particle concentrations per milliliter: (a) 4.1×10^7, (b) 4.1×10^6, (c) 4.1×10^5. (From Ref. 14. Copyright 1995 American Chemical Society.)

aged in feedback mode. Recently the opposite effect, local activation, has been demonstrated in which the bulk solution was maintained at pH 6 below the optimum (pH 9) for alcohol dehydrogenase immobilized on agarose beads. The tip was used to locally raise the pH by reduction of oxygen and water and the enzymatic activity observed by monitoring the reduced cofactor, NADH, via fluorescence microscopy (67).

III. CONCLUSION AND OUTLOOK

A. Information Content of SECM Measurements and Spatial Resolution

SECM images contain information about both local reactivity and the topography of the sample surface. However, it is our opinion that the greatest utility of SECM in biological and biochemical investigations is its ability to map chemical species and reactivity (i.e., processes) rather than topography. Since the theoretical models of SECM are based on established electrochemical principles, the data can often be interpreted in a quantitative manner, and kinetic investigations at fixed lateral tip position are possible with a time resolution of ms or better.

Typical SECM tips have radii in the μm range, hence the lateral resolution of the feedback images is in the μm scale on a routine basis and the vertical resolution (sample height variations) is better than 1 μm. There may be some advantages to the use of SECM feedback imaging for recording the topography of biological samples because there is no physical contact between probe and sample. Molecular scale images of sample topography by SECM have been demonstrated in a few cases where a nonconducting sample like DNA can be fixed on a flat insulator such as mica and imaged in humid air with an STM-type of tip (3). However, the majority of investigations have been devoted to systems with spatial heterogeneity on the μm scale. Whole cells, immobilized enzyme layers, and immunoassay systems have all been studied. At this level of spatial resolution, the flux of chemical species at individual cells is easily resolved, but mapping of chemical or enzyme activity at the subcellular level is limited to the largest cells. It can be expected that current methods of tip fabrication will be improved and that higher resolution experiments will be feasible. It should be noted that the spatial resolution of the positioning systems is typically much greater than that of the tips, which are therefore the limiting factor at present. Strategies for the formation of very small tips are discussed in Chapter 3.

Several proof-of-concept studies have demonstrated how various metabolic processes can be utilized for contrast formation in SECM imaging. From a methodological point of view, two possibilities exist. The scanning microelectrode may be used to convert a mediator to a redox active cofactor of an enzyme at the specimen surface and to detect the regeneration of the mediator in the course of the enzymatic reaction (feedback mode). Alternatively, the tip can monitor amperometrically or potentiometrically the diffusional flux caused by a metabolic reaction at the specimen in a similar manner to the investigation of localized mass transport phenomena (Chapter 9). If one is interested in studying selective binding events at biorecognition layers, enzyme-labeling techniques can be used to enable detection by the SECM.

Current efforts are devoted to expand the range of accessible enzymatic reactions and to apply the technique as a day-to-day research tool in biochemical and physiological problem solving. This tendency will certainly become more evident as commercial instrumentation enters the market and will be available to those interested in biological applications, but reluctant to devote their efforts to instrument design, programming, and testing. The combination of SECM operation principles with reactions catalyzed by dissolved enzymes may be used to broaden the range of SECM applications as redox inactive metabolites may be converted in a rapid homogeneous enzyme reaction to a redox active product detectable by SECM. Homogeneous enzymatic reaction may be used for selective signal quenching, thus providing a tool for image verification similar to that discussed in Sec. II.B. Strategies for image verification will attain more importance in the future as more complex specimens are investigated.

B. Applications in Biotechnology

Some of the most promising applications of SECM were reported in the area of biotechnology, i.e., optimization of immobilized enzyme systems for use in synthesis and analysis and novel immunoassay methodologies. The two strategies for the imaging of immobilized enzyme activity, feedback, and generation collection provide complementary methods of investigating the spatial distribution of enzymes. The well-established theory enables extraction of quantitative information and the range of electrochemical probes that are available gives SECM wide application and great flexibility. A particular problem that does remain is the necessity to trade off image resolution and spatial control of the diffusion fields in the feedback mode against the greater sensitivity of the generation-collection mode. In some cases spatial confinement of diffusion fields has been achieved by means of microstructured specimens (29,31,39) or by a second solution-phase reaction that limits the lifetime of the species detected at the tip (66,93). Feedback imaging with tip-position modulation (TPM) and phase-selective detection (94) may also enhance the sensitivity to the feedback flux from the immobilized enzyme compared to the contribution from hindered diffusion from the bulk solution.

One application is the use of SECM to make sequential readings from a sensing array. This might be done as demonstrated by Shiku et al. (31) for a dual electrochemical enzyme immunoassay. Such methods are rapid compared to conventional assays because there is no need to incubate the enzyme-labeled antigens to produce enough product for the detection step. The assay speed might be further enhanced if the microstructured samples are combined with likewise precise sample-holding facilities, such that the time-consuming scanning of the tip may be replaced by a direct movement to the sensing region to record a single value. A problem of sensitivity remains, although electrochemical transduction can achieve low limits of detection (LOD), the very small currents (<1 pA) involved in direct amperometric measurement with UMEs means that the full potential has not yet been realized. The use of stripping methods or enzyme amplification to decrease the LOD may be valuable.

Characterization of enzyme electrodes for synthesis, which have recently been considered as a clean manufacturing technology (95), is another area where SECM may play a role in investigating the factors limiting electrode performance. For such work, determination of species (pH, enzymatic reaction products such as H_2O_2) in the diffusion layer of these technological enzyme electrodes and monitoring the local enzyme activity is likely to be vital. Serious issues in the optimization of biosensor manufacturing processes can be addressed by SECM. An incomplete list of examples includes entrapment of microbubbles above recessed microelectrodes during immersion (96) and the question whether enzyme immobilization extends beyond the active electrode surface and covers also the support and encapsulation of the sensors (97). Restricting the enzyme modification to the electrode surface will be one of the crucial prerequisites in miniaturized multianalyte sensors to prevent cross-talk between neighboring sensing regions (53).

C. Application in Cell Biology and Physiology

The application of SECM to the study of living cells offers several unique opportunities owing to the wide range of analytes that may be detected by suitable choices of tip electrodes. The technique is noninvasive, sensitive, quantitative, and capable of good time and spatial resolution. To date comparatively little work has been carried out. The recent literature indicates that strong efforts are being made to perform electrochemical measurements at single cells. This goal has been approached by fabricating submicroliter electrochemical cells in which a single living cell may be contained (98,99). Bratten et al. (98) introduced an array of microvials with a set of individual working electrodes, reference electrodes, and counterelectrodes at the bottom of each cell. As demonstrated previously in the context of enzyme immunoassay (100), electrochemical detection of substances originating from a limited number of active centers leads to very low detection limits if carried out in a restricted volume. This restricted solution volume prevents the dilution of the analytes due to diffusion into the bulk solution found in conventional SECM with cells of some milliliter volume. Furthermore, Bratten et al. (98) demonstrated an electrochemical feedback between the working electrode and the counterelectrode and discussed possibilities to use this phenomenon for signal amplification. However, since the relative position of the single cell versus the working and counterelectrode is not exactly controlled and cannot be varied once the cell grows in the microvials, these devices will probably not allow the experimenter to couple the heterogeneous electrochemical reactions and the biochemical reactions with the same degree of precision and quantitative theoretical understanding found in SECM. The concept of Clark et al. (99) also aims at the investigation of single cells contained in microvials that are micromachined from polystyrene. An assembly consisting of a 5 μm carbon fiber and a 1 μm Ag/AgCl (1 M KCl) reference electrode together with a capillary for dispensing the solution was used to fill and carry out voltammetric measurements in each microvial consecutively. The movement of the electrodes from vial to vial is a concept bearing some similarity to SECM. Instead of reading a current value at each point of an orthogonal grid in SECM, a cyclic voltammogram is recorded in each microvial. The approaches of Bratten et al. (98) and Clark et al. (99) aim to analyze the amount of a metabolite after it has been released by the cells, while SECM would try to couple the process of substance production and SECM detection, i.e., measuring the release in real time.

An entirely different approach to the study of single living cells by means of SECM has been reported by Schuhmann and coworkers employing the damping of the mechanical vibration of a microelectrode. The tip is positioned close to the sample and vibrated in the plane parallel to the specimen surface with an amplitude of less than 50 nm and the damping of this vibration as the tip approaches the surface is detected optically. Using the amplitude of the vibration as the signal in the feedback loop of the positioning system, a topographic image of an immobilized nerve cell can be obtained independent of the electrochemical process occurring at the tip (28). Using this topographic image, the tip electrode can be moved to features indicative of an upcoming exocystosis event. With this technology it is conceivable to investigate the possibility of triggering the exocystosis by chemical species generated at the tip or by an electrical stimulus provided by the metal surface onto which the cell is immobilized. Since the microelectrode is placed close to the location of an expected exocystosis event, the dilution of analyte by diffusion to the bulk is less important, and instead it may be possible to monitor the kinetics of such an event. Yasukawa et al. (27) used a dual amperometric microelectrode to detect simultaneously O_2 (indicative of cellular activity) and $[Fe(CN)_6]^{4-}$ (reflecting the topography) around single cultivated protoplasts.

As the size of readily available tips is decreased, current-independent positioning schemes may become very important because the working distance has to be decreased in proportion to the tip radius. As the tip radius approaches the height variations of a living cell (μm scale), a mechanical contact between the insulating shielding and the living cell becomes very likely. Once a current-independent positioning scheme is routinely available, new experiments will become possible at the subcellular level such as the mapping of membrane-bound enzyme activity across a cell surface and intracellular measurements with precise localization of the tip microelectrode. It would also open the door for the use of probes that are not electrodes at all. This concept has been demonstrated with a capillary filled with glucose oxidase–loaded and glucose dehydrogenase–loaded gels (101). The enzymatic production of H_2O_2 or NADH at the scanning capillary was detected at a Pt microelectrode or poly(methylene blue)–modified Pt microelectrode used as the sample in the SECM cell (tip generation–substrate collection mode). Such arrangements might be useful if living cells are grown onto microstructured electrodes, or in microvials, and their reaction towards an external excitation introduced by the enzyme-loaded capillary can be tested under the condition of controlled distance between specimen and probe capillary.

The possibility to excite cells and tissues locally will certainly be a very interesting future application of hyphenated techniques such as the combinations SECM–near field scanning optical microscopy (NSOM) and SECM–scanning force microscopy (SFM). A very recent application uses the TG/SC mode to generate H_2O_2 at the tip. In a solution of sodium luminol immobilized horseradish peroxidase caused chemiluminescence. The light intensity plotted as a function of the microelectrode position gave an image of HRP activity (102). The reader will find an assessment of the current stage of development of these combinations in Chapter 14 of this volume.

Mass transport phenomena in biological systems can be investigated with SECM if the species of interest can be detected either by an potentiometric or amperometric microelectrode. Theory and selected studies of localized mass transport are covered in Chapter 9 of this volume. SECM is particularly appropriate in these studies in view of the intimate connection of the imaging mechanism to mass transport effects. The investigation of oxygen and ion transport in various tissues and under a variety of driving forces (concentration gradient, electric field, convection) has been demonstrated with quantification of the transport process on a μm length scale. It is likely that the number of applications of the SECM in this area will grow.

D. Recording Speed

When evaluating ranges of application, the time scale of SECM experiments should be considered. Although SECM allows the investigation of very rapid kinetics under steady-state conditions (μs–ms time resolution) at laterally fixed microelectrode position, images can be obtained only with a rather limited frequency. It would take about 34 minutes to record a 100 $\times$ 100 μm^2 image composed of 100 line scans, each 100 μm long with a typical lateral scan rate of 10 $\mu m\ s^{-1}$. If the experimenter wishes to follow the activity of biological specimen as a function of time, such frame acquisition rates will turn out to be unacceptably long in many cases. In fact, limited recording speed represents a fundamental problem for all scanning probe techniques, but it is particularly evident for SECM. The upper limit of the lateral scan rate is imposed by the time to establish a steady-state response at each grid point for amperometric tips in the feedback mode, or to attain the equilibrium potential for potentiometric tips. This characteristic time becomes smaller as the active tip electrode sizes and tip-sample distances shrink (see Chapter 5). Undesired stirring also sets an upper limit of the lateral scan rate. The problem of stirring will be less severe as the total tip diameter is reduced. Smaller active tip diameters also require a reduced tip-sample distance. Since biological specimens are rarely flat, a distance-control mechanism will be required if microelectrodes with sizes of 1 μm and below are used. These distance-control mechanisms tend to decrease the recording speed further.

To overcome the limited image acquisition rate, the tip electrode may be positioned above a region of particular interest once a complete image has been recorded. At the fixed tip position the signal is followed as a function of time and external stimuli as has been shown by Tsionsky et al. (23). Another concept has been demonstrated by Meyer et al. (103). On one chip of 1 $\times$ 1 cm^2 an array of 20 $\times$ 20 individually addressable electrochemical cells were integrated each measuring 500 $\times$ 500 μm^2 in size. Each cell contained a platinum working electrode of 36 $\times$ 36 μm^2, a readout amplifier, and a sensor control unit. By sequentially reading the current from the working electrodes, images of O_2 and H_2O_2 distributions were obtained within 2 minutes. After covering the array with a glucose containing polypyrrole membrane, glucose distributions could also be measured. Since there is no need for mechanical displacement of the probe array, the restriction imposed by stirring effects or the time to establish a steady-state current at each grid point is circumvented. The necessity to design and manufacture new arrays for each spacing of the array elements (corresponding to grid points of the image) is likely to be one of the main reasons why this scheme has not been used for other studies. Furthermore, as electrode size and in particular interelectrode distances are reduced to achieve the spatial resolution routinely obtained with SECM, crosstalk, and shielding between the different microelectrodes of the probe array will become a significant problem.

An interesting compromise between time-dependent studies at fixed-probe position (no spatial resolution) and complete two-dimensional imaging (low time resolution) might be "smart" imaging. The image information is mainly contained in the regions where signal transitions between low and high levels occur. Quite often systems are investigated that contain one or a few interesting regions embedded in an indifferent matrix covering the largest area of the image frame. Recording images with high grid point density at the borders between the different regions and with a low point density above indifferent areas would decrease the information content only marginally but would enhance the recording speed significantly. Hyphenated techniques may greatly assist in the preselection of interesting sample regions.

IV. ABBREVIATIONS, ACRONYMS, AND SYMBOLS

a	tip radius (active area)
Ab	antibody
Ab:Ag	antibody-antigen complex
AFM	atomic force microscopy/microscope
AC	alternating current
Ag	antigen
Ag-E	enzyme-labeled antigen
ATR-IR	attenuated total reflectance-infrared (spectroscopy)
c	concentration; subscript = species
c'	detection limit of the tip
cAMP	cyclic adenosine monophosphate
CEA	carcinoembryonic antigen
d	tip-sample distance in micrometers

D	diffusion coefficient
ELISA	enzyme-labeled immunosorbent assay
E_S, E_T	specimen, tip potential
f	flux
FMA/FMA^+	hydroxymethyl ferrocene/hydroxymethyl ferrocinium
FWHM	full width at half maximum
G	conductance between tip and external electrode
G_∞	conductance between tip and external electrode at quasi-infinite distance
Γ_{enz}	surface concentration of enzyme
GC	generation collection
GOx	glucose oxidase
H_2Q	hydroquinone
HCG	human choronic gonadotropin
HPL	human placental lactogen
HRP	horseradish peroxidase
ISE	ion-selective electrode
i_T	tip current
$i_{T\infty}$	tip current at quasi-infinite distance
k_{cat}	reaction rate constant between SECM mediator and enzyme
K_M	Michaelis-Menten constant
L	normalized tip-sample distance = distance/sample radius
LOD	limit of detection
MV^{2+}	methyl viologen
MW	molecular weight
NAD^+/NADH	oxidized/reduced forms of nicotinamide adenine dinucleotide

$NADP^+$/NADPH	oxidized/reduced forms of nicotinamide adenine dinucleotide phosphate
PAP	4-aminophenolate
PAPP	4-aminophenyl phosphate
PPy	polypyrrole
PTH	parathyroid hormone
QRE	"quasi"-reference electrode (metal wire inserted in solution)
r	radial coordinate in polar coordinate system
r_S	sample radius
RG	ratio between the total tip diameter (active area plus insulator) and the active tip diameter
SAM	self-assembled monolayer
SECM	scanning electrochemical microscopy/microscope
STM	scanning tunneling microscopy/microscope
TMPD	tetramethyl-*p*-phenylene diamine
u	normalized radial distance in polar coordinates = r/r_S
v_t	translation speed of tip vs. sample

ACKNOWLEDGMENTS

B.R.H. would like to thank Action Research, NATO, The Royal Society, and the University of Newcastle upon Tyne for supporting research in the area of SECM. G.W. acknowledges financial support for ongoing research on SECM by Deutsche Forschungsgemeinschaft (Wi 1617/1-1 . . . 2 and Wi 1617/2-1 . . . 4). The authors are grateful to A. J. Bard (Figs. 4, 5, 21, 22) and C. Kranz (Fig. 11) for providing original printouts. G.W. thanks Gerhild Wiedeman for technical support in preparing the manuscript.

REFERENCES

1. J Wang, L-H Wu, R Li. Scanning electrochemical microscopic monitoring of biological processes. J Electroanal Chem 272:285–292, 1989.

2. AJ Bard, G Denuault, C Lee, D Mandler, DO Wipf. Scanning electrochemical microscopy: a new technique for characterization and modification of surfaces. Acc Chem Res 23:357–368, 1990.
3. R Guckenberger, M Heim G Cevc, HF Knapp, W Wiegräbe, A Hillebrand. Scanning tunneling microscopy of insulators and biological specimens based on lateral conductivity of ultrathin water films. Science 266:1538–1540, 1994.
4. R Guckenberger, T Hartmann, W. Wiegräbe, W. Baumeister. Imaging and conduction mechanisms. In: R Wiesendanger, H-J Güntherodt, eds. Scanning Tunneling Microscopy II. 2nd ed. Berlin: Springer, 1995, pp 84–92.
5. F-RF Fan, AJ Bard. STM on wet insulators: electrochemistry or tunneling? Science 270:1849–1851, 1995.
6. R Guckenberger, M Heim. Response to [5]. Science 270:1851–1852, 1995.
7. M Patel, MC Davies, M Lomas, CJ Roberts, SJB Tendler, PM Williams. STM of Insulators with the probe in contact with an aqueous layer. J Phys Chem B 101:5138–5142, 1997.
8. F Forouzan, AJ Bard. Evidence for faradaic processes in scanning probe microscopy on mica in humid air. J Phys Chem B 101:10876–10879, 1997.
9. J Lipkowski. Ion and electron transfer across monolayers of organic surfactants. In: BE Conway, JOM Bockris, RE White, eds. Modern Aspects of Electrochemistry, Vol. 23. New York: Plenum Press, 1992, pp 1–99.
10. DT Pierce, PR Unwin, AJ Bard. Scanning electrochemical microscopy. 17. Studies of enzyme-mediator kinetics for membrane- and surface-immobilized glucose oxidase. Anal Chem 64:1795–1804, 1992.
11. AEG Cass, G Davis, GD Francis, HAO Hill, WJ Aston, IJ HIggins, EV Plotkin, LDL Scott, APF Turner. Ferrocene-mediated enzyme electrode for amperometric determination of glucose. Anal Chem 56:667–671, 1984.
12. DT Pierce, AJ Bard. Scanning electrochemical microscopy. 23. Reaction localization of artificially patterned and tissue-bound enzymes. Anal Chem 65: 3598–3604, 1993.
13. C Kranz, G Wittstock H Wohlschläger, W Schuhmann. Imaging of microstructured biochemically active surfaces by means of scanning electrochemical microscopy. Electrochim Acta 42:3105–3111, 1997.
14. CA Wijayawardhana, G Wittstock, HB Halsall, WR Heineman. Spatially addressed deposition and imaging of biochemically active bead microstructures by scanning electrochemical microscopy. Anal Chem 72:333–338, 2000.
15. H Shiku, T Takeda H Yamada, T Matsue, I Uchida. Microfabrication and characterization of diaphorase-patterned surfaces by scanning electrochemical microscopy. Anal Chem 67:312–317, 1995.
16. H Shiku, T Matsue, I Uchida. Detection of microspotted carcinoembryonic antigen on a glass substrate by scanning electrochemical microscopy. Anal Chem 68:1276–1278, 1996.
17. J Zaumseil, G Wittstock, S Bahrs, P Steinrücke. Imaging the activity of nitrate reductase by means of a scanning electrochemical microscope. Fresenius J Anal Chem 367:352–355, 2000.
18. I Katakis, E Domínguez. Catalytic electrooxidation of NADH for dehydrogenase amperometric biosensors. Mikrochim Acta 126:11–32, 1997.

19. CEM Berger, BR Horrocks, HK Datta. cAMP-dependent inhibition is dominant in regulating superoxide production in the bone-resorbing osteoclasts. J Endocrin 158:311–318, 1998.
20. K Borgwarth, C Ricken, DG Ebling, J Heinze. Surface-analysis by scanning electrochemical microscopy—resolution studies and applications to polymer samples. Fresenius J Anal Chem 356:288–294, 1996.
21. G Wittstock, W Schuhmann. Formation and imaging of microscopic enzymatically active spots on an alkanethiolate-covered gold electrode by scanning electrochemical microscopy. Anal Chem 69:5059–5066, 1997.
22. BR Horrocks, D Schmidtke, A Heller, AJ Bard. Scanning electrochemical microscopy. 24. Enzyme ultramicroelectrodes for the measurement of hydrogen peroxide at surfaces. Anal Chem 65:3605–3614, 1993.
23. M Tsionsky, ZG Cardon, AJ Bard, RB Jackson. Photosynthetic electron transport in single guard cells as measured by scanning electrochemical microscopy. Plant Physiol 113:895–901, 1997.
24. CM Lee, JY Kwak, AJ Bard. Application of scanning electrochemical microscopy to biological samples. Proc Natl Acad Sci USA 87:1740–1743, 1990.
25. T Yasukawa, Y Kondo, I Uchida, T Matsue. Imaging of cellular activity of single cultured cells by scanning electrochemical microscopy. Chem Lett 767–768, 1998.
26. T Yasukawa, T Kaya, T Matsue. Imaging of photosynthetic and respiratory activity of a single algal protoplast by scanning electrochemical microscopy. Chem Lett 975–976, 1999.
27. T Yasukawa, T Kaya, T Matsue. Dual imaging of topography and photosynthetic activity of a single protoplast by scanning electrochemical microscopy. Anal Chem 71:4637–4641, 1999.
28. A Hengstenberg W Schuhmann, I Dietzel, A Blöchl. Zell-Zell-Kommunikationsprozesse mittels elektrochemischer Rastermikroskopie. BIOforum 22: 595–599, 1999.
29. BR Horrocks, MV Mirkin, DT Pierce, AJ Bard, G Nagy, K Toth. Scanning electrochemical microscopy. 19. Ion-selective potentiometric microscopy. Anal Chem 65:1213–1224, 1993.
30. BR Horrocks, MV Mirkin. Evidence for a potential dependent reversible inactivation of urease adsorbed on a gold electrode. J Chem Soc Faraday Trans 94:1115–1118, 1998.
31. H Shiku, Y Hara, T Matsue, I Uchida, T Yamauchi. Dual immunoassay of human chorionic gonadotropin and human placental lactogen at a microfabricated substrate by scanning electrochemical microscopy. J Electroanal Chem 438:187–190, 1997.
32. G Wittstock, KJ Yu, HB Halsall, TH Ridgway, WR Heineman. Imaging of immobilized antibody layers with scanning electrochemical microscopy. Anal Chem 67:3578–3582, 1995.
33. YN Antonenko, P Pohl, E Rosenfeld. Visualization of the reaction layer in the immediate membrane vicinity. Arch Biochem Biophys 333:225–232, 1996.

34. S Nugues, G Denuault. Scanning electrochemical microscopy—amperometric probing of diffusional ion fluxes through porous membranes and human dentine. J Electroanal Chem 408:125–140, 1996.
35. JL Gilbert, SM Smith, EP Lautenschlager. Scanning electrochemical microscopy of metallic biomaterials—reaction-rate and ion release imaging modes. J Biomed Mater Res 27:1357–1366, 1993.
36. PS Cahill, QD Walker, JM Finnegan, GE Mickelson, ER Travis, RM Wightman. Microelectrodes for the measurement of catecholamines in biological systems. Anal Chem 68:3180–3186, 1996.
37. D Ammann, P Caroni. Preparation and use of microelectrodes and macroelectrodes for measurement of transmembrane potentials and ion activities. Methods Enzymol 172:136–155, 1989.
38. CJ McNeil, KA Smith, P Bellavite, JV Bannister. Application of the electrochemistry of cytochrome-c to the measurement of superoxide radical production. Free Radical Res Commun 7:89–96, 1989.
39. C Wei, AJ Bard, G Nagy, K Toth. Scanning electrochemical microscopy. 28. Ion-selective neutral carrier-based microelectrode potentiometry. Anal Chem 67:1346–1356, 1995.
40. E Klusmann, JW Schultze. pH-Microscopy—theoretical and experimental investigations. Electrochim Acta 42:3123–3134, 1997.
41. C Wei, AJ Bard, I Kapui, G Nagy, K Toth. Scanning electrochemical microscopy. 32. Gallium ultramicroelectrodes and their application in ion-selective probes. Anal Chem 68:2651–2655, 1996.
42. M Ludwig, C Kranz, W Schuhmann, HE Gaub. Topography feedback mechanism for the scanning electrochemical microscope based on hydrodynamic-forces between tip and sample. Rev Sci Instrum 66:2857–2860, 1995.
43. C Kranz, HE Gaub, W Schuhmann. Polypyrrole towers grown with the scanning electrochemical microscope. Adv Mater 8:634–637, 1996.
44. K Karrai, RD Grober. Piezoelectric tip-sample distance control for near-field optical microscopes. Appl Phys Lett 66:1842–1844, 1995.
45. R Wiesendanger. Scanning Probe Microscopy and Spectroscopy—Methods and Applications. Cambridge: Cambridge University Press, 1994, pp 525–534.
46. H Yamada, H Shiku, T Matsue, I Uchida. Microvoltammetric characterization of diaphorase monolayer at glass surfaces. Bioelectrochem Bioenerg 33:91–93, 1994.
47. T Sawaguchi, T Matsue, I Uchida. Catalytic capability of diaphorase bound to a self-assembled thiol monolayer at a gold electrode. Bioelectrochem Bioenerg 29:127–133, 1992.
48. C Wei, AJ Bard, MV Mirkin. Scanning electrochemical microscopy. 31. Application of SECM to the study of charge-transfer processes at the liquid-liquid interface. J Phys Chem 99:16033–16042, 1995.
49. MJ Waner, M Gilchrist, M Schindler, M Dantos. Imaging the molecular dimensions and oligomerization of proteins at liquid/solid interfaces. J Phys Chem 102:1649–1657, 1998.

50. A Ball, RAL Jones. Conformational changes in adsorbed proteins. Langmuir 11:3542–3548, 1995.
51. FW Scheller, F Schubert, J Fedrowitz, eds. Frontiers in Biosensorics. Vols. I and II. Basel: Birkhäuser Verlag, 1997.
52. LC Clark Jr, C Lyons. Electrode system for continuous monitoring in cardiovascular surgery. Ann NY Acad Sci 102:29–45, 1962.
53. DJ Strike, A Hengstenberg, M Quinto, C Kurzawa, M Koudelka-Hep, W Schuhmann. Mikrochim Acta 131:47–55, 1999.
54. DL Allara. Critical issues in application of self-assembled monolayers. Biosens Bioelectron 10:771–783, 1995.
55. T Wink, SJ van Zwuilen, A Bult, WP van Bennekom. Self-assembled monolayers for biosensors. Analyst 122:43R–50R, 1997.
56. BB Ratcliff, JW Klancke, MD Koppang, RC Engstrom. Microderivatization of anodic glassy carbon. Anal Chem 68:2010–2014, 1996.
57. WB Nowall, N Dontha, WG Kuhr. Electron transfer kinetics at a biotin/avidin patterned glassy carbon electrode. Biosensors Bioelectronics 13:1237–1244, 1998.
58. G Wittstock, H Emons, M Kummer, JR Kirchhoff, WR Heineman. Application of scanning electrochemical microscopy and scanning electron microscopy for the characterisation of carbon-spray modified electrodes. Fresenius J Anal Chem 348:712–718, 1994.
59. G Wittstock. Elektrochemische Rastermikroskopie zur Grenzflächencharakterisierung. CLB Chem Labor Biotechnol 46:166–169, 1995.
60. C Kranz, M Ludwig, HE Gaub, W Schuhmann. Lateral deposition of polypyrrole lines by means of the scanning electrochemical microscope. Adv Mater 7:38–40, 1995.
61. C Kranz, M Ludwig, HE Gaub, W Schuhmann. Lateral deposition of polypyrrole lines on insulating gaps. Towards the development of polymer-based electronic devices. Adv Mater 7:568–571, 1995.
62. T Schalkhammer, E Mann-Buxbaum, F Pittner, G Urban. Electrochemical glucose sensors on permselevtive non-conducting substituted pyrrole polymers. Sensors Actuators B4:273–281, 1991.
63. G Wittstock, R Hesse, W Schuhmann. Patterned self-assembled alkanethiolate monolayers on gold. Patterning and imaging by means of scanning electrochemical microscopy. Electroanalysis 9:746–750, 1997.
64. H Emons, T Schmidt. Properties of protein layers at electrodes. In: F Scheller, RD Schmidt, eds. Biosensors: Fundamentals, Technologies and Application. Weinheim: VCH, 1992, pp 287–290.
65. ZW Tian, ZD Fen, ZQ Tian, XD Zhuo, JQ Mu, CZ Li, HS Lin, B Ren, ZX Xie, WL Hu. Confined etchant layer technique for 2-dimensional lithography at high resolution using electrochemical scanning-tunneling-microscopy. Faraday Discuss 94:37–44, 1992.
66. RC Engstrom, B Small, L Kattan. Observation of microscopically local electron-transfer kinetics with scanning electrochemical microscopy. Anal Chem 64:241–244, 1992.

67. JC O'Brien, J Shumakerparry, RC Engstrom. Microelectrode control of surface-bound enzymatic activity. Anal Chem 70:1307–1311, 1998.
68. I Shohat, D Mandler. Deposition of nickel hydroxide structures using scanning electrochemical microscopy. J Electrochem Soc 141:995–999, 1994.
69. K Borgwarth, C Ricken, DG Ebling, J Heinze. Surface characterization and modification by the scanning electrochemical microscope (SECM). Ber Bunsenges Phys Chem 99:1421–1426, 1995.
70. G Wittstock, H Emons, TH Ridgway, EA Blubaugh, WR Heineman. Development and experimental evaluation of a simple system for scanning electrochemical microscopy. Anal Chim Acta 298:285–302, 1994.
71. MD Hawley, SV Tatawawadi, S Piekarski, RN Adams. Electrochemical studies of the oxidation pathways of catecholamines. J Am Chem Soc 89:447–450, 1967.
72. RC Engstrom, T Meaney, R Tople, RM Wightman. Spatiotemporal description of the diffusion layer with a microelectrode probe. Anal Chem 59:2005–2010, 1987.
73. H Shiku, Y Hara, T Takeda, T Matsue, I Uchida. Microfabrication and characterization of solid surfaces patterned with enzymes or antigen-antibodies by scanning electrochemical microscopy. In: G Jerkiewiecz, MP Soriaga, K Uosaki, A Wieckowski, eds. Solid-Liquid Electrochemical Interfaces. Washington, DC: ACS, 1997, pp 202–209.
74. CA Wijayawardhana, G Wittstock, HB Halsall, WR Heineman. Electrochemical immunoassay with microscopic immunomagnetic bead domains and scanning electrochemical microscopy. Electroanalysis 12:640–644, 2000.
75. CEM Berger, BR Horrocks, HK Datta. Application of ion-selective microelectrodes to the detection of calcium release during bone resorption. Electrochim Acta 44:2677–2683, 1999.
76. SA Nesbitt, MA Horton. Trafficking of matrix collagens through bone-resorbing osteoclasts. Science 276:266–269, 1997.
77. HK Datta, P Manning, H Rathod, CJ McNeil. Effect of calcitonin, elevated calcium and extracellular matrices on superoxide anion production by rat osteoclasts. Exp Physiol 80:713–719, 1995.
78. HK Datta, H Rathod, P Manning, Y Turnbull, CJ McNeil. Parathyroid-hormone induces superoxide anion burst in the osteoclast—evidence for the direct instantaneous activation of the osteoclast by the hormone. J Endocrinol 149:269–275, 1996.
79. N Agarwala, CV Gay. Specific binding of parathyroid hormone to living osteoclasts. J Bone Min Res 7:531–539, 1992.
80. JV Macpherson, D O'Hare, PR Unwin, CP Winlove. Quantitative spatially resolved measurements of mass transfer through laryngeal cartilage. Biophys J 73:2771–2781, 1997.
81. JV Macpherson, MA Beeston, PR Unwin, NP Hughes, D Littlewood. Scanning electrochemical microscopy as a probe of local fluid-flow through porous solids—application to the measurement of convective rates through a single dentinal tubule. J Chem Soc Faraday Trans 91:1407–1410, 1995.

82. ER Scott, JP Phipps, HS White. Direct imaging of molecular-transport through skin. J Invest Dermatol 104:142–145, 1995.
83. SM Dzekunov, YN Antonenko. Dynamics of formation and dissipation of local pH gradients in the unstirred layers near bilayer lipid membranes. Bioelectrochem Bioenerg 41:187–190, 1996.
84. P Pohl, SM Saparov, YN Antonenko. The effect of a transmembrane osmotic flux on the ion concentration distribution in the immediate membrane vicinity measured by microelectrodes. Biophys J 72:1711–1718, 1997.
85. YN Antonenko, P Pohl, GA Denisov. Permeation of ammonia across bilayer lipid membranes studied by ammonium ion selective microelectrodes. Biophys J 72:2187–2195, 1997.
86. W Schuhmann. Conducting polymer based amperometric enzyme electrodes. Mikrochim Acta 121:1–29, 1995.
87. W Schuhmann, C Kranz, H Wohlschläger, J Strohmeier. Pulse technique for the electrochemical deposition of polymer films on electrode surfaces. Biosensors Bioelectron 12:1157–1167, 1997.
88. DE Weisshaar, MM Walczak, MD Porter. Electrochemically induced transformation of monolayers formed by self-assembly of mercaptoethanol on gold. Langmuir 9:323–329, 1993.
89. DF Yang, CP Wilde, M Morin. Studies of the electrochemical and efficient reformation of a monolayer of hexadecanethiol self-assembled at a Au(111) single crystal in aqueous solutions. Langmuir 13:243–249, 1997.
90. T Wilhelm, G Wittstock. Localized electrochemical desorption of gold alkanethiolate monolayers by means of scanning electrochemical microscopy (SECM). Mikrochim Acta 133:1–9, 2000.
91. H Shiku, I Uchida, T Matsue. Microfabrication of alkylsilanized glass substrate by electrogenerated hydroxyl radical using scanning electrochemical microscopy. Langmuir 13:7329–7244, 1997.
92. WB Wilbur, DO Wipf, WG Kuhr. Localized avidin/biotin derivatization of glassy carbon electrodes using SECM. Anal Chem 70:2601–2606, 1998.
93. C Hess, K Borgwarth, C Ricken, DG Ebling, J Heinze. Scanning electrochemical microscopy: study of silver deposition on non-conducting substrates. Electrochim Acta 42:3065–3073, 1997.
94. DO Wipf, AJ Bard. Scanning electrochemical microscopy. 15. Improvements in imaging via tip-position modulation and lock-in detection. Anal Chem 64: 1362–1367, 1992.
95. PN Bartlett, D Pletcher, J Zeng. Approaches to the integration of electrochemistry and biotechnology. 1. Enzyme-modified reticulated vitreous carbon electrodes. J Electrochem Soc 144:3705–3710, 1997.
96. G Wittstock, B Gründig, B Strehlitz, K Zimmer. Evaluation of microelectrode arrays for amperometric detection by scanning electrochemical microscopy. Electroanalysis 10:526–531, 1998.
97. T Wilhelm, G Wittstock, R Szargan. Scanning electrochemical microscopy of enzymes immobilized on structured glass-gold substrates. Fresensius J Anal Chem 365:163–167, 1999.

98. CDT Bratten, PH Cobbold, JM Cooper. Micromachining sensors for electrochemical measurement in subnanoliter volumes. Anal Chem 69:252–258, 1997.
99. RA Clark, PB Hietpas, AG Ewing. Electrochemical analysis in picoliter microvials. Anal Chem 69:259–263, 1997.
100. G Wittstock, SH Jenkins, HB Halsall, WR Heineman. Continuing challenges for the immunoassay field. Nanobiology 4:153–162, 1998.
101. A Hengstenberg, C Kranz, W Schuhmann. Facilitated tip-positioning and application of non-electrode tips in scanning electrochemical microscopy using a shear force-based constant-distance mode. Chem Eur J 6:1547–1554, 2000.
102. H Zhou, S Kasai, T Yasukawa, T Matsue. Imging the activity of immobilized horseradish peroxidase with scanning electrochemical/chemiluminescence microscopy. Electrochemistry 67:1135–1137, 1999.
103. H Meyer, H Drewer, B Gründig, K Cammann, R Kakerow, Y Manoli, W Mokwa, M Rospert. Two-dimensional imaging of O_2, H_2O_2 and glucose distribution by an array of 400 individually addressable microelectrodes. Anal Chem 67:1164–1170, 1995.

12

PROBING REACTIONS AT SOLID/LIQUID INTERFACES

Julie V. Macpherson and Patrick R. Unwin

University of Warwick, Coventry, England

I. INTRODUCTION

The aim of this chapter is to demonstrate that scanning electrochemical microscopy (SECM) is ideally suited to probing solid/liquid interfacial processes, overcoming many of the deficiencies inherent in conventional methodologies by providing new quantitative insights into rates and mechanisms. In general, this chapter is concerned only with processes that involve the net phase transfer of material, specifically (1) adsorption/desorption kinetics at hydrous metal oxides (1), (2) the dissolution of ionic crystals (2–8), and (3) corrosion phenomena (9–19). These processes are of considerable fundamental and practical importance (20–26). Related applications of SECM as a probe of heterogeneous electron transfer reactions at electrode/electrolyte interfaces and in the modification of solid surfaces are described in Chapters 5 and 13, respectively.

To appreciate the impact of SECM on the study of phase transfer kinetics, it is useful to briefly review the basic steps in reactions at solid/liquid interfaces. Processes of dissolution (growth) or desorption (adsorption), which are of interest herein, may be described in terms of some, or all, of the series of events shown in Figure 1. Although somewhat simplistic, this schematic identifies the essential elements in addressing the kinetics of interfacial processes. In one limit, when any of the surface processes in Figure 1 (e.g., the detachment of ions or molecules from an active site, surface diffusion of a species across the surface, or desorption) are slow compared to the mass transport step between the bulk solution and the interface, the reaction is kinetically surface-controlled. In the other limit, if the surface events are fast compared to mass transport, the overall process is in a mass transport–controlled regime.

The view of solid/liquid interfacial reactivity in Figure 1 is further complicated when the nature of a typical solid surface is considered. For many materials, the surface morphology is highly complex with a variety of active

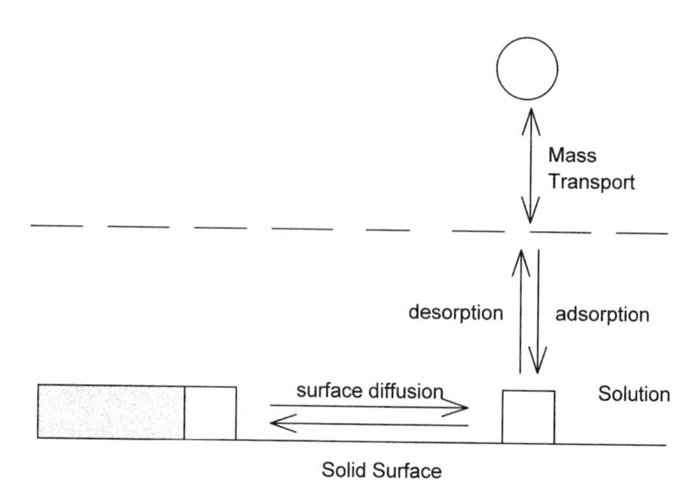

FIG. 1 The elementary steps involved in dissolution/growth and adsorption/desorption processes.

sites exposed. Even on a single crystal there is a considerable variety of surface microstructure and defects, such as steps, terraces, kink sites, vacancies, inclusion sites, adatoms, along with screw and edge dislocations (23,24,27), all of which can play a role in determining interfacial reactivity. Thus, at an elementary level, a number of surface mechanisms (and rates) may operate in parallel over a solid surface.

It follows from this brief introduction, that in the quantitative investigation of reactions at solid/liquid interfaces, it is key that the role of local mass transport in controlling the rate is understood, together with the relationship between surface structure and reactivity. As described herein, SECM is well suited to addressing these two questions, since mass transfer is defined and calculable (28,29) and can be varied over a wide dynamic range simply by changing the size of the ultramicroelectrode (UME) and the tip/substrate separation. In particular, for very close tip/sample distances and small electrode dimensions, high rates of mass transfer can be generated, which is extremely advantageous for studying fast reactions (29). Coupled with this, the local nature of SECM opens up the possibility of directly monitoring the reactivity of a targeted spot on an interface (down to micrometer dimensions), enabling surface activity to be directly correlated with structural features on this scale. Measurements at higher spatial resolution are possible by combining the local electrochemical capabilities of SECM with the topographical imaging features of the atomic force microscope (AFM). Preliminary steps in this direction through the development of an integrated electrochemical (IE)-AFM (8) will be discussed in Sec. III.C.

II. MEASUREMENT OF ADSORPTION/DESORPTION KINETICS AND SURFACE DIFFUSION RATES

A. Principles

For the investigation of adsorption/desorption kinetics and surface diffusion rates, SECM is employed to locally perturb adsorption/desorption equilibria and measure the resulting flux of adsorbate from a surface. In this application, the technique is termed scanning electrochemical induced desorption (SECMID) (1), but historically this represents the first use of SECM in an equilibrium perturbation mode of operation. Later developments of this mode are highlighted towards the end of Sec. II.C. The principles of SECMID are illustrated schematically in Figure 2, with specific reference to proton adsorption/desorption at a metal oxide/aqueous interface, although the technique should be applicable to any solid/liquid interface, provided that the adsorbate of interest can be detected amperometrically.

For this type of investigation, the tip UME is placed close to the surface of the substrate, such that the tip/substrate separation, d, is of the order of the electrode radius, a, or less. The substrate is bathed in a solution of the electroactive adsorbate and the adsorption/desorption process is allowed to come to equilibrium. Upon stepping the electrode potential from a value where no electrode reactions occur to one where the solution component of the adsorbate (H^+ in Fig. 2) is removed at a diffusion-controlled rate, H^+ is locally depleted in the solution between the tip and substrate surface. This process perturbs the adsorption/desorption equilibrium, inducing the desorption of protons from the surface and promoting the diffusion of protons from the surrounding solution into the tip/substrate gap. Additionally since the desorption process depletes the concentration of adsorbed protons on the substrate directly under the UME, a radial concentration gradient of adsorbate develops across the surface, thereby promoting surface diffusion as a pathway for the transport of protons into the gap. In principle, the resulting tip current-time behavior provides quantitative information on the rates of desorption, adsorption, and surface diffusion.

B. Theory

In order to obtain quantitative information from SECMID measurements, it is key that mass transport in the SECM coordinate system can be addressed theoretically. Although it is beyond the scope of this chapter to provide a complete overview of the treatment of SECM problems, since this is covered in detail in Chapter 5, some discussion of the elements involved in the formulation of the SECM problem is warranted.

To calculate the tip current response, the diffusion equation appropriate to the axisymmetric SECM geometry must be solved subject to boundary

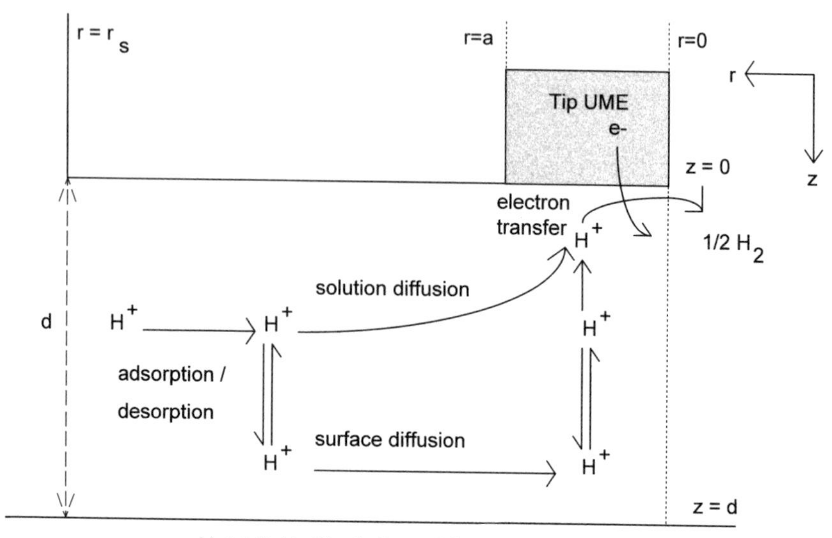

FIG. 2 Principles of SECMID using H^+ as a model adsorbate. Schematic of the transport processes in the tip/substrate domain for a reversible adsorption/desorption process at the substrate following the application of a potential step to the tip UME where the reduction of H^+ is diffusion-controlled. The coordinate system and notation for the axisymmetric cylindrical geometry is also shown. Note that the diagram is not to scale as the tip/substrate separation is typically $\leq 0.01\ r_s$.

conditions that define the chemical processes occurring at the tip UME and the substrate surface.

$$\frac{\partial C}{\partial \tau} = \frac{\partial^2 C}{\partial R^2} + \frac{1}{R}\frac{\partial C}{\partial R} + \frac{\partial^2 C}{\partial Z^2} \tag{1}$$

Equation (1) is the relevant time-dependent diffusion equation expressed in the form of dimensionless variables:

$$C = c/c^* \tag{2}$$

$$\tau = tD_{soln}/a \tag{3}$$

$$R = r/a \tag{4}$$

$$Z = z/a \tag{5}$$

where r and z are the coordinates, respectively, in the directions radial and normal to the electrode surface (Fig. 2), starting at its center and t is time; c and D_{soln} are the concentration and solution diffusion coefficient of the species (adsorbate) of interest and c^* is the concentration of adsorbate in bulk solution.

Prior to the potential step, the concentration of adsorbate in solution is at the bulk value, over all space:

$$\tau = 0, \text{ all } R, \text{ all } Z: \qquad C = 1 \tag{6}$$

Following the potential step ($\tau > 0$) the boundary conditions at the surface of the UME,

$$Z = 0, 0 \leq R \leq 1: \qquad C = 0 \tag{7}$$

$$Z = 0, 1 \leq R \leq RG: \qquad \partial C/\partial Z = 0 \tag{8}$$

denote that electrolysis of the species of interest occurs at a diffusion-controlled rate [Eq. (7)] and the adsorbate is inert on the insulating sheath surrounding the electrode [Eq. (8)]. Additional conditions define zero radial flux at the axis of symmetry [Eq. (9)] and the recovery of the bulk concentration of the species beyond the radial edge of the tip/substrate domain [Eq. (10)]. This latter assumption is generally valid provided that $RG \geq 10$ (28,30).

$$R = 0, 0 < Z < L: \qquad \partial C/\partial R = 0 \tag{9}$$

$$R > RG, 0 < Z < L: \qquad C = 1 \tag{10}$$

where:

$$L = d/a \tag{11}$$

$$RG = r_s/a \tag{12}$$

and r_s is the radial distance from the center of the disk to the edge of the insulating sheath surrounding the electrode (Fig. 2).

The boundary condition at the substrate depends upon the nature of the adsorption/desorption process and whether or not adsorbate surface diffusion needs to be considered. For the induced desorption of ions, there is an accompanying change in the local surface charge density. A model taking this into account (1), based on the diffuse layer (Gouy-Chapman) surface charge density-potential relationship (31), has been developed with specific reference to the deprotonation of hydrous metal oxides (32). A simpler limiting case, involving Langmuirian adsorption/desorption, has also been described, and we consider this here to illustrate the general features of SECMID. For this case, the substrate boundary condition is:

$$Z = L, 0 \leq R \leq RG: \qquad \partial C/\partial Z = -K_d\theta + K_aC(1 - \theta) \tag{13}$$

where

$$K_a = k_a a / D_{soln} \tag{14}$$

$$K_d = k_d a / D_{soln} c^* \tag{15}$$

k_a (cm s^{-1}) and k_d (mol cm^{-2} s^{-1}) are the apparent adsorption and desorption rate constants and θ is the fractional surface coverage.

The time dependence of θ is:

$$Z = L,\ 0 \le R \le RG: \qquad \lambda \left(\frac{\partial \theta}{\partial \tau} \right) = -K_d \theta + K_a C(1 - \theta) \tag{16}$$

where

$$\lambda = N / c^* a \tag{17}$$

and N is the density of adsorption sites (mol cm^{-2}). Prior to the potential step, the adsorption/desorption process is at equilibrium and the corresponding initial substrate condition is:

$$\tau = 0\text{: substrate: } 0 \le R \le RG: \qquad \theta = [1 + (K_a/K_d)]^{-1} \tag{18}$$

When surface diffusion of the adsorbate operates in addition to adsorption/desorption, the following transport equation applies to the substrate rather than Eq. (16):

$$Z = L,\ 0 \le R \le RG: \qquad \lambda \left(\frac{\partial \theta}{\partial \tau} \right) = \lambda \beta \left[\frac{\partial^2 \theta}{\partial R^2} + \frac{1}{R} \frac{\partial \theta}{\partial R} \right] - K_d \theta + K_a C_{Z=L}(1 - \theta) \tag{19}$$

where

$$\beta = D_{sur} / D_{soln} \tag{20}$$

D_{sur} is the surface diffusion coefficient of the adsorbate, which is assumed to be direction-independent. Equation (20) has the following associated conditions:

$$\text{substrate: } R = 0: \qquad \partial \theta / \partial R = 0 \tag{21}$$

$$\text{substrate: } R > RG: \qquad \theta = [1 + (K_a/K_d)]^{-1} \tag{22}$$

The flux of material from the surface of the substrate for this case is again given by Eq. (13).

The quantity to be determined from the calculations, the tip current, i, normalized with respect to the steady-state current which flows at an UME in bulk solution, $i(\infty)$, is:

$$i/i(\infty) = (\pi/2) \int_0^1 (\partial C/\partial R)_{Z=0}\, dR \tag{23}$$

This problem was solved numerically using the alternating direction implicit (ADI) finite difference method (30). It follows from above that the current-time behavior depends on five parameters: K_a, K_d, λ, L, and β, all of which influence the shape of a chronoamperogram.

Figure 3 illustrates the effect of the adsorption/desorption kinetics on the transient profile, in the absence of surface diffusion, where $K_a/K_d = 1$, i.e., half the surface sites are occupied at equilibrium. The results are presented as $i/i(\infty)$ versus $\tau^{-1/2}$ in order to emphasize the short-time behavior. At very short times, i.e., the largest $\tau^{-1/2}$, the UME response is identical for all values of K_a, since under these conditions, the diffusion field adjacent to the electrode is much smaller than the tip/sample separation and so does not sense the presence of the substrate (30,33). At times sufficient for the dif-

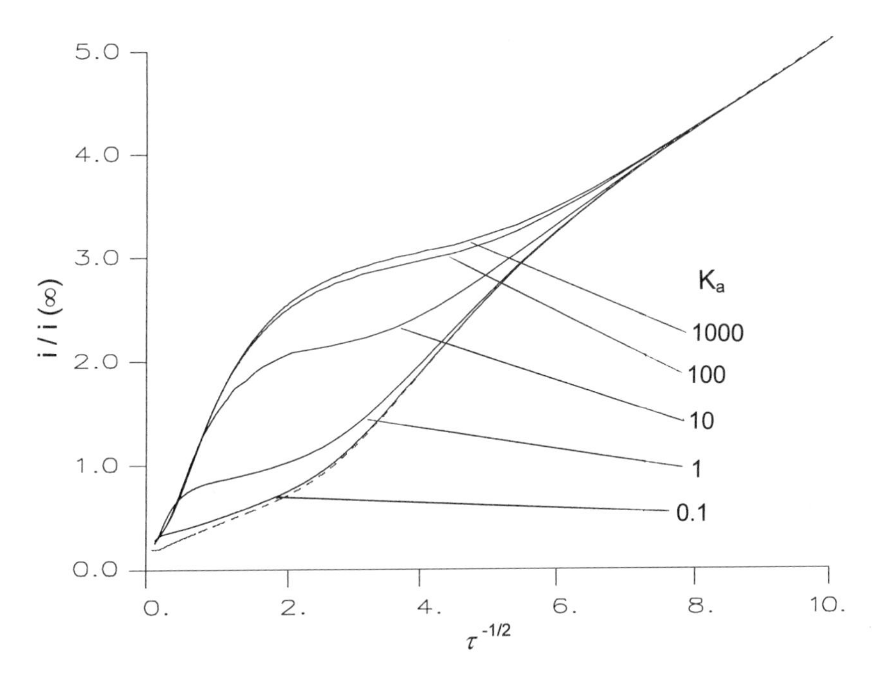

FIG. 3 Theoretical $i/i(\infty)$ versus $\tau^{-1/2}$ characteristics for a SECMID adsorption/desorption process characterized by $\lambda = 8$, $K_a/K_d = 1$, $L = 0.32$, and K_a values (solid lines) of 0.1 (lower), 1, 10, 100, and 1000 (upper). The dashed line indicates the behavior for an inert substrate.

fusion field to intercept the interface, the resulting chronoamperometric behavior becomes sensitive to the magnitude of K_a and K_d.

When the adsorption/desorption kinetics are slow compared to the rate of diffusional mass transfer through the tip/substrate gap, the system responds sluggishly to depletion of the solution component of the adsorbate close to the interface and the current-time characteristics tend towards those predicted for an inert substrate. As the kinetics increase, the response to the perturbation in the interfacial equilibrium is more rapid and, at short to moderate times, the additional source of protons from the induced-desorption process increases the current compared to that for an inert surface. This occurs up to a limit where the interfacial kinetics are sufficiently fast that the adsorption/desorption process is essentially always at equilibrium on the time scale of SECM measurements. For the case shown in Figure 3 this is effectively reached when $K_a = K_d = 1000$. In the absence of surface diffusion, at times sufficiently long for the system to attain a true steady state, the UME currents for all kinetic cases approach the value for an inert substrate. In this situation, the adsorption/desorption process reaches a new equilibrium (governed by the local solution concentration of the target species adjacent to the substrate/solution interface) and the tip current depends only on the rate of (hindered) diffusion through solution.

The effective adsorbate surface coverage, governed by the occupancy (θ) and density of adsorption sites (λ) has a significant effect on the current-time characteristics. It is particularly interesting to note that much lower surface coverages than considered in Figure 3 can readily be detected using SECMID, provided that the kinetics are sufficiently fast. This is demonstrated in Figure 4 for the case where $K_a/K_d = 0.1$, i.e., $\theta = 0.091$. In this particular case, the lower initial surface coverage leads to currents that more rapidly approach the value for an inert surface. However, it is still possible to distinguish the various kinetic cases in the short-time region. Theoretical simulations indicate that kinetic resolution is feasible down to θ values of ca. 0.01, i.e., $K_a/K_d = 0.01$ (1). Figure 5 shows the effect of λ on the SECMID current-time behavior. As expected, increasing λ (increasing the surface concentration relative to that in solution) serves to increase the overall charge passed during the transient.

The UME current response becomes increasingly sensitive to the surface process as the tip/substrate separation is decreased, since this serves both to maximize the ratio of effective surface area to solution volume probed by SECMID and to hinder solution diffusion in the gap. This effect is illustrated by Figures 6a and b for various values of K_a and an adsorption/desorption system characterized by $K_a/K_d = 1$ and $\lambda = 8$ for $L = 0.1$ and 1.0, respectively. The closer the tip is positioned with respect to the substrate, the greater the depletion of the solution component of the adsorbate adjacent to

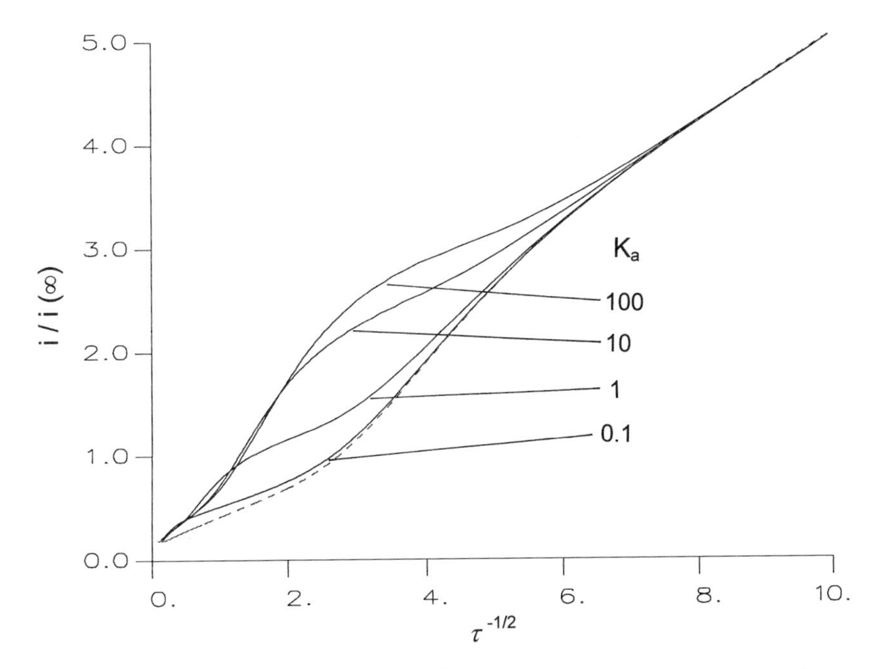

FIG. 4 Theoretical $i/i(\infty)$ versus $\tau^{-1/2}$ SECMID characteristics for a substrate adsorption/desorption process characterized by $\lambda = 8$, $K_a/K_d = 0.1$, $L = 0.32$, and K_a values (solid lines) of 0.1 (lower), 1, 10, 100, and 1000 (upper). The dashed line indicates the behavior for an inert substrate.

the substrate and thus the greater the perturbation of the overall equilibrium. This perturbation of the interfacial equilibrium can be very extensive; for example, at extremely close tip/substrate separations ($L = 0.1$) the local solution concentration adjacent to the substrate can be depleted by over five orders of magnitude compared with the bulk concentration value (1). Under these conditions, the spatial resolution of the technique (at short to moderate times) can approach the size of the electrode.

Surface diffusion provides an additional pathway for the transport of adsorbate into the tip/substrate domain, becoming most important when a significant radial concentration gradient develops in surface-bound adsorbate. The main effect is to increase the magnitude of the current flowing in the longer time region of the transient and, in particular, enhance the final steady-state current compared to that for an inert surface (or the case where adsorption/desorption, in the absence of surface diffusion, comes to equilibrium). Clearly, the larger the value of the surface diffusion coefficient compared to that in solution, the larger the steady-state current. This is illustrated

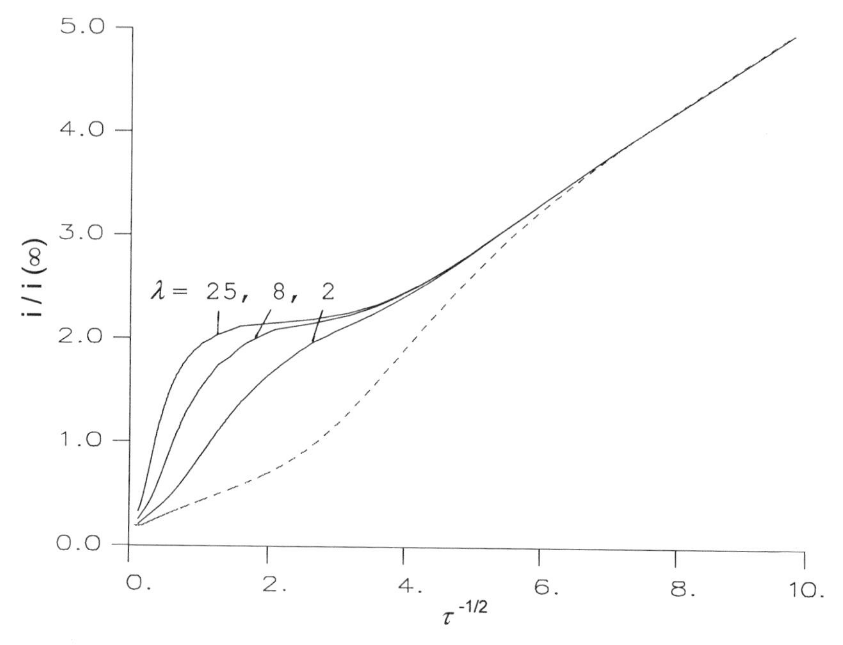

FIG. 5 The effect of λ on the SECMID $i/i(\infty)$ versus $\tau^{-1/2}$ characteristics for an adsorption/desorption process characterized by $K_a = K_d = 10$ at a tip-substrate separation of $L = 0.32$ for $\lambda = 2$, 8, and 25. The dashed line represents the behavior for an inert substrate.

in Figure 7 for β values of 0, 0.01, 0.1, and 1.0. The data are presented as $i/i(\infty)$ versus τ in order to emphasize the long time portion of the transient, as the system approaches a steady state. It has been shown that the larger the effective surface concentration of adsorbate, governed by the values of θ and λ, the more pronounced the effect of surface diffusion on the chronoamperometric behavior (1). Moreover, the steady-state current becomes increasingly sensitive to the surface diffusion process as the tip/substrate separation is decreased for the reasons discussed above.

Thus, the most suitable route for obtaining information on the adsorption/desorption kinetics is from the short-time transient behavior. Under these conditions, surface diffusion effects are negligible and the short-time current response depends only on K_a, K_d, and λ for a given tip/substrate separation. Provided that an independent measurement of λ can be made, an absolute assignment of the interfacial kinetics is possible. Furthermore, analysis of the long-time current allows the importance, and magnitude, of surface diffusion to be determined.

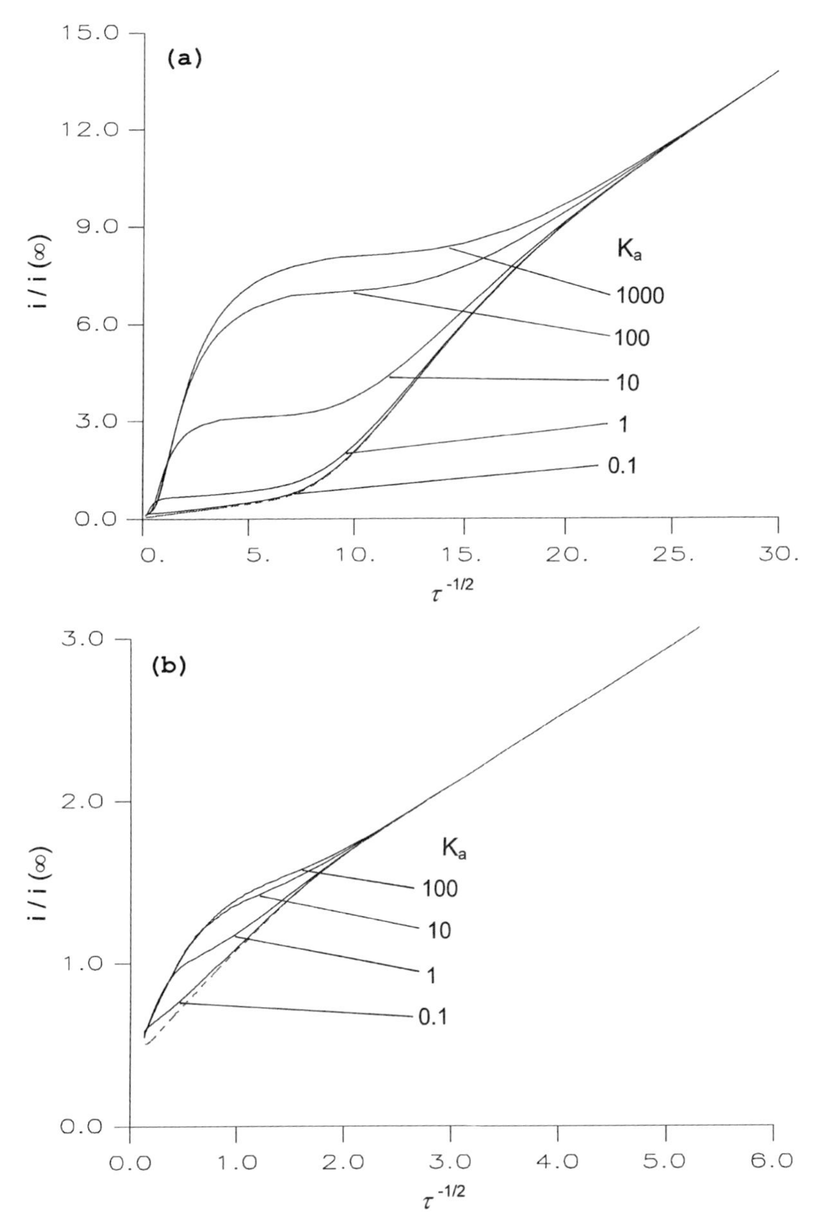

FIG. 6 The effect of tip-substrate separation on the SECMID response for an adsorption/desorption process characterized by $\lambda = 8$, $K_a/K_d = 1$. Data are given for (a) $L = 0.1$ with K_a values (solid lines) of 0.1 (lower), 1, 10, 100, and 1000 (upper) and (b) $L = 1.0$ with K_a values (solid lines) of 0.1 (lower), 1, 10, and 100 (upper). The behavior for an inert substrate is indicated by the dashed line.

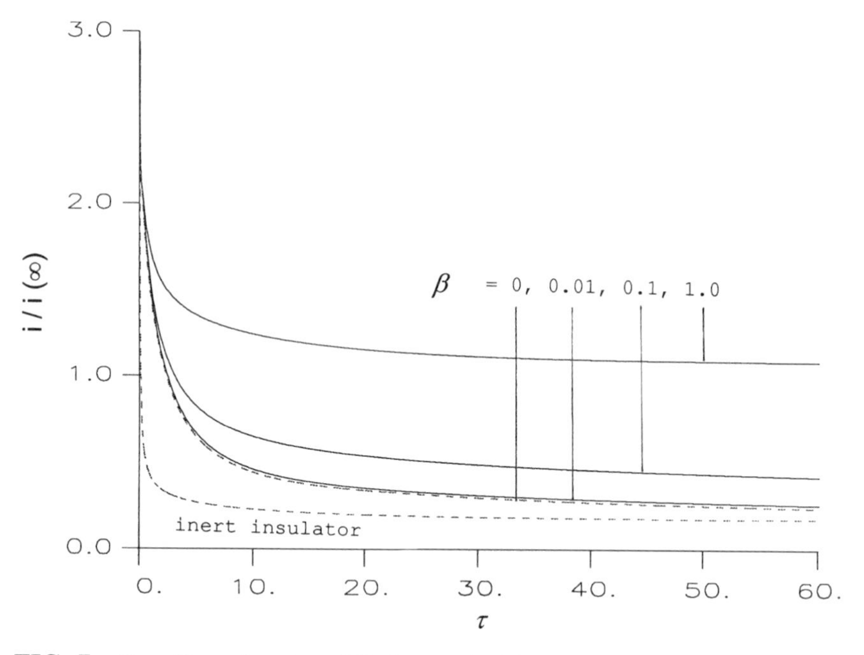

FIG. 7 The effect of surface diffusion on the SECMID current-time response for a substrate adsorption/desorption process characterized by $K_a = K_d = 10$, $\lambda = 8$, and $\beta = 0$, 0.01, 0.1, and 1.0, at a tip-substrate separation, $L = 0.32$. The behavior for an inert substrate is also given.

C. Applications

SECMID has been applied to investigate the adsorption/desorption kinetics of protons on the (001) face of TiO_2 (rutile) and (010) face of $NaAlSi_3O_8$ (albite) single crystals in dilute acid solutions (2×10^{-4} mol dm^{-3} HCl) (1). The interfacial proton concentration was perturbed locally, and the resulting current-time behavior measured, by positioning a 25 μm diameter Pt disc electrode close to the crystal surface of interest and stepping the tip potential to a value to reduce protons at a diffusion-controlled rate. The importance of surface diffusion in the interfacial process was determined by comparing steady-state approach curves ($i/i(\infty)$ as a function of d) for the diffusion-limited reduction of H^+ and the oxidation of aqueous $Fe(CN)_6^{4-}$ (a distance calibrant) against the two target substrates, with the theoretical behavior predicted for an inert surface (solution diffusion only). For each substrate, there was excellent agreement between the experimental data and the theory

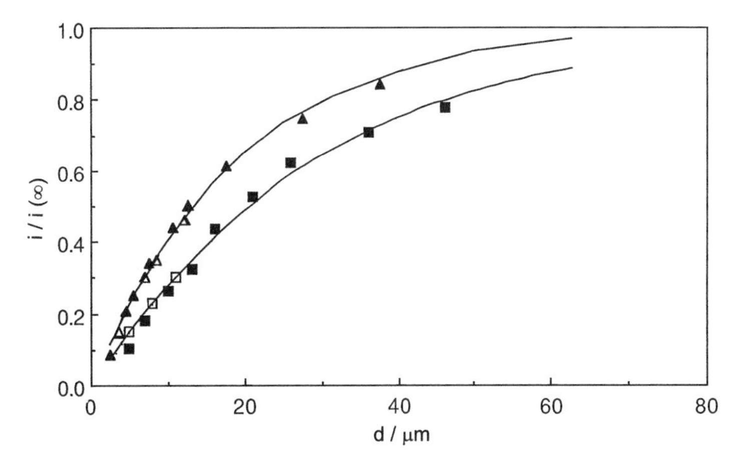

FIG. 8 SECM steady-state current-distance approach curves with substrates of rutile (001) (▲, △), and albite (010) (■, □). In each case the filled symbols are data for H^+ reduction and open symbols are tip-substrate distance calibration data for ferrocyanide oxidation. The solid lines show the negative feedback behavior for an inert substrate and an electrode geometry characterized by $RG = 10$ (rutile experiments) and $RG = 20$ (albite experiments).

for an inert interface (28), as shown in Figure 8, indicating that surface diffusion of adsorbed H^+ was essentially negligible for both crystal surfaces.

At close tip/substrate separations, the short-time currents for H^+ reduction above both crystal surfaces were larger than those predicted for an inert surface, as shown in Figures 9 and 10. This was attributed to H^+ desorption from the surface providing an additional source of protons, thereby increasing the overall flux at the UME. Using known values of K_a/K_d and λ for rutile (32) and albite (34), it was possible to assign values for the adsorption and desorption rate constants, taking into account surface potential effects for the former system, which were found to be important. For rutile, the intrinsic adsorption and desorption rate constants were $k_a = 0.4 \pm 0.1$ cm s^{-1} and $k_d = 5.5 \pm 1.4 \times 10^{-8}$ mol cm^{-2} s^{-1}. Under the experimental conditions, the initial surface concentration of H^+ was only $\sim 8 \times 10^{-11}$ mol cm^{-2}. Figure 9 thus illustrates admirably the effectiveness of SECMID as a probe of low surface coverages, when the adsorption/desorption characteristics are fast. For albite, the measured rate constants were much lower; $k_a = 0.003 \pm 0.001$ cm s^{-1} and $k_d = 0.84 \pm 0.37 \times 10^{-9}$ mol cm^{-2} s^{-1}. Figure 10 demonstrates that for sluggish interfacial kinetics, the measured chronoamperometric responses at close tip/substrate separations are only slightly larger than for an inert surface.

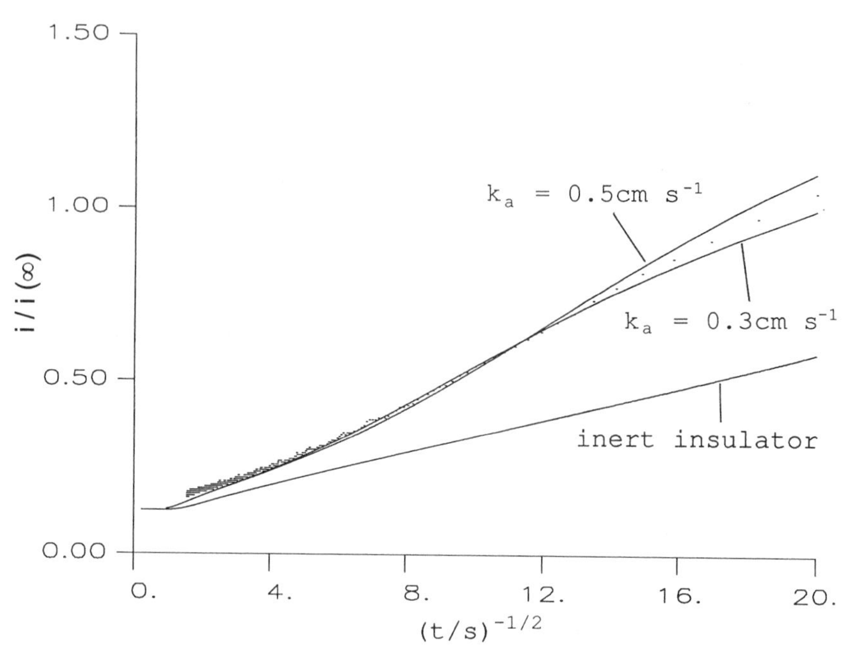

FIG. 9 SECMID chronoamperometric characteristics for H^+ reduction against a rutile (001) surface at a tip-substrate separation of 2.6 μm, with a 12.5 μm radius Pt UME (RG = 10). The solid lines show the behavior for an inert substrate and an adsorption/desorption process characterized by $K_a/K_d = 1.49$, $\lambda = 6.6$, and $k_a = 0.3$ and 0.5 cm s^{-1}. The adsorption/desorption fit includes surface potential effects that have not been considered in this chapter.

Although only two applications of SECMID have been described at solid/liquid interfaces, the technique should be more widely applicable to (reversible) adsorption/desorption processes involving electroactive adsorbates. Irreversible adsorption can be studied using a modification of SECMID—reactant injection chronoamperometry—provided that the adsorbate can be electrogenerated and is part of a reversible electrode couple (35). This technique has been used to investigate the interaction of Ag^+ with the (100) face of pyrite (35,36). This is a complex process, involving adsorption (possibly by ion exchange) and Ag^+ reduction, for which the elementary kinetics and mechanisms remain elusive (37). The basis of the SECM approach is to use an Ag UME, positioned close to the surface of pyrite, to electrogenerate Ag^+ potentiostatically. The current response determines the flux of Ag^+, which is governed by mass transport and interfacial reactivity (Fig. 11). In essence, when adsorption is important, Ag^+ is depleted more

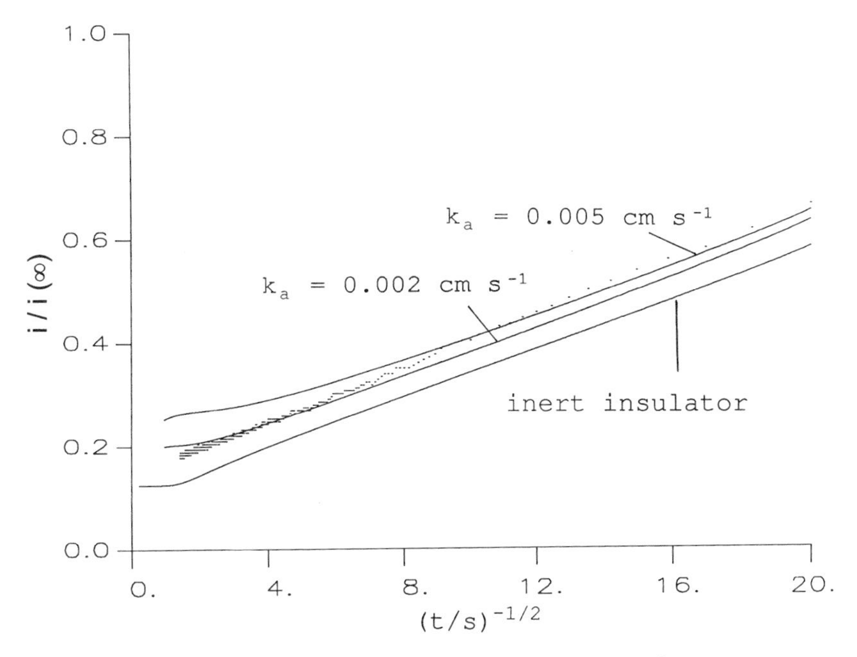

FIG. 10 SECMID chronoamperometric characteristics for H^+ reduction against an albite (010) surface at a tip-substrate separation of 2.8 μm, with a 12.5 μm radius Pt UME (*RG* = 20). The solid lines show the behavior for an inert substrate and an adsorption/desorption process characterized by $K_a/K_d = 0.86$, $\lambda = 28.4$, and k_a = 0.002 and 0.005 cm s^{-1}.

rapidly in the tip/substrate gap than when the interface is inert. To maintain the equilibrium concentration of Ag^+ adjacent to the UME, a larger current thus has to flow, the form of which is governed by the adsorption kinetics. Preliminary results indicate a value of 0.25 cm s^{-1} for the initial adsorption rate constant of Ag^+ on the (100) face of pyrite (36).

For completeness, it is worth mentioning that the equilibrium perturbation mode is not just limited to solid/liquid interfaces. The technique is applicable, in principle, to adsorption/desorption processes at liquid/liquid interfaces. Moreover, with the introduction of a novel electrode design for performing SECM measurements at air/liquid interfaces (38,39), SECMID can readily be extended to this environment. A variant of SECMID, currently finding much use in the study of liquid/liquid and biomaterial/liquid interfaces, is scanning electrochemical induced transfer (SECMIT) (40,41). In this case, the electroactive species partitions throughout the two phases forming the target interface. Electrochemical depletion of the species of interest

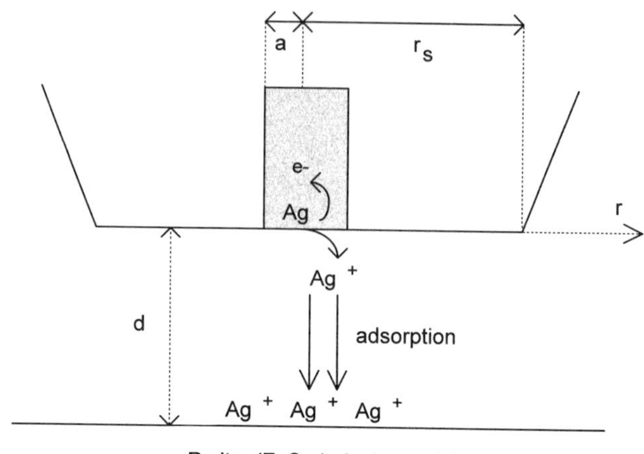

FIG. 11 Schematic illustration of the application of SECM to investigate Ag^+ adsorption at a pyrite/aqueous solution interface.

using the tip UME positioned in one phase, close to the interface boundary, induces transfer from the second phase into the first. The technique can be used to probe transfer kinetics, permeabilities, and to measure concentrations and/or diffusion coefficients of the species in the second phase, without the probe electrode having to directly enter that phase.

The use of SECMID to map spatial variations in the adsorption/desorption characteristics of a substrate has not yet been attempted. In principle, however, it should be possible to correlate adsorption/desorption kinetics with surface structural features, using the imaging capabilities of SECM.

III. DISSOLUTION KINETICS OF IONIC SINGLE CRYSTALS

A. Principles

One of the first uses of the SECM was as a fabrication tool to form dissolution channels and growth hillocks on materials at the micrometer to submicrometer level. This area is dealt with in greater detail in Chapter 13. For these studies, the feedback mode of the device was predominantly utilized. Because the purpose was only to alter the topographical features of a substrate, little information was provided on the rates or nature of the dissolution processes. In this section we describe how the equilibrium perturbation mode can be employed to initiate, and quantitatively monitor, dissolution reactions, providing unequivocal information on the kinetics and mechanism of the process under conditions of controlled mass transport and with high spatial resolution.

The SECM arrangement for the study of ionic crystal dissolution is illustrated in Figure 12, using copper sulfate pentahydrate as an example substrate. The surrounding solution is saturated with respect to $CuSO_4 \cdot 5H_2O$, so that initially the system is at equilibrium and there is no net dissolution or growth. With the UME positioned close to the substrate, the tip potential is stepped from a value where no electrochemical reactions occur to one where electrolysis of one type of the lattice ion in solution (in this case the reduction of Cu^{2+} to Cu) occurs at a diffusion-controlled rate. The resulting electrode process depletes the concentration of Cu^{2+} in the tip-substrate gap, creating a local undersaturation at the crystal/solution interface. This perturbs the interfacial equilibrium and provides the thermodynamic force for the dissolution reaction. Dissolving ions diffuse from the crystal across the gap and are detected at the tip UME, thus producing a current flow, the magnitude of which depends on the rate and mechanism of the dissolution process. Using the imaging mode it is possible to initiate the dissolution reaction over wider areas of the substrate. In this case the UME, which is held at a potential necessary to promote dissolution, is scanned over a selected area of the surface. The current, measured as a

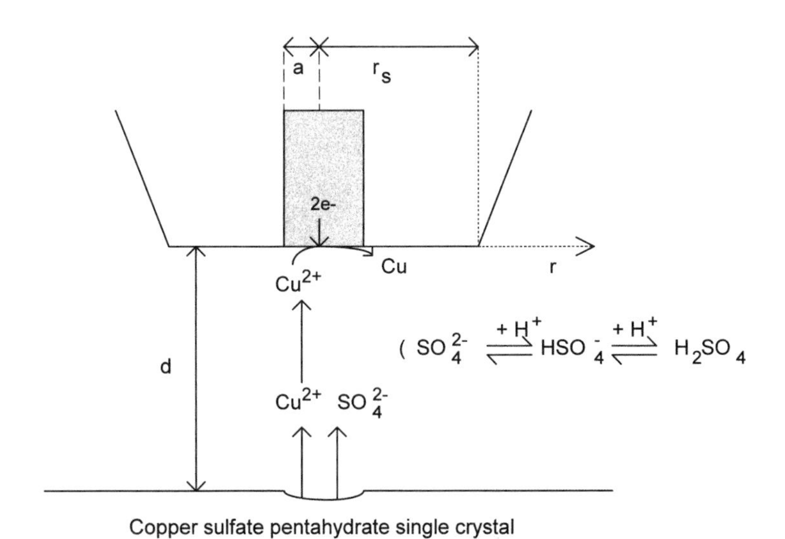

FIG. 12 Principles of SECM-induced dissolution using copper sulfate pentahydrate as an example substrate. Note that the diagram is not to scale as the tip-crystal separation is typically $\leq 0.01\ r_s$.

function of probe position in the plane of the surface, can be used to provide a map of the dissolution activity.

The key features of this technique, which result in considerable advantages over conventional methodologies, are as follows:

1. Mass transfer in the tip-substrate gap is well defined, variable, and calculable, enabling the role of interfacial undersaturation in the dissolution process to be quantified (2–8).
2. Extremely high mass transfer rates can be generated using close tip-substrate separations and small-radii UMEs (42,43) facilitating the study of fast reactions (2–8).
3. The extent of dissolution can be controlled by varying the period of electrolysis (down to the millisecond timescale) with a stationary probe, the scan speed for a moving probe in the imaging mode, and/or the UME potential. This enables the study of materials with a very wide range of solubilities (presently 10^{-5} to 10^{-1} mol dm^{-3}) (5,6).
4. As the reactivity is probed at a targeted microscopic region of the surface (often a single crystal face), it is possible to establish the relationship between surface structure and dissolution activity on the micrometer scale.
5. Since UME measurements can readily be made with and without supporting electrolyte (44), the role of added electrolytes on dissolution processes can be identified.

B. Theory

Although a large number of rate equations, both empirical and mechanistic in origin, have been proposed to describe the dissolution of ionic crystals (45–47), we consider here just two simple kinetic cases, in which the dissolution rate is controlled by first- and second-order terms in the interfacial undersaturation. These cases arise as limits to many dissolution mechanisms, including the classical Burton, Cabrera, and Frank (BCF) spiral dissolution model (48), which treats the dissolution process in terms of the retreat of monatomic steps arising from screw dislocations. This model was originally developed for crystal growth from the vapor phase, but it is often used to describe crystal dissolution in liquids. In this context, the first-order rate law applies at high undersaturation where desorption of ions or molecules from the crystal surface limits the rate, while the second-order rate law holds for low undersaturation where surface diffusion becomes rate limiting.

For a crystal such as copper sulfate pentahydrate, these rate laws are embodied in the following equation:

$$j_{Cu^{2+}} = j_{SO_4^{2-}} = k_n \sigma^n \tag{24}$$

where j_i denotes the dissolution flux of species i (in this example Cu^{2+} or SO_4^{2-}), k_n is a dissolution rate constant, and the dissolution process is defined as being either first- or second-order, n, in the local undersaturation, σ, at the crystal/solution interface:

$$\sigma = 1 - S \tag{25}$$

In Eq. (25), S is the saturation ratio, given by:

$$S = (I_p/K_s)^{1/v} \tag{26}$$

where I_p is the ionic activity product, K_s is the solubility product, and v is the number of ions in the formula unit of the ionic crystal. Hence, for copper sulfate pentahydrate:

$$S = \left(\frac{a_{Cu^{2+}} a_{SO_4^{2-}}}{a_{Cu^{2+}}^{sat} a_{SO_4^{2-}}^{sat}} \right)^{1/2} \tag{27}$$

where a_i denotes the activity of species i and the superscript *sat* refers to saturated solution.

The dissolution problem can readily be reduced to the consideration of one species (Cu^{2+} in this particular case) by experimentally setting the conditions such that the counterion (SO_4^{2-}) concentration remains essentially constant, i.e., is buffered, during a measurement. For the $CuSO_4 \cdot 5H_2O$ dissolution experiments, this was achieved by making measurements in aqueous sulfuric acid solutions (carefully avoiding concentrations that would promote the dehydration of the crystal), which additionally serves as a supporting electrolyte.

The UME response during the SECM chronoamperometric experiment is calculated by solving Eq. (1) with respect to the boundary conditions defining the processes occurring at the electrode and substrate. The initial conditions and boundary conditions are as previously stated in Eqs. (6)–(10), except C now corresponds to the normalized concentration of Cu^{2+}. The substrate boundary condition, reflecting the dissolution rate law is:

$$\tau > 0;\ Z = L;\ 0 \leq R \leq RG: \qquad \frac{\partial C_{Cu^{2+}}}{\partial Z} = -K_n[1 - C_{Cu^{2+}}^{1/2}]^n \tag{28}$$

where K_n is the normalized dissolution rate constant given by

$$K_n = \frac{k_n a}{\gamma_{Cu^{2+}} c_{Cu^{2+}}^{sat} D_{Cu^{2+}}} \tag{29}$$

and $\gamma_{Cu^{2+}}$ is the activity coefficient of Cu^{2+} in the solution of interest.

The UME tip current was calculated from Eq. (23). The problem has been solved numerically using the ADI method. The results of theoretical simulations allow the effect of K_n and n on the chronoamperometric response to be directly assessed and the spatial resolution of the technique to be identified. Figure 13 shows typical current-time transients for various values of K_1 (first-order dissolution) at a normalized tip-substrate separation, $L = 0.1$. At times sufficient for the diffusion field to intercept the substrate surface, the chronoamperometric behavior reflects the dissolution kinetics. As the rate constant decreases, the UME current approaches that for an inert surface while an increase in the dissolution rate constant serves to increase the flux of material from the crystal surface. This results in an increase in the tip current, up to a point where the UME response is limited by mass transfer between the tip and the substrate, i.e., the dissolution process is diffusion-controlled. Under these conditions the kinetics are sufficiently fast to maintain the concentration at the substrate at the saturated solution value. It is also clear that the time taken for the current to approach a steady value decreases with increasing K_1.

The effect of n on the tip current behavior can best be understood by considering the steady-state current response, obtained from the long time portion of simulated transients. Figure 14 shows typical working curves of normalized current versus log K_n for n values of 1 and 2, at normalized

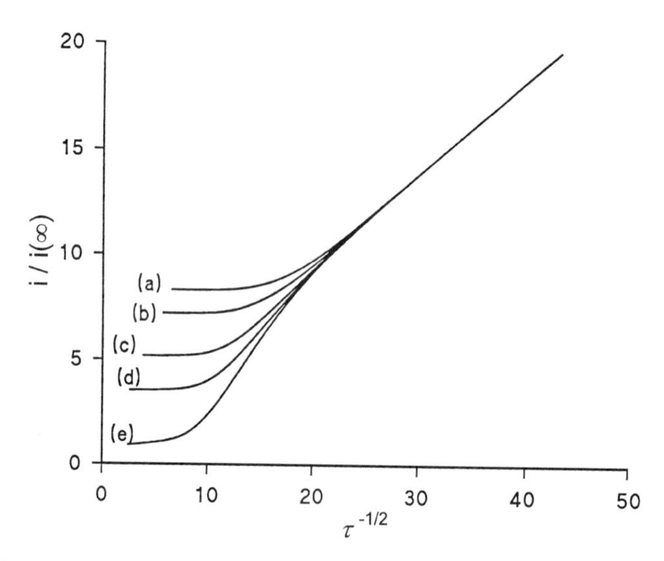

FIG. 13 SECM chronoamperometric characteristics for a dissolution process [Eq. (24)] of order $n = 1$ with a tip/substrate separation $L = 0.1$. The data shown are for normalized rate constants K_1 = (a) 1000, (b) 100, (c) 25, (d) 10, and (e) 1.

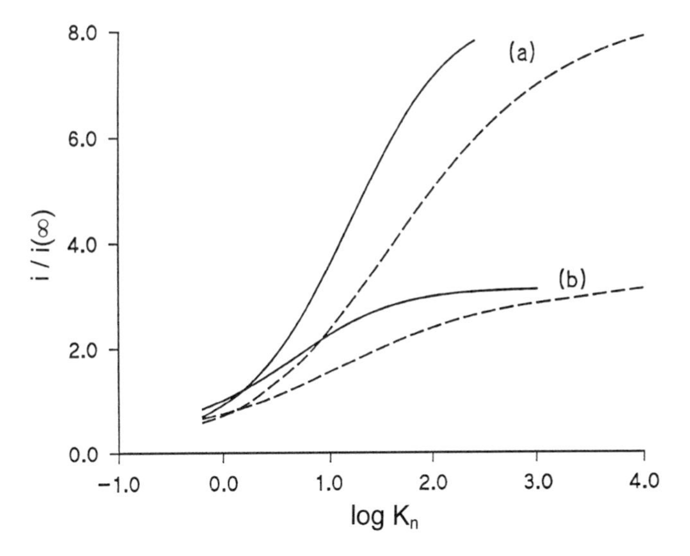

FIG. 14 Working curves of $i/i(\infty)$ versus log K_n for a measurement time $\tau = 1$ and tip-substrate distances L of (a) 0.1 and (b) 0.32. For each distance, curves are shown for $n = 1$ (——) and 2 (– – –).

UME-substrate separations of 0.1 and 0.32. The data relate to a τ value of 1.0 which, for the range of kinetics considered, is sufficient for a steady-state to be established at the UME. Figure 14 demonstrates that the shape of the working curves, at constant L, are significantly distinct to allow first- and second-order dissolution processes to be resolved experimentally. Moreover, as observed for other kinetic applications of SECM, dissolution kinetics are measurable with greatest sensitivity at the closest tip-substrate separations (1,30,42,49).

The tip-generated interfacial undersaturation is governed by the interplay between mass transport in the tip/substrate gap and the dissolution kinetics. This concept is illustrated in Figures 15 and 16. Figure 15a and b shows the radial dependence of the steady-state concentration and flux at the crystal/solution interface for a first-order dissolution process characterized by $K_1 = 1$, 10, and 100. For rapid kinetics ($K_1 = 100$), the dissolution process is able to maintain the interfacial concentration close to the saturated value and only a small depletion in the concentration adjacent to the crystal is observed over a radial distance of about one electrode dimension. Under these conditions, diffusion in the z-direction dominates over radial diffusion. As the rate constant decreases, diffusion is able to compete with the interfacial kinetics and consequently the undersaturation at the crystal surface

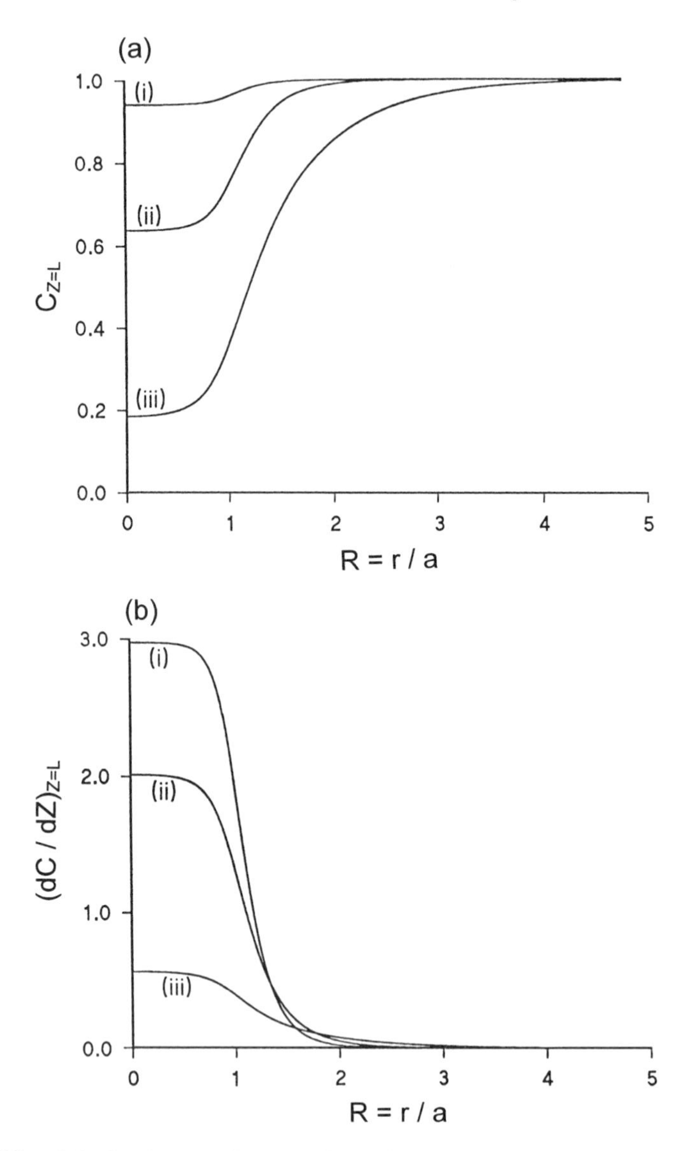

FIG. 15 Calculated normalized profiles of (a) concentration and (b) flux of a target species at a substrate/solution boundary for SECM-induced dissolution with L = 0.32 and K_1 = (i) 100, (ii) 10, and (iii) 1. The data relate to a normalized measurement time, τ, of 1.0.

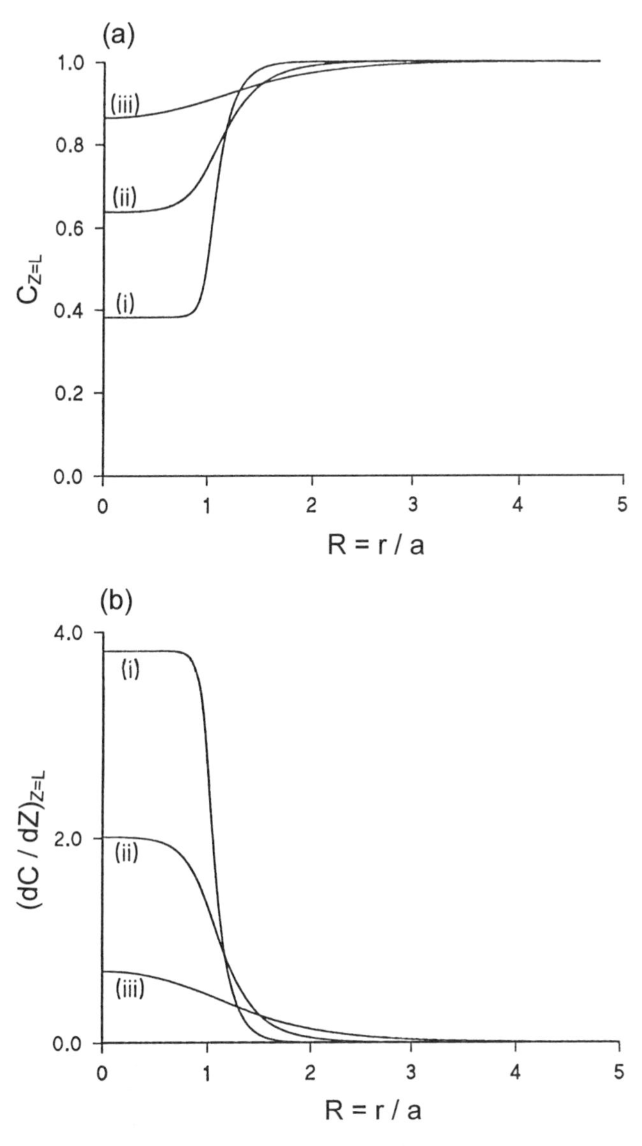

FIG. 16 Calculated normalized profiles of (a) concentration and (b) flux of a target species at a substrate/solution boundary for SECM-induced dissolution with $K_1 = 10$ and $L =$ (i) 0.1, (ii) 0.32, and (iii) 1.0. The data relate to a normalized measurement time, τ, of 1.0.

increases (surface concentration decreases). The dissolution process provides less material for diffusion towards the UME, as verified by the magnitude of the interfacial flux (Fig. 15b), and the diffusion field around the UME extends further in the radial direction decreasing the spatial resolution of the technique (Fig. 15a).

Figure 16a and b shows the effect of L on the radial dependence of the steady-state concentration and flux at the substrate/solution interface for a first-order dissolution process characterized by $K_1 = 10$ and $L = 0.1$, 0.32, and 1.0. As the tip-substrate separation decreases, the effective rate of diffusion between the probe and the surface increases, forcing the crystal/solution interface to become more undersaturated. Conversely, as the UME is retracted from the substrate, the interfacial undersaturation approaches the saturated value, since the solution mass transfer coefficient decreases compared to the first-order dissolution rate constant. Movement of the tip electrode away from the substrate also has the effect of promoting radial diffusion, and consequently the area of the substrate probed by the UME increases.

It follows from the above discussion that the greatest spatial and kinetic resolutions are achieved at the closest tip-substrate separations. Moreover, Figure 14 demonstrates that a wide range of rate constants are accessible to characterization, for example, when $L = 0.1$, log K_1 values in the range -0.2 to 2.8 should be determinable. Additionally, Eq. (29) indicates that this range of K_1 values depends on the activity of Cu^{2+} in the saturated solution and the values of a and $D_{Cu^{2+}}$. Each of these parameters can be tuned to some extent (particularly the value of a) in order to enhance the sensitivity of the technique when studying a particular system.

C. Applications

1. Effect of Dislocation Density on Dissolution Characteristics

The ability to induce dissolution for brief periods in the chronoamperometric mode has enabled the first comprehensive studies of dissolution from various regions of the (100) face of copper sulfate pentahydrate single crystals, grown from aqueous solution (2–4). Dissolution was initiated through the chronoamperometric reduction of Cu^{2+} to Cu at a 25 μm diameter Pt disc UME, over selected spots on the crystal surface, as a function of tip-substrate separation. Dissolution was induced for only fractions of a second, in order to prevent the accumulation of Cu at the tip electrode and to avoid extensive dissolution of the solid, which might otherwise change the geometry of the tip-substrate gap and the integrity of the exposed surface. Experiments were carried out with solutions containing a range of sulfuric acid concentrations

(2.8–10.3 mol dm^{-3}), which served both to suppress migration effects and to buffer the sulfate concentration during the course of a measurement.

In studies of the (100) face, two distinct types of dissolution behavior were observed. In the center of the face, dislocation etching revealed that the average dislocation spacing was considerably less than the size of the UME probe (50), i.e., in the SECM configuration the electrode would sit over a constant source of dissolution sites once the process had been initiated. This was reflected in the shape of experimental SECM chronoamperometric transients (3), such as those depicted in Figure 17 at different tip-substrate separations. In all cases, the current rapidly attained a steady state, indicative of a uniform, constant rate of dissolution. As the UME was moved closer to the crystal surface, the limiting current increased, consistent with an increase in the flux of Cu^{2+} from the crystal surface to the tip electrode.

Steady-state current-distance approach curves, derived from the long-time portion of transients, are shown in Figure 18 for dissolution experiments involving saturated copper sulfate solutions with sulfuric acid at concentrations of 10.2, 7.3, 6.4, 3.6, and 2.8 mol dm^{-3}. As the concentration of sulfuric acid decreased, the dissolution characteristics moved away from a position of diffusion-control (dashed line in Fig. 18). This was due to the

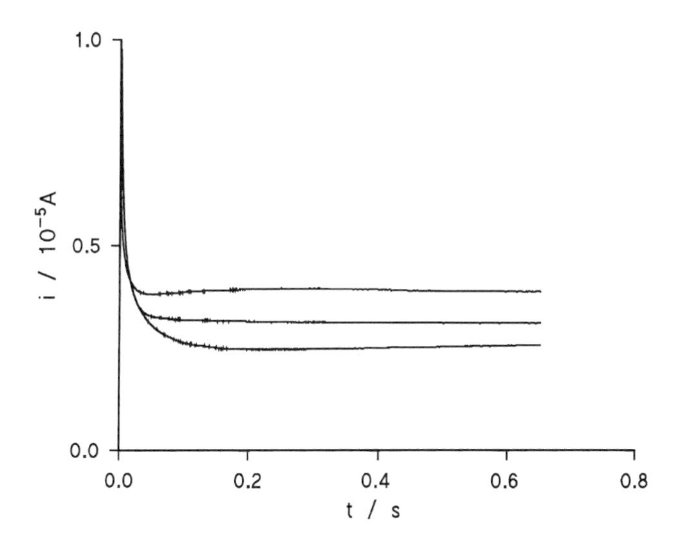

FIG. 17 SECM chronoamperometric transients for the reduction of Cu^{2+} with a 25 μm diameter Pt UME positioned at distances of 2 (upper curve), 5, 10 μm (lowest curve) from the (100) face of a copper sulfate pentahydrate single crystal. The initial solution conditions were 3.6 mol dm^{-3} H_2SO_4 and saturated copper sulfate. In each case the potential was stepped from −0.2 to −0.65 V versus AgQRE.

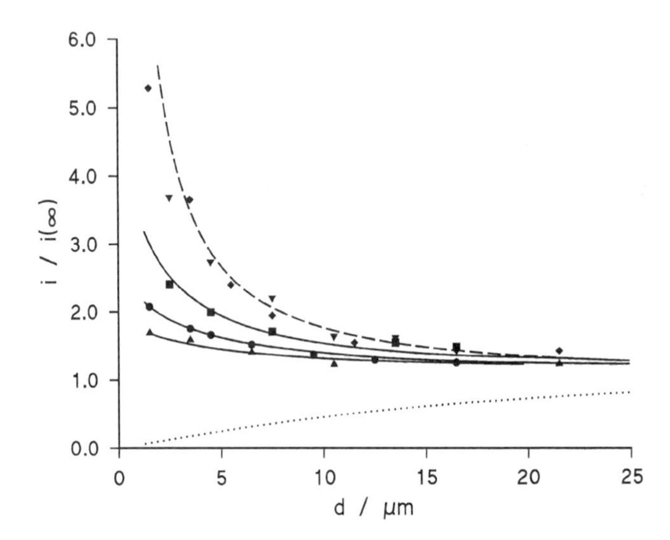

FIG. 18 SECM approach curves (measurement time 0.6 s after a potential step) of normalized current for the reduction of Cu^{2+} versus separation between a 25 μm diameter Pt tip and the (100) face of $CuSO_4 \cdot 5H_2O$. Data are shown for saturated copper sulfate solutions containing 2.3 (▲), 3.6 (●), 6.4 (■), 7.3 (▼), and 10.2 (♦) mol dm^{-3} sulfuric acid. The solid lines through each data set represent the best fits for a first-order dissolution process characterized by log K_1 = 0.46 (▲), log K_1 = 0.60 (●), and log K_1 = 0.91 (■). The dashed line shows the theoretical behavior predicted for a diffusion-controlled dissolution process, while the dotted line shows the behavior expected for an inert surface.

increase in the activity and diffusion coefficient of Cu^{2+} in solution (reducing K_1 [(Eq. 29)], which increased mass transfer in the UME/crystal domain and pushed the dissolution reaction towards surface control.

An excellent fit to the experimental data was found assuming a first-order process (Fig. 18). This is further illustrated by Figure 19 for a saturated copper sulfate solution containing 3.6 mol dm^{-3} sulfuric acid, which clearly shows that discrimination between $n = 1$ and 2 is possible. The dissolution flux from this region of the surface was described by the following rate law:

$$j_{Cu^{2+}}/\text{mol cm}^{-2}\ \text{s}^{-1} = 2.0 \times 10^{-6}\sigma \qquad (30)$$

suggesting mechanistically that surface diffusion was not kinetically significant (48). Furthermore, it provided indirect evidence for the operation of the BCF mechanism under conditions where dissolution occurred in an environment that was nonstoichiometric with respect to the concentrations of Cu^{2+} and SO_4^{2-} ions in solution.

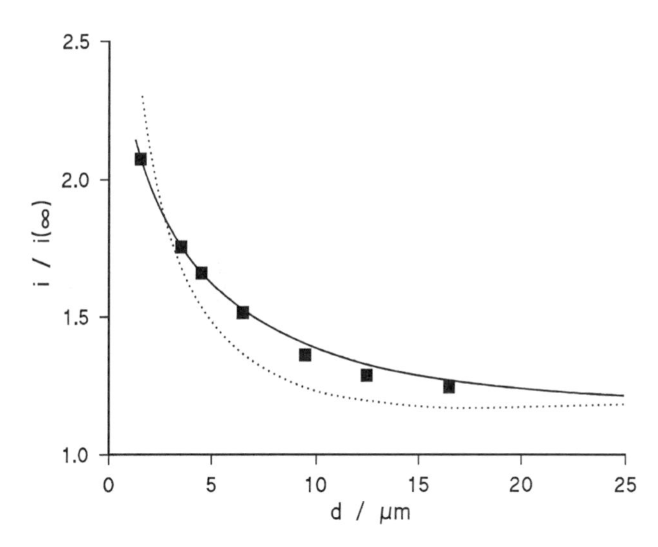

FIG. 19 Comparison of the best fits for a first-order (——) and second-order (– – –) dissolution process, for the rate constants log K_1 = 0.60 and log K_2 = 1.08, with the experimental current-distance data (■) obtained with a 25 μm diameter Pt UME in a saturated copper sulfate solution containing 3.6 mol dm^{-3} sulfuric acid.

A differential interference contrast (DIC) micrograph of a typical pit resulting from SECM-induced dissolution, with an initial solution containing 3.6 mol dm^{-3} sulfuric acid, is shown in Figure 20a. In this case, the probe was positioned about 2 μm from the crystal surface and dissolution was induced via potential step chronoamperometry for a period of 0.6 s. It is interesting to compare the form of the pit with the simulated steady-state interfacial flux and concentration profiles, for the conditions of the experiment. As shown in Figure 20b, the profiles indicate that there is an essentially constant zone of interfacial undersaturation on the area of the crystal surface directly under the tip electrode. This zone is approximately the same size of the central flat portion of the pit in Figure 20a, confirming that the experimental and theoretical results correlate well.

Further out in the radial direction ($1 < R < 2$), the interfacial concentration rises (undersaturation falls) steeply, while the flux decreases, again consistent with the shallow inclining walls of the pit in Figure 20a. For $R > 2$ the calculations indicate that the solution is only slightly undersaturated with respect to Cu^{2+}. It is clear from Figure 20a that this is still sufficient to induce dissolution, as the microtopography in this region shows many closely spaced elongated etch pits. This confirms that the SECM measurements relate to regions of the (100) surface with a high dislocation density.

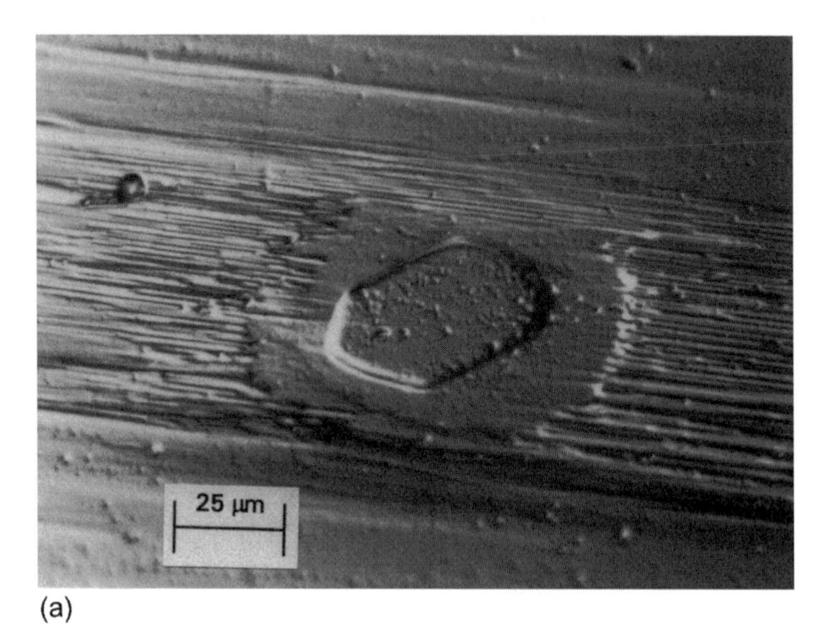

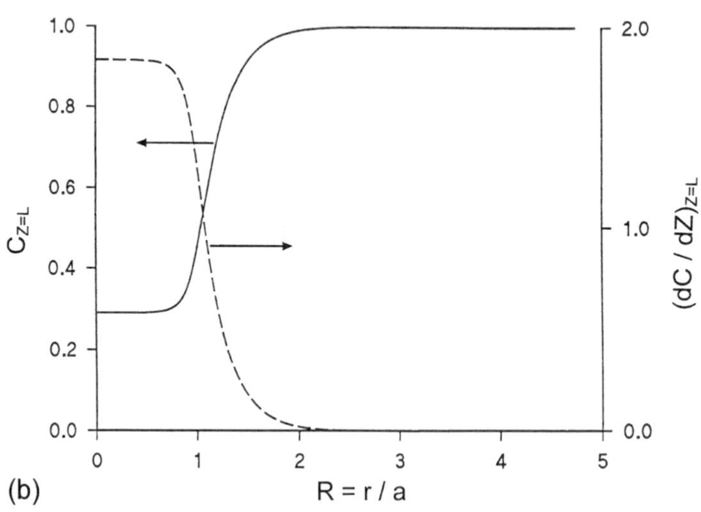

FIG. 20 (a) Differential interference contrast (DIC) micrograph of a typical SECM-induced dissolution pit in the center of the (100) face of copper sulfate pentahydrate. The 25 μm diameter Pt UME was located about 2 μm from the surface and dissolution was effected through SECM diffusion-controlled chronoamperometry for a period of 0.6 s. (b) Calculated normalized concentration (——) and flux (–––) at the crystal/solution boundary for conditions pertaining to those for the formation of the SECM-induced dissolution pit shown in (a). A value of log K_1 = 0.60, appropriate to this case, was used in the simulations.

At the edges of the (100) face of copper sulfate pentahydrate single crystals, etching revealed a mean interdislocation spacing in excess of 100 μm, i.e., much greater than the size of the UME employed (50). This resulted in a high probability of inducing dissolution in an area effectively devoid of dislocation cores with a UME positioned above this region of the surface. Coupled with the above studies, this provided an opportunity to compare the dissolution activity in dislocated and dislocation-free areas of the same crystal surface, so allowing an assessment of the role of dislocations in controlling the rate and mechanism of the reaction.

Figure 21 shows a typical example of the chronoamperometric response observed when dissolution is initiated in a region towards the edge of the crystal face. In stark contrast to the behavior described above (Fig. 17), SECM-induced dissolution, in the absence of dislocations, occurs via an oscillatory mechanism (4). After an initial period of approximately one unit

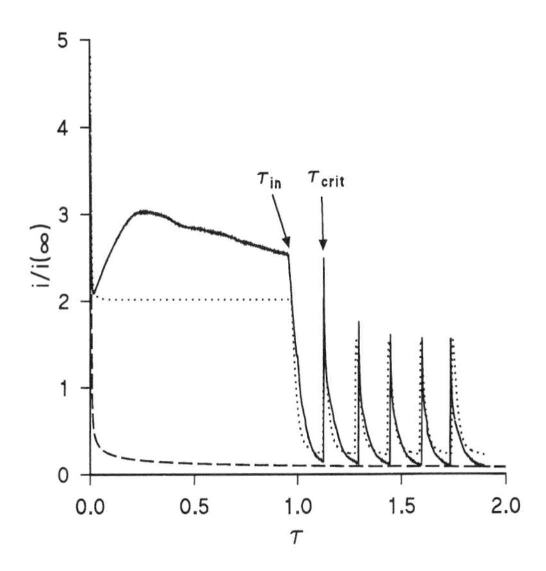

FIG. 21 Experimental current-time characteristics (——) for the reduction of Cu^{2+} at a 25 μm diameter Pt UME located 1.0 μm from a copper sulfate pentahydrate (100) face in an area of the crystal with a low dislocation density. The experimental characteristics have been normalized using the steady-state current measured at effectively infinite probe-crystal separation, $i(\infty) = 1.88\ \mu$A. $D_{Cu^{2+}} = 4.56 \times 10^{-6}$ cm^2 s^{-1}, and $a = 12.5\ \mu$m. The best fit of the theoretical model (••••••) to the experimental data was obtained with the following parameters: $\tau_{in} = 0.967$, $K = 3.5$, $C_{crit} = 1.25 \times 10^{-6}$, and $N = 12.5 \times 10^{-9}$ mol cm^{-2} (50). The theoretical behavior predicted for an inert surface (–––) is shown for comparison.

of normalized time (~0.3 s) of high dissolution activity, the rate—reflected by the UME current—rapidly decreases and thereafter proceeds in a series of periodic bursts, each lasting ~0.05 s.

The initial dissolution activity (at the shortest times) was attributed to dissolution from steps that occur in abundance on the surfaces of crystals grown from solution. Dissolution causes the steps to retreat out of the microscopic zone probed by the SECM. In the absence of dislocation cores in the targeted region, it has been postulated that the surface becomes depleted of active sites (4,23,50), as shown schematically in Figure 22. The area probed is essentially wiped clean, leaving an almost perfect surface below. This can be thought of as "electrochemically induced sputtering and annealing" of the crystal surface. Consequently, the dissolution rate decreases to a value indicative of zero dissolution activity. This is reflected in a drop

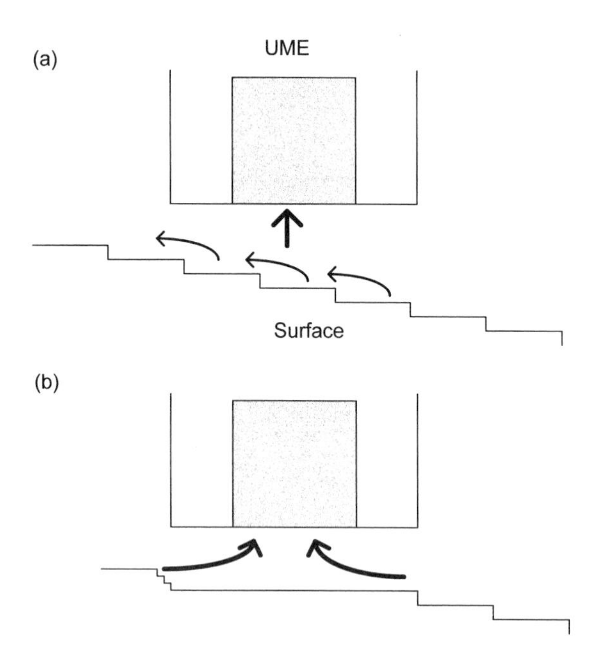

FIG. 22 Schematic illustration of SECM-induced dissolution. Steps initially provide a source of sites for dissolution, which is detected at the UME (a). As dissolution proceeds, steps move out of the probe/crystal domain and the activity of the surface decreases. In this case (b) mass transfer tends towards that for an inert surface (hindered diffusion). Note that this diagram is not to scale. The height of and separation between the steps has been significantly exaggerated. The surface is effectively flat on the scale of the probe/crystal geometry: the probe-crystal separation is ~1 μm, the UME diameter is 25 μm, and the overall probe diameter is ~250 μm.

in the current flowing at the tip electrode, at the time marked τ_{in} on Figure 21, to a value that corresponds to mass transfer by hindered diffusion only in the tip/substrate gap (28).

Dissolution in the initial period, $0 \rightarrow \tau_{crit}$, was modeled successfully (4,23,50) in terms of the following rate law, which describes the flux of Cu^{2+} from the crystal surface:

$$Z = L, 0 \leq R \leq RG: \qquad \frac{\partial C_{\mathrm{Cu}^{2+}}}{\partial Z} = -K_1[1 - C_{\mathrm{Cu}^{2+}}^{1/2}]\theta \tag{31}$$

coupled with the following conditions on the fraction of active dissolution sites, θ:

$$0 < \tau < \tau_{in}: \qquad \theta = 1 \tag{32}$$

$$\tau > \tau_{in}: \mathrm{C}_{Z=L,0\leq R\leq RG} > C_{crit}: \qquad d\theta/d\tau = (K_1/\lambda)[C^{1/2} - 1]\theta \tag{33}$$

C_{crit} is a critical concentration (see later) below which active sites can be nucleated on the crystal surface and

$$\lambda = N/a_{\mathrm{Cu}^{2+}}^{sat} a \tag{34}$$

where N is the effective surface density of Cu^{2+} ions involved in dissolution during this period (related to the number of layers of copper sulfate pentahydrate removed in each burst of dissolution activity).

The decrease in crystal activity at τ_{in}, coupled with strongly hindered diffusion in solution in the SECM configuration, results in a rapid depletion in the local concentration of electroactive material in the tip-substrate gap and crucially at the crystal/solution interface. Eventually a value for the interfacial undersaturation is reached that exceeds the threshold required to spontaneously nucleate fresh dissolution sites from the perfect surface and dissolution is initiated again at τ_{crit}. This is reflected in the sudden surge of current. To model this simply, the entire surface was allowed to become active once the undersaturation at any point on the surface exceeded a critical value (50). Thus, the following condition applied, in conjunction with Eqs. (31) and (33):

$$C_{Z=L,0\leq R\leq RG} < C_{crit}: \qquad \theta = 1 \tag{35}$$

As the freshly generated steps, associated with nucleation sites, traverse from the tip-crystal domain, as a consequence of dissolution, the activity of the surface decreases and the current falls accordingly. The process outlined above is thus repeated, leading to the observed oscillatory current response shown in Figure 21. This sequence of events is illustrated schematically in Figure 23.

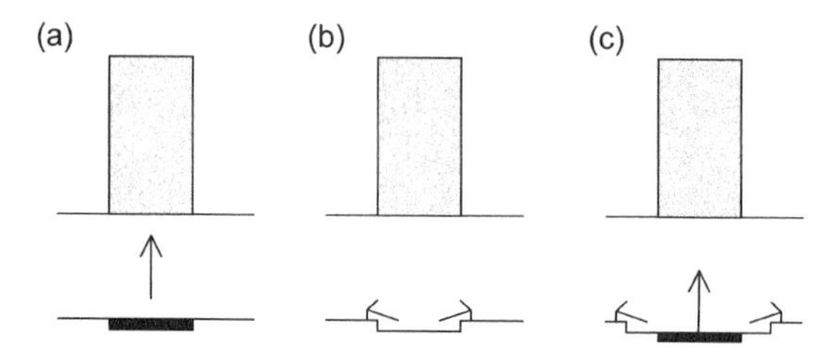

FIG. 23 Schematic illustration of the formation of an SECM-induced dissolution pit with terraced walls. The UME induces dissolution from the area directly under it (a), resulting in a current surge. The flux of material from this area then rapidly declines and slow dissolution occurs from the edge of the resulting pit (b), resulting in expansion. Concurrently, the area directly under the center of the UME becomes increasingly undersaturated, until the critical value for the creation of fresh dissolution sites is attained, when there is a subsequent burst of material from the surface (c), producing a terraced structure within the pit. The overall process may then be repeated.

The theoretical model developed to take account of these factors, Eqs. (31)–(35), is consistent with the experimental data presented in Figure 21. This model indicates that ~20 lattice layers of the crystal surface are removed in each current surge [given the deduced value of $N = 12.5 \times 10^{-9}$ mol cm^{-2}, and the density of Cu^{2+} in the (100) surface of $\sim 5.3 \times 10^{-10}$ mol cm^{-2} (51)]. The value of $C_{crit} = 1.25 \times 10^{-6}$, is some five orders of magnitude smaller than the value typically required for the nucleation of observable dissolution etch pits at dislocation sites (3,52,53). Intuitively, this indicates that the measured value of C_{crit} is consistent with dissolution from a dislocation-free area.

Further evidence for the validity of the proposed dissolution model was revealed through microscopy of the crystal surface in areas of SECM measurements. Figure 24 is a DIC light micrograph of the dissolution pit corresponding to the transient in Figure 21. A series of approximately five ledges are visible inside the pit, consistent with the number of current surges in Figure 21. The formation of ledges of this type concurs with dissolution occurring in bursts, in a layer-by-layer fashion, as shown schematically in Figure 23. The morphology of the pit in Figure 24 contrasts markedly with that due to induced dissolution in the central region of the (100) face (Fig. 20a). As discussed above, dissolution in the latter case resulted in a central pit, with additional significant pitting of the surrounding surface, diagnostic of a high dislocation density (3). In contrast, the lack of any pitting, or roughening, of the surface around the SECM induced pit in Figure 24 is consistent with a lack of dislocations in the area of the SECM measurement.

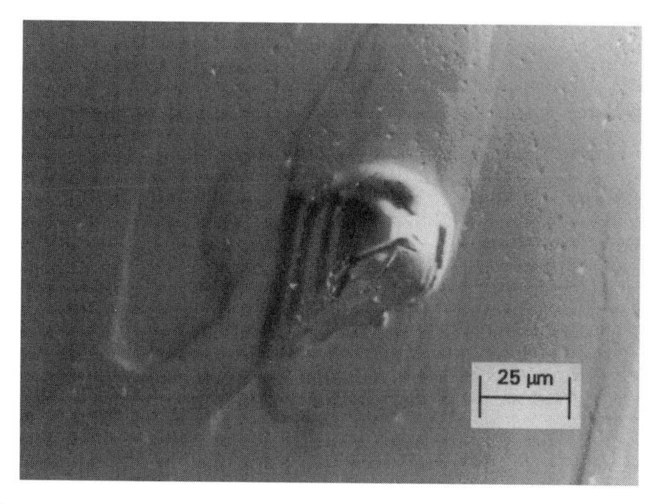

FIG. 24 DIC micrograph of the SECM-induced dissolution pit corresponding to the current-time characteristics in Figure 21.

2. *Dissolution Rate Imaging*

Dissolution from the (010) face of monoclinic potassium ferrocyanide trihydrate in solutions containing 3.5 mol dm^{-3} potassium chloride saturated with respect to potassium ferrocyanide trihydrate ($[Fe(CN)_6^{4-}]$ = 0.133 mol dm^{-3}) was investigated using SECM chronoamperometry at Pt disc UMEs (5). The reaction was induced by stepping the potential of the tip, positioned close to the crystal surface, to a value sufficient to oxidize $Fe(CN)_6^{4-}$ to $Fe(CN)_6^{3-}$ at a diffusion-controlled rate. Dissolution was induced for millisecond time periods only in order to prevent the significant build up of the tip electrolysis product, $Fe(CN)_6^{3-}$, in the electrode-crystal gap, as well as to avoid extensive dissolution of the solid, which might otherwise change the geometry of the tip-substrate gap. Steady-state current versus tip-substrate separation (approach) curves were obtained as a function of electrode radius and are shown in Figure 25.

As the size of the electrode was decreased from 25 to 5 μm in diameter, the dissolution reaction was shifted from a position where the initial rate was predominantly diffusion-controlled to a time scale where surface-kinetic limitations became apparent. The procedure for analyzing the rate data was similar to that described above, with boundary conditions that reflected first- and second-order dissolution rate laws under the defined conditions. The dissolution fluxes were found to be governed by a second-order dependence on the interfacial undersaturation:

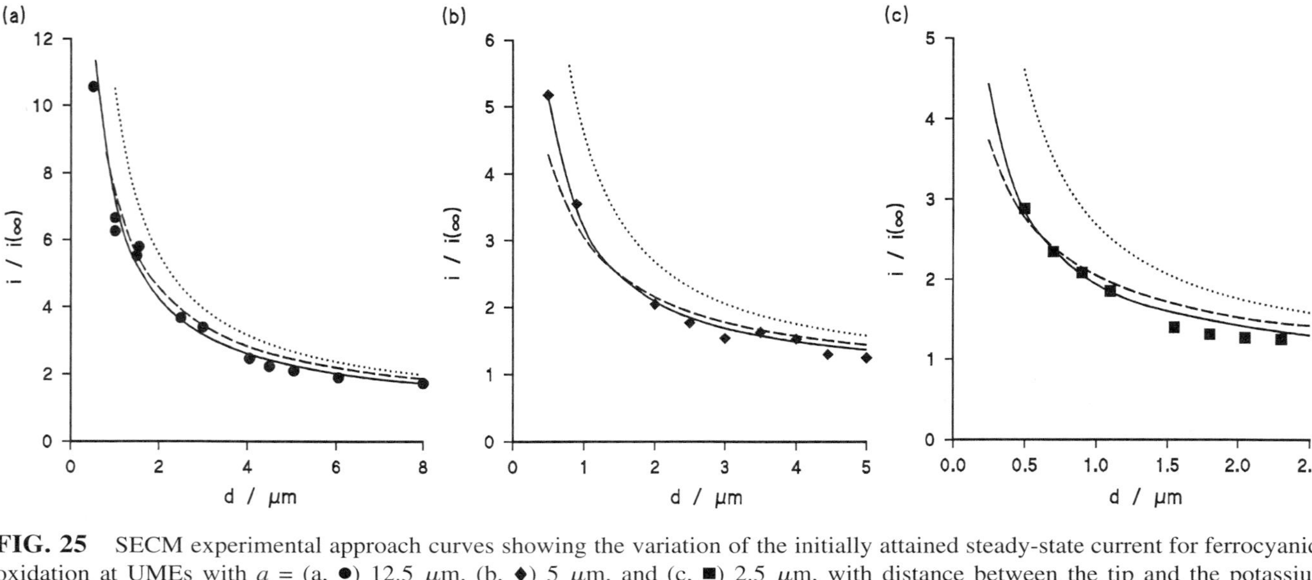

FIG. 25 SECM experimental approach curves showing the variation of the initially attained steady-state current for ferrocyanide oxidation at UMEs with a = (a, ●) 12.5 μm, (b, ◆) 5 μm, and (c, ■) 2.5 μm, with distance between the tip and the potassium ferrocyanide trihydrate (010) surface. The data were derived from chronoamperometric measurements. For comparison, the theoretical behavior for a diffusion-controlled dissolution process (•••••) is also shown along with the best fits to the experimental data for first- and second-order dissolution models, characterized by (a) log K_1 = 2.1 (–––) and log K_2 = 3.2 (——), (b) log K_1 = 1.54 (–––) and log K_2 = 2.77 (——), and (c) log K_1 = 1.35 (–––) and log K_2 = 2.51 (——).

$$j_{Fe(CN)_6^{4-}} = k_2' a_{Fe(CN)_6^{4-}} \sigma^2 \tag{36}$$

and $k_2' = 6.9\ (\pm 0.3)\ \text{cm s}^{-1}$. This provided supporting evidence for the operation of the BCF spiral dissolution model at low interfacial undersaturation (48) in an environment that was nonstoichiometric with respect to the concentrations of lattice ions. Mechanistically, this pointed to surface diffusion as the rate-limiting step in the dissolution process.

Given that the product of the tip reaction, $Fe(CN)_6^{3-}$, did not deposit on the UME surface (unlike the Cu^{2+}/Cu procedure discussed above), this system represented a useful model for testing the abilities of SECM to image the dissolution activity over selected areas of a crystal surface. For these experiments, the current for the diffusion-limited oxidation of $Fe(CN)_6^{4-}$ was recorded as a function of position in the x, y plane, as the probe UME was scanned over the surface at a constant height. The solution conditions were as defined above. Problems that might have arisen from the accumulation of the electrolysis product, $Fe(CN)_6^{3-}$, in the tip-substrate gap, with the possible deposition of potassium ferricyanide, were largely alleviated by careful choice of a relatively fast tip scan speed and moderate initial UME/crystal separation.

A typical dissolution rate image is shown in Figure 26a of a pit that was electrochemically pre-etched in the surface, using a 50 μm diameter Pt UME. A tip electrode (a = 5 μm) was scanned at 50 $\mu\text{m s}^{-1}$ at an initial height of 5.3 μm over a square region 300 μm $\times$ 300 μm. Over the area surrounding the pit, the currents were relatively constant and had values of ~140–150 nA, consistent with the initially attained steady-state currents derived from chronoamperometric kinetic measurements under similar conditions. This confirmed that the imaging experiment monitored the dissolution activity of the crystal surface, while the uniformity of the current with tip position, in the area of the crystal surrounding the pit, indicated that this part of the surface was of constant activity with respect to dissolution.

As the probe encountered the edge of the pit, there was a slight increase in the current, corresponding to an increase in the dissolution rate, from which it was deduced that macroscopic edges dissolve more rapidly than a planar surface (5). Similar (qualitative) effects have been seen in ex situ microscopy studies on powders, which have been subjected to partial dissolution, where corners and edges show enhanced reactivity (54). Once the UME was above the base of the pit, the current fell due to the increase in the tip-crystal separation and thus a lower flux of $Fe(CN)_6^{4-}$ ions from the substrate to the tip.

A DIC optical micrograph of the crystal surface in the region of the SECM image is shown in Figure 26b. Although there is some precipitated material on the surface, which resulted from evaporation when the crystal

(a)

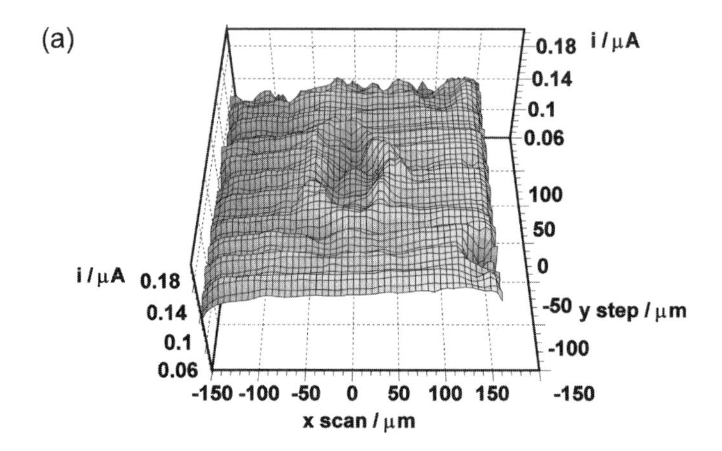

(b)

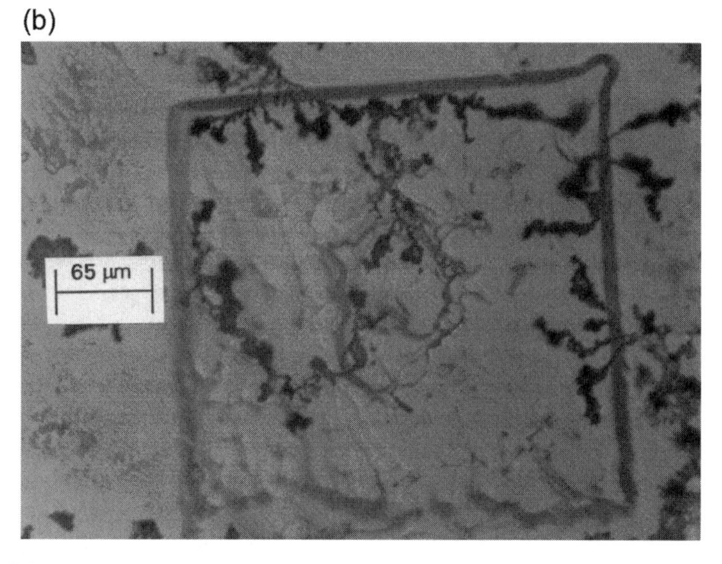

FIG. 26 (a) Dissolution rate image of a single pit on the (010) surface of potassium ferrocyanide trihydrate. The 300 μm × 300 μm image was recorded by scanning a probe UME (a = 5.0 μm) at a height of 5.3 μm above the crystal surface and recording the diffusion-limited current for the oxidation of ferrocyanide as a function of tip position. The step size between line scans employed in the raster scan was 5 μm. (b) DIC micrograph of the area of the crystal surface in which part (a) was recorded. An etched square of 300 μm length, resulting from induced dissolution during the imaging process, is clearly evident.

was removed from the saturated solution, the image clearly shows a square etched region, where the crystal surface was dissolved during the course of the scan. The etched region is the same size as the scan employed to record the dissolution rate image in Figure 26a. The outline of the pit imaged in Figure 26a is clearly evident in the center of the scanned region in Figure 26b. Comparison of Figures 26a and b establishes that the location and dimensions of the pit, measured by dissolution rate imaging and optical microscopy are well correlated.

Chapter 4 describes how SECM can be employed as a tool for imaging surface topography. This facet of the technique can be exploited to obtain SECM images of both dissolution activity and topography, thereby avoiding the need to make ex situ topographical characterizations. For the system discussed here, a small concentration (2×10^{-3} mol dm^{-3}) of potassium ferricyanide was added to a saturated potassium ferrocyanide trihydrate (3.5 mol dm^{-3} potassium chloride) solution. The former acted as a calibrant of the tip-crystal separation, since the current depended only on the hindered diffusion of $Fe(CN)_6^{3-}$ (28). Thus, for imaging experiments, the current flowing for the diffusion-controlled reduction of $Fe(CN)_6^{3-}$ in an initial scan, provided topographical information. In a second scan, the oxidation of $Fe(CN)_6^{4-}$ probed the dissolution activity.

Typical results from this type of experiment are displayed in Figures 27 (topography) and 28 (dissolution reactivity). To emphasize the topographical imaging capabilities of SECM, the crystal surface was deliberately misorientated by $\sim 1°$, and a 50 μm diameter pit was electrochemically pre-etched into the surface. Figure 27a represents the raw steady-state $i/i(\infty)$ data for the reduction of $Fe(CN)_6^{3-}$ at the tip UME, as it is scanned over the substrate in the x, y plane, reflecting changes in the topography of the surface. The current ratio can be related to changes in the tip-substrate distance by (55):

$$i/i(\infty) = [0.292 + 1.515/L + 0.655 \exp(-2.4035/L)]^{-1} \tag{37}$$

Although Eq. (37) strictly applies only for a planar surface, it can provide semiquantitative information on the topography of the crystal surface. Transformation of the current data displayed in Figure 27a to a topography map, using this equation, is shown in Figure 27b. The technique is clearly able to pick out the key features of the surface, namely the slope and the pit. Figure 28 is the corresponding dissolution activity image for the area shown in Figure 27. The current ratio over the entire area scanned is greater than unity due to the induced dissolution process.

3. *Effect of Background Electrolyte*

The solid materials considered above were characterized by very high solubilities, and dissolution was considered under conditions with a supporting

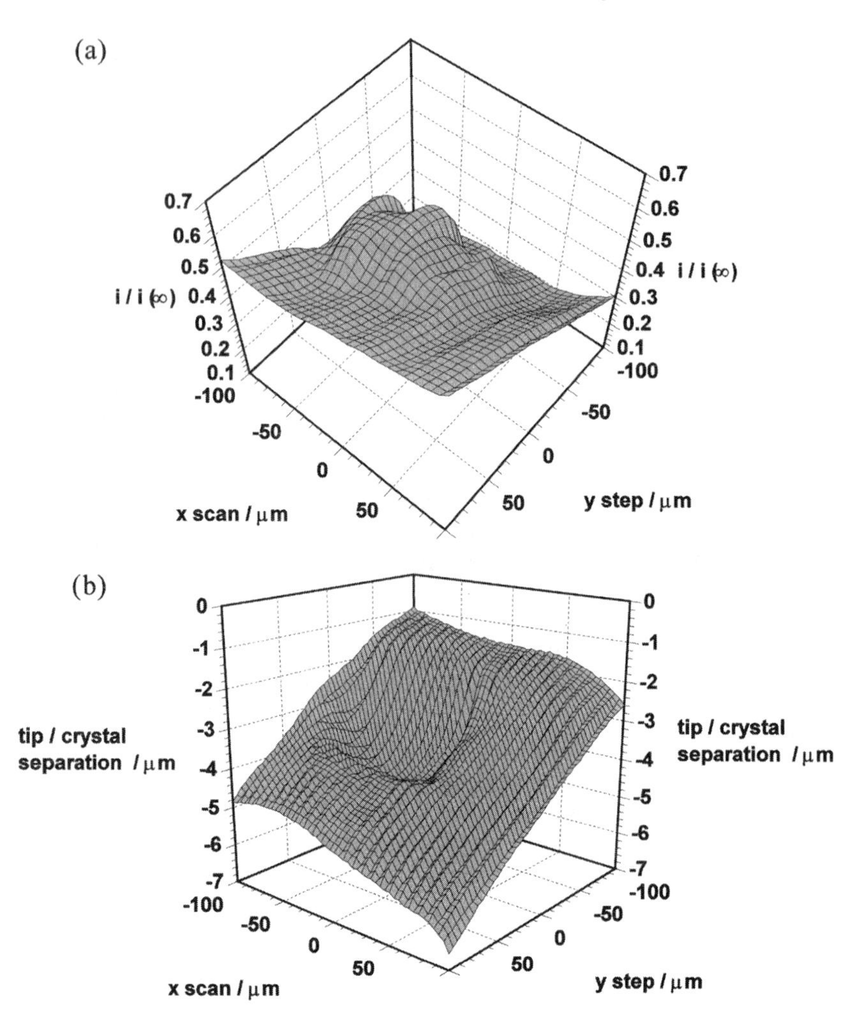

FIG. 27 SECM topographical image of a pit on the (010) surface of potassium ferrocyanide trihydrate, which was misorientated by $\sim 1°$ with respect to the base of the cell. The image is displayed in terms of (a) $i/i(\infty)$ versus tip position in the x,y plane, for the diffusion-limited reduction of ferricyanide, and (b) tip to crystal separation, d, as a function of tip position in the x,y plane, obtained by transforming the current data using Eq. (37).

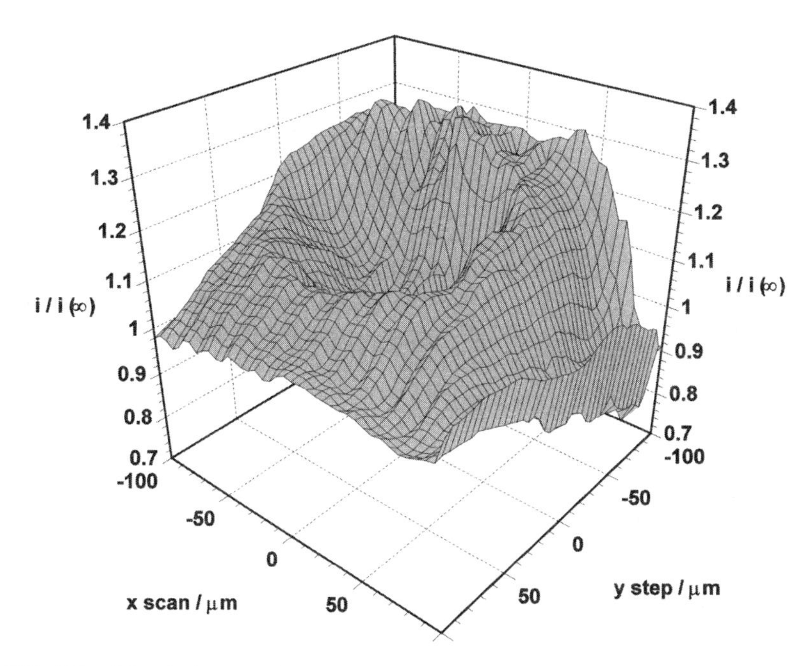

FIG. 28 Dissolution rate image of the (010) surface of potassium ferrocyanide trihydrate recorded in the same area of the crystal as the topographic image shown in Figure 27. The tip was held at a potential to establish the diffusion-controlled oxidation of ferrocyanide and scanned at a speed of 50 $\mu m\ s^{-1}$.

electrolyte present, which contained a common ion with the substrate of interest. For SECM-induced dissolution of sparingly soluble ionic materials, it is not always possible to select a supporting electrolyte of this type, as this might reduce the saturated concentration of the electroactive ion to a level where UME currents are impractically low. More generally, it is not always possible to select an appropriate background electrolyte to buffer the concentration of the counterion during the course of an experiment. The effects of (1) a supporting electrolyte that does not contain a common ion with the dissolving material (6) and (2) no added supporting electrolyte (7) have thus been considered, with the aim of significantly diversifying the range of dissolution systems open to study with SECM.

To examine these effects, investigations were carried out on the dissolution of electrochemically grown films and pressed pellets of silver chloride in 0.1 mol dm^{-3} aqueous potassium nitrate and in water (6,7). Silver chloride was an ideal system since the kinetics and mechanisms controlling dissolution were unresolved by conventional approaches, despite a number of studies. Dissolution was initiated using SECM chronoamperometry, at 50, 25, and 10 μm diameter Pt disc UMEs, by stepping the potential of a tip, positioned close to an AgCl surface, to a value where the reduction of Ag^+ occurred at a diffusion-controlled rate.

In the presence of 0.1 mol dm^{-3} potassium nitrate solution, electrochemical depletion of Ag^+ in the tip-substrate gap induces the dissolution of AgCl. As Cl^- is electroinactive at the tip potential of interest, its concentration increases in the tip-substrate domain during the period of the chronoamperometric measurement. This results in dissolution into an environment, which is spatially and temporally nonstoichiometric with respect to the lattice ions of the solid material. Theoretically, the dissolution behavior is now a two species problem, which is described by the following tip electrode and substrate boundary conditions (6):

$$\tau > 0,\ Z = 0,\ 0 \leq R \leq 1: \qquad C_{Ag^+} = 0,\ \frac{\partial C_{Cl^-}}{\partial Z} = 0 \tag{38}$$

$$\tau > 0,\ Z = 0,\ 1 \leq R \leq RG: \qquad \frac{\partial C_{Ag^+}}{\partial Z} = 0,\ \frac{\partial C_{Cl^-}}{\partial Z} = 0 \tag{39}$$

$$\tau > 0,\ Z = L,\ 0 \leq R \leq RG: \qquad \frac{\partial C_i}{\partial Z} = -\frac{K_n}{\gamma_i}\left[1 - (C_{Ag^+} C_{Cl^-})^{1/2}\right]^n \tag{40}$$

where i denotes Ag^+ or Cl^- and the parameter

$$\gamma_i = \frac{D_i}{D_{Ag^+}} \tag{41}$$

reflects the fact that, under the conditions of the experiment, Ag^+ and Cl^- have different diffusion coefficients. Values of $n = 1$ and 2 were considered, for the reasons outlined in Sec. III.B.

The concentration profiles displayed in Figure 29 demonstrate the buildup of Cl^- in the tip-substrate gap, as a function of time, for a first-order dissolution process characterized by $K_1 = 100$ and $L = 0.16$ (approaching diffusion-controlled conditions). As the Cl^- levels increase, there is a concomitant decrease in the concentration of Ag^+ to prevent the solution from exceeding the saturated value. Ultimately, this leads to a decrease in the flux of Ag^+ and Cl^- from the interface, which is reflected in a diminution in the current flowing at the tip electrode towards long times. The overall effect of this process is to suppress the attainment of high interfacial undersaturations, as shown by Figure 30. This is evident even for dissolution reactions characterized by fairly slow kinetics. For example when $K_1 = 1$ (Fig. 30iii), there is only a small undersaturation at the substrate/solution interface in the

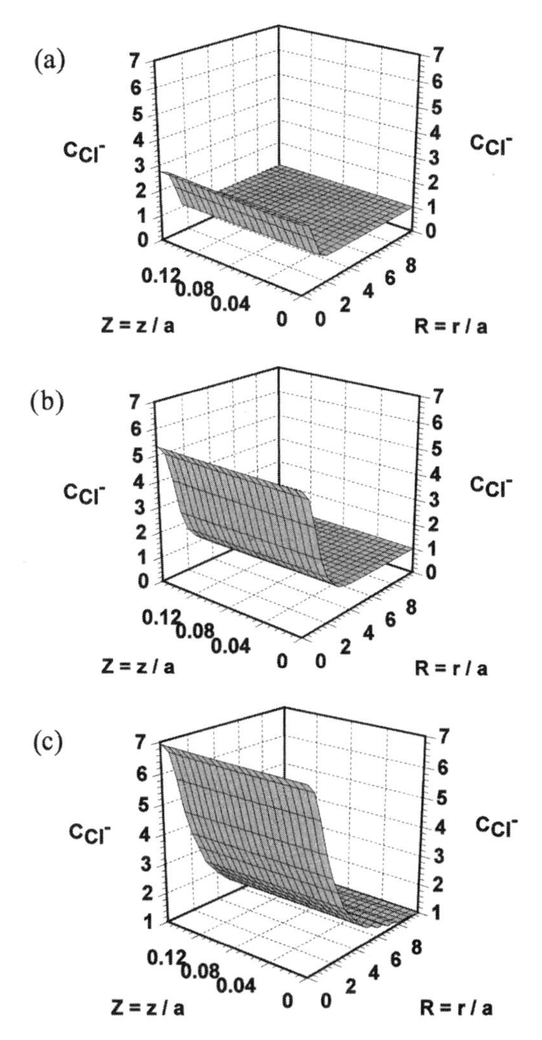

FIG. 29 Chloride ion concentration profiles in the tip/substrate gap at times, τ, of (a) 0.1, (b) 1.0, and (c) 10.0 during a first-order SECM-induced dissolution process, characterized by $K_1 = 100$ at a tip/substrate separation, $L = 0.16$, where the supporting electrolyte does not contain a common ion with the dissolving material.

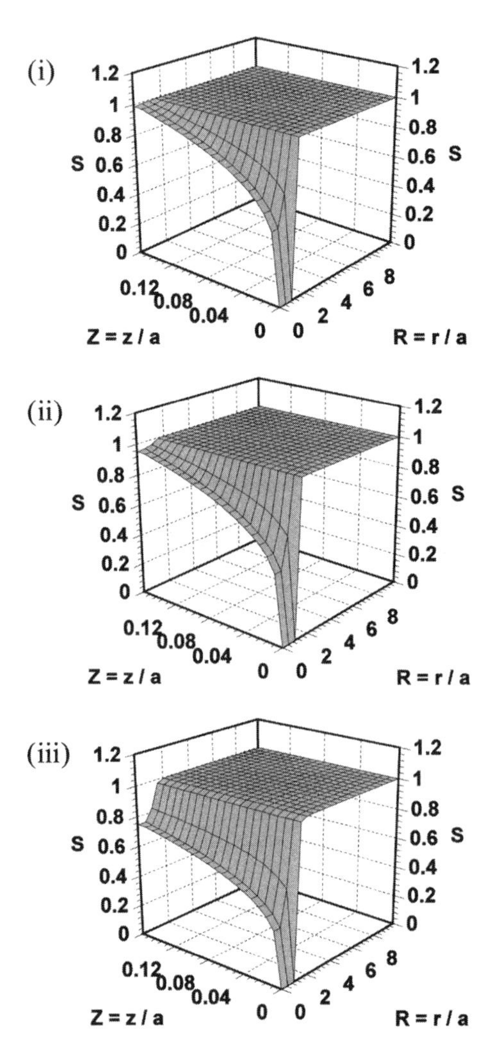

FIG. 30 Steady-state profiles of the saturation ratio in the tip-substrate gap ($L = 0.16$) for a first-order induced dissolution process, characterized by K_1 = (i) 100, (ii) 10, and (iii) 1. The data are for an induced dissolution reaction in the presence of a supporting electrolyte, which does not contain a common ion with the dissolving material.

zone directly under the tip, and consequently the dissolution kinetics are merely slightly surface-limited.

Approach curves of long-time normalized current for the diffusion-controlled reduction of Ag^+ versus tip-substrate separation, with pressed pellets of AgCl as the substrate, are shown in Figure 31. The best fit of the experimental data to the theory outlined above, for these conditions, was obtained for a diffusion-controlled dissolution process. Also shown in Figure 31 is the theory for a diffusion-controlled dissolution process if the concentration of the electroinactive Cl^- ion were to remain constant—i.e., the conditions considered earlier for $CuSO_4 \cdot 5H_2O$ dissolution. Comparison of the responses clearly demonstrates that the accumulation of Cl^-, which occurs during SECM-induced dissolution with potassium nitrate as the supporting electrolyte, has a dramatic effect on the current response. Similar results were obtained when electrochemically grown films of AgCl were used as the substrate.

The scanning mode of the SECM can also be used to provide kinetic information via dissolution rate imaging (5,6). In these particular experiments, the tip was scanned at a series of constant heights above the surface of an Ag UME coated with a AgCl layer, while recording the diffusion-controlled current for the reduction of Ag^+ as a function of tip position. To avoid prolonged deposition of Ag on the UME surface, small-diameter AgCl substrates (discs of 125 μm) were employed. Typical dissolution rate images, recorded as a function of tip-substrate distance, with a tip scan speed of 10 $\mu m\ s^{-1}$, are presented in Figure 32. The position of the AgCl film in each image is clearly evident as the peak in the current when the probe passes over the film and initiates dissolution, which provides an additional source of Ag^+ for reduction at the tip. Taking account of the residence time of the tip over the AgCl disc, it was deduced that the peak current in each image tended to a steady-state value. Approach curves constructed from such dissolution activity images were in good agreement with those recorded from transient measurements (Fig. 31) again suggesting diffusion-controlled dissolution.

In the absence of supporting electrolyte, the electroneutrality principle demands that the local concentrations of the electroactive and electroinactive ions remain equal (56) in the tip-substrate gap when the concentration of the former is depleted by electrolysis at the tip UME. In the case of AgCl dissolution, the mass transport problem was shown to reduce to the consideration of a single species (7). Figure 33 shows steady-state profiles that illustrate the interfacial undersaturations, obtainable for a range of first-order dissolution rate constants, with no added supporting electrolyte. Although the saturation ratio at the substrate/solution interface is close to unity for $K_1 = 100$ (Fig. 33i), i.e., the dissolution kinetics are close to the diffusion-

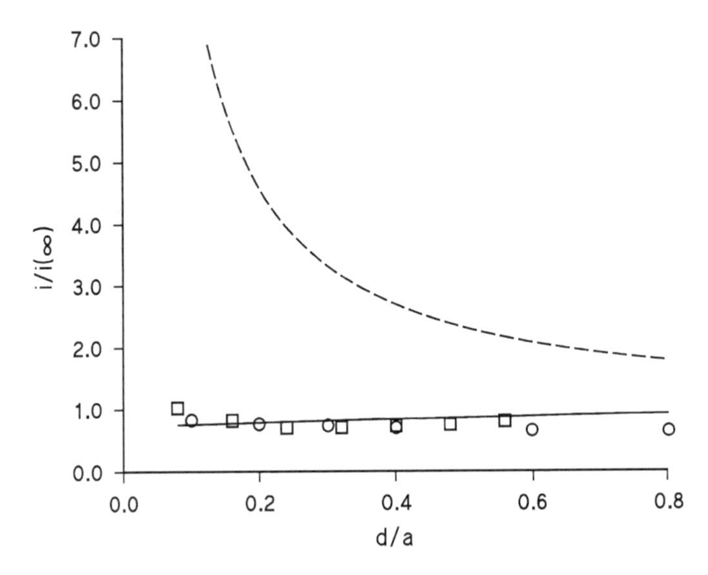

FIG. 31 Experimental approach curves of normalized long-time current for Ag^+ reduction versus tip-substrate separation measured over a AgCl pellet with tips characterized by a = 12.5 μm (□) and 5 μm (○). Also shown is the theoretical behavior for a diffusion-controlled process in the presence of a supporting electrolyte, which does not contain a common ion with the dissolving material (——), together with the characteristics for the situation where the concentration of the electroinactive dissolving counterion remains constant (---) during SECM measurement.

controlled limit, there is a moderate undersaturation at the portion of the substrate directly under the tip when K_1 = 10 (Fig. 33ii) and a large undersaturation in this zone when K_1 = 1 (Fig. 33iii). This contrasts markedly with the situation depicted in Figure 30. The absence of supporting electrolyte clearly favors the establishment of large interfacial undersaturations, compared to the case where the added supporting electrolyte does not contain a common ion with the dissolving material.

Dissolution measurements in the absence of supporting electrolyte were made on both electrochemically grown films and pressed pellets of AgCl, with similar results (7). Approach curves for tips characterized by a = 5, 12.5, and 25 μm, constructed by plotting the normalized long-time currents, from chronoamperometric measurements, as a function of normalized tip-substrate distance are shown in Figure 34a and b. The curves cover different values of L since, in each case, measurements were made over a range of distances up to 2 μm from the substrate surface. For comparison with the experimental data, theoretical approach curves are also shown for second-

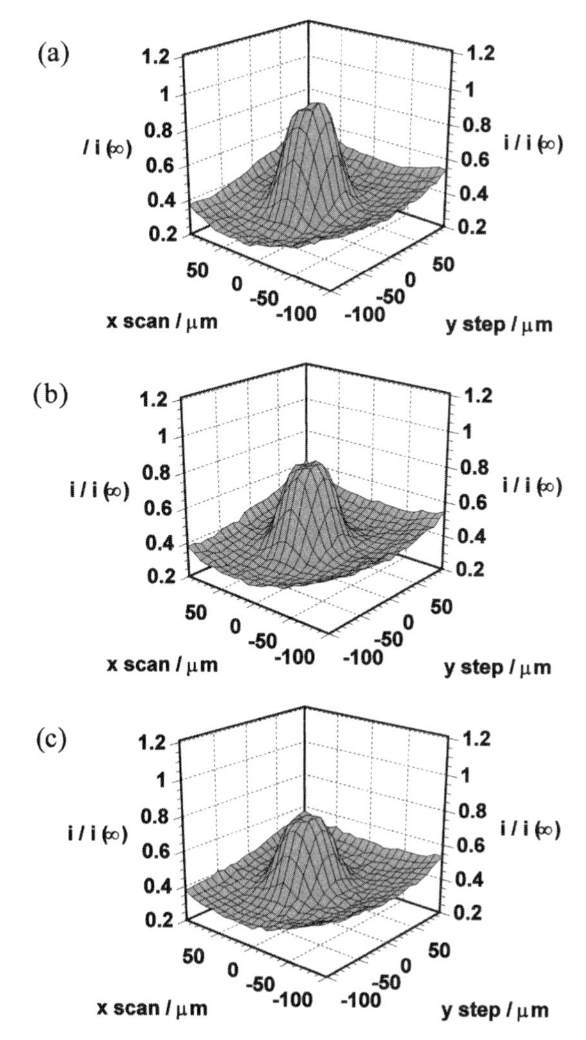

FIG. 32 Selection of typical dissolution rate images for the diffusion-controlled reduction of Ag^+ at a tip UME (a = 12.5 μm) scanned at heights of (a) 3.0 μm, (b) 4.0 μm, and (c) 5.0 μm over a silver chloride disc (125 μm diameter) surrounded by a glass sheath.

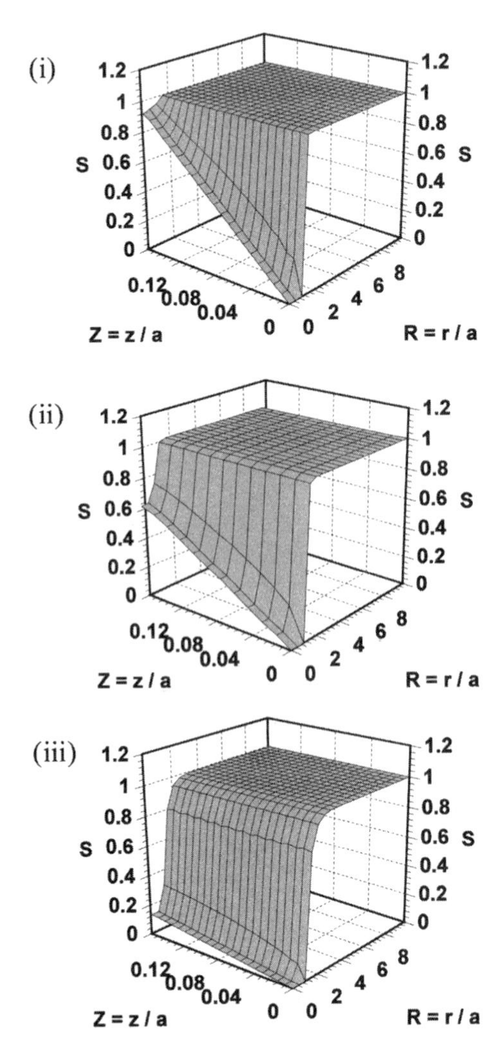

FIG. 33 Steady-state profiles of the saturation ratio in the tip/substrate gap (L = 0.16) for a first-order induced dissolution process, characterized by K_1 = (i) 100, (ii) 10, and (iii) 1. The data are for an induced dissolution reaction in the absence of a supporting electrolyte.

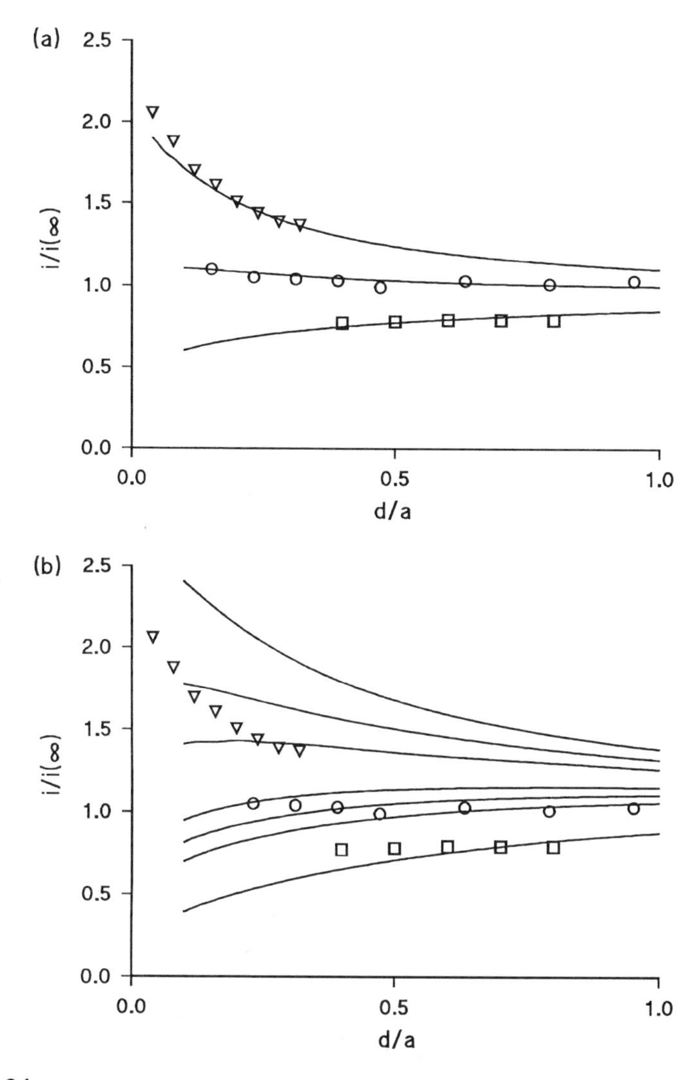

FIG. 34 Normalized approach curves of steady-state current for Ag^+ reduction versus separation between the tip and AgCl substrate. Data relate to experiments with Pt tips with a = 5.0 μm (□), 12.5 μm (○), and 25.0 μm (▽). Corresponding theory is shown for (a) second-order process with log K_2 = 0.4 (upper curve), 0.1, and −0.3 (lower curve), (b) a first-order process with log K_1 = 0.5 (upper curve), 0.3, 0.15, −0.1, −0.2, −0.3, and −0.7 (lower curve).

and first-order dissolution processes with a range of rate constants (Fig. 34a and b, respectively). Decreasing a has the effect of increasing mass transport in the system, which serves to push the process towards surface control, and consequently lower current ratios are measured.

It was found that the second-order rate law provided an excellent fit to each set of experimental data (Fig. 34a). Moreover, the normalized rate constants derived, log $K_2 = -0.3$ ($a = 5$ μm), 0.1 ($a = 12.5$ μm), and 0.4 ($a = 25$ μm), yielded a consistent rate constant, $k_2/c^{sat}_{\mathrm{AgCl}} = 0.016$ cm s^{-1}. In contrast, the first-order dissolution theory did not provide a good description of any of the sets of experimental data, and even the most consistent fits of this model did not yield a constant value of $k_1/c^{sat}_{\mathrm{AgCl}}$. These results thus reemphasize the ability of SECM to distinguish between candidate dissolution rate laws.

4. *High-Resolution Dissolution Imaging: The Integrated Electrochemical–Atomic Force Microscope*

Although many of the above studies have provided kinetic evidence for the BCF dissolution model (3,5), SECM lacks the (atomic-level) resolution as a topographical imaging tool to provide direct structural evidence for the operation of this mechanism. In order to significantly increase the spatial resolution of electrochemically induced dissolution imaging, an integrated electrochemical–atomic force microscope (IE-AFM) probe has been introduced (8). The device consists of a Pt-coated AFM tip which, in this application, is used to measure the topography of a dissolving crystal surface while simultaneously inducing the dissolution process electrochemically under conditions that mimic those employed for SECM kinetic measurements.

The approach has been illustrated through SECM and IE-AFM studies of dissolution from the (100) cleavage face of potassium bromide single crystals (8) in acetonitrile solutions, containing 0.05 mol dm^{-3} $LiClO_4$ serving as a supporting electrolyte, and saturated with respect to potassium bromide. Dissolution was initiated by depleting the local concentration of bromide ions in the direct vicinity of the crystal/solution interface via oxidation to tribromide or bromine at the electrode surface. The current, monitored in the SECM configuration as a function of time, provided kinetic information on the dissolution characteristics, while in situ topographical measurements with the IE-AFM allowed the structural changes that accompanied the reaction to be identified.

Figure 35 shows a typical approach curve of long-time current, obtained in an SECM configuration, for the diffusion-controlled oxidation of Br^- to Br_3^- as a function of tip-crystal separation for a 5 μm diameter UME. The measured current has been normalized with respect to the steady-state current that flowed when the UME was placed far from the crystal surface. The

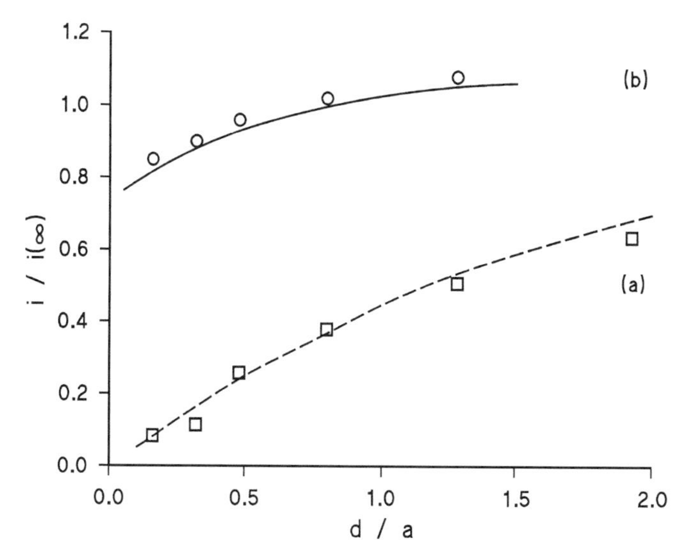

FIG. 35 Steady-state approach curves of the diffusion-limited current for the oxidation of Br^- to Br_3^- at a Pt tip (a = 2.5 μm), as a function of distance from a glass surface (□) and the (100) face of KBr (○), in an acetonitrile solution containing 0.05 mol dm^{-3} $LiClO_4$ and saturated with respect to potassium bromide. The theoretical characteristics for (a) negative feedback (---) and (b) a diffusion-controlled dissolution process (——) are also shown.

experimental and theoretical $i/i(\infty)$ versus d behavior for the same electrode against a glass surface is also shown. The dissolution limiting currents are significantly greater than those due only to hindered diffusion in the tip-substrate gap, clearly illustrating the ability of SECM to induce dissolution from the potassium bromide surface.

For comparison with experiment, the behavior predicted for diffusion-controlled dissolution (K_1 = 100) of a binary (1:1) crystal under conditions where only one ion type is depleted and none of the lattice ions are buffered during the course of a measurement is shown. It is clear that even on the fastest time scale (a = 2.5 μm was the smallest electrode employed in this study), the dissolution process is effectively diffusion-controlled even at a probe-crystal separation as close as 400 nm (L = 0.16). This was not unexpected given that, under the experimental conditions, K^+ levels in the gap between the tip and the crystal surface increase [as described above for Cl^- in the dissolution of AgCl in KNO_3 solutions (6)], ultimately suppressing the attainment of high undersaturations and thereby holding the dissolution process in the mass-transfer control regime.

The experimental arrangement for the IE-AFM induced-dissolution measurements is shown in Figure 36. Silicon nitride probes were used that were sputter-coated with Pt (300 Å) deposited on a Cr anchor layer (100 Å). After the Pt-coated AFM probe had been secured in the cell, the underside and the part of the tip holder that came into contact with the solution were electrically insulated by coating with a thin film of polystyrene, leaving Pt exposed only on the tip and a significant fraction of the cantilever. With an imaging tip, most of the electrochemically active part of the probe (apart from the apex of the tip) was at a sufficiently great distance from the crystal surface (3–40 μm) to ensure that dissolution images related to diffusion-controlled dissolution based on the earlier SECM results (8).

Dissolution with the IE-AFM was induced by pulsing the potential at the probe for 1 s from a value where no electrode reactions occurred to various potentials on the Br^-/Br_3^- or Br_3^-/Br_2 waves. Images of the surface topography were recorded in a specific region of the surface in the AFM mode, prior to and after each pulse, as a function of time. A typical sequence of images is shown in Figure 37, in which the potential was pulsed to a value approximately at the half-wave potential for the Br^-/Br_3^- couple. Prior

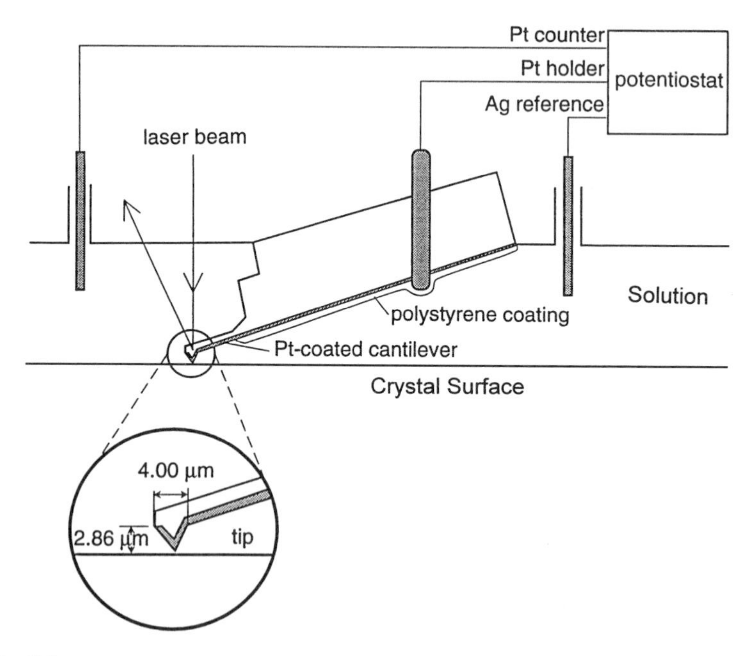

FIG. 36 Schematic of the cell and the probe for IE-AFM induced dissolution experiments.

to the potential step, Figure 37a clearly shows the presence of a dissolution spiral, which was found to slowly rotate (due to dissolution) over a period of several minutes. The wide spacing of the steps is diagnostic of only a very slightly undersaturated solution (57). There were a number of possible explanations as to why, for an initially saturated solution, this might occur: small thermal changes in the sample chamber (58), local frictional heating effects from the scanning tip, or as a result of the very slight local electrochemical depletion of Br^- with the potential of the electrode held at the foot of the Br^- to Br_3^- wave.

Figure 37b, recorded immediately after the potential pulse clearly demonstrates that under conditions of diffusion-controlled dissolution, the predominant mechanism is the unwinding of screw dislocations consistent with the BCF mechanism for a rapid dissolution process where the interfacial undersaturation is small (48). These results provided the first in situ evidence for the validity of this mechanism in describing fast dissolution processes of ionic single crystal surfaces. The local electrochemically controlled undersaturation provides the driving force, which enables many more dislocations to be activated and hence revealed, compared with the situation shown by Figure 37a. It is interesting to note that the breaks in the steps in the spiral depicted in Figure 37a coincide with the cores of the dislocations that are exposed in Figure 37b. Heterogeneities in the shape and depth of the dislocations in Figure 37b suggest differences in the associated strain energies of these defects (27).

After application of the 1 s pulse, the solution in the probe-substrate gap returns to a saturated state via dissolution of the crystal surface. The images in Figure 37c and d provide vital information on how steps of unit cell height retreat and interact during dissolution. Evidently, further dissolution normal to the surface slows considerably, and continued dissolution occurs via the motion of steps away from the dislocation cores, resulting in the step spacing becoming wider towards the center of the spiral and closer at a distance from the core. Spirals with this morphology have been predicted theoretically for dissolution processes close to saturation (59). Figure 37c and d also demonstrate elegantly the interactions between closely spaced spirals at the points where they intercept. It is particularly interesting to note the step pattern that results from the interception of spirals that rotate in opposite directions, labeled A (anticlockwise rotation, moving outwards from the center) and C (clockwise rotation) in Figure 37c. The steps emerging from dislocations A and C travel towards one another and annihilate when they intersect, leaving a single step connecting the dislocation cores, the morphology of which is very similar to that predicted theoretically for this situation by Burton, Cabrera, and Frank (48). After several minutes the sur-

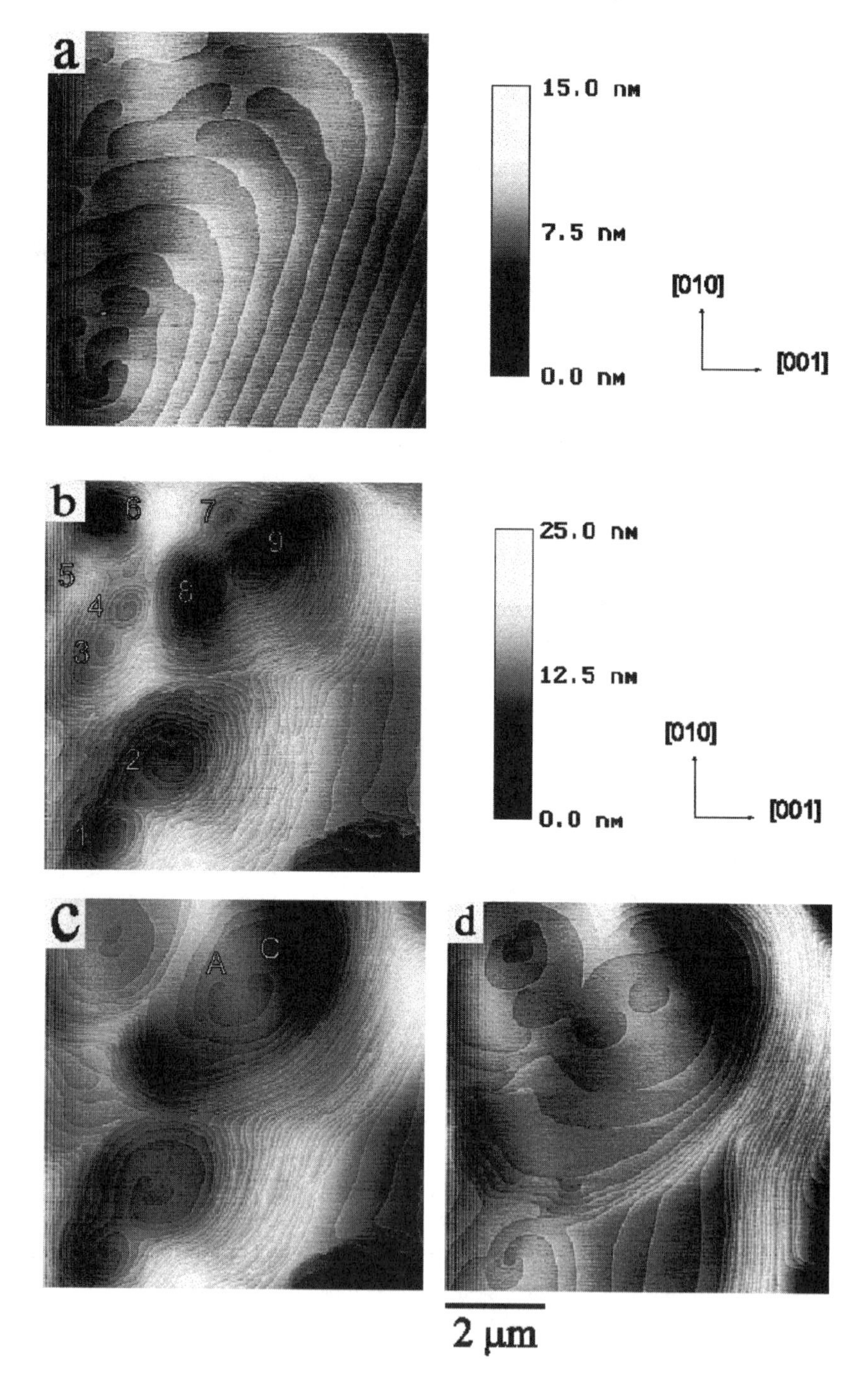
a
15.0 nm
7.5 nm
0.0 nm
[010]
[001]
b
6
7
9
5
4
8
3
2
1
25.0 nm
12.5 nm
0.0 nm
[010]
[001]
c
A
C
d
2 μm

face was found to reconstruct to a form similar to that which prevailed prior to the induction of dissolution.

To increase the driving force exerted on the dissolution process, the potential of the probe was stepped to a value where Br_3^- was oxidized to Br_2 at a diffusion-limited rate. Under these conditions, the following equilibrium becomes important in acetonitrile solution:

$$Br^- + Br_2 \Leftrightarrow Br_3^- \tag{42}$$

for which the equilibrium constant is 1×10^7 mol^{-1} dm^3 (60). With the tip potential held in this region, Br^- in the probe/substrate zone is locally depleted in two ways, by (1) direct electrolysis at the IE-AFM probe and (2) titration of Br^- with electrogenerated Br_2 to produce Br_3^-. The image in Figure 38 corresponds to the area highlighted in Figure 37, 120 s after application of the potential pulse described (50). The centers of seven dislocations are observed, which spiral down to hollow cores (cylindrical holes at the center of each dislocation), with radii ranging from 100 to 1000 nm. Such features were absent from Figure 37 and all images recorded at lower driving potentials (8,50).

These results are the first dynamic and in situ observations of cores at the center of dislocations characterized by small step spacings. The appearance of hollow cores on the dissolving surface is diagnostic of dissolution occurring with a moderate interfacial undersaturation (57,59,61–63), with the surface kinetics at least partly rate-limiting. Hollow cores were not observed for dissolution under diffusion-controlled conditions (Fig. 37), since the interfacial concentrations of K^+ and Br^- were maintained at a level very close to the saturated value (8).

IV. CORROSION STUDIES

SECM and related methodologies, such as scanning photoelectrochemical microscopy (SPECM) (16,17), are ideally suited to investigating the phenomena underpinning corrosion processes. This is because (1) they are

←

FIG. 37 AFM images of a 7 μm × 7 μm area of the (100) face of potassium bromide in contact with an acetonitrile solution containing 0.05 mol dm^{-3} $LiClO_4$ and saturated with respect to potassium bromide: (a) Prior to and (b–d) immediately after the application of a 1 s potential pulse to the electrochemically active probe (from the foot to half-way up the Br^-/Br_3^- wave; +0.5 V to +1.1 V vs. a silver quasi-reference electrode [AgQRE]). Images (b) to (d) were recorded sequentially at a rate of 21 s/frame.

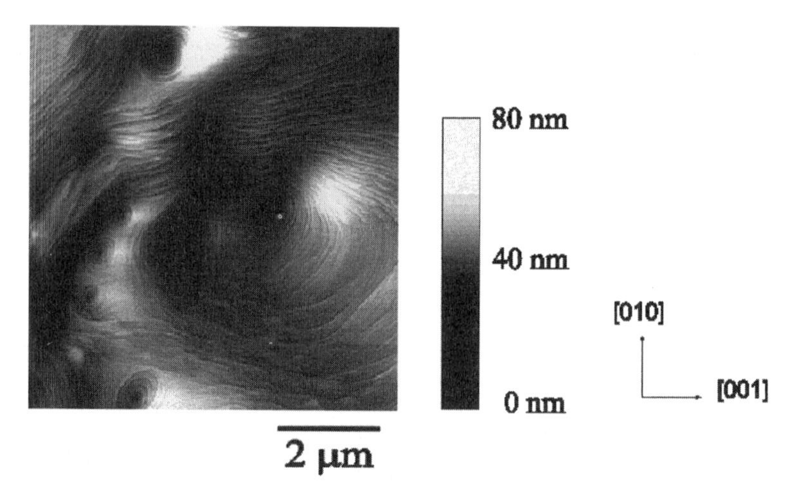

FIG. 38 AFM image of the same area of the KBr (100) face as in Figure 37 120 s after the application of a 1 s potential pulse (+0.5 V to +1.7 V vs. AgQRE) to the electrochemically active probe.

equally applicable to insulating and conducting surfaces (28,29), (2) the spatial resolution is high (3,4), enabling surface structural effects to be correlated with activity, and (3) the tip UME can be used both to initiate the reaction and to detect the corrosion products, with the tip (and substrate) current providing quantitative information on the process. To date, SECM and SPECM have been used to explore three fundamental areas, with a view to providing more detailed information on the nature of corrosion (Secs. IV.A–IV.C).

A. Identification of Pitting Precursor Sites

The sites on a corroding surface that are susceptible to pit initiation are referred to as pitting precursor sites (PPS). An important question in corrosion research is why some surface sites lead to pit formation while others do not. White and coworkers have used SECM to investigate the relationship between electroactivity and pitting precursor sites on the surface of Ti foil covered with a 50 Å thin film of TiO_2 (9,10). The Ti/TiO_2 surface was bathed in Br^- ions (from a solution containing 1 mol dm^{-3} KBr and 0.05 mol dm^{-3} H_2SO_4). Initial studies were carried out with the surface biased at a potential between +1.3 and +1.6 V versus a saturated calomel electrode (SCE), sufficient to convert Br^- to Br_2 at regions of the surface that were electrochemically active. An 8 μm diameter carbon fiber UME, poised at a potential to reconvert any substrate-generated Br_2 back to Br^-, was scanned over the surface at an initial tip-sample separation of ~20 μm. Thus, areas of high electroactivity corresponded with increases in the cathodic tip current, measured as a function of tip position in the *x,y* plane.

This substrate generation-tip collection type experiment (SG-TC) is illustrated schematically in Figure 39. SECM images revealed a number of microscopic sites where the local Br_2 concentration was high. Typical images of tip current as a function of tip position and substrate potential for a single active site are shown in Figure 40. The large differences between the maximum and minimum currents and the increase in the maximum tip current with applied substrate potential demonstrates that Br^- oxidation occurs heterogeneously across the surface and is also a potential dependent process (9,10,64). There is also, however, an underlying slow passivation of the surface with this mediator in this potential region, as indicated by the decrease in the current in Figure 40d, recorded 40 minutes after the image in Figure 40c. From low-resolution SECM scans, the number of active surface sites was estimated as ~30 ± 10 sites cm^{-2}, with site sizes ranging from 10 to 50 μm.

Increasing the substrate potential to +3.0 V initiated pitting of the surface. As the pits rapidly grew to microscopic dimensions, their positions were identified using video microscopy. Comparison of SECM and video images revealed a close correspondence between the sites of pit formation and the locations of electroactivity observed at lower potentials. These studies provided the first experimental evidence of a correlation between electrochemical activity of the surface and the subsequent occurrence of oxide

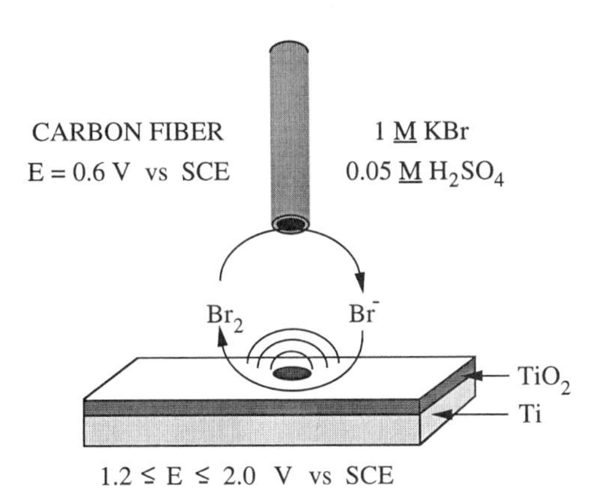

FIG. 39 Identification of pitting precursor sites using the SECM. Schematic illustration demonstrating the localized oxidation of Br^- at a Ti/TiO_2 electrode and the reduction of electrogenerated Br_2 at an SECM tip.

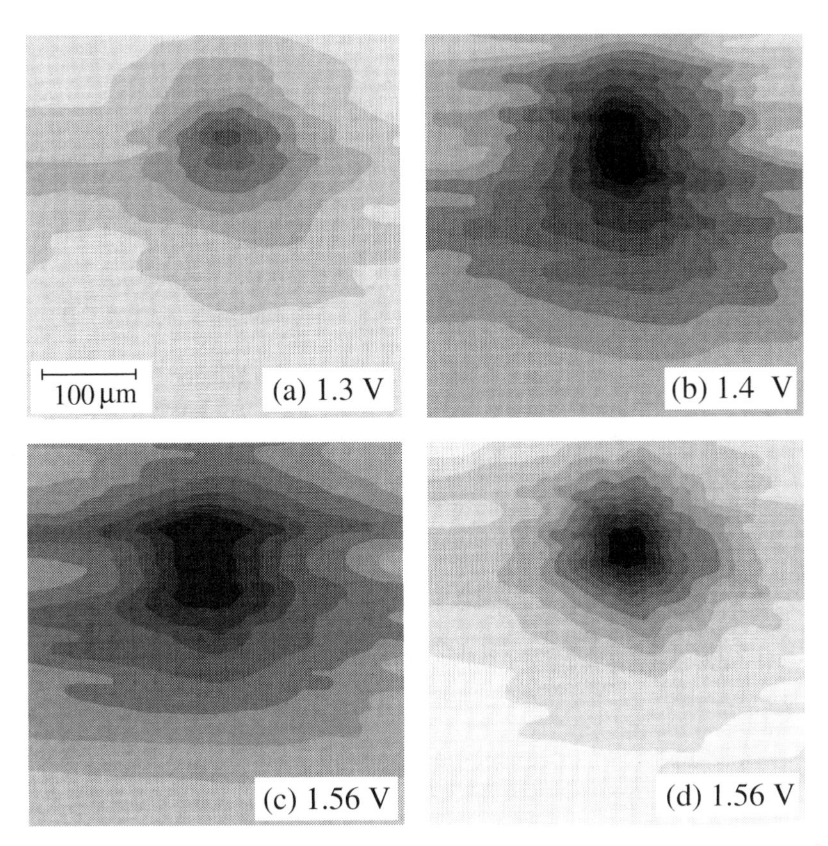

FIG. 40 SECM images of an individual electroactive site on a Ti/TiO_2 (50 Å) electrode biased at potentials versus SCE of (a) +1.3 V, (b) +1.4 V, (c) +1.56 V, and (d) +1.56 V 40 minutes later. All images were obtained at a tip-substrate separation of ~20 μm in a 1 mol dm^{-3} KBr, 0.05 mol dm^{-3} H_2SO_4 (pH 1.05) solution. The SECM tip potential was +0.6 V. The gray scale contrast corresponds to an absolute maximum-minimum tip current range of (a) 0.16–0.09 nA, (b) 0.18–0.1 nA, (c) 0.24–0.1 nA and (d) 0.18–0.09 nA.

breakdown. The observations are of fundamental interest but also demonstrate a great practical advantage of SECM in allowing precursor sites for oxide breakdown to be identified nondestructively (i.e., prior to pit initiation), which facilitates subsequent examination of PPS by complementary surface analysis techniques.

To obtain further insights, both qualitative (64) and quantitative (65), into the electrochemical activity of oxide-covered metals, the same generation/collection approach was used by Basame and White to map redox processes of a range of mediators at Ti electrodes covered by a thin oxide layer. The oxidations of iodide, bromide, and ferrocyanide were shown to occur at the same randomly distributed microscopic sites, from which it was concluded that the activity of the Ti/TiO_2 surface was due to the local properties of the oxide layer. Spatially localized processes were associated with reactions occurring at potentials positive of the conduction band edge of TiO_2. The corresponding sites were metal-like in that they were electrochemically active at potentials where the depletion layer structure of the oxide film should have inhibited the oxidation processes. In contrast, the reduction of ferricyanide at potentials negative of the conduction band edge of TiO_2 occurred uniformly over the electrode surface, indicating that, in this case, the heterogeneous electron-transfer reaction was spatially delocalized at the SECM level (64).

Investigations by Garfias-Mesias et al. (11) also employed the SG/TC mode to identify pitting precursor sites on the surface of polycrystalline Ti covered with a 50 Å thin film of TiO_2. The Ti/TiO_2 surface contained particles (inclusions) composed of Al and Si. Areas of electrochemical activity were located using a tip UME to detect the local concentrations of either Br_2 or $Fe(CN)_6^{3-}$, generated from the oxidation of Br^- or $Fe(CN)_6^{4-}$ at the substrate, with the latter considered to be the most reliable mediator (11). Subsequent studies of active sites with scanning electron microscopy and energy dispersive x-ray analysis revealed, in most cases, that the areas of activity corresponded with surface particles (typically a few micrometers in size, but occasionally as large as 20 μm) consisting mainly of Al and Si. Pitting corrosion tests in Br^- solutions, with the substrate biased at +2.5 V (vs. SCE), in conjunction with in situ confocal microscopy, demonstrated that most of the localized attack began in the areas where particles were located.

The photoelectrochemical activity of pitting precursor sites on the surface of inclusion-free Ti, covered with a 50 Å TiO_2 film, has also been investigated by Smyrl and coworkers using SPECM (16,17), which combines SECM and scanning photoelectrochemical microscopy (PEM) (66). A schematic of the probe, which consists of an optical fiber coated with gold and a further insulating polymer layer, is shown in Figure 41, along with the experimental arrangement employed to study photoelectrochemical reactions at a Ti/TiO_2 surface. In contrast to the probe design in Figure 39, a gold ring electrode surrounds an optical fiber, which facilitates local illumination of the surface during which both the tip (ring) and substrate currents can be monitored.

A typical sequence of images recorded using SECM, PEM, and SPECM are shown in Figure 42, centered at the location of a particularly active precursor site. The initial SECM image (Fig. 42a) was recorded using an

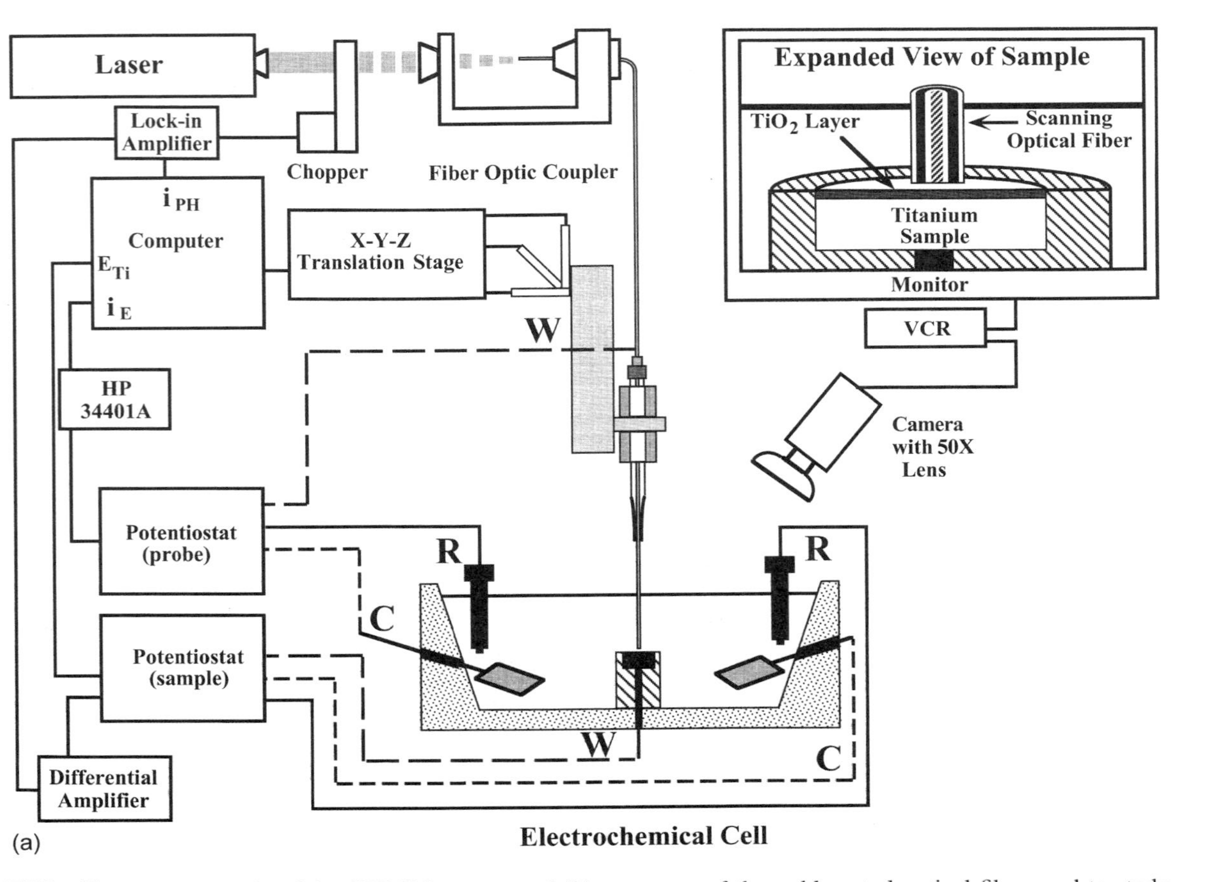

FIG. 41 (a) Schematic of the SPECM system and (b) geometry of the gold-coated optical fiber used to study reactions at the surface of a Ti/TiO_2 substrate.

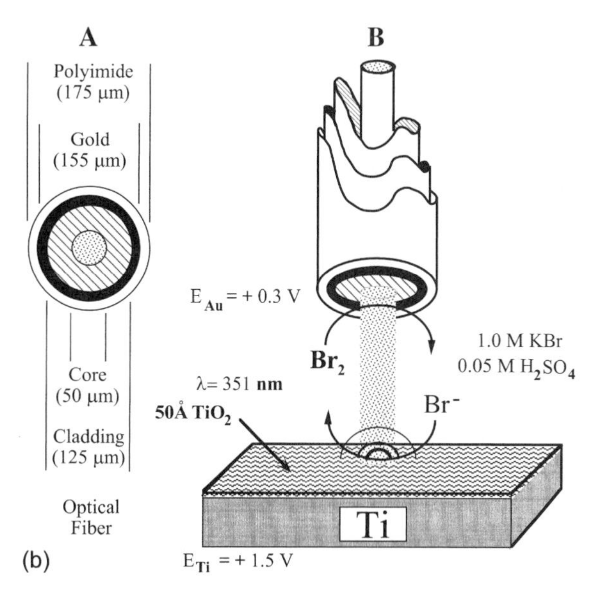

FIG. 41 Continued.

8 μm diameter C fiber electrode, in the same manner as described above, with the substrate biased at +1.0 V (vs. SCE) to promote the oxidation of Br^- to Br_2, and the tip held at +0.6 V (vs. SCE) to detect Br_2 via reduction to Br^-. The area of high current indicates the site of high electroactivity. The C fiber electrode was then replaced with the SPECM probe. With no illumination and the Au ring biased at +0.3 V to reduce substrate-generated Br_2 to Br^-, the PPS is still evident, although the image is slightly elongated in the x direction (Fig. 42b). This was attributed to stirring effects caused by the large size of the probe. The currents detected are higher than in Figure 42a due to the larger area of the Au ring electrode.

The resolution of the image in Figure 42b is largely a function of the probe-sample separation and the ring electrode size. One problem with probes of the size shown in Figure 41 and those employed to obtain the data in Figure 42b is that at close probe-substrate separations the spatial resolution is governed by the ring size (17). For microscopic electroactive regions, an image of the ring electrode itself is obtained. This artefact was eliminated by employing a large tip-substrate separation (d = 200 μm for Fig. 42b), but the resulting spatial resolution was lower than the SECM image obtained with a conventional electrode (Fig. 42a). Improvements in the imaging capabilities of SPECM have been made by Shi et al. (67), who etched the tips of optical fibers in concentrated hydrofluoric acid solutions,

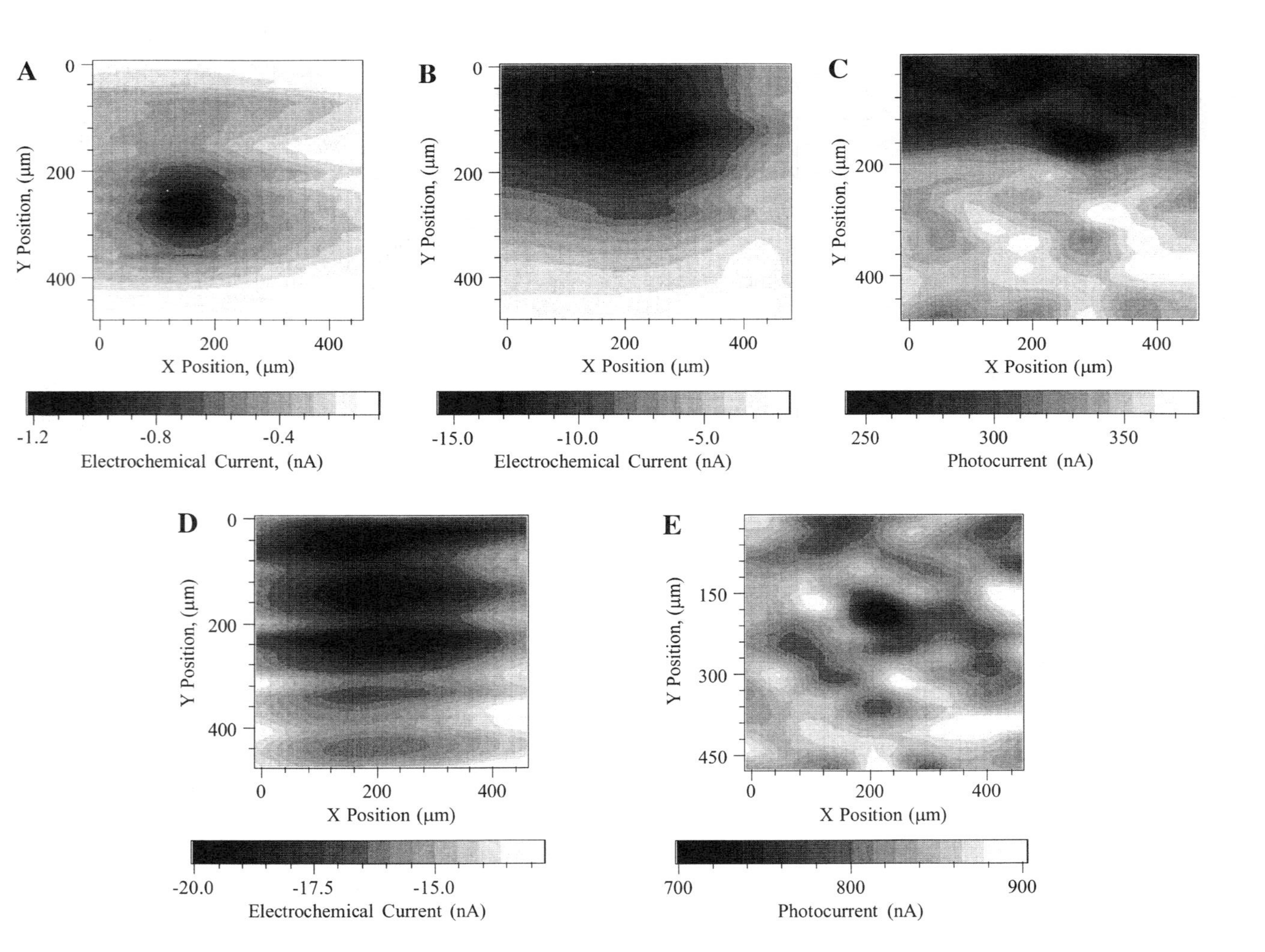
A
Y Position, (μm)
X Position, (μm)
Electrochemical Current, (nA)
-1.2
-0.8
-0.4
B
Y Position, (μm)
X Position (μm)
Electrochemical Current (nA)
-15.0
-10.0
-5.0
C
Y Position, (μm)
X Position (μm)
Photocurrent (nA)
250
300
350
D
Y Position, (μm)
X Position (μm)
Electrochemical Current (nA)
-20.0
-17.5
-15.0
E
Y Position, (μm)
X Position (μm)
Photocurrent (nA)
700
800
900

sputter-coated with gold, and then finally coated with an insulating varnish, to produce probes with an overall diameter of less than 5 μm.

Figure 42c shows the PEM (substrate photocurrent) image recorded close to the area of Figure 42a and b. The Ti/TiO_2 substrate was biased at a potential of +0.5 V, where only the photoassisted turnover of Br^- to Br_2 was possible. Taking into account a shift in the position of the imaged area towards the right by 100 μm, Figure 42c demonstrates that the PPS appears to have a lower photocurrent than adjacent surface sites. This was further confirmed from SPECM images, which show ring (Br_2 reduction; Fig. 42d) and substrate (Br^- oxidation; Fig. 42e) current maps recorded simultaneously under local illumination of the surface. Although the ring current image is blurred, it is possible to directly correlate the region of high electrochemical activity with decreased substrate photocurrent.

In a different application, Williams and coworkers were interested in using the SECM to identify a precursor state to the pitting corrosion of stainless steel (12) and also to elucidate the mechanism by which a pit could be maintained and propagate further (13), once it had been initiated. For these studies, a commercial AFM (Quesant 'Resolver') was adapted for use as an SECM. The tip electrode was a Pt-Ir wire electrolytically sharpened and insulated, as for electrochemical STM. A two-electrode mode was used with a Pt counterelectrode of much larger area than the stainless steel working electrode. The potential of the counterelectrode was standardized against an SCE. The probe UME was maintained at the same potential as the Pt counterelectrode. Typical tip-substrate separations of 0.1–0.5 μm were employed for imaging purposes, with the distance carefully established by ap-

FIG. 42 Gray scale images generated by performing SECM, PEM, and SPECM over a pitting precursor site on Ti/TiO_2 in 0.05 mol dm^{-3} H_2SO_4 and 1.0 mol dm^{-3} KBr. The probe/sample separation was ~30 μm for the carbon fiber probe and 200 μm for the optical fiber. The carbon UME (a = 4 μm) was biased at +0.6 V (vs. SCE) and the gold ring (inner diameter 125 μm, outer diameter 155 μm) at +0.3 V (vs. SCE). The illumination used for (C), (D), and (E) was λ = 351 nm provided by an Ar ion laser. Images (D) and (E) were obtained simultaneously in the SPECM configuration, with the tip substrate biased at various potentials (vs. SCE). (A) SECM tip current with the carbon fiber electrode, Ti/TiO_2 biased at +1.0 V. (B) SECM tip current with the gold ring electrode, Ti/TiO_2 biased at +1.0 V. (C) PEM (substrate current) performed with the optical fiber, Ti/TiO_2 biased at +0.5 V. (D) SECM tip current image obtained during SPECM with the Au ring electrode, Ti/TiO_2 biased at +1.0 V. (E) PEM (substrate current) image obtained during SPECM with the optical fiber, Ti/TiO_2 biased at +1.0 V.

proaching the tip to within tunneling distance and then retracting a designated distance.

Initial studies were carried out on type 304L stainless steel surfaces, effectively biased at +0.430 V versus SCE in a solution containing 0.3 mol dm^{-3} $NaClO_4$ and 0.3 mol dm^{-3} NaCl. Figure 43 shows typical current fluctuations, which were observed as the tip was scanned over the surface. The current spikes, of the order of 100 pA, were heterogeneously distributed across the surface, typically with a duration less than the time taken for the tip to complete a line scan (0.25 s). As the passive layer thickened with time, the number of fluctuations decreased.

In some instances, the currents observed were nontransitory in nature. Indeed, small local enhancements of passive current (as low as 1 pA) were found to mark the site at which a pit nucleated, as demonstrated by Figure 44. These occurrences were thought to be linked to pitting precursor events and the formation of metastable pits. In most cases (Fig. 43) the local environment was unable to support continued propagation of the pit and the spiked current signal indicated repassivation. Where pit growth was possible (Fig. 44), it was postulated that this was due to the presence of manganese sulfide inclusions in the surface.

Further work by Williams et al. (13) was carried out to confirm this hypothesis using 316F stainless steel (a high-sulfur steel with a rich inclusion

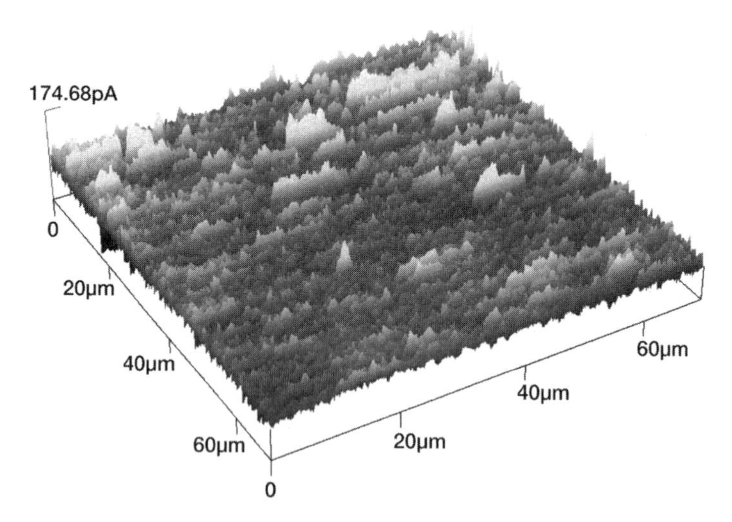

FIG. 43 Current fluctuations observed with an SECM tip upon initial polarization of type 304L stainless steel at +0.430 V versus SCE in a solution containing 0.3 mol dm^{-3} $NaClO_4$ and 0.3 mol dm^{-3} NaCl. The image is composed of 310 lines recorded at a tip scan rate of 0.25 s/line.

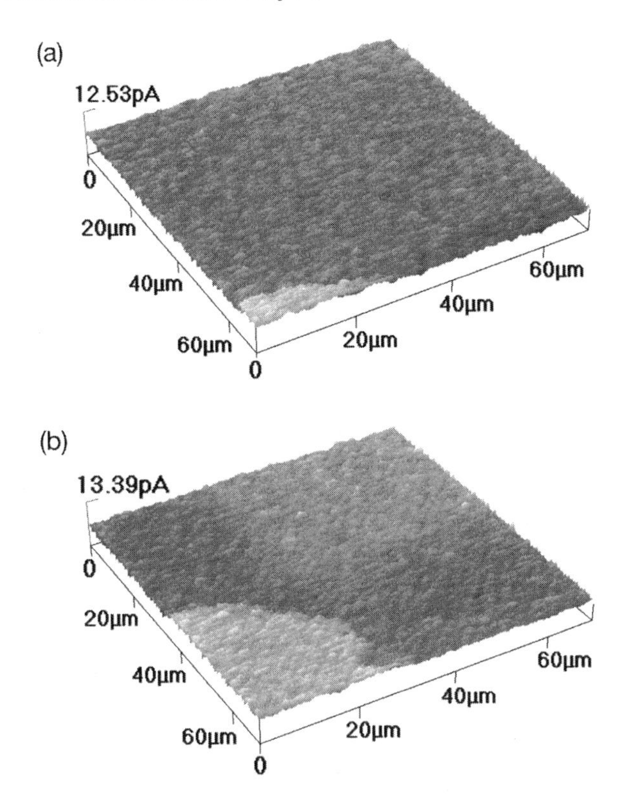

FIG. 44 (a, b) SECM images demonstrating localized minor perturbation of passive current density with the substrate, 304L stainless steel biased at +0.03 V versus SCE, in a solution containing 0.3 mol dm^{-3} $NaClO_4$ and 0.3 mol dm^{-3} NaCl: (a) 5 min and (b) 70 min after application of the substrate potential. (c, d) SECM images demonstrating the nucleation of the pit in the area (a, b) of perturbed passive current immediately (c) and 5 min (d) after the potential of the substrate was raised to +0.230 V versus SCE.

content) as the test substrate. Figure 45a shows a typical SECM image of a zone of enhanced current attributed to dissolution over and around a MnS inclusion. The substrate was biased at +0.330 V versus SCE and the bathing solution contained 0.03 mol dm^{-3} NaCl and 0.3 mol dm^{-3} $NaClO_4$. Figure 45b shows an AFM image of the topography of the site depicted in Figure 45a. The presence of a deposit partly covering the area where increased current flow was detected is clearly evident. Electron microprobe analysis of this region revealed a high sulfur and chloride content.

The evidence collected in these experiments and others, including complementary PEM studies, enabled a mechanism to be elucidated for the pit-

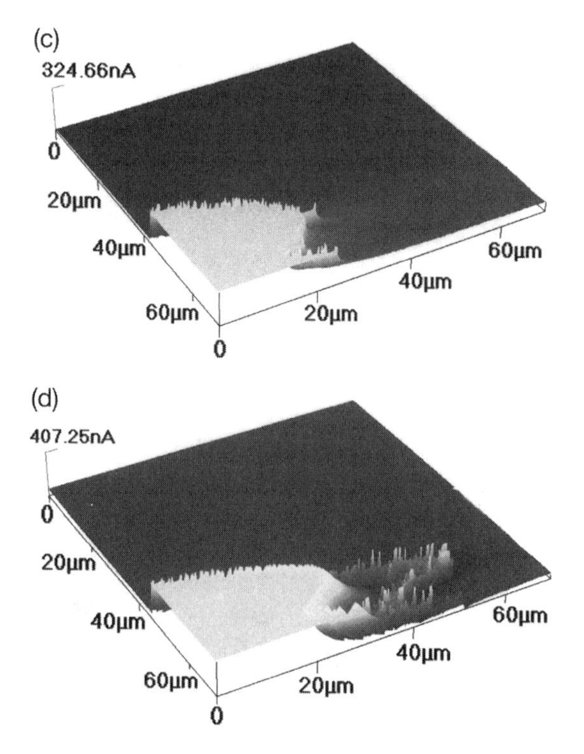

FIG. 44 Continued.

ting corrosion of stainless steels (13). It was conjectured that chloride-catalyzed dissolution of MnS inclusions, at extremely high current densities, resulted in the formation of a sulfur-rich crust which could extend over the inclusion and surrounding metal (Fig. 45b). Solution trapped in the small volume beneath the crust was subject to extreme conditions of local pH and Cl^-, the former resulting from the dissolution of MnS and the latter due to the migration of Cl^- into this region to support the dissolution current. Both served to promote the necessary conditions for the depassivation of stainless steel, with the formation of a pit and its continued growth.

B. Passive Film Breakdown and Pit Initiation

An attractive feature of the SECM is the ability to use the tip UME, positioned close to a target interface, to electrogenerate a local, known concentration of a solution species. For corrosion studies, this mode of operation has been exploited by Wipf and Still to generate local concentrations of the aggressive Cl^- ion in close proximity to passivated iron (15), stainless steel

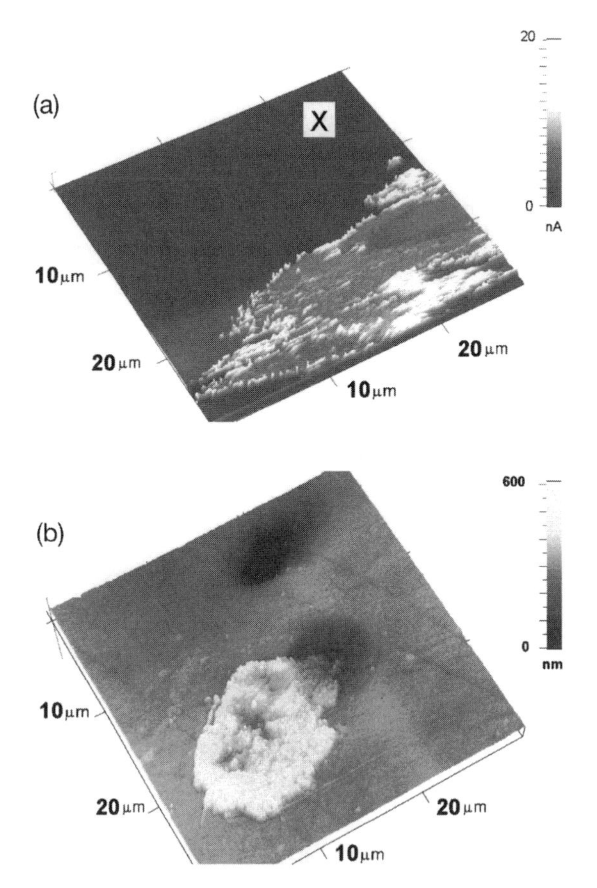

FIG. 45 (a) SECM image showing the current distribution over the region on and around an inclusion on the surface of 316F stainless steel. The substrate potential was +0.330 V versus SCE and the bathing solution contained 0.03 mol dm^{-3} NaCl and 0.3 mol dm^{-3} $NaClO_4$. X shows the position where the surface was marked to locate this site for ex situ characterization. (b) Subsequent ex situ AFM image of the region shown in (a). The identification marker is the circular depression at the top right.

(14), and aluminum (14) surfaces. Subsequent breakdown of the passive layer and pit initiation were detected by monitoring both the substrate current as the sample dissolved and the tip current as the dissolution products were detected cathodically.

Typical current-time traces for an SECM-initiated iron corrosion experiment are shown in Figure 46. In this experiment, the potential of an initially

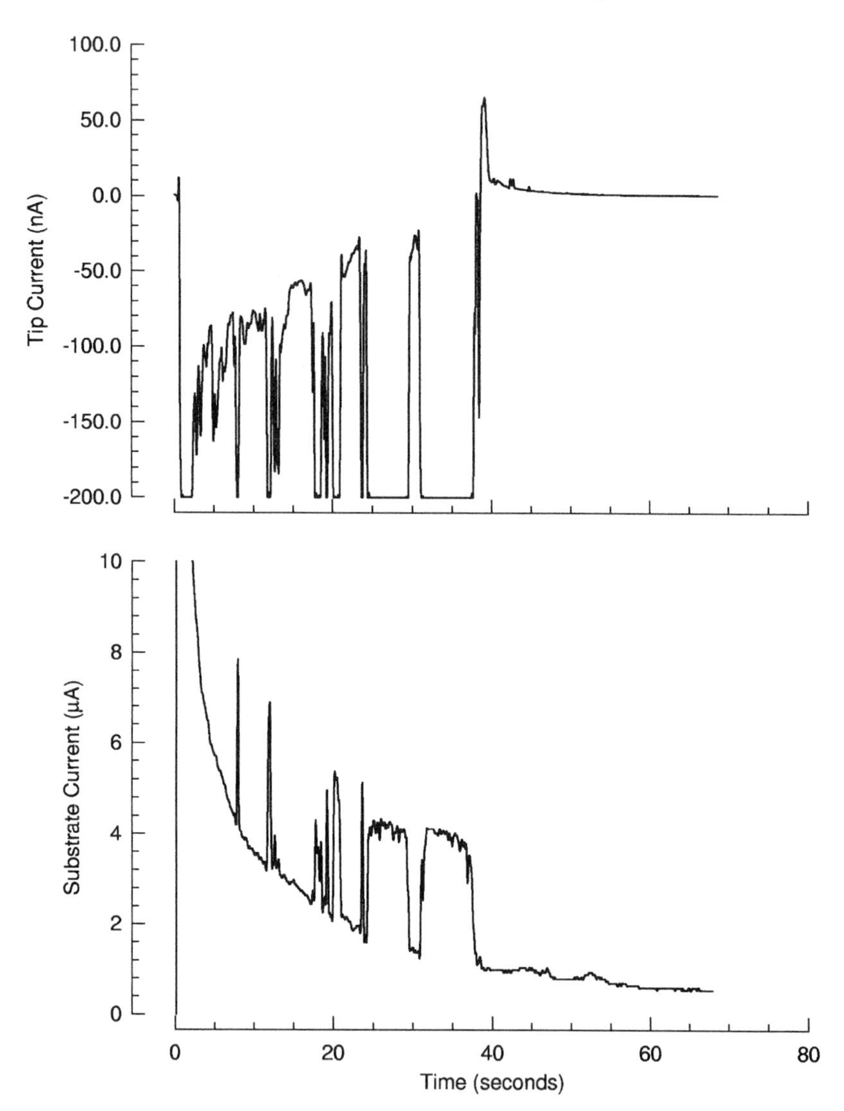

FIG. 46 Typical current-time traces for an SECM-initiated iron corrosion experiment. Traces (a) and (b) are, respectively, the current at the tip (12.7 μm diameter Au UME) and substrate (1.0 mm diameter iron electrode). The tip (biased at -1.1 V vs. MSE) was positioned ~6.3 μm from the iron surface (biased at -0.1 V vs. MSE). A 0.5 s passive layer growth time was used. The tip potential was returned to 0.0 V (vs. MSE) at t = 38 s.

oxide-free, 1.0 mm diameter iron electrode was stepped from −1.1 to −0.1 V versus a mercurous sulfate reference electrode (MSE), promoting passive layer growth. After a 0.5 s time delay, the potential of a 12.7 μm diameter gold tip UME, positioned approximately one electrode radius from the surface, was stepped from 0.0 to −1.1 V versus MSE. This initiated Cl^- production due to the reduction of 0.03 mol dm^{-3} trichloroacetic acid (TCA) in the solution (buffered with pH 6.0 phosphate/citrate). Fluctuations in the tip and substrate current were indicative of iron corrosion. In cases where the localized production of Cl^- failed to initiate the reaction, the current at the substrate showed a normal decay containing only charging current and passive layer growth contributions, while the tip current maintained a steady state due to the diffusion-controlled reduction of TCA.

The tip and substrate current spikes in Figure 46 are generally well correlated (particularly at times greater than 8 s), suggesting that the breakdown of the passive layer (substrate current) involves the release of Fe^{2+} from the iron surface, which was detected by reduction to Fe(0) at the tip UME. Evidence for the presence of Fe(0) at the tip came from the visual observation of a reddish-brown film at the electrode surface after such measurements and cyclic voltammograms (CVs) recorded with the tip positioned close to the iron surface, before and after a corrosion experiment. Prior to corrosion measurements, the tip CV displayed features consistent only with the reduction of TCA, while after corrosion the CV also showed a cathodic wave, possibly due to the reduction of Fe^{2+} to Fe and a corresponding anodic stripping peak. The latter occurred at the same potential as the anodic dissolution of iron, and was thus attributed to the reoxidation of Fe(0). Denuault and Tan (68,69) used a similar approach to identify the dissolution products for mild steel subjected to an acidic corrosive environment. In contrast to the work of Wipf and Still, the tip electrode was used only as a detector and not as an initiator of the corrosion process. CVs recorded with the tip placed close to the substrate detected the presence of Fe^{2+} and H_2.

A further interesting feature of Figure 46 occurs at $t = 38$ s, where the tip potential was adjusted to 0.0 V, causing the cessation of Cl^- production. At this point, the substrate current also stopped fluctuating, providing further evidence for the importance of Cl^- in the breakdown of the passive layer of iron. Light microscopy of the substrate revealed a small (9 μm × 14 μm) pit on the iron surface where the tip had been located, which was similar in size to that predicted based on the charge passed in the current spikes on the substrate.

Further studies were carried out to examine the effect, on the corrosion process, of the time lag between the start of passivation and Cl^- generation. For small-tip UMEs ($a = 6.4$ μm) positioned in the center of the iron substrate, the longer Cl^- production was delayed, the lower the probability of passive film breakdown. Moreover, Cl^- production at times greater than 1.5 s into the growth of the passive layer very rarely resulted in pitting corrosion. This suggested the existence of a critical thickness of the oxide layer, which, if formed before Cl^- ions were generated, protected the substrate from further attack by Cl^-, unless the site was associated with a grain boundary or defect structure, as discussed below.

For a fully passivated iron surface, which had been biased at 0.0 V for 1 hour to form a thick oxide layer, the probability of pit initiation was found to be much higher when larger (100 μm) diameter UMEs were used, which probed a much larger area of the substrate, particularly if the tip was positioned close to the edge of the iron electrode. Subsequent etching of the iron disc revealed a radial distribution in the grain boundaries of the surface. Smaller grains, with higher number densities, were located at the edge of the iron electrode, while larger grains, with lower number densities, were in the center. The results supported earlier studies (70), which established a link between grain boundaries (and the defects associated with them) and the location of pitting sites on iron.

C. Imaging of Active Pitting Corrosion

The substrate generation/tip collection mode of the SECM has also been used by Wipf to image active pitting corrosion on the surface of 304 stainless steel (14). A typical image of an active corrosion pit is shown in Figure 47. The steel sample was biased at +0.5 V versus Ag/AgCl and bathed in a solution containing 0.01 mol dm^{-3} Cl^- at pH 3.0, conditions severe enough to initiate pitting corrosion (substrate generation). A 12.5 μm diameter Au tip electrode was scanned above the surface of a corrosion pit at an initial tip-substrate separation of 20 μm. The potential of the tip was held at +1.0 V such that the image in Figure 47 was considered to represent a map of the distribution of oxidizable corrosion products emanating from the corrosion pit (tip collection). The current distribution in the pit can be seen to be heterogeneous in nature, which was attributed to a nonuniform rate of corrosion. With the tip UME placed close to the corroding pit, a broad CV was recorded with a half-wave potential of +0.86 V, suggesting that the current map was due to the oxidation of Fe^{2+}.

V. CONCLUSIONS

SECM techniques have made many valuable contributions to the investigation of phase transfer processes at solid/liquid interfaces by facilitating quantitative measurements under conditions of well-defined and high mass transfer coupled with high spatial resolution. At present there are effectively

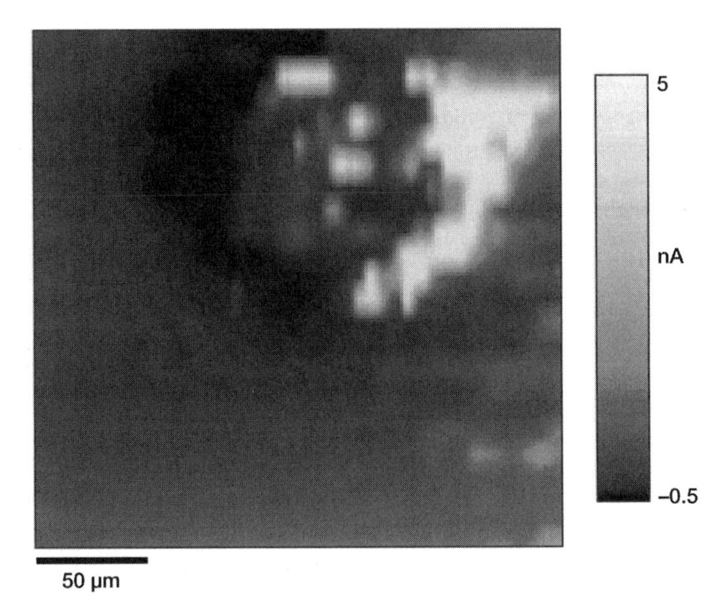

FIG. 47 A gray-scale SECM image (230 μm × 230 μm) of an active corrosion pit on 304 stainless steel. The image was acquired with a 12.5 μm diameter Au tip UME biased at +1.0 V vs. Ag/AgCl scanned above the surface of the substrate at an initial tip-substrate separation of ~20 μm at a scan rate of 40 $\mu m\ s^{-1}$. The steel sample was biased at +0.5 V in a solution containing 0.01 mol dm^{-3} total chloride concentration at pH 3.0.

two broad strategies for investigating reactivity: (1) the tip can be used as a (passive) detector of species produced or consumed at a reacting surface, and (2) the tip can be employed to both initiate and monitor the reaction of interest. The continued expansion in the types of UMEs available, as exemplified by the introduction of liquid membrane UMEs—both amperometric and potentiometric—which prove to be suitable as SECM tips, will lead to a diversification in the range of interfacial processes that can be studied using the tip as a detector. For cases where a UME reaction drives the interfacial process, there is also scope for extending the range of reactions that can be studied through the use of other modes. For example, double potential step chronoamperometry in the SECM configuration allows the electrogeneration of reactive species, which are subsequently recollected when the electrode potential is reversed in a second step [38]. This mode opens up the possibility of studying irreversible adsorption and absorption processes, as well as chemical reactions involving tip-generated species, thereby complementing the equilibrium perturbation mode described extensively in this chapter.

SECM has provided major new insights into localized corrosion in a relatively short period of time. The ability of SECM to quantitatively map the current distribution across complex heterogeneous surfaces has proven particularly valuable for identifying and characterizing pitting precursor sites, prior to pit initiation and growth. The possibility of characterizing the same microscopic regions of a sample with SECM, in combination with a range of other high-resolution structural and chemical probes, is a particularly exciting prospect for understanding the behavior of complex solid surfaces. Further developments in the capabilities of hybrid SECM techniques such as IE-AFM and SPECM (among others) are expected to contribute to this area.

Although this chapter has focused on phase transfer reactions at solid/liquid interfaces, many of the techniques and principles are generally applicable to such processes at liquid/liquid and air/liquid interfaces. Studies of adsorption/desorption, absorption, dissolution, and lateral interfacial diffusion at these types of interface are of considerable fundamental and practical importance, and SECM studies in these areas are already appearing.

ACKNOWLEDGMENTS

We thank the EPSRC (UK) for support of much of our research with SECM. We are grateful to Andy Hillier, Bill Smyrl, Henry White, David Williams, and David Wipf for providing us with figures.

REFERENCES

1. PR Unwin, AJ Bard. J Phys Chem 96:5035, 1992.
2. JV Macpherson, PR Unwin. J Chem Soc Faraday Trans 89:1883, 1993.
3. JV Macpherson, PR Unwin. J Phys Chem 98:1704, 1994.
4. JV Macpherson, PR Unwin. J Phys Chem 98:11764, 1994.
5. JV Macpherson, PR Unwin. J Phys Chem 99:3338, 1995.
6. JV Macpherson, PR Unwin. J Phys Chem 99:14824, 1995.
7. JV Macpherson, PR Unwin. J Phys Chem 100:19475, 1996.
8. JV Macpherson, PR Unwin, AC Hillier, AJ Bard. J Am Chem Soc 118:6445, 1996.
9. N Casillas, SJ Charlesbois, WH Smyrl, HS White. J Electrochem Soc 140: L142, 1993.
10. N Casillas, SJ Charlesbois, WH Smyrl, HS White. J Electrochem Soc 141:636, 1994.
11. LF Garfias-Mesias, M Alodan, PI James, WH Smyrl. J Electrochem Soc 145: 2005, 1998.

12. Y Zhu, DE Williams. J Electrochem Soc 144:L43, 1997.
13. DE Williams, TF Mohiuddin, Y Zhu. J Electrochem Soc 145:2664, 1998.
14. DO Wipf. Colloids Surfaces A Physicochem Eng Aspects 93:251, 1994.
15. JW Still, DO Wipf. J Electrochem Soc 144:2657, 1997.
16. N Casillas, P James, WH Smyrl. J Electrochem Soc 142:L16, 1995.
17. P James, N Casillas, WH Smyrl. J Electrochem Soc 143:3853, 1996.
18. JL Luo, YC Lu, MB Ives. J Electroanal Chem 326:51, 1992.
19. T Misawa, H Tanabe. ISIJ Int 36:787, 1996.
20. AC Lasaga. In: MF Hochella Jr, AF White, eds. Reviews in Mineralogy: Mineral-Water Interface Geochemistry. Vol. 23. Washington: Mineralogical Society of Washington, 1990, pp 17–85.
21. W Stumm, G Furrer. In: W Stumm, ed. Aquatic Surface Chemistry. New York: Wiley, 1989, pp 197–219.
22. RB Heimann. Crystals: Growth, Properties and Applications. Berlin: Springer-Verlag, 1982.
23. PR Unwin, JV Macpherson. Chem Soc Rev 24:109, 1995.
24. JV Macpherson, PR Unwin. Prog React Kinet 20: 185, 1995.
25. PR Unwin. J Chem Soc Faraday Trans 94:3183, 1998.
26. Interface 20, 1997.
27. K. Sangwal. Etching of Crystals: Theory, Experiment and Application. Amsterdam: Elsevier Science, 1987.
28. J Kwak, AJ Bard, Anal Chem 61:1221, 1989.
29. AJ Bard, F-RF Fan, MV Mirkin. In: AJ Bard, ed. Electroanalytical Chemistry. Vol. 18. New York: Marcel Dekker, 1993, pp 243–373.
30. PR Unwin, AJ Bard. J Phys Chem 95:7814, 1991.
31. AJ Bard, LR Faulkner. Electrochemical Methods. New York: Wiley, 1980, p 507.
32. KF Hayes, G Redden, W Ela, JO Leckie. J Colloid Interface Sci 142:448, 1991.
33. AJ Bard, G Denuault, RA Freisner, BC Dornblaser, LS Tuckerman. Anal Chem 63:1282, 1991.
34. PR Unwin, AJ Bard. Anal Chem 64:113, 1992.
35. PR Unwin, AL Barker, JV Macpherson, RD Martin, CJ Slevin. Abstracts of the 193rd Meeting of The Electrochemical Society, San Diego, CA, Spring 1998, abstract no. 982.
36. JV Macpherson, RD Martin, CF McConville, PR Unwin. In: SR Taylor, AC Hillier, and M Seo, eds. Localised In-situ Methods for Investigating Electrochemical Surfaces. New Jersey: The Electrochemical Society, 2000, p. 104–121.
37. GM Bancroft, ME Hyland. Rev Mineral 23:511, 1990.
38. CJ Slevin, JV Macpherson, PR Unwin. J Phys Chem 101:10851, 1997.
39. CJ Slevin, S Ryley, DJ Walton, PR Unwin. Langmuir 14:5331, 1998.
40. JV Macpherson, D O'Hare, CP Winlove, PR Unwin. Biophys J 73:2771, 1997.
41. AL Barker, JV Macpherson, CJ Slevin, PR Unwin. J Phys Chem B 102:1586, 1998.
42. MV Mirkin, TC Richards, AJ Bard. J Phys Chem 97:7672, 1993.

43. MV Mirkin, LOS Bulhões, AJ Bard. J Am Chem Soc 115:201, 1993.
44. KB Oldham. In: MI Montenegro, MA Queirós, JL Dashbach, eds. Microelectrodes, Theory and Applications. Vol. 197. Dordrecht, The Netherlands: NATO ASI Ser E, Kluwer, 1991, pp 3–16.
45. JW Zhang, GH Nancollas. Rev Mineral 23:365, 1990.
46. HG Linge. Adv Colloid Interface Sci 14:239, 1981.
47. M Ohara, RC Reid. Modeling Crystal Growth Rates from Solution. Englewood Cliffs, NJ: Prentice-Hall, 1973.
48. WK Burton, N Cabrera, FC Frank. Phil Trans R Soc London. A243:299, 1951.
49. AJ Bard, MV Mirkin, PR Unwin, DO Wipf. J Phys Chem 96:1861, 1992.
50. JV Macpherson. PhD thesis, University of Warwick, 1996.
51. WH Baur, JL Rolin. Acta Cryst B28:1448, 1972.
52. WJP Van Enckevort, WH Van Der Linden. J Cryst Growth 47:196, 1979.
53. K Onuma, K Tsukamoto, I Sunagawa. J Cryst Growth 110:724, 1991.
54. RA Berner, JW Morse. Am J Sci 274:108, 1974.
55. MV Mirkin, F-RF Fan, AJ Bard. J Electroanal Chem 328:47, 1992.
56. JS Newman. Electrochemical Systems. Englewood Cliffs, NJ: Prentice Hall, 1991.
57. N Cabrera, MM Levine. Philos Mag 1:450, 1951.
58. S Kipp, R Lacmann, MA Schneeweiss. Ultramicroscopy 57:333, 1995.
59. B Van Der Hoek, JP Van Der Eerden, P Bennema, I Sunagawa, J Cryst Growth 58:365, 1982.
60. JC Marchon. CR Acad Sci Paris Ser C 267:1123, 1968.
61. FC Frank. Acta Cryst 4:497, 1951.
62. B Van Der Hoek, JP Van Der Eerden, P Bennema. J Cryst Growth 56:261, 1982.
63. P Bennema. J Cryst Growth 69:182, 1984.
64. SB Basame, HS White. J Phys Chem 99:16430, 1995.
65. SB Basame, HS White. J Phys Chem 102:9812, 1998.
66. R Peat, A Riley, DE Williams, LM Peter. J Electrochem Soc 136:3352, 1989.
67. G Shi, LF Garfias-Mesias, WH Smyrl. J Electrochem Soc 145:2011, 1998.
68. YM Tan, G Denuault. Results presented at Electrochem 97, University College London, August 1997.
69. YM Tan. MRes Electrochemistry Report. Department of Chemistry, University of Southampton, 1997.
70. LF Lin, CY Chao, DD Macdonald. J Electrochem Soc 128:1194, 1981.

13

MICRO- AND NANOPATTERNING USING THE SCANNING ELECTROCHEMICAL MICROSCOPE

Daniel Mandler

The Hebrew University of Jerusalem
Jerusalem, Israel

Early in the development of the scanning electrochemical microscope (SECM) it was recognized that when an ultramicroelectrode (UME) is brought near a conducting surface, electron transfer is confined to a small area on the surface. This realization led to the development of a number of methods for using the SECM as a tool for surface modification (1). The term "microelectrochemistry" was coined to denote the production of micropatterns by means of electrochemical techniques. The surface reactions that have been explored in this context range from metal deposition and etching to the patterning of surfaces by enzymes, and they have not all been electron-transfer processes. Moreover, work has been done to increase the resolution of the fabricated patterns, extending the capability of the SECM into the nanometer domain.

The aim of this chapter is to review systematically the various systems and methods that have been used in conjunction with the SECM for fabricating small patterns on surfaces. Most of these can be divided into two main categories based on the mode of operation: the direct mode and the feedback mode. In the direct mode the substrate serves as the auxiliary electrode, while in the feedback mode the substrate is unbiased (in most cases) and a mediator that shuttles between the UME and surface is used (Fig. 1). This difference is significant because, as will be shown, it dictates most of intrinsic parameters, such as the type of reactions that can be carried out, as well as the extrinsic parameters, i.e., the resolution and rate of patterning. In this chapter, the direct mode is considered first, followed by the feedback mode. The chapter concludes with a discussion of some technological issues that are an integral part of micro- and nanopatterning techniques and presents several approaches that either are being currently undertaken or are likely to be demonstrated in the future.

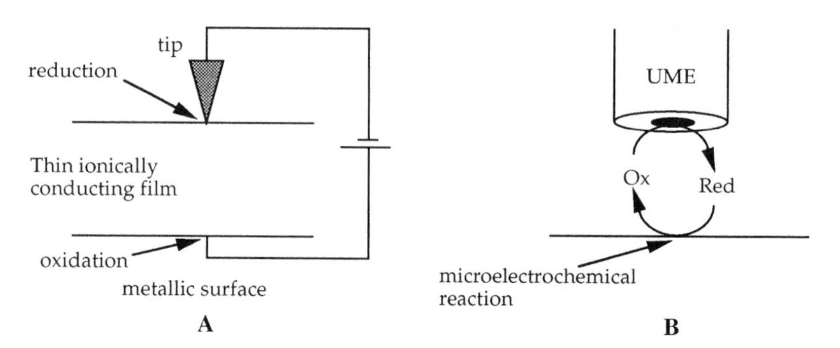

FIG. 1 Schematic representation of the direct (A) and the feedback mode (B) of the SECM.

I. PATTERNING BY THE DIRECT MODE OF THE SECM

This mode of operation was primarily developed by Bard and coworkers in the early days of SECM. The direct mode is based on approaching a *conducting* substrate with a biased UME. The substrate acts as the auxiliary electrode, which means that if a reduction process occurs at the UME, an oxidation reaction must take place at the substrate. The latter reduction is the driving force for the patterning process. Because the whole substrate is biased, localization of the process on the substrate is caused by the distribution of the electrical field between the UME and the substrate (2). Accordingly, the resolution of the patterns chiefly depends on the distance between the UME and the substrate. Therefore, in most cases where the direct mode has been applied, care has been taken to minimize the distance between the UME and the substrate.

A. Semiconductor Etching

The only system that employs the direct mode for semiconductor etching was reported by Lin et al. (3). n-GaAs was photoetched by holding a sharp metallic tip, insulated in glass, in close proximity (<1 μm) to a positively biased (4 V) n-GaAs while illuminating the substrate. The formation of etched patterns with line widths of 0.3–2 μm was attributed to the nonuniform spatial current distribution. Nonetheless, some etching also occurred across the whole surface because there was no attempt to focus the illumination on the area beneath the UME.

B. Metal Deposition and Etching

One of the critical problems of the previous system was the lack of a mechanism for controlling the distance between the UME and the substrate using a feedback loop. The current measured at the tip was insensitive to the distance, and therefore the latter was determined by scanning the tip across the predefined pattern in a STM mode in order to "learn the topography" of the path. This problem was resolved by Bard and colleagues (4) in the following way: an exposed STM-like tip approached a conducting substrate that was coated with a thin film made of an ionically conducting polymer, e.g., Nafion. The sample was kept in air and the tip was biased versus the substrate. As soon as the tip touched the polymeric film, a faradaic current passed across the film. The magnitude of the current was a function of the area of the tip that penetrated the polymer. Therefore, by keeping the current constant with an electronic feedback loop, the degree of tip penetration, and thus the tip-substrate distance, could be controlled.

Local metal deposition in the polymer was accomplished by incorporating metal ions inside the polymer and applying a negative potential to the tip (Fig. 2). The etching of metals was, however, accomplished by applying a positive potential to the substrate in the absence of metal ions in the polymer. Specifically, Craston et al. (4) used this approach for the high-resolution deposition of silver in Nafion, while Hüsser et al. (5) reported on the deposition of silver, gold, copper, and palladium in a variety of ionically conducting polymers and high-resolution etching of copper, silver, and gold. For example, Figure 3 shows a scanning electron micrograph of a pattern of silver lines deposited in Nafion. The width of the lines was as thin as 0.3 μm. Carrying out etching as well as deposition processes on the substrate/polymer interface was less successful and more difficult to detect. The films had to be removed in order to detect the patterns on the substrate, and in

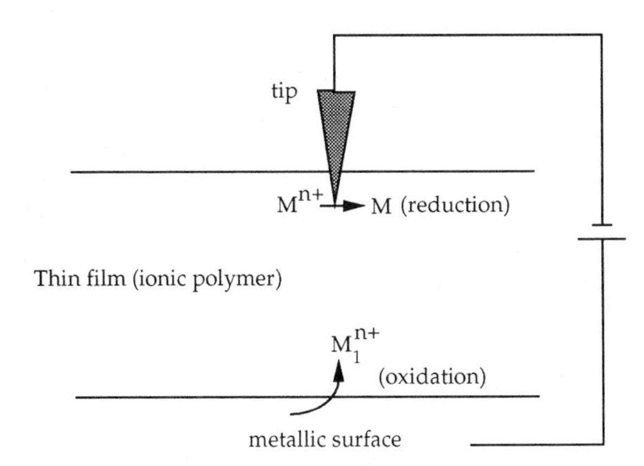

FIG. 2 Schematic representation of the direct mode for deposition of metal ions in polymeric films.

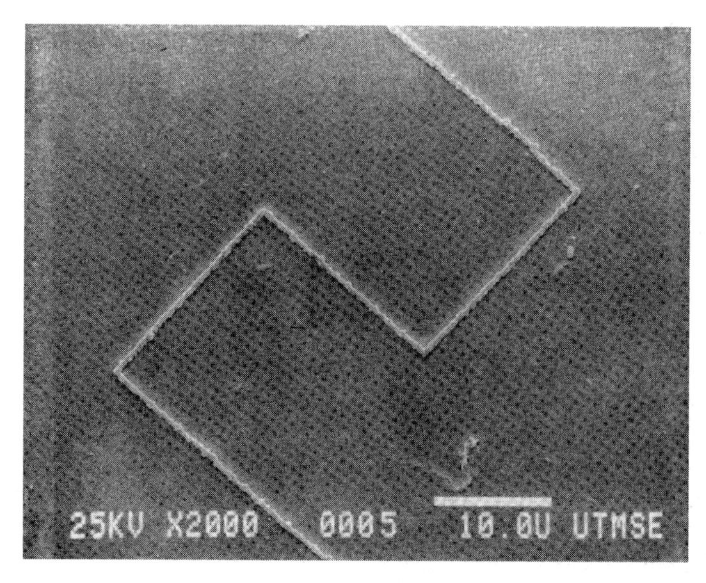

FIG. 3 SEM picture of a pattern of silver lines deposited in a Nafion film. (From Ref. 5b.)

most cases the resolution was lower as compared with that of patterns formed at the UME/polymer interface. In summary, this approach offered several advantages over the wet etching of GaAs: the tip-substrate distance could be nicely controlled by the electronic feedback, the tips used were conventional STM tips that did not have to be insulated, and the deposits did not stick to the tip. Even so, the speed of patterning was relatively low, i.e., 0.1 $\mu m \cdot s^{-1}$, because of the movement inside a very thin polymer and the response of the electronic feedback.

More recently, a remarkable example of the capabilities of the direct mode of the SECM has been demonstrated by Forouzan and Bard (6). It basically uses the same concept of controlling the distance between the tip and the substrate by its degree of penetration into an ionically conducting matrix. However, a very thin water layer, on the order of 10–15 Å, on an insulator is used instead of a polyelectrolyte film (Fig. 4). The reduction of silver ions and the formation of submicrometer rectangular patterns were driven by an STM tip that was biased negatively versus a gold contact on the mica. The thin water layer, which was formed by maintaining the relative humidity of the environment between 80–95%, made it possible for a faradaic current to flow between the tip and the gold contact. As long as the applied potential between the tip and the substrate was kept sufficiently

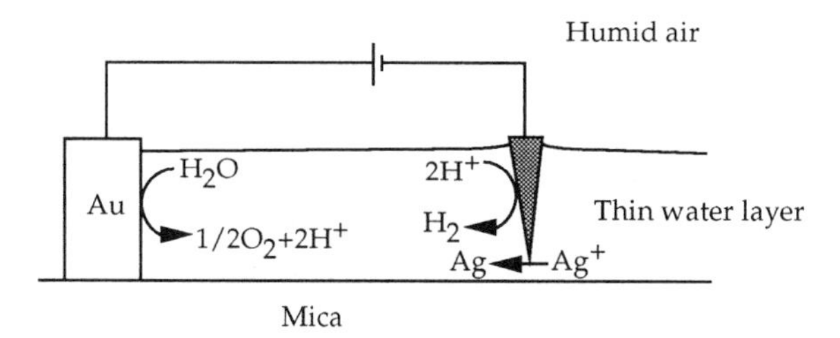

FIG. 4 Schematic representation of high resolution silver deposition on mica by the STM used in the direct mode of the SECM.

negative to reduce silver ions and water, the faradaic current prevented crashing of the tip into the mica.

C. Deposition of Conducting Polymers

The fabrication of micro- and nanopatterns of conducting polymers has been the subject of continuous research. Since the best method for depositing thin films of conducting polymers involves electropolymerization, it is only logical that continuous efforts have been made to apply the SECM as a tool for driving the local deposition of conducting polymers. The first report by Wuu et al. (7) exploits the direct mode for the deposition of polyaniline. More specifically, anilinium ions were incorporated into a thin Nafion film and the Pt substrate was biased positively, which caused the electropolymerization of aniline. Scanning the negatively biased tip across the film resulted in the simultaneous reduction of protons at the UME and the deposition of submicrometer-width patterns of polyaniline. As before, the resolution of the deposit was controlled by the tip size, its depth of penetration, and the thickness of the Nafion.

Although this sort of approach in the use of the direct mode has seen no further development, the lateral deposition of conducting polymers has been the subject of continuous and very successful research, in particular by Schuhmann and coworkers (8). Schuhmann realized that in order to deposit a conducting polymer locally, high concentrations of the radical cation of the monomer must be maintained within the volume between the UME and the substrate. Obviously, a constant potential applied to the substrate will cause the depletion of the monomer in the gap between the UME and the substrate and presumably the deposition of the polymer over the entire substrate. The solution was in the form of a series of potentials applied to the gold substrate. The first pulse was of 1.2 V versus Ag/AgCl for a time interval of 2 seconds, which caused the oxidation of nearly all the pyrrole monomers in the volume between the substrate and a 10 μm Pt UME. Then a second, lower potential of 0.85 V for 1 second was applied in order to oxidize the soluble oligomers. Finally a third relaxation pulse of 0.35 V for 2 seconds brought the system back to its initial stage. During this series of pulses the UME was scanned with a speed of a few $\mu m \cdot s^{-1}$ back and forth to obtain a sufficiently thick polypyrrole line of 50–60 μm width. The width of the deposited lines correlated with the total diameter of the UME, while no polypyrrole was detected outside the patterns. Other polymers such as polythiophene have been deposited as well (Fig. 5).

The lateral deposition of polypyrrole lines over insulating gaps has been accomplished by Kranz et al. using the same approach (9). The motivation of this work was to develop a polymer-based transistor. Hence, a galvanostatic pulse profile was applied between gold electrodes separated by 100 μm and a 25 μm Pt UME that served as an auxiliary electrode. This three-pulse series resulted in the deposition of a continuous spot of polypyrrole in the gap between the two electrodes. As a result, the potential applied to the polypyrrole (the gate) controlled the current between the two gold electrodes, which resembled a source and a drain. Moreover, the combination of this direct mode with the "topography feedback mode" developed by Kranz (10) paved the way to deposition of polypyrrole towers as high as 400 μm, with diameters of about 80 μm (Fig. 6).

In a more recent report, Schuhmann applied the same concept to the deposition of poly-*N*-(ω-amino-alkyl)pyrrole (11). The functionalized patterns were used for covalently attaching periodate-oxidized glucose oxidase. Instead of applying a series of potential pulses, a series of galvanostatic pulses was used. The SECM was used in this study to verify the formation of the patterns of the functionalized polypyrrole as well as the enzymatic activity. Scanning a UME across the functionalized polypyrrole while generating a redox couple resulted in a negative feedback current due to the low conductivity of the polymer as compared with the bare gold surface. The activity of the enzyme was detected with the SECM by comparing the feedback current (obtained over the enzyme patterns) of a mediator, i.e., ferrocene carboxylic acid, in the presence and absence of the substrate of the enzyme, β-D-glucose.

D. Organic and Biomolecule Patterning

The formation of miniature structures of organic and biological molecules is of particular importance, since conventional photolithographic techniques are not suited for these materials. Hence, efforts have been made to develop novel approaches for micro- and nanopatterning of surfaces with thin films and monolayers of organic and biological molecules.

In this area, the first report that took advantage of the direct mode of the SECM was presented by Masuhara et al. (12). Fluorescent patterns of

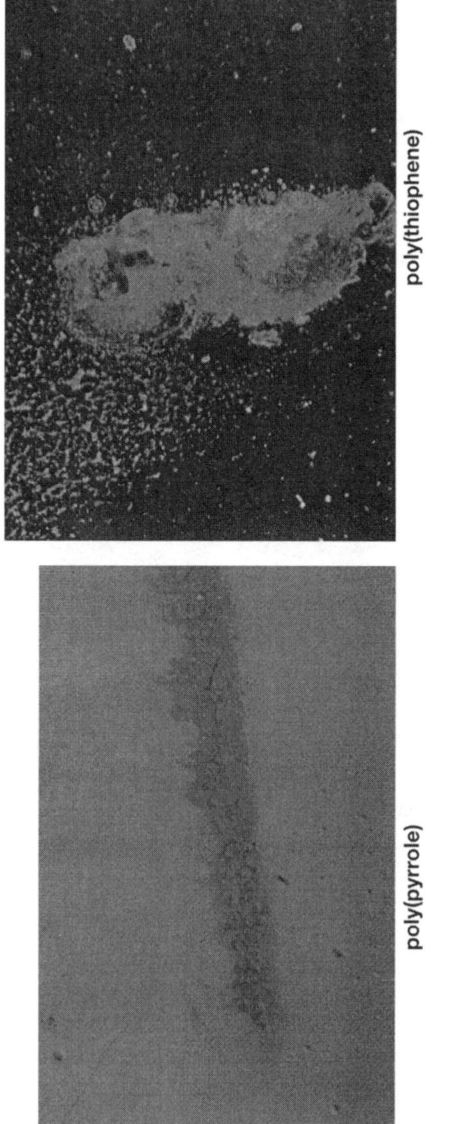

FIG. 5 Scanning electron micrographs of lines of polypyrrole and polythiophene deposited with the direct mode of the SECM. (From Ref. 8.)

FIG. 6 A scanning electron microscope image of a polypyrrole tower deposited by the direct mode of the SECM using a 10 μm UME. (From Ref. 10b.)

Rhodamine 6G in an ionically conducting polymer were formed as a result of the local decomposition of the quencher, methyl viologen. The latter was incorporated inside the film with the dye and quenched the fluorescence of the Rhodamine 6G. However, when a tungsten tip biased at -4 V versus the platinum substrate was scanned at 1 $\mu m \cdot s^{-1}$, the quencher was decomposed, probably due to its reaction with hydroxyl ions that were generated at the tip, and the fluorescence of the dye was recovered. The authors concluded that the factors governing the resolution of the patterns were the moisture in the film and the shape of the tip. The mechanism of the fluorescent micropattern formation in the Rhodamine 6G-methyl viologen polymer film was further studied using the direct mode in a three-electrode configuration (13). An additional reference electrode was constructed on the surface and was covered by the polymer. This made it possible to control the potential of the tip and as a result to correlate between the tip potential and the fluorescent micropattern formation. In essence, the tip had to be biased more negatively than a certain threshold potential, which depended on the tip material, in order to form the micropatterns. This potential was substantially more negative than that needed to electrogenerate reduced methyl viologen, indicating that the decomposition of the fluorescence quencher, i.e., methyl viologen, was driven by the generation of hydroxyl anions.

Recent efforts to pattern surfaces with organic and biomolecules using the direct mode have been carried out by Wittstock et al. (14). The localized desorption of alkanethiol monolayers has been induced by the direct mode of the SECM using, again, the UME as the auxiliary electrode and applying either negative or positive potentials to a gold substrate coated with dodecanethiol. The reductive and oxidative desorption of alkanethiols has been studied extensively (e.g., Ref. 15) and is best carried out in an alkaline solution. Accordingly, Schuhmann found that while the oxidative and the reductive desorption of the thiols in phosphate buffer resulted in only partial desorption, a significant improvement has been achieved using 0.2 M KOH. The width of the patterns was two to three times larger than that of the active microelectrode diameter. Although the resolution of the patterns was not comparable with that obtained by other methods, such as the microcontact printing, the SECM could be used in the feedback mode for imaging the patterns as soon as they were formed.

Local desorption of alkanethiols was further exploited by Schuhmann for the formation of enzymatic patterns of glucose oxidase (16). Cystamine, $(SCH_2CH_2NH_2)_2$, was adsorbed in the exposed patterns of the gold surface, and periodate-oxidized glucose oxidase was covalently attached to the amino groups of cystamine. The activity of the enzyme was followed by the generation-collection mode of the SECM. The conventional feedback mode, in which ferrocenium carboxylic acid was generated at the UME, could not be used for detecting the enzymatic activity because the mediator was partially regenerated on unmodified areas. On the other hand, the enzyme patterns could be detected upon introducing oxygen and glucose and applying a positive potential (0.4 V vs. SCE) at the UME that caused the oxidation of hydrogen peroxide. The latter was formed in the modified patterns through a glucose oxidase–catalyzed reaction. Improved images were obtained by adding catalase, which decreased the diffusion layer of the H_2O_2. The major advantage of this patterning approach, according to the authors, stems from the possibility of repeating the procedure at different spots. By using different enzymes, multienzyme structures could be produced.

The main advantage of the direct mode approach originates from the submicrometer distance that the tip is held close to and above the substrate. Calculations, as well as experimental results, indicate that the electrical field between the tip and the substrate is relatively focused, which results in submicrometer resolution of the fabricated patterns. The resolution is also a function of the ionically conducting matrix (in the case where a noninsulated STM tip is used) as well as the shape of the tip and its lateral speed.

On the other hand, the direct mode has a number of significant limitations. The reduction or oxidation at the tip must be accompanied by a reversed reaction at the substrate interface, or vice versa. For example, the deposition of silver was followed by the oxidation of water or substrate etching. This reaction can sometimes, e.g., when gas bubbles are formed, result in degradation of the pattern formed at the substrate. The resolution is also a function of the tip shape. Very small, preferably needle-like tip electrodes are essential for establishing the focused electrical field that is the key for obtaining high resolution. However, such tips are likely to be damaged as a result of their lateral movement or by contact with the substrate. The lateral speed of the tip in the direct mode varies and depends on whether the UME scans inside a polymer or in solution. In the former case, the UME speed is of the order of hundreds of Å per second, which makes the formation of large patterns tedious. Nevertheless, possible practical applications of this technique have been suggested, such as the fabrication of lithographic masks and the repair of broken contacts in integrated circuits.

Finally, it must be realized that the SECM in the direct mode is basically related to STM probe used for microfabrication. Therefore, additional studies, which are usually categorized under STM or scanning probes in general, must not be neglected. In many STM studies distinguishing between patterning through a tunneling or a faradaic mechanism is not evident. Numerous studies on STM-based electrochemical modification of materials have been demonstrated so far. For example, nanostructures with dimensions of 10 nm have been formed on gold surfaces by Schneir et al. (17). Penner and coworkers have fabricated metal pillars with 10–30 nm diameters on graphite surfaces (18). Sugimura et al. (19) fabricated oxide patterns of titanium with a spatial resolution of 20 nm using the STM in the presence of adsorbed water. Sugimura also used this approach, i.e., scanning probe anodization, in order to locally decompose trimethylsilyl monolayers on a Si substrate (20). Crooks termed this mode of operation scanning probe lithography and realized the similarity of his approach to the SECM direct mode (21). A number of reports (22) have speculated that an electrochemical mechanism could be responsible for STM-induced surface modifications in air. No attempt will be made here to review these and related contributions. However, there is no clear border between the direct mode of the SECM and STM-based electrochemical mode carried out under nondry conditions.

II. PATTERNING BY THE FEEDBACK MODE OF THE SECM

The versatility of this mode of operation has made it extremely powerful for fabrication of microstructures. In the feedback mode an ultramicroelectrode is held close above a substrate in a solution containing one form of electroactive species, either reduced or oxidized, that serves as a mediator (Fig. 1). The latter is usually used both as a means of controlling the distance between the UME and the surface and to drive the microelectrochemical process on the surface. This poses a number of requirements that must be taken into account when configuring the system. The basic limitation stems from the requirement that the electrochemical reaction be confined only to the surface. This means that the electroactive species generated at the UME will react with the surface or with other species attached to it. In addition, it is preferable in most cases that the redox couple used should exhibit chemical and electrochemical reversibility, so that it is effectively regenerated on the surface. The regeneration of the redox couple on the surface is required for controlling the UME-substrate distance. Finally, the thermodynamics and kinetics of the electrochemical process on the surface will dictate the choice of the redox couple introduced.

A. Metal Deposition and Etching

The first system that was based on the feedback mode aimed to deposit metals (Fig. 7). Metal ions were incorporated in a polymer layer, e.g., protonated polyvinyl pyridine, as a means of attaching them to the surface. Hexaaminoruthenium(III), $Ru(NH_3)_6^{3+}$, was reduced at an UME, diffused to the substrate, and drove the reduction of $AuCl_4^-$ and $PdCl_4^{2-}$ to their respective metals (23). The different factors that determined the size and pattern of deposited metal were examined. Although this initial system was expected to be simple and straightforward, it concealed complicated chemistry predominantly due to the diffusion of the mediator into the polymer. While gold ions were efficiently reduced by $Ru(NH_3)_6^{2+}$ inside the polymer to form gold patterns, $PdCl_4^{2-}$ ions formed different structures upon reduction. A detailed study of this system revealed that the reduction of palladium ions inside the polymer exhibited significantly more sluggish kinetics than in homogeneous solutions and thus required a stronger reductant than $Ru(NH_3)_6^{2+}$.

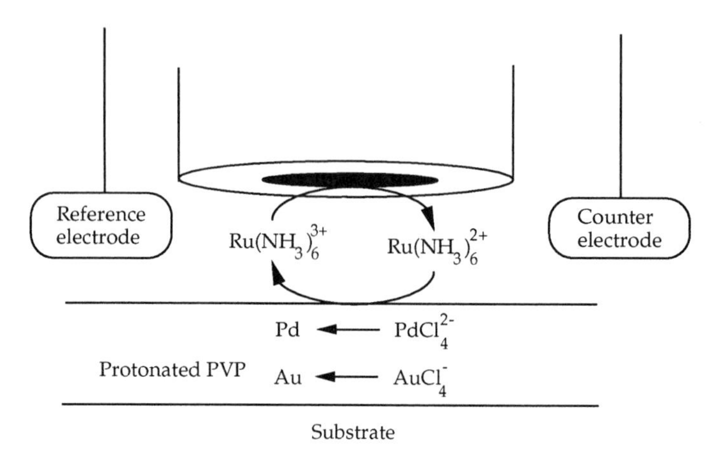

FIG. 7 Schematic representation of the feedback mode for the deposition of metal in polymeric films.

This system made the investigators change their concept. It was clear that if diffusion inside polymers could be abandoned, the interfacial chemistry involved would likely be faster and simpler. One possible approach has been the use of reactive surfaces that react directly with the electrogenerated mediators. The etching of copper (24) as well as of a variety of semiconductors (see below) represent the implementation of this concept.

The etching of copper using the feedback mode has been driven by electrogenerating a strong oxidant, e.g., $Fe(phen)_3^{2+}$ or $Os(bpy)_3^{2+}$ (phen = 1,10-phenanthroline, bpy = 2,2′-bipyridine) at a UME held close above a copper surface. The positive feedback current that was observed while approaching a copper surface indicated that the electroactive species were regenerated via electron transfer at the surface. As a result, copper dissolution occurred, which was strictly limited to the diffusion range of the oxidized mediator. Figure 8 shows a scanning electron micrograph of three etching spots made by holding a 25 μm Pt UME at a constant distance above a Cu thin layer for various periods of time. The profiles of these spots are also shown.

This system has further been studied by Unwin et al. (25) using the SECM in an ingenious configuration, in which two UMEs were made to approach each other. In this configuration the contribution of the lateral charge transport to the feedback current was totally eliminated. Thus, any deviation from an absolutely negative feedback current was due to the chemical regeneration of the mediator on the surface. Accordingly, tris(2,2′-bipyridine)ruthenium(III) and bromine was electrochemically generated at a

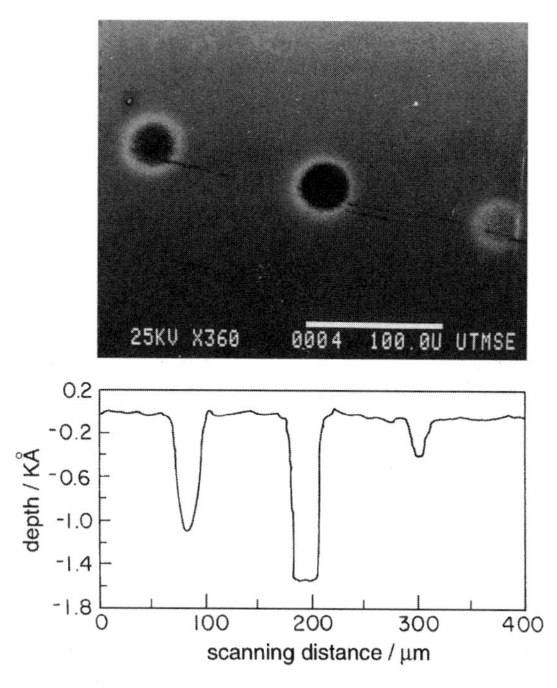

FIG. 8 Scanning electron micrograph and profiles of etching pits on a thin copper layer. The etching pits were formed as a result of leaving a 25 μm diameter Pt UME (biased at 0.7 V vs. SCE) close above (9 μm) the surface for 5, 10, and 20 minutes. (From Ref. 24.)

Pt UME and feedback currents, as the Pt UME and an unbiased Cu UME approached one another, were recorded. Investigations of Cu etching by both strong oxidants revealed that the metal dissolution process is diffusion-controlled under all the conditions examined, suggesting that the heterogeneous rate constant controlling the etching process is large.

Etching of iron by the feedback mode of the SECM has recently been reported by Still and Wipf (26). They produced localized corrosion at passivating iron surfaces by generating chloride ions at an SECM tip. Here is a case in which the metal is covered with an oxide layer that needs to be removed in order to facilitate metal dissolution. These experiments are described in great detail in Chapter 12.

Etching of surfaces in the feedback mode is rather straightforward. It requires the generation of a strong oxidant that is capable of oxidizing the surface. (The electrolyte, however, must be carefully chosen to favor the dissolution of the metal ions, a problem that is discussed in the following section.) On the other hand, the deposition of metal poses an additional challenge, namely, that of ensuring it is deposited on the targeted surface rather than the UME. A few methods have been described that address this challenge. The local deposition of gold patterns, so-called microwriting, has been reported by Meltzer and Mandler (27). The basic idea has been to dissolve anodically a gold UME held in close proximity to a negatively biased surface (Fig. 9). In the presence of bromide ions, the Au UME generated a constant flux of $AuBr_4^-$ ions that diffused to an indium tin oxide substrate and were further reduced to form micro-crystalline structures of gold. The bromide ions, besides acting as a promoter for gold dissolution, were responsible for the positive feedback current detected upon approach to the surface. This was due to their regeneration on the surface, which increased their effective concentration and thus the dissolution of the gold UME. The resolution of the gold patterns, which were made of $1 \pm 0.1\ \mu m$ gold crystals, was determined by the diameter of the UME (Fig. 10).

On the other hand, Heinze and coworkers (28) developed a method for deposition of silver lines, initially on conducting substrates, e.g., gold, and later on insulators. The deposition of perfect silver lines was accomplished by adopting the method of Mandler et al. (37) (see below), in which the pH in the volume between the UME and the substrate is altered. Specifically, silver was deposited on a negatively biased gold substrate as a result of shifting the dissociation of a silver complex [Eqs. (1) and (2)] by decreasing the pH. The latter was caused by oxidizing nitrite ions at the tip [Eq. (3)]. The resolution of the deposit was a function of the distance between the UME and the substrate. There was no broadening of the pattern, because ammonia diffused from the solution into the reaction zone and shifted the equilibrium to the silver complex. This latter phenomenon represents the

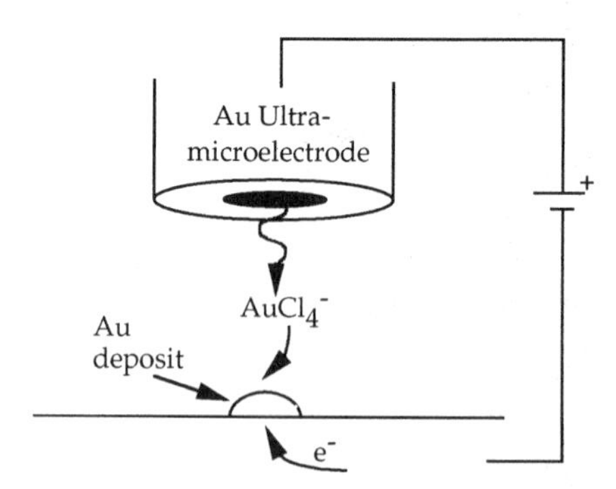

FIG. 9 Schematic representation of the system for "microwriting" of gold patterns with the SECM.

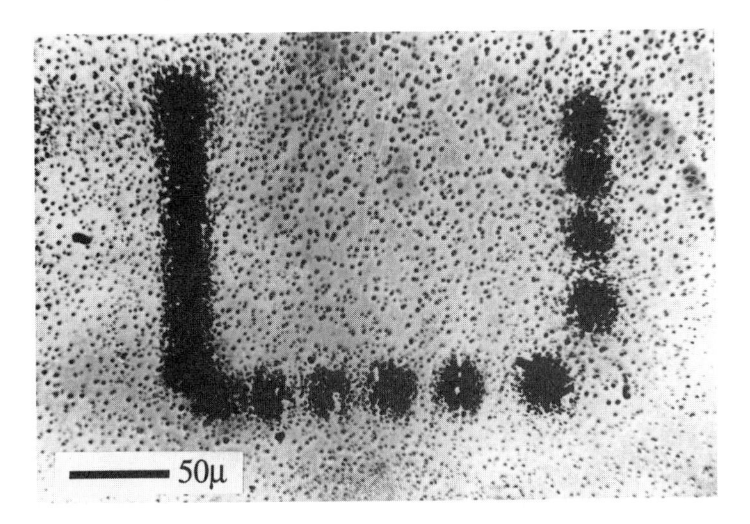

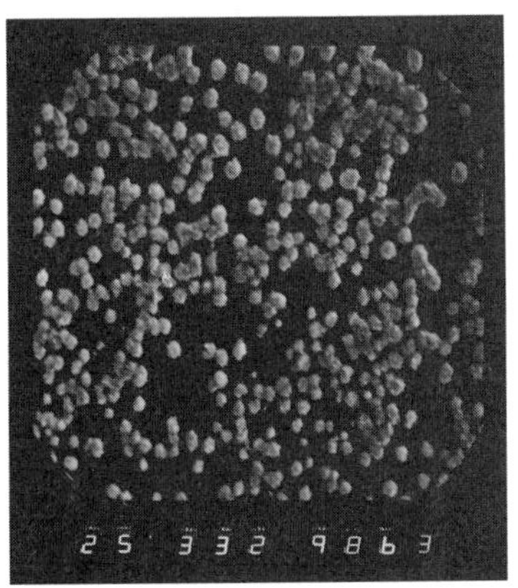

FIG. 10 (Top) A scanning electron micrograph of the gold particles deposited on ITO. (Bottom) A SEM micrograph of the gold particles deposited, the average size of the particles is $1.0 \pm 0.1\ \mu m$.

first example of a "chemical lens" that focuses the flux of the reactive species on the surface (see below).

$$[Ag(NH_3)_2]^+ + 2H^+ \rightleftarrows Ag^+ + 2NH_4^+ \quad (1)$$

$$Ag^+ + e^- \rightarrow Ag \quad (2)$$

$$NO_2^- + H_2O \rightarrow NO_3^- + 2H^+ + 2e^- \quad (3)$$

Recently, Heinze et al. (29) studied the local deposition of silver on a nonconducting substrate. Silver lines were formed by the local reduction of a silver chloride thin film using the feedback mode of the SECM (Fig. 11). AgCl thin films of the thickness of 1–2 μm were vapor deposited on smooth Teflon surfaces. Hydroquinone, electrochemically generated from benzoquinone in a mixture of ethylene glycol and water, diffused to the silver chloride film and caused initially the formation of square-shaped well-ordered silver crystals via "chemical development." After an induction period, these nuclei grew through "physical development" and merged to form fine silver patterns. This caused a two-dimensional propagation of the silver until a limit of lateral extension, which was probably controlled by the mixed potential in the periphery of the deposit, was reached. Then the growth continued in a vertical direction. One of the major accomplishments here was the development of a mechanism that resulted in the formation of patterns with smaller dimensions than the UME. That is, under appropriate conditions, the width of the silver lines was thinner than the metallic diameter of the UME. This "chemical lens," which was further used in other systems (see below) and was the result of the reaction of silver ions (present in the solution to some extent) with the hydroquinone flux, "focused" the flux on the AgCl surface. Accordingly, the minimum line width formed with a 10 μm diameter UME was 7.5 μm.

Finally, the local etching of nickel has recently been demonstrated by the same group (30). The electrochemical reduction of nitrile [Eq. (3)] at a Pt tip close to a thin Ni layer that had been deposited on gold caused the local etching of the nickel and the exposure of the gold underneath. The

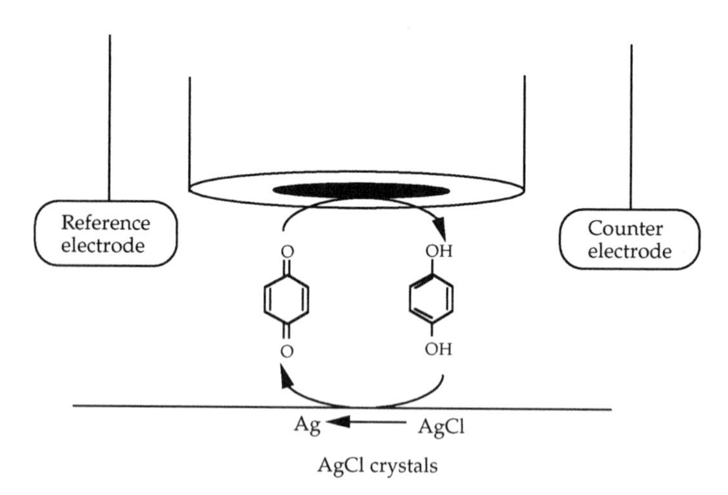

FIG. 11 Schematic representation of the feedback mode for the formation of silver patterns on AgCl.

addition of hydroxyl ions to the solution was successfully used as the "chemical lens" and increased substantially the resolution of the etching patterns.

B. Semiconductor Etching

High-resolution etching of semiconductor surfaces is of technological importance in the fabrication of microelectronic devices and is usually carried out by photolithography through several consecutive steps that include coating with photoresist, x-ray or UV irradiation, etching and stripping. Since the wet etching of semiconductors is an oxidation process, whereby a reactive surface, namely, the semiconductor, is subjected to strong oxidants, it was only natural to exploit the advantages of the SECM for this process.

Indeed, the few studies in which the SECM has been used for etching semiconductors represent one of the most impressive examples of the capabilities of the SECM as a modifying tool. The first system was developed by Mandler and Bard (31) and involved the electrochemical generation of bromine (Fig. 12). This strong oxidant has been widely applied for etching III-V and II-VI semiconductors. The positive feedback that was observed while approaching a GaAs wafer with an UME generating bromine was followed by the formation of clear and deep etching pits (Fig. 13). The conditions for continuous etching required efficiently dissolving the oxide layer that was formed. Oxide layers of GaAs are insoluble in aqueous solutions only at pHs of 3–7, which caused the positive feedback current to decrease rapidly while approaching in this range of pH. On the other hand, a continuous and isotropic etching was observed at either acidic or basic media. The parameters that affected the etching size and shape were examined and included the UME size, the distance between the UME and the

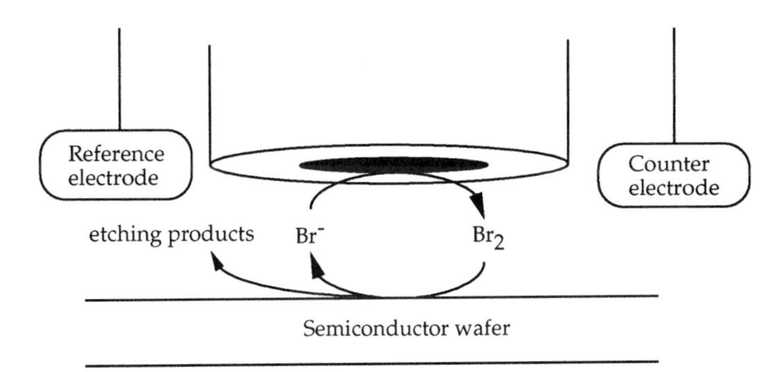

FIG. 12 Overall scheme for the etching of semiconductors using the feedback mode of the SECM.

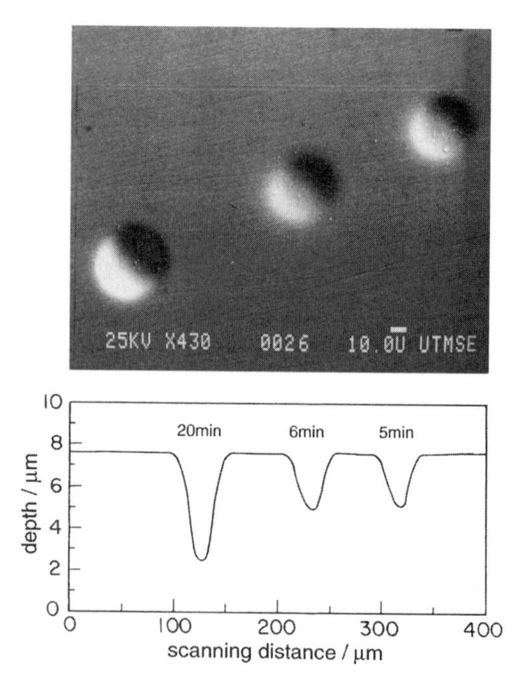

FIG. 13 Scanning electron micrograph and profiles of etching pits made on a GaAs wafer as a result of leaving a 25 μm diameter Pt UME (biased at 1.0 V vs. SCE) above the surface in a 0.02 M HBr/0.1 M HCl solution for different duration. (From Ref. 31.)

surface and the electrolysis time. In a succeeding study (32), the feedback current was carefully studied and used for collecting information about the surface processes involved in the etching of GaAs. Assignment of the energy of the valence band edge of the semiconductor was made by studying the behavior of the feedback current at various pHs and with various redox couples. Changing the pH shifted the band edges of the semiconductor, thus decreasing the overlap between the energy states of the oxidized redox species and the valence band. The feedback current was sensitive to the degree of the overlap and could therefore be used for determining the energy of the edge of the valence band. Moreover, the unique characteristics of the SECM to inject charge locally resulted in the selective etching of n-GaAs versus p-GaAs. That is, n-GaAs was effectively etched by one-electron redox couples, e.g., $Fe(phen)_3^{3+}$, while no etching patterns could be detected on p-GaAs even after prolonged generation of such oxidants close to its surface. This behavior was explained by the band bending of the semiconductor, i.e., while locally injected holes remained on the surface of an n-GaAs leading to efficient etching, fast diffusion of injected holes into the bulk of p-GaAs totally blocked etching. This general phenomenon, which was demonstrated also in the case of silicon etching (see below) and has recently been observed in charge injection into tungsten oxide thin films (33), suggests that in order to drive a local chemical reaction with the feedback mode of the SECM the charge carries responsible for the reaction ought to be trapped on the surface.

Other semiconductors, e.g., CdTe, $Hg_{1-x}Cd_xTe$, and GaP, have also been etched using the SECM. Nevertheless, the latter could be etched only with a stronger oxidant, such as $Ru(phen)_3^{3+}$. Accordingly, a negative feedback current was detected upon approaching GaP while generating bromine, whereas an increase of the steady-state current followed the approach of GaP using the ruthenium complex.

Silicon, the major substrate material used in the fabrication of solid-state electronic devices, was naturally the next semiconductor to be etched with the SECM. Its etching has been accomplished in acidic fluoride solutions by electrogenerating bromine (34). The different parameters that affected the etching process, such as the nature of the oxidant and its concentration, the fluoride concentration, the acidity and the type of silicon, were examined and a detailed mechanism of the overall process was proposed. The most interesting finding was the fact that only bromine etched silicon (with close to 100% current efficiency) whereas a one-electron redox couple had no effect on the silicon surface. The explanation given by the authors is that it is necessary to trap the injected charge on the silicon surface in order to produce etching. The formation of Si-Br bonds (when bromine is generated) scavenges the holes on the Si surface. On the other hand, the local injection of holes by one-electron redox couples, e.g., $Fe(phen)_3^{3+}$, is insufficient since the holes are not trapped on the surface and therefore diffuse away. Micropatterning was demonstrated as well. Figure 14 shows an optical micrograph of a pattern that was formed using a 10 μm diameter platinum UME that was scanned laterally at a rate of 0.1 $\mu m \cdot s^{-1}$.

This system has been the subject of a subsequent study (35) that aimed to improve the etching resolution using a special focusing technique. The so-called confined etchant layer technique (CELT) involves the introduction of a soluble substance that reacts with the electrogenerated species, thus limiting the diffusion layer of the latter around the UME. The CELT technique is somewhat similar to the above-mentioned "chemical lens" reported by Heinze and coworkers (29). While Heinze used this method to increase the resolution of deposited metal patterns as well as metal etching, Tian applied the approach for increasing the resolution of semiconductor etching. Specifically, the formation of etching pits on silicon that matched closely to the UME size were formed by adding H_3AsO_3 into the solution and electro-

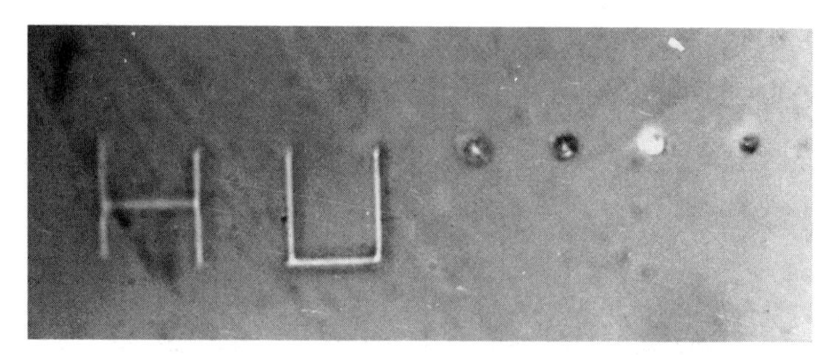

FIG. 14 A "HU" etching micropattern formed on silicon ⟨111⟩ surface via the feedback mode of the SECM using a 10 μm diameter Pt UME with a lateral scanning speed of 0.1 $\mu m \cdot s^{-1}$. The total width of the "H" is ~150 μm.

generating bromine, as in the previous study. The diffusion layer of bromine, μ, could be calculated from the pseudo-first-order reaction rate constant, k_s, of the arsenic acid (that was in access) with bromine [Eq. (4)] and the diffusion coefficient.

$$\mu = (D_{Br_2}/k_s)^{0.5} \tag{4}$$

Nevertheless, there is a significant difference between the "chemical lens" and the CELT. In the former approach the additive, hydroxyl ions, is electrochemically inactive and therefore its concentration is not affected by the electrochemical reaction at the tip. This causes the additive to react solely with the electrogenerated species and therefore its flux is focused even more than the UME diameter. On the other hand, the reagent added in the CELT is electrochemically active and thus is also consumed at the UME. Obviously, this will limit the "focusing" effect to the size of the UME.

C. Inorganic Materials—Deposition and Dissolution

Most of the SECM work involving inorganic materials has been carried out by Unwin and colleagues and focused on local dissolution kinetics. In a series of excellent papers they have developed a novel approach and modeling capabilities for studying the kinetics of dissolution of inorganic crystal. These studies are described in detail in Chapter 12 and only the concept will be presented here (Fig. 15). The principle of Unwin's method is to induce the dissolution of an ionic crystal by locally depleting the concentration of one (or more) of the solute ions in the solution zone between the crystal and the UME from an initial saturated value. This has been accomplished by stepping the UME potential from a value where no electrode reaction occurs to a potential at which the ion is electrolyzed at a diffusion-

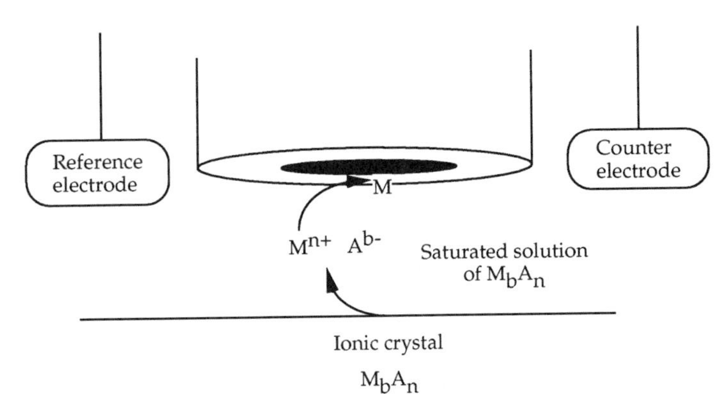

FIG. 15 Schematic representation of the induced dissolution of an ionic crystal by the feedback mode of the SECM.

limited rate. This causes the crystal to dissolve, and the resulting ion flux from the crystal is detected electrochemically at the UME, providing quantitative information on the kinetics of dissolution. Although no attempts have been made to pattern etching structures on crystals, the etching pits could be studied with atomic resolution as a result of the successful combination of the AFM and the SECM (36) (see Chapter 4).

Shohat and Mandler (37), and more recently Turyan et al. (33), have reported additional studies of patterning inorganic materials. The goal of the first report was to develop an approach for driving local acid-base reactions on surfaces. The motivation stems from previous work in which stable metal oxides, e.g., nickel hydroxide, served as an anchor for attaching organic and biological molecules onto surfaces in specific patterns.

The basic idea was to utilize an electrochemical reaction on the surface, which would be coupled with proton consumption. This configuration would be characterized by a pH gradient between the surface and the UME. The developed approached is depicted schematically in Figure 16. Specifically, the reduction of protons on the surface was driven by reduced methyl viologen, $MV^{+\bullet}$, a process that is catalyzed by platinum. A mercury UME had to be used to prevent the reduction of protons at the UME surface. As a result of the local increase of the pH on the surface beneath the UME, $Ni(OH)_2$ was irreversibly deposited. Interestingly, the $Ni(OH)_2$ patterns could be imaged by the SECM. Moving the microelectrode across the deposited patterns in a Ni^{2+}-free solution that contained only MV^{2+} showed that the feedback current above the $Ni(OH)_2$ deposits was less positive than over the bare platinum surface. A detailed study suggested that the difference in the feedback current was due to the different conductivity of the Pt versus

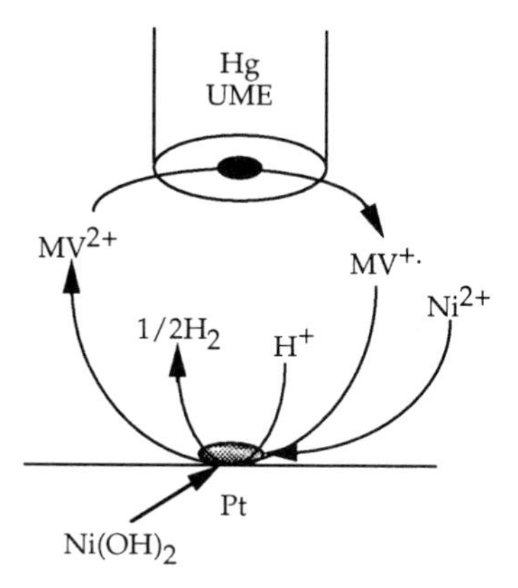

FIG. 16 Schematic representation of the approach for deposition of $Ni(OH)_2$ by the feedback mode of the SECM.

$Ni(OH)_2$ and to the catalytic effect that regenerated MV^{2+} on the metallic surface. This approach of altering the pH at the electrode surface was used further by Heinze et al. for driving the deposition of silver (29).

The second study involved the local reduction or oxidation of an electrochromic material, tungsten oxide (33). This system is of particular interest since it has been the first system where the local injection of charge could be followed by an instantaneous change of color, making it possible to follow the process with an optical microscope. The stable form of tungsten oxide is WO_3, which is colorless and poorly conductive. The reduction of WO_3 [Eq. (5)] results in the formation of deep blue tungsten bronze that is substantially more conducting.

$$WO_3 \text{ (colorless)} + xe^- + xM^+ \rightarrow M_xWO_3 \text{ (blue)} \tag{5}$$

Two systems were studied using the feedback mode of the SECM, the local reduction of oxidized tungsten oxide by electrochemically generated methylviologen radical cation and the local oxidation of reduced tungsten oxide by iron(III) generated at the UME (Fig. 17). The tungsten oxide films were prepared by the sol-gel method, which made it possible to control the thickness of the films as well as the substrates on which they were dip-coated. Observation of the feedback current as the concentration of the mediator, the thickness of the film, and the potential of the substrate were changed enabled the authors to determine whether the feedback current was governed by the diffusion of the mediator or by the kinetics of charge transport inside the film. Moreover, the development of a "writing-reading-erasing" system based on the local change of the color of the tungsten oxide

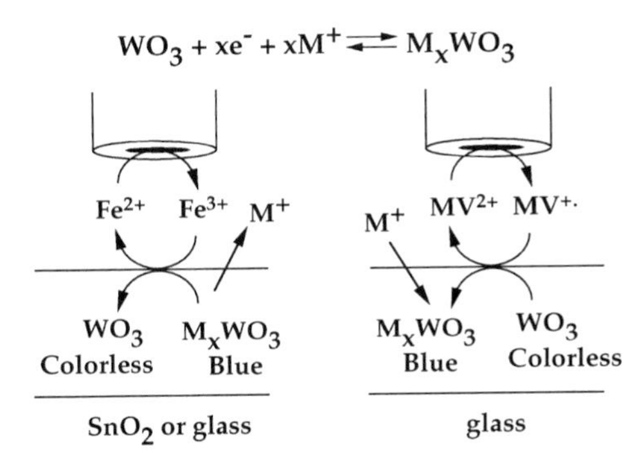

FIG. 17 Schematic representation of the "writing-reading-erasing" system based on tungsten oxide.

wad demonstrated. Remarkably, the "writing" process was extremely fast and the local injected charge was very stable.

D. Deposition of Conducting Polymers

The major obstacle to a much wider application of the feedback mode for microfabricating conducting polymer patterns is the basic requirement that the monomer be attached to the surface. Nevertheless, the development of strategies for patterning conducting polymers has attracted much interest, as conducting polymers may one day supplant solid-state devices. Moreover, the SECM could be used for the local doping and undoping of conducting polymers as well. Indeed, Bard, Mirkin, Heinze and others have used the SECM for studying surface processes at conducting polymers (see Chapter 6). In these investigations the feedback current was studied upon approach to biased surfaces coated with conducting polymers. The behavior of the feedback current as a function of the film thickness, the concentration of the mediator in the solution, the potential of the surface, and the nature of the electrolyte provided qualitative and quantitative information about the film and the electron-transfer processes inside. Only those studies that aimed to microfabricate patterns of conducting polymers will be discussed here.

The first successful attempt to deposit conducting polymers using the feedback mode of the SECM was reported by Heinze and coworkers (28). The system has subsequently been refined (39). The essence of their method was to take advantage of the different solubility properties of the monomer versus the polymer in different solvents. The monomer was 2,5-bis(1-methyl-pyrrol-2yl)-thiophene (NSN) (Fig. 18), which is very soluble in organic solvents but insoluble in aqueous solutions. On the other hand, the respective polymer is insoluble in either aqueous or nonaqueous solutions.

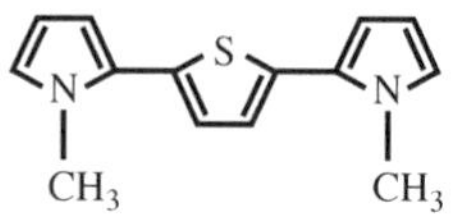

FIG. 18 2,5-Bis(1-methyl-pyrrol-2yl)-thiophene (NSN).

A film of the monomer on different substrates was deposited by thermal evaporation, and micropatterning was performed in an aqueous solution consisting of a reduced redox couple—Br^-, $Ru(bpy)_3^{2+}$, and $IrCl_6^{3-}$. The oxidant that was electrogenerated at the UME caused the polymerization of the monomer to form patterns of poly-NSN. Then the unreacted monomer was washed away by an organic solvent that did not remove the polymer. This resulted in the formation of freestanding polymeric structures (Fig. 19) on either ITO or on an insulating substrate such as polymethylmethacrylate. It was found that the mediator, as well as the nature of the substrate and its interaction with the monomer and oligomers formed in the course of the polymerization, had a significant effect on the polymer.

Recently, Zhou and Wipf reported on the deposition of polyaniline patterns on gold, platinum, and carbon surfaces using the feedback mode of the SECM (40). The essence of their method, termed "microreagent," has been to locally increase the pH in the volume between the UME and the surface upon the reduction of protons. The increase of alkalinity shifted the oxidation potential of aniline cathodically. Hence, polyaniline was electrodeposited beneath the UME as a result of applying a constant positive potential to the substrate, a potential that caused aniline oxidation under the pH conditions generated by the UME. The different parameters that affected the width and thickness of the deposited films were examined and included the UME dimension and potential, substrate potential, deposition time, and UME-substrate separation. Features as small as 3 μm in diameter were formed using a 3 μm diameter Pt UME. The UME and substrate potentials had to be adjusted to cause a large local decrease in proton concentration (required negative potentials at the UME) and polyaniline deposition beneath the UME only (positive potentials at the substrate). Smaller tips required more negative potentials at the UME in order to get good polyaniline deposits. Figure 20 shows an optical micrograph of a structure formed by scanning the UME laterally across the substrate surface. The effect of UME-substrate separation, UME potential, and deposition time were strongly coupled. Close separations allowed using shorter deposition times or lower UME potentials. On the other hand, thickness of the film could be controlled by these parameters. The polyaniline structures could be imaged by the SECM. The fact that positive feedback currents, which were larger than

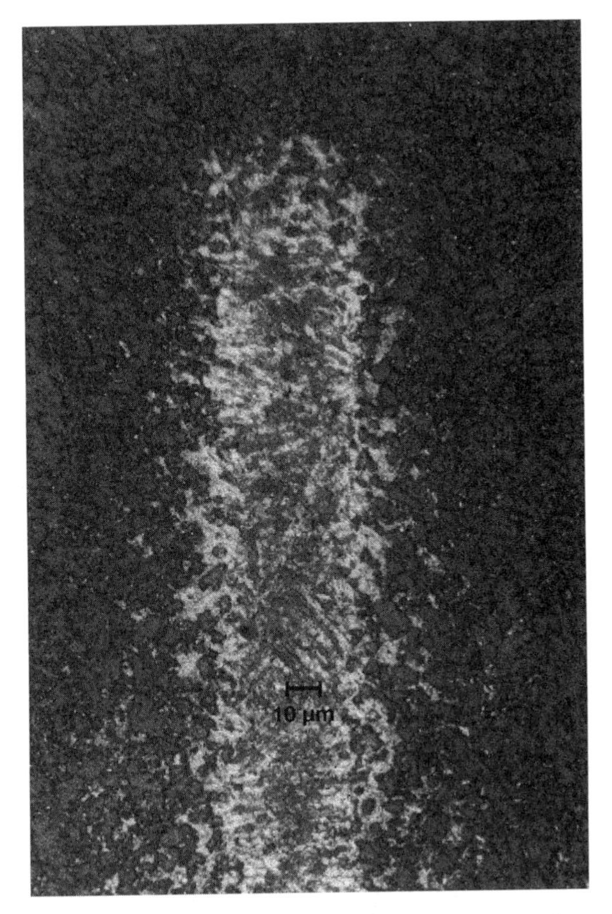

FIG. 19 A pattern of poly-NSN on polymethylmethylmethacrylate.

those detectd on the bare substrate, were measured over the structures of polyaniline (while the potential of the substrate was held at 0.6 V vs. Ag/AgCl) verified their high conductivity. Finally, measuring the shorting current between the UME and the substrate upon touching the polyaniline provided a rough estimate of the conductivity of ~0.18 $S \cdot cm^{-1}$.

E. Organic and Biomolecule Patterning

The exploitation of the SECM in the feedback mode for microfabrication of organic and biological structures has only recently been reported and lags substantially behind its application in the direct mode as described above.

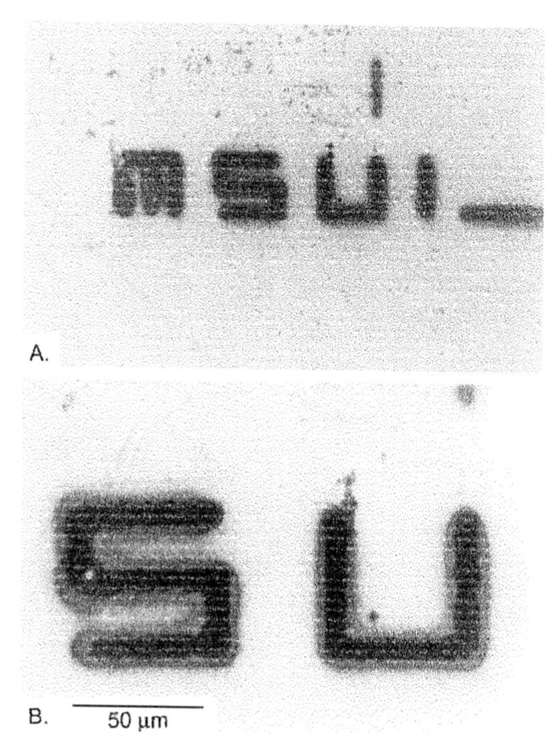

FIG. 20 An optical microscope image of a polyaniline pattern deposited on a gold substrate using a 10 μm Pt UME in the "microreagent" mode. (From Ref. 40.) Figure B is a magnification of part of Figure A.

Presumably, the reason for this is that most organic and biological molecules do not exhibit simple one-electrode redox behavior and thus are not suitable as mediators. This means that in order to use the feedback mode of the SECM for micropatterning organic and biomolecules, one has first to attach them onto surfaces and then develop ways to locally desorb/deactivate or trigger/activate them by an electrochemical reaction driven by the mediator. Although methods for attaching organic and biomolecules onto conducting as well as insulating substrates have been elaborated for many years, the second stage of altering their reactivity by a redox mediator is not straightforward. Accordingly, and as will be described, most contributions to this area have used a similar concept of locally deactivating sites on organic and biological thin film substrates.

The first study was reported by Matsue and coworkers (41) and aimed at creating micropatterns of diaphorase—a flavin enzyme on a glass substrate. The approach is based on the local electrogeneration of highly reactive species, e.g., bromine, that deactivated immobilized enzyme molecules forming nonreactive patterns. The formation of these patterns was verified by the SECM used as an imaging tool. Diaphorase catalyzes the oxidation of the reduced nicotinamide cofactor NADH by an artificial mediator such as ferrocenylmethanol. Therefore, the latter was introduced into the solution together with NADH and the diaphorase-nonreactive patterns were detected by the negative feedback current observed upon the approach of an UME biased at a positive potential that caused the oxidation of the mediator (Fig. 21). The size of the deactivated zones reflected the diffusion length of the electrogenerated species, and therefore their diameter varied roughly with the square root of the pulse period of deactivation. Smaller patterns were formed as a result of electrogenerating chlorine instead of bromine. The higher resolution could be explained by the "chemical lens" (or CELT), which suggested that the diffusion layer of chlorine was thinner due to its reaction with water.

Although the SECM has been used by the same group as a means of characterizing enzyme and antigen/antibody patterns (42) of various types, a glass capillary pen was used to create the patterns. Recently Matsue et al. (43) presented a slightly different approach in which, instead of deactivating an enzyme, silane monolayers attached onto glass surfaces were modified

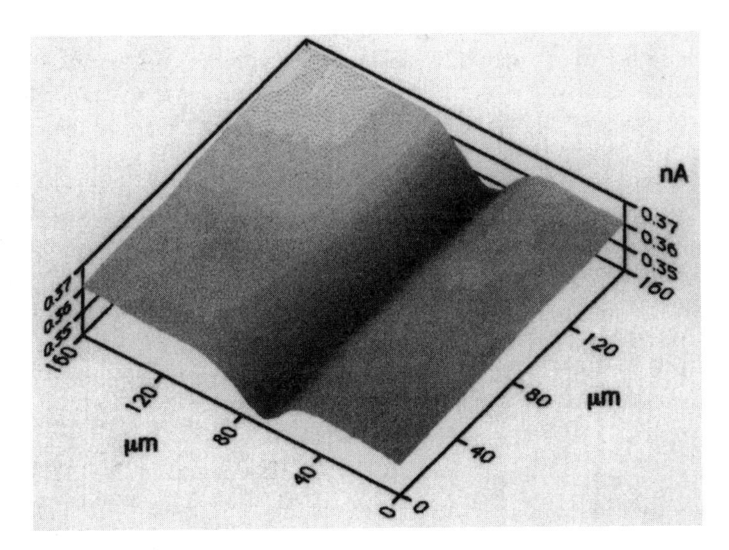

FIG. 21 SECM image of a diaphorase-immobilized substrate with a deactivated line formed by scanning an UME while generating bromine across the surface. (From Ref. 41.)

locally by reactive species, i.e., hydroxyl radicals. Three procedures for the preparation of diaphorase micropatterns have been demonstrated (Fig. 22). Through all these approaches, hydroxyl radicals were electrochemically generated by Fenton's reaction, where Fe^{3+} was reduced at the UME in the presence of H_2O_2 [Eqs. (6) and (7)]:

$$Fe^{3+} + e^- \rightarrow Fe^{2+} \tag{6}$$

$$Fe^{2+} + H_2O_2 \rightarrow Fe^{3+} + HO^{\bullet} + HO^- \tag{7}$$

The radicals reacted with the end-functional groups of the monolayers, altering their chemical and physical properties in terms of their ability to adsorb or covalently attach the enzyme. The SECM itself was used, as previously described, for verifying the presence of attached enzymes. In the first approach (Fig. 22A) hydroxyl radicals were used to locally degrade octadecylsilyl chains. The hydrophobic chains that are capable of adsorbing diaphorase became significantly more hydrophilic, preventing diaphorase adsorption and creating enzyme-free zones. The second concept (Fig. 22B), also producing enzyme-free zones, involved the degradation of (3-aminopropyl)trimethoxysilane-derived monolayers by hydroxyl radicals. The enzyme was covalently attached through the amino groups of the unaffected molecules using glutardialdehyde as a cross-linker. The resolution of the resulting patterns depended on the different reaction paths that the hydroxyl radicals could follow. The concentration of these highly reactive species depended on the concentration of the oxidized and reduced iron ions, as well as on that of hydrogen peroxide. While increasing the concentration of Fe^{3+} increased the levels of the radicals, both Fe^{2+} and H_2O_2 reacted with $OH^{\bullet}$. In other words, the diffusion layer of the hydroxyl radicals became thinner by the "chemical lens" mechanism.

The third approach (Fig. 22C) resulted in "positive" patterns, i.e., in the formation of defined areas where diaphorase was selectively attached. This approach consisted of three steps. Initially, a mercapto-terminated monolayer was locally exposed to hydroxyl radicals formed via the Fenton's reaction. Two simultaneous processes occurred: the monolayer in the areas exposed to the radicals was removed to form hydroxyl-terminated surface and, at the same time, the mercapto groups in the remaining area were

FIG. 22 Schematic representation of the different approaches for micropatterning glass with diaphorase via electrogenerating hydroxyl radicals. (A) Diaphorase is physically adsorbed on hydrophobic zones. (B) Diaphorase is covalently bound through nonattacked amino groups. (C) Diaphorase is covalently attached to hydroxyl-activated sites.

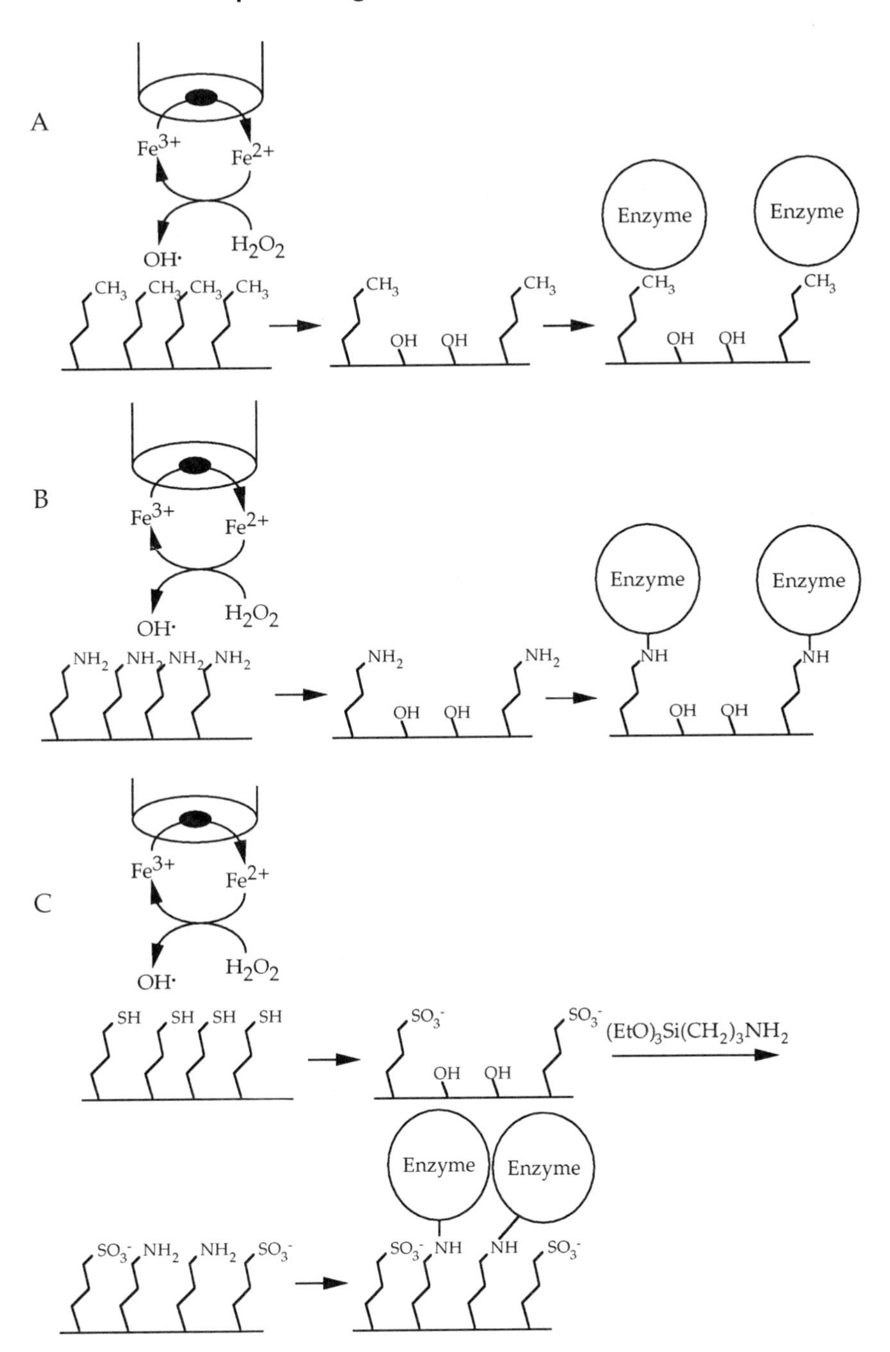
A
Fe^{3+}
Fe^{2+}
$OH\cdot$
H_2O_2
CH_3 CH_3 CH_3 CH_3
CH_3 CH_3
OH OH
Enzyme
Enzyme
B
NH_2 NH_2 NH_2 NH_2
NH_2 NH_2
NH NH
C
SH SH SH SH
SO_3^- SO_3^-
$(EtO)_3Si(CH_2)_3NH_2$
SO_3^- NH_2 NH_2 SO_3^-

converted to sulfonate groups in an oxidation process driven by the H_2O_2/Fe^{2+} solution. Diaphorase was immobilized onto the hydroxylated sites through the glutardialdehyde linker and verified by the SECM using the same enzymatic catalyzed reaction as before (42) (Fig. 23).

The last example of micropatterning surfaces with biological molecules by the feedback mode of the SECM has been demonstrated by Wipf and coworkers (44) and involves the attachment and detachment of biotin on a glassy carbon electrode (GCE). Biotin/avidin chemistry has been used extensively as a means of modifying surfaces by protein immobilization. The high specificity of the biotin/avidin interaction permitted selective attachment of avidin-labeled biomolecules. Two complementary approaches have been developed for either locally attaching biotin hydrazide or locally removing attached biotin. In the former case, an active product was formed at the UME upon oxidizing biotin hydrazine that diffused to the GCE and was irreversibly attached through an unknown mechanism. The biotinylated regions were further reacted with a fluorescent avidin in order to form fluorescent spots. A number of parameters affected the formation and resolution of the patterns, namely, the potentials of the UME and the substrate, as well as the distance and the speed at which the UME was moved. The potential of the UME had to be sufficiently positive (1.1 V vs. Ag/AgCl) to generate

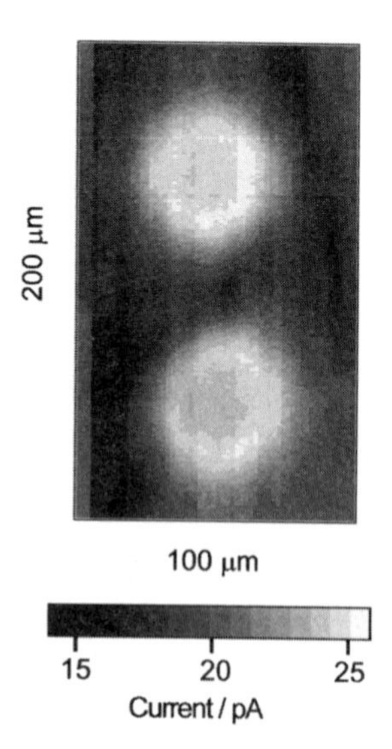

FIG. 23 SECM image of diaphorase attached through the local degradation of a mercapto-terminated monolayer that was subsequently treated with (3-aminopropyl)trimethoxysilane. (From Ref. 43.)

a large but local increase in the activated form of the biotin, while the potential of the surface had to be controlled to avoid generation of active biotin all over the GCE (which occurred at both negative and positive potentials). Moreover, because of the high reactivity of the oxidized biotin, the UME had to be positioned within the thin diffusion layer of the generated species, and its scan speed had to be rather slow (between 0.1–0.5 $\mu m \cdot s^{-1}$). Optimization of these parameters resulted in clear and reproducible patterns of attached biotin.

On the other hand, the local removal of an attached biotin/avidin complex has been accomplished by generating hydroxide ions electrochemically. Investigating the mechanism of the reaction between hydroxide ions and the various constituents of the system led the authors to conclude that the reaction occurring at the electrode was a hydrolysis of all the amide bonds involved in the immobilization reaction.

III. PERSPECTIVE APPROACHES

In order to evaluate the prospects of the SECM as a tool for micro- and nanopatterning, its advantages and disadvantages need to be considered. In this last section we measure the capabilities of the SECM against the characteristics of an effective patterning technique, namely, speed, resolution, and scope of materials that can be patterned.

A. Speed of Patterning

As in any other scanning probe microscopy technique, the speed of patterning is limited by the movement of the probe across the substrate as well as by the rate of the patterning reaction. The latter is sometimes diffusion controlled, which suggests that the UME can be scanned relatively quickly when it is held very close to the surface. This obviously requires a reliable and fast feedback mechanism in order to assure that the UME does not crash onto the surface. Examples of micropatterning surfaces with speeds exceeding micrometers per second have been demonstrated in the case of copper etching (24), tungsten oxide reduction (33), and deposition of a conducting polymer (39). At the same time, increasing the speed of the UME causes the steady state current to increase because the UME outruns the growth of the diffusion layer, as has been shown by Borgwarth et al. (38) and termed the picking mode. The picking mode is of particular use when the nature of the surface is not investigated and the importance is to keep the tip at a constant distance above it. Under such conditions the diffusion layer is substantially smaller than the microelectrode diameter (including the insulating sheath), which results in patterns with the width of that of the conducting part of the UME. Increasing the UME speed will make it possible to apply smaller tips, which will maintain also a reasonable aspect ratio of the patterns that are formed.

On the other hand, there are many cases where the chemical or electrochemical reactions taking place on the surface are the rate-limiting steps in the patterning process. This will limit the patterning rate or will require development of other approaches whereby the whole pattern is made at the same time. At present, the common approaches used in SECM, i.e., the direct and feedback modes, cannot compete with the conventional photolithography methods. Nonetheless, future approaches such as multitip configuration may dramatically enhance the SECM capabilities in terms of speed of surface modification.

B. Resolution

If speed is an obstacle to the wide application of scanning probe techniques as surface modification tools, their resolution keeps them in the running. The resolution of the patterns formed by the SECM depends on the size and geometry of the UME, as well as the distance it is held above the surface. At the same time, the resolution of the patterns can easily be controlled (as compared to other SPM techniques) by approaches such as the "chemical lens." Pattern width can be controlled merely by varying the electrolyte and the surface-UME distance. The UME need not be altered.

In the direct mode the resolution is governed by the electric field distribution, which therefore requires that the UME be brought very close to the surface. At the same time, the direct mode does not require a UME with a disc shape insulated in an inert matrix, as long as the active part of the UME is small and close to the surface. As described above, high-resolution modification of surfaces by the direct mode of the SECM is basically an STM-driven approach. On the other hand, the resolution of the feedback mode is presently limited by our ability to produce micro- and nanoelectrodes with well-defined shapes. Recently, Mirkin and Lewis reported (45) the preparation and characterization of nanoelectrodes that are suitable for SECM (see Chapter 3). This development may pave the way for a substantial increase in feedback mode resolution.

C. Scope of Materials and Approaches

The major advantage of the SECM as a patterning tool lies in its ability to drive a wide range of chemical and electrochemical surface reactions. As compared with other scanning probe techniques, e.g., STM and AFM, in an SECM experiment the mechanism of the surface process is usually known and involves electrochemical and chemical processes. Moreover, the conditions for patterning are mild, i.e., the UME is biased moderately and, therefore, processes are very easily studied and controlled. This has made it possible to exploit the SECM in the feedback mode for driving almost all type of processes including oxidation-reduction, acid-base, and precipitation reactions which resulted in a variety of surface patterning. These accomplishments stand in contrast with the fact that there are very few straightforward approaches that can be used for patterning surfaces with metals, inorganic materials, organic and biochemicals. The conventional photolithographic methods are capable of depositing metals and inorganic materials, i.e., oxides, whereas novel methods such as micro-contact printing enable one to pattern surfaces mostly with organic materials. The fact that the SECM (in the feedback mode) does not require that the surface be conductive makes it an attractive method for the modification (and characterization) of inorganic materials such as salt crystals, amorphous materials such as glass, and, in the future, biological membranes and tissues.

Various approaches have been developed based on the feedback mode. Besides driving irreversible electrochemical surface processes such as the dissolution of metals and semiconductors, the SECM has been used for driving acid-base reactions, generating radicals that deactivate biological and organic substances, and shifting precipitation reactions that affected the dissolution of salts.

Finally, the SECM is not limited to electroactive materials. Novel approaches based on other modes, currently under investigation, will enable us to modify surfaces with organic monolayers, polymers, and biological molecules. This is the area in which the future of SECM nanofabrication looks most bright. However, the fulfillment of its promise depends on the development of generic techniques that will increase the speed at which it can carry out modification.

The SECM was born of the union of scanning probe microscopy with the ultramicroelectrode. Its extraordinary versatility, which has been enhanced by the ongoing miniaturization of electrodes, is likely to make it an increasingly useful tool for surface nanopatterning.

ACKNOWLEDGMENT

J. Heinze and K. Borgwarth from the University of Freiburg are warmly acknowledged for fruitful discussions.

REFERENCES

1. Mandler D, Meltzer S, Shohat I. Isr J Chem 36:73, 1996.
2. Trachtenberg I. Unpublished results.

3. Lin CW, Fan F-R, Bard AJ. J Electrochem Soc 134:1038, 1987.
4. Craston DH, Lin CW, Bard AJ. J Electrochem Soc 135:785, 1988.
5. (a) Hüsser OE, Craston DH, Bard AJ, J Vac Sci Technol B 6:1873, 1988. (b) Hüsser OE, Craston DH, Bard AJ. J Electrochem Soc 136:3222, 1989.
6. Forouzan F, Bard AJ. J Phys Chem B 101:10876, 1997.
7. Wuu Y-M, Fan F-R, Bard AJ. J Electrochem Soc 136:885, 1989.
8. Kranz C, Ludwig M, Gaub HE, Schuhmann W. Adv Mater 7:38, 1995.
9. Kranz C, Ludwig M, Gaub HE, Schuhmann W. Adv Mater 7:568, 1995.
10. (a) Ludwig C, Kranz W, Schuhmann W, Gaub HE. Rev Sci Instrum 66:2857, 1995. (b) Kranz W, Gaub HE, Schuhmann W. Adv Mater 8:634, 1996.
11. Kranz C, Wittstock G, Wohlschläger H, Schuhmann W. Electrochim Acta 42: 3105, 1997.
12. Sugimura H, Uchida T, Shimo N, Kitamura N, Masuhara H. Ultramicroscopy 42–44:468, 1992.
13. Sugimura H, Uchida T, Shimo N, Kitamura N, Shimo N, Masuhara HJ. Electroanal Chem 361:57, 1993.
14. Wittstock G, Hesse R, Schuhmann W. Electroanalysis 9:746, 1997.
15. (a) Widrig CA, Chung C, Porter MD. J Electroanal Chem 310:335, 1991. (b) Yang D-F, Wilde CP, Morin M. Langmuir 12:6570, 1996. (c) Yang D-F, Wilde CP, Morin M. Langmuir 13:243, 1997.
16. Wittstock G, Schuhmann W. Anal Chem 69:5059, 1997.
17. Schneir J, Sonnenfeld R, Marti O, Hansma PK, Demuth JE, Hamers RJ. J Appl Phys 63:717, 1988.
18. (a) Li W, Virtanen JA, Penner RM. Appl Phys Lett 60:1181, 1992. (b) Li W, Virtanen JA, Penner RM. J Phys Chem 96:6529, 1992.
19. Sugimura H, Uchida T, Kitamura N, Masuhara H. J Phys Chem 98:4352, 1994.
20. Sugimura H, Nakagiri N. Langmuir 11:3623, 1995.
21. Schoer JK, Zamborini FP, Crooks RM. J Phys Chem 100:11086, 1996.
22. (a) Corbitt TS, Crooks RM, Ross CB, Hampden-Smith MJ, Schoer JK. Adv Mater 5:935, 1993. (b) McCarley RL, Hendricks SA, Bard AJ. J Phys Chem 96:10089, 1992. (c) Staufer U. In Wiesendanger R, Güntherodt H-J, eds. Scanning Tunneling Microscopy II. New York: Springer-Verlag, 1992, pp 273–302. (d) Lebreton C, Wang ZZ. Scanning Microsc 8:441, 1994.
23. Mandler D, Bard AJ. J Electrochem Soc 137:1079, 1990.
24. Mandler D, Bard AJ. J Electrochem Chem 136:3143, 1989.
25. Macpherson JV, Slevin CJ, Unwin PR. J Chem Soc Faraday Trans 92:3799, 1996.
26. Still JW, Wipf DO. J Electrochem Soc 144:2657, 1997.
27. Meltzer S, Mandler D. J Electrochem Soc 142:L82, 1995.
28. Borgwarth K, Ricken C, Ebling DG, Heinze J. Ber Bunsenges Phys Chem 99: 1421, 1995.
29. Heß C, Borgwarth K, Ricken C, Ebling DG, Heinze J. Electrochim Acta 42: 3065, 1997.
30. Ufheil Y, Borgwarth K, Heinze J. In preparation.
31. Mandler D, Bard AJ. J Electrochem Soc 137:2468, 1990.
32. Mandler D, Bard AJ. Langmuir 6:1489, 1990.

33. Turyan I, Krasovec UO, Orel B, Reisfeld R, Mandler D. Adv Mater 12:330, 2000.
34. Meltzre S, Mandler D. J Chem Soc Faraday Trans 91:1019, 1995.
35. (a) Zu YB, Xie L, Mao BW, Mu JO, Xie ZX, Tian ZW. Chi Sci Bull 42:1318, 1997. (b) Zu YB, Xie L, Mao BW, Tian ZW. Electrochim Acta 43:1683, 1998.
36. Macpherson JV, Unwin PR, Hillier AC, Bard AJ. J Am Chem Soc 118:6445, 1996.
37. Shohat I, Mandler D. J Electrochem Soc 141:995, 1994.
38. Borgwarth K, Ebling DG, Heinze J. Ber Bunsenges Phys Chem 98:1317, 1994.
39. Borgwarth K, Rohde N, Ricken C, Hallensleben ML, Mandler D, Heinze J. Adv Mater 11:1221, 1999.
40. Zhou J, Wipf DO. J Electrochem Soc 144:1202, 1997.
41. Shiku H, Takeda T, Yamada H, Matsue T, Uchida I. Anal Chem 67:312, 1995.
42. Shiku H, Hara Y, Takeda T, Matsue T, Uchida I. ACS Symp Ser 656, Solid-Liquid Electrochem Interfaces. 1997.
43. Shiku H, Uchida I, Matsue T. Langmuir 13:7239, 1997.
44. Nowall WB, Wipf DO, Kuhr WG. Anal Chem 70:2601, 1998.
45. Shao YH, Mirkin MV, Fish G, Kokotov S, Palanker D, Lewis A. Anal Chem 69:1627, 1997.

14

CONCLUSIONS AND PROSPECTS

Allen J. Bard

University of Texas at Austin
Austin, Texas

The preceding chapters have provided a detailed exposition of the principles and applications of scanning electrochemical microscopy (SECM). Although SECM has already shown its applicability in a wide range of problems, from fast heterogeneous kinetics at interfaces to imaging of biological molecules, and from characterization to fabrication, the technique will continue to develop. New instrumentation and types of tips will be introduced and the theoretical treatments appropriate to these will be developed. New applications will follow. In this concluding chapter we will give an overview of some of the work on the horizon and will project possible paths for future development of SECM.

I. COMBINING SECM WITH OTHER TECHNIQUES

Carrying out SECM experiments in concert with other techniques, such as atomic force microscopy (AFM) or near-field scanning optical microscopy (NSOM), greatly increases the difficulty of the experiment and presents challenging problems in tip fabrication. However, obtaining information from two different techniques at the same time and at exactly the same location can greatly increase the power of the SECM technique, e.g., by providing independent topographical or optical data during the SECM scan. Initial experiments have been reported along these lines, and these should continue in the coming years.

A. AFM/SECM

AFM is based on a measurement of the force between a scanning probe tip and a surface. This force is frequently measured by noting the deflection of a Si cantilever in its approach to the surface, as revealed by changes in the deflection of a laser beam reflected from the cantilever. Alternatively, the tip can be made to oscillate in the x-y plane (called "dithering") and the effect of the shear force caused by the substrate on the magnitude of the oscillation used as a measure of lateral force. This is carried out by a reflected laser beam or by an attached tuning fork. AFM measurements made during an SECM experiment can be useful, allowing an independent determination of the zero distance point ($d = 0$) and aiding in the positioning of the tip—especially important when tips of diameters below a few μm are employed, since with these SECM feedback effects only become important at very small distances. Moreover, this combination of techniques opens the way to experiments in which changes in topography are studied by AFM as a reactant is generated electrochemically at the tip.

Several studies of this type have already been reported. For example, Ludwig and coworkers (1) used a piezo element attached to the tip shaft for dithering and measured this modulation by noting changes in the Fresnel diffraction pattern created with an illuminating laser beam focused on the lower part of the electrode on a split photodiode. This type of detection is also sometimes used with NSOM tips, but it is a little cumbersome, especially if the tip is largely immersed in solution. It has rarely been used in SECM studies. An alternative is to use a cantilever-based AFM and modify the cantilever so that the tip behaves as an SECM tip. On conventional cantilevers the tip is an integral part of the Si cantilever and is nonconductive (e.g., Si_3N_4). Thus the tip and bottom of the cantilever must be metallized to establish a conducting path to the tip, e.g., by evaporating or sputtering a thin film (~300 Å) of Pt, and then all of the conductive surface except the very end of the tip insulated from the contacting solution. Such an arrangement was used to study the AFM of a KBr crystal undergoing dissolution, where the dissolution was promoted by oxidation of Br^- in the solution at the tip (2). While the AFM imaging was not affected significantly by the metallization, the SECM performance of the tip was not good, mainly because of difficulties with simultaneously providing good tip insulation and tip configuration. A different approach, where special cantilever-type SECM tips were constructed, showed better SECM performance (3). The principle of construction of these tips is shown in Figure 1. A 50-μm-diameter Pt wire is bent and electrochemically etched to a small point by the same procedure as that used for STM tips, then flattened and insulated using electrophoretic paint. It was then epoxied to a commercial AFM Si probe replacing the usual cantilever that had been removed. This could then be used at the probe in a commercial AFM, with contact to the tip allowing for SECM measurements.

An alternative approach is the use of a tuning fork glued to the tip (4–6). In this work a commercial NSOM instrument employing tuning fork positioning was modified to use a tungsten tip insulated with varnish. So far

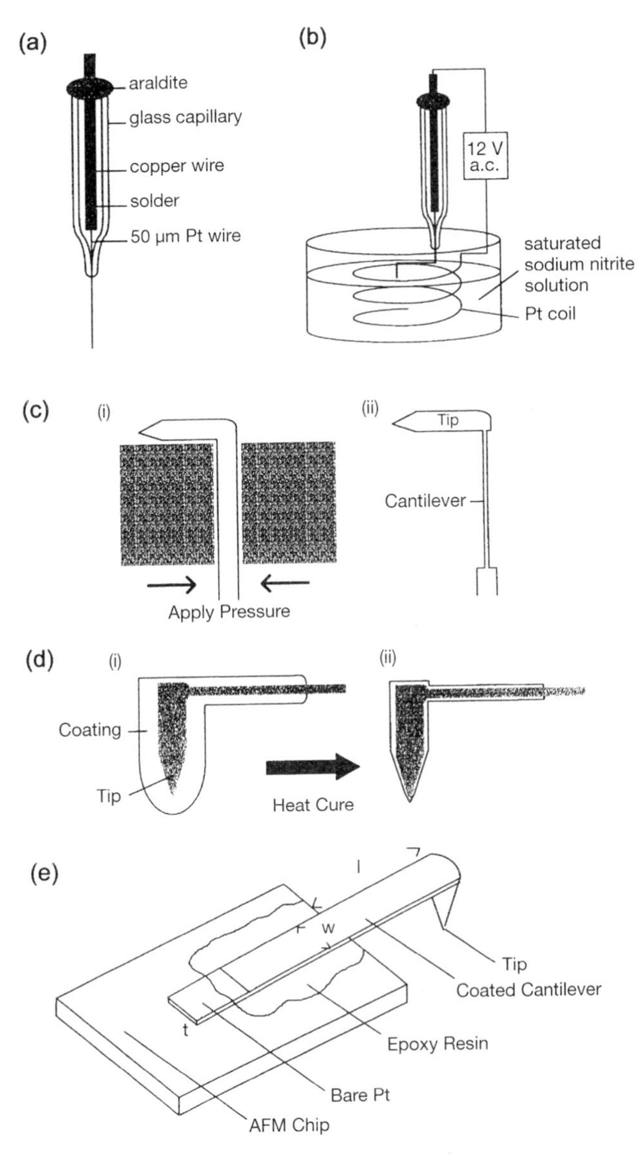

FIG. 1 Steps involved in constructing an SECM/AFM tip: (a) Pt microwire inserted in holder; (b) electrochemical etching; (c) compression to form cantilever component; (d) insulation of electrode and cantilever with electrophoretic paint followed by heating to expose tip; (e) attachment to commercial AFM chip. (From Ref. 3.)

only SECM images have been reported. A key issue in the application of AFM/SECM is the ability to obtain good quantitative approach curves for both insulators and conductors that agree with the theoretical treatments.

B. NSOM/SECM

NSOM is a scanning probe technique for obtaining high-resolution optical images and performing spectroscopic measurements. By using a fiberoptic probe that has been pulled down to a very small tip (50–100 nm) and scanned in the x-y plane very close to the sample, one can obtain images whose resolution is governed by the tip diameter and placement rather than the usual diffraction limits that govern conventional microscope optics (7). Typical fiberoptic tips have coatings of aluminum, which help minimize loss of light in passage from the laser source to the tip. For a tip that is also useful in SECM, the metal coating should be of a more noble metal, such as gold, and an insulating layer must be added so that only the conducting end ring of metal is exposed to the solution. This has been difficult to accomplish with a true NSOM tip, maintaining the metal layer pinhole-free, so that all light comes from the tip, with the insulating layer producing a good ring geometry with no leaks to other locations on the metal. One useful method of judging the integrity of the insulating layer, used with tips where insulation was accomplished with electrophoretic paint deposition, was to lower the tip from the air into the test solution and to note the current as the tip was immersed deeper into the solution (8).

There have been reports of the use of commercial NSOM with a modified tip coated with varnish for SECM and photoelectrochemical experiments (9,10). So far only far field optical imaging has been reported with resolution in the μm regime. No SECM approach curves for these tips have been shown. Because of the power of optical imaging and spectroscopic measurements, there is strong motivation to perfect NSOM/SECM techniques, and further developments in this area appear to be on the horizon. Since positioning of the NSOM tip frequently involves shear force measurements, e.g., with a tuning fork, techniques for making simultaneous SECM, NSOM, and AFM measurements should be possible, given the proper tips.

C. ECL/SECM

An alternative approach to obtaining optical imaging with the SECM is through light generation via an electrogenerated chemiluminescence (ECL) process at the tip. Light generation at ultramicroelectrodes can be accomplished either by the radical ion annihilation approach (in aprotic solvents) (11,12) or by a coreactant route (also applicable in aqueous solutions)

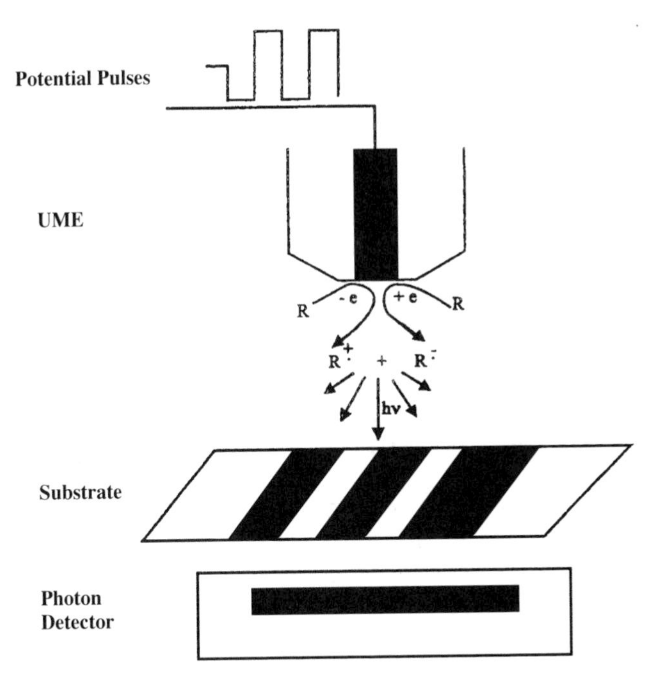

FIG. 2 Schematic diagram for ECL generation at an SECM tip for optical imaging. (From Ref. 13.)

(13,14). A schematic diagram showing the basic concept is shown in Fig. 2. In the only optical imaging study reported, the resolution was at the μm level, far below the resolution needed to be competitive with NSOM (13). However, approach curves in which the current or the emission intensity varied as a function of tip-substrate spacing, d, were demonstrated. This suggests that monitoring tip current could be useful in approaching the tip to the substrate and maintaining d during optical imaging. For this technique to be successful in optical imaging, it will be necessary to show that sufficient light levels for detection can be generated at tips of the 100-nm-size scale. It will also be necessary to demonstrate ECL systems that produce constant light levels over the time it takes to scan a sample (several minutes). For these very small tips, tuning fork or other sensing methods will probably be required to allow scanning at close distances.

D. EQCM/SECM

SECM has been used to monitor a substrate in contact with a quartz crystal microbalance (QCM) and detect species generated at the substrate at the

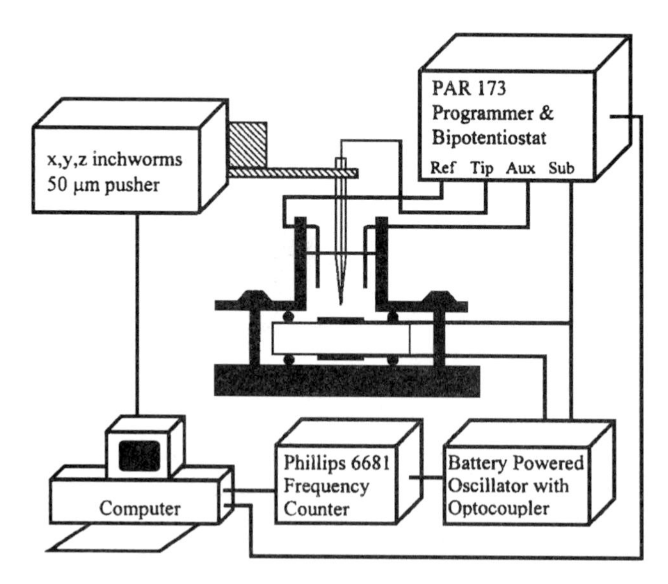

FIG. 3 Block diagram of combined SECM/QCM instrument. (From Ref. 15.)

same time mass changes are monitored by QCM (15,16). The setup used is shown in Figure 3. It may be necessary to make the QCM oscillator battery-driven to avoid coupling with the SECM bipotentiostat. When the SECM tip is close to the crystal, tip movement causes changes in the QCM resonant frequency. However, these studies show that combined measurements could be used to monitor simultaneously both mass changes in a C_{60} film and the production of a soluble reduced species during reduction in acetonitrile solutions (15).

II. NOVEL INTERFACES

Most research with SECM has dealt with the solid/liquid interface and involved solid samples in contact with an electrolyte containing the tip and the electroactive species. However, more recently various types of samples and interfaces have been probed by SECM, and further work in this area is expected (17).

Liquid/liquid interfaces have been discussed in Chapter 8. A related approach involves using an expanding droplet of a nonaqueous solvent positioned above a stationary microelectrode (microelectrochemical measurements at expanding droplets, or MEMED) (18,19). More recent work has probed bilayer lipid membranes (BLMs). A BLM is produced by placing a small amount of a lipid, such as lecithin, on a small orifice (~600 μm diameter) that separates two aqueous solutions. The BLM is a good model for biological membranes (much better than the interface between an aqueous and nonaqueous solvent), and much research is being carried out on transport through BLMs and on the nature of channel-forming proteins for selective transport of ions and other species. The study of transport by means of scanning amperometric and potentiometric tips has excellent potential in imaging and in quantitative measurements on BLMs.

Ion-selective microelectrodes have been used earlier to study concentration gradients in unstirred layers near BLMs (20–22). The SECM has the advantage, however, of allowing precise placement of the electrode near the BLM and making it possible to scan across the BLM and image ion flux. Initial studies in this general area were made by Matsue and coworkers (23–25). In most studies with BLMs, the membrane is vertical, which allows ready maintenance of equal pressure on both sides of the membrane. This configuration is less convenient for SECM studies, however. Recently a cell was devised for use with a horizontal BLM, and SECM approach curves and scans were demonstrated (26). For example, the ion flux with iodine in the membrane was studied, and the membrane thickness across the containing orifice was determined. Ion selective microprobes for potassium ion with tip diameters of the order of 1 μm have also been used to study the fluxes through gramicidin channels in a BLM (27). Further work in this area should be fruitful, and one can dream of imaging single protein channels with SECM. Related work on biological membranes and actual biological systems, including live cells, has excellent potential. Preliminary studies have been described in Chapter 11, and recent studies on cells demonstrate that such an approach is very promising for studying the transport of materials across a living cell membrane (28).

Little work has been done with SECM at the gas/liquid interface. Unwin and coworkers (29,30) devised an "inverted tip" that can approach an interface from below (Fig. 4). This configuration was used to measure the transport of bromine and oxygen across the electrolyte/air interface. A particularly interesting aspect of this study was that the oxygen transport was examined on a Langmuir trough with a monolayer of surfactant (1-ocatadecanol) on the surface. The approach curves could be determined as a function of the surface pressure on the monolayer (Fig. 5), and these could be used to find an interfacial transport rate constant as a function of monolayer structure. The ability to combine the SECM with the Langmuir trough opens up many new possibilities. These include not only measurements at the liquid/air interface but also studies of the monolayers themselves, as has been previously been accomplished by the "horizontal touch" technique from the air phase (31).

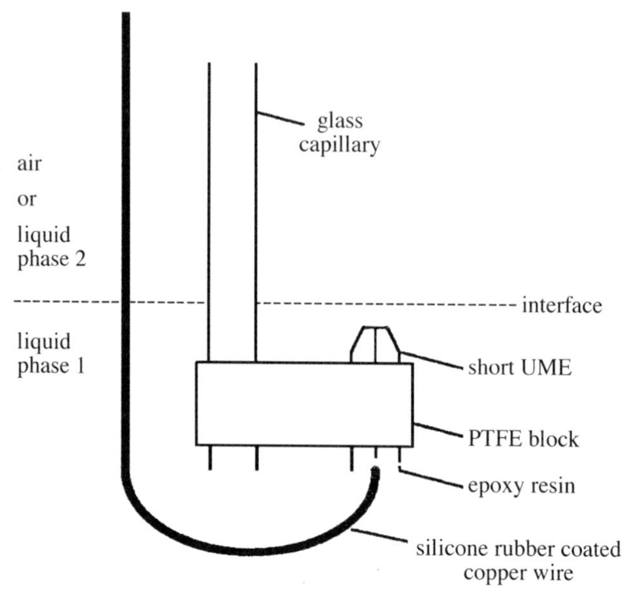

FIG. 4 The inverted tip configuration to investigate the liquid/gas interface. The glass capillary is connected to the piezo positioners and the insulated copper wire makes contact to the tip. (From Ref. 29.)

III. INSTRUMENTATION IMPROVEMENTS

Two key issues in SECM are speed and resolution. These are tied to the nature of the available tips and the techniques used to move them in the vicinity of the sample. One approach to greater speed is the use of multiple tips. There are a number of approaches to the fabrication of an array of tips. A problem with these arrays is alignment of the array with the sample so that all of the tips are at the same or a known distance from the sample. This can be accomplished by having each tip on its own controllable positioning device, such as an array of cantilevers with individual piezoelectric control (32,33). For example, 50 cantilevers with a 200-μm period (to cover a span of 1 cm) were fabricated with integrated piezo sensors and zinc oxide actuators. Such an array provided 35 Å resolution and a 20 kHz bandwidth. A difficulty with such arrays is the complexity of the instrumentation needed to use them. Each sensor in the array essentially needs its own driver and potentiostat. An alternative would be an array of individually addressable tips on a single chip that all move in the x, y, and z directions in unison. This would require some additional means to orient and align the array chip

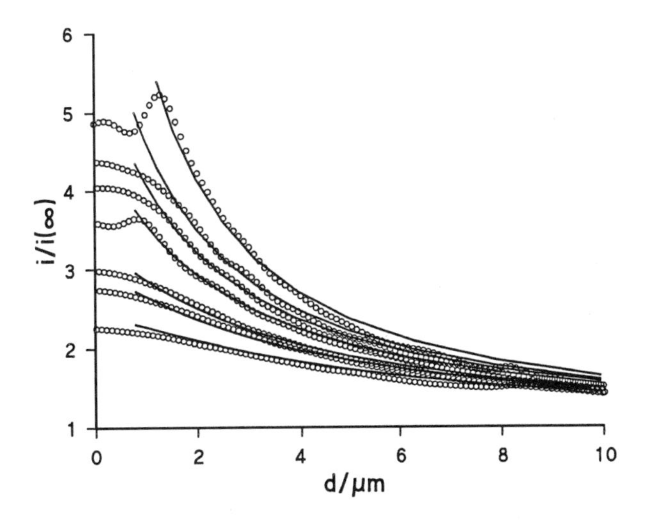

FIG. 5 Approach curves for oxygen reduction at an inverted silver ultramicroelectrode approaching the air/water interface with a 1-octadecanol monolayer at the interface at different surface pressures: Top to bottom, uncompressed, 5, 10, 20, 30, 40, and 50 mN/m. (From Ref. 28.)

with respect to the sample, e.g., with piezoelectric pushers, and would still require means for individually addressing each electrode in the array.

The highest resolution in SECM has been accomplished by imaging in very thin layers of water (i.e., SECM "in air" as described in Chapter 4). This technique is still not routine and requires careful control of conditions and tip configuration. Ultimately, high-resolution imaging will require the use of an alternative approach, such as those discussed in Sec. I, to maintain the tip at a small (nm scale), fixed distance from the sampling during scanning. One also expects progress in new approaches to tip fabrication, e.g., by the use of focused ion beam apparatus and new methods of tip insulation.

REFERENCES

1. M Ludwig, C Kranz, W Schuhmann, HE Gaub. Rev Sci Instr 66:2857, 1995.
2. JV Macpherson, PR Unwin, AC Hillier, AJ Bard. J Am Chem Soc 118:6445, 1996.
3. JV Macpherson, PR Unwin. Anal Chem 72:276, 2000.
4. PL James, LF Garfias-Mesias, PJ Moyer, WH Smyrl. J Electrochem Soc 145: L64, 1998.
5. M Buchler, SC Kelley, WH Smyrl. Electrochem Solid-State Lett 3:35, 2000.
6. FR Fan, AJ Bard. Unpublished experiments.

7. (a) MA Paesler, PJ Moyer. Near-Field Optics: Theory, Instrumentation and Applications. New York: Wiley, 1996; (b) E Betzig, JK Trautman. Science 257: 189, 1992.
8. Y-M Lee, AJ Bard. Unpublished experiments.
9. G Shi, LF Garfias-Mesias, WH Smyrl. J Electrochem Soc 145:3011, 1998.
10. P James, N Casillas, WH Smyrl. J Electrochem Soc 143:3853, 1996.
11. MA Collinson, RM Wightman. Anal Chem 65:2576, 1993.
12. RG Maus, EM McDonald, RM Wightman. Anal Chem 71:4944, 1999.
13. F-R F Fan, D Cliffel, AJ Bard. Anal Chem 70:2941, 1998.
14. Y-B Zu, AJ Bard. Unpublished experiments.
15. DE Cliffel, AJ Bard. Anal Chem 70:1993, 1998.
16. AC Hillier, MD Ward. Anal Chem 64:2539, 1992.
17. AL Barker, M Gonsalves, JV Macpherson, CJ Slevin, PR Unwin. Anal Chim Acta 385:223, 1999.
18. CJ Slavin, PR Unwin. Langmuir 13:4799, 1997.
19. J Zhang, PR Unwin. Phys Chem Chem Phys 2:1267, 2000.
20. YN Antonenko, AA Bulychev. Biochim Biophys Acta 1070:474, 1991.
21. YN Antonenko, P Pohl, GA Denisov. Biophys J 72:2187, 1997.
22. P Pohl, SM Saparov, YN Antonenko. Biophys J 75:1403, 1998.
23. H Yamada, T Matsue, I Uchida. Biochem Biophys Res Commun 180:1330, 1991
24. T Matsue, H Shiku, H Yamada, I Uchida. J Phys Chem 98:11001, 1994.
25. T Yasukawa, I Uchida, T Matsue. Biochim Biophys Acta 1369:152, 1998.
26. M Tsionsky, J Zhou, S Amemiya, F-R F Fan, AJ Bard, RAW Dryfe. Anal Chem 71:4300, 1999.
27. S Amemiya, AJ Bard. Anal Chem 72:4940, 2000.
28. B Liu, SA Rotenberg, MV Mirkin. Proc Natl Acad Sci USA 97:9855, 2000.
29. CJ Slevin, S Ryley, DJ Walton, PR Unwin. Langmuir 14:5331, 1998.
30. AL Barker, M Gonsalves, JV Macpherson, CJ Slevin, PR Unwin. Anal Chim Acta 385:223, 1999.
31. X Zhang, AJ Bard. J Am Chem Soc 111:8098, 1989.
32. SC Minne, JD Adams, G Yaralioglu, SR Manalis, A Atalar, CF Quate. Appl Phys Lett 73:1742, 1998.
33. SC Minne, G Yaralioglu, SR Manalis, JD Adams, J Zesch, A Atalar, CF Quate. Appl Phys Lett 72:2340, 1998.

INDEX